JOHNSON/EVINRUDE

Outbo
1958-72 REPAIR MANUAL
1.5-125 HP, 1-3 CYLINDER & V4 MODELS

SELOC®

©2018 SeaStar Solutions
1-877-663-8396
Rev. May/22/2018

www.sierramarine.com

CONTENTS

CONTENTS

POWERHEAD   6

LOWER UNIT   7

TRIM & TILT   8

REMOTE CONTROLS   9

HAND REWIND STARTER   10

MASTER INDEX

SAFETY NOTICE

Proper service and repair procedures are vital to the safe, reliable operation of all marine engines, as well as the personal safety of those performing repairs. This manual outlines procedures for servicing and repairing engines and drive systems using safe, effective methods. The procedures contain many NOTES, CAUTIONS and WARNINGS which should be followed, along with standard procedures, to minimize the possibility of personal injury or improper service which could damage the vehicle or compromise its safety.

It is important to note that repair procedures and techniques, tools and parts for servicing these engines, as well as the skill and experience of the individual performing the work, vary widely. It is not possible to anticipate all of the conceivable ways or conditions under which the engine may be serviced, or to provide cautions as to all possible hazards that may result. Standard and accepted safety precautions and equipment should be used during cutting, grinding, chiseling, prying, or any other process that can cause material removal or projectiles.

Some procedures require the use of tools specially designed for a specific task. Before substituting another tool or procedure, you must be completely satisfied that neither your personal safety, nor the performance of the vessel, will be endangered. All procedures covered in this manual requiring the use of special tools will be noted at the beginning of the procedure by means of an **OEM symbol** ⬭OEM⬭

Additionally, any procedure requiring the use of an electronic tester or scan tool will be noted at the beginning of the procedure by means of a **DVOM symbol**

Although information in this manual is based on industry sources and is complete as possible at the time of publication, the possibility exists that some manufacturers made later changes which could not be included here. While striving for total accuracy, Seloc Publishing cannot assume responsibility for any errors, changes or omissions that may occur in the compilation of this data. We must therefore warn you to follow instructions carefully, using common sense. If you are uncertain of a procedure, seek help by inquiring with someone in your area who is familiar with these motors before proceeding.

PART NUMBERS

Part numbers listed in this reference are not recommendations by Seloc Publishing for any particular product brand name, simply iterations of the manufacturer's suggestions. They are also references that can be used with interchange manuals and aftermarket supplier catalogs to locate each brand supplier's discrete part number.

SPECIAL TOOLS

Special tools are recommended by the manufacturers to perform a specific job. Use has been kept to a minimum, but, where absolutely necessary, they are referred to in the text by the part number of the manufacturer if at all possible; and also noted at the beginning of each procedure with one of the following symbols: **OEM** or **DVOM.**

The **OEM** symbol usually denotes the need for a unique tool purposely designed to accomplish a specific task, it will also be used, less frequently, to notify the reader of the need for a tool that is not commonly found in the average tool box.

The **DVOM** symbol is used to denote the need for an electronic test tool like an ohmmeter, multi-meter or, on certain later engines, a scan tool.

These tools can be purchased, under the appropriate part number, from your local dealer or regional distributor, or an equivalent tool can be purchased locally from a tool supplier or parts outlet. Before substituting any tool for the one recommended, read the SAFETY NOTICE at the top of this page.

Providing the correct mix of service and repair procedures is an endless battle for any publisher of "How-To" information. Users range from first time do-it yourselfers to professionally trained marine technicians, and information important to one is frequently irrelevant to the other. The editors at Seloc Publishing strive to provide accurate and articulate information on all facets of marine engine repair, from the simplest procedure to the most complex. In doing this, we understand that certain procedures may be outside the capabilities of the average DIYer. Conversely we are aware that many procedures are unnecessary for a trained technician.

SKILL LEVELS

In order to provide all of our users, particularly the DIYers, with a feeling for the scope of a given procedure or task before tackling it we have included a rating system denoting the suggested skill level needed when performing a particular procedure. One of the following icons will be included at the beginning of most procedures:

① EASY

EASY. These procedures are aimed primarily at the DIYer and can be classified, for the most part, as basic maintenance procedures; battery, fluids, filters, plugs, etc. Although certainly valuable to any experience level, they will generally be of little importance to a technician.

② MODERATE

MODERATE. These procedures are suited for a DIYer with experience and a working knowledge of mechanical procedures. Even an advanced DIYer or professional technician will occasionally refer to these procedures. They will generally consist of component repair and service procedures, adjustments and minor rebuilds.

③ DIFFICULT

DIFFICULT. These procedures are aimed at the advanced DIYer and professional technician. They will deal with diagnostics, rebuilds and internal engine/drive components and will frequently require special tools.

④ SKILLED

SKILLED. These procedures are aimed at highly skilled technicians and should not be attempted without previous experience. They will usually consist of machine work, internal engine work and gear case rebuilds.

Please remember one thing when considering the above ratings—they are a guide for judging the complexity of a given procedure and are subjective in nature. Only you will know what your experience level is, and only you will know when a procedure may be outside the realm of your capability. First time DIYer, or life-long marine technician, we all approach repair and service differently so an easy procedure for one person may be a difficult procedure for another, regardless of experience level. All skill level ratings are meant to be used as a guide only! Use them to help make a judgement before undertaking a particular procedure, but by all means read through the procedure first and make your own decision—after all, our mission at Seloc is to make boat maintenance and repair easier for everyone whether you are changing the oil or rebuilding an engine. Enjoy boating!

ALL RIGHTS RESERVED

ACKNOWLEDGMENTS

Seloc Publishing expresses appreciation to the following companies who supported the production of this book:
- Marine Mechanics Institute—Orlando, FL
- Belks Marine—Holmes, PA

Thanks to John Hartung and Judy Belk of Belk's Marine for for their assistance, guidance, patience and access to some of the motors photographed for this manual.

Seloc Publishing would like to express thanks to the fine companies who participate in the production of all our books:
- Hand tools supplied by Craftsman are used during all phases of our vehicle teardown and photography.
- Many of the fine specialty tools used in our procedures were provided courtesy of Lisle Corporation.
- Much of our shop's electronic testing equipment was supplied by Universal Enterprises Inc. (UEI).

1

GENERAL INFORMATION, SAFETY AND TOOLS

HOW TO USE THIS MANUAL

This service is designed to be a handy reference guide to maintaining and repairing your Johnson/Evinrude Outboard. We strongly believe that regardless of how many or how few year's experience you may have, there is something new waiting here for you.

This service covers the topics that a factory service manual (designed for factory trained mechanics) and a manufacturer owner's manual (designed more by lawyers than boat owners these days) covers. It will take you through the basics of maintaining and repairing your outboard, step-by-step, to help you understand what the factory trained mechanics already know by heart. By using the information in this service, any boat owner should be able to make better informed decisions about what they need to do to maintain and enjoy their outboard.

Even if you never plan on touching a wrench (and if so, we hope that we can change your mind), this service will still help you understand what a mechanic needs to do in order to maintain your engine.

Can You Do It?

If you are not the type who is prone to taking a wrench to something, NEVER FEAR. The procedures provided here cover topics at a level virtually anyone will be able to handle. And just the fact that you purchased this service shows your interest in better understanding your outboard.

You may even find that maintaining your outboard yourself is preferable in most cases. From a monetary standpoint, it could also be beneficial. The money spent on hauling out to a marina and paying a tech to service the engine could buy you fuel for a whole weekend of boating. And, if you are really that unsure of your own mechanical abilities, at the very least you should fully understand what a marine mechanic does to your boat. You may decide that anything other than maintenance and adjustments should be performed by a mechanic (and that's your call), but if so you should know that every time you board your boat, you are placing faith in the mechanic's work and trusting him or her with your well-being, and maybe your life.

It should also be noted that in most areas a factory-trained mechanic will command a hefty hourly rate for off site service. If the tech comes to you this hourly rate is often charged from the time they leave their shop to the time that they return home. When service is performed at a boat yard, the clock usually starts when they go out to get the boat and bring it into the shop and doesn't end until it is tested and put back in the yard. The cost savings in doing the job yourself might be readily apparent at this point.

Of course, if even you're already a seasoned Do-It-Yourselfer or a Professional Technician, you'll find the procedures, specifications, special tips as well as the schematics and illustrations helpful when tackling a new job on a motor.

■ To help you decide if a task is within your skill level, procedures will often be rated using a wrench symbol in the text. When present, the number of wrenches designates how difficult we feel the procedure to be on a 1-4 scale. For more details on the wrench icon rating system, please refer to the information under Skill Levels at the beginning of this service.

Where to Begin

Before spending any money on parts, and before removing any nuts or bolts, read through the entire procedure or topic. This will give you the overall view of what tools and supplies will be required to perform the procedure or what questions need to be answered before purchasing parts. So read ahead and plan ahead. Each operation should be approached logically and all procedures thoroughly understood before attempting any work.

Avoiding Trouble

Some procedures in this service may require you to "label and disconnect . . . " a group of lines, hoses or wires. Don't be lulled into thinking you can remember where everything goes - you won't. If you reconnect or install a part incorrectly, the motor may operate poorly, if at all. If you hook up electrical wiring incorrectly, you may instantly learn a very expensive lesson.

A piece of masking tape, for example, placed on a hose and another on its fitting will allow you to assign your own label such as the letter "A", or a short name. As long as you remember your own code, you can reconnect the lines by matching letters or names. Do remember that tape will dissolve when saturated in some fluids (especially cleaning solvents). If a component is to be washed or cleaned, use another method of identification. A permanent felt-tipped marker can be very handy for marking metal parts; but remember that some solvents will remove permanent marker. A scribe can be used to carefully etch a small mark in some metal parts, but be sure NOT to do that on a gasket-making surface.

SAFETY is the most important thing to remember when performing maintenance or repairs. Be sure to read the information on safety in this service.

Maintenance or Repair?

Proper maintenance is the key to long and trouble-free engine life, and the work can yield its own rewards. A properly maintained engine performs better than one that is neglected. As a conscientious boat owner, set aside a Saturday morning, at least once a month, to perform a thorough check of items that could cause problems. Keep your own personal log to jot down which services you performed, how much the parts cost you, the date, and the amount of hours on the engine at the time. Keep all receipts for parts purchased, so that they may be referred to in case of related problems or to determine operating expenses. As a do-it-yourselfer, these receipts are the only proof you have that the required maintenance was performed. In the event of a warranty problem (on new motors), these receipts can be invaluable.

It's necessary to mention the difference between maintenance and repair. Maintenance includes routine inspections, adjustments, and replacement of parts that show signs of normal wear. Maintenance compensates for wear or deterioration. Repair implies that something has broken or is not working. A need for repair is often caused by lack of maintenance.

For example: draining and refilling the gearcase oil is maintenance recommended by all manufacturers at specific intervals. Failure to do this can allow internal corrosion or damage and impair the operation of the motor, requiring expensive repairs. While no maintenance program can prevent items from breaking or wearing out, a general rule can be stated: MAINTENANCE IS CHEAPER THAN REPAIR.

Directions & Locations

◆ See Figure 1

Two basic rules should be mentioned here. First, whenever the Port side of the engine (or boat) is referred to, it is meant to specify the left side of the engine when you are sitting at the helm. Conversely, the Starboard means your right side. The Bow is the front of the boat and the Stern or Aft is the rear.

Most screws and bolts are removed by turning counterclockwise, and tightened by turning clockwise. An easy way to remember this is: righty-

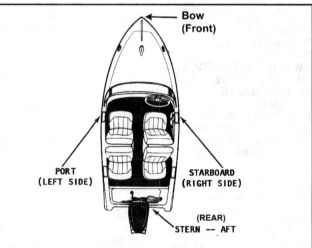

Fig. 1 Common terminology used for reference designation on boats of all size. These terms are used through out the text

tighty; lefty-loosey. Corny, but effective. And if you are really dense (and we have all been so at one time or another), buy a ratchet that is marked ON and OFF (like Snap-on® ratchets), or mark your own. This can be especially helpful when you are bent over backwards, upside down or otherwise turned around when working on a boat-mounted component.

Professional Help

Occasionally, there are some things when working on an outboard that are beyond the capabilities or tools of the average Do-It-Yourselfer (DIYer). This shouldn't include most of the topics of this service, but you will have to be the judge. Some engines require special tools or a selection of special parts, even for some basic maintenance tasks.

Talk to other boaters who use the same model of engine and speak with a trusted marina to find if there is a particular system or component on your engine that is difficult to maintain.

You will have to decide for yourself where basic maintenance ends and where professional service should begin. Take your time and do your research first (starting with the information contained within) and then make your own decision. If you really don't feel comfortable with attempting a procedure, DON'T DO IT. If you've gotten into something that may be over your head, don't panic. Tuck your tail between your legs and call a marine mechanic. Marinas and independent shops will be able to finish a job for you. Your ego may be damaged, but your boat will be properly restored to its full running order. So, as long as you approach jobs slowly and carefully, you really have nothing to lose and everything to gain by doing it yourself.

On the other hand, even the most complicated repair is within the ability of a person who takes their time and follows the steps of a procedure. A rock climber doesn't run up the side of a cliff, he/she takes it one step at a time and in the end, what looked difficult or impossible was conquerable. Worry about one step at a time.

Purchasing Parts

◆ See Figure 2

When purchasing parts there are two things to consider. The first is quality and the second is to be sure to get the correct part for your engine. To get quality parts, always deal directly with a reputable retailer. To get the proper parts always refer to the model number from the information tag on your engine prior to calling the parts counter. An incorrect part can adversely affect your engine performance and fuel economy, and will cost you more money and aggravation in the end.

Just remember a tow back to shore will cost plenty. That charge is per hour from the time the towboat leaves their home port, to the time they return to their home port. Get the picture. . .$$$?

So whom should you call for parts? Well, there are many sources for the parts you will need. Where you shop for parts will be determined by what kind of parts you need, how much you want to pay, and the types of stores in your neighborhood.

Your marina can supply you with many of the common parts you require. Using a marina as your parts supplier may be handy because of location (just walk right down the dock) or because the marina specializes in your particular brand of engine. In addition, it is always a good idea to get to know the marina staff (especially the marine mechanic).

The marine parts jobber, who is usually listed in the yellow pages or whose name can be obtained from the marina, is another excellent source for parts. In addition to supplying local marinas, they also do a sizeable business in over-the-counter parts sales for the do-it-yourselfer.

Almost every boating community has one or more convenient marine chain stores. These stores often offer the best retail prices and the convenience of one-stop shopping for all your needs. Since they cater to the do-it-yourselfer, these stores are almost always open weeknights, Saturdays, and Sundays, when the jobbers are usually closed.

The lowest prices for parts are most often found in discount stores or the auto department of mass merchandisers. Parts sold there are name and private brand parts bought in huge quantities, so they can offer a competitive price. Private brand parts are made by major manufacturers and sold to large chains under a store label. And, of course, more and more large automotive parts retailers are stocking basic marine supplies.

Avoiding the Most Common Mistakes

There are 3 common mistakes in mechanical work:

1. Following the Incorrect order of assembly, disassembly or adjustment. When taking something apart or putting it together, performing steps in the wrong order usually just costs you extra time; however, it CAN break something. Read the entire procedure before beginning disassembly. Perform everything in the order in which the instructions say you should, even if you can't immediately see a reason for it. When you're taking apart something that is very intricate, you might want to draw a picture of how it looks when assembled at one point in order to make sure you get everything back in its proper position. When making adjustments, perform them in the proper order; often, one adjustment affects another, and you cannot expect satisfactory results unless each adjustment is made only when it cannot be changed by subsequent adjustments.

■ **Digital cameras are handy. If you've got access to one, take pictures of intricate assemblies during the disassembly process and refer to them during assembly for tips on part orientation.**

2. Over-torquing (or under-torquing). While it is more common for over-torquing to cause damage, under-torquing may allow a fastener to vibrate loose causing serious damage. Especially when dealing with plastic and aluminum parts, pay attention to torque specifications and utilize a torque wrench in assembly. If a torque figure is not available, remember that if you are using the right tool to perform the job, you will probably not have to strain yourself to get a fastener tight enough. The pitch of most threads is so slight that the tension you put on the wrench will be multiplied many times in actual force on what you are tightening.

Fig. 2 By far the most important asset in purchasing parts is a knowledgeable and enthusiastic parts person

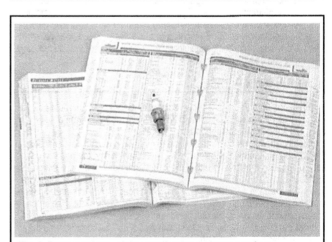

Fig. 3 Parts catalogs, giving application and part number information, are provided by manufacturers for most replacement parts

3. Cross-threading. This occurs when a part such as a bolt is screwed into a nut or casting at the wrong angle and forced. Cross-threading is more likely to occur if access is difficult. It helps to clean and lubricate fasteners, then to start threading with the part to be installed positioned straight inward. Always start a fastener, etc. with your fingers. If you encounter resistance, unscrew the part and start over again at a different angle until it can be inserted and turned several times without much effort. Keep in mind that some parts may have tapered threads, so that gentle turning will automatically bring the part you're threading to the proper angle, but only if you don't force it or resist a change in angle. Don't put a wrench on the part until it has been tightened a couple of turns by hand. If you suddenly encounter resistance, and the part has not seated fully, don't force it. Pull it back out to make sure it's clean and threading properly.

BOATING SAFETY

In 1971 Congress ordered the U.S. Coast Guard to improve recreational boating safety. In response, the Coast Guard drew up a set of regulations.

Aside from these federal regulations, there are state and local laws you must follow. These sometimes exceed the Coast Guard requirements. This section discusses only the federal laws. State and local laws are available from your local Coast Guard. As with other laws, "Ignorance of the boating laws is no excuse." The rules fall into two groups: regulations for your boat and required safety equipment on your boat.

Regulations For Your Boat

DOCUMENTING OF VESSELS

Most boats on waters within Federal jurisdiction must be registered or documented. These waters are those that provide a means of transportation between two or more states or to the sea. They also include the territorial waters of the United States.

A vessel of five or more net tons may be documented as a yacht. In this process, papers are issued by the U.S. Coast Guard as they are for large ships. Documentation is a form of national registration. The boat must be used solely for pleasure. Its owner must be a citizen of the U.S., a partnership of U.S. citizens, or a corporation controlled by U.S. citizens. The captain and other officers must also be U.S. citizens. The crew need not be.

If you document your yacht, you have the legal authority to fly the yacht ensign. You also may record bills of sale, mortgages, and other papers of title with federal authorities. Doing so gives legal notice that such instruments exist. Documentation also permits preferred status for mortgages. This gives you additional security, and it aids in financing and transfer of title. You must carry the original documentation papers aboard your vessel. Copies will not suffice.

REGISTRATION OF BOATS

If your boat is not documented, registration in the state of its principal use is probably required. If you use it mainly on an ocean, a gulf, or other similar water, register it in the state where you moor it.

If you use your boat solely for racing, it may be exempt from the requirement in your state. Some states may also exclude dinghies, while others require registration of documented vessels and non-power driven boats.

All states, except Alaska, register boats. In Alaska, the U.S. Coast Guard issues the registration numbers. If you move your vessel to a new state of principal use, a valid registration certificate is good for 60 days. You must have the registration certificate (certificate of number) aboard your vessel when it is in use. A copy will not suffice. You may be cited if you do not have the original on board.

NUMBERING OF VESSELS

A registration number is on your registration certificate. You must paint or permanently attach this number to both sides of the forward half of your boat. Do not display any other number there.

The registration number must be clearly visible. It must not be placed on the obscured underside of a flared bow. If you can't place the number on the bow, place it on the forward half of the hull. If that doesn't work, put it on the superstructure. Put the number for an inflatable boat on a bracket or fixture. Then, firmly attach it to the forward half of the boat. The letters and numbers must be plain block characters and must read from left to right. Use a space or a hyphen to separate the prefix and suffix letters from the numerals. The color of the characters must contrast with that of the background, and they must be at least three inches high.

In some states your registration is good for only one year. In others, it is good for as long as three years. Renew your registration before it expires. At that time you will receive a new decal or decals. Place them as required by state law. You should remove old decals before putting on the new ones. Some states require that you show only the current decal or decals. If your vessel is moored, it must have a current decal even if it is not in use.

If your vessel is lost, destroyed, abandoned, stolen, or transferred, you must inform the issuing authority. If you lose your certificate of number or your address changes, notify the issuing authority as soon as possible.

SALES AND TRANSFERS

Your registration number is not transferable to another boat. The number stays with the boat unless its state of principal use is changed.

HULL IDENTIFICATION NUMBER

A Hull Identification Number (HIN) is like the Vehicle Identification Number (VIN) on your car. Boats built between November 1, 1972 and July 31, 1984 have old format HINs. Since August 1, 1984 a new format has been used.

Your boat's HIN must appear in two places. If it has a transom, the primary number is on its starboard side within two inches of its top. If it does not have a transom or if it was not practical to use the transom, the number is on the starboard side. In this case, it must be within one foot of the stern and within two inches of the top of the hull side. On pontoon boats, it is on the aft crossbeam within one foot of the starboard hull attachment. Your boat also has a duplicate number in an unexposed location. This is on the boat's interior or under a fitting or item of hardware.

LENGTH OF BOATS

For some purposes, boats are classed by length. Required equipment, for example, differs with boat size. Manufacturers may measure a boat's length in several ways. Officially, though, your boat is measured along a straight line from its bow to its stern. This line is parallel to its keel.

The length does not include bowsprits, boomkins, or pulpits. Nor does it include rudders, brackets, outboard motors, outdrives, diving platforms, or other attachments.

CAPACITY INFORMATION

◆ **See Figure 4**

In order to receive maximum enjoyment, with safety and performance, from your boat, take care not to exceed the load capacity given by the manufacturer.

Manufacturers must put capacity plates on most recreational boats less than 20 feet long. Sailboats, canoes, kayaks, and inflatable boats plus non-mono hull boats (tri-hulls, catamaran's etc) are usually exempt. Outboard boats must display the maximum recommended horsepower of their engines. The plates must also show the recommended maximum weights of the people on board. And they must show the recommended maximum combined weights of people, engine(s), and gear. Inboards and stern drives need not show the weight of their engines on their capacity plates. The capacity plate must appear where it is clearly visible to the operator when underway. This information serves to remind you of the capacity of your boat under normal circumstances. You should ask yourself, "Is my boat loaded above its recommended capacity" and, "Is my boat overloaded for the present sea and wind conditions?" If you are stopped by a legal authority, you may be sent back to port if you are overloaded or cited if something occurs.

■ **Keep in mind that state law can affect how coast guard regulations are enforced, potentially making it illegal to operate a vessel while Loading**

If the capacity plate states the maximum person capacity to be 750 pounds and you assume each person to weigh an average of 150 lbs., then the boat could carry five persons safely. If you add another 250 lbs. for motor

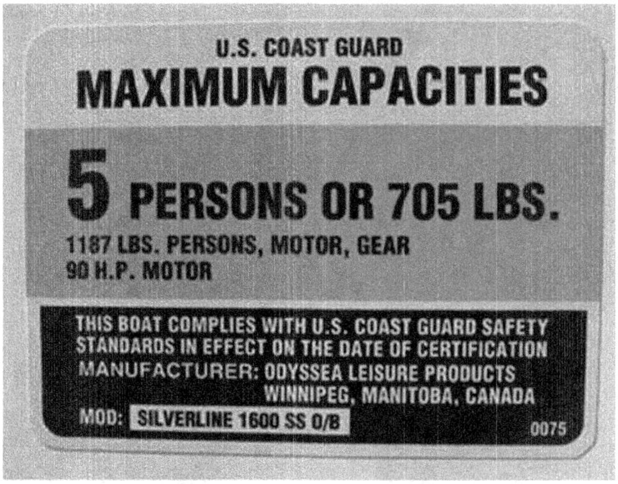

U.S. COAST GUARD

MAXIMUM CAPACITIES

5 PERSONS OR 705 LBS.

1187 LBS. PERSONS, MOTOR, GEAR

90 H.P. MOTOR

THIS BOAT COMPLIES WITH U.S. COAST GUARD SAFETY
STANDARDS IN EFFECT ON THE DATE OF CERTIFICATION
MANUFACTURER: ODYSSEA LEISURE PRODUCTS
WINNIPEG, MANITOBA, CANADA

MOD: SILVERLINE 1600 SS O/B   0075

Fig. 4 A U.S. Coast Guard certification plate indicates the amount of occupants and gear appropriate for safe operation of the vessel (and allowable engine size for outboard boats)

and gear, and the maximum weight capacity for persons and gear is 1,000 lbs. or more, then the five persons and gear would be within the limit.

Try to load the boat evenly port and starboard. If you place more weight on one side than on the other, the boat will list to the heavy side and make steering difficult. You will also get better performance by placing heavy supplies aft of the center to keep the bow light for more efficient planning.

Clarification

Much confusion arises from the terms, certification, requirements, approval, regulations, etc. Perhaps the following may clarify a couple of these points:

1. The Coast Guard does not approve boats in the same manner as they "Approve" life jackets. The Coast Guard applies a formula to inform the public of what is safe for a particular craft.

2. If a boat has to meet a particular regulation, it must have a Coast Guard certification plate. The public has been led to believe this indicates approval of the Coast Guard. Not so.

3. The certification plate means a willingness of the manufacturer to meet the Coast Guard regulations for that particular craft. The manufacturer may recall a boat if it fails to meet the Coast Guard requirements.

4. The Coast Guard certification plates, see accompanying illustration, may or may not be metal. The plate is a regulation for the manufacturer. It is only a warning plate and the public does not have to adhere to the restrictions set forth on it. Again, the plate sets forth information as to the Coast Guard's opinion for safety on that particular boat.

5. Coast Guard Approved equipment is equipment which has been approved by the Commandant of the U.S. Coast Guard and has been determined to be in compliance with Coast Guard specifications and regulations relating to the materials, construction, and performance of such equipment.

Horsepower

The maximum horsepower engine for each individual boat should not be increased by any great amount without checking requirements from the Coast Guard in your area. The Coast Guard determines horsepower requirements based on the length, beam, and depth of the hull. TAKE CARE NOT to exceed the maximum horsepower listed on the plate or the warranty and possibly the insurance on the boat may become void.

CERTIFICATE OF COMPLIANCE

◆ See Figure 4

Manufacturers are required to put compliance plates on motorboats greater than 20 feet in length. The plates must say, "This boat," or "This equipment complies with the U. S. Coast Guard Safety Standards in effect on the date of certification." Letters and numbers can be no less than one-eighth of an inch high. At the manufacturer's option, the capacity and compliance plates may be combined.

VENTILATION

A cup of gasoline spilled in the bilge has the potential explosive power of 15 sticks of dynamite. This statement, commonly quoted over 20 years ago, may be an exaggeration; however, it illustrates a fact. Gasoline fumes in the bilge of a boat are highly explosive and a serious danger. They are heavier than air and will stay in the bilge until they are vented out.

Because of this danger, Coast Guard regulations require ventilation on many powerboats. There are several ways to supply fresh air to engine and gasoline tank compartments and to remove dangerous vapors. Whatever the choice, it must meet Coast Guard standards.

■ **The following is not intended to be a complete discussion of the regulations. It is limited to the majority of recreational vessels. Contact your local Coast Guard office for further information.**

General Precautions

Ventilation systems will not remove raw gasoline that leaks from tanks or fuel lines. If you smell gasoline fumes, you need immediate repairs. The best device for sensing gasoline fumes is your nose. Use it! If you smell gasoline in a bilge, engine compartment, or elsewhere, don't start your engine. The smaller the compartment, the less gasoline it takes to make an explosive mixture.

Ventilation for Open Boats

In open boats, gasoline vapors are dispersed by the air that moves through them. So they are exempt from ventilation requirements.

To be "open," a boat must meet certain conditions. Engine and fuel tank compartments and long narrow compartments that join them must be open to the atmosphere." This means they must have at least 15 square inches of open area for each cubic foot of net compartment volume. The open area must be in direct contact with the atmosphere. There must also be no long, unventilated spaces open to engine and fuel tank compartments into which flames could extend.

Ventilation for All Other Boats

Powered and natural ventilation are required in an enclosed compartment with a permanently installed gasoline engine that has a cranking motor. A compartment is exempt if its engine is open to the atmosphere. Diesel powered boats are also exempt.

VENTILATION SYSTEMS

◆ See Figure 5

There are two types of ventilation systems. One is "natural ventilation." In it, air circulates through closed spaces due to the boat's motion. The other type is "powered ventilation." In it, air is circulated by a motor-driven fan or fans.

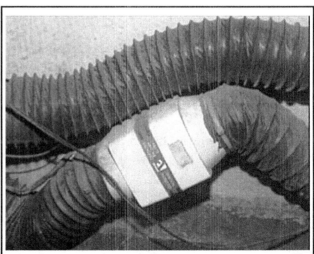

Fig. 5 Typical blower and duct system to vent fumes from the engine compartment

Natural Ventilation System Requirements

A natural ventilation system has an air supply from outside the boat. The air supply may also be from a ventilated compartment or a compartment open to the atmosphere. Intake openings are required. In addition, intake ducts may be required to direct the air to appropriate compartments.

The system must also have an exhaust duct that starts in the lower third of the compartment. The exhaust opening must be into another ventilated compartment or into the atmosphere. Each supply opening and supply duct, if there is one, must be above the usual level of water in the bilge. Exhaust openings and ducts must also be above the bilge water. Openings and ducts must be at least three square inches in area or two inches in diameter. Openings should be placed so exhaust gasses do not enter the fresh air intake. Exhaust fumes must not enter cabins or other enclosed, non-ventilated spaces. The carbon monoxide gas in them is deadly.

Intake and exhaust openings must be covered by cowls or similar devices. These registers keep out rain water and water from breaking seas. Most often, intake registers face forward and exhaust openings aft. This aids the flow of air when the boat is moving or at anchor since most boats face into the wind when properly anchored.

Power Ventilation System Requirements

Powered ventilation systems must meet the standards of a natural system, but in addition, they must also have one or more exhaust blowers. The blower duct can serve as the exhaust duct for natural ventilation if fan blades do not obstruct the air flow when not powered. Openings in engine compartment, for carburetion are in addition to ventilation system requirements.

Required Safety Equipment

MINIMUM REQUIREMENTS

Coast Guard regulations require that your boat have certain equipment aboard. These requirements are minimums. Exceed them whenever you can.

TYPES OF FIRES

There are four common classes of fires:
- Class A - fires are of ordinary combustible materials such as paper or wood.
- Class B - fires involve gasoline, oil and grease.
- Class C - fires are electrical.
- Class D - fires involve ferrous metals

One of the greatest risks to boaters is fire. This is why it is so important to carry the correct number and type of extinguishers onboard.

The best fire extinguisher for most boats is a Class B extinguisher. Never use water on Class B or Class C fires, as water spreads these types of fires. Additionally, you should never use water on a Class C fire as it may cause you to be electrocuted.

FIRE EXTINGUISHERS

◆ **See Figure 6**

If your boat meets one or more of the following conditions, you must have at least one fire extinguisher aboard. The conditions are:
- Inboard or stern drive engines
- Closed compartments under seats where portable fuel tanks can be stored
- Double bottoms not sealed together or not completely filled with flotation materials
- Closed living spaces
- Closed stowage compartments in which combustible or flammable materials are stored
- Permanently installed fuel tanks
- Boat is 26 feet or more in length.

Contents of Extinguishers

Fire extinguishers use a variety of materials. Those used on boats usually contain dry chemicals, Halon, or Carbon Dioxide (CO2). Dry chemical extinguishers contain chemical powders such as Sodium Bicarbonate - baking soda.

Carbon dioxide is a colorless and odorless gas when released from an extinguisher. It is not poisonous but caution must be used in entering

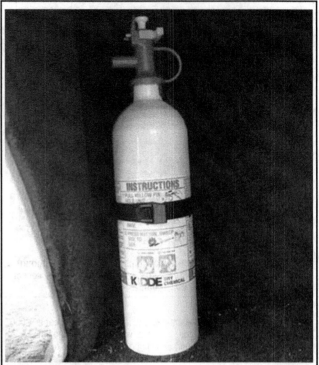

Fig. 6 An approved fire extinguisher should be mounted close to the operator for emergency use

compartments filled with it. It will not support life and keeps oxygen from reaching your lungs. A fire-killing concentration of Carbon Dioxide can be lethal. If you are in a compartment with a high concentration of CO2, you will have no difficulty breathing. But the air does not contain enough oxygen to support life. Unconsciousness or death can result.

Halon Extinguishers

Some fire extinguishers and "built-in" or "fixed" automatic fire extinguishing systems contain a gas called Halon. Like carbon dioxide it is colorless and odorless and will not support life. Some Halons may be toxic if inhaled.

To be accepted by the Coast Guard, a fixed Halon system must have an indicator light at the vessel's helm. A green light shows the system is ready. Red means it is being discharged or has been discharged. Warning horns are available to let you know the system has been activated. If your fixed Halon system discharges, ventilate the space thoroughly before you enter it. There are no residues from Halon but it will not support life.

Although Halon has excellent fire fighting properties; it is thought to deplete the earth's ozone layer and has not been manufactured since January 1, 1994. Halon extinguishers can be refilled from existing stocks of the gas until they are used up, but high federal excise taxes are being charged for the service. If you discontinue using your Halon extinguisher, take it to a recovery station rather than releasing the gas into the atmosphere. Compounds such as FE 241, designed to replace Halon, are now available.

Fire Extinguisher Approval

Fire extinguishers must be Coast Guard approved. Look for the approval number on the nameplate. Approved extinguishers have the following on their labels: "Marine Type USCG Approved, Size..., Type..., 162.208/," etc. In addition, to be acceptable by the Coast Guard, an extinguisher must be in serviceable condition and mounted in its bracket. An extinguisher not properly mounted in its bracket will not be considered serviceable during a Coast Guard inspection.

Care and Treatment

Make certain your extinguishers are in their stowage brackets and are not damaged. Replace cracked or broken hoses. Nozzles should be free of obstructions. Sometimes, wasps and other insects nest inside nozzles and make them inoperable. Check your extinguishers frequently. If they have pressure gauges, is the pressure within acceptable limits? Do the locking pins and sealing wires show they have not been used since recharging?

Don't try an extinguisher to test it. Its valves will not reseat properly and the remaining gas will leak out. When this happens, the extinguisher is useless.

Weigh and tag carbon dioxide and Halon extinguishers twice a year. If their weight loss exceeds 10 percent of the weight of the charge, recharge them. Check to see that they have not been used. They should have been inspected by a qualified person within the past six months, and they should have tags showing all inspection and service dates. The problem is that they can be partially discharged while appearing to be fully charged.

Some Halon extinguishers have pressure gauges the same as dry chemical extinguishers. Don't rely too heavily on the gauge. The extinguisher can be partially discharged and still show a good gauge reading. Weighing a Halon extinguisher is the only accurate way to assess its contents.

If your dry chemical extinguisher has a pressure indicator, check it frequently. Check the nozzle to see if there is powder in it. If there is, recharge it. Occasionally invert your dry chemical extinguisher and hit the base with the palm of your hand. The chemical in these extinguishers packs and cakes due to the boat's vibration and pounding. There is a difference of opinion about whether hitting the base helps, but it can't hurt. It is known that caking of the chemical powder is a major cause of failure of dry chemical extinguishers. Carry spares in excess of the minimum requirement. If you have guests aboard, make certain they know where the extinguishers are and how to use them.

Using a Fire Extinguisher

A fire extinguisher usually has a device to keep it from being discharged accidentally. This is a metal or plastic pin or loop. If you need to use your extinguisher, take it from its bracket. Remove the pin or the loop and point the nozzle at the base of the flames. Now, squeeze the handle, and discharge the extinguisher's contents while sweeping from side to side. Recharge a used extinguisher as soon as possible.

If you are using a Halon or carbon dioxide extinguisher, keep your hands away from the discharge. The rapidly expanding gas will freeze them. If your fire extinguisher has a horn, hold it by its handle.

Legal Requirements for Extinguishers

You must carry fire extinguishers as defined by Coast Guard regulations. They must be firmly mounted in their brackets and immediately accessible.

A motorboat less than 26 feet long must have at least one approved hand-portable, Type B-1 extinguisher. If the boat has an approved fixed fire extinguishing system, you are not required to have the Type B-1 extinguisher. Also, if your boat is less than 26 feet long, is propelled by an outboard motor, or motors, and does not have any of the first six conditions described at the beginning of this section, it is not required to have an extinguisher. Even so, it's a good idea to have one, especially if a nearby boat catches fire, or if a fire occurs at a fuel dock.

A motorboat 26 feet to less than 40 feet long, must have at least two Type B-1 approved hand-portable extinguishers. It can, instead, have at least one Coast Guard approved Type B-2. If you have an approved fire extinguishing system, only one Type B-1 is required.

A motorboat 40 to 65 feet long must have at least three Type B-1 approved portable extinguishers. It may have, instead, at least one Type B-1 plus a Type B-2. If there is an approved fixed fire extinguishing system, two Type B-1 or one Type B-2 is required.

WARNING SYSTEM

Various devices are available to alert you to danger. These include fire, smoke, gasoline fumes, and carbon monoxide detectors. If your boat has a galley, it should have a smoke detector. Where possible, use wired detectors. Household batteries often corrode rapidly on a boat.

There are many ways in which carbon monoxide (a by-product of the combustion that occurs in an engine) can enter your boat. You can't see, smell, or taste carbon monoxide gas, but it is lethal. As little as one part in 10,000 parts of air can bring on a headache. The symptoms of carbon monoxide poisoning - headaches, dizziness, and nausea - are like seasickness. By the time you realize what is happening to you, it may be too late to take action. If you have enclosed living spaces on your boat, protect yourself with a detector.

PERSONAL FLOTATION DEVICES

◆ See Figures 7, 8 and 9

Personal Flotation Devices (PFDs) are commonly called life preservers or life jackets. You can get them in a variety of types and sizes. They vary with their intended uses. To be acceptable, PFDs must be Coast Guard approved.

Type I PFDs

A Type I life jacket is also called an offshore life jacket. Type I life jackets will turn most unconscious people from facedown to a vertical or slightly backward position. The adult size gives a minimum of 22 pounds of buoyancy. The child size has at least 11 pounds. Type I jackets provide more protection to their wearers than any other type of life jacket. Type I life jackets are bulkier and less comfortable than other types. Furthermore, there are only two sizes, one for children and one for adults.

Type I life jackets will keep their wearers afloat for extended periods in rough water. They are recommended for offshore cruising where a delayed rescue is probable.

Type II PFDs

A Type II life jacket is also called a near-shore buoyant vest. It is an approved, wearable device. Type II life jackets will turn some unconscious people from facedown to vertical or slightly backward positions. The adult size gives at least 15.5 pounds of buoyancy. The medium child size has a minimum of 11 pounds. And the small child and infant sizes give seven pounds. A Type II life jacket is more comfortable than a Type I but it does not have as much buoyancy. It is not recommended for long hours in rough water. Because of this, Type IIs are recommended for inshore and inland cruising on calm water. Use them only where there is a good chance of fast rescue.

Type III PFDs

Type III life jackets or marine buoyant devices are also known as flotation aids. Like Type IIs, they are designed for calm inland or close offshore water where there is a good chance of fast rescue. Their minimum buoyancy is 15.5 pounds. They will **not** turn their wearers face up.

Type III devices are usually worn where freedom of movement is necessary. Thus, they are used for water skiing, small boat sailing, and fishing among other activities. They are available as vests and flotation coats. Flotation coats are useful in cold weather. Type IIIs come in many sizes from small child through large adult.

Fig. 7 Type III PFDs are recommended for inshore/inland use on calm water (where there is a good chance of fast rescue)

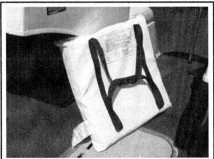
Fig. 8 Type IV buoyant cushions are thrown to people in the water. If you can squeeze air out of the cushion, it should be replaced

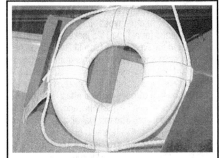
Fig. 9 Type IV throwables, such as this ring life buoy, are not designed for unconscious people, non-swimmers, or children

Life jackets come in a variety of colors and patterns - red, blue, green, camouflage, and cartoon characters. From purely a safety standpoint, the best color is bright orange. It is easier to see in the water, especially if the water is rough.

Type IV PFDs

Type IV ring life buoys, buoyant cushions and horseshoe buoys are Coast Guard approved devices called throwables. They are made to be thrown to people in the water, and should not be worn. Type IV cushions are often used as seat cushions. But, keep in mind that cushions are hard to hold onto in the water, thus, they do not afford as much protection as wearable life jackets.

The straps on buoyant cushions are for you to hold onto either in the water or when throwing them, they are **NOT** for your arms. A cushion should never be worn on your back, as it will turn you face down in the water.

Type IV throwables are not designed as personal flotation devices for unconscious people, non-swimmers, or children. Use them only in emergencies. They should not be used for, long periods in rough water.

Ring life buoys come in 18, 20, 24, and 30 in. diameter sizes. They usually have grab lines, but you will need to attach about 60 feet of polypropylene line to the grab rope to aid in retrieving someone in the water. If you throw a ring, be careful not to hit the person. Ring buoys can knock people unconscious

Type V PFDs

Type V PFDs are of two kinds, special use devices and hybrids. Special use devices include boardsailing vests, deck suits, work vests, and others. They are approved only for the special uses or conditions indicated on their labels. Each is designed and intended for the particular application shown on its label. They do not meet legal requirements for general use aboard recreational boats.

Hybrid life jackets are inflatable devices with some built-in buoyancy provided by plastic foam or kapok. They can be inflated orally or by cylinders of compressed gas to give additional buoyancy. In some hybrids the gas is released manually. In others it is released automatically when the life jacket is immersed in water.

The inherent buoyancy of a hybrid may be insufficient to float a person unless it is inflated. The only way to find this out is for the user to try it in the water. Because of its limited buoyancy when deflated, a hybrid is recommended for use by a non-swimmer only if it is worn with enough inflation to float the wearer.

If they are to count against the legal requirement for the number of life jackets you must carry, hybrids manufactured before February 8, 1995 must be worn whenever a boat is underway and the wearer must not go below decks or in an enclosed space. To find out if your Type V hybrid must be worn to satisfy the legal requirement, read its label. If its use is restricted it will say, "REQUIRED TO BE WORN" in capital letters.

Hybrids cost more than other life jackets, but this factor must be weighed against the fact that they are more comfortable than Types I, II or III life jackets. Because of their greater comfort, their owners are more likely to wear them than are the owners of Type I, II or III life jackets.

The Coast Guard has determined that improved, less costly hybrids can save lives since they will be bought and used more frequently. For these reasons, a new federal regulation was adopted effective February 8, 1995. The regulation increases both the deflated and inflated buoyancys of hybrids, makes them available in a greater variety of sizes and types, and reduces their costs by reducing production costs.

Even though it may not be required, the wearing of a hybrid or a life jacket is encouraged whenever a vessel is underway. Like life jackets, hybrids are now available in three types. To meet legal requirements, a Type I hybrid can be substituted for a Type I life jacket. Similarly Type II and III hybrids can be substituted for Type II and Type III life jackets. A Type I hybrid, when inflated, will turn most unconscious people from facedown to vertical or slightly backward positions just like a Type I life jacket. Type II and III hybrids function like Type II and III life jackets. If you purchase a new hybrid, it should have an owner's manual attached that describes its life jacket type and its deflated and inflated buoyancys. It warns you that it may have to be inflated to float you. The manual also tells you how to don the life jacket and how to inflate it. It also tells you how to change its inflation mechanism, recommended testing exercises, and inspection or maintenance procedures. The manual also tells you why you need a life jacket and why you should wear it. A new hybrid must be packaged with at least three gas cartridges. One of these may already be loaded into the inflation mechanism. Likewise, if it has an automatic inflation mechanism, it must be packaged with at least

three of these water sensitive elements. One of these elements may be installed.

Legal Requirements

A Coast Guard approved life jacket must show the manufacturer's name and approval number. Most are marked as Type I, II, III, IV or V. All of the newer hybrids are marked for type.

You are required to carry at least one wearable life jacket or hybrid for each person on board your recreational vessel. If your vessel is 16 feet or more in length and is not a canoe or a kayak, you must also have at least one Type IV on board. These requirements apply to all recreational vessels that are propelled or controlled by machinery, sails, oars, paddles, poles, or another vessel. Sailboards are not required to carry life jackets.

You can substitute an older Type V hybrid for any required Type I, II or III life jacket provided:

1. Its approval label shows it is approved for the activity the vessel is engaged in

2. It's approved as a substitute for a life jacket of the type required on the vessel

3. It's used as required on the labels

and

4. It's used in accordance with any requirements in its owner's manual (if the approval label makes reference to such a manual.)

A water skier being towed is considered to be on board the vessel when judging compliance with legal requirements.

You are required to keep your Type I, II or III life jackets or equivalent hybrids readily accessible, which means you must be able to reach out and get them when needed. All life jackets must be in good, serviceable condition.

General Considerations

The proper use of a life jacket requires the wearer to know how it will perform. You can gain this knowledge only through experience. Each person on your boat should be assigned a life jacket. Next, it should be fitted to the person who will wear it. Only then can you be sure that it will be ready for use in an emergency. This advice is good even if the water is calm, and you intend to boat near shore.

Boats can sink fast. There may be no time to look around for a life jacket. Fitting one on you in the water is almost impossible. Most drownings occur in inland waters within a few feet of safety. Most victims had life jackets, but they weren't wearing them.

Keeping life jackets in the plastic covers they came wrapped in, and in a cabin, assure that they will stay clean and unfaded. But this is no way to keep them when you are on the water. When you need a life jacket it must be readily accessible and adjusted to fit you. You can't spend time hunting for it or learning how to fit it.

There is no substitute for the experience of entering the water while wearing a life jacket. Children, especially, need practice. If possible, give your guests this experience. Tell them they should keep their arms to their sides when jumping in to keep the life jacket from riding up. Let them jump in and see how the life jacket responds. Is it adjusted so it does not ride up? Is it the proper size? Are all straps snug? Are children's life jackets the right sizes for them? Are they adjusted properly? If a child's life jacket fits correctly, you can lift the child by the jacket's shoulder straps and the child's chin and ears will not slip through. Non-swimmers, children, handicapped persons, elderly persons and even pets should always wear life jackets when they are aboard. Many states require that everyone aboard wear them in hazardous waters.

Inspect your lifesaving equipment from time to time. Leave any questionable or unsatisfactory equipment on shore. An emergency is no time for you to conduct an inspection.

Indelibly mark your life jackets with your vessel's name, number, and calling port. This can be important in a search and rescue effort. It could help concentrate effort where it will do the most good.

Care of Life Jackets

Given reasonable care, life jackets last many years. Thoroughly dry them before putting them away. Stow them in dry, well-ventilated places. Avoid the bottoms of lockers and deck storage boxes where moisture may collect. Air and dry them frequently.

Life jackets should not be tossed about or used as fenders or cushions. Many contain kapok or fibrous glass material enclosed in plastic bags. The bags can rupture and are then unserviceable. Squeeze your life jacket gently. Does air leak out? If so, water can leak in and it will no longer be safe to use. Cut it up so no one will use it, and throw it away. The covers of some life jackets are made of nylon or polyester. These materials are plastics. Like many plastics, they break down after extended exposure to the ultraviolet light in sunlight. This process may be more rapid when the materials are dyed with bright dyes such as "neon" shades.

Ripped and badly faded fabrics are clues that the covering of your life jacket is deteriorating. A simple test is to pinch the fabric between your thumbs and forefingers. Now try to tear the fabric. If it can be torn, it should definitely be destroyed and discarded. Compare the colors in protected places to those exposed to the sun. If the colors have faded, the materials have been weakened. A life jacket covered in fabric should ordinarily last several boating seasons with normal use. A life jacket used every day in direct sunlight should probably be replaced more often.

SOUND PRODUCING DEVICES

All boats are required to carry some means of making an efficient sound signal. Devices for making the whistle or horn noises required by the Navigation Rules must be capable of a four-second blast. The blast should be audible for at least one-half mile. Athletic whistles are not acceptable on boats 12 meters or longer. Use caution with athletic whistles. When wet, some of them come apart and loose their "pea." When this happens, they are useless.

If your vessel is 12 meters long and less than 20 meters, you must have a power whistle (or power horn) and a bell on board. The bell must be in operating condition and have a minimum diameter of at least 200mm (7.9 in.) at its mouth.

VISUAL DISTRESS SIGNALS

◆ **See Figures 10 and 11**

Visual Distress Signals (VDS) attract attention to your vessel if you need help. They also help to guide searchers in search and rescue situations. Be sure you have the right types, and learn how to use them properly.

It is illegal to fire flares improperly. In addition, they cost the Coast Guard and its Auxiliary many wasted hours in fruitless searches. If you signal a distress with flares and then someone helps you, please let the Coast Guard or the appropriate Search And Rescue (SAR) Agency know so the distress report will bo cancoled.

Recreational boats less than 16 feet long must carry visual distress signals on coastal waters at night. Coastal waters are:

*   The ocean (territorial sea)
*   The Great Lakes
*   Bays or sounds that empty into oceans
*   Rivers over two miles across at their mouths upstream to where they narrow to two miles.

Recreational boats 16 feet or longer must carry VDS at all times on coastal waters. The same requirement applies to boats carrying six or fewer passengers for hire. Open sailboats less than 26 feet long without engines are exempt in the daytime as are manually propelled boats. Also exempt are boats in organized races, regattas, parades, etc. Boats owned in the United States and operating on the high seas must be equipped with VDS.

A wide variety of signaling devices meet Coast Guard regulations. For pyrotechnic devices, a minimum of three must be carried. Any combination can be carried as long as it adds up to at least three signals for day use and at least three signals for night use. Three day/night signals meet both requirements. If possible, carry more than the legal requirement.

■ **The American flag flying upside down is a commonly recognized distress signal. It is not recognized in the Coast Guard regulations, though. In an emergency, your efforts would probably be better used in more effective signaling methods.**

Types of VDS

VDS are divided into two groups; daytime and nighttime use. Each of these groups is subdivided into pyrotechnic and non-pyrotechnic devices.

Daytime Non-Pyrotechnic Signals

A bright orange flag with a black square over a black circle is the simplest VDS. It is usable, of course, only in daylight. It has the advantage of being a continuous signal. A mirror can be used to good advantage on sunny days. It can attract the attention of other boaters and of aircraft from great distances. Mirrors are available with holes in their centers to aid in "aiming." In the absence of a mirror, any shiny object can be used. When another boat is in sight, an effective VDS is to extend your arms from your sides and move them up and down. Do it slowly. If you do it too fast the other people may think you are just being friendly. This simple gesture is seldom misunderstood, and requires no equipment.

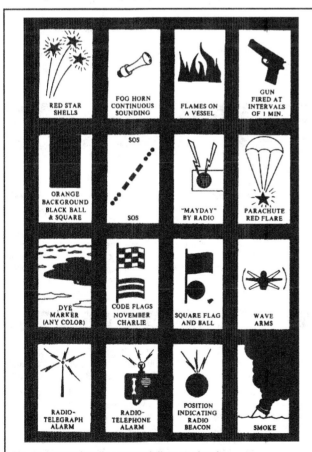

Fig. 10 Internationally accepted distress signals

Fig. 11 Moisture-protected flares should be carried onboard any vessel for use as a distress signal

Daytime Pyrotechnic Devices

Orange smoke is a useful daytime signal. Hand-held or floating smoke flares are very effective in attracting attention from aircraft. Smoke flares don't last long, and are not very effective in high wind or poor visibility. As with other pyrotechnic devices, use them only when you know there is a possibility that someone will see the display.

To be usable, smoke flares must be kept dry. Keep them in airtight containers and store them in dry places. If the "striker" is damp, dry it out before trying to ignite the device. Some pyrotechnic devices require a forceful "strike" to ignite them.

All hand-held pyrotechnic devices may produce hot ashes or slag when burning. Hold them over the side of your boat in such a way that they do not burn your hand or drip into your boat.

Nighttime Non-Pyrotechnic Signals

An electric distress light is available. This light automatically flashes the international morse code SOS distress signal (••• --- •••). Flashed four to six times a minute, it is an unmistakable distress signal. It must show that it is approved by the Coast Guard. Be sure the batteries are fresh. Dated batteries give assurance that they are current.

Under the Inland Navigation Rules, a high intensity white light flashing 50-70 times per minute is a distress signal. Therefore, use strobe lights on inland waters only for distress signals.

Nighttime Pyrotechnic Devices

Aerial and hand-held flares can be used at night or in the daytime. Obviously, they are more effective at night.

Currently, the serviceable life of a pyrotechnic device is rated at 42 months from its date of manufacture. Pyrotechnic devices are expensive. Look at their dates before you buy them. Buy them with as much time remaining as possible.

Like smoke flares, aerial and hand-held flares may fail to work if they have been damaged or abused. They will not function if they are or have been wet. Store them in dry, airtight containers in dry places. But store them where they are readily accessible.

Aerial VDSs, depending on their type and the conditions they are used in, may not go very high. Again, use them only when there is a good chance they will be seen.

A serious disadvantage of aerial flares is that they burn for only a short time; most burn for less than 10 seconds. Most parachute flares burn for less than 45 seconds. If you use a VDS in an emergency, do so carefully. Hold hand-held flares over the side of the boat when in use. Never use a road hazard flare on a boat; it can easily start a fire. Marine type flares are specifically designed to lessen risk, but they still must be used carefully.

Aerial flares should be given the same respect as firearms since they are firearms! Never point them at another person. Don't allow children to play with them or around them. When you fire one, face away from the wind. Aim it downwind and upward at an angle of about 60 degrees to the horizon. If there is a strong wind, aim it somewhat more vertically. Never fire it straight up. Before you discharge a flare pistol, check for overhead obstructions that might be damaged by the flare. An obstruction might deflect the flare to where it will cause injury or damage.

Disposal of VDS

Keep outdated flares when you get new ones. They do not meet legal requirements, but you might need them sometime, and they may work. It is illegal to fire a VDS on federal navigable waters unless an emergency exists. Many states have similar laws.

Emergency Position Indicating Radio Beacon (EPIRB)

There is no requirement for recreational boats to have EPIRBs. Some commercial and fishing vessels, though, must have them if they operate beyond the three-mile limit. Vessels carrying six or fewer passengers for hire must have EPIRBs under some circumstances when operating beyond the three-mile limit. If you boat in a remote area or offshore, you should have an EPIRB. An EPIRB is a small (about 6 to 20 in. high), battery-powered, radio transmitting buoy-like device. It is a radio transmitter and requires a license or an endorsement on your radio station license by the Federal Communications Commission (FCC). EPIRBs are either automatically activated by being immersed in water or manually by a switch.

Courtesy Marine Examinations

One of the roles of the Coast Guard Auxiliary is to promote recreational boating safety. This is why they conduct thousands of Courtesy Marine Examinations each year. The auxiliarists who do these examinations are well-trained and knowledgeable in the field.

These examinations are free and done only at the consent of boat owners. To pass the examination, a vessel must satisfy federal equipment requirements and certain additional requirements of the coast guard auxiliary. If your vessel does not pass the Courtesy Marine Examination, no report of the failure is made. Instead, you will be told what you need to correct the deficiencies. The examiner will return at your convenience to redo the examination.

If your vessel qualifies, you will be awarded a safety decal. The decal does not carry any special privileges, it simply attests to your interest in safe boating.

BOATING EQUIPMENT (NOT REQUIRED BUT RECOMMENDED)

Oar/Paddle (Second Means of Propulsion)

All boats less than 16 feet long should carry a second means of propulsion. A paddle or oar can come in handy at times. For most small boats, a spare trolling or outboard motor is an excellent idea. If you carry a spare motor, it should have its own fuel tank and starting power. If you use an electric trolling motor, it should have its own battery.

Bailing Devices

All boats should carry at least one effective manual bailing device in addition to any installed electric bilge pump. This can be a bucket, can, scoop, hand-operated pump, etc. If your battery "goes dead" it will not operate your electric pump.

First Aid Kit

◆ See Figure 12

All boats should carry a first aid kit. It should contain adhesive bandages, gauze, adhesive tape, antiseptic, aspirin, etc. Check your first aid kit from time to time. Replace anything that is outdated. It is to your advantage to know how to use your first aid kit. Another good idea would be to take a Red Cross first aid course.

Anchors

◆ See Figure 13

All boats should have anchors. Choose one of suitable size for your boat. Better still, have two anchors of different sizes. Use the smaller one in calm water or when anchoring for a short time to fish or eat. Use the larger one when the water is rougher or for overnight anchoring.

Carry enough anchor line, of suitable size, for your boat and the waters in which you will operate. If your engine fails you, the first thing you usually should do is lower your anchor. This is good advice in shallow water where you may be driven aground by the wind or water. It is also good advice in windy weather or rough water, as the anchor, when properly affixed, will usually hold your bow into the waves.

VHF-FM Radio

Your best means of summoning help in an emergency or in case of a breakdown is a VHF-FM radio. You can use it to get advice or assistance from the Coast Guard. In the event of a serious illness or injury aboard your boat, the Coast Guard can have emergency medical equipment meet you ashore.

■ **Although the VHF radio is the best way to get help, in this day and age, cell phones are a good backup source, especially for boaters on inland waters. You probably already know where you get a signal when boating, keep the phone charged, handy and off (so it doesn't bother you when boating right?). Keep phone numbers for a local dockmaster, coast guard, tow service or maritime police unit handy on board or stored in your phone directory.**

Fig. 12 Always carry an adequately stocked first aid kit on board for the safety of the crew and guests

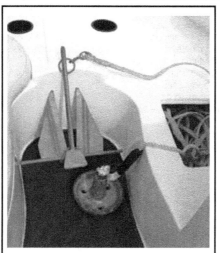

Fig. 13 Choose an anchor of sufficient weight to secure the boat without dragging

Fig. 14 Don't hesitate to spend a few extra dollars for a reliable compass

Compass

SELECTION

◆ See Figure 14

The safety of the boat and her crew may depend on her compass. In many areas, weather conditions can change so rapidly that, within minutes, a skipper may find himself socked in by a fog bank, rain squall or just poor visibility. Under these conditions, he may have no other means of keeping to his desired course except with the compass. When crossing an open body of water, his compass may be the only means of making an accurate landfall.

During thick weather when you can neither see nor hear the expected aids to navigation, attempting to run out the time on a given course can disrupt the pleasure of the cruise. The skipper gains little comfort in a chain of soundings that does not match those given on the chart for the expected area. Any stranding, even for a short time, can be an unnerving experience.

Fig. 15 The compass is a delicate instrument which should be mounted securely in a position where it can be easily observed by the helmsman

A pilot will not knowingly accept a cheap parachute. By the same token, a good boater should not accept a bargain in lifejackets, fire extinguishers, or compass. Take the time and spend the few extra dollars to purchase a compass to fit your expected needs. Regardless of what the salesman may tell you, postpone buying until you have had the chance to check more than one make and model.

Lift each compass, tilt and turn it, simulating expected motions of the boat. The compass card should have a smooth and stable reaction.

The card of a good quality compass will come to rest without oscillations about the lubber's line. Reasonable movement in your hand, comparable to the rolling and pitching of the boat, should not materially affect the reading.

INSTALLATION

◆ See Figure 15

Proper installation of the compass does not happen by accident. Make a critical check of the proposed location to be sure compass placement will permit the helmsman to use it with comfort and accuracy. First, the compass should be placed directly in front of the helmsman, and in such a position that it can be viewed without body stress as he sits or stands in a posture of relaxed alertness. The compass should be in the helmsman's zone of comfort. If the compass is too far away, he may have to bend forward to watch it; too close and he must rear backward for relief.

Second, give some thought to comfort in heavy weather and poor visibility conditions during the day and night. In some cases, the compass position may be partially determined by the location of the wheel, shift lever and throttle handle.

Third, inspect the compass site to be sure the instrument will be at least two feet from any engine indicators, bilge vapor detectors, magnetic instruments, or any steel or iron objects. If the compass cannot be placed at least 2 ft. (6 ft. would be better but on a small craft, let's get real - even 2 ft. is usually pushing it) from one of these influences, then either the compass or the other object must be moved, if first order accuracy is to be expected.

Once the compass location appears to be satisfactory, give the compass a test before installation. Hidden influences may be concealed under the cabin top, forward of the cabin aft bulkhead, within the cockpit ceiling, or in a wood-covered stanchion.

Move the compass around in the area of the proposed location. Keep an eye on the card. A magnetic influence is the only thing that will make the card turn. You can quickly find any such influence with the compass. If the influence cannot be moved away or replaced by one of non-magnetic material, test to determine whether it is merely magnetic, a small piece of iron or steel, or some magnetized steel. Bring the north pole of the compass near the object, then shift and bring the south pole near it. Both the north and south poles will be attracted if the compass is demagnetized. If the object attracts one pole and repels the other, then the compass is magnetized. If your compass needs to be demagnetized, take it to a shop equipped to do the job PROPERLY.

After you have moved the compass around in the proposed mounting area, hold it down or tape it in position. Test everything you feel might affect the compass and cause a deviation from a true reading. Rotate the wheel from hard over-to-hard over. Switch on and off all the lights, radios, radio direction finder, radio telephone, depth finder and, if installed, the shipboard intercom. Sound the electric whistle, turn on the windshield wipers, start the engine (with water circulating through the engine), work the throttle, and move the gear shift lever. If the boat has an auxiliary generator, start it.

If the card moves during any one of these tests, the compass should be relocated. Naturally, if something like the windshield wipers causes a slight deviation, it may be necessary for you to make a different deviation table to use only when certain pieces of equipment are operating. Bear in mind, following a course that is off only a degree or two for several hours can make considerable difference at the end, putting you on a reef, rock or shoal.

Check to be sure the intended compass site is solid. Vibration will increase pivot wear.

Now, you are ready to mount the compass. To prevent an error on all courses, the line through the lubber line and the compass card pivot must be exactly parallel to the keel of the boat. You can establish the fore-and-aft line of the boat with a stout cord or string. Use care to transfer this line to the compass site. If necessary, shim the base of the compass until the stile-type lubber line (the one affixed to the case and not gimbaled) is vertical when the boat is on an even keel. Drill the holes and mount the compass.

COMPASS PRECAUTIONS

◆ See Figures 16, 17 and 18

Many times an owner will install an expensive stereo system in the cabin of his boat. It is not uncommon for the speakers to be mounted on the aft bulkhead up against the overhead (ceiling). In almost every case, this position places one of the speakers in very close proximity to the compass, mounted above the ceiling.

You probably already know that a magnet is used in the operation of the speaker. Therefore, it is very likely that the speaker, mounted almost under

the compass in the cabin will have a very pronounced effect on the compass accuracy.

Consider the following test and the accompanying photographs as proof:

First, the compass was read as 190° while the boat was secure in her slip.

Next, a full can of soda in an aluminum can was placed on one side and the compass read as 204°, a good 14° off.

Next, the full can was moved to the opposite side of the compass and again a reading was observed, this time as 189°, 11° off from the original reading.

Finally, the contents of the can were consumed, the can placed on both sides of the compass with NO effect on the compass reading.

Two very important conclusions can be drawn from these tests.

• Something must have been in the contents of the can to affect the compass so drastically.

• Keep even innocent things clear of the compass to avoid any possible error in the boat's heading.

■ **Remember, a boat moving through the water at 10 knots on a compass error of just 5° will be almost 1.5 miles off course in only ONE hour. At night, or in thick weather, this could very possibly put the boat on a reef, rock or shoal with disastrous results.**

Tools & Spare Parts

◆ See Figures 19 and 20

Carry a few tools and some spare parts, and learn how to make minor repairs. Many search and rescue cases are caused by minor breakdowns that boat operators could have repaired. Carry spare parts such as propellers, fuses or basic ignition components (like spark plugs, wires or even ignition coils) and the tools necessary to install them.

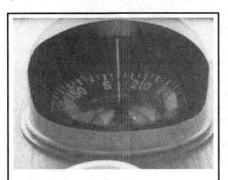

Fig. 16 This compass is giving an accurate reading, right?

Fig. 17 . . .well think again, as seemingly innocent objects may cause serious problems. . .

Fig. 18 . . .a compass reading off by just a few degrees could lead to disaster

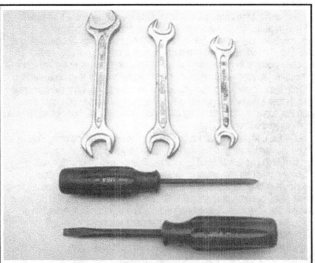

Fig. 19 A few wrenches, a screwdriver and maybe a pair of pliers can be very helpful to make emergency repairs

Fig. 20 A flashlight with a fresh set of batteries is handy when repairs are needed at night. It can also double as a signaling device

SAFETY IN SERVICE

It is virtually impossible to anticipate all of the hazards involved with maintenance and service, but care and common sense will prevent most accidents.

The rules of safety for mechanics range from "don't smoke around gasoline," to "use the proper tool(s) for the job." The trick to avoiding injuries is to develop safe work habits and to take every possible precaution. Whenever you are working on your boat, pay attention to what you are doing. The more you pay attention to details and what is going on around you, the less likely you will be to be hurt yourself or damage your boat.

Do's

• Do keep a fire extinguisher and first aid kit handy.
• Do wear safety glasses or goggles when cutting, drilling, grinding or prying, even if you have 20-20 vision. If you wear glasses for the sake of vision, wear safety goggles over your regular glasses.
• Do shield your eyes whenever you work around the battery. Batteries contain sulfuric acid. In case of contact with the eyes or skin, flush the area with water or a mixture of water and baking soda; then seek immediate medical attention.
• Do use adequate ventilation when working with any chemicals or hazardous materials.
• Do disconnect the negative battery cable when working on the electrical system. The secondary ignition system contains EXTREMELY HIGH VOLTAGE. In some cases it can even exceed 50,000 volts. Furthermore, an accidental attempt to start the engine could cause the propeller or other components to rotate suddenly causing a potentially dangerous situation.
• Do follow manufacturer's directions whenever working with potentially hazardous materials. Most chemicals and fluids are poisonous if taken internally.
• Do properly maintain your tools. Loose hammerheads, mushroomed punches and chisels, frayed or poorly grounded electrical cords, excessively worn screwdrivers, spread wrenches (open end), cracked sockets, or slipping ratchets can cause accidents.
• Likewise, keep your tools clean; a greasy wrench can slip off a bolt head, ruining the bolt and often harming your knuckles in the process.
• Do use the proper size and type of tool for the job at hand. Do select a wrench or socket that fits the nut or bolt. The wrench or socket should sit straight, not cocked.

• Do, when possible, pull on a wrench handle rather than push on it, and adjust your stance to prevent a fall.
• Do be sure that adjustable wrenches are tightly closed on the nut or bolt and pulled so that the force is on the side of the fixed jaw. Better yet, avoid the use of an adjustable if you have a fixed wrench that will fit.
• Do strike squarely with a hammer; avoid glancing blows.
• Do use common sense whenever you work on your boat or motor. If a situation arises that doesn't seem right, sit back and have a second look. It may save an embarrassing moment or potential damage to your beloved boat.

Don'ts

• Don't run the engine in an enclosed area or anywhere else without proper ventilation - EVER! Carbon monoxide is poisonous; it takes a long time to leave the human body and you can build up a deadly supply of it in your system by simply breathing in a little every day. You may not realize you are slowly poisoning yourself.
• Don't work around moving parts while wearing loose clothing. Short sleeves are much safer than long, loose sleeves. Hard-toed shoes with neoprene soles protect your toes and give a better grip on slippery surfaces. Jewelry, watches, large belt buckles, or body adornment of any kind is not safe working around any craft or vehicle. Long hair should be tied back under a hat.
• Don't use pockets for toolboxes. A fall or bump can drive a screwdriver deep into your body. Even a rag hanging from your back pocket can wrap around a spinning shaft.
• Don't smoke when working around gasoline, cleaning solvent or other flammable material.
• Don't smoke when working around the battery. When the battery is being charged, it gives off explosive hydrogen gas. Actually, you shouldn't smoke anyway, it's bad for you. Instead, save the cigarette money and put it into your boat!
• Don't use gasoline to wash your hands; there are excellent soaps available. Gasoline contains dangerous additives that can enter the body through a cut or through your pores. Gasoline also removes all the natural oils from the skin so that bone dry hands will suck up oil and grease.
• Don't use screwdrivers for anything other than driving screws! A screwdriver used as a prying tool can snap when you least expect it, causing injuries. At the very least, you'll ruin a good screwdriver.

TROUBLESHOOTING

Troubleshooting can be defined as a methodical process during which one discovers what is causing a problem with engine operation. Although it is often a feared process to the uninitiated, there is no reason to believe that you cannot figure out what is wrong with a motor, as long as you follow a few basic rules.

To begin with, troubleshooting must be systematic. Haphazardly testing one component, then another, **might** uncover the problem, but it will more likely waste a lot of time. True troubleshooting starts by defining the problem and performing systematic tests to eliminate the largest and most likely causes first.

Start all troubleshooting by eliminating the most basic possible causes. Begin with a visual inspection of the boat and motor. If the engine won't crank, make sure that the kill switch or safety lanyard is in the proper position. Make sure there is fuel in the tank and the fuel system is primed before condemning the carburetor or fuel injection system. On electric start motors, make sure there are no blown fuses, the battery is fully charged, and the cable connections (at both ends) are clean and tight before suspecting a bad starter, solenoid or switch.

The majority of problems that occur suddenly can be fixed by simply identifying the one small item that brought them on. A loose wire, a clogged passage or a broken component can cause a lot of trouble and are often the cause of a sudden performance problem.

The next most basic step in troubleshooting is to test systems before components. For example, if the engine doesn't crank on an electric start motor, determine if the battery is in good condition (fully charged and properly connected) before testing the starting system. If the engine cranks, but doesn't start, you know already know the starting system and battery (if it cranks fast enough) are in good condition, now it is time to look at the ignition or fuel systems. Once you've isolated the problem to a particular

system, follow the troubleshooting/testing procedures in the section for that system to test either subsystems (if applicable, for example: the starter circuit) or components (starter solenoid).

Basic Operating Principles

UNDERSTANDING ENGINE OPERATION

◆ **See Figures 21 and 22**

Before attempting to troubleshoot a problem with your motor, it is important that you understand how it operates. Once normal engine or system operation is understood, it will be easier to determine what might be causing the trouble or irregular operation in the first place. System descriptions are found throughout this service, but the basic mechanical operating principles for both 2-stroke engines (like most of the outboards covered here) and 4-stroke engines (like some outboards and like your car) are given here. A basic understanding of both types of engines is useful not only in understanding and troubleshooting your outboard, but also for dealing with other motors in your life.

All motors covered by this service (and probably MOST of the motors you own) operate according to the Otto cycle principle of engine operation. This means that all motors follow the stages of intake, compression, power and exhaust. But, the difference between a 2- and 4-stroke motor is in how many times the piston moves up and down within the cylinder to accomplish this. On 2-stroke motors (as the name suggests) the four cycles take place in 2 movements (one up and one down) of the piston. Again, as the name suggests, the cycles take place in 4 movements of the piston for 4-stroke motors.

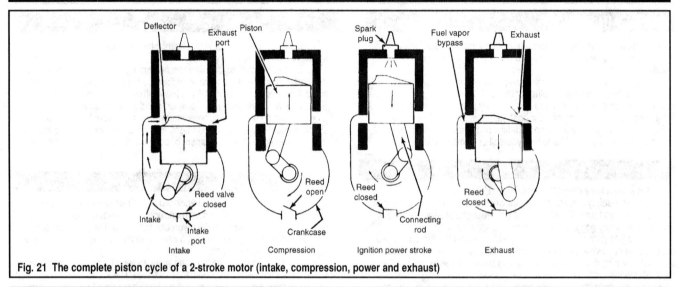

Fig. 21 The complete piston cycle of a 2-stroke motor (intake, compression, power and exhaust)

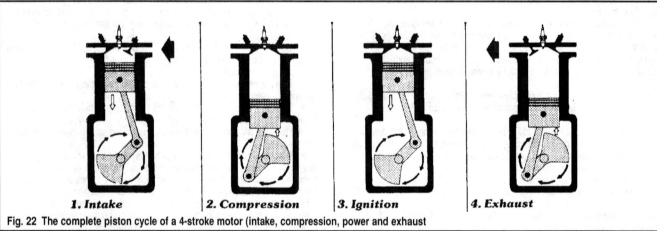

Fig. 22 The complete piston cycle of a 4-stroke motor (intake, compression, power and exhaust

2-STROKE MOTORS

The 2-stroke engine differs in several ways from a conventional four-stroke (automobile or marine) engine.

1. The intake/exhaust method by which the fuel-air mixture is delivered to the combustion chamber.
2. The complete lubrication system.
3. The frequency of the power stroke.

Let's discuss these differences briefly (and compare 2-stroke engine operation with 4-stroke engine operation.)

Intake/Exhaust

◆ **See Figures 23 thru 26**

Two-stroke engines utilize an arrangement of port openings to admit fuel to the combustion chamber and to purge the exhaust gases after burning has been completed. The ports are located in a precise pattern in order for them to be open and closed off at an exact moment by the piston as it moves up and down in the cylinder. The exhaust port is located slightly higher than the fuel intake port. This arrangement opens the exhaust port first as the piston starts downward and therefore, the exhaust phase begins a fraction of a second before the intake phase.

Actually, the intake and exhaust ports are spaced so closely together that both open almost simultaneously. For this reason many OMC (Johnson/Evinrude) 2-stroke engines utilize deflector-type pistons. This design of the piston top serves two purposes very effectively.

First, it creates turbulence when the incoming charge of fuel enters the combustion chamber. This turbulence results in a more complete burning of the fuel than if the piston top were flat. The second effect of the deflector-type piston crown is to force the exhaust gases from the cylinder more rapidly. Although this configuration is used in many OMC outboards, a few of the larger twins (the 50 hp, 55 hp and 60 hp, as well as the 1983 and later 40 hp motors) covered here are Loop (L) charged.

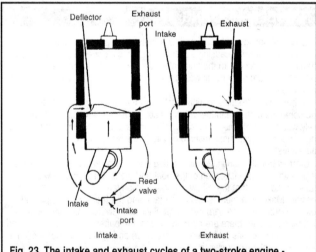

Fig. 23 The intake and exhaust cycles of a two-stroke engine - Cross flow (CV) design shown

Loop charged motors, or as they are commonly called "loopers", differ in how the air/fuel charge is introduced to the combustion chamber. Instead of the charge flowing across the top of the piston from one side of the cylinder to the other (CV) the use a looping action on top of the piston as the charge is forced through irregular shaped openings cut in the piston's skirt. In a LV motor, the charge is forced out from the crankcase by the downward motion of the piston, through the irregular shaped openings and transferred upward by long, deep grooves in the cylinder wall. The charge completes its looping action by entering the combustion chamber, just above the piston, where the

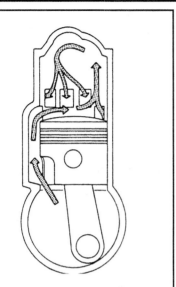

Fig. 24 Cross-sectional view of a typical loop-charged cylinder, showing charge flow while piston is moving downward

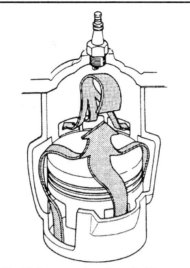

Fig. 25 Cutaway view of a typical loop-charged cylinder, depicting exhaust leaving the cylinder as the charge enters through 3 ports in the piston

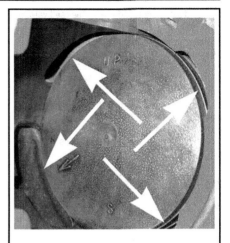

Fig. 26 The combustion chamber of a typical looper, notice the piston is far enough down the cylinder bore to reveal intake and exhaust ports

upward motion of the piston traps it in the chamber and compresses it for optimum ignition power.

Unlike the knife-edged deflector top pistons used in CV motors, the piston domes on Loop motors are relatively flat.

These systems of intake and exhaust are in marked contrast to individual intake and exhaust valve arrangement employed on 4-stroke engines (and the mechanical methods of opening and closing these valves).

■ It should be noted here that there are some 2-stroke engines that utilize a mechanical valve train, though it is very different from the valve train employed by most 4-stroke motors. Rotary 2-stroke engines use a circular valve or rotating disc that contains a port opening around part of one edge of the disc. As the engine (and disc) turns, the opening aligns with the intake port at and for a predetermined amount of time, closing off the port again as the opening passes by and the solid portion of the disc covers the port.

Lubrication

A 2-stroke engine is lubricated by mixing oil with the fuel. Therefore, various parts are lubricated as the fuel mixture passes through the crankcase and the cylinder. In contrast, four-stroke engines have a crankcase containing oil. This oil is pumped through a circulating system and returned to the crankcase to begin the routing again.

Power Stroke

◆ **See Figures 21 thru 27**

The combustion cycle of a 2-stroke engine has four distinct phases.
1. Intake
2. Compression
3. Power
4. Exhaust

The 4 phases of the cycle are accomplished with each up and down stroke of the piston, and the power stroke occurs with each complete revolution of the crankshaft. Compare this system with a 4-stroke engine. A separate stroke of the piston is required to accomplish each phase of the cycle and the power stroke occurs only every other revolution of the crankshaft. Stated another way, 2 revolutions of the 4-stroke engine crankshaft are required to complete 1 full cycle, the 4 phases.

Physical Laws

The 2-stroke engine is able to function because of 2 very simple physical laws.

One: Gases will flow from an area of high pressure to an area of lower pressure. A tire blowout is an example of this principle. The high-pressure air escapes rapidly if the tube is punctured.

Two: If a gas is compressed into a smaller area, the pressure increases, and if a gas expands into a larger area, the pressure is decreased.

If these 2 laws are kept in mind, the operation of the 2-stroke engine will be easier understood.

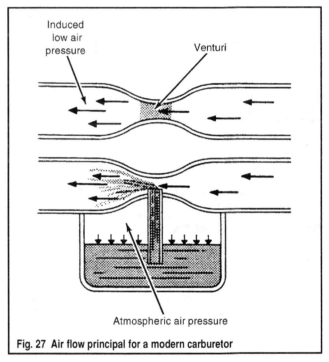

Fig. 27 Air flow principal for a modern carburetor

Actual Operation

Beginning with the piston approaching top dead center on the compression stroke: the intake and exhaust ports are physically closed (blocked) by the piston. During this stroke, the reed valve is open (because as the piston moves upward, the crankcase volume increases, which reduces the crankcase pressure to less than the outside atmosphere (creates a vacuum under the piston). The spark plug fires; the compressed fuel-air mixture is ignited; and the power stroke begins.

As the piston moves downward on the power stroke, the combustion chamber is filled with burning gases. As the exhaust port is uncovered, the gases, which are under great pressure, escape rapidly through the exhaust ports. The piston continues its downward movement. Pressure within the crankcase (again, under the piston) increases, closing the reed valves against their seats. The crankcase then becomes a sealed chamber so the air-fuel mixture becomes compressed (pressurized) and ready for delivery to the combustion chamber. As the piston continues to move downward, the intake port is uncovered. The fresh fuel mixture rushes through the intake port into the combustion chamber striking the top of the piston where it is deflected along the cylinder wall. The reed valve remains closed until the piston moves upward again.

When the piston begins to move upward on the compression stroke, the reed valve opens because the crankcase volume has been increased, reducing crankcase pressure to less than the outside atmosphere. The intake and exhaust ports are closed and the fresh fuel charge is compressed inside the combustion chamber.

Pressure in the crankcase (beneath the piston) decreases as the piston moves upward and a fresh charge of air flows through the carburetor picking up fuel. As the piston approaches top dead center, the spark plug ignites the air-fuel mixture, the power stroke begins and 1 complete Otto cycle has been completed.

4-STROKE MOTORS

◆ See Figure 22

The 4-stroke motor may be easier to understand for some people either because of its prevalence in automobile and street motorcycle motors today or perhaps because each of the four strokes corresponds to one distinct phase of the Otto cycle. Essentially, a 4-stroke engine completes 1 Otto cycle of intake, compression, ignition/power and exhaust using 2 full revolutions of the crankshaft and f4 distinct movements of the piston (down, up, down and up).

Intake

The intake stroke begins with the piston near the top of its travel. As crankshaft rotation begins to pull the piston downward, the exhaust valve closes and the intake opens. As volume of the combustion chamber increases, a vacuum is created that draws in the air/fuel mixture from the intake manifold.

Compression

Once the piston reaches the bottom of its travel, crankshaft rotation will begin to force it upward. At this point the intake valve closes. As the piston rises in the bore, the volume of the sealed combustion chamber (both intake and exhaust valves are closed) decreases and the air/fuel mixture is compressed. This raises the temperature and pressure of the mixture and increases the amount of force generated by the expanding gases during the Ignition/Power stroke.

Ignition/Power

As the piston approaches top dead center (TDC, the highest point of travel in the bore), the spark plug will fire, igniting the air/fuel mixture. The resulting combustion of the air/fuel mixture forces the piston downward, rotating the crankshaft (causing other pistons to move in other phases/strokes of the Otto cycle on multi-cylinder motors).

Exhaust

As the piston approaches the bottom of the Ignition/Power stroke, the exhaust valve opens. When the piston begins its upward path of travel once again, any remaining unburned gasses are forced out through the exhaust valve. This completes 1 Otto cycle, which begins again as the piston passes top dead center, the intake valve opens and the Intake stroke starts.

COMBUSTION

Whether we are talking about a 2- or 4-stroke engine, all Otto cycle, internal combustion engines require three basic conditions to operate properly,

1. Compression
2. Ignition (Spark)
3. Fuel

A lack of any one of these conditions will prevent the engine from operating. A problem with any one of these will manifest itself in hard-starting or poor performance.

Compression

An engine that has insufficient compression will not draw an adequate supply of air/fuel mixture into the combustion chamber and, subsequently, will not make sufficient power on the power stroke. A lack of compression in just 1 cylinder of a multi-cylinder motor will cause the motor to stumble or run irregularly.

But, keep in mind that a sudden change in compression is unlikely in 2-stroke motors (unless something major breaks inside the crankcase, but that would usually be accompanied by other symptoms such as a loud noise when it occurred or noises during operation). On 4-stroke motors, a sudden change in compression is also unlikely, but could occur if the timing belt or chain was to suddenly break. Remember that the timing belt/chain is used to synchronize the valve train with the crankshaft. If the valve train suddenly ceases to turn, some intake and some exhaust valves will remain open, relieving compression in that cylinder.

Ignition (Spark)

Traditionally, the ignition system is the weakest link in the chain of conditions necessary for engine operation. Spark plugs may become worn or fouled, wires will deteriorate allowing arcing or misfiring, and poor connections can place an undue load on coils leading to weak spark or even a failed coil. The most common question asked by a technician under a no-start condition is: "Do I have spark and fuel?" (as they've already determined that they have compression).

A quick visual inspection of the spark plug(s) will answer the question as to whether or not the plug(s) is/are worn or fouled. While the engine is shut **OFF** a physical check of the connections could show a loose primary or secondary ignition circuit wire. An obviously physically damaged wire may also be an indication of system problems and certainly encourages one to inspect the related system more closely.

If nothing is turned up by the visual inspection, perform the Spark Test provided in the Ignition System section to determine if the problem is a lack of or a weak spark. If the problem is not compression or spark, it's time to look at the fuel system.

Fuel

If compression and spark is present (and within spec), but the engine won't start or won't run properly, the only remaining condition to fulfill is fuel. As usual, start with the basics. Is the fuel tank full? Is the fuel stale? If the engine has not been run in some time (a matter of months, not weeks) there is a good chance that the fuel is stale and should be properly disposed of and replaced.

■ **Depending on how stale or contaminated (with moisture) the fuel is, it may be burned in an automobile or in yard equipment, though it would be wise to mix it well with a much larger supply of fresh gasoline to prevent moving your driveability problems to that motor. But it is better to get the lawn tractor stuck on stale gasoline than it would be to have your boat motor quit in the middle of the bay or lake.**

For hard starting motors, is the choke or primer system operating properly. Remember that the choke/prime should only be used for **cold** starts. A true cold start is really only the first start of the day, but it may be applicable to subsequent starts on cooler days, if the engine sat for more than a few hours and completely cooled off since the last use. Applying the primer to the motor for a hot start may flood the engine, preventing it from starting properly. One method to clear a flood is to crank the motor while the engine is at wide-open throttle (allowing the maximum amount of air into the motor to compensate for the excess fuel). But, keep in mind that the throttle should be returned to idle immediately upon engine start-up to prevent damage from over-revving.

Fuel delivery and pressure should be checked before delving into the carburetor(s) or fuel injection system. Make sure there are no clogs in the fuel line or vacuum leaks that would starve the motor of fuel.

Make sure that all other possible problems have been eliminated before touching the carburetor. It is rare that a carburetor will suddenly require an adjustment in order for the motor to run properly. It is much more likely that an improperly stored motor (one stored with untreated fuel in the carburetor) would suffer from one or more clogged carburetor passages sometime after shortly returning to service. Fuel will evaporate over time, leaving behind gummy deposits. If untreated fuel is left in the carburetor for some time (again typically months more than weeks), the varnish left behind by evaporating fuel will likely clog the small passages of the carburetor and cause problems with engine performance. If you suspect this, remove and disassemble the carburetor following procedures under Fuel System.

SHOP EQUIPMENT

Safety Tools

WORK GLOVES

◆ See Figure 28

Unless you think scars on your hands are cool, enjoy pain and like wearing bandages, get a good pair of work gloves. Canvas or leather gloves are the best. And yes, we realize that there are some jobs involving small parts that can't be done while wearing work gloves. These jobs are not the ones usually associated with hand injuries.

A good pair of rubber gloves (such as those usually associated with dish washing) or vinyl gloves is also a great idea. There are some liquids such as solvents and penetrants that don't belong on your skin. Avoid burns and rashes. Wear these gloves.

And lastly, an option. If you're tired of being greasy and dirty all the time, go to the drug store and buy a box of disposable latex gloves like medical professionals wear. You can handle greasy parts, perform small tasks, wash parts, etc. all without getting dirty! These gloves take a surprising amount of abuse without tearing and aren't expensive. Note however, that some people are allergic to the latex or the powder used inside some gloves, so pay attention to what you buy.

EYE & EAR PROTECTION

◆ See Figures 29 and 30

Don't begin any job without a good pair of work goggles or impact resistant glasses! When doing any kind of work, it's all too easy to avoid eye injury through this simple precaution. And don't just buy eye protection and leave it on the shelf. Wear it all the time! Things have a habit of breaking, chipping, splashing, spraying, splintering and flying around. And, for some reason, your eye is always in the way!

If you wear vision-correcting glasses as a matter of routine, get a pair made with polycarbonate lenses. These lenses are impact resistant and are available at any optometrist.

Often overlooked is hearing protection. Engines and power tools are noisy! Loud noises damage your ears. It's as simple as that! The simplest and cheapest form of ear protection is a pair of noise-reducing ear plugs. Cheap insurance for your ears! And, they may even come with their own, cute little carrying case.

More substantial, more protection and more money is a good pair of noise reducing earmuffs. They protect from all but the loudest sounds. Hopefully those are sounds that you'll never encounter since they're usually associated with disasters.

WORK CLOTHES

Everyone has "work clothes." Usually these consist of old jeans and a shirt that has seen better days. That's fine. In addition, a denim work apron is a nice accessory. It's rugged, can hold some spare bolts, and you don't feel bad wiping your hands or tools on it. That's what it's for.

When working in cold weather, a one-piece, thermal work outfit is invaluable. Most are rated to below freezing temperatures and are ruggedly constructed. Just look at what local marine mechanics are wearing and that should give you a clue as to what type of clothing is good.

Chemicals

TYPES OF CHEMICALS

There is a whole range of chemicals that you'll find handy for maintenance and repair work. The most common types are: lubricants, penetrants and sealers. Keep these handy. There are also many chemicals that are used for detailing or cleaning.

When a particular chemical is not being used, keep it capped, upright and in a safe place. These substances may be flammable, may be irritants or might even be caustic and should always be stored properly, used properly and handled with care. Always read and follow all label directions and be sure to wear hand and eye protection!

LUBRICANTS & PENETRANTS

◆ See Figure 31

Anti-seize is used to coat certain fasteners prior to installation. This can be especially helpful when two dissimilar metals are in contact (to help prevent corrosion that might lock the fastener in place). This is a good practice on a lot of different fasteners, BUT, NOT on any fastener that might vibrate loose causing a problem. If anti-seize is used on a fastener, it should be checked periodically for proper tightness.

Lithium grease, chassis lube, silicone grease or a synthetic brake caliper grease can all be used pretty much interchangeably. All can be used for coating rust-prone fasteners and for facilitating the assembly of parts that are a tight fit. Silicone and synthetic greases are the most versatile.

■ Silicone dielectric grease is a non-conductor that is often used to coat the terminals of wiring connectors before fastening them. It may sound odd to coat metal portions of a terminal with something that won't conduct electricity, but here is it how it works. When the connector is fastened the metal-to-metal contact between the terminals will displace the grease (allowing the circuit to be completed). The grease that is displaced will then coat the non-contacted surface and the cavity around the terminals, SEALING them from atmospheric moisture that could cause corrosion.

Silicone spray is a good lubricant for hard-to-reach places and parts that shouldn't be gooped up with grease.

Penetrating oil may turn out to be one of your best friends when taking something apart that has corroded fasteners. Not only can they make a job easier, they can really help to avoid broken and stripped fasteners. The most familiar penetrating oils are Liquid Wrench® and WD-40®. A newer penetrant, PB Blaster® works very well (and has become a mainstay in our shops). These products have hundreds of uses. For your purposes, they are vital!

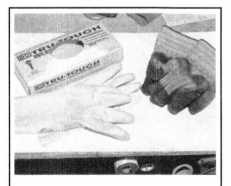

Fig. 28 Three different types of work gloves. The box contains latex gloves

Fig. 29 Don't begin major repairs without a pair of goggles for your eyes and earmuffs to protect your hearing

Fig. 30 Things have a habit of, splashing, spraying, splintering and flying around during repairs

Before disassembling any part, check the fasteners. If any appear rusted, soak them thoroughly with the penetrant and let them stand while you do something else (for particularly rusted or frozen parts you may need to soak them a few days in advance). This simple act can save you hours of tedious work trying to extract a broken bolt or stud.

SEALANTS

◆ See Figures 32 and 33

Sealants are an indispensable part for certain tasks, especially if you are trying to avoid leaks. The purpose of sealants is to establish a leak-proof bond between or around assembled parts. Most sealers are used in conjunction with gaskets, but some are used instead of conventional gasket material.

The most common sealers are the non-hardening types such as Permatex® No.2 or its equivalents. These sealers are applied to the mating surfaces of each part to be joined, then a gasket is put in place and the parts are assembled.

■ A sometimes overlooked use for sealants like RTV is on the threads of vibration prone fasteners.

One very helpful type of non-hardening sealer is the "high tack" type. This type is a very sticky material that holds the gasket in place while the parts are being assembled. This stuff is really a good idea when you don't have enough hands or fingers to keep everything where it should be.

The stand-alone sealers are the Room Temperature Vulcanizing (RTV) silicone gasket makers. On some engines, this material is used instead of a gasket. In those instances, a gasket may not be available or, because of the shape of the mating surfaces, a gasket shouldn't be used. This stuff, when used in conjunction with a conventional gasket, produces the surest bonds.

RTV does have its limitations though. When using this material, you will have a time limit. It starts to set-up within 15 minutes or so, so you have to assemble the parts without delay. In addition, when squeezing the material out of the tube, don't drop any glops into the engine. The stuff will form and set and travel around a cooling passage, possibly blocking it. Also, most types are not fuel-proof. Check the tube for all cautions.

CLEANERS

◆ See Figures 34 and 35

There are two basic types of cleaners on the market today: parts cleaners and hand cleaners. The parts cleaners are for the parts; the hand cleaners are for you. They are **not** interchangeable.

There are many good, non-flammable, biodegradable parts cleaners on the market. These cleaning agents are safe for you, the parts and the environment. Therefore, there is no reason to use flammable, caustic or toxic substances to clean your parts or tools.

As far as hand cleaners go; the waterless types are the best. They have always been efficient at cleaning, but they used to all leave a pretty smelly odor. Recently though, most of them have eliminated the odor and added stuff that actually smells good. Make sure that you pick one that contains lanolin or some other moisture-replenishing additive. Cleaners not only remove grease and oil but also skin oil.

■ Most women already know to use a hand lotion when you're all cleaned up. It's okay. Real men DO use hand lotion too! Believe it or not, using hand lotion *before* your hands are dirty will actually make them easier to clean when you're finished with a dirty job. Lotion seals your hands, and keeps dirt and grease from sticking to your skin.

Fig. 31 Keep a supply of anti-seize, penetrating oil, lithium grease, electronic cleaner4 and silicone spray

Fig. 32 Sealants are essential for preventing leaks

Fig. 33 On some engines, RTV is used instead of gasket material to seal components

Fig. 34 Citrus hand cleaners not only work well, but they smell pretty good too. Choose one with pumice for added cleaning power

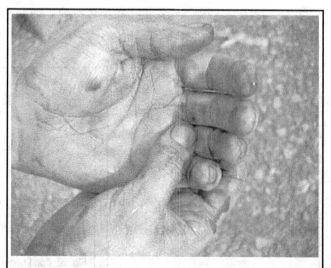

Fig. 35 The use of hand lotion seals your hands and keeps dirt and grease from sticking to your skin

TOOLS

◆ See Figure 36

Tools; this subject could fill a completely separate manual. The first thing you will need to ask yourself, is just how involved do you plan to get. If you are serious about maintenance and repair you will want to gather a quality set of tools to make the job easier, and more enjoyable. BESIDES, TOOLS ARE FUN!!!

Almost every do-it-yourselfer loves to accumulate tools. Though most find a way to perform jobs with only a few common tools, they tend to buy more over time, as money allows. So gathering the tools necessary for maintenance or repair does not have to be an expensive, overnight proposition.

When buying tools, the saying "You get what you pay for ..." is absolutely true! Don't go cheap! Any hand tool that you buy should be drop forged and/or chrome vanadium. These two qualities tell you that the tool is strong enough for the job. With any tool, go with a name that you've heard of before, or, that is recommended buy your local professional retailer. Let's go over a list of tools that you'll need.

Even though most of the world today uses the metric system, many parts for these motors and some American-built aftermarket accessories use standard fasteners. So, accumulate your tools accordingly. Any good DIYer should have a decent set of both U.S. and metric measure tools.

■ **Don't be confused by terminology. Most advertising refers to "SAE and metric", or "standard and metric." Both are misnomers. The Society of Automotive Engineers (SAE) did not invent the English system of measurement; the English did. The SAE likes metrics just fine. Both English (U.S.) and metric measurements are SAE approved. Also, the current "standard" measurement IS metric. So, if it's not metric, it's U.S. measurement.**

Hand Tools

SOCKET SETS

◆ See Figures 37 and 43

aSocket sets are the most basic hand tools necessary for repair and maintenance work. For our purposes, socket sets come in three drive sizes: 1/4 in., 3/8 in. and 1/2 in. Drive size refers to the size of the drive lug on the ratchet, breaker bar or speed handle.

A 3/8 in. set is probably the most versatile set in any mechanic's toolbox. It allows you to get into tight places that the larger drive ratchets can't and gives you a range of larger sockets that are still strong enough for heavy-duty work. The socket set that you'll need should range in sizes from 1/4 in. through 1 in. for standard fasteners, and a 6mm through 19mm for metric fasteners.

You'll need a good 1/2 in. set since this size drive lug assures that you won't break a ratchet or socket on large or heavy fasteners. Also, torque wrenches with a torque scale high enough for larger fasteners are usually 1/2 in. drive.

Plus, 1/4 in. drive sets can be very handy in tight places. Though they usually duplicate functions of the 3/8 in. set, 1/4 in. drive sets are easier to use for smaller bolts and nuts.

As for the sockets themselves, they come in shallow (standard) and deep lengths as well as 6 or 12 point. The 6 and 12 points designation refers to how many sides are in the socket itself. Each has advantages. The 6 point socket is stronger and less prone to slipping which would strip a bolt head or nut. 12 point sockets are more common, usually less expensive and can operate better in tight places where the ratchet handle can't swing far.

Standard length sockets are good for just about all jobs, however, some stud-head bolts, hard-to-reach bolts, nuts on long studs, etc., require the deep sockets.

Most marine manufacturers use recessed hex-head fasteners to retain many of the engine parts. These fasteners require a socket with a hex shaped driver or a large sturdy hex key. To help prevent torn knuckles, we would recommend that you stick to the sockets on any tight fastener and leave the hex keys for lighter applications. Hex driver sockets are available individually or in sets just like conventional sockets.

More and more, manufacturers are using Torx® head fasteners, which were once known as tamper resistant fasteners (because many people did not have tools with the necessary odd driver shape). Since Torx® fasteners have become commonplace in many DIYer tool boxes, manufacturers designed newer tamper resistant fasteners that are essentially Torx® head bolts that contain a small protrusion in the center (requiring the driver to contain a small hole to slide over the protrusion. Tamper resistant fasteners are often used where the manufacturer would prefer only knowledgeable mechanics or advanced Do-It-Yourselfers (DIYers) work.

TORQUE WRENCHES

◆ See Figures 44 and 48

In most applications, a torque wrench can be used to ensure proper installation of a fastener. Torque wrenches come in various designs and most stores will carry a variety to suit your needs. A torque wrench should be used any time you have a specific torque value for a fastener. Keep in mind that because there is no worldwide standardization of fasteners, so charts or figure found in each repair section refer to the manufacturer's fasteners. Any general guideline charts that you might come across based on fastener size (they are sometimes included in a repair manual or with torque wrench packaging) should be used with caution. Just keep in mind that if you are using the right tool for the job, you should not have to strain to tighten a fastener.

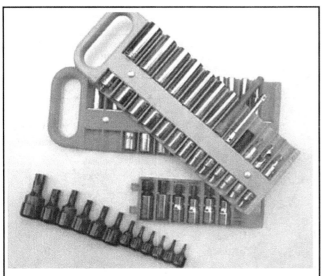

Fig. 36 Socket holders, especially the magnetic type, are handy items to keep tools in order

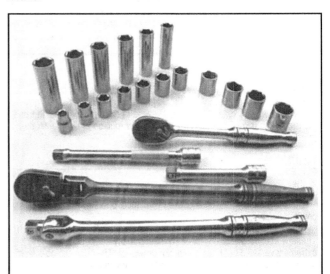

Fig. 37 A 3/8 in. socket set is probably the most versatile tool in any mechanic's tool box

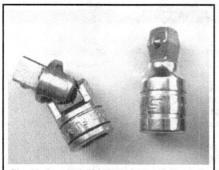

Fig. 38 A swivel (U-joint) adapter (left), a wobble-head adapter (center) and a 1/2 in.-to-3/8 in. adapter (right)

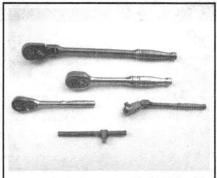

Fig. 39 Ratchets come in all sizes and configurations from rigid to swivel-headed

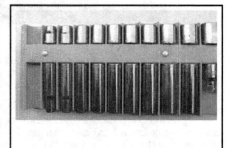

Fig. 40 Shallow sockets (top) are good for most jobs. But, some bolts require deep sockets (bottom)

Fig. 41 Hex-head fasteners require a socket with a hex shaped driver

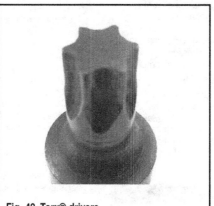

Fig. 42 Torx® drivers . . .

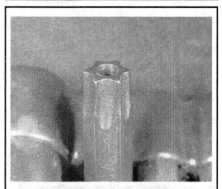

Fig. 43 . . . and tamper resistant drivers are required to remove special fasteners

Beam Type

The beam type torque wrench is one of the most popular styles in use. If used properly, it can be the most accurate also. It consists of a pointer attached to the head that runs the length of the flexible beam (shaft) to a scale located near the handle. As the wrench is pulled, the beam bends and the pointer indicates the torque using the scale.

Click (Breakaway) Type

Another popular torque wrench design is the click type. The clicking mechanism makes achieving the proper torque easy and most use a ratcheting head for ease of bolt installation. To use the click type wrench you pre-adjust it to a torque setting. Once the torque is reached, the wrench has a reflex signaling feature that causes a momentary breakaway of the torque wrench body, sending an impulse to the operator's hand. But be careful, as continuing the turn the wrench after the momentary release will increase torque on the fastener beyond the specified setting.

BREAKER BARS

◆ See Figure 49

Breaker bars are long handles with a drive lug. Their main purpose is to provide extra turning force when breaking loose tight bolts or nuts. They come in all drive sizes and lengths. Always take extra precautions and use the proper technique when using a breaker bar (pull on the bar, don't push, to prevent skinned knuckles).

WRENCHES

◆ See Figures 50 and 54

Basically, there are 3 kinds of fixed wrenches: open end, box end, and combination.

Open-end wrenches have 2-jawed openings at each end of the wrench. These wrenches are able to fit onto just about any nut or bolt. They are extremely versatile but have one major drawback. They can slip on a worn or rounded bolt head or nut, causing bleeding knuckles and a useless fastener.

■ Line wrenches are a special type of open-end wrench designed to fit onto more of the fastener than standard open-end wrenches, thus reducing the chance of rounding the corners of the fastener.

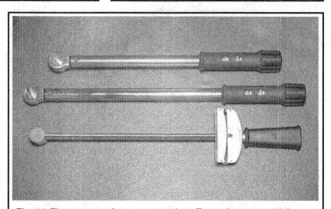

Fig. 44 Three types of torque wrenches. Top to bottom: a 3/8 in. drive beam type that reads in inch lbs., a 1/2 in. drive clicker type and a 1/2 in. drive beam type

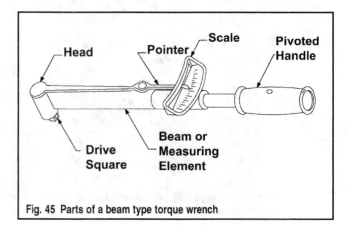

Fig. 45 Parts of a beam type torque wrench

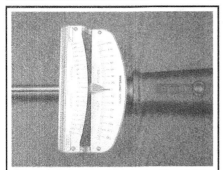

Fig. 46 A beam type torque wrench consists of a pointer attached to the head that runs the length of the flexible beam (shaft) to a scale located near the handle

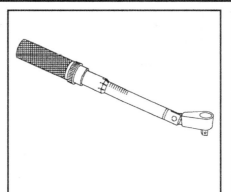

Fig. 47 A click type or breakaway torque wrench - note this one has a pivoting head

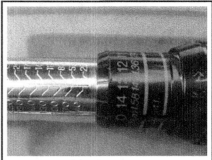

Fig. 48 Setting the torque on a click type wrench involves turning the handle until the specification appears on the dial

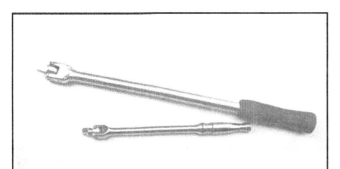

Fig. 49 Breaker bars are great for loosening large or stuck fasteners

Box-end wrenches have a 360° circular jaw at each end of the wrench. They come in both 6 and 12 point versions just like sockets and each type has some of the same advantages and disadvantages as sockets.

Combination wrenches have the best of both. They have a 2-jawed open end and a box end. These wrenches are probably the most versatile.

As for sizes, you'll probably need a range similar to that of the sockets, about 1/4 in. through 1 in. for standard fasteners, or 6mm through 19mm for metric fasteners. As for numbers, you'll need 2 of each size, since, in many instances, one wrench holds the nut while the other turns the bolt. On most fasteners, the nut and bolt are the same size so having two wrenches of the same size comes in handy.

■ Although you will typically just need the sizes we specified, there are some exceptions. Occasionally you will find a nut that is larger. For these, you will need to buy ONE expensive wrench or a very large adjustable. Or you can always just convince the spouse that we are talking about *safety* here and buy a whole (read expensive) large wrench set.

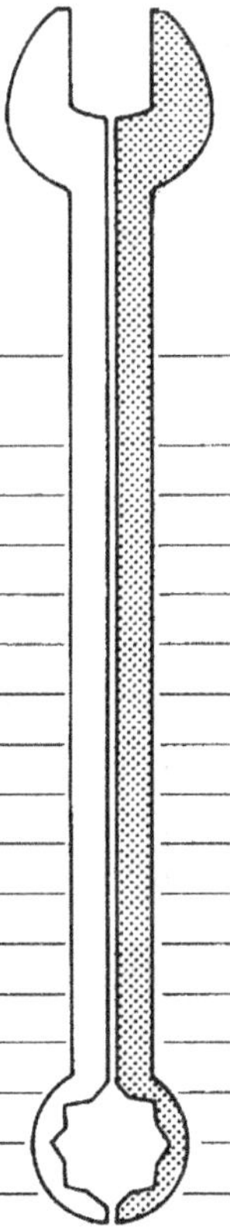

INCHES	DECIMAL	DECIMAL	MILLIMETERS
1/8″	.125	.118	3mm
3/16″	.187	.157	4mm
1/4″	.250	.236	6mm
5/16″	.312	.354	9mm
3/8″	.375	.394	10mm
7/16″	.437	.472	12mm
1/2″	.500	.512	13mm
9/16″	.562	.590	15mm
5/8″	.625	.630	16mm
11/16″	.687	.709	18mm
3/4″	.750	.748	19mm
13/16″	.812	.787	20mm
7/8″	.875	.866	22mm
15/16″	.937	.945	24mm
1″	1.00	.984	25mm

Fig. 50 Comparison of U.S. measure and metric wrench sizes

One extremely valuable type of wrench is the adjustable wrench. An adjustable wrench has a fixed upper jaw and a moveable lower jaw. The lower jaw is moved by turning a threaded drum. The advantage of an adjustable wrench is its ability to be adjusted to just about any size fastener.

The main drawback of an adjustable wrench is the lower jaw's tendency to move slightly under heavy pressure. This can cause the wrench to slip if it is not facing the right way. Pulling on an adjustable wrench in the proper direction will cause the jaws to lock in place. Adjustable wrenches come in a large range of sizes, measured by the wrench length.

PLIERS

◆ **See Figure 55**

Pliers are simply mechanical fingers. They are, more than anything, an extension of your hand. At least 3 pairs of pliers are an absolute necessity - standard, needle nose and slip joint.

In addition to standard pliers there are the slip-joint, multi-position pliers such as ChannelLock® pliers and locking pliers, such as Vise Grips®.

Slip joint pliers are extremely valuable in grasping oddly sized parts and fasteners. Just make sure that you don't use them instead of a wrench too often since they can easily round off a bolt head or nut.

Locking pliers are usually used for gripping bolts or studs that can't be removed conventionally. You can get locking pliers in square jawed, needle-nosed and pipe-jawed. Locking pliers can rank right up behind duct tape as the handy-man's best friend.

SCREWDRIVERS

You can't have too many screwdrivers. They come in 2 basic flavors, either standard or Phillips. Standard blades come in various sizes and thickness for all types of slotted fasteners. Phillips screwdrivers come in sizes with number designations from 1 on up, with the lower number designating the smaller size. Screwdrivers can be purchased separately or in sets.

HAMMERS

◆ **See Figure 56**

You need a hammer for just about any kind of work. You need a ball-peen hammer for most metal work when using drivers and other like tools. A plastic hammer comes in handy for hitting things safely. A soft-faced dead-blow hammer is used for hitting things safely and hard. Hammers are also VERY useful with non air-powered impact drivers.

Other Common Tools

There are a lot of other tools that every DIYer will eventually need (though not all for basic maintenance). They include:

- Funnels
- Chisels
- Punches
- Files
- Hacksaw
- Portable Bench Vise
- Tap and Die Set
- Flashlight
- Magnetic Bolt Retriever
- Gasket scraper
- Putty Knife

Fig. 51 Always use a backup wrench to prevent rounding flare nut fittings

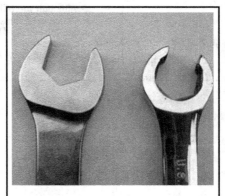

Fig. 52 Note how the flare wrench jaws are extended to grip the fitting tighter and prevent rounding

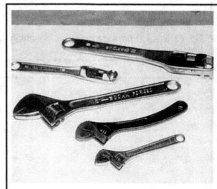

Fig. 53 Several types and sizes of adjustable wrenches

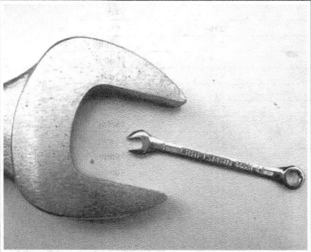

Fig. 54 You may find a nut that requires a particularly large or small wrench (that is usually available at your local tool store)

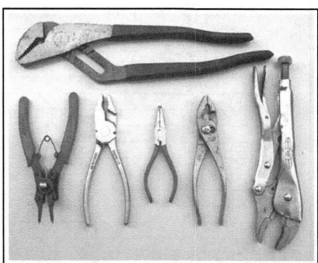

Fig. 55 Pliers come in many shapes and sizes. You should have an assortment on hand

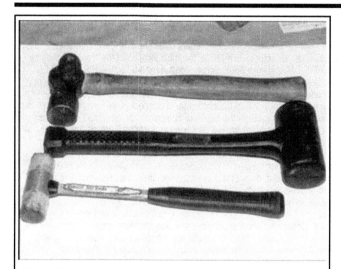

Fig. 56 Three types of hammers. Top to bottom: ball peen, rubber dead-blow, and plastic

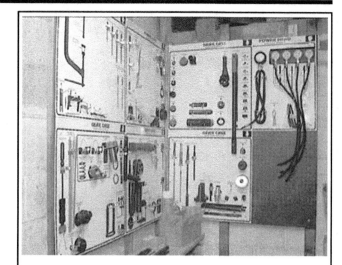

Fig. 57 Almost every marine engine around today requires at least one special tool to perform a certain task

- Screw/Bolt Extractors
- Prybars

Hacksaws have just one use - cutting things off. You may wonder why you'd need one for something as simple as maintenance or repair, but you never know. Among other things, guide studs to ease parts installation can be made from old bolts with their heads cut off.

A tap and die set might be something you've never needed, but you will eventually. It's a good rule, when everything is apart, to clean-up all threads, on bolts, screws or threaded holes. Also, you'll likely run across a situation in which you will encounter stripped threads. The tap and die set will handle that for you.

Gasket scrapers are just what you'd think, tools made for scraping old gasket material off of parts. You don't absolutely need one. Old gasket material can be removed with a putty knife or single edge razor blade. However, putty knives may not be sharp enough for some really stubborn gaskets and razor blades have a knack of breaking just when you don't want them to, inevitably slicing the nearest body part! As the old saying goes, "always use the proper tool for the job". If you're going to use a razor to scrape a gasket, be sure to always use a blade holder.

Putty knives really do have a use in a repair shop. Just because you remove all the bolts from a component sealed with a gasket doesn't mean it's going to come off. Most of the time, the gasket and sealer will hold it tightly. Lightly inserting a putty knife at various points between the two parts will break the seal without damage to the parts.

A small - 8-10 in. (20-25cm) long - prybar is extremely useful for removing stuck parts.

■ **Never use a screwdriver as a prybar! Screwdrivers are not meant for prying. Screwdrivers, used for prying, can break, sending the broken shaft flying!**

Screw/bolt extractors are used for removing broken bolts or studs that have broken off flush with the surface of the part.

Special Tools

◆ **See Figure 57**

Almost every marine engine around today requires at least one special tool to perform a certain task. In most cases, these tools are specially designed to overcome some unique problem or to fit on some oddly sized component.

When manufacturers go through the trouble of making a special tool, it is usually necessary to use it to ensure that the job will be done right. A special tool might be designed to make a job easier, or it might be used to keep you from damaging or breaking a part.

Don't worry, MOST maintenance procedures can either be performed without any special tools OR, because the tools must be used for such basic things, they are commonly available for a reasonable price. It is usually just

the low production, highly specialized tools (like a super thin 7-point star-shaped socket capable of 150 ft. lbs. (203 Nm) of torque that is used only on the crankshaft nut of the limited production what-dya-callit engine) that tend to be outrageously expensive and hard to find. Hopefully, you will probably never need such a tool.

Special tools can be as inexpensive and simple as an adjustable strap wrench or as complicated as an ignition tester. A few common specialty tools are listed here, but check with your dealer or with other boaters for help in determining if there are any special tools for YOUR particular engine. There is an added advantage in seeking advice from others, chances are they may have already found the special tool you will need, and know how to get it cheaper (or even let you borrow it).

Electronic Tools

BATTERY TESTERS

The best way to test a non-sealed battery is using a hydrometer to check the specific gravity of the acid. Luckily, these are usually inexpensive and are available at most parts stores. Just be careful because the larger testers are usually designed for larger batteries and may require more acid than you will be able to draw from the battery cell. Smaller testers (usually a short, squeeze bulb type) will require less acid and should work on most batteries.

Electronic testers are available and are often necessary to tell if a sealed battery is usable. Luckily, many parts stores have them on hand and are willing to test your battery for you.

BATTERY CHARGERS

◆ **See Figure 57**

If you are a weekend boater and take your boat out every week, then you will most likely want to buy a battery charger to keep your battery fresh. There are many types available, from low amperage trickle chargers to electronically controlled battery maintenance tools that monitor the battery voltage to prevent over or undercharging. This last type is especially useful if you store your boat for any length of time (such as during the severe winter months found in many Northern climates).

Even if you use your boat on a regular basis, you will eventually need a battery charger. The charger should be used anytime the boat is going to be in storage for more than a few weeks or so. Never leave the dock or loading ramp without a battery that is fully charged.

Also, some smaller batteries are shipped dry and in a partial charged state. Before placing a new battery of this type into service it must be filled and properly charged. Failure to properly charge a battery (which was shipped dry) before it is put into service will prevent it from ever reaching a fully charged state.

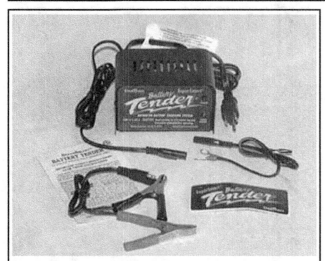

Fig. 58 The Battery Tender® is more than just a battery charger, when left connected, it keeps your battery fully charged

MULTI-METERS (DVOMS)

◆ See Figure 59

Multi-meters or Digital Volt Ohmmeter (DVOMs) are an extremely useful tool for troubleshooting electrical problems. They can be purchased in either analog or digital form and have a price range to suit any budget. A multi-meter is a voltmeter, ammeter and ohmmeter (along with other features) combined into one instrument. It is often used when testing solid state circuits because of its high input impedance (usually 10 mega-ohms or more). A brief description of the multi-meter main test functions follows:

• Voltmeter - the voltmeter is used to measure voltage at any point in a circuit or to measure the voltage drop across any part of a circuit. Voltmeters usually have various scales and a selector switch to allow the reading of different voltage ranges. The voltmeter has a positive and a negative lead. To avoid the possibility of damage to the meter, whenever possible, connect the negative lead to the negative (-) side of the circuit (to ground or nearest the ground side of the circuit) and connect the positive lead to the positive (+) side of the circuit (to the power source or the nearest power source). Luckily, most quality DVOMs can adjust their own polarity internally and will indicate (without damage) if the leads are reversed. Note that the negative voltmeter lead will always be black and that the positive voltmeter will always be some color other than black (usually red).

• Ohmmeter - the ohmmeter is designed to read resistance (measured in ohms) in a circuit or component. Most ohmmeters will have a selector switch which permits the measurement of different ranges of resistance (usually the selector switch allows the multiplication of the meter reading by 10, 100, 1,000 and 10,000). Some ohmmeters are "auto-ranging" which means the meter itself will determine which scale to use. Since the meters are powered by an internal battery, the ohmmeter can be used like a self-powered test light. When the ohmmeter is connected, current from the ohmmeter flows through the circuit or component being tested. Since the ohmmeter's internal resistance and voltage are known values, the amount of current flow through the meter depends on the resistance of the circuit or component being tested. The ohmmeter can also be used to perform a continuity test for suspected open circuits. In using the meter for making continuity checks, do not be concerned with the actual resistance readings. Zero resistance, or any ohm reading, indicates continuity in the circuit. Infinite resistance indicates an opening in the circuit. A high resistance reading where there should be little or none indicates a problem in the circuit. Checks for short circuits are made in the same manner as checks for open circuits, except that the circuit must be isolated from both power and normal ground. Infinite resistance indicates no continuity, while zero resistance indicates a dead short.

✳✳ WARNING

Never use an ohmmeter to check the resistance of a component or wire while there is voltage applied to the circuit.

• Ammeter - an ammeter measures the amount of current flowing through a circuit in units called amperes or amps. At normal operating voltage, most circuits have a characteristic amount of amperes, called "current draw" which can be measured using an ammeter. By referring to a specified current draw rating, then measuring the amperes and comparing the two values; one can determine what is happening within the circuit to aid in diagnosis. An open circuit, for example, will not allow any current to flow, so the ammeter reading will be zero. A damaged component or circuit will have an increased current draw, so the reading will be high. The ammeter is always connected in series with the circuit being tested. All of the current that normally flows through the circuit must also flow through the ammeter; if there is any other path for the current to follow, the ammeter reading will not be accurate. The ammeter itself has very little resistance to current flow and, therefore, will not affect the circuit, but, it will measure current draw only when the circuit is closed and electricity is flowing. Excessive current draw can blow fuses and drain the battery, while a reduced current draw can cause motors to run slowly, lights to dim and other components to not operate properly.

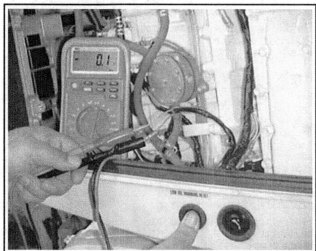

Fig. 59 Multi-meters, such as this one from UEI, are an extremely useful tool for troubleshooting electrical problems

Fig. 60 Cylinder compression test results are extremely valuable indicators of internal engine condition

GAUGES

Compression Gauge

◆ See Figure 60

An important element in checking the overall condition of your engine is to check compression. This becomes increasingly more important on outboards with high hours. Compression gauges are available as screw-in types and hold-in types. The screw-in type is slower to use, but eliminates the possibility of a faulty reading due to pressure escaping by the seal. A compression reading will uncover many problems that can cause rough running. Normally, these are not the sort of problems that can be cured by a tune-up.

Vacuum/Pressure Gauge/Pump

◆ See Figures 61, 62 and 63

Vacuum gauges are handy for discovering air leaks, late ignition or valve timing, and a number of other problems. A hand-held vacuum/pressure pump can be purchased at many automotive or marine parts stores and can be used for multiple purposes. The gauge can be used to measure vacuum or pressure in a line, while the pump can be used to manually apply vacuum or pressure to a solenoid or fitting. The hand-held pump can also be used to power-bleed trailer and tow vehicle brakes.

Measuring Tools

MICROMETERS & CALIPERS

Micrometers and calipers are devices used to make extremely precise measurements. The simple truth is that you really won't have the need for many of these items just for routine maintenance. But, measuring tools, such as an outside caliper can be handy during repairs. And, if you decide to tackle a major overhaul, a micrometer will absolutely be necessary.

Should you decide on becoming more involved in boat engine mechanics, such as repair or rebuilding, then these tools will become very important. The success of any rebuild is dependent, to a great extent on the ability to check the size and fit of components as specified by the manufacturer. These measurements are often made in thousandths and ten-thousandths of an inch.

Micrometers

A micrometer is an instrument made up of a precisely machined spindle that is rotated in a fixed nut, opening and closing the distance between the end of the spindle and a fixed anvil. When measuring using a micrometer, don't over-tighten the tool on the part as either the component or tool may be damaged, and either way, an incorrect reading will result. Most micrometers are equipped with some form of thumbwheel on the spindle that is designed

Fig. 61 Vacuum gauges are useful for troubleshooting including testing some fuel pumps

Fig. 62 You can also use the gauge on a hand-operated vacuum pump for tests

Fig. 63 Hand-held vacuum/pressure pumps are available at most parts stores

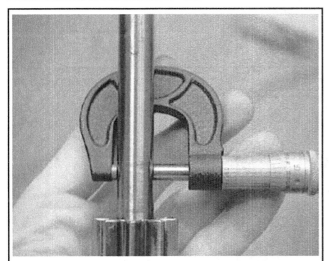

Fig. 64 Outside micrometers measure thickness, like shims or a shaft diameter

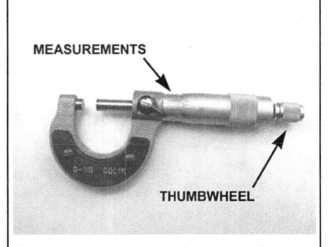

Fig. 65 Be careful not to over-tighten the micrometers always use the thumbwheel

to freewheel over a certain light touch (automatically adjusting the spindle and preventing it from over-tightening).

Outside micrometers can be used to check the thickness of parts such shims or the outside diameter of components like the crankshaft journals. They are also used during many rebuild and repair procedures to measure the diameter of components such as the pistons. The most common type of micrometer reads in 1/1000 of an inch. Micrometers that use a vernier scale can estimate to 1/10 of an inch.

Inside micrometers are used to measure the distance between two parallel surfaces. For example, in powerhead rebuilding work, the "inside mike" measures cylinder bore wear and taper. Inside mikes are graduated the same way as outside mikes and are read the same way as well.

Remember that an inside mike must be absolutely perpendicular to the work being measured. When you measure with an inside mike, rock the mike gently from side to side and tip it back and forth slightly so that you span the widest part of the bore. Just to be on the safe side, take several readings. It takes a certain amount of experience to work any mike with confidence.

Metric micrometers are read in the same way as inch micrometers, except that the measurements are in millimeters. Each line on the main scale equals 1mm. Each fifth line is stamped 5, 10, 15 and so on. Each line on the thimble scale equals 0.01 mm. It will take a little practice, but if you can read an inch mike, you can read a metric mike.

Calipers

◆ **See Figures 66, 67 and 68**

Inside and outside calipers are useful devices to have if you need to measure something quickly and absolute precise measurement is not necessary. Simply take the reading and then hold the calipers on an accurate steel rule. Calipers, like micrometers, will often contain a thumbwheel to help ensure accurate measurement.

DIAL INDICATORS

◆ **See Figure 69**

A dial indicator is a gauge that utilizes a dial face and a needle to register measurements. There is a movable contact arm on the dial indicator. When the arm moves, the needle rotates on the dial. Dial indicators are calibrated to show readings in thousandths of an inch and typically, are used to measure end-play and run-out on various shafts and other components.

Dial indicators are quite easy to use, although they are relatively expensive. A variety of mounting devices are available so that the indicator can be used in a number of situations. Make certain that the contact arm is always parallel to the movement of the work being measured.

TELESCOPING GAUGES

◆ **See Figure 70**

A telescope gauge is really only used during rebuilding procedures (NOT during basic maintenance or routine repairs) to measure the inside of bores. It can take the place of an inside mike for some of these jobs. Simply insert the gauge in the hole to be measured and lock the plungers after they have contacted the walls. Remove the tool and measure across the plungers with an outside micrometer.

DEPTH GAUGES

◆ **See Figure 71**

A depth gauge can be inserted into a bore or other small hole to determine exactly how deep it is. One common use for a depth gauge is measuring the distance the piston sits below the deck of the block at top dead center. Some outside calipers contain a built-in depth gauge so you can save money and buy just one tool.

Fig. 66 Calipers are the fast and easy way to make precise measurements

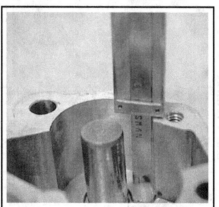
Fig. 67 Calipers can also be used to measure depth . . .

Fig. 68 . . . and inside diameter measurements, to 0.001 in. accuracy

Fig. 69 This dial indicator is measuring the end-play of a crankshaft during a powerhead rebuild

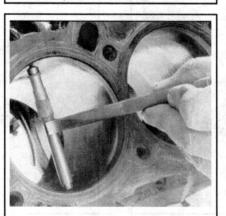
Fig. 70 Telescoping gauges are used during powerhead rebuilding procedures to measure the inside diameter of bores

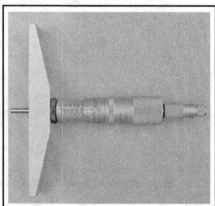
Fig. 71 Depth gauges are used to measure the depth of bore or other small holes

FASTENERS, MEASUREMENTS AND CONVERSIONS

Bolts, Nuts and Other Threaded Retainers

◆ See Figure 72 and 73

Although there are a great variety of fasteners found in the modern boat engine, the most commonly used retainer is the threaded fastener (nuts, bolts, screws, studs, etc). Most threaded retainers may be reused, provided that they are not damaged in use or during the repair.

■ Some retainers (such as stretch bolts or torque prevailing nuts) are designed to deform when tightened or in use and should not be reused.

Whenever possible, we will note any special retainers which should be replaced during a procedure. But you should always inspect the condition of a retainer when it is removed and you should replace any that show signs of damage. Check all threads for rust or corrosion that can increase the torque necessary to achieve the desired clamp load for which that fastener was originally selected. Additionally, be sure that the driver surface itself (on the fastener) is not compromised from rounding or other damage. In some cases a driver surface may become only partially rounded, allowing the driver to catch in only one direction. In many of these occurrences, a fastener may be installed and tightened, but the driver would not be able to grip and loosen the fastener again. (This could lead to frustration down the line should that component ever need to be disassembled again).

If you must replace a fastener, whether due to design or damage, you must always be sure to use the proper replacement. In all cases, a retainer of the same design, material and strength should be used. Markings on the heads of most bolts will help determine the proper strength of the fastener. The same material, thread and pitch must be selected to assure proper installation and safe operation of the motor afterwards.

Thread gauges are available to help measure a bolt or stud's thread. Most part or hardware stores keep gauges available to help you select the proper size. In a pinch, you can use another nut or bolt for a thread gauge. If the bolt you are replacing is not too badly damaged, you can select a match by finding another bolt that will thread in its place. If you find a nut that will thread properly onto the damaged bolt, then use that nut as a gauge to help select the replacement bolt. If however, the bolt you are replacing is so badly damaged (broken or drilled out) that its threads cannot be used as a gauge, you might start by looking for another bolt (from the same assembly or a similar location) which will thread into the damaged bolt's mounting. If so, the other bolt can be used to select a nut; the nut can then be used to select the replacement bolt.

In all cases, be absolutely sure you have selected the proper replacement. Don't be shy, you can always ask the store clerk for help.

WARNING

Be aware that when you find a bolt with damaged threads, you may also find the nut or tapped bore into which it was threaded has also been damaged. If this is the case, you may have to drill and tap the hole, replace the nut or otherwise repair the threads. Never try to force a replacement bolt to fit into the damaged threads.

Torque

Torque is defined as the measurement of resistance to turning or rotating. It tends to twist a body about an axis of rotation. A common example of this would be tightening a threaded retainer such as a nut, bolt or screw. Measuring torque is one of the most common ways to help assure that a threaded retainer has been properly fastened.

When tightening a threaded fastener, torque is applied in three distinct areas, the head, the bearing surface and the clamp load. About 50% of the measured torque is used in overcoming bearing friction. This is the friction between the bearing surface of the bolt head, screw head or nut face and the base material or washer (the surface on which the fastener is rotating). Approximately 40% of the applied torque is used in overcoming thread friction. This leaves only about 10% of the applied torque to develop a useful clamp load (the force that holds a joint together). This means that friction can account for as much as 90% of the applied torque on a fastener.

Standard and Metric Measurements

Specifications are often used to help you determine the condition of various components, or to assist you in their installation. Some of the most common measurements include length (in. or cm/mm), torque (ft. lbs., inch lbs. or Nm) and pressure (psi, in. Hg, kPa or mm Hg).

In some cases, that value may not be conveniently measured with what is available in your toolbox. Luckily, many of the measuring devices that are available today will have two scales so U.S. or Metric measurements may easily be taken. If any of the various measuring tools that are available to you do not contain the same scale as listed in your specifications, use the conversion factors that are provided in the Specifications section to determine the proper value.

The conversion factor chart is used by taking the given specification and multiplying it by the necessary conversion factor. For instance, looking at the first line, if you have a measurement in inches such as "free-play should be 2 in." but your ruler reads only in millimeters, multiply 2 in. by the conversion factor of 25.4 to get the metric equivalent of 50.8mm. Likewise, if a specification was given only in a Metric measurement, for example in Newton Meters (Nm), then look at the center column first. If the measurement is 100 Nm, multiply it by the conversion factor of 0.738 to get 73.8 ft. lbs.

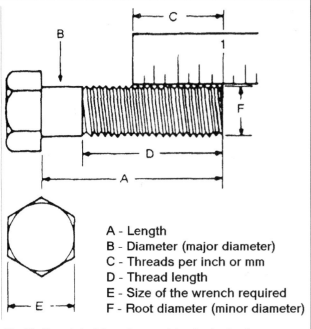

A - Length
B - Diameter (major diameter)
C - Threads per inch or mm
D - Thread length
E - Size of the wrench required
F - Root diameter (minor diameter)

Fig. 72 Threaded retainer sizes are determined using these measurements

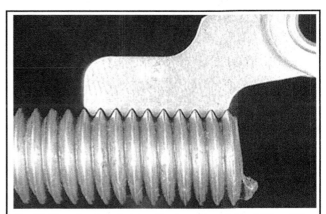

Fig. 73 Thread gauges measure the threads-per-inch and the pitch of a bolt or stud's threads

CONVERSION FACTORS

LENGTH–DISTANCE

Inches (in.)	x 25.4	= Millimeters (mm)	x .0394	= Inches
Feet (ft.)	x .305	= Meters (m)	x 3.281	= Feet
Miles	x 1.609	= Kilometers (km)	x .0621	= Miles

VOLUME

Cubic Inches (in3)	x 16.387	= Cubic Centimeters	x .061	= in3
IMP Pints (IMP pt.)	x .568	= Liters (L)	x 1.76	= IMP pt.
IMP Quarts (IMP qt.)	x 1.137	= Liters (L)	x .88	= IMP qt.
IMP Gallons (IMP gal.)	x 4.546	= Liters (L)	x .22	= IMP gal.
IMP Quarts (IMP qt.)	x 1.201	= US Quarts (US qt.)	x .833	= IMP qt.
IMP Gallons (IMP gal.)	x 1.201	= US Gallons (US gal.)	x .833	= IMP gal.
Fl. Ounces	x 29.573	= Milliliters	x .034	= Ounces
US Pints (US pt.)	x .473	= Liters (L)	x 2.113	= Pints
US Quarts (US qt.)	x .946	= Liters (L)	x 1.057	= Quarts
US Gallons (US gal.)	x 3.785	= Liters (L)	x .264	= Gallons

MASS–WEIGHT

Ounces (oz.)	x 28.35	= Grams (g)	x .035	= Ounces
Pounds (lb.)	x .454	= Kilograms (kg)	x 2.205	= Pounds

PRESSURE

Pounds Per Sq. In. (psi)	x 6.895	= Kilopascals (kPa)	x .145	= psi
Inches of Mercury (Hg)	x .4912	= psi	x 2.036	= Hg
Inches of Mercury (Hg)	x 3.377	= Kilopascals (kPa)	x .2961	= Hg
Inches of Water (H_2O)	x .07355	= Inches of Mercury	x 13.783	= H_2O
Inches of Water (H_2O)	x .03613	= psi	x 27.684	= H_2O
Inches of Water (H_2O)	x .248	= Kilopascals (kPa)	x 4.026	= H_2O

TORQUE

Pounds–Force Inches (in–lb)	x .113	= Newton Meters (N·m)	x 8.85	= in–lb
Pounds–Force Feet (ft–lb)	x 1.356	= Newton Meters (N·m)	x .738	= ft–lb

VELOCITY

Miles Per Hour (MPH)	x 1.609	= Kilometers Per Hour (KPH)	x .621	= MPH

POWER

Horsepower (Hp)	x .745	= Kilowatts	x 1.34	= Horsepower

FUEL CONSUMPTION*

Miles Per Gallon IMP (MPG)	x .354	= Kilometers Per Liter (Km/L)
Kilometers Per Liter (Km/L)	x 2.352	= IMP MPG
Miles Per Gallon US (MPG)	x .425	= Kilometers Per Liter (Km/L)
Kilometers Per Liter (Km/L)	x 2.352	= US MPG

*It is common to covert from miles per gallon (mpg) to liters/100 kilometers (1/100 km), where mpg (IMP) x 1/100 km = 282 and mpg (US) x 1/100 km = 235.

TEMPERATURE

Degree Fahrenheit (°F)   = (°C x 1.8) + 32

Degree Celsius (°C)   = (°F – 32) x .56

Metric Bolts						
Relative Strength Marking	4.6, 4.8			8.8		
Bolt Markings						
Usage	Frequent			Infrequent		
Bolt Size	Maximum Torque			Maximum Torque		
Thread Size x Pitch (mm)	Ft-Lb	Kgm	Nm	Ft-Lb	Kgm	Nm
6 x 1.0	2–3	.2–.4	3–4	3–6	.4–.8	5–8
8 x 1.25	6–8	.8–1	8–12	9–14	1.2–1.9	13–19
10 x 1.25	12–17	1.5–2.3	16–23	20–29	2.7–4.0	27–39
12 x 1.25	21–32	2.9–4.4	29–43	35–53	4.8–7.3	47–72
14 x 1.5	35–52	4.8–7.1	48–70	57–85	7.8–11.7	77–110
16 x 1.5	51–77	7.0–10.6	67–100	90–120	12.4–16.5	130–160
18 x 1.5	74–110	10.2–15.1	100–150	130–170	17.9–23.4	180–230
20 x 1.5	110–140	15.1–19.3	150–190	190–240	26.2–46.9	160–320
22 x 1.5	150–190	22.0–26.2	200–260	250–320	34.5–44.1	340–430
24 x 1.5	190–240	26.2–46.9	260–320	310–410	42.7–56.5	420–550

Typical Torque Values - Metric Bolts

	SAE Bolts								
SAE Grade Number	1 or 2			5			6 or 7		
Bolt Markings Manufacturers' marks may vary—number of lines always two less than the grade number.									
Usage	Frequent			Frequent			Infrequent		
Bolt Size (inches)—(Thread)	Maximum Torque			Maximum Torque			Maximum Torque		
	Ft-Lb	kgm	Nm	Ft-Lb	kgm	Nm	Ft-Lb	kgm	Nm
1/4—20	5	0.7	6.8	8	1.1	10.8	10	1.4	13.5
—28	6	0.8	8.1	10	1.4	13.6			
5/16—18	11	1.5	14.9	17	2.3	23.0	19	2.6	25.8
—24	13	1.8	17.6	19	2.6	25.7			
3/8—16	18	2.5	24.4	31	4.3	42.0	34	4.7	46.0
—24	20	2.75	27.1	35	4.8	47.5			
7/16—14	28	3.8	37.0	49	6.8	66.4	55	7.6	74.5
—20	30	4.2	40.7	55	7.6	74.5			
1/2—13	39	5.4	52.8	75	10.4	101.7	85	11.75	115.2
—20	41	5.7	55.6	85	11.7	115.2			
9/16—12	51	7.0	69.2	110	15.2	149.1	120	16.6	162.7
—18	55	7.6	74.5	120	16.6	162.7			
5/8—11	83	11.5	112.5	150	20.7	203.3	167	23.0	226.5
—18	95	13.1	128.8	170	23.5	230.5			
3/4—10	105	14.5	142.3	270	37.3	366.0	280	38.7	379.6
—16	115	15.9	155.9	295	40.8	400.0			
7/8— 9	160	22.1	216.9	395	54.6	535.5	440	60.9	596.5
—14	175	24.2	237.2	435	60.1	589.7			
1— 8	236	32.5	318.6	590	81.6	799.9	660	91.3	894.8
—14	250	34.6	338.9	660	91.3	849.8			

Typical Torque Values - U.S. Standard Bolts

2

MAINTENANCE & TUNE-UP

GENERAL INFORMATION (WHAT EVERYONE SHOULD KNOW ABOUT MAINTENANCE)

At Seloc, we estimate that 75% of engine repair work can be directly or indirectly attributed to lack of proper care for the engine. This is especially true of care during the off-season. There is no way on this green earth for a mechanical engine, particularly an outboard motor, to be left sitting idle for an extended period of time, say for 6 months, and then be ready for instant, satisfactory service.

Imagine, if you will, leaving your car or truck for 6 months, and then expecting to turn the key, having it roar to life, and being able to drive off in the same manner as a daily occurrence. Not likely, eh?

Therefore it is critical for an outboard engine to either be run (at least once a month), preferably, in the water and properly maintained between uses or for it to be specifically prepared for storage and serviced again immediately before the start of the season.

Only through a regular maintenance program can the owner expect to receive long life and satisfactory performance at minimum cost.

Many times, if an outboard is not performing properly, the owner will "nurse" it through the season with good intentions of working on the unit once it is no longer being used. As with many New Year's resolutions, the good intentions are not completed and the outboard may lie for many months before the work is begun or the unit is taken to the marine shop for repair.

Imagine, if you will, the cause of the problem being a blown head gasket. And let us assume water has found its way into a cylinder. This water, allowed to remain over a long period of time, will do considerably more damage than it would have if the unit had been disassembled and the repair work performed immediately. Therefore, if an outboard is not functioning properly, do not stow it away with promises to get at it when you get time, because the work and expense will only get worse, the longer corrective action is postponed. In the example of the blown head gasket, a relatively simple and inexpensive repair job could very well develop into major overhaul and rebuild work.

Maintenance Equals Safety

OK, perhaps no one thing that we do as boaters will protect us from risks involved with enjoying the wind and the water on a powerboat. But, each time we perform maintenance on our boat or motor, we increase the likelihood that we will find a potential hazard before it becomes a problem. Each time we service and inspect our boat and motor, we decrease the possibility that it could leave us stranded on the water.

In this way, performing boat and engine service is one of the most important ways that we, as boaters, can help protect ourselves, our boats, and the friends and family that we bring aboard.

Outboards On Sail Boats

Owners of sailboats pride themselves in their ability to use the wind to clear a harbor or for movement from Port A to Port B, or maybe just for a day sail on a lake. For some, the outboard is carried only as a last resort - in case the wind fails completely, or in an emergency situation or for ease of docking.

Therefore, in some cases, the outboard is stowed below, usually in a very poorly ventilated area, and subjected to moisture and stale air - in short, an excellent environment for "sweating" and corrosion.

If the owner could just take the time at least once every month, to pull out the outboard, clean it up, and give it a short run, not only would he/she have "peace of mind" knowing it will start in an emergency, but also maintenance costs will be drastically reduced.

Maintenance Coverage In This Manual

At Seloc, we strongly feel that every boat owner should pay close attention to this section. We also know that it is one of the most frequently used portions of our service. The material in this section is divided into sections to help simplify the process of maintenance. Be sure to read and thoroughly understand the various tasks that are necessary to keep your outboard in tip-top shape.

Topics covered in this section include:

1. General Information (What Everyone Should Know About Maintenance) - an introduction to the benefits and need for proper maintenance; a guide to tasks that should be performed before, and after, each use.

2. Lubrication Service - after the basic inspections that you should perform each time the motor is used, the most frequent form of periodic maintenance you will conduct will be the Lubrication Service. This section takes you through each of the various steps you must take to keep corrosion from slowly destroying your motor before your very eyes.

3. Engine Maintenance - the various procedures that must be performed on a regular basis in order to keep the motor and all of its various systems operating properly.

4. Boat Maintenance - the various procedures that must be performed on a regular basis in order to keep the boat hull and its accessories looking and working like new.

5. Tune-Up - also known as the pre-season tune-up, but don't let the name fool you. A complete tune-up is the best way to determine the condition of your outboard while also preparing it for hours and hours of hopefully trouble-free enjoyment.

6. Winter Storage and Spring Commissioning Checklists - use these sections to guide you through the various parts of boat and motor maintenance that protect your valued boat through periods of storage and return it to operating condition when it is time to use it again.

7. Specification Charts - located at the end of the section are quick-reference, easy to read charts that provide you with critical information such as General Engine Specifications, Maintenance Intervals, Lubrication Service (intervals and lubricant types) and Capacities.

Engine Identification

Johnson and Evinrude produced an large number of models with regards to horsepower ratings, as well a large number of trim and option variances on each of those models. In this service guide, we've included the 1-, 2-, 3-cylinder and V4 models (all of which are 2-stroke motors). We chose to do this because of the many similarities these motors have to each other. But, enough differences exist that many procedures will apply only to a sub-set of these motors. When this occurs, we'll either refer to the differences within a procedure or, if the differences are significant, we'll break the motors out and give separate procedures. In order to prevent confusion, we try to sort and name the models in a way that is most easily understood.

The **absolute best** method of engine identification is to start by referring to the engine serial number tag. For all models covered by this service guide an ID tag is affixed to the engine. On some (if not most) models, the serial number and model number were stamped on a plate mounted between the 2 swivel brackets underneath the hood (top engine case/cover). On other model engines, the plate is mounted on the port side of the engine on the front or side of the swivel bracket. The hp and rpm range may also be found on the plate.

ENGINE SERIAL NUMBERS

◆ See Figures 1, 2 and 3

The engine serial numbers are the manufacturer's key to engine changes. These alpha-numeric codes identify the year of manufacture, the horsepower rating, lower unit shaft length and various model/option differences (such as tiller electric, remote electric or commercial models). If any correspondence or parts are required, the engine serial number must be used for proper identification.

Remember that the serial number establishes the model-year for which the engine was produced, and is often *not* the year of first installation.

The engine serial number tag contains information such as the plant in which the motor was produced, the model number or code, the serial number (a unique sequential identifier given ONLY to that one motor) as well as other useful information in some cases such as such as weight (mass) in Kilograms (kg).

Prior to 1980 Johnson/Evinrude did not use a single specific formula to decipher model codes and as a result, by 1979 had 32 pages of a specific model decoding book dedicated to the subject on their outboards. If you need to decipher a model code on an engine you believe to be 1972 or earlier please contact a dealer or the Seloc offices directly with the code.

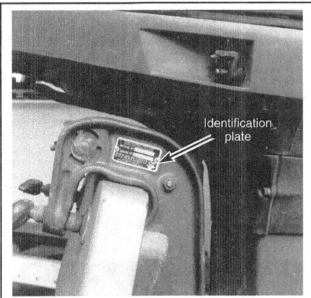

Fig. 1 There should be an identification plate installed between the transom brackets. . .

Identification plate

Identification plate

Fig. 2 . . .or on the port side of the bracket like this. . .

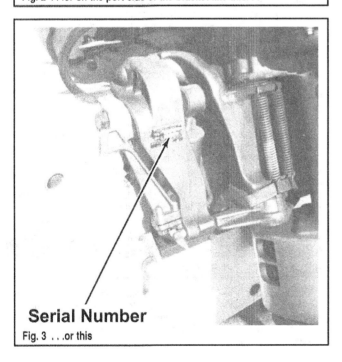

Serial Number

Fig. 3 . . .or this

Before/After Each Use

VISUALLY INSPECTING THE BOAT AND MOTOR

◆ See Figures 4 and 5

As stated earlier, the best means of extending engine life and helping to protect yourself while on the water is to pay close attention to boat/engine maintenance. This starts with an inspection of systems and components before and after each time you use your boat.

Before each launch and immediately after each retrieval, visually inspect the boat and motor as follows:

1. **Check the fuel and oil levels** according to the procedures in this service. Do NOT launch a boat without properly topped off fuel and oil tanks (on oil injected motors). It is not worth the risk of getting stranded or of damage to the motor. Likewise, upon retrieval, check the oil and fuel levels while it is still fresh in your mind. This is a good way to track fuel consumption (one indication of engine performance). Compare the fuel consumption to the oil consumption (a dramatic change in proportional use may be an early sign of trouble).

2. **Check for signs of fuel or oil leakage.** Probably as important as making sure enough fuel and oil is onboard, is the need to make sure that no dangerous conditions might arise due to leaks. Thoroughly check all hoses, fittings and tanks for signs of leakage. Oil leaks may prevent proper oiling of the powerhead and, although all VRO systems have warning systems, reduced oiling could damage the powerhead or, if the system fails completely, could strand the boat. Fuel leaks can cause a fire hazard, or worse, an explosive condition. This check is not only about properly maintaining your boat and motor, but about helping to protect your life.

3. **Inspect the boat hull and engine cases** for signs of corrosion or damage. Don't launch a damaged boat or motor. And don't surprise yourself dockside or at the launch ramp by discovering damage that went unnoticed last time the boat was retrieved. Repair any hull or case damage now.

4. **Check the battery** connections (if equipped) to make sure they are clean and tight. A loose or corroded connection will cause charging problems (damaging the system or preventing charging). There's only one thing worse than a dead battery dockside or on the launch ramp and that's a dead battery in the middle of a bay, river or heavens, the ocean. Whenever possible, make a quick visual check of battery electrolyte levels (keeping an eye on the level will give some warning of overcharging problems). This is especially true if the engine is operated at high speeds for extended periods of time.

5. **Check the propeller and lower unit.** Make sure the propeller shows no signs of damage. A broken or bent propeller may allow the engine to over-rev and it will certainly waste fuel. The lower unit should be checked before and after each use for signs of leakage. Check the lower unit oil for signs of contamination if any leakage is noted. Also, visually check behind the propeller for signs of entangled rope or fishing lines that could cut through the lower gearcase propeller shaft seal. This is a common cause of lower unit lubricant leakage, and eventually, water contamination that can lead to lower unit failure. Even if no lower unit leakage is noted when the boat is first retrieved, check again next time before launching. A nicked seal might not seep fluid right away when still swollen from heat immediately after use, but might begin seeping over the next day, week or month as it sat, cooled and dried out.

6. **Check all accessible fasteners for tightness.** Make sure all easily accessible fasteners appear to be tight. This is especially true for the propeller nut, any anode retaining bolts, all steering or throttle linkage fasteners and the engine mounting bolts. Don't risk loosing control or becoming stranded due to loose fasteners. Perform these checks before heading out, and immediately after you return (so you'll know if anything needs to be serviced before you want to launch again.)

7. **Check operation of all controls including the throttle/shifter, steering and emergency stop/start switch and/or safety lanyard.** Before launching, make sure that all linkage and steering components operate properly and move smoothly through their range of motion. All electrical switches (such as power trim/tilt) and especially the emergency stop system(s) must be in proper working order. While underway, watch for signs that a system is not working or has become damaged. With the steering,

Fig. 4 Rope and fishing line entangled behind the propeller can cut through the seal, allowing water to enter and lubricant to escape

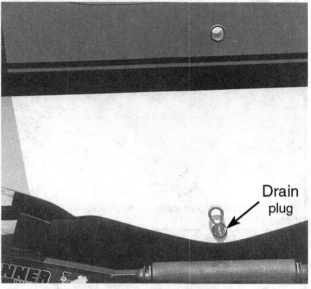

Fig. 5 Always make sure the transom plug is installed and tightened securely before a launch

shifter or throttle, keep a watchful eye out for a change in resistance or the start of jerky/notchy movement.

8. **Check the water pump intake grate and water indicator.** The water pump intake grate should be clean and undamaged before setting out. Remember that a damaged grate could allow debris into the system that could destroy the impeller or clog cooling passages. Once underway, make sure the cooling indicator stream (or spray on some early models) is visible at all times. Make periodic checks, including one final check before the motor is shut down each time. If a cooling indicator stream (or spray) is not present at any point, troubleshoot the problem before further engine operation.

9. **If used in salt, brackish or polluted waters thoroughly rinse the engine (and hull), then flush the cooling system** according to the procedure in this section.

■ Even if used in fresh water, it is never a bad idea to flush the system with fresh clean water from a garden hose. Keep in mind that sand, silt or other debris may be picked up by the cooling system during normal motor operation. Removing this debris before it can build-up and clog the engine is wise service.

10. **Inspect all anodes** after each use for signs of wear, damage or to make sure they just plain didn't fall off (especially if you weren't careful about checking all the accessible fasteners the last time you launched).

11. **For Pete's sake, make sure the plug is in!** We shouldn't have to say it, but unfortunately we do. If you've been boating for any length of time, you've seen or heard of someone whose backed a trailer down a launch ramp, forgetting to check the transom drain plug before literally submerging the boat. Always make sure the transom plug is installed and tight before a launch.

LUBRICATION SERVICE

General Information

An outboard motor's greatest enemy is corrosion. Face it, oil and water just don't mix and, as anyone who has visited a junkyard knows, metal and water aren't the greatest of friends either. To expose an engine to a harsh marine environment of water and wind is to expect that these elements will take their toll over time. But, there is a way to fight back and help prevent the natural process of corrosion that will destroy your beloved boat motor.

Various marine grade lubricants are available that serve two important functions in preserving your motor. Lubricants reduce friction on metal-to-metal contact surfaces, and they also displace air and moisture, therefore slowing or preventing corrosion damage. Periodic lubrication services are your best method of preserving an outboard motor.

Lubrication takes place through various forms. For all engines, internal moving parts are lubricated by 2-stroke engine oil, through oil contained in the fuel/oil mixture. Pay close attention to the oil level in the tank on oil injected models, or to the oil/fuel mixing process on pre-mix motors. Also, the lower unit is filled with gear oil that lubricates the driveshaft, propshaft, gears and other internal gearcase components. The gear oil should be periodically checked and replaced following the appropriate Engine Maintenance procedures.

✳✳ WARNING

If equipped with power trim/tilt, maintaining proper fluid level is necessary for the built-in impact protection system. Incorrect fluid level could lead to significant lower unit damage in the event of an impact.

On motors equipped with power trim/tilt (unlikely), the fluid level and condition in the reservoir should be checked periodically to ensure proper operation. Also, on these motors, correct fluid level is necessary to ensure operation of the motor impact protection system.

Most other forms of lubrication occur through the application of grease (Johnson/Evinrude Triple-Guard, Johnson/Evinrude EP/Wheel bearing grease, Johnson/Evinrude Starter Pinion Lube, or their equivalents) to various points on the motor. These lubricants are either applied by hand (an old toothbrush can be helpful in preventing a mess) or using a grease gun to pump the lubricant into grease fittings (also known as zerk fittings). When using a grease gun, do not pump excessive amounts of grease into the fitting. Unless otherwise directed, pump until either the rubber seal (if used) begins to expand or until the grease just begins to seep from the joints of the component being lubricated (if no seal is used).

To ensure your motor is getting the protection it needs, perform a visual inspection of the various lubrication points at least once a week during regular seasonal operation (this assumes that the motor is being used at least once a week). Most manually lubricated (greased) items should be serviced at least every 60 days when the boat is operated in fresh water or every 30 days when the boat is operated in salt, brackish or polluted waters. We said *at least*, meaning you should perform these services more often, as discovered by your weekly inspections.

Electric Starter Motor Pinion

RECOMMENDED LUBRICANT

Check your owner's manual. On most motors a periodic lubrication of the starter pinion is necessary to keep the pinion from sticking. When it doubt, lubricate this component (but do so lightly, as excessive amounts of grease could cause problems). Use Johnson/Evinrude Starter Pinion Lube, General Electric Versalube or an equivalent lubricant.

LUBRICATION

◆ **See Figures 6, 7 and 8**

Periodic lubrication of the starter motor pinion is required on most electric start models.

The starter pinion is the gear and slider assembly located on the top of the starter motor as it is mounted to the engine. When power is applied to the starter, the gear on the pinion assembly slides upward to contact and mesh with the gear teeth on the outside of the flywheel. Periodically, apply a small amount of lubricant to the sliding surface of the starter pinion in order to prevent excessive wear or possible binding on the shaft.

■ **On models that require periodic lubrication, easy access is normally provided to the starter pinion. Though, it is possible that a flywheel cover may need to be removed on a few models.**

Engine Cover Latches

RECOMMENDED LUBRICANT

For some reason, not all Johnson/Evinrude factory literature mentions the periodic lubrication of the engine cover latches (including one of the author's own motors and owner's manual). But, most motors are equipped with a grease fitting for each cover latch and/or exposed metal-to-metal contact surfaces that will benefit from periodic lubrication).

Use Johnson/Evinrude Triple-Guard, or equivalent water-resistant marine grease for lubrication.

LUBRICATION

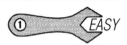

◆ **See Figures 9 and 10**

For some reason, not all Johnson/Evinrude factory literature mentions the periodic lubrication of the engine cover latches (including one of the author's own motors and owner's manual). But, *most* motors are equipped with a grease fitting for each cover latch and/or exposed metal-to-metal contact surfaces that will benefit from periodic lubrication).

Although the sliding surfaces of all cover latches can benefit from an application of grease, the design of the latches used on some motors (those equipped with grease fittings) makes periodic greasing necessary to prevent the latches from binding or wearing. Some models are equipped with 2 or 3 grease fittings on the engine case, while other models have a single grease fitting on each latch (located facing upward, inside of the engine covers).

Depending on the latch type, either apply a small amount of grease to the metal surfaces using an applicator brush or use a grease gun to pump grease into the zerk fitting facing outward from the latch assembly.

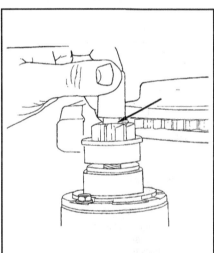

Fig. 6 Apply lubricant to the sliding surface of the electric starter pinion

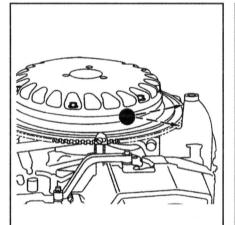

Fig. 7 In most cases, models that require periodic lubrication provide easy access to the starter pinion

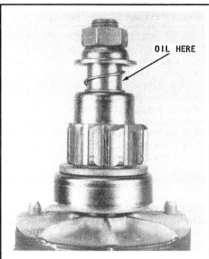

Fig. 8 A good look at the pinion

Fig. 9 If equipped, use the grease fittings often found at the front . . .

Fig. 10 . . . and rear of the engine covers to lubricate the engine cover latches

Power Trim/Tilt Reservoir

RECOMMENDED LUBRICANT

✳✳ WARNING

When equipped with power trim/tilt (and this is unlikely with the models here), proper fluid level is necessary for the built-in impact protection system. Incorrect fluid level could lead to significant lower unit damage in the event of an impact.

The power trim/tilt reservoir must be kept full of Johnson/Evinrude Power Trim/Tilt and Power Steering Fluid.

CHECKING FLUID LEVEL/CONDITION

✳✳ WARNING

When equipped with power trim/tilt, proper fluid level is necessary for the built-in impact protection system. Incorrect fluid level could lead to significant lower unit damage in the event of an impact.

The fluid in the power trim/tilt reservoir should be checked periodically to ensure it is full and is not contaminated. To check the fluid, tilt the motor upward to the full tilt position and manually engage the tilt support, for safety and to prevent damage. Remove the filler cap and make a visual inspection of the fluid. The cap is usually threaded in position on the outboard side of the reservoir and equipped with a flat to accept a bladed screwdriver. The fluid, should seem clear and not milky. The level is correct if, with the motor at full tilt, the level is even with the bottom of the filler cap hole (but only when the motor is tilted upward, as the oil level will rise above this point when the motor is tilted downward).

The total system capacity is usually 25 fl. oz. (740mL).

Power Steering Fluid Reservoir

RECOMMENDED LUBRICANT

◆ **See Figure 11**

Although unlikely, a few models may equipped with a power steering system consisting of a belt, pump and reservoir. The fluid level should be checked periodically (ideally with each outing, but at **minimum** at least every 30 days.

✳✳ CAUTION

Remember, steering a boat is a matter of safety, don't risk poor performance or failure of the system due to something as silly a low fluid level.

✳✳ WARNING

Should the pump lose pressure or become inoperative on the water, shut the engine off and cut the drive belt free of the pulleys. This will allow normal "non-power" steering operation (which might be a little slow and heavy, but predictable) and will prevent serious and permanent damage which can occur to the pump if it is run low on fluid/pressure.

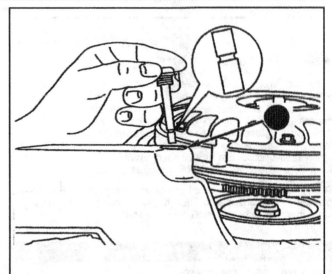

Fig. 11 Check the power steering fluid level periodically to ensure safe, trouble-free system operation

The reservoir must be kept full of Johnson/Evinrude Power Trim/Tilt and Power Steering Fluid or with Dexron II (or latest superceding) automatic transmission fluid.

CHECKING FLUID LEVEL/CONDITION

◆ **See Figure 11**

■ **The system should be completely drained (by disconnecting a fluid fitting or line, especially at the filter, if equipped) and the fluid should be completely changed every 500 hours.**

The fluid in the power steering reservoir should be checked periodically to ensure it is full and is not contaminated. To check the fluid, the motor must be in the normal, fully vertical position (trimmed level with the ground, NOT the gauge). Remove the engine top case for access to the reservoir (usually located near the top of the motor), then unthread the dipstick. Wipe the dipstick off and insert it back into the reservoir, but DO NOT thread it back into position, instead let it sit for a second on top of the threads, then withdraw the dipstick. Hold the dipstick vertically, with the bottom downward, to prevent a false high reading by fluid running up the stick if it was tilted past vertical. Then read the dipstick by looking at the highest wet line across the surface of the dipstick. Fluid should be kept at the full mark. If not, add **just** enough fluid to top it off to the full mark, NOT above. While checking the level, also take note of the fluid condition. It should seem clear and not milky.

■ **If the fluid level is low, thoroughly inspect the system for signs of leakage and repair, as necessary).**

CHANGING FLUID & FILTER

◆ **See Figure 12**

Every 500 hours of operation or anytime the fluid inspection shows signs of contamination; the system should be completely emptied and refilled using fresh fluid. some systems are also equipped with an inline power steering fluid filter to help keep the fluid free of particles and contamination. The filter should be changed anytime the fluid is changed. When equipped, the filter is a convenient way of draining the system, as it is normally mounted inline, beneath the reservoir. Once the lines are disconnected they can be pointed downward to ensure thorough system draining. Keep in mind that the bottom filter line must be positioned downward at a point lower than the lowest point in the system in order to ensure draining, this might necessitate placing an additional length of line on it, or following it downward to disconnect it at the other end.

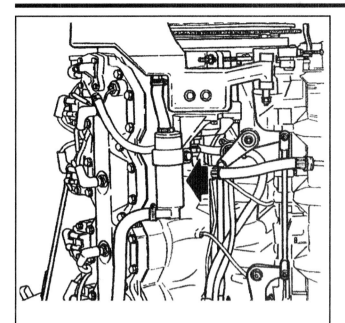

Fig. 12 Most power steering systems utilize an inline filter that must be changed whenever the fluid is replaced

Linkage, Cables and Shafts (Shift, Carburetor and/or Throttle Shaft)

RECOMMENDED LUBRICANT

Use Johnson/Evinrude Triple-Guard, or equivalent water-resistant marine grease for lubrication.

LUBRICATION

◆ See Figures 13, 14 and 15

Every Johnson and Evinrude outboard uses some combination of cables and/or linkage in order to actuate the throttle plate (of the carburetors or throttle bodies) and the lower unit shifter. Because linkage and cables contain moving parts that work in contact with other moving parts, the contact points can become worn and loose if proper lubrication is not maintained. These small parts are also susceptible to corrosion and breakage if they are not protected from moisture by light coatings of grease. Periodically apply a light coating of suitable water-resistant marine grease on each of these surfaces where either two moving parts meet or where a cable end enters a housing. For more details on grease points refer to the accompanying illustrations.

■ On some models, the lower engine covers must be removed for access to some of the cable/linkage greasing points.

Steering Arm (Cable Ram/Tiller Arm)

RECOMMENDED LUBRICANT

Use Johnson/Evinrude Triple-Guard, or equivalent water-resistant marine grease for lubrication.

LUBRICATION

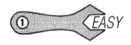

◆ See Figures 16 and 17

All motors covered within are equipped with a tiller control and/or a remote control assembly. On models equipped with a tiller, the arm's pivot

Fig. 13 All engines contain cable throttle and shift linkage. . .

Fig. 14 . . . whose metal-to-metal contact points should be periodically coated with grease

Fig. 15 Typical carburetor, throttle and shift linkage lubrication points

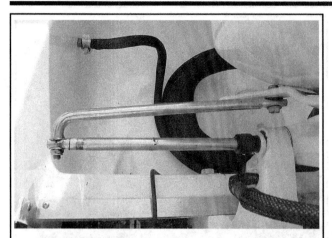

Fig. 16 On remote models, the steering arm (cable ram) must be greased periodically to prevent corrosion and ensure smooth operation

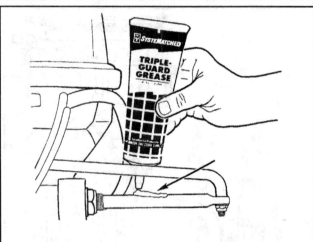

Fig. 17 Apply a light coating of water-resistance marine grease to the steering cable ram

point (where it attaches to the engine) should be lubricated periodically. On models with remote controls, the steering arm should be given a light coating of fresh lubricant to prevent corrosion or scoring.

Swivel Bracket & Tilt Support

RECOMMENDED LUBRICANT

Use Johnson/Evinrude Triple-Guard, or equivalent water-resistant marine grease for lubrication.

LUBRICATION

◆ **See Figures 18 and 19**

Most Johnson/Evinrude 1- and 2-cylinder motors are equipped with a grease fitting on the lower portion of the swivel bracket. Use a grease gun to apply fresh water-resistant marine grease until a small amount of lubricant begins to seep from the swivel bracket. It is important to keep this system corrosion free in order to prevent corrosion that would lead to excessive resistance or even binding that might cause dangerous operational conditions.

■ **The grease fitting for the swivel bracket is sometimes located behind the tilt support (trailering) bracket, when equipped. In these cases, the fitting may be hidden when the bracket is stowed and accessible when the bracket is engaged to hold the motor in the full tilt position.**

When equipped, the pivot points of the integral support (trailering) bracket should also be lubricated periodically to ensure smooth operation and to prevent corrosion. Since they are normally not equipped with a grease fitting, pump a small amount of grease out of the grease gun and spread it by hand or using an old toothbrush.

Tilt Tube Assembly

RECOMMENDED LUBRICANT

Use Johnson/Evinrude Triple-Guard, or an equivalent water-resistant marine grease for lubrication.

LUBRICATION

◆ **See Figure 20**

The tilt tube assembly must be greased periodically to prevent corrosion or binding, ensuring reliable and trouble-free operation.

Most Johnson/Evinrude motors have 2 grease fittings on the front of the tilt tube, facing the boat's transom. Apply a water-resistant marine grade grease to the fitting(s) until a small amount of grease seeps from the joints.

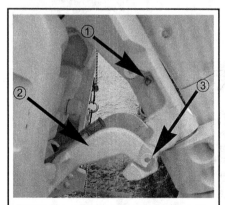

Fig. 18 The swivel bracket grease fitting (1), is sometimes hidden by the tilt bracket (2). When equipped, be sure to grease the pivot points of the tilt bracket (3). . .

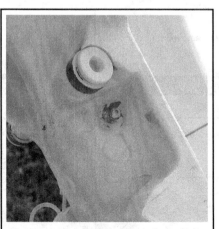

Fig. 19 . . . then apply grease to the swivel bracket through the fitting

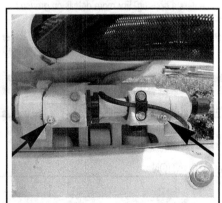

Fig. 20 Using a grease gun, lubricate both tilt tube assembly zerk fittings (normally there are 2 on the tilt tube, facing the transom

ENGINE MAINTENANCE

Engine Covers (Top & Lower Cases)

REMOVAL & INSTALLATION

◆ See Figures 21, 22 and 23

Removal of the top cover is necessary for the most basic of maintenance and inspection procedures. The cover should come off before and after each use in order to perform basic safety checks. The lower covers do not need to be removed nearly as often, but on models where they are easily removed, they should be removed at least seasonally for service and inspection procedures, especially linkage/cable lubrication procedures.

■ **On many models, the lower engine covers must be removed for access to some of the cable/linkage greasing points.**

On all models, the top cover is attached by some type of lever and latch assembly. No tools are usually necessary to remove the cover itself (except on a few of the smallest outboards). The exact shape and design of the levers vary somewhat from model-to-model, though they are usually located on the engine cover at the front and the aft portions of the split line between the top cover and the lower cases.

On some models, the cover latches must be pulled outward slightly or otherwise removed from a bushing or snap fixture that holds them in the locked position when closed. Once the end of the lever is freed, it is rotated 45-90° from the locked position to a top cover released position. With all of the levers released, most top covers will simply lift straight off the outboard.

No matter what design is used, once installed be certain that the cover is fully seated and mounted tightly to the lower cases in order to prevent the possibility of it coming loose in service. Make sure that the levers are secured once they are returned to the locked position.

■ **Cover screws on most Johnson/Evinrude V-motors are usually retained by various hex-head bolts, but some models may use Phillips, slotted head, or even star-headed Torx® screws. Be sure to use only the proper-sized socket or driver on fastener heads.**

The lower covers of most motors are screwed or bolted together by fasteners found around the perimeter of one or both sides of the cover. However, some motors are equipped with 1-piece covers that are not designed for easy removal. On the other hand for models so equipped, this cover is a low-rise component that should not interfere with service procedures. For this reason, the cover is not usually removed except during a complete overhaul where the powerhead is removed from the lower unit.

In most cases, some of the engine wiring and the fuel or oil lines must be disconnected in order to remove the lower case(s) completely from the outboard. But, some models may be equipped with removable panels or covers that allow most lines and wiring to remain connected and intact. Some lower cover designs utilize cutouts at the cover split-lines through which cables are passed.

✳✳ WARNING

It is especially important that you take note how each hose and wire is routed before disconnecting or moving them during service. Unless the person who worked on the motor previously made a mistake (which could cause damage and the need for repairs), all hoses and wires should already be routed in a manner that will prevent interference with and damage from moving components. Unless there are signs of damage from contact with components wires and hoses should be returned to the exact same positions as noted during disassembly. Don't be afraid to grab a digital camera and take pictures as you're disassembling. If you are unsure how a wire or hose was routed, work slowly, checking the positioning as the covers are installed to prevent damage.

Cooling System

FLUSHING THE COOLING SYSTEM

◆ See Figures 24, 25 and 26

The most important service that you can perform on your motor's cooling system is to flush it periodically using fresh, clean water. This should be done immediately following any use in salt, brackish or polluted waters in order to prevent mineral deposits or corrosion from clogging cooling passages. Even if you do not always boat in salt or polluted waters, get used to the flushing procedure and perform it often (ideally, immediately following every outing) to ensure no silt or debris clogs your cooling system over time.

■ **Flush the cooling system after any use in which the motor was operated through suspended/churned-up silt, debris or sand.**

Although the flushing procedure should take place right away (dockside or on the trailer), be sure to protect the motor from damage due to possible thermal shock. If the engine has just been run under high load or at continued high speeds, allow time for it to cool to the point where the powerhead can be touched. Do not pump very cold water through a very hot engine, or you are just asking for trouble. If you trailer your boat short distances, the flushing procedure can probably wait until you arrive home or wherever the boat is stored, but ideally it should occur within an hour of use in salt water. Remember that the corrosion process begins as soon as the motor is removed from the water and exposed to air.

The flushing procedure is not used only for cooling system maintenance, but it is also a tool with which a technician can provide a source of cooling water to protect the engine (and water pump impeller) from damage anytime the motor needs to be run out of the water. **Never** start or run the engine out of the water, even for a few seconds, for any reason. Water pump impeller

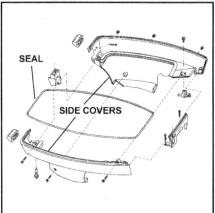

Fig. 21 Most Johnson/Evinrude motors utilize a 2-piece lower port and lower starboard cover assembly. . .

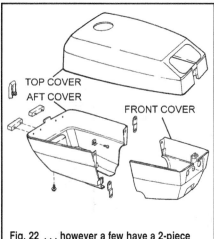

Fig. 22 . . . however a few have a 2-piece front and aft cover assembly

Fig. 23 Covers are secured using internal (shown) and external fasteners

damage can occur instantly and damage to the engine from overheating can follow shortly thereafter. If the engine must be run out of the water for tuning or testing, always connect an appropriate flushing device **before** the engine is started and leave it turned on until **after** the engine is shut off.

✳✳ WARNING

ANYTIME the engine is run, the first thing you should do is check the cooling water indicator (a stream on most, but a mist on some early motors). All models covered by this service are equipped with some form of a cooling stream indicator towards the aft portion of the lower engine cover. Anytime the engine is operating, either a steady stream of water or a constant mist should come from the indicator, showing that the pump is supplying water to the engine for cooling. If the stream or mist is ever absent, stop the motor and determine the cause before restarting.

As we stated earlier, flushing the cooling system consists of supplying fresh, clean water to the system in order to clean deposits from the internal passages. If the engine is running, the water does not normally have to be pressurized, as it is delivered through the normal water intake passages and the water pump (the system can self flush if supplied with clean water). If your engine can be placed in a test tank that is filled with fresh, clean water, then in theory, simply running it will self-flush the motor. But, some larger 2-cylinder motors are of a high enough horsepower application that their size may make this impractical. For this reason, you may have to come up with another flushing method (i.e. using a garden hose and, in most cases, an adapter, to deliver the fresh water).

All Johnson/Evinrude engines will either accept flush fittings or adapters. Most marina's or boat supply shops will carry adapters of the generic type that are designed to fit over the engine water intakes on the lower unit (and resemble a pair of strange earmuffs with a hose fitting on one side). But, some models require special adapters (available from the manufacturer) that thread into special flushing fittings on the powerhead. When using the later type adapter, follow the manufacturer's instructions closely regarding flushing conditions. In some cases, flushing with this type of adapter should occur only with the motor turned off, so as to prevent damage to the water pump impeller or other engine components. This varies with each motor, so be sure to check with your dealer regarding these direct to the powerhead adapters when you purchase one.

■ **When running the engine on a flushing adapter and a garden hose, make sure the hose delivers about 20-40 psi (140-300 kPa) of pressure.**

✳✳ CAUTION

For safety, the propeller should be removed ANYTIME the motor is run on the trailer or on an engine stand. We realize that this is not always practical when flushing the engine on the trailer, but cannot emphasize enough how much caution must be exercised to prevent injury to you or someone else. Either take the time to remove the propeller or take the time to make sure no one or nothing comes close enough to it to become injured. Serious personal injury or death could result from contact with the spinning propeller.

When using a flushing device and a pressurized water source, most motors can be flushed in either a tilted or a vertical position, BUT, the manufacturer specifically warns against flushing most motors in the tilted position IF the engine running. Some models can be seriously damaged by attempting to flush them with the engine running in the full tilt position. If the motor must be flushed tilted (dockside) then your best bet it to do so with the engine shut off.

1. Check the engine top case and, if necessary remove it to check the powerhead, to ensure it is cooled enough to flush without causing thermal shock.

2. Prepare the engine for flushing by attaching the adapter following the instructions that came with the adapter. Then, attach the garden hose to the adapter.

■ **When using a clamp-type adapter, position the suction cup(s) over water intake grate(s) in such a way that they form tight seals. A little pressure seepage should not be a problem, but look to the water indicator once the motor is running to be sure that sufficient water is reaching the powerhead.**

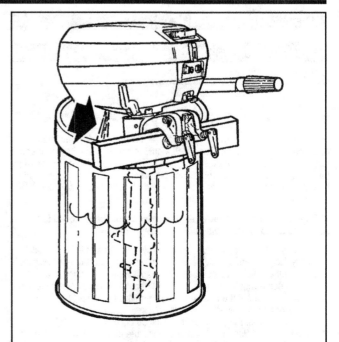

Fig. 24 All models must be flushed or run in a test tank. Smaller ones, in a garbage can

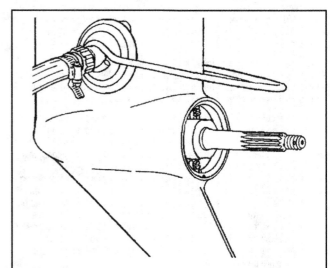

Fig. 25 The easiest way to flush some models is using a clamp-type adapter. . .

Flush attachment

Fig. 26 A water source must be used ANYTIME the engine is started

3. If the motor is to be run (during flushing or for testing), position the outboard vertically and remove the propeller, for safety. Also, be sure to position the water hose so it will not contact with moving parts (tie the hose out of the way with mechanic's wire or wire ties, as necessary).

4. Unless using a test tank, turn the water on, making sure that pressure does not exceed 45 psi (300 kPa).

5. If the motor must be run for testing/tuning procedures, start the engine and run in neutral until the motor reaches operating temperature. For most motors, the motor will continue to run at fast idle until warmed.

✳✳ WARNING

Unless you are flushing a motor through the water indicator fitting (which is the case on some large late-model Johnson/Evinrude motors) check the cooling system indicator as SOON as the engine is started. It must be present and strong as long as the motor is operated. If not, stop the motor and rectify the problem before proceeding. Common problems could include insufficient water pressure or incorrect flush adapter installation. When you are flushing a motor through the cooling indicator fitting(s), water should exit through the other passages at the bottom of the drive unit.

6. Flush the motor for at least 5-10 minutes or until the water exiting the engine is clear. When flushing while running the motor, check the engine temperature (using a gauge or carefully by touch) and stop the engine immediately if steam or overheating starts to occur. Make sure that carbureted motors slow to low idle for the last few minutes of the flushing procedure.

7. Stop the engine (if running), **then** shut the water off.

8. Remove the adapter from the engine or the engine from the test tank, as applicable.

9. If flushing did not occur with the motor running (so the motor would already by vertical), be sure to place it in the full vertical position allowing the cooling system to drain. This is especially important if the engine is going to be placed into storage and could be exposed to freezing temperatures. Water left in the motor could freeze and crack the powerhead or lower unit.

2-Stroke Engine Oil

OIL RECOMMENDATIONS

◆ See Figure 27

■ Johnson/Evinrude recommends the use of Carbon Guard fuel additive to help prevent the build-up of harmful carbon deposits in the combustion chambers. The manufacturer also recommends de-carboning the pistons, twice as often, if Carbon Guard is NOT used.

Use only an NMMA (National Marine Manufacturers Association) certified TC-W3 or equivalent 2-stroke lubricant. Of course, the manufacturer recommends using Johnson/Evinrude brand oils, since they are specially formulated to match the needs of Johnson/Evinrude motors. In all cases, a high quality TC-W3 oils are proprietary lubricants designed to ensure optimal engine performance and to minimize combustion chamber deposits, to avoid

detonation and prolong spark plug life. Use only 2-stroke type outboard oil. Never use automotive motor oil.

■ Remember, it is this oil, mixed with the gasoline that lubricates the internal parts of the 2-stroke engine. Lack of lubrication due to the wrong mix or improper type of oil can cause catastrophic powerhead failure.

FUEL:OIL RATIO

The proper fuel:oil ratio is 50:1 for **normal** operating conditions. Most manufacturers define normal as a motor operated under varying conditions from idle to wide open throttle, without excessive amounts of use at either. Unfortunately, no one seems to put a definition to "excessive amount" either, so you'll have to use common sense. We don't think an hour of low speed trolling mixed in with some high speed operation or an hour or two of pulling a skier constitutes "excessive amounts," but you'll have to make your own decision. Also necessary for defining normal operating conditions is the ambient and seawater temperatures. The seawater temperatures should be above 32° F (0° C) and below 68° F (20° C). Ambient air conditions should be above freezing and below the point of extreme discomfort (90-100° F).

■ **The fuel/oil ratios listed here are Johnson/Evinrude recommendations given in service literature. Because your engine may differ slightly from service manual specification, refer to your Owner's manual or a reputable dealer to be certain that your mixture meets your conditions of use.**

If an outboard is to be used under severe conditions including, long periods of idle, long periods of heavy load, used in severe ambient temperatures (outside the range of normal use) or under high performance (constant wide-open throttle or racing conditions) some adjustment may be necessary to the fuel:oil ratio. Most carbureted outboards require a 25:1 ratio for severe and high performance conditions.

■ **No additional oiling is necessary for these engines when used in commercial, rental or extended severe service other than high performance applications.**

All motors covered here require a 25:1 ratio during some portion of break-in (the first 20 hours of operation).

PRE-MIX

◆ See Figures 28 and 29

Mixing the engine lubricant with gasoline before pouring it into the tank is by far the simplest method of lubrication for 2-stroke outboards. However, this method is the messiest and potentially causes the most amount of harm to our environment.

The most important part of filling a pre-mix system is to determine the proper fuel/oil ratio. Most operating conditions require a 50:1 ratio (that is 50 parts of fuel to 1 part of oil). Consult the information in this section on Fuel:Oil Ratio and your owner's manual to determine what the appropriate ratio should be for your engine.

Fig. 27 This scuffed piston is an example of the damage caused by improper 2-stroke oil or mixture

Fig. 28 Add the oil and gasoline at the same time, or add the oil first, then add the gasoline to ensure proper mixing

Fig. 29 Portable tanks are nice because you can agitate them to mix the oil further

The procedure itself is uncomplicated, but you've got a couple options depending on how the fuel tank is set-up for your boat. To fill an empty portable tank, add the appropriate amount of oil to the tank, then add gasoline and close the cap. Rock the tank from side-to-side to gently agitate the mixture, thereby allowing for a thorough mixture of gasoline and oil. When just topping off built-in or larger portable tanks, it is best to use a separate 3 or 6 gallon (11.4 or 22.7 L) mixing tank in the same manner as the portable tank noted earlier. In this way a more exact measurement of fuel occurs in 3 or 6 gallon increments (rather than just adding fuel directly to the tank and possibly realizing that you've just added 2.67 gallons of gas and need to add, uh, a little less than 8 oz of oil, but exactly how many ounces would that be?) Use of a separate mixture tank will prevent the need for such mathematical equations. Of course, the use of a mixing tank may be inconvenient or impossible under certain circumstances, so the next best method for topping off is to take a good guess (but be a little conservative to prevent an excessively rich oil ratio). Add the oil and gasoline at the same time, or add the oil first, then add the gasoline to ensure proper mixing. For measurement purposes, it would obviously be more exact to add the gasoline first and then add a suitable amount of oil to match it. The problem with adding gasoline first is that unless the tank could be thoroughly agitated afterward (and that would be **really** difficult on built-in tanks), the oil might not mix properly with the gasoline. Don't take that unnecessary risk.

To determine the proper amount of oil to add to achieve the desired fuel:oil ratio, refer to the Fuel:Oil Ratio chart at the end of this section.

Lower Unit (Gearcase) Oil

◆ See Figures 4 and 30

Regular maintenance and inspection of the lower unit is critical for proper operation and reliability. A lower unit can quickly fail if it becomes heavily contaminated with water or excessively low on oil. The most common cause of a lower unit failure is water contamination.

Water in the lower unit is usually caused by fishing line or other foreign material, becoming entangled around the propeller shaft and damaging the seal. If the line is not removed, it will eventually cut the propeller shaft seal and allow water to enter the lower unit. Fishing line has also been known to cut a groove in the propeller shaft if left neglected over time. This area should be checked frequently.

Fig. 30 This lower unit was destroyed because the bearing carrier froze due to lack of lubrication

OIL RECOMMENDATIONS

◆ See Figure 31

Use only Johnson/Evinrude Hi-Vis or Ultra-HPF gearcase lubricant (or another suitable equivalent high quality, marine gearcase lubricant that meets GL5 specifications). In both cases, these oils are proprietary lubricants designed to ensure optimal performance and to minimize corrosion in the lower unit.

■ Remember, it is this lower unit lubricant that prevents corrosion and lubricates the internal parts of the drive gears. Lack of lubrication due to water contamination or the improper type of oil can cause catastrophic lower unit failure.

CHECKING LOWER UNIT OIL LEVEL & CONDITION

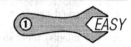

◆ See Figures 32, 33 and 34

✳✳ CAUTION

On certain lower units, the Phillips screw securing the shift fork in place may be located very close to the vent or drain screw. On some units the Phillips screw is located on the other side. Either way, the DRAIN and VENT SCREWS SHOULD BE FLAT HEAD (slotted) NOT PHILLIPS. If the wrong screw is removed, BAD NEWS, VERY BAD NEWS. The lower unit will have to be disassembled in order to return the shift fork to its proper location.

Visually inspect the lower unit before and after each use for signs of leakage. At least monthly, or as needed, remove the lower unit level plug in order to check the lubricant level and condition as follows:

■ Certain very low hp motors do not have separate drain and vent screws.

1. Position the engine in the upright position with the motor shut off for at least 1 hour. Whenever possible, check the level overnight cold in order to

Fig. 31 Use Johnson/Evinrude Ultra-HPF, Johnson/Evinrude Hi-Vis or equivalent marine gear oil

Fig. 32 On most motors, the vent/level plug is on top, while the drain/fill plug is at the bottom of the gearcase (newer models shown)

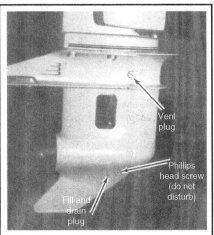

Fig. 33 On most motors, the vent/level plug is on top, while the drain/fill plug is at the bottom of the gearcase (port or starboard)

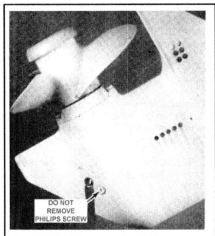

Fig. 34 DO NOT remove the Phillips screw securing the shift fork on models so equipped!

get the best indication of the level without having to account for heat expansion.

2. Disconnect the negative battery cable and/or remove the propeller for safety.

✳✳ CAUTION

Always observe extreme care when working anywhere near the propeller. Take steps to ensure that no accidental attempt to start the engine occurs while work is being performed or remove the propeller completely to be safe.

3. Position a small rag or drain pan under the lower unit, then unthread and remove the vent/level plug (remember it's a slotted head screw, NOT a Phillips). The level of lubricant should come up just even with the bottom of the threads (if the gearcase was filled properly). If lubricant does not come all the way up, check for signs of leakage. If lubricant leaks out, either the case was overfilled or water has entered the case as well. In all cases, it is best to take a small sample of lubricant from the drain plug in order to examine it further.

■ If a large amount of lubricant escapes when the vent/level plug is removed, either the lower unit was seriously overfilled on the last service, the crankcase is still too hot from the last use (and the fluid is expanded) or a large amount of water has entered the lower unit. If the later is true, some water should escape before the oil and/or the oil will be a milky white in appearance (showing the moisture contamination).

4. To take small sample, unthread the drain/fill plug at the bottom of the housing and allow a small sample (a teaspoon or less) to drain from the lower unit. Quickly install the drain/filler plug and tighten securely.

5. Examine the gear oil as follows:
   a. Check the oil for obvious signs of water. A small amount of moisture may be present from condensation, especially if a motor has been stored for some time, but a milky appearance indicates that either the fluid has not been changed in ages or the lower unit allowing some water to intrude. If significant water contamination is present, the first suspect is the propeller shaft seal.
   b. Dip an otherwise clean finger into the oil and then rub a small amount of the fluid between your finger and your thumb to check for the presence of debris. The lubricant should feel smooth. A **very** small amount of metallic shavings may be present, but should not really be felt. Large amounts of grit or metallic particles indicate a probable need to overhaul the lower unit looking for damaged/worn gears, shafts, bearings or thrust surfaces.

6. If it is necessary to add fluid, a very small amount of fluid may be added through the level plug, but larger amounts of fluid should be added through the drain/filler plug opening to make certain that the case is properly filled. If necessary, add gear oil until fluid flows from the level/vent opening. If much more than 1 oz. (29 ml) is required to fill the lower unit, check the case carefully for leaks. Install the drain/filler plugs and/or the level/vent plug, tightening both securely.

■ One trick that makes adding lower unit oil less messy is to install the vent/level plug BEFORE removing the pump from the drain/fill opening and threading the drain/fill plug back into position. This creates a partial vacuum, which will slow the leakage of gearcase oil out of the drain/fill opening while you are attempting to rethread the plug.

7. Once fluid is pumped into the lower unit, let the unit sit in a shaded area for at least 1 hour for the fluid to settle. Recheck the fluid level and, if necessary, add more lubricant.

8. Install the propeller and/or connect the negative battery cable, as applicable.

DRAINING & FILLING

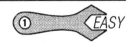

◆ See Figures 31 and 35 thru 40

Avoid removing the vent/level or drain/fill plugs when the lower unit is hot as expanded lubricant will be forced through the hole.

✳✳ WARNING

The EPA warns that prolonged contact with oils may cause a number of skin disorders, including cancer! You should make every effort to minimize your exposure to used engine oil. Protective gloves should be worn when changing the oil. Wash your hands and any other exposed skin areas as soon as possible after exposure to used engine oil. Soap and water or waterless hand cleaner should be used.

1. Place a suitable container under the lower unit.

✳✳ CAUTION

On certain lower units, the Phillips screw securing the shift fork in place may be located very close to the vent or drain screw. On some units the Phillips screw is located on the other side. Either way, the DRAIN and VENT SCREWS SHOULD BE FLAT HEAD (slotted) NOT PHILLIPS. If the wrong screw is removed, BAD NEWS, VERY BAD NEWS. The lower unit will have to be disassembled in order to return the shift fork to its proper location.

2. Loosen the oil vent/level plug on the lower unit. This step is important! If the oil vent/level cannot be loosened or removed, you will have a VERY difficult time adding oil.

■ Certain very low hp motors do not have separate drain and vent screws.

3. Remove the drain/fill plug from the lower end of the gear housing followed by the oil vent/level plug.

4. Allow the lubricant to completely drain from the lower unit.

5. If applicable, check the magnet end of the drain screw for metal particles. Some amount of metal is considered normal wear is to be expected but if there are signs of metal chips or excessive metal particles, the lower unit needs to be disassembled and inspected.

6. Inspect the lubricant for the presence of a milky white substance, water or metallic particles. If any of these conditions are present, the lower unit should be serviced immediately.

■ **Certain low hp models do not have a vent screw. Therefore, this unit must be laid in a horizontal position for filling and time taken to allow the lubricant to work into the lower unit cavity by raising the skeg slightly from time-to-time.**

7. Place the outboard in the proper position for filling the lower unit. The lower unit should not list to either port or starboard and should be completely vertical.

8. Insert the lubricant tube into the oil drain hole at the bottom of the lower unit and inject lubricant until the excess begins to come out the oil level hole.

■ **The lubricant must be filled from the bottom to prevent air from being trapped in the lower unit. Air could temporarily displace lubricant and causes an improperly full reading that would lead to a lack of lubrication in the lower unit.**

9. Oil should be squeezed in using a tube or with the larger quantities, by using a pump kit to fill the lower unit through the drain plug.

■ **One trick that makes adding lower unit oil less messy is to install the vent/level plug BEFORE removing the pump from the drain/fill opening and threading the drain/fill plug back into position.**

10. Using new gaskets (washers), install the oil level/vent plug first and then install the oil fill plug.

11. Wipe the excess oil from the lower unit and inspect the unit for leaks.

12. Place the used lubricant in a suitable container for transportation to an authorized recycling facility.

Fuel Filter

A fuel filter is designed to keep particles of dirt and debris from entering the carburetors and clogging the tiny internal passages of either. A small speck of dirt or sand can drastically affect the ability of the fuel system to deliver the proper amount of air and fuel/oil to the engine. If a filter becomes clogged, the flow of gasoline will be impeded. This could cause lean fuel mixtures, hesitation and stumbling and idle problems in carburetors.

Regular cleaning or replacement of the fuel filter (depending on the type or types used) will decrease the risk of blocking the flow of fuel to the engine, which could leave you stranded on the water. It will also decrease

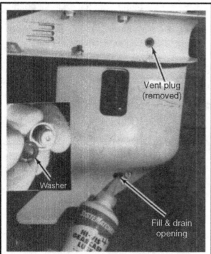

Fig. 35 Lower unit oil is pumped or squeezed into the lower unit through the filler opening, while the vent opening is removed to let air escape

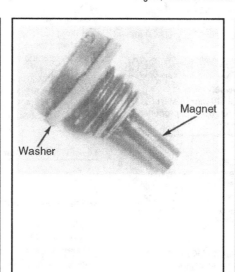

Fig. 36 Later model units will utilize a magnetic drain plug

Fig. 37 When draining a unit with no shifting, you needn't worry about the shift screw

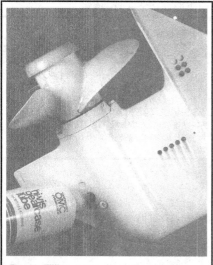

Fig. 38 Filling a propeller exhaust unit

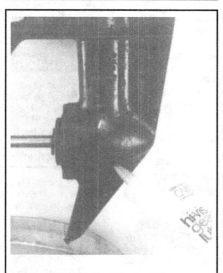

Fig. 39 Filling a non-shifting unit

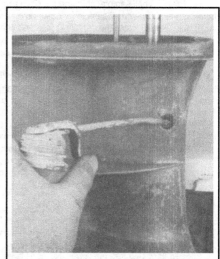

Fig. 40 A small squirt can be used to top off the last little bit

the risk of damage to the small passages of a carburetor that could require more extensive and expensive replacement. Keep in mind that fuel filters are usually pretty inexpensive (at lease when compared to a tow) and replacement is a simple task. Service your fuel filter on a regular basis to avoid fuel delivery problems.

The type of fuel filter used on your engine will vary not only with the year and model, but also with the accessories and rigging. Because of the number of possible variations it is impossible to accurately give instructions based on model. Instead, we will provide instructions for the different types of filters the manufacturer used on various families of motors or systems with which they are equipped. To determine what filter(s) are utilized by your boat and motor rigging, trace the fuel line from the tank to the fuel pump and then from the pump to the carburetors.

In addition to the fuel filter mounted on the engine, a filter is usually found inside or near the fuel tank. Because of the large variety of differences in both portable and fixed fuel tanks, it is impossible to give a detailed procedure for removal and installation. Most in-tank filters are simply a screen on the pick-up line inside the fuel tank. Filters of this type usually only need to be cleaned and returned to service (assuming they are not torn or otherwise damaged). Fuel filters on the outside of the tank are typically of the inline type and are replaced by simply removing the clamps, disconnecting the hoses and installing a new filter. When installing the new filter, make sure the arrow on the filter points in the direction of fuel flow.

FUEL PUMP FILTERS

◆ See Figures 41 thru 44

✳✳ CAUTION

Observe all applicable safety precautions when working around fuel. Whenever servicing the fuel system, always work in a well-ventilated area. Do not allow fuel spray or vapors to come in contact with a spark or open flame. Do not smoke while working around gasoline. Keep a dry chemical fire extinguisher near the work area. Always keep fuel in a container specifically designed for fuel storage; also, always properly seal fuel containers to avoid the possibility of fire or explosion.

Many models will be equipped with only a small, flat, vacuum (pulse) driven fuel pump mounted somewhere on the powerhead. Although the exact shape and design of this pump may vary slightly from model-to-model, for this discussion, they are serviced virtually the same way. The various mechanical fuel pumps normally contain a serviceable fuel filter screen mounted just underneath the fuel inlet cover. On all versions, the cover is connected to the fuel inlet hose from the fuel tank. Additionally, on many models the cover is usually round and is retained by a single bolt at the center.

To service the fuel inlet screen on these fuel pumps, remove the inlet cover screw(s), then carefully separate the inlet cover, gasket or O-ring and screen from the fuel pump body. Clean the screen using a suitable solvent and blow dry with low pressure compressed air or allow it to air dry. Once the screen is dry, check it carefully for clogs or tears and replace, if necessary. Depending on the gasket material and condition it may be reused, but it is normally best (and safest) to simply replace the gasket(s).

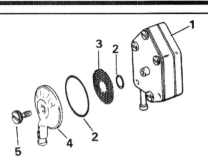

① FUEL PUMP BODY
② O-RING or GASKET
③ FILTER ELEMENT
④ INLET COVER
⑤ SCREW

Fig. 41 Versions of a fuel pump filter are found on most models covered here

✳✳ CAUTION

To prevent the danger of fire of explosion, pressurize the fuel system after service by slowly squeezing the fuel primer bulb until firm. Then, once the system is pressurized, inspect it carefully for leaks, especially around the fuel pump/cover.

■ Some fuel pump motors will also be equipped with an inline filter. When present, be sure to replace all inline filters at least annually.

INLINE FILTERS

◆ See Figure 45

✳✳ CAUTION

Observe all applicable safety precautions when working around fuel. Whenever servicing the fuel system, always work in a well-ventilated area. Do not allow fuel spray or vapors to come in contact with a spark or open flame. Do not smoke while working around gasoline. Keep a dry chemical fire extinguisher near the work area. Always keep fuel in a container specifically designed for fuel storage; also, always properly seal fuel containers to avoid the possibility of fire or explosion.

Johnson/Evinrude has used different types of inline filters on these models. Some of them are sealed and therefore cannot be cleaned and reused. A few of them however utilize 2-piece housings that can be opened

Fig. 42 Remove the bolt(s) securing the pump cover. . .

Fig. 43 . . .then remove the inlet cover to access the filter element

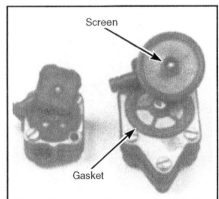

Fig. 44 Two separate fuel pumps and their internal filters

to access the filter element. Generally speaking though, most models are equipped with the non-serviceable filter design.

■ Because of the relative ease and relatively low expense of an inline filter (when compared with the time and hassle of a carburetor overhaul or the expense of a tow) we encourage you to replace the filter at least annually.

To replace a non-serviceable inline filter, release the hose clamps (they are usually equipped with spring-type clamps that are released by squeezing the tabs using a pair of pliers) and slide them back on the hose, past the raised portion of the filter inlet/outlet nipples. Once a clamp is released, position a small drain pan or a shop towel under the filter and carefully pull the hose from the nipple. Allow any fuel remaining in the filter and fuel line to drain into the drain pan or catch fuel with the shop towel. Repeat on the other side, noting which fuel line connects to which portion of the filter (for assembly purposes). Inline filters are usually marked with an arrow indicating fuel flow. The arrow should point towards the fuel line that runs to the motor (not the fuel tank).

Before installation of the new filter, make sure the hoses are in good condition and not brittle, cracking and otherwise in need of replacement. During installation, be sure to fully seat the hoses and then place the clamps over the raised portions of the nipples to secure them. Spring clamps will weaken over time, so replace them if they've lost their tension. If wire ties or adjustable clamps were used, be careful not to over-tighten the clamp. If the clamp cuts into the hose, it's too tight; loosen the clamp or cut the wire tie (as applicable) and start again.

Fig. 45 Typical disposable Johnson/Evinrude inline filter

SEDIMENT BOWL FILTERS

◆ See Figures 46, 47 and 48

✳✳ CAUTION

Observe all applicable safety precautions when working around fuel. Whenever servicing the fuel system, always work in a well-ventilated area. Do not allow fuel spray or vapors to come in contact with a spark or open flame. Do not smoke while working around gasoline. Keep a dry chemical fire extinguisher near the work area. Always keep fuel in a container specifically designed for fuel storage; also, always properly seal fuel containers to avoid the possibility of fire or explosion.

With some rags handy, loosen the bowl yoke screw and slide off the yoke and bowl.

Remove the filter element and gasket. Clean the bowl thoroughly with solvent and dry it completely.

Install a new element and gasket, position the bowl and then tighten the yoke screw securely.

Propeller

◆ See Figures 49, 50 and 51

The propeller is mounted to the lower unit propeller shaft using a nut that is in turn secured either by a shear pin, cotter pin through the castellations on the nut itself, or through a separate nut keeper. The propeller is driven by a splined connection to the shaft and the rubber drive hub found inside the propeller. The rubber hub provides a cushioning that allows softer shifts, but more importantly, provides some measure of protection for the lower unit components in the event of an impact. The amount of force necessary to break the hub is supposed to be just less than the amount of force necessary to cause lower unit component damage. In this way, the hope is that the propeller and hub will be sacrificed in the event of a collision, but the more expensive lower unit components will survive unharmed. Although these systems do supply a measure of protection, this, unfortunately, is not always the case and lower unit component damage will still occur with the right impact or with a sufficient amount of force.

✳✳ WARNING

Do not use excessive force when removing the propeller from the shaft as excessive force can result in damage to the propeller, shaft and, even other lower unit components. If the propeller cannot be removed by normal means, consider having a reputable marine shop remove it. Using heat or impact wrenches to free the propeller will likely lead to damage.

Fig. 46 Some models use a sediment bowl filter on the pump

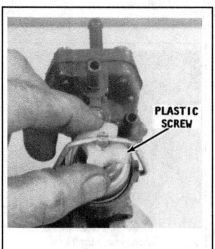

Fig. 47 Loosen the plastic screw. . .

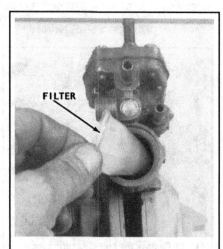

Fig. 48 . . .and lift out the filter

■ Clean and lubricate the propeller and shaft splines using a high-quality, water-resistant, marine grease every time the propeller is removed from the shaft. This will help keep the hub from seizing to the shaft due to corrosion (which would require special tools to remove without damage to the shaft or gearcase.)

Many outboards are equipped with aftermarket propellers. Because of this, the attaching hardware may differ slightly from what is shown. Contact a reputable propeller shop or marine dealership for parts and information on other brands of propellers.

INSPECTION

◆ See Figures 49, 50 and 51

The propeller is secured to the gearcase propshaft either by a drive pin on smaller motors, or by a castellated hex nut on larger motors.

On models secured by a hex nut, the propeller is driven by a splined connection to the shaft and the rubber drive hub found inside the propeller. The rubber hub provides a cushioning that allows softer shifts, but more importantly, it provides some measure of protection for the gearcase components in the event of an impact.

On motors where the propeller is retained by a drive pin, impact protection is provided by the drive pin itself (also known as a shear pin for this reason). The pin is designed to break or shear when a specific amount of force is applied because the propeller hits something. In both cases (rubber hubs or shear pins) the amount of force necessary to break the hub or shear the pin is supposed to be just less than the amount of force necessary to cause gearcase component damage. In this way, the hope is that the propeller and hub or shear pin will be sacrificed in the event of a collision, but the more expensive gearcase components will survive unharmed. Although these systems do supply a measure of protection, this, unfortunately, is not always the case and gearcase component damage will still occur with the right impact or with a sufficient amount of force.

The propeller should be inspected before and after each use to be sure the blades are in good condition. If any of the blades become bent or nicked, this condition will set up vibrations in the motor. Remove and inspect the propeller. Use a file to trim nicks and burrs. Take care not to remove any more material than is absolutely necessary.

✳✳ CAUTION

Never run the engine with serious propeller damage, as it can allow for excessive engine speed and/or vibration that can damage the motor. Also, a damaged propeller will cause a reduction in boat performance and handling.

Also, check the rubber and splines inside the propeller hub for damage. If there is damage to either of these, take the propeller to your local marine dealer or a "prop shop". They can evaluate the damaged propeller and determine if it can be saved by rehubbing.

Additionally, the propeller should be removed every 100 hours of operation or at the end of each season, whichever comes first for cleaning, greasing and inspection.

Whenever the propeller is removed, apply a fresh coating of Johnson/Evinrude Triple-Guard or an equivalent water-resistant, marine grease to the propeller shaft and the inner diameter of the propeller hub. This is necessary to prevent possible propeller seizure onto the shaft that could lead to costly or troublesome repairs. Also, whenever the propeller is removed, any material entangled behind the propeller should be removed before any damage to the shaft and seals can occur. This may seem like a waste of time at first, but the small amount of time involved in removing the propeller is returned many times by reduced maintenance and repair, including the replacement of expensive parts.

■ Propeller shaft greasing and debris inspection should occur more often depending upon motor usage. Frequent use in salt, brackish or polluted waters would make it advisable to perform greasing more often. Similarly, frequent use in areas with heavy marine vegetation, debris or potential fishing line would necessitate more frequent removal of the propeller to ensure the gearcase seals are not in danger of becoming cut.

CLEARING THE FISH LINE TRAP

Some motors by this service are equipped with a special propeller thrust washer. It contains an integral fishing line trap to keep line that becomes entangled from cutting the propeller shaft seal. For models so equipped, the manufacturer recommends removing the propeller every 15-20 hours (or anytime fishing line may become entangled) in order to check and clean the trap. Whenever the propeller and thrust washer are removed, the washer must be positioned with the line trap groove facing the gearcase in order to work properly. Always note the direction of the trap groove during removal.

REMOVAL & INSTALLATION

Propellers w/o A Shear Pin (Castellated Nut Or Keeper Retained)

◆ See Figures 52 thru 56

There are essentially 2 slightly different ways that propellers are secured to the propshaft on these outboards. Both place a thrust washer over the shaft, followed by the propeller, a spacer and a nut that is tightened to secure the assembly. The difference comes in the size of the nut and the method that the nut is secured to keep it from loosening in service. Some of the Johnson/Evinrude motors use a slotted or castellated nut (so named because, when viewed from the side, it appears similar to the upper walls or tower of a castle.) On these models a cotter pin is placed through the slots in the nut in order to lock it in place. The rest of the models use a larger, standard flat-sided nut and a separate keeper that is fitted in place over the nut and then secured using a cotter pin.

You'll notice that in both cases, the nut is locked in place by a cotter pin to ensure that it cannot loosen while the motor is running. The pin passes through a hole in the propeller shaft, as well as through the notches in the sides of the castellated nut or the keeper. Install a new cotter pin anytime the propeller is removed and, perhaps more importantly, make sure the cotter pin is of the correct size and is made of materials designed for marine use.

Fig. 49 This propeller is long overdue for repair or replacement

Fig. 50 Although minor damage can be dressed with a file. . .

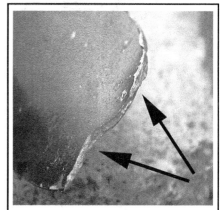

Fig. 51 . . . a propeller specialist should repair large nicks or damage

Make sure that you include the cotter pin in all pre- and post-launch checks.

Whenever working around the propeller, check for the presence of black rubber material in the drive hub and spline grease. Presence of this material normally indicates that the hub has turned inside the propeller bore (have the propeller checked by a propeller repair shop). Keep in mind that a spun hub will not allow proper torque transfer from the motor to the propeller and will allow the engine to over-rev in order to produce thrust (or will just over-rev producing little or no thrust). If the propeller has spun on the hub it has been weakened and is more likely to fail completely in use.

1. For safety, disconnect the negative cable (if so equipped) and/or disconnect the spark plug leads from the plugs (ground the leads to prevent possible ignition damage should the motor be cranked at some point before the leads are reconnected to the spark plugs).

✳✳ CAUTION

Don't ever take the risk of working around the propeller if the engine could accidentally be cranked or started. Always take precautions such as disconnecting the spark plug leads and, if equipped, the negative battery cable.

2. Cut the ends off the cotter pin (as that is easier than trying to straighten them in most cases). Next, free the pin by grabbing the head with a pair of needle nose pliers. Either tap on the pliers gently with a hammer to help free the pin from the nut or carefully use the pliers as a lever by prying back against the castellated nut. Discard the cotter pin once it is removed.

3. On models with a separate nut keeper, pull the keeper free and place it aside for installation purposes.

4. Place a block of wood between the propeller and the anti-ventilation housing to lock the propeller and shaft from turning, then loosen and remove the castellated nut. Note the orientation, then remove the splined spacer from the propeller shaft.

5. Slide the propeller from the shaft. If the prop is stuck, use a block of wood to prevent damage and carefully drive the propeller from the shaft.

■ **If the propeller is completely seized on the shaft, refer to Frozen Propeller in the Lower Unit section for details on how it can be removed. But use care, don't risk damage to the propeller or gearcase by applying excessive force.**

6. Note the direction in which the thrust washer is facing (the shoulder is normally positioned to the aft, facing the propeller). Remove the thrust washer from the propshaft (if the washer appears stuck, tap lightly to free it from the propeller shaft).

7. Clean the thrust washer, propeller and shaft splines of any old grease. Small amounts of corrosion can be removed carefully using steel wool or fine grit sandpaper.

8. Inspect the shaft for signs of damage including twisted splines or excessively worn surfaces. Rotate the shaft while looking for any deflection. Replace the propeller shaft if these conditions are found. Inspect the thrust washer for signs of excessive wear or cracks and replace, if found.

To Install:

9. Apply a light coat of Johnson/Evinrude Triple-Guard or equivalent high-quality, water-resistant, marine grease to all surfaces of the propeller shaft and to the splines inside the propeller hub.

10. Position the thrust washer over the propshaft in the direction noted during removal. On all models, the shoulder should normally face the propeller.

11. Carefully slide the propeller onto the propshaft, rotating the propeller to align the splines. Push the propeller forward until it seats against the thrust washer.

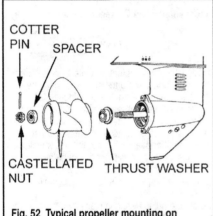

Fig. 52 Typical propeller mounting on models utilizing a castellated nut

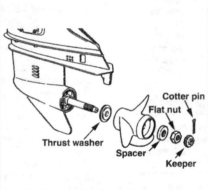

Fig. 53 Typical propeller mounting on models using a flat nut with a separate keeper

Fig. 54 Although designs vary, propeller nuts are secured with a cotter pin

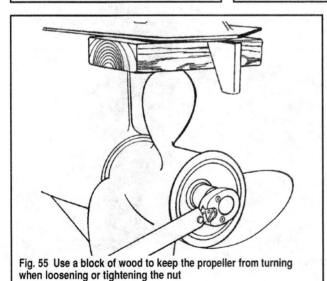

Fig. 55 Use a block of wood to keep the propeller from turning when loosening or tightening the nut

Fig. 56 Notice the cotter pin is gently spread, NOT bent 90° or more

12. Install the splined spacer onto the propeller shaft, as noted during removal.

13. Place a block of wood between the propeller and housing to hold the prop from turning, then thread the castellated nut onto the shaft with the cotter pin grooves facing outward.

14. Tighten the nut securely.

15. If used, install the keeper over the flat-nut.

16. Install a new cotter pin through the grooves in the nut or the keeper (as applicable) that align with the hole in the propshaft. If the cotter pin hole and the grooves do not align, tighten the nut additionally, just enough to align them (**do not** loosen the nut to achieve alignment.) Once the cotter pin is inserted, **spread** the ends sufficiently to lock the pin in place. Do not bend the ends over at 90° or greater angles as the pin will lose tension and rattle in the slot.

17. Connect the spark plug leads and/or the negative battery cable, as applicable.

Propellers w/Shear Pin

◆ See Figures 57 thru 62

❊❊ CAUTION

Check to be sure the high-tension leads to the spark plugs have been removed from the plugs BEFORE starting to remove the propeller. This simple safety task will prevent the engine from accidentally "firing" while the propeller is being removed. Such action could cause SEVERE PERSONAL INJURY.

Most units covered here have a shear pin located between the propeller nut and the propeller. The propeller nut should be removed and the shear pin checked.

1. Remove the cotter key, and then remove the propeller nut.

2. Remove the shear pin and washer. Because the shear pin is not a tight fit, the propeller is able to move on the pin and cause burrs on the hole. The propeller may be difficult to remove because of these burrs. To overcome this problem, the propeller hub has 2 grooves running the full length of the hub. Hold the shaft from turning, and then rotate the propeller 1/4 turn to position the grooves over the drive pin holes. The propeller can then be pulled straight off the shaft.

3. After the propeller has been removed, file the drive pin holes on both sides of the shaft to remove the burrs.

To Install:

The propeller washer and shear pin, play an extremely important role. When shifting gears during normal operation, or if the propeller should hit an underwater obstacle, the propeller is subjected to considerable shock. A washer is installed between the propeller and drive pin. This washer MUST always be in place for proper operation. If the hub should slip, the propeller will move back towards the propeller nut and lock against the drive pin. The washer is designed to stop propeller movement so the drive pin can be easily removed for service. Now, on with the installation.

4. Coat the propeller shaft with anti-corrosion grease.

5. Install the propeller with the drive pin holes aligned. Install the washer and shear pin.

6. Slide the propeller cap into place and secure it with the cotter pin.

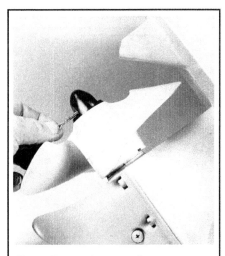

Fig. 57 Remove the cotter pin. . .

Fig. 58 . . .and then pull off the cap/nut

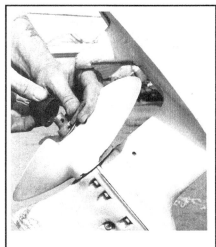

Fig. 59 Remove the shear pin and prop nut. . .

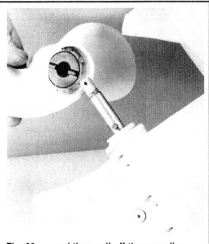

Fig. 60 . . .and then pull off the propeller

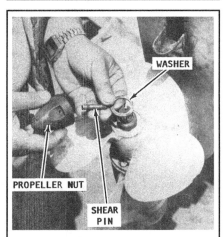

Fig. 61 A good look at the hardware on a shear pin propeller

Fig. 62 Note that the prop will have 2 grooves running the length of the hub

Exhaust Propellers

◆ **See Figures 63 thru 71**

Propellers with the exhaust passing through the hub MUST be removed more frequently than the standard propeller. Removal after each weekend use, or outing, is not considered excessive.

These propellers do not have a shear pin. The shaft and propeller have splines which MUST be coated with anti-corrosion lubricant prior to installation as an aid to removal the next time the propeller is pulled. Even with the lubricant applied to the shaft splines, the propeller may be difficult to remove. If this type propeller is "frozen" to the shaft, see the Frozen Propeller procedures in the Lower Unit section for special instructions.

The propeller with the exhaust hub is more expensive than the standard propeller and therefore, the cost of rebuilding the unit, if the hub is damaged, is justified.

A replaceable diffuser ring on the backside of the propeller disperses the exhaust away from the propeller blades as the boat moves through the water. If the ring becomes broken or damaged "ventilation" would be created pulling the exhaust gases back into the negative pressure area behind the propeller. This condition would create considerable air bubbles and reduce the effectiveness of the propeller.

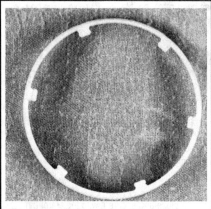

Fig. 63 A good look at the diffuser ring on its own. . .

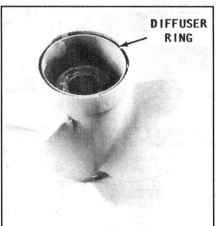

Fig. 64 . . .and installed

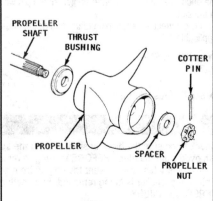

Fig. 65 Exploded view of the exhaust propeller

Fig. 66 Don't forget to coat the shaft splines

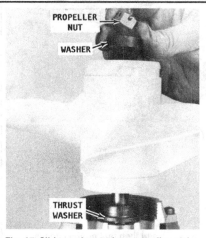

Fig. 67 Slide on the washer, propeller and nut. . .

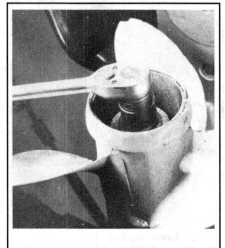

Fig. 68 . . .tighten the nut. . .

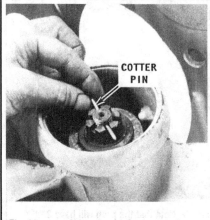

Fig. 69 . . .and then insert the cotter pin

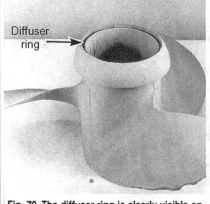

Fig. 70 The diffuser ring is clearly visible on this unit

Fig. 71 Installing the thrust washer

✳✳ CAUTION

Check to be sure the high-tension leads to the spark plugs have been removed from the plugs BEFORE starting to remove the propeller. This simple safety task will prevent the engine from accidentally "firing" while the propeller is being removed. Such action could cause SEVERE PERSONAL INJURY.

1. Pull the cotter pin from the propeller nut.
2. Wedge a piece of wood between one of the propeller blades and the cavitation plate to prevent the propeller from rotating. Back off the castellated propeller nut and remove the splined washer.
3. Pull the propeller straight off the shaft. It may be necessary to carefully tap on the front side of the propeller with a soft headed mallet to jar it loose.
4. Remove the thrust washer from the propeller shaft.

To Install:

5. Slide the thrust washer onto the propeller shaft.
6. Coat the propeller shaft with Perfect Seal No.4, Triple Guard Grease, or similar good grade of lubricant to prevent the propeller from becoming "frozen" to the shaft.
7. Slide the propeller onto the shaft with the splines in the propeller indexing with the splines on the shaft.
8. Slide the splined washer onto the shaft.
9. Thread the castellated nut onto the shaft. Jam a piece of board between 1 of the propeller blades and the cavitation plate to prevent the propeller from turning. Tighten the propeller nut securely and then a bit more to align the hole through the nut with the hole through the propeller shaft. NEVER back the nut off in an effort to align the hole, ALWAYS tighten the nut for alignment. Install the cotter pin through the nut and propeller shaft.

FROZEN PROPELLER

◆ **See Figures 72, 73 and 74**

If an exhaust propeller is "frozen" to the propeller shaft and the usual methods of removal fail, using a mallet to beat on the propeller blades in an effort to dislodge the propeller will have little if any affect. This is due to the cushioning affect the rubber hub has on the blow being struck.

A "frozen" propeller is caused by the inner sleeve becoming corroded and stuck to the propeller shaft. Therefore, special procedures are required to free the propeller and remove it from the shaft.

The following detailed procedures are presented as the only practical method to remove a "frozen" propeller from the shaft without damaging other more expensive parts. In almost all cases it is successful.

1. Remove the cotter pin, and then the castellated nut and splined washer from the propeller shaft.

2. Heat the inside diameter of the propeller with a torch. Do not apply the heat to the outside surface of the propeller. Concentrate the heat in the area of the hub and shaft where the nut was removed and as far into the hub as possible. Continue applying heat, and at the same time have an assistant use a piece of 2" x 4" wooden block wedged between one of the blades and the lower unit housing. Use a prying force on the propeller while the heat is being applied. As the heat melts the inner rubber hub, the propeller will come free.

✳✳ WARNING

As the force and heat are applied, the propeller may "pop" loose suddenly and without warning. Therefore, stand to one side while applying the heat as a precaution against personal injury.

3. After the propeller has been removed, the sleeve and what's left of the rubber hub will still be stuck to the propeller shaft. Attach a puller to the thrust washer. Apply more heat to the sleeve and rubber hub, and at the same time take up on the puller. When the sleeve reaches the proper temperature, it will be released and come free.

4. If a suitable puller is not available use a sharp knife and cut the rubber hub from the sleeve. An alternate method is to use canned heat, or the equivalent, and set fire to the rubber hub. Allow the hub to burn away from the sleeve. When the fire burns out, only a small amount may be left on the sleeve. Heat the sleeve again, and then while it is still hot, use a chisel, punch, or similar tool with a hammer, and drive the sleeve free of the shaft. Allow the propeller shaft to cool, and then clean the splines thoroughly. Take time to remove any corrosion.

Anodes (Zincs)

◆ **See Figure 75**

The idea behind anodes (also known as sacrificial anodes) is simple: When dissimilar metals are dunked in water and a small electrical current is leaked between or amongst them, the less-noble metal (galvanically speaking) is sacrificed (corrodes).

The zinc alloy of which the anodes are made is designed to be less noble than the aluminum alloy of which your outboard is constructed. If there's any electrolysis, and there almost always is, the inexpensive zinc anodes are consumed in lieu of the expensive outboard motor.

These zincs require a little attention in order to make sure they are capable of performing their function. Anodes must be solidly attached to a clean mounting site. Also, they must not be covered with any kind of paint, wax or marine growth.

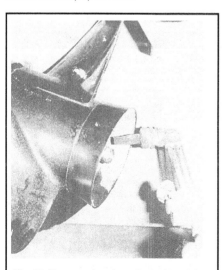

Fig. 72 Use a torch to heat the inside of the bore

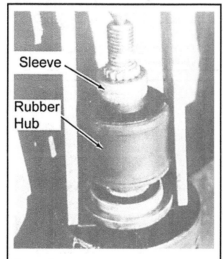

Fig. 73 You're going to have to use a puller to get it out. . .

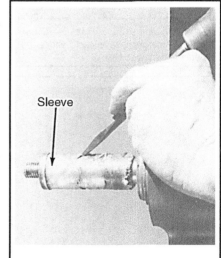

Fig. 74 . . or cut it away

Fig. 75 Extensive corrosion of an anode suggests a problem or a complete disregard for maintenance

zincs will insulate them and prevent them from performing their function properly. They must be left bare and must be installed onto bare metal of the motor. If the zincs are installed properly and not painted or waxed, inspect around them for sings of corrosion. If corrosion is found, strip it off immediately and repaint with a rust inhibiting paint. If in doubt, replace the zincs.

On the other hand, if your zinc seems to erode in no time at all, this may be a symptom of the zincs themselves. Each manufacturer uses a specific blend of metals in their zincs. If you are using zincs with the wrong blend of metals, they may erode more quickly or leave you with diminished protection.

At least annually or whenever an anode has been removed or replaced, check the mounting for proper electrical contact using a multi-meter. Set the multi-meter to check resistance (ohms), then connect 1 meter lead to the anode and the other to a good, unpainted or corroded ground on the motor. Resistance should be very low or zero. If resistance is high or infinite, the anode is insulated and cannot perform its function properly.

INSPECTION

◆ **See Figures 75 thru 81**

Visually inspect the anodes, especially gearcase mounted ones, before and after each use. You'll want to know right away if it has become loose or fallen off in service. Periodically inspect them closely to make sure they haven't eroded too much. At a certain point in the erosion process, the mounting holes start to enlarge, which is when the zinc might fall off. Obviously, once that happens your engine no longer has any protection. Generally, a zinc anode is considered worn if it has shrunken to 2/3 or less than the original size. To help judge this, buy a spare and keep it handy (in the boat or tow vehicle for comparison).

If you use your outboard in salt water or brackish water, and your zincs never seem to wear, inspect them carefully. Paint, wax or marine growth on

SERVICING

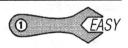

◆ **See Figures 75 thru 81**

Depending on your boat, motor and rigging, you may have anywhere from 1-4 (or even more) anodes. Regardless of the number, there are some fundamental rules to follow that will give your boat and motor's sacrificial anodes the ability to do the best job protecting your boat's underwater hardware that they can.

All motors covered by this service are equipped with at least 1 gearcase anode, normally mounted in, on, or near the anti-ventilation plate. Some of the motors covered by this service also have a powerhead mounted anode and/or an engine clamp bracket anode. When present, the location of the powerhead zincs will vary slightly from motor-to-motor including mounting bosses specifically cast in the motor or on the manifold assemblies. Some of the largest 2-cylinder motors and most 3 cyl/V4 motors may be equipped with 1 or more anodes on the engine mount clamp bracket.

Fig. 76 All motors have at least 1 anode mounted on the gearcase . . .

Fig. 77 . . .though some are under the anti-ventilation plate and . . .

Fig. 78 . . .others are mounted inside the gearcase (often accessed from underneath)

Fig. 79 All motors use at least one (1) anode on the bottom of the transom bracket. .

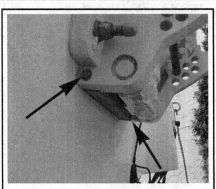

Fig. 80 . . .this one is secured by a bolt on either side

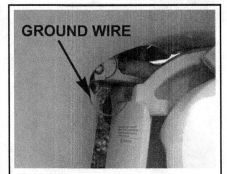

Fig. 81 Some motors may use lead wires used to connect bracketed parts and assist in corrosion resistance

Some people replace zincs annually. This may or may not be necessary, depending on the type of waters in which you boat and depending on

whether or not the boat is hauled with each use or left in for the season. Either way, it is a good idea to remove zincs at least annually in order to make sure the mounting surfaces are still clean and free of corrosion.

The first thing to remember is that zincs are electrical components and like all electrical components, they require good clean connections. So after you've undone the mounting hardware you want to get the zinc mounting sites clean and shiny.

Get a piece of coarse emery cloth or some 80-grit sandpaper. Thoroughly rough up the areas where the zincs attach (there's often a bit of corrosion residue in these spots). Make sure to remove every trace of corrosion.

Zincs are attached with stainless steel machine screws that thread into the mounting for the zincs. Over the course of a season, this mounting hardware is inclined to loosen. Mount the zincs and tighten the mounting hardware securely. Tap the zincs with a hammer hitting the mounting screws squarely. This process tightens the zincs and allows the mounting hardware to become a bit loose in the process. Now, do the final tightening. This will insure your zincs stay put for the entire season.

BOAT MAINTENANCE

Batteries

◆ See Figures 82 and 83

Batteries require periodic servicing, so a definite maintenance program will help ensure extended life.

A failure to maintain the battery in good order can prevent it from properly charging or properly performing its job even when fully charged. Low levels of electrolyte in the cells, loose or dirty cable connections at the battery terminals or possibly an excessively dirty battery top can all contribute to an improperly functioning battery. So battery maintenance, first and foremost, involves keeping the battery full of electrolyte, properly charged and keeping the casing/connections clean of corrosion or debris.

If a battery charges and tests satisfactorily but still fails to perform properly in service, 1 of 3 problems could be the cause.

1. An accessory left on overnight or for a long period of time can discharge a battery.

2. Using more electrical power than the stator assembly or lighting coil can replace would slowly drain the battery during motor operation, resulting in an undercharged condition.

3. A defect in the charging system. A faulty stator assembly or lighting coil, defective regulator or rectifier or high resistance somewhere in the system could cause the battery to become undercharged.

MAINTENANCE

Electrolyte Level

The most common and important procedure in battery maintenance is checking the electrolyte level. On most batteries this is accomplished by removing the cell caps and visually observing the level in the cells. The bottom of each cell has a split vent that will cause the surface of the electrolyte to appear distorted when it makes contact. When the distortion first appears at the bottom of the split vent, the electrolyte level is correct. Smaller marine batteries are sometimes equipped with translucent cases that are printed or embossed with high and low level markings on the side. On

some of these, shining a flashlight through the battery case will help make it easier to determine the electrolyte level.

During hot weather and periods of heavy use, the electrolyte level should be checked more often than during normal operation. Add distilled water to bring the level of electrolyte in each cell to the proper level. Take care not to overfill, because adding an excessive amount of water will cause loss of electrolyte and any loss will result in poor performance, short battery life and will contribute quickly to corrosion.

■ Never add electrolyte from another battery. Use only distilled water. Even tap water may contain minerals or additives that will promote corrosion on the battery plates, so distilled water is always the best solution.

Although less common in marine applications than other uses today, sealed maintenance-free batteries also require electrolyte level checks, through the window built into the tops of the cases. The problem for marine applications is the tendency for deep cycle use to cause electrolyte evaporation and electrolyte cannot be replenished in a sealed battery.

The second most important procedure in battery maintenance is periodically cleaning the battery terminals and case.

Cleaning

◆ See Figures 83 thru 86

Dirt and corrosion should be cleaned from the battery as soon as it is discovered. Any accumulation of acid film or dirt will permit a small amount of current to flow between the terminals. Such a current flow will drain the battery over a period of time.

Clean the exterior of the battery with a solution of diluted ammonia or a paste made from baking soda and water. This is a base solution to neutralize any acid that may be present. Flush the cleaning solution off with plenty of clean water.

■ Take care to prevent any of the neutralizing solution from entering the cells as it will quickly neutralize the electrolyte (ruining the battery).

Fig. 82 Explosive hydrogen gas is released from the batteries in a discharged state. This one exploded when the gas ignited from someone smoking with a cap. Explosions can also be caused by a spark from the battery terminals or jumper cables

Fig. 83 Ignoring a battery (and corrosion) to this extent is asking for it to fail

Fig. 84 Place a battery terminal tool over posts, then rotate back and forth . . .

Fig. 85 . . . until the internal brushes expose a fresh, clean surface on the post

Fig. 86 Clean the insides of cable ring terminals using the tool's wire brush

Poor contact at the terminals will add resistance to the charging circuit. This resistance will cause the voltage regulator to register a fully charged battery and thus cut down on the stator assembly or lighting coil output adding to the low battery charge problem.

At least once a season, the battery terminals and cable clamps should be cleaned. Loosen the clamps and remove the cables, negative cable first. On batteries with top mounted posts, if the terminals appear stuck, use a puller specially made for this purpose to ensure the battery casing is not damaged. NEVER pry a terminal off a battery post. These are inexpensive and available in most parts stores.

Clean the cable clamps and the battery terminal with a wire brush until all corrosion, grease, etc., is removed and the metal is shiny. It is especially important to clean the inside of the clamp thoroughly (a wire brush or brush part of a battery post cleaning tool is useful here), since a small deposit of foreign material or oxidation there will prevent a sound electrical connection and inhibit either starting or charging. It is also a good idea to apply some dielectric grease to the terminal, as this will aid in the prevention of corrosion.

After the clamps and terminals are clean, reinstall the cables, negative cable last, do not hammer the clamps onto battery posts. Tighten the clamps securely but do not distort them. To help slow or prevent corrosion, give the clamps and terminals a thin external coating of grease after installation.

Check the cables at the same time that the terminals are cleaned. If the insulation is cracked or broken or if its end is frayed, that cable should be replaced with a new one of the same length and gauge.

TESTING

◆ See Figure 87

A quick check of the battery is to place a voltmeter across the terminals. Although this is by no means a clear indication, it gives you a starting point when trying to troubleshoot an electrical problem that could be battery related. Most marine batteries will be of the 12 volt DC variety. They are constructed of 6 cells, each of which is capable of producing slightly more than 2 volts, wired in series so that total voltage is 12 and a fraction. A fully charged battery will normally show more than 12 and slightly less than 13 volts across its terminals. But keep in mind that just because a battery reads 12.6 or 12.7 volts does NOT mean it is fully charged. It is possible for it to have only a surface charge with very little amperage behind it to maintain that voltage rating for long under load. A discharged battery will read some value less than 12 volts, but can be brought back to 12 volts through recharging. Of course a battery with 1 or more shorted or un-chargeable cells will also read less than 12, but it cannot be brought back to 12+ volts after charging. For this reason, the best method to check battery condition on most marine batteries is through a specific gravity check.

A hydrometer is a device that measures the density of a liquid when compared to water (specific gravity). Hydrometers are used to test batteries by measuring the percentage of sulfuric acid in the battery electrolyte in terms of specific gravity. When the condition of the battery drops from fully

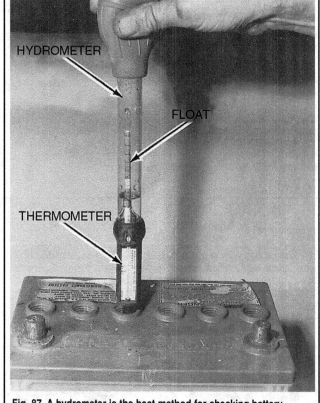

Fig. 87 A hydrometer is the best method for checking battery condition

charged to discharged, the acid is converted to water as electrons leave the solution and enter the plates, causing the specific gravity of the electrolyte to drop.

It may not be common knowledge but hydrometer floats are calibrated for use at 80°F (27°C). If the hydrometer is used at any other temperature, hotter or colder, a correction factor must be applied.

■ Remember, a liquid will expand if it is heated and will contract if cooled. Such expansion and contraction will cause a definite change in the specific gravity of the liquid, in this case the electrolyte.

A quality hydrometer will have a thermometer/temperature correction table in the lower portion, as illustrated in the accompanying illustration. By measuring the air temperature around the battery and from the table, a correction factor may be applied to the specific gravity reading of the hydrometer float. In this manner, an accurate determination may be made as to the condition of the battery.

When using a hydrometer, pay careful attention to the following points:

1. Never attempt to take a reading immediately after adding water to the battery. Allow at least 1/4 hour of charging at a high rate to thoroughly mix the electrolyte with the new water. This time will also allow for the necessary gases to be created.

2. Always be sure the hydrometer is clean inside and out as a precaution against contaminating the electrolyte.

3. If a thermometer is an integral part of the hydrometer, draw liquid into it several times to ensure the correct temperature before taking a reading.

4. Be sure to hold the hydrometer vertically and suck up liquid only until the float is free and floating.

5. Always hold the hydrometer at eye level and take the reading at the surface of the liquid with the float free and floating.

6. Disregard the slight curvature appearing where the liquid rises against the float stem. This phenomenon is due to surface tension.

7. Do not drop any of the battery fluid on the boat or on your clothing, because it is extremely caustic. Use water and baking soda to neutralize any battery liquid that does accidentally drop.

8. After drawing electrolyte from the battery cell until the float is barely free, note the level of the liquid inside the hydrometer. If the level is within the charged (usually green) band range for all cells, the condition of the battery is satisfactory. If the level is within the discharged (usually white) band for all cells, the battery is in fair condition.

9. If the level is within the green or white band for all cells except one, which registers in the red, the cell is shorted internally. No amount of charging will bring the battery back to satisfactory condition.

10. If the level in all cells is about the same, even if it falls in the red band, the battery may be recharged and returned to service. If the level fails to rise above the red band after charging, the only solution is to replace the battery.

STORAGE

If the boat is to be laid up (placed into storage) for the winter or for more than a few weeks, special attention must be given to the battery to prevent complete discharge and/or possible damage to the terminals and wiring. Before putting the boat in storage, disconnect and remove the batteries. Clean them thoroughly of any dirt or corrosion and then charge them to full specific gravity readings. After they are fully charged, store them in a clean cool dry place where they will not be damaged or knocked over, preferably on a couple blocks of wood. Storing the battery up off the deck, will permit air to circulate freely around and under the battery and will help to prevent condensation.

Never store the battery with anything on top of it or cover the battery in such a manner as to prevent air from circulating around the filler caps. All batteries, both new and old, will discharge during periods of storage, more so if they are hot than if they remain cool. Therefore, the electrolyte level and the specific gravity should be checked at regular intervals. A drop in the specific gravity reading is cause to charge them back to a full reading.

In cold climates, care should be exercised in selecting the battery storage area. A fully charged battery will freeze at about 60°F (17°C) below zero. The electrolyte of a discharged battery, almost dead, will begin forming ice at about 19°F (-7°C) above zero.

■ For more information on batteries and the engine electrical systems, please refer to the Ignition and Electrical section of this service.

Fiberglass Hull

INSPECTION & CARE

◆ See Figures 88, 89 and 90

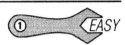

Fiberglass reinforced plastic hulls are tough, durable and highly resistant to impact. However, like any other material they can be damaged. One of the advantages of this type of construction is the relative ease with which it may be repaired.

A fiberglass hull has almost no internal stresses. Therefore, when the hull is broken or stove-in, it retains its true form. It will not dent to take an out-of-shape set. When the hull sustains a severe blow, the impact will be either absorbed by deflection of the laminated panel or the blow will result in a definite, localized break. In addition to hull damage, bulkheads, stringers and other stiffening structures attached to the hull may also be affected and therefore, should be checked. Repairs are usually confined to the general area of the rupture.

■ The best way to care for a fiberglass hull is to wash it thoroughly, immediately after hauling the boat while the hull is still wet.

A foul bottom can seriously affect boat performance. This is one reason why racers, large and small, both powerboat and sail, are constantly giving attention to the condition of the hull below the waterline.

In areas where marine growth is prevalent, a coating of vinyl, anti-fouling bottom paint should be applied if the boat is going to be left in the water for extended periods of time such as all or a large part of the season. If growth has developed on the bottom, it can be removed with a diluted solution of muriatic acid applied with a brush or swab and then rinsed with clear water. Always use rubber gloves when working with Muriatic acid and take extra care to keep it away from your face and hands. The fumes are toxic. Therefore, work in a well-ventilated area, or if outside, keep your face on the windward side of the work.

■ If marine growth is not too severe you may avoid the unpleasantness of working with muriatic acid by trying a power washer instead. Most marine vegetation can be removed with pressurized water and a little bit of scrubbing using a rough sponge (don't use anything that will scratch or damage the surface).

Barnacles have a nasty habit of making their home on the bottom of boats that have not been treated with anti-fouling paint. Actually they will not harm the fiberglass hull but can develop into a major nuisance.

If barnacles or other crustaceans have attached themselves to the hull, extra work will be required to bring the bottom back to a satisfactory condition. First, if practical, put the boat into a body of fresh water and allow it to remain for a few days. A large percentage of the growth can be removed in this manner. If this remedy is not possible, wash the bottom thoroughly with a high-pressure fresh water source and use a scraper. Small particles of hard shell may still hold fast. These can be removed with sandpaper.

Fig. 88 The best way to care for a fiberglass hull is to wash it thoroughly

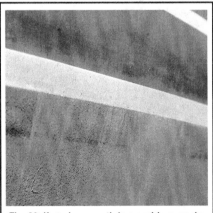

Fig. 89 If marine growth is a problem, apply a coating of anti-foul bottom paint

Fig. 90 Fiberglass, vinyl and rubber care products, like those from Meguiar's, protect your boat

TUNE-UP

Introduction

A proper tune-up is the key to long and trouble-free outboard life and the work can yield its own rewards. Studies have shown that a properly tuned and maintained outboard can achieve better fuel economy than an out-of-tune engine. As a conscientious boater, set aside a Saturday morning, say once a month, to check or replace items that could cause major problems later. Keep your own personal log to jot down which services you performed, how much the parts cost you, the date and the number of hours on the engine at the time. Keep all receipts for such items as oil and filters, so that they may be referred to in case of related problems or to determine operating expenses. These receipts are the only proof you have that the required maintenance was performed. In the event of a warranty problem on newer engines, these receipts will be invaluable.

The efficiency, reliability, fuel economy and enjoyment available from boating are all directly dependent on having your outboard tuned properly. The importance of performing service work in the proper sequence cannot be over emphasized. Before making any adjustments, check the specifications. Never rely on memory when making critical adjustments.

Before tuning any outboard, insure it has satisfactory compression. An outboard with worn or broken piston rings, burned pistons or scored cylinder walls, will not perform properly no matter how much time and expense is spent on the tune-up. Poor compression must be corrected or the tune-up will not give the desired results.

The extent of the engine tune-up is usually dependent on the time lapse since the last service. In this section, a logical sequence of tune-up steps will be presented in general terms. If additional information or detailed service work is required, refer to the section of this service containing the appropriate instructions.

Tune-Up Sequence

◆ See Figures 91 and 92

A tune-up can be defined as pre-determined series of procedures (adjustments, tests and replacement of worn components) that are performed to bring the engine operating parameters back to original condition (or as near original as possible). The series of steps are important, as the later procedures (especially adjustments) are dependent upon the earlier procedures. In other words, a procedure is performed only when subsequent steps would not change the result of that procedure (this is mostly for adjustments or settings that would be incorrect after changing another part or setting). For instance, fouled or excessively worn spark plugs may affect engine idle. If adjustments were made to the idle speed or mixture *before* these plugs were cleaned or replaced, the idle speed or mixture might be wrong after replacing the plugs. The possibilities of such an effect become much greater when dealing with multiple adjustments such as timing, idle speed and/or idle mixture. Therefore, be sure to follow each of the steps given here. Since many of the steps listed here are full procedures in themselves, refer to the procedures of the same name in this section for details.

A complete pre-season tune-up should be performed at the beginning of each season or anytime a motor is removed from storage. Operating conditions, amount of use and the frequency of maintenance required by your motor may make one or more additional tune-ups necessary during the season. Perform additional tune-ups as use dictates.

■ **Under normal conditions a tune-up is expected about every 100 hours of operation. Excessive idle or wide-open throttle operation, use of poor quality engine oil or fuels, or other variables may necessitate shortening that timeframe.**

Burned piston

Fig. 91 No amount of tune-ups can restore an engine to satisfactory performance when the pistons have been damaged this badly

Fig. 92 A lower unit covered with this much marine growth can be a serious hindrance to satisfactory performance

If twenty different mechanics were asked the question, "What constitutes a major and minor tune-up?" it is entirely possible twenty different answers would be given. As the terms are used in this service, the following work is normally performed for a minor and major tune-up.

Minor Tune-up
- Lubricate engine.
- Drain and replace gear oil.
- Adjust (or replace) points.
- Adjust carburetor (if applicable).
- Clean exterior surface of engine.
- Tank test engine for fine adjustments.
- Check synchronization and timing.

Major Tune-up
- Remove head.
- Clean carbon from pistons and cylinders.
- Clean and overhaul carburetor.
- Clean and overhaul fuel pump.
- Test ignition system.
- Lubricate engine.
- Drain and replace gear oil.
- Clean exterior surface of engine.
- Tank test engine for fine adjustments.
- Check synchronization and timing

During a major tune-up, a definite sequence of service work should be followed to return the engine to the maximum performance desired. This type of work should not be confused with attempting to locate problem areas of "why" the engine is not performing satisfactorily. This work is classified as "troubleshooting". In many cases, these two areas will overlap, because many times a minor or major tune-up will correct the malfunction and return the system to normal operation.

The following list is a suggested sequence of tasks to perform during the tune-up service work. The tasks are merely listed here. Generally procedures are given in subsequent sections of this section or other sections of this guide.

1. Before starting, inspect the motor thoroughly for signs of obvious leaks, damage and loose or missing components. Make repairs, as necessary.

2. Perform a complete Lubrication service, as detailed in this section.

3. If Johnson/Evinrude Carbon Guard of equivalent is not used consistently with each fill-up, remove carbon from the pistons and combustion chamber after every 50 hours of operation. Refer to the De-carboning the Pistons in this section.

4. Perform a compression check to make sure the motor is mechanically ready for a tune-up. An engine with low compression on 1 or more cylinder should be overhauled, not tuned. A tune-up will not be successful without sufficient engine compression. Refer to the Compression Test in this section.

5. Since the spark plugs must be removed for the compression check, take the opportunity to inspect them thoroughly for signs of oil fouling, carbon fouling, damage due to detonation, etc. Clean and re-gap the plugs or, better yet, install new plugs, as no amount of cleaning will precisely match the performance and life of new plugs. Refer to Spark Plugs, in this section.

6. Inspect all ignition system components for signs of obvious defects. Look for signs of burnt, cracked or broken insulation. Replace wires or components with obvious defects. If spark plug condition suggests weak or no spark on 1 or more cylinders, perform ignition system testing to eliminate possible worn or defective components. Refer to the Ignition System Inspection procedures in this section and the Ignition & Electrical System section.

7. Inspect the fuel system. Remove and clean (on serviceable filters) or replace the inline filter and/or fuel pump filter, as equipped. Refer to the Fuel Filter procedures in this section. Perform a thorough inspection of the fuel system, hoses and components. Replace any cracked or deteriorating hoses.

8. Remove the propeller in order to inspect for leaks thoroughly at the shaft seal. Inspect the propeller or rotor condition, look for nicks, cracks or other signs of damage and repair or replace, as necessary. If available, install a test wheel to run the motor in a test tank after completion of the tune-up. If no test wheel is available, lubricate the shaft/splines, then install the propeller. Refer to the procedure for Propeller, in this section.

9. Change the lower unit oil as directed under the Lower Unit Oil procedures in this section. If you are conducting a pre-season tune-up and the oil was changed immediately prior to storage this is not necessary. But, be sure to check the oil level and condition. Drain the oil anyway if significant contamination is present.

■ **Anytime large amounts of water or debris is present in the lower unit oil, be sure to troubleshoot and repair the problem before returning the lower unit to service. The presence of water may indicate problems with the seals, while debris could a sign that overhaul is required.**

10. Check all accessible bolts and fasteners and tighten any that are loose.

11. Pressurize the fuel system using the primer bulb, then check carefully for leaks.

12. Perform engine Timing & Synchronization adjustments as described in this section.

13. Perform a test run of the engine to verify proper operation of the starting, fuel, oil and cooling systems. Although this can be performed using a flush/test adapter, or even on the boat itself (if operating with a normal load/passengers), the preferred method is the use of a test tank. If possible, run the engine, in a test tank using the appropriate test wheel. Monitor the cooling system indicator stream or mist to ensure the water pump is working properly. Once the engine is fully warmed, slowly advance the engine to wide-open throttle, then note and record the maximum engine speed. Refer to the Tune-Up Specifications chart to compare engine speeds with the test propeller minimum rpm specifications. If engine speeds are below specifications, yet engine compression was sufficient at the beginning of this procedure, recheck the fuel and ignition system adjustments.

De-Carboning the Pistons 1Wrench

A by-product of the normal combustion process; carbon will build-up on the pistons and in the combustion chambers of a motor over time. Engine tuning and condition will affect this process, as a properly tuned engine running high-quality fuels under proper conditions will reduce the amount of build-up, but not stop it completely. Generally speaking an out-of-tune motor, a motor running too rich or a motor run under extended idle conditions will increase the rate at which carbon deposits are formed. Carbon, when its presence becomes significant enough, will increase the compression ratio (by decreasing effective combustion chamber size) and will lead to detonation. Also, over time, carbon may cause piston rings to stick, which would lead to blow-by. For this reason, Johnson/Evinrude recommends the use of Johnson/Evinrude Carbon Guard fuel additive with each fill up in order to help slow this process.

The manufacturer also warns that, if this fuel additive is not used, the pistons and combustion chambers should be cleaned of deposits using Johnson/Evinrude Engine Tuner after, at least every 50 hours of engine operation. As noted earlier, variables such as types of fuel used and patterns of usage (wide-open throttle vs. extensive idle) will also have an effect upon how often this procedure should be followed. Let your own experience (and the amount of carbon found on your spark plugs) be your guide.

1. Provide the engine with a cooling water source (either and engine flushing adapter, a test tank, or if necessary, perform this procedure with the boat and motor in the water, attached to a sturdy dock).

2. Start and run the engine at normal idle until it reaches normal operating temperature.

3. Set the engine to fast idle. On most engines a fast idle of around 1200 rpm is sufficient.

4. For severe cases of carbon build-up, run the engine to normal operating temperature, then shut the engine off and remove the spark plugs. Lay the engine into a horizontal position and peer through the spark plug holes as you slowly turn the motor over by hand (do so by turning the engine clockwise when viewed from above the flywheel). With the pistons leveled so as to best block of the ports, cover the tops of the pistons with engine tuner and let sit for approximately 1 hour. After at least that amount of time, rotate the engine a couple of revolutions by hand to begin removing the cleaner. Then, proceed with the next step to finish the can of Engine Tuner.

■ **Johnson/Evinrude service literature gives 2 conflicting recommendations for the amount of time Engine Tuner should be left in the motor. Materials published before 1996 tell technicians to leave the Engine Tuner in for no MORE than 1 hour. But, literature published after 1996 instructs the technician to allow the engine tuner to soak for 3-16 hours. There does not appear to be any change in design or materials on later motors which would affect this AND, perhaps more importantly, the cans of Engine Tuner available while this text was being written do not give any cautions against letting the motor soak longer than an hour. You'll have to make up your own mind, but we can't find the harm in allowing the motor to soak longer.**

5. For less severe (typical cases of carbon build-up), spray the entire contents of the Johnson/Evinrude Engine Tuner can with the engine still running from Step 3 above. Spray the Tuner through the carburetor throats, moving the spray nozzle from carburetor-to-carburetor in a back and forth sequence until the can is emptied. Once all of the Engine Tuner has been sprayed, shut the engine **Off** and allow the cleaner to penetrate for at least 15 minutes (but more time is permissible).

6. If removed, reconnect the flushing device or place the engine back in the water (test tank or dockside), then start the engine again and warm it to normal operating temperature. When warmed, run the engine above 1/2 throttle for at least 3-5 minutes.

7. Shut the engine **Off**, then remove and inspect the spark plugs. Shine a small light through each spark plug bore to examine the tops of the pistons and compare the visual evidence of carbon build-up to that before the procedure. If necessary, repeat the procedure using a second can of engine tuner and following the step for severe cases.

Compression Check

The quickest (but not necessarily most accurate) way to gauge the condition of an internal combustion engine is through a compression check. In order for an internal combustion engine to work properly, it must be able to generate sufficient compression in the combustion chamber to take advantage of the explosive force generated by the expanding gases after ignition. This is true on all motors whether they are of the 2- or 4-stroke design.

If the combustion chambers or ports (or any mating surfaces like cylinder heads and gaskets) are worn or damaged in some fashion as to allow pressure to escape, the engine cannot develop sufficient horsepower. Under these circumstances, combustion will not occur properly, air/fuel mixtures cannot be set to maximize power and minimize emissions. An engine with poor compression on 1 or more cylinders cannot be given a proper tune-up, it should be overhauled.

There are 2 types of compression checks generally conducted by technicians. The first, which is included here, is called a compression check or sometimes a tune-up compression check. It is a quick-test used during a tune-up to determine if you should continue or stop and overhaul the motor. This test is what technicians think of when you say compression check as it measures the ability of a motor to create compression.

A compression check requires a compression gauge and a spark plug port adapter that matches the plug threads of your motor.

Some technicians, during deeper diagnostic work or to verify a rebuild before returning it to service, will perform a second compression check known as a leakage or leak-down check. This test, which uses special gauges, adapters and a pressurized air supply, measures the ability of an engine to hold pressure (as opposed to create it).

PERFORMING A COMPRESSION CHECK

◆ See Figure 93

When analyzing the results of a compression check, generally the actual amount of pressure measured during a compression check is not AS important as the variation from cylinder-to-cylinder on multi-cylinder motors. However, it appears that the manufacturer changed its recommendations to Johnson/Evinrude field technicians over the years. Through much of the fifties, sixties, seventies and eighties Johnson/Evinrude recommended that the variation between cylinders not exceed 5 psi (34.5 kPa). Then sometime in the eighties (with no apparent significant change in design or construction), Johnson/Evinrude raised that limit, advising that the variations between cylinders should **not** exceed 15 psi (100 kPa) or more and kept advising this limit through about 1995. Finally, starting in 1996 (with still no apparent change in design or construction), Johnson/Evinrude eased even further on the compression specifications. Beginning in 1996, the manufacturer instructed technicians that there should be no more than a 20% variation between the lowest and highest cylinders (i.e. that the lowest cylinder reading must be 80% or more of the highest cylinder reading).

■ It does not appear that Johnson/Evinrude made significant changes in the motor designs, materials or manufacturing processes between years they made these changes in specifications for compression checks. It seems that they instead changed their mind set, adopting a new, and somewhat less strict, set of standards each time. Whether this was from practical testing results or for other reasons we cannot say. Which one you decide to follow is your own choice. Following the stricter standard, however, could lead to overhauling a motor that otherwise is performing properly. We'd recommend that a motor that is out of the 5 psi (34.5 kPa) or 15 psi (100 kPa) spec, but STILL within the 80% spec, be tuned. If it runs properly, then an overhaul is not necessary. . .yet.

■ A good rule of thumb on single cylinder engines is that compression should be in the 90-100 psi range.

Of course, the MOST important specification when it comes to compression checks is how much has the spec changed from the last test (especially on single cylinder motors, for which there is otherwise nothing with which to compare your reading). The first thing you should do with a new motor is to take a compression reading for each cylinder and mark it down. The same should be done with each successive tune-up thereafter. In this way, you can track the internal wear in the motor over time, possibly even predicting at what point an overhaul might be necessary (unless a component failure necessitates one sooner). Even for a used motor, a compression check is the first step in knowing where you stand.

Ok, for the point of argument's sake let's say you bought the engine used and the last owner didn't have any information regarding previous compression checks, or let's say you never checked compression the first season or so, assuming it wasn't something you needed to worry about. You're not alone. Although Johnson/Evinrude does not publish a specification for the exact amount of compression each of their engines should generate, a general rule of thumb that can be applied is that internal combustion engines should generate at least 100 psi (690 kPa).

When taking readings during the compression check, repeat the procedure a few times for each cylinder, recording the highest reading for that cylinder. Then, compare the readings. The compression reading on the lowest cylinder should within 5 psi (34.5 kPa), 15 psi (100 kPa) or 80% (depending on the standard that you wish to apply) of the highest reading.

■ When using the 80% standard, the compression reading on the low cylinder should be equal to 80% or more of the reading from the high cylinder (or from the last test on singles). In other words, the low reading should be the equal to or greater than the high reading multiplied by 0.8. For example, if the high reading was 150 psi (1034 kPa), then the low reading must be equal to or more than 150 psi x 0.8 (1035 kPa x 0.8) or 120 psi (827 kPa).

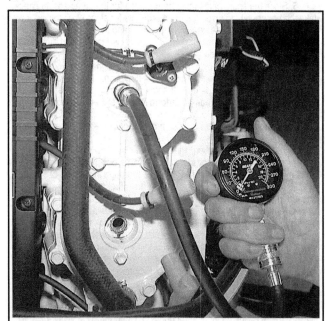

Fig. 93 Compression check on a typical multi-cylinder powerhead

■ If the powerhead has been in storage for an extended period, the piston rings may have relaxed. This will often lead to initially low and misleading readings. Always run an engine to normal operating temperature to ensure that the readings are accurate.

■ If you've never removed the spark plugs from this cylinder head before, break each one loose and retighten them, to make sure they will not seize in the head once it is warmed. Better yet, remove each one and coat the threads very lightly with some fresh anti-seize compound.

1. Using a test tank, flush fitting adapter or other water supply, start and run the engine until it reaches normal operating temperature, then shut the engine **off**.

2. Disable the ignition system by connecting each spark plug wire to a good engine ground (using a jumper wire from the ground to the wire inside each spark plug boot). Never simply disconnect plug wires.

✳✳ CAUTION

Removing both of the spark plugs from twins and cranking the powerhead can lead to an explosion if raw fuel/oil sprays out of the plug holes. A plug wire could spark and ignite this mix outside of the combustion chamber if it isn't grounded to the engine. Also, on many of the ignition systems covered, cranking the engine and firing the coil without allowing the coils to discharge through the spark plug leads can lead to severe damage to the ignition system.

3. Remove the spark plug(s). On twins, be sure to keep them in order. Carefully inspect the plug(s), looking for any inconsistency in coloration and for any sign of water or rust near the tip. Refer to the procedures on Spark Plugs in this section for more details.

4. Thread the compression gauge into the No. 1 spark-plug hole, taking care to not cross-thread the fitting.

5. Open the throttle to the wide open throttle position and hold it there.

■ Some engines allow only minimal opening if the gearshift is in neutral, to guard against over-revving.

6. Crank over the engine an equal number of times for each cylinder you test, zeroing the gauge for each cylinder.

7. If you have electric start, count the number of seconds you crank. On manual start, pull the starter rope 4-5 times for each cylinder you are testing. (And, if you have manual start, and are about to try a compression check, eat your spinach first, especially on the larger twins, because you are going to be ONE TIRED PUPPY when you're finished).

■ On manual start twin motors, it really does make sense to remove both of the spark plugs before attempting to check compression. Removing the plugs on the cylinder not being checked will relieve compression in that cylinder making it easier to turn the motor using the rope.

8. Record your readings from the cylinder. When both cylinders are tested on twins, compare the readings and determine if pressures are within the 5 psi (34.5 kPa), 15 psi (100 kPa) or 80% criterion, as applicable.

9. If compression readings are lower than normal for any cylinder, try a "wet" compression test, which will temporarily seal the piston rings and determine if they are the cause of the low reading. Using a can of fogging oil, fog the cylinder with a circular motion to distribute oil spray all around the perimeter of the piston. Retest the cylinder:

   a. If the compression rises noticeably in a wet test, the piston rings are sticking. You may be able to cure the problem by de-carboning the powerhead.

   b. If the dry compression test was really low and no change is evident during the wet test, the cylinder is dead. The piston and/or cylinder are worn beyond specification (possibilities include damaged pistons, broken or stuck pistons rings, scored cylinder walls or a blown head gasket) and a powerhead overhaul or replacement is necessary.

10. If the 2 cylinders on a twin engine give a similarly low reading then the problem may be a faulty head gasket. This should be suspected especially if there is evidence of water or rust on the spark plugs from these cylinders.

11. If the engine has compression within specification (on both cylinders for twins), yet is hard to start and runs poorly, there may still be damage to the powerhead, suspect the possibilities of scored cylinder walls, damaged pistons and/or stuck or worn piston rings.

Spark Plugs

■ On older outboards such as those we are discussing here, spark plugs have a relatively short life. The output of the magneto capacitor discharge ignition is not nearly as high as that of modern electronic ignitions. This means that as spark plug gap widens during use it will quickly reach a point where the gap becomes troublesome for the spark to jump. This will inevitably lead to engine misfiring and stumbling.

The spark plug performs 4 main functions:
- First and foremost, it provides spark for the combustion process to occur.
- It also removes heat from the combustion chamber.
- Its removal provides access to the combustion chamber (for inspection or testing) through a hole in the cylinder head.
- It acts as a dielectric insulator for the ignition system.

It is important to remember that spark plugs do not create heat, they help remove it. Anything that prevents a spark plug from removing the proper amount of heat can lead to pre-ignition, detonation, premature spark plug failure and even internal engine damage, especially in 2-stroke engines.

In the simplest of terms, the spark plug acts as the thermometer of the engine. Much like a doctor examining a patient, this "thermometer" can be used to effectively diagnose the amount of heat present in each combustion chamber.

Spark plugs are valuable tuning tools, when interpreted correctly. They will show symptoms of other problems and can reveal a great deal about the engine's overall condition. Evaluating the appearance of the spark plug's firing tip, gives visual cues to determine the engine's overall operating condition, in order to get a feel for air/fuel ratios and even diagnose driveability problems.

As spark plugs grow older, they lose their sharp edges and material from the center and ground electrodes slowly erodes away. As the gap between these 2 points grows, the voltage required to bridge this gap increases proportionally. The ignition system must work harder to compensate for this higher voltage requirement and hence there is a greater rate of misfires or incomplete combustion cycles. Each misfire means lost horsepower, reduced fuel economy and higher emissions. Replacing worn out spark plugs with new ones (that have sharp new edges) effectively restores the ignition system's efficiency and reduces the percentage of misfires, restoring power, economy and reducing emissions.

■ Although spark plugs can typically be cleaned and re-gapped if they are not excessively worn, no amount of cleaning or re-gapping will return most spark plugs to original condition and it is usually best to just go ahead and replace them.

How long spark plugs last will depend on a variety of factors, including engine compression, fuel used, gap, center/ground electrode material and the conditions in which the outboard is operated.

SPARK PLUG HEAT RANGE

◆ See Figure 94

Spark plug heat range is the ability of the plug to dissipate heat from the combustion chamber. The longer the insulator (or the farther it extends into the engine), the hotter the plug will operate; the shorter the insulator (the closer the electrode is to the engine's cooling passages) the cooler it will operate.

Selecting a spark plug with the proper heat range will ensure that the tip maintains a temperature high enough to prevent fouling, yet cool enough to prevent pre-ignition. A plug that absorbs little heat and remains too cool will quickly accumulate deposits of oil and carbon since it won't be able to burn them off. This leads to plug fouling and consequently to misfiring. A plug that absorbs too much heat will have no deposits but, due to the excessive heat, the electrodes will burn away quickly and might also lead to pre-ignition or other ignition problems.

Pre-ignition takes place when plug tips get so hot that they glow sufficiently to ignite the air/fuel mixture before the actual spark occurs. This early ignition will usually cause a pinging during heavy loads and if not corrected, will result in severe engine damage. While there are many other things that can cause pre-ignition, selecting the proper heat range spark plug will ensure that the spark plug itself is not a hot-spot source.

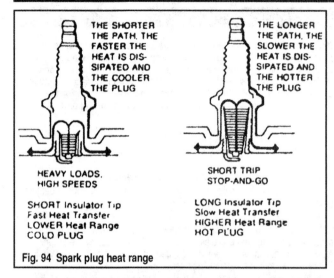

THE SHORTER THE PATH. THE FASTER THE HEAT IS DIS-SIPATED AND THE COOLER THE PLUG

HEAVY LOADS. HIGH SPEEDS

SHORT Insulator Tip
Fast Heat Transfer
LOWER Heat Range
COLD PLUG

THE LONGER THE PATH. THE SLOWER THE HEAT IS DIS-SIPATED AND THE HOTTER THE PLUG

SHORT TRIP STOP-AND-GO

LONG Insulator Tip
Slow Heat Transfer
HIGHER Heat Range
HOT PLUG

Fig. 94 Spark plug heat range

■ The manufacturer recommended spark plugs for carbureted motors are listed in the Tune-Up Specifications chart. When provided, alternate plugs for extended idle and/or extended wide-open throttle service are also listed.

REMOVAL & INSTALLATION

◆ See Figures 95, 96 and 97

■ On older outboards such as those we are discussing here, spark plugs have a relatively short life. The output of the magneto points or capacitor discharge ignition is not nearly as high as that of modern electronic ignitions. This means that as spark plug gap widens during use it will quickly reach a point where the gap becomes troublesome for the spark to jump. This will inevitably lead to engine misfiring and stumbling.

Typically, spark plugs will require replacement once a season, but this varies greatly with engine condition and usage. The electrode on a new spark plug has a sharp edge but with use, this edge becomes rounded by wear, causing the plug gap to increase. As the gap increases, the plug's voltage requirement also increases. It requires a greater voltage to jump the wider gap and about 2-3 times as much voltage to fire a plug at high speeds then at idle.

■ Fouled plugs can cause hard-starting, engine mis-firing or other problems. You don't want that happening on the water. Take time, at least once a month to remove and inspect the spark plugs. Early signs of other tuning or mechanical problems may be found on the plugs that could save you from becoming stranded or even allow you to address a problem before it ruins the motor.

Tools needed for spark plug replacement include: a ratchet, short extension, spark plug socket (there are 2 types; either 13/16 in. or 5/8 in., depending upon the type of plug), a combination spark plug gauge and gapping tool and a can of anti-seize type compound.

1. When removing spark plugs from multi-cylinder motors, work on one at a time. Don't start by removing all plug wires at once, because unless you number them, they may become mixed up. Take a minute before you begin and number the wires with tape.

2. For safety, disconnect the negative battery cable or turn the battery switch **OFF**.

3. If the engine has been run recently, allow the engine to thoroughly cool (unless performing a compression check). Attempting to remove plugs from a hot cylinder head could cause the plugs to seize and damage the threads in the cylinder head, especially on aluminum heads!

■ To ensure an accurate reading during a compression check, the spark plugs must be removed from a hot engine. But, DO NOT force a plug if it feels like it is seized. Instead, wait until the engine has cooled, remove the plug and coat the threads lightly with anti-seize then reinstall and tighten the plug, then back off the tightened position a little less than 1/4 turn. With the plug(s) installed in this manner, re-warm the engine and conduct the compression check.

4. Carefully twist the spark plug wire boot to loosen it, then pull the boot using a twisting motion to remove it from the plug. Be sure to pull on the boot and not on the wire, otherwise the connector located inside the boot may become separated from the high-tension wire.

■ A spark plug wire removal tool is recommended as it will make removal easier and help prevent damage to the boot and wire assembly. Most tools have a wire loom that fits under the plug boot so the force of pulling upward is transmitted directly to the bottom of the boot.

5. Using compressed air (and safety glasses), blow debris from the spark plug area to assure that no harmful contaminants are allowed to enter the combustion chamber when the spark plug is removed. If compressed air is not available, use a rag or a brush to clean the area. Compressed air is available from both an air compressor or from compressed air in cans available at photography stores. In a pinch, blow up a balloon by hand and use the escaping air to blow debris from the spark plug port(s).

■ Remove the spark plugs when the engine is cold, if possible, to prevent damage to the threads. If plug removal is difficult, apply a few drops of penetrating oil to the area around the base of the plug and allow it a few minutes to work.

6. Using a spark plug socket that is equipped with a rubber insert to properly hold the plug, turn the spark plug counterclockwise to loosen and remove the spark plug from the bore.

⁕⁕ WARNING

Avoid the use of a flexible extension on the socket. Use of a flexible extension may allow a shear force to be applied to the plug. A shear force could break the plug off in the cylinder head, leading to costly and/or frustrating repairs. In addition, be sure to support the ratchet with your other hand - this will also help prevent the socket from damaging the plug.

Fig. 95 With the lead removed, loosen the plug using a plug socket...

Fig. 96 ... then remove the spark plug from the cylinder head

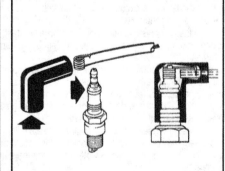

Fig. 97 To prevent corrosion, apply a small amount of grease to the plug and boot during installation

7. Evaluate each cylinder's performance by comparing the spark condition. Check each spark plug to be sure they are from the same plug manufacturer and have the same heat range rating. Inspect the threads in the spark plug opening of the block and clean the threads before installing the plug.

8. When purchasing new spark plugs, always ask the dealer if there has been a spark plug change for the engine being serviced. Sometimes manufacturers will update the type of spark plug used in an engine to offer better efficiency or performance.

9. Inspect the spark plug boot for tears or damage. If a damaged boot is found, the spark plug boot and possibly the entire wire will need replacement.

To install:

10. Check the spark plug gap prior to installing the plug. Most spark plugs do not come gapped to the proper specification. On most motors, set the spark plug gap to 0.030 in. (0.75mm). On 1971-72 50 hp 2 cyl. motors, set the spark plug gap to 0.040 in. (1.02mm). To be certain though, please check the Tune-Up Specifications chart at the end of this section

■ **All UL76V and UL77V plugs are not adjustable.**

11. Apply a thin coating of anti-seize on the thread of the plug. This is extremely important on aluminum head engines to prevent corrosion and heat from seizing the plug in the threads (which could lead to a damaged cylinder head upon removal).

12. Carefully thread the plug into the bore by hand. If resistance is felt before the plug completely bottoms, back the plug out and begin threading again.

✳✳ WARNING

Do not use the spark plug socket to thread the plugs. Always carefully thread the plug by hand or using an old plug wire/boot to prevent the possibility of cross-threading and damaging the cylinder head bore. An old plug wire/boot can be used to thread the plug if you turn the wire by hand. Should the plug begin to cross-thread, the wire will twist before the cylinder head would be damaged. This trick is useful when accessories or a deep cylinder head design prevents you from easily keeping fingers on the plug while it is threaded by hand.

13. Carefully tighten the spark plug. If the plug you are installing is equipped with a crush washer, tighten the plug until the washer seats, then turn it 1/4 turn to crush the washer.

14. Apply a small amount of Johnson/Evinrude Triple-Guard or a silicone dielectric grease to the ribbed, ceramic portion of the spark plug lead and inside the spark plug boot to prevent sticking, then install the boot to the spark plug and push until it clicks into place. The click may be felt or heard. Gently pull back on the boot to assure proper contact.

15. Connect the negative battery cable or turn the battery switch **ON**.

16. Test run the outboard (using a test tank or flush fitting) and insure proper operation.

READING SPARK PLUGS

◆ **See Figures 98 thru 103**

Reading spark plugs can be a valuable tuning aid. By examining the insulator firing nose color, you can determine much about the engine's overall operating condition.

In general, a light tan/gray color tells you that the spark plug is at the optimum temperature and that the engine is in good operating condition.

Dark coloring, such as heavy black wet or dry deposits usually indicate a fouling problem. Heavy, dry deposits can indicate an overly rich condition, too cold a heat range spark plug, possible vacuum leak, low compression, overly retarded timing or too large a plug gap.

■ **Note, carbon fouling can also occur from excessive idling conditions. If you put through a lot of no wake zones, for hours at a time, then you either need a hotter plug, or you need to balance that use by running the motor up at or near wide-open throttle too for periods of time. If you can do this for a while on the way back to the dock or ramp, after those idling conditions, you may alleviate the need to change to a different type of plug.**

If the deposits are wet, it can be an indication of a breached head gasket (water) or an extremely rich condition (fuel/oil), depending on what liquid is present at the firing tip.

Also look for signs of detonation, such as silver specs, black specs or melting or breakage at the firing tip.

Compare your plugs to the illustrations shown to identify the most common plug conditions.

Fouled Spark Plugs

A spark plug is "fouled" when the insulator nose at the firing tip becomes coated with a foreign substance, such as fuel, oil or carbon. This coating makes it easier for the voltage to follow along the insulator nose and leach back down into the metal shell, grounding out, rather than bridging the gap normally.

Fuel, oil and carbon fouling can all be caused by different things but in any case, once a spark plug is fouled, it will not provide voltage to the firing tip and that cylinder will not fire properly. In many cases, the spark plug cannot be cleaned sufficiently to restore normal operation. It is therefore recommended that fouled plugs be replaced.

Signs of fouling or excessive heat must be traced quickly to prevent further deterioration of performance and to prevent possible engine damage.

Overheated Spark Plugs

When a spark plug tip shows signs of melting or is broken, it usually means that excessive heat and/or detonation was present in that particular combustion chamber or that the spark plug was suffering from thermal shock.

Since spark plugs do not create heat by themselves, one must use this visual clue to track down the root cause of the problem. In any case, damaged firing tips most often indicate that cylinder pressures or temperatures were too high. Left unresolved, this condition usually results in more serious engine damage.

Detonation refers to a type of abnormal combustion that is usually preceded by pre-ignition. It is most often caused by a hot spot formed in the combustion chamber.

As air and fuel is drawn into the combustion chamber during the intake stroke, this hot spot will "pre-ignite" the air fuel mixture without any spark from the spark plugs.

Detonation

Detonation exerts a great deal of downward force on the pistons as they are being forced upward by the mechanical action of the connecting rods. When this occurs, the resulting concussion, shock waves and heat can be severe. Spark plug tips can be broken or melted and other internal engine components such as the pistons or connecting rods themselves can be damaged.

Left unresolved, engine damage is almost certain to occur, with the spark plug usually suffering the first signs of damage.

■ **When signs of detonation or pre-ignition are observed, they are symptom of another problem. You must determine and correct the situation that caused the hot spot to form in the first place.**

INSPECTION & GAPPING

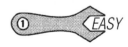

◆ **See Figures 104 and 105**

A particular spark plug might fit hundreds of powerheads and although the factory will typically set the gap to a pre-selected setting, this gap may not be the right one for your particular powerhead.

Insufficient spark plug gap can cause pre-ignition, detonation, even engine damage. Too much gap can result in a higher rate of misfires, a noticeable loss of power, plug fouling and poor economy.

■ **Most Johnson/Evinrude motors from this time period require a 0.030 in. (0.8mm) or 0.040 in. (1.0mm) gap. Remember also that a number of larger engines from the late 60's on use a plug that is not adjustable.**

Check the spark plug gap before installation. The ground electrode (the L-shaped one connected to the body of the plug) must be parallel to the center electrode and the specified size wire gauge must pass between the electrodes with a slight drag.

Do not use a flat feeler gauge when measuring the gap on a used plug, because the reading may be inaccurate. A round-wire type gapping tool is the best way to check the gap. The correct gauge should pass through the electrode gap with a slight drag. If you're in doubt, try a wire that is one size smaller and one larger; the smaller gauge should go through easily, while the larger one shouldn't go through at all.

Wire gapping tools usually have a bending tool attached. **USE IT!** This tool greatly reduces the chance of breaking off the electrode and is much more accurate. Never attempt to bend or move the center electrode. Also, be careful not to bend the side electrode too far or too often as it may weaken and break off within the engine, requiring removal of the cylinder head to retrieve it.

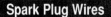

Spark Plug Wires

TESTING

◆ See Figures 106 and 107

■ All Johnson/Evinrude motors of this time frame are equipped with secondary spark leads or spark plug wires to carry ignition voltage from the coils to the spark plugs. Overtime the insulation on these wires will break down, allowing arcing (voltage leakage or shorts to ground) and/or corrosion (causing higher resistance). The wires must be inspected periodically and replaced when worn or damaged in order to ensure optimum ignition performance.

Fig. 98 A normally worn spark plug should have light tan or gray deposits on the firing tip (electrode)

Fig. 99 A carbon-fouled plug, identified by soft, sooty black deposits, may indicate an improperly tuned powerhead

Fig. 100 This spark plug has been left in the powerhead too long, as evidenced by the extreme gap. Plugs with such an extreme gap can cause misfiring and stumbling accompanied by a noticeable lack of power

Fig. 101 An oil-fouled spark plug indicates a powerhead with worn piston rings or a malfunctioning oil injection system that allows excessive oil to enter the combustion chamber

Fig. 102 A physically damaged spark plug may be evidence of severe detonation in that cylinder. Watch the cylinder carefully between services, as a continued detonation will not only damage the plug but will most likely damage the powerhead

Fig. 103 A bridged or almost bridged spark plug, identified by the build-up between the electrodes caused by excessive carbon or oil build up on the plug

Each time you remove the engine cover, visually inspect the spark plug wires for burns, cuts or breaks in the insulation. Check the boots on the coil and at the spark plug end. Replace any wire that is damaged.

If the spark plug wire is not integral with the ignition coil, about once a year (this should probably be performed when you change your spark plugs during a pre-season tune-up), check the resistance of the spark plug wires with an ohmmeter. Wires with excessive resistance will cause misfiring and may make the engine difficult to start. In addition worn wires will allow arcing and misfiring in humid conditions.

Remove the spark plug wire from the engine. Test the wires by connecting 1 lead of an ohmmeter to the coil end of the wire and the other lead to the spark plug end of the wire. Typically resistance for spark plug leads would measure approximately 7000 ohms per foot of wire. However, on carbureted Johnson/Evinrude motors, the manufacturer calls for a reading very close to or equal to zero ohms resistance. If a spark plug wire is found to have excessive resistance the entire set should be replaced.

■ Keep in mind that just because a spark plug wire passes a resistance test doesn't mean that it is in good shape. Cracked or deteriorated insulation will allow the circuit to misfire under load, especially when wet. Always visually check wires to cuts, cracks or breaks in the insulation. If found, run the engine in a test tank or on a flush device either at night (looking for a bluish glow from the wires that would indicate arcing) or while spraying water (from a spray bottle, NOT a garden hose) on them while listening for an engine stumble.

Regardless of resistance tests and visual checks, it is never a bad idea to replace spark plug leads at least every couple of years, and to keep the old ones around for spares. Think of spark plug wires as a relatively low cost item that whose replacement can also be considered maintenance.

REMOVAL & INSTALLATION

■ All Johnson/Evinrude motors covered here are equipped with secondary spark leads or spark plug wires to carry ignition voltage from the coils to the spark plugs. Overtime the insulation on these wires will break down, allowing arcing (voltage leakage or shorts to ground) and/or corrosion (causing higher resistance). The wires must be inspected periodically and replaced when worn or damaged in order to ensure optimum ignition performance.

When installing a new set of spark plug wires, replace the wires 1 at a time so there will be no confusion. Coat the inside of the boots with Johnson/Evinrude Triple-Guard or dielectric grease to prevent sticking. Install the boot firmly over the spark plug until it clicks into place. The click may be felt or heard. Gently pull back on the boot to assure proper contact. Repeat the process for each wire.

■ It is important to route the new spark plug wire the same as the original and install it in a similar manner on the powerhead. Improper routing of spark plug wires may cause powerhead performance problems.

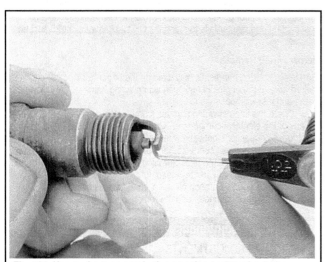
Fig. 104 Use a wire-type spark plug gapping tool to check the distance between center and ground electrodes

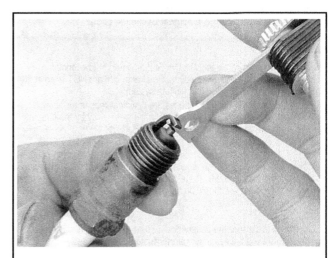
Fig. 105 Most plug gapping tools have an adjusting fitting used to bend the ground electrode

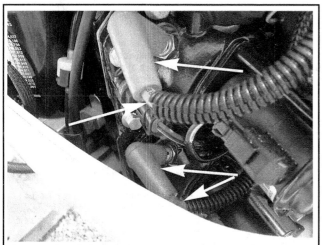
Fig. 106 Visually inspect the spark plug boot and wire (under the protective cover). . .

Fig. 107 . . .all the way back to the ignition coils for signs of wear or damage (typical engine shown)

Ignition & Electrical System Maintenance

INSPECTION

◆ See Figure 108

A number of different ignition systems are used on outboard engines covered in this service: mainly a flywheel magneto and versions of a capacitor discharge (CD) system. There is only a little maintenance involved in the operation of these ignition systems and even less to repair if they fail. Most systems are sealed and there is no option other than to replace failed components.

1. However, just as a tune-up is pointless on an engine with no compression, installing new spark plugs will not do much for an engine with a damaged ignition system. At each tune-up, visually inspect all ignition system components for signs of obvious defects. Look for signs of burnt, cracked or broken insulation. Replace wires or components with obvious defects. If spark plug condition suggests weak or no spark on 1 or more cylinders, perform ignition system testing to eliminate possible worn or defective components.

If trouble is suspected, it is very important to narrow down the problem to the ignition system and replace the correct components rather than just replace parts hoping to solve the problem. Electronic components can be very expensive and are usually not returnable.

■ **For more details on ignition system component testing and service, please refer to the Ignition & Electrical System section.**

Breaker Points

Engines equipped with either the flywheel magneto or low-tension magneto systems utilize breaker points. Breaker points are NOT used in the magneto capacitor discharge (CD) ignition system.

Rough or discolored contact surfaces are sufficient reason for replacement. The cam follower will usually have worn away by the time the points have become unsatisfactory for efficient service.

Check the resistance across the contacts. If the test indicates ZERO resistance, the points are serviceable. A slight resistance across the points will affect idle operation. A high resistance may cause the ignition system to malfunction and loss of spark. Therefore, if any resistance across the points is indicated, the point set should be replaced.

Starter Motor Test

1. Check to be sure the battery has a 70-ampere rating and is fully charged. Would you believe, many starter motors are needlessly disassembled, when the battery is actually the culprit?

2. Lubricate the pinion gear and screw shaft with No. 10 oil or Johnson/Evinrude Starter Pinion Lubricant.

Fig. 108 The fuel and ignition systems on any engine must be properly synchronized before maximum performance can be obtained.

3. Connect 1 lead of a voltmeter to the positive terminal of the starter motor. Connect the other meter lead to a good ground on the engine. Check the battery voltage under load by turning the ignition switch to the **START** position and observing the voltmeter reading.

4. If the reading is 9 1/2 volts or greater, and the starter motor fails to operate, repair or replace the starter motor. Refer to the Ignition & Electrical section.

Solenoid Test

An ohmmeter is the only instrument required to effectively test a solenoid. Test the ohmmeter by connecting the red and black leads together. Adjust the pointer to the right side of the scale.

On all Johnson/Evinrude engines the case of the solenoid does NOT provide a suitable ground to the engine. Hundreds of solenoids have been discarded because of the erroneous belief the case is providing a ground and the unit should function when 12-volts is applied. Not so! One terminal of the solenoid is connected to a 12-volt source. The other terminal is connected via a white wire to a cut-out switch on top of the engine. This cut-out switch provides a safety to break the ground to the solenoid in the event the engine starts at a high rpm. Therefore, the solenoid ground is made and broken by the cut-out switch.

■ **NEVER connect the battery leads to the large terminals of the solenoid, or the test meter will be damaged. Connect each lead of the test meter to each of the large terminals on the solenoid.**

Using battery jumper leads, connect the positive lead from the positive terminal of the battery to the small **S** terminal of the solenoid. Connect the negative lead to the negative battery terminal and the **I** terminal of the solenoid. If the meter pointer hand moves into the OK block, the solenoid is serviceable. If the pointer fails to reach the OK block, the solenoid must be replaced.

Internal Wiring Harness

An internal wiring harness is only used on the larger horsepower engines covered here. If the engine is equipped with a wiring harness, the following checks and test will apply.

1. Check the internal wiring harness if problems have been encountered with any of the electrical components. Check for frayed or chafed insulation and/or loose connections between wires and terminal connections.

2. Check the harness connector for signs of corrosion. If the harness shows any evidence of damage or corrosion, the problem must be corrected before proceeding with any harness testing.

3. Convince yourself a good electrical connection is being made between the harness connector and the remote control harness.

Fuel System Maintenance

CARBURETOR ADJUSTMENTS

High-Speed Adjustment

◆ See Figures 109 thru 112

■ **Carburetor mixture adjustments are usually not performed for the sake of periodic maintenance, but are only required after repairs, rebuilds or after many hundreds of hours of operation, in order to compensate for changes in mechanical condition of the motor, carburetor and fuel system. In other words, don't tinker un-necessarily.**

The high-speed needle valve is adjustable on some models covered here through 1972. On a few 2 cyl. motors, and almost all 3 cyl. and V4 motors, the high-speed orifice is fixed at the factory and is NOT adjustable. However, larger or smaller orifices may be installed for different elevations.

On all Johnson/Evinrude engines, the high-speed needle valve, or orifice, is the lower valve on the carburetor. The upper needle valve is always the idle adjustment.

A beginning "rough" adjustment for the high-speed needle valve is 3/4 turn out (counterclockwise) from the lightly seated (closed) position. TAKE CARE not to seat the valve firmly to prevent damage to the valve or the carburetor.

To make the high-speed adjustment:

1. Mount the engine in a test tank or body of water, preferably with a test wheel. Engines up to 40 hp may be operated in the high rpm range in a test tank without sustaining damage.

NEVER, AGAIN NEVER, operate the engine at high speed with a flush device attached. The engine, operating at high speed with such a device attached, would RUN-AWAY from lack of a load on the propeller, causing extensive damage.

2. Connect a tachometer to the engine.

Water must circulate through the lower unit to the engine any time the engine is run to prevent damage to the water pump in the lower unit. Just 5 seconds without water will damage the water pump.

3. Start the engine and allow it to warm to operating temperature.
4. Shift the engine into forward gear.
5. With the engine running in forward gear, advance the throttle to the wide open position, and then very SLOWLY turn the high-speed needle valve inward (CLOCKWISE) until the engine begins to lose rpm. Now, SLOWLY rotate the needle valve outward (COUNTERCLOCKWISE) until the engine peaks out at the highest rpm.
6. If the high-speed needle valve adjustment is too lean, the low-speed adjustment will be affected. Under certain conditions it may be necessary to adjust the high-speed needle valve just a bit richer in order to obtain a satisfactory idle adjustment.
7. After the high-speed needle adjustment has been obtained, proceed with the idle adjustment as outlined in this section.

Fig. 109 Adjustment points on a 35 hp engine. Note the pressure-type fuel connector

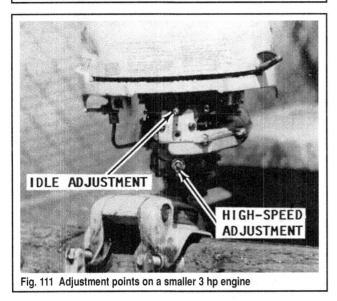

Fig. 111 Adjustment points on a smaller 3 hp engine

■ For detailed procedures to disassemble, clean, assemble, and adjust the carburetor, see the appropriate section under Fuel System for the carburetor type on the engine being serviced.

Idle Adjustment

◆ See Figures 109 thru 112

■ Carburetor mixture adjustments are usually not performed for the sake of periodic maintenance, but are only required after repairs, rebuilds or after many hundreds of hours of operation, in order to compensate for changes in mechanical condition of the motor, carburetor and fuel system. In other words, don't tinker un-necessarily.

It is necessary to adjust the carburetor while the engine is running in a test tank or with the boat in a body of water. For maximum performance, the idle mixture and the idle rpm should be adjusted under actual operating conditions.

1. Set the idle mixture screw at the specified number of turns open from a lightly seated position. In most cases this is from 1 to 1 1/2 turns open from close (but for more details, please refer to the Carburetor overhaul information in the Fuel System section.

NEVER, AGAIN NEVER, operate the engine at high speed with a flush device attached. The engine, operating at high speed with such a device attached, would RUN-AWAY from lack of a load on the propeller, causing extensive damage.

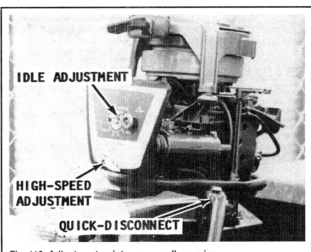

Fig. 110 Adjustment points on a smaller engine

Fig. 112 Adjustment points on a larger engine

2. Start the engine and allow it to warm to operating temperature.

✳✳ CAUTION

Water must circulate through the lower unit to the engine any time the engine is run to prevent damage to the water pump in the lower unit. Just 5 seconds without water will damage the water pump.

3. With the engine running in forward gear, slowly turn the idle mixture screw COUNTERCLOCKWISE until the affected cylinders start to load up or fire unevenly, due to an over-rich mixture.

4. Slowly turn the idle mixture screw CLOCKWISE until the cylinders fire evenly and engine rpm increases.

5. Continue to slowly turn the screw CLOCKWISE until too lean a mixture is obtained and the rpm falls off and the engine begins to misfire.

6. Now, set the idle mixture screw one-quarter (1/4) turn out (counterclockwise) from the lean-out position. This adjustment will result in an approximate true setting. A too-lean setting is a major cause of hard starting a cold engine. It is better to have the adjustment on the rich side rather than on the lean side. Stating it another way, do not make the adjustment any leaner than necessary to obtain a smooth idle.

■ **If the engine hesitates during acceleration after adjusting the idle mixture, the mixture is too lean. Enrich the mixture slightly, by turning the adjustment screw inward until the engine accelerates correctly.**

7. With the engine running in forward gear, rotate the nylon idle adjustment screw, located on the port-side of the engine, until the engine idles at the recommended rpm, as given in the Specifications. This idle adjustment screw is always exposed on the outside of the shroud.

■ **For detailed procedures to disassemble, clean, assemble, and adjust the carburetor, see the appropriate section under Fuel System for the carburetor type on the engine being serviced.**

FUEL & FUEL TANKS

◆ **See Figures 113 and 114**

Take time to check the fuel tank and all of the fuel lines, fittings, couplings, valves, flexible tank fill and vent. Turn on the fuel supply valve at the tank, if the engine is equipped with a self-contained fuel tank. If the fuel was not drained at the end of the previous season, make a careful inspection for gum formation. When gasoline is allowed to stand for long periods of time, particularly in the presence of copper, gummy deposits form. This gum can clog the filters, lines, and passageway in the carburetor.

If the condition of the fuel is in doubt, drain, clean, and fill the tank with fresh fuel.

Fuel pressure at the carburetor should be checked whenever a lack of fuel volume at the carburetor is suspected.

FUEL PUMPS

Many times, a defective fuel pump diaphragm is mistakenly diagnosed as a problem in the ignition system. The most common problem is a tiny pin-hole in the diaphragm. Such a small hole will permit gas to enter the crankcase and wet foul the spark plug of the cylinder to which the fuel pump is attached at idle. During high-speed operation, gas quantity is limited; the plug is not foul and will therefore fire in a satisfactory manner.

If the fuel pump fails to perform properly, an insufficient fuel supply will be delivered to the carburetor. This lack of fuel will cause the engine to run lean, lose rpm or cause piston scoring.

When a fuel pressure gauge is added to the system, it should be installed at the end of the fuel line leading to the upper carburetor. To ensure maximum performance, the fuel pressure must be 2 psi or more at full throttle.

Tune-up Task

Most fuel pumps are equipped with a fuel filter. The filter may be cleaned by first removing the cap, then the filter element, cleaning the parts and drying them with compressed air, and finally installing them in their original position.

A fuel pump pressure test should be made any time the engine fails to perform satisfactorily at high speed.

■ **NEVER use liquid Neoprene on fuel line fittings. Always use Permatex when making fuel line connections. Permatex is available at almost all marine and hardware stores.**

Cooling System Maintenance

WATER PUMP CHECK

■ **Please refer to the Cooling System or Lower Unit sections for complete details on the water pump and impeller.**

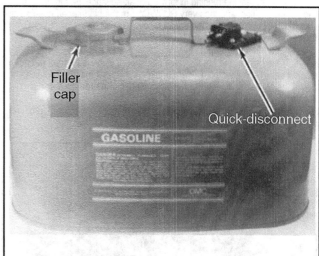

Fig. 113 A typical 6 gal. fuel tank

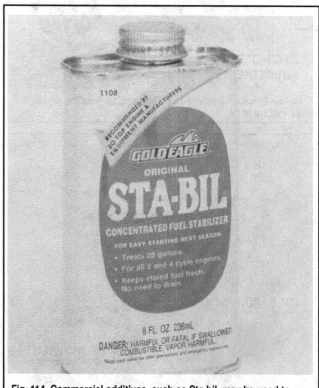

Fig. 114 Commercial additives, such as Sta-bil, may be used to keep gasoline fresh

The water pump MUST be in very good condition for the engine to deliver satisfactory service. The pump performs an extremely important function by supplying enough water to properly cool the engine. Therefore, in most cases, it is advisable to replace the complete water pump assembly at least once a year, or anytime the lower unit is disassembled for service.

Sometimes during adjustment procedures, it is necessary to run the engine with a flush device attached to the lower unit. NEVER operate the engine over 1000 rpm with a flush device attached, because the engine may "RUNAWAY" due to the no-load condition on the propeller. A "runaway" engine could be severely damaged. As the name implies, the flush device is primarily used to flush the engine after use in salt water or contaminated fresh water. Regular use of the flush device will prevent salt or silt deposits

from accumulating in the water passageway. During and immediately after flushing, keep the motor in an upright position until all of the water has drained from the drive shaft housing. This will prevent water from entering the powerheads by way of the drive shaft housing and the exhaust ports, during the flush. It will also prevent residual water from being trapped in the driveshaft housing and other passageways.

To test the water pump, the lower unit MUST be placed in a test tank or the boat moved into a body of water. The pump must now work to supply a volume to the engine.

Lack of adequate water supply from the water pump thru the engine will cause any number of powerhead failures, such as stuck rings, scored cylinder walls, burned pistons, etc.

TIMING & SYNCHRONIZATION - 1 & 2 CYLINDER ENGINES

General Information

◆ See Figure 115

The fuel and ignition systems MUST be carefully synchronized to achieve maximum performance from the engine. In simple terms, synchronization is timing the carburetion to the ignition. This means, as the throttle is advanced to increase engine rpm, the carburetor and ignition systems are both advanced equally and at the same rate.

Therefore, any time the fuel system or the ignition system is serviced to replace a faulty part, or any adjustments are made for any reason, the engine synchronization must be carefully checked and verified.

Before making any adjustments with the synchronization, the ignition system should be thoroughly checked according to the procedures outlined in this section (and the Ignition & Electrical System section, as necessary) as well as the fuel system checked according to the procedures outlined under Fuel System.

The timing on small Johnson/Evinrude outboard engines (models which are NOT equipped with a CD ignition) is controlled through adjustment of the points and therefore covered within the specific Ignition system section. On the 40, 50, 55 and 60 hp engines, and most 3 cyl/V4 models (models equipped with a type of CD ignition), the timing is adjustable through the synchronization.

When equipped, if the points are adjusted too closely, the spark plugs will fire early; if adjusted with excessive gap, the plugs will fire too late, for efficient operation.

Therefore, correct point adjustment and synchronization are essential for proper engine operation. An engine may be in apparent excellent mechanical condition, but perform poorly, unless the points and synchronization have been adjusted precisely, according to the Specifications. The correct point setting for ALL engines covered in this service is 0.020 in. (0.5mm).

Timing & Synchronization

■ Due to the nature of their individual systems, all timing procedures are included with the specific Ignition System procedures in the Ignition & Electrical section.

PRIMARY PICK-UP ADJUSTMENTS & LOCATIONS

To properly adjust the synchronization, locations are assigned letters and primary adjustments are assigned numbers for every powerhead covered in this service. These letters and numbers are referenced in the Tune-up Specifications chart at the end of this section.

The locations are lettered **A** thru **E** and the adjustments are numbered from 1 thru 5. The information, a letter and a number, to be used is taken from the chart by following across from the first column for the engine being serviced to the column titled Primary Pick-up Location and the column titled Primary Pick-Up Adjustment. Therefore, from the chart, a letter will be obtained for the location of the primary adjustment and a number, indicating the method of making the adjustment.

EXAMPLE

If the powerhead being serviced is a 1971 25 hp, then the primary pick-up location letter obtained from the chart is **D** and the primary pick-up adjustment note number is **3**.

In this case, perform the first few steps, valid for all powerheads with the **D** location. DO NOT perform D1 or D5. However, you WOULD perform Step D3 which describes you engine's procedure.

Fig. 115 Typical Johnson/Evinrude flywheel timing marks, aligned with a timing pointer

When making the synchronization adjustment, it is well to know and understand exactly what to look for and why. The critical time when the throttle shaft in the carburetor begins to move is of the utmost importance.

First, realize that the time the cam follower makes contact with the cam is not the time the throttle shaft starts to move. Instead, the critical time is when the follower hits the designated position (as described in the next paragraphs) and the throttle shaft AT THE CARBURETOR begins to move.

A considerable amount of play exists between the followers at the top of the carburetor through the linkage to the actual throttle shaft. Therefore, the most important consideration is to watch for movement of the THROTTLE SHAFT, and not the follower. Movement of the shaft can be exaggerated by attaching a short piece of stiff wire to an alligator clip; grinding down the teeth on one side of the clip; and then attaching the clip to the throttle shaft, as shown. The wire jiggling will instantly indicate movement of the shaft.

Almost all of the photographs were taken with the flywheel removed for clarity. Normally the synchronization is set with the flywheel installed.

The following paragraphs describe the location in detail and give specific instructions as to exactly how the synchronization is to be made.

The "A" Location

◆ See Figures 116, 117 and 118

The pick-up for the **A** location is Port side of the mark. This means that the pick-up on the carburetor arm should be just to the Port side of the mark on the cam. To obtain this position, the cam, or the cam follower, is to be adjusted until the pick-up is at the proper location and the throttle shaft just begins to move.

Movement of the shaft can be exaggerated by attaching a short piece of stiff wire to an alligator clip; grinding down the teeth on one side of the clip, and then attaching the clip to the throttle shaft, as shown. The wire jiggling will instantly indicate movement of the shaft. The actual adjustment is accomplished by ONLY ONE method, depending on the engine being serviced. Check the Tune-Up Specifications chart for the adjustment

Fig. 116 Pick-up location for 'A' is Port side of mark

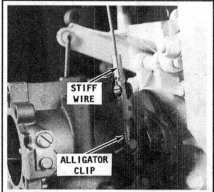

Fig. 117 Use a clip to help with the adjustment

Fig. 118 Adjustment A1

procedure to be performed. The reference numbers are listed in the Primary Pick-up Location and Adjustment columns.

A1 Loosen the 2 screws under the armature plate and move the primary pick-up inward or outward to meet the follower.

The "B" Location

◆ See Figures 119 and 120

Powerheads referenced to the **B** Primary Pick-up Location are to be adjusted to the Starboard side of the mark. To obtain this position, the cam, or the cam follower, (depending the type of powerhead being serviced), is to

Fig. 119 Pick-up location for 'B' is Starboard side of mark. Use a clip to help with the adjustment

Fig. 120 Adjustment 'B1'

be adjusted until the pick-up is at the proper location and the throttle shaft just begins to move. Movement of the shaft can be exaggerated by attaching a short piece of stiff wire to an alligator clip; grinding down the teeth on one side of the clip; and then attaching the clip to the throttle shaft, as shown. The wire jiggling will instantly indicate movement of the shaft.

The actual adjustment is accomplished by ONLY ONE method, regardless of the powerhead being serviced. Check the Tune-Up Specifications chart for the adjustment procedure to be performed. The reference numbers are listed in the Primary Pick-up Location and Adjustment columns.

B1 Loosen the 2 screws under the armature plate and move the primary pick-up inward or outward to meet the follower.

The "C" Location

◆ See Figures 121 thru 126

Powerheads referenced to the **C** Primary Pick-up Location are to be adjusted to the CENTER of the mark. To obtain this position, the cam, or the cam follower, (depending the type of engine being serviced), is to be adjusted until pick-up is at the proper location and the throttle shaft just begins to move.

Movement of the shaft can be exaggerated by attaching a short piece of stiff wire to an alligator clip; grinding down the teeth on one side of the clip; and then attaching the clip to the throttle shaft, as shown. The wire jiggling will instantly indicate movement of the shaft.

The actual adjustment is accomplished by ONLY ONE of two means, depending on the powerhead being serviced. Check the Tune-Up Specifications chart for the adjustment procedure to be performed. The reference numbers are listed in the Primary Pick-up Location and Adjustment columns.

AFTER the pick-up adjustment note has been obtained from the Chart, perform ONE of the following numbered procedures with the matching note number from the Chart.

C1 Loosen the 2 screws under the armature plate and move the primary pick-up inward or outward to meet the follower.

C2 Loosen the center screw on the throttle lever and move the lever inward or outward to match the line on the cam,

C3 On the starboard side of the engine, loosen the clamps on the throttle shaft and move the roller to make contact with the armature cam.

C4 On the starboard side, loosen the throttle arm screw and move the roller in or out to align with the mark on the cam.

The "D" Location

◆ See Figures 127 thru 132

Powerheads referenced to the **D** Primary Pick-up Location are to be adjusted with the cam follower midway between the 2 marks on the cam, at the moment the throttle shaft at the carburetor begins to move. The marks on the cam are about 1/4 in. apart.

Movement of the shaft can be exaggerated by attaching a short piece of stiff wire to an alligator clip; grinding down the teeth on one side of the clip; and then attaching the clip to the throttle shaft, as shown. The wire jiggling will instantly indicate movement of the shaft.

Fig. 121 Pick-up location for 'C' is center of the mark

Fig. 122 Use a clip to help with the adjustment

Fig. 123 Adjustment 'C1'

Fig. 124 Adjustment 'C2'

Fig. 125 Adjustment 'C3'

Fig. 126 Adjustment 'C4'

Fig. 127 Pick-up location for 'D' is between the marks on 1967-70 motors. . .

Fig. 128 . . .and also 1971-72 motors

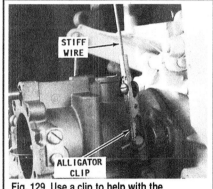

Fig. 129 Use a clip to help with the adjustment

Fig. 130 Adjustment 'D1'

Fig. 131 Adjustment 'D3'

Fig. 132 Adjustment 'D5'

The actual adjustment is accomplished by ONLY ONE of two means, depending on the powerhead being serviced. Check the Tune-Up Specifications chart for the adjustment procedure to be performed. The reference numbers are listed in the Primary Pick-up Location and Adjustment columns.

AFTER the pick-up adjustment note has been obtained from the Chart, perform ONE of the following numbered procedures with the matching note number from the Chart.

D1 Loosen the 2 screws under the armature plate and move the primary pick-up inward or outward to meet the follower.

D3 On the starboard side of the engine, loosen the clamps on the throttle shaft and move the roller to make contact with the armature cam.

D5 Loosen the eccentric lock screw on the Port side on the throttle shaft and turn the eccentric to move the roller to make contact with the armature cam.

The "E" Location

◆ **See Figures 133 thru 136**

Powerheads referenced to the **E** Primary Pick-up Location have a pointer attached to the intake manifold. Synchronization is made by advancing the magneto until the mark on the cam is aligned with the pointer on the intake manifold. At this point the throttle shaft should just begin to move.

Movement of the shaft can be exaggerated by attaching a short piece of stiff wire to an alligator clip; grinding down the teeth on one side of the clip; and then attaching the clip to the throttle shaft, as shown. The wire jiggling will instantly indicate movement of the shaft.

The actual adjustment is accomplished by ONLY ONE of two means, depending on the powerhead being serviced. Check the Tune-Up Specifications chart for the adjustment procedure to be performed. The reference numbers are listed in the Primary Pick-up Location and Adjustment columns.

AFTER the pick-up adjustment note has been obtained from the Chart, perform ONE of the following numbered procedures with the matching note number from the Chart.

E3 On the starboard side of the engine, loosen the clamps on the throttle shaft and move the roller to make contact with the armature cam.

E5 Loosen the eccentric lock screw on the Port side on the throttle shaft and turn the eccentric to move the roller to make contact with the armature cam.

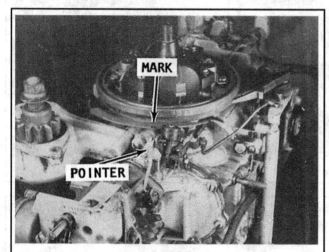
Fig. 133 Pick-up location for 'E' is center of the mark

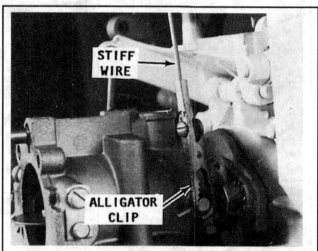
Fig. 134 Use a clip to help with the adjustment

Fig. 135 Adjustment' E3

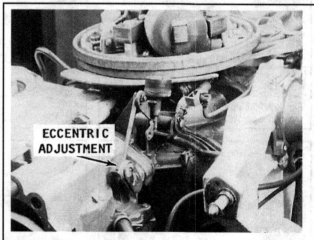
Fig. 136 Adjustment 'E5'

TIMING AND SYNCHRONIZATION - 3 CYLINDER & V4 ENGINES

General Information

The fuel and ignition systems MUST be carefully synchronized to achieve maximum performance from the engine. In simple terms, synchronization is timing the carburetion to the ignition. This means, as the throttle is advanced to increase engine rpm, the carburetor and ignition systems are both advanced equally and at the same rate.

Therefore, any time the fuel system or the ignition system is serviced to replace a faulty part, or any adjustments are made for any reason, the engine synchronization must be carefully checked and verified.

Before making any adjustments with the synchronization, the ignition system should be thoroughly checked according to the procedures outlined in this section and the fuel checked according to the procedures outlined in the Fuel System section.

■ Like most adjustments, the results of the Timing & Synchronization procedures build on one another. Do not just jump in at the middle adjustment, perform it and think you've addressed the problem. The procedures are presented in the order that they must be performed, starting at the first and ending at the last. You may find that the adjustment for a given procedure is already in spec, and that is fine, but if you don't check it you cannot be sure.

Carburetor Adjustments

HIGH-SPEED ADJUSTMENT

◆ See Figure 137

The high-speed needle valve is adjustable on most models covered through 1965. After 1965, the high speed orifice is fixed at the factory and is NOT adjustable. However, larger or smaller orifices may be installed for different elevations.

On all Johnson/Evinrude engines, the high speed needle valve is the upper valve on the carburetor. The upper needle is always the idle adjustment. The 3-cylinder units do have a high speed adjustment. A beginning "rough" adjustment for the high speed needle valve is 3/4 turn out (counterclockwise) from the lightly seated (closed) position. TAKE CARE not to seat the valve firmly to prevent damage to the valve or the carburetor.

To make the high-speed adjustment:
1. Mount the engine in a body of water.

✳✳ CAUTION

NEVER, AGAIN NEVER, operate the engine at high speed with a flush device attached. The engine, operating at high speed with such a device attached, would RUN-AWAY from lack of a load on the propeller, causing extensive damage.

2. Connect a tachometer to the engine.

✳✳ CAUTION

Water must circulate through the lower unit to the engine any time the engine is run to prevent damage to the water pump in the lower unit. Just 5 seconds without water will damage the water pump.

3. Start the engine and allow it to warm to operating temperature.
4. Shift the engine into forward gear.
5. With the engine running in forward gear, advance the throttle to the wide open position (WOT), and then very SLOWLY turn the high speed needle valve inward (CLOCKWISE) until the engine begins to lose rpm.
6. Now, SLOWLY rotate the needle valve outward (COUNTERCLOCKWISE) until the engine peaks out at the highest rpm.

If the high speed needle valve adjustment is too lean, the low speed adjustment will be affected. Under certain conditions it may be necessary to adjust the high speed needle valve just a bit richer in order to obtain a satisfactory idle adjustment.

7. After the high-speed needle adjustment has been obtained, proceed with the idle adjustment as outlined in the next paragraphs.

IDLE ADJUSTMENT

◆ See Figures 138 and 139

Due to local conditions, it may be necessary to adjust the carburetor while the engine is running in a test tank or with the boat in a body of water. For maximum performance, the idle mixture and the idle rpm should be adjusted under actual operating conditions.

1. Set the idle mixture screw at the specified number of turns open from a lightly seated position. In most cases this is from 1-1 1/2 turns open from closed position. Refer to the Tune-Up Specifications chart.
2. Start the engine and allow it to warm to operating temperature.

✳✳ CAUTION

Water must circulate through the lower unit to the engine any time the engine is run to prevent damage to the water pump in the lower unit. Just 5 seconds without water will damage the water pump.

✳✳ CAUTION

NEVER, AGAIN NEVER, operate the engine at high speed with a flush device attached. The engine, operating at high speed with such a device attached, would RUN-AWAY from lack of a load on the propeller, causing extensive damage.

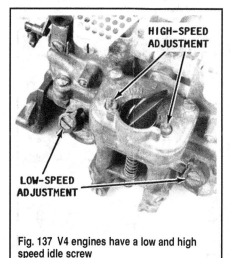

Fig. 137 V4 engines have a low and high speed idle screw

Fig. 138 A good look at the idle speed screws on a typical V4. . .

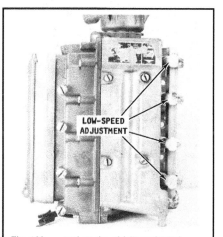

Fig. 139 . . .and on the old 90 and 100 hp engines

3. With the engine running in forward gear, slowly turn the idle mixture screw COUNTERCLOCKWISE until the affected cylinders start to load up or fire unevenly due to an over-rich mixture. Slowly turn the idle mixture screw CLOCKWISE until the cylinders fire evenly and engine rpm increases.

4. Continue to slowly turn the screw CLOCKWISE until too lean a mixture is obtained and the rpm falls off and the engine begins to misfire.

5. Now, set the idle mixture screw 1/4 turn out (counterclockwise) from the lean-out position. This adjustment will result in an approximate true setting. A too-lean setting is a major cause of hard starting a cold engine. It is better to have the adjustment on the rich side rather than on the lean side. Stating it another way, do not make the adjustment any leaner than necessary to obtain a smooth idle.

6. If the engine hesitates during acceleration after adjusting the idle mixture, the mixture is too lean. Enrich the mixture slightly, by turning the adjustment screw inward until the engine accelerates correctly.

7. With the engine running in forward gear, rotate the nylon idle adjustment screw, located on the portside of the engine, until the engine idles at the recommended rpm, as given in the Specifications. This idle adjustment screw is always exposed on the outside of the shroud.

■ **For detailed procedures to disassemble, clean, assemble, and adjust the carburetor, see the appropriate procedures in the Fuel Systems section.**

Synchronization

1958-59 50 HP MODELS

■ **This procedure is highly important - synchronizing the ignition system with the fuel system to achieve maximum performance from the outboard unit. It might be best to read through the complete synchronizing procedure before actually performing the work to obtain a good overall idea of exactly what needs to be done, why, and how it is accomplished.**

Tower Shaft To Distributor

◆ **See Figures 140, 141 and 142**

1. Place the advance arm from the distributor on the bottom of the advance arm on the tower shaft. Start the 2 screws that secure the arm in this position; but DO NOT tighten them at this time.

2. Rotate the flywheel CLOCKWISE until the mark on the distributor pulley is aligned with the mark on the lower housing. Grasp the distributor housing and move it until the mark on the breaker plate is in between the pulley mark and the timing mark.

3. When the 3 marks are aligned, advance the throttle by moving the tower shaft arm until it is 90° from the embossed mark on the bracket on the port side attached to the tower shaft (this is all viewed from above the engine.) Check again to be sure the distributor pulley mark is aligned with the timing mark on the distributor base.

If the marks are still aligned and the tower shaft arm is still at 90° to the tower shaft bracket - you are a winner. Tighten the 2 screws securing the advance arm between the distributor and the tower shaft.

Tower Shaft To Carburetor

◆ **See Figures 143 thru 146**

1. Remove the rod between the tower shaft and the throttle advance cam.

■ **Some engines have a double set screw arrangement. One setscrew locks in the second. The first setscrew must be completely removed before the second can be backed off.**

2. SLOWLY advance the throttle cam and at the same time watch the mark on the throttle cam and the mark on the nylon cam roller. The nylon roller should just begin to move (open) the throttle shaft arm as the center of the nylon roller comes into alignment with the throttle cam timing mark. Now, if this setting requires adjustment, hold the nylon roller in contact with the throttle cam and at the same time loosen the throttle arm screw and push the throttle arm *towards* the throttle cam. After the adjustment has been made, tighten the screw securely.

3. Shift the engine into forward gear and at the same time, rotate the propeller to verify the engine is fully engaged in the forward gear. Grasp the magneto and turn it to the full advance position. Hold the magneto in the full advance position with a large rubber band.

4. Insert a 0.020 in. (0.51mm) feeler gauge between the throttle shaft arm and its stop.

5. At the carburetor, insert the rod from the tower shaft into the throttle advance cam. Move the throttle advance cam to the wide open throttle position, and then tighten the set screw. Remove the rubber band and allow the magneto to return to the idle position. Grasp the arm at the bottom of the tower shaft and rotate the tower shaft to the full advance position. Hold the tower shaft in this position and check the following 3 places:

    a. The throttle advance cam at the carburetor is against its stop.

    b. Tower shaft advance arm is against its stop (without the feeler gauge.)

    c. The distributor is fully advanced against its stop.

6. Mount the engine in a large test tank or body of water and check the completed work.

■ **The engine MUST be mounted in a large test tank or body of water to adjust the timing. NEVER attempt to make this adjustment with a flush attachment connected to the lower unit or in a small, confined test tank. The no-load condition on the propeller would cause the engine to RUNAWAY resulting in serious damage or destruction of the unit.**

✳✳ CAUTION

Water must circulate through the lower unit to the engine any time the engine is run to prevent damage to the water pump in the lower unit. Just 5 seconds without water will damage the water pump.

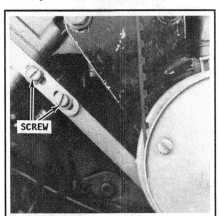

Fig. 140 Start the 2 screws, but do not tighten them

Fig. 141 Line up all the marks. . .

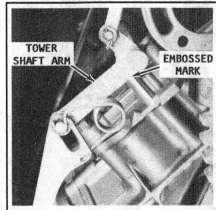

Fig. 142 . . .before moving the tower shaft arm

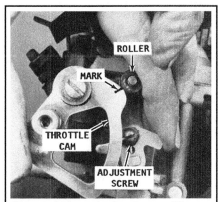
Fig. 143 Advance the throttle cam while watching the mark

Fig. 144 Run it all the way up against the stop

Fig. 145 Check the clearance between the throttle shaft arm and stop

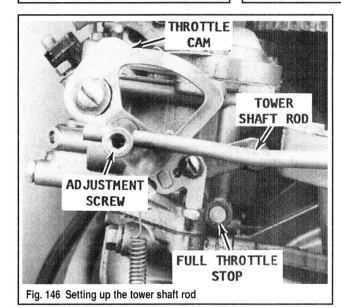
Fig. 146 Setting up the tower shaft rod

Fig. 147 Run the housing fully against the stop

1964-66 60 HP, 1960-65 75 HP, 1964-65 90 HP & 1966-67 80/100 HP MODELS

Tower Shaft To Distributor

MODERATE

◆ See Figures 147 and 148

1. Grasp the top of the distributor and rotate the distributor COUNTERCLOCKWISE to the fully advanced position until the tab on the distributor housing rests firmly against the rubber bumper. Use a large strong rubber band to hold the distributor at this position.

2. Check to be sure the inside edge of the control shaft arm is aligned with the outer edge of the triangular shaped boss on the control shaft bracket (when viewed from above).

3. If the setting is not correct as described, loosen the 2 control link adjusting screws and either lengthen or shorten the control shaft until the edges of the arm and boss are aligned. Tighten the adjusting screws and remove the band.

Tower Shaft To Carburetor

MODERATE

◆ See Figures 149 thru 152

1. Remove the rod between the tower shaft and the throttle advance cam.

■ Some engines have a double set screw arrangement to hold the rod in place. One set screw locks in the second. The first set screw must be completely removed before the second can be backed off. On some other engines, a nylon yoke and pin arrangement may be used instead of the set screws.

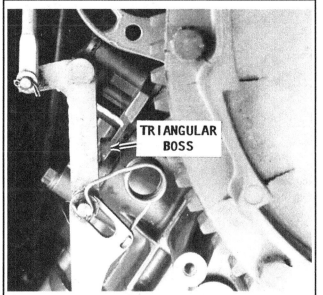
Fig. 148 The control arm shaft must be aligned with the edge of the boss

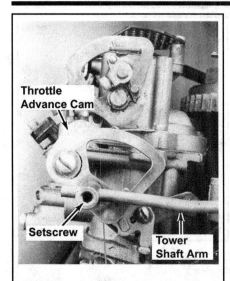

Fig. 149 Loosen the set screw and remove the arm. . .

Fig. 150 . . .although some models may use a yoke set-up

Fig. 151 Adjustment screw locations may be different

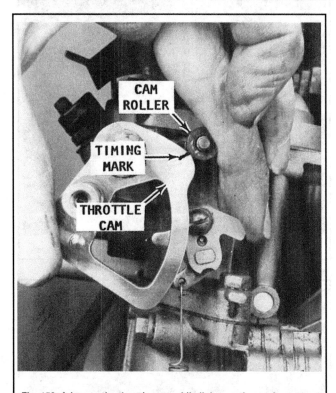

Fig. 152 Advance the throttle cam while lining up the mark

■ On the 90 and 100 hp models, the throttle pick-up and cam mentioned in the following step is located on TOP of the carburetor. All other engines covered here have these items on the SIDE of the carburetor

2. SLOWLY advance the throttle cam and at the same time watch the mark on the throttle cam and the mark on the nylon cam roller. The nylon roller should just begin to move (open) the throttle shaft arm as the center of the nylon roller comes into alignment with the throttle cam timing mark.

3. Now, if this setting requires adjustment, hold the nylon roller in contact with the throttle cam and at the same time loosen the throttle arm screw and push the throttle arm TOWARDS the throttle cam.

4. After the adjustment has been made, tighten the screw securely.

Checking The Synchronizing Work

MODERATE

◆ See Figures 153 and 154

1. Shift the engine into forward gear and at the same time rotate the propeller to verify the engine is fully engaged in the forward gear. Grasp the distributor and turn it to the full advance position. Hold the distributor in the full advance position with a large rubber band.

2. Insert a 0.020 in. (0.51mm) feeler gauge between the throttle shaft arm and its stop. At the carburetor: insert the rod from the tower shaft into the throttle advance cam. Move the throttle advance cam to the wide open throttle position, and then tighten the set screw.

3. If the engine being serviced has the nylon yoke and pin arrangement, adjust the yoke on the rod until the pin can be inserted through the yoke and the cam follower. Insert the pin and hold it in place with a cotter pin. Remove the rubber band.

4. Hold the tower shaft in the full advance position and check the following 3 places:

a. The throttle advance cam at the carburetor is against its stop.

b. Tower shaft advance arm is against its stop, but disregard the feeler gauge.

c. The distributor is fully advanced against its stop, illustration.

5. Mount the engine in a large test tank or body of water and check the completed work.

■ The engine MUST be mounted in a large test tank or body of water to adjust the timing. NEVER attempt to make this adjustment with a flush attachment connected to the lower unit or in a small, confined test tank. The no-load condition on the propeller would cause the engine to RUNAWAY resulting in serious damage or destruction of the unit.

✳✳ CAUTION

Water must circulate through the lower unit to the engine any time the engine is run to prevent damage to the water pump in the lower unit. Just 5 seconds without water will damage the water pump.

1968-69 55 HP & 1970-71 60 HP MODELS

Adjust Cam Follower

◆ See Figure 155

On the starboard side of the engine, loosen the screw on the roller arm and move the roller until the mark on the throttle cam is aligned with the center of the cam follower roller *just* as the roller makes contact with the cam. Tighten the screw securely to hold the adjustment.

Fig. 153 Turn the housing until it is fully advanced and against the stop

Fig. 154 Measure the clearance between the arm and the stop

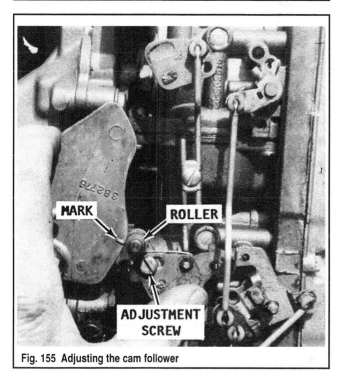

Fig. 155 Adjusting the cam follower

Adjust Carburetor & Distributor Linkage

◆ See Figures 156 and 157

1. On the starboard side of the engine, move the distributor base and throttle lever to the full throttle position against their stops. The carburetor throttle shaft MUST also be against its full throttle stop. To adjust, turn the throttle cam yoke on the throttle control rod until the conditions just described are satisfied.

2. If the throttle cam yoke has been removed from the throttle control rod, install the yoke onto the rod to a rough adjustment of 5 in. (12.7 cm) from the end of the throttle link to the face of the yoke, and then make the linkage adjustment as just described.

3. Set the timing.

Fig. 156 Adjusting the throttle cam yoke

Fig. 157 If the rod had been removed from the yoke, set the rough adjustment to 5 in.

1968 65, 85 & 100 HP MODELS

Adjust Cam Follower

◆ **See Figure 158**

On the starboard side of the engine alongside the carburetor, position the throttle cam so the scribe mark is aligned with the center of the cam follower roller shaft. To make an adjustment, loosen the adjusting screw on the throttle cam and move the throttle arm to its limit and the throttle valves are closed. Check to be sure the cam follower roller is contacting the cam. Tighten the screw to hold the adjustment.

Adjust Carburetor & Distributor Linkage

◆ **See Figure 159**

■ **The flywheel has been removed in the illustration ONLY for photographic clarity in order to see the marks.**

On the starboard side toward the rear of the engine, move the distributor to the full advance position. The mark on the distributor base plate arm should align with the mark on the distributor cap.

With the control shaft at the full throttle position and against its stop, the throttle shaft arm must also be against its stop. To adjust, turn the throttle cam yoke on the throttle control rod until the throttle shaft arm is against its stop when the control shaft is in the full throttle position.

1969 85 HP MODELS

■ **The cam follower has 2 marks, one above the other. The top mark is used for the cam-to-carburetor adjustment, and the lower mark is used for the carburetor-to-distributor adjustment.**

Adjust Cam Follower

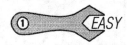

◆ **See Figure 160**

On the starboard side of the carburetor, check to be sure the throttle valves are closed. Work the adjustment screw until the center of the cam follower roller is aligned with the upper mark on the throttle cam *just* as the throttle begins to open.

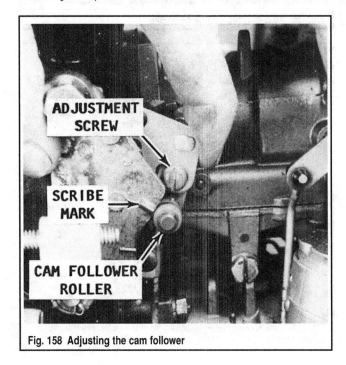

Fig. 158 Adjusting the cam follower

Adjust Carburetor & Distributor Linkage

◆ **See Figure 161**

■ **The flywheel has been removed in the illustration ONLY for photographic clarity to see the marks.**

1. On the starboard side of the engine, adjust the spark advance stop screw until the marks on the distributor base plate and the distributor cap are aligned.

2. Still on the starboard side of the engine, adjust the throttle cam yoke until the lower mark on the throttle cam is aligned with the center of the cam follower roller. Check to be sure the distributor base and the distributor cap are still aligned on their marks.

1970 85 HP & 1969-70 115 HP MODELS

Adjust Cam Follower

◆ **See Figure 160**

1. On the starboard side of the carburetor, check to be sure the throttle valves are closed.

2. Work the adjustment screw until the center of the cam follower roller is aligned with the upper mark on the throttle cam *just* as the throttle begins to open.

Adjust Carburetor And Distributor Linkage

◆ **See Figure 161**

■ **The flywheel has been removed in the illustration ONLY for photographic clarity to see the marks.**

1. On the starboard side of the engine, adjust the spark advance stop screw until the mark on the distributor base plate aligns with the mark on the distributor cap.

2. On the starboard side of the carburetor, hold the throttle lever against the stop on the crankcase and adjust the throttle cam yoke until the throttle valves are wide open.

3. Check to be sure the mark on the distributor base plate still aligns with the mark on the distributor cap.

Fig. 159 Linkage adjustment

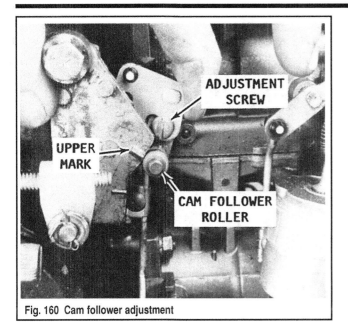

Fig. 160 Cam follower adjustment

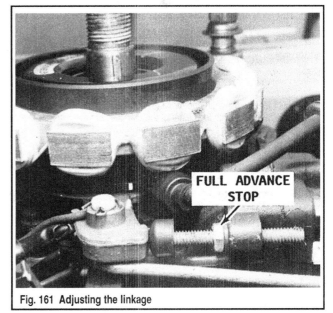

Fig. 161 Adjusting the linkage

1971-72 85,100 & 125 HP MODELS

Adjust Cam Follower

◆ See Figures 162 and 163

1. On the starboard side of the carburetor, check to be sure the throttle valves are closed. Work the adjustment screw until the center of the cam follower roller is aligned with the upper mark on the throttle cam *just* as the throttle begins to open.

✳✳ CAUTION

The engine MUST be mounted in a body of water with a test wheel to make the following adjustments. NEVER attempt to run an engine at above idle speed with a flush attachment connected or in a confined test tank. Such practice, with a no-load condition on the propeller, would cause the engine to RUNAWAY causing serious damage or destruction of the unit. The test cannot be performed with the boat moving through the water.

✳✳ CAUTION

Water must circulate through the lower unit to the engine any time the engine is run to prevent damage to the water pump and an overheating condition. Just 5 seconds without water will damage the water pump and cause the engine to overheat.

2. Connect a timing light to the No. 1 (top, starboard bank) cylinder. Start the motor and allow it to warm to operating temperature at 500 rpm.

3. Adjust the idle speed adjustment screw to 5° advance timing and then turn off the engine.

4. Adjust the throttle cam yoke until the upper embossed mark on the throttle cam is aligned with the center of the throttle cam roller.

Adjust Carburetor & Distributor Linkage

◆ See Figure 164

1. Start the engine and shift into FORWARD gear.

✳✳ CAUTION

The engine MUST be in gear during the following adjustment to prevent a RUNAWAY condition and possible destruction of the unit.

2. Advance the throttle to the wide open position. In the wide open position the upper carburetor throttle shaft roll pin should be against its stop.

3. Insert a thin piece of paper (approximately 0.003 in. {0.08mm}) between the roll pin and the stop on the intake manifold.

4. Now, adjust the full throttle advance screw on the throttle advance arm until the paper can be withdrawn with just a slight amount of drag. This procedure will prevent a strain on the throttle shaft.

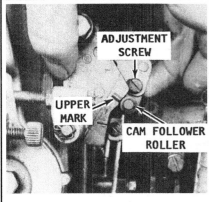

Fig. 162 Set up the cam follower...

Fig. 163 ...and then use a timing light

Fig. 164 Make sure the roll pin is against its stop

1972 65 HP MODELS

Pick-Up Timing Adjustment

◆ **See Figure 165**

1. On the starboard side of the power-head: loosen the screw on the roller arm of the throttle shaft and move the roller until the short mark on the throttle cam is aligned with the center of the cam follower roller *just* as the roller makes contact with the cam. Tighten the screw securely to hold the adjustment.

2. Mount the engine in a test tank, or move the boat into a body of water. NEVER use a flush device while making the primary pick-up adjustments. Connect a timing light to the No. 1 (top) spark plug lead.

✳✳ CAUTION

Water must circulate through the lower unit to the powerhead anytime the power-head is operating to prevent damage to the water pump in the lower unit. Just 5 seconds without water will damage the water pump impeller.

3. Start the powerhead and allow it to warm to operating temperature.

4. With the outboard in FORWARD gear and running at idle speed, aim the timing light at the timing decal on the flywheel. The primary pick-up timing is 2° ATDC to TDC.

5. If an adjustment is required, lengthen the throttle control rod to increase the timing degrees or shorten the rod to decrease the timing degrees.

Maximum Spark Advance

◆ **See Figure 166**

1. Shut down the powerhead (after Pick-Up Timing Adjustment). Manually advance the throttle lever to the WOT position. Adjust the WOT stop screw until all 3 roll pins on the carburetor shafts are in the vertical position, as indicated in the accompanying illustration.

2. Start the powerhead. Shift the unit into Forward gear, and then increase engine speed to 5,000 rpm. Aim the timing light at the timing decal. The maximum spark advance is 18-20° BTDC.

3. If an adjustment is necessary, shut down the powerhead and loosen the lock nut on the timing adjustment screw located port side under the flywheel just aft of the cranking motor. Rotate the screw one turn CLOCKWISE to retard the timing 1° and COUNTERCLOCKWISE one turn to advance the timing 1°.

Idle Speed Adjustment

◆ **See Figure 167**

1. With the powerhead running (after Maximum Spark Advance), adjust the idle speed screw to obtain the correct idle speed listed in the Tune-Up Specifications chart for the unit being serviced. Rotating the screw COUNTERCLOCKWISE will decrease rpm and rotating the screw will increase rpm.

Timing

TIMING POINTER ADJUSTMENT - 1971-72 85, 100 & 125 HP MODELS

◆ **See Figures 168 and 169**

■ **A special tool (OMC #384887, or a dial indicator) is required to accurately make this adjustment.**

1. Remove all spark plugs from the block. Install the special tool or the dial indicator into the No.1 cylinder.

Fig. 165 Pick-up timing adjustment

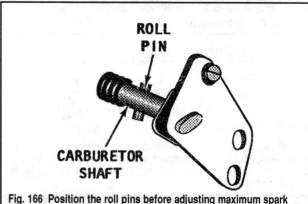

Fig. 166 Position the roll pins before adjusting maximum spark advance

Fig. 167 Idle speed adjustment

2. Rotate the flywheel CLOCKWISE until the piston crown makes contact with the dial indicator or the special tool at 120° TDC (top dead center). Lock the tool or the dial indicator and scribe a mark on the flywheel in line with the pointer.

3. *Slowly* rotate the flywheel COUNTERCLOCKWISE (under normal conditions we say NEVER to rotate the flywheel counterclockwise but here it is permissible), until the piston again makes contact with the dial indicator or the special tool.

4. Scribe another mark on the flywheel in line with the pointer. Measure the distance between the 2 marks just scribed on the flywheel. The midway point between the 2 marks is top dead center. Loosen the screw and adjust the pointer to align with the midway point between the 2 marks.

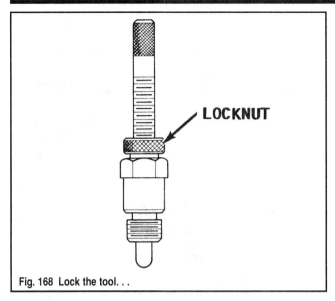

Fig. 168 Lock the tool. . .

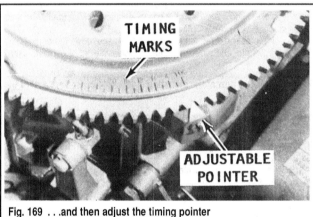

Fig. 169 . . .and then adjust the timing pointer

TIMING - 1968-69 55 HP & 1970-71 60 HP MODELS

◆ See Figures 170 and 171

The timing should not require adjustment unless the spark advance screw has been moved, or the amplifier has been changed and an updated one installed.

The timing should be checked any time the ignition system is serviced.

1. Note the square timing mark on the flywheel ring guard. Observe the safety bar installed directly above the carburetors (the only purpose of the bar is to prevent a person's hand from becoming caught in the flywheel while the engine is operating).

2. Look behind the safety bar and observe the adjustment screw with a locknut and a rubber cap. The timing adjustment is made by loosening the nut slightly, and then threading the screw inward or outward.

■ On some engines, the timing mark is an embossed mark on the flywheel and the degree marks are embossed on the safety guard just above the carburetor. On other engines, the opposite is true. The timing mark is embossed on the safety guard and the degree marks are on the flywheel. *However*, some engines may also have a triangular mark embossed on the flywheel. When timing the engines covered in this section, DISREGARD the triangular mark.

3. Connect a timing light to the No. 1 (top) cylinder. Start the engine.

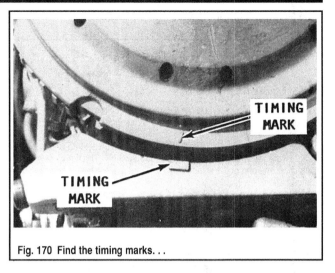

Fig. 170 Find the timing marks. . .

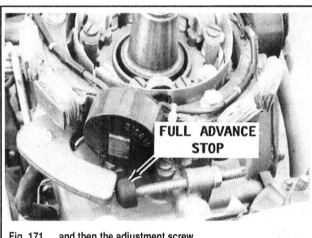

Fig. 171 . .and then the adjustment screw

■ The engine MUST be mounted in a body of water with a test wheel to make the following adjustment. NEVER attempt to run an engine at above idle speed with a flush attachment connected or in a confined test tank. Such practice, with a no-load condition on the propeller, would cause the engine to RUNAWAY causing serious damage or destruction of the unit. The test cannot be performed with the boat moving through the water.

■ Water must circulate through the lower unit to the engine any time the engine is run to prevent damage to the water pump and an overheating condition. Just 5 seconds without water will damage the water pump and cause the engine to overheat.

4. Allow the engine to warm to operating temperature at 500 rpm. Now increase engine speed to full throttle, approximately 4500 rpm. The degree timing mark on the flywheel should align with the mark on the flywheel guard. If the marks do not align, shut down the engine and move the adjusting screw inward or outward just a wee bit.

■ The reason for shutting the engine down is SAFETY. The adjusting screw is so close to the flywheel, the adjustment should NOT be attempted while the engine is running. Keep ten fingers and both hands, they'll come in handy the rest of your life.

5. Start the engine and repeat the check until the marks do align when the engine is operating at 4500 rpm. Once the marks align at 4500 rpm, tighten the locknut to hold the adjustment and replace the rubber cap.

TIMING - 1968 65 HP, 1968-70 85 HP, 1968-69 100 HP &
1969-70 115 HP MODELS

◆ See Figures 172, 173 and 174

The timing should not require adjustment unless the spark advance screw has been moved, or the amplifier has been changed and an updated one installed.

The timing should be checked any time the ignition system is serviced.

1. Note the timing degree marks on the flywheel. The timing pointer is located at one end of the lifting bracket attached to the power head opposite the carburetor.

2. Look for the adjustment screw attached to the power head underneath the flywheel and just behind the throttle advance arm on the starboard side of the engine. Except on 1968 65, 85 and 100 hp models, where the adjustment screw on these engines is located on the port side and

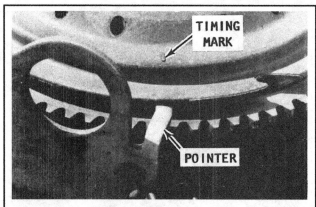

Fig. 172 Find the timing marks. . .

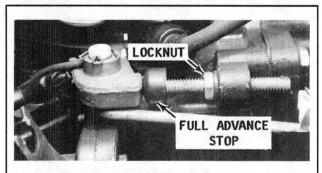

Fig. 173 . .and then the adjustment screw

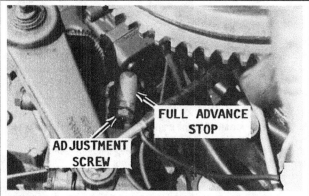

Fig. 174 The adjustment screw is in a different location on a few 1968 models

is attached to the powerhead. The screw is mounted in such a manner that it acts as a stop for a flange attached to the distributor advance arm as this arm rotates on the end of the tower shaft.

3. Connect a timing light to the No. 1 (top, starboard bank) cylinder. Start the engine.

✳✳ CAUTION

The engine MUST be mounted in a body of water with a test wheel to make the following adjustment. NEVER attempt to run an engine at above idle speed with a flush attachment connected or in a confined test tank. Such practice, with a no-load condition on the propeller, would cause the engine to RUNAWAY causing serious damage or destruction of the unit. The test cannot be performed with the boat moving through the water.

✳✳ CAUTION

Water must circulate through the lower unit to the engine any time the engine is run to prevent damage to the water pump and an overheating condition. Just 5 seconds without water will damage the water pump and cause the engine to overheat.

4. Allow the engine to warm to operating temperature at 500 rpm. Now increase engine speed to full throttle, approximately 4500 rpm. The pointer on the bracket should align with the proper degree mark on the flywheel as detailed in the Tune-Up Specifications chart.

5. If the proper degree mark does not align; shut down the engine, loosen the locknut, and move the adjusting screw inward or outward just a wee bit.

■ **One complete turn of the adjusting screw will advance or retard the timing one full degree.**

✳✳ CAUTION

The reason for shutting the engine down is one of SAFETY. The adjusting screw is very close to the flywheel. The adjustment can be made with a screwdriver while the engine is running, but we *strongly* recommend the engine be shut down. The few extra minutes involved will help you keep ten fingers and both hands - they'll come in handy the rest of your life.

6. Start the engine and repeat the check until the pointer and the proper degree mark do align when the engine is operating at 4500 rpm. Once the marks align at 4500 rpm, tighten the locknut to hold the adjustment.

TIMING - CDI/FLYWHEEL W/SENSOR (1967 100 HP)

Idle Timing Check

◆ See Figure 175

1. Connect one lead of a timing light to the No. 1 (top, starboard bank) cylinder. Connect the other lead to a good ground on the engine.

2. Start and operate the engine at 500 rpm.

■ **In the illustration, the distributor cap and belt have been removed ONLY for photographic clarity to show the timing marks.**

✳✳ CAUTION

Water must circulate through the lower unit to the engine any time the engine is run to prevent damage to the water pump and an overheating, condition. Just 5 seconds without water will damage the water pump and cause the engine to overheat.

3. At 500 rpm, the mark on the distributor pulley *must* align with the mark on the distributor housing. If the marks do not align, the keyway in the flywheel, the Woodruff key and rotor wheel may be damaged; or the sensor may not be adjusted properly; or the sensor may not be located in the correct pin. Actually, there is no adjustment. The timing marks must align at 500 rpm. If they do not, the defective part or condition must be isolated and corrected.

Fig. 175 Find the timing marks

Full-Advance Timing Check

◆ See Figure 176   OEM ② MODERATE

✳✳ CAUTION

The engine MUST be mounted in a body of water with a test wheel to make the following full-advance timing check. NEVER attempt to run an engine at above idle speed with a flush attachment connected or in a confined test tank. Such practice, with a no-load condition on the propeller, would cause the engine to RUNAWAY causing serious damage or destruction of the unit. The test cannot be performed with the boat moving through the water.

　　1. With the timing light connected to the No. 1 cylinder, observe the timing mark on top of the flywheel and the square mark on the aft starboard side of the flywheel guard.
　　2. Start the engine and allow it to warm to operating temperature at 500 rpm. Now increase engine speed to 4500 rpm and observe the timing marks, If the marks do not align, loosen the nut on the starboard side of the

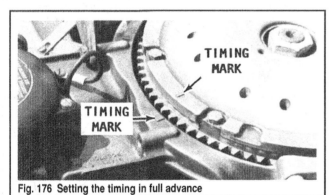

Fig. 176 Setting the timing in full advance

Fig. 177 Adjusting the idle speed

distributor housing. Adjust the screw inward or outward until the timing marks are aligned.

TIMING - 1971-72 125 HP MODELS

✳✳ CAUTION

The engine MUST be mounted in a body of water with a test wheel to make the following timing check. NEVER attempt to run an engine at above idle speed with a flush attachment connected or in a confined test tank. Such practice, with a no-load condition on the propeller, would cause the engine to RUNAWAY causing serious damage or destruction of the unit. The test cannot be performed with the boat moving through the water.

✳✳ CAUTION

Water must circulate through the lower unit to the engine any time the engine is run to prevent damage to the water pump and an overheating condition. Just 5 seconds without water will damage the water pump and cause the engine to overheat.

Idle Adjustment

◆ See Figure 177   OEM ② MODERATE

　　First, check the Specifications to determine the proper number of degrees of timing advance.
　　1. Connect one lead of a timing light to the No.1 (top, starboard bank) cylinder. Connect the other lead to a good ground on the engine.
　　2. Start and operate the engine at 500 rpm until it reaches normal operating temperature. The flywheel ring gear has degree marks clearly visible. Observe the pointer located on the forward side of the engine.
　　3. Start and operate the engine at 500 rpm until it reaches normal operating temperature. With the engine operating at 500 rpm, the pointer should be aligned with the 5° mark on the flywheel.
　　4. If the pointer is not aligned, adjust the top screw on the throttle advance arm to bring the pointer into alignment with the 5° mark.

High Speed Adjustment

◆ See Figure 178   OEM ② MODERATE

　　With the engine still running from the last adjustment, increase engine speed to 4500 rpm. The pointer should be directed to the proper degree mark (as determined from the Specifications).
　　If the proper degree mark is not indicated at 4500 rpm, loosen the nut on the throttle stop on the starboard side of the engine. Move the screw inward or outward until the proper degree mark is indicated at 4500 rpm, then tighten the nut securely to hold the adjustment.

■ The timing pointer is secured to the bracket with attaching screws. Now, it is possible that the pointer could have been changed intentionally or accidentally. Therefore, a Timing Pointer Adjustment may be necessary as outlined in the following procedure.

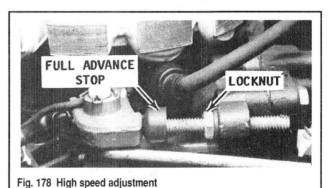

Fig. 178 High speed adjustment

STORAGE (WHAT TO DO BEFORE AND AFTER)

Winterization

◆ See Figure 179

Taking extra time to store the boat and motor properly at the end of each season or before any extended period of storage will greatly increase the chances of satisfactory service at the next season. Remember, that next to hard use on the water, the time spent in storage can be the greatest enemy of an outboard motor. Ideally, outboards should be used regularly. If weather in your area allows it, don't store the motor, enjoy it. Use it, at least on a monthly basis. It's best to enjoy and service the boat's steering and shifting mechanism several times each month. If a small amount of time is spent in such maintenance, the reward will be satisfactory performance, increased longevity and greatly reduced maintenance expenses.

But, in many cases, weather or other factors will interfere with time for enjoying a boat and motor. If you must place them in storage, take time to properly winterize the boat and outboard. This will be your best shot at making time stand still for them.

For many years there was a widespread belief that simply shutting off the fuel at the tank and then running the powerhead until it stops constituted prepping the motor for storage. Right? Well, WRONG!

First, it is not possible to remove all fuel in the carburetor by operating the powerhead until it stops. Considerable fuel will remain trapped in the float chamber and other passages, especially in the lines leading to carburetors. The only guaranteed method of removing all fuel is to take the physically drain the carburetors from the float bowls.

■ On VRO equipped motors, disconnecting the fuel line to run the engine out of fuel may cause the consumption of excessive amounts of oil. This can lead to hard starting problems later, from deposits formed in the combustion chamber.

Proper storage involves adequate protection of the unit from physical damage, rust, corrosion and dirt. The following steps provide an adequate maintenance program for storing the unit at the end of a season.

PREPPING FOR STORAGE

◆ See Figure 180

Where to Store Your Boat and Motor

Ok, a well lit, locked, heated garage and work area is the best place to store you precious boat and motor, right? Well, we're probably not the only

ones who wish we had access to a place like that, but if you're like most of us, we place our boat and motor wherever we can.

Of course, no matter what storage limitations are placed by where you live or how much space you have available, there are ways to maximize the storage site.

If possible, select an area that is dry. Covered is great, even if it is under a carport or sturdy portable structure designed for off-season storage. Many people utilize canvas and metal frame structures for such purposes. If you've got room in a garage or shed, that's even better. If you've got a heated garage, God bless you, when can we come over? If you do have a garage or shed that's not heated, an insulated area will help minimize the more extreme temperature variations and an attached garage is usually better than a detached for this reason. Just take extra care to make sure you've properly inspected the fuel system before leaving your boat in an attached garage for any amount of time.

If a storage area contains large windows, mask them to keep sunlight off the boat and motor otherwise, use a high-quality, canvas cover over the boat, motor and if possible, the trailer too. A breathable cover is best to avoid the possible build-up of mold or mildew, but a heavy duty, non-breathable cover will work too. If using a non-breathable cover, place wooden blocks or length's of 2 x 4 under various reinforced spots in the cover to hold it up off the boat's surface. This should provide enough room for air to circulate under the cover, allowing for moisture to evaporate and escape.

■ Marine supply stores normally sell various types of desiccant (water absorbent) products. These mesh bags or small plastic pails are filled with a material that tends to draw moisture from the air and hold it in suspension. They can be very helpful in the prevention of mildew when insufficient airflow is present to naturally remove moisture from underneath a cover (or shrink wrapping). Follow the product instructions closely when using such products (and keep them away from small children).

Whenever possible, avoid storing your boat in industrial buildings or parks areas where corrosive emissions may be present. The same goes for storing your boat too close to large bodies of saltwater. Hey, on the other hand, if you live in the Florida Keys, we're jealous again, just enjoy it and service the boat often to prevent corrosion from causing damage.

Finally, when picking a place to store your motor, consider the risk or damage from fire, vandalism or even theft. Check with your insurance agent regarding coverage while the boat and motor is stored.

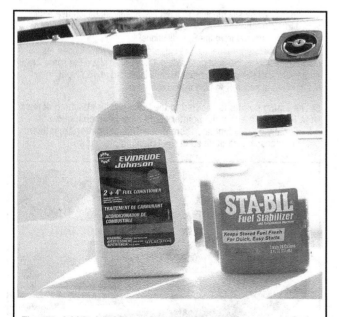

Fig. 179 Add fuel stabilizer to the system anytime it will be stored without being completely drained

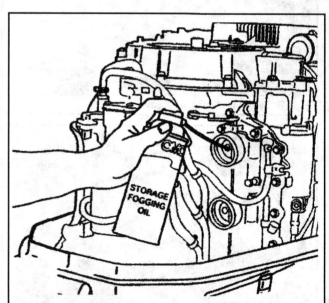

Fig. 180 Fogging oil can be sprayed down the throats of the carburetors, but it should still be added directly to the cylinders through the spark plug ports

Storage Checklist (Preparing the Boat and Motor)

◆ See Figures 181 thru 184

■ Due to availability, we have used a newer model engine in the pictures shown.

The amount of time spent and number of steps followed in the storage procedure will vary with factors such as the length of planed storage time, the conditions under which boat and motor are to be stored and your personal decisions regarding storage.

But, even considering the variables, plans can change, so be careful if you decide to perform only the minimal amount of preparation. A boat and motor that has been thoroughly prepared for storage can remain so with minimum adverse affects for as short or long a time as is reasonably necessary. The same cannot be said for a boat or motor on which important winterization steps were skipped.

■ Always store your Johnson/Evinrude motor vertically on the boat or on a suitable engine stand.

1. Thoroughly wash the boat motor and hull. Be sure to remove all traces of dirt, debris or marine life. Check the water stream fitting (if equipped) and water inlet(s) for debris. If equipped, inspect the speedometer opening at the leading edge of the lower unit or any other lower unit drains for debris (clean debris with a compressed air or a piece of thin wire).

■ The manufacturer recommends the use of Johnson/Evinrude 2+4 Fuel Conditioner when treating the fuel systems on Johnson/Evinrude motors. In the past, Johnson/Evinrude recommended using 2+4 in a ratio of 1.0 oz. (30 ml) for every gallon (3.8 L), but products may change, so be sure to follow the directions on the bottle if they differ. On these motors, you always have the work intensive option of draining the fuel system instead, either using the float bowl drains on the carburetors or by removing the carburetors completely from the motor. For more details on carburetor service, please refer to the Fuel System section, but keep in mind that the manufacturer currently recommends using the fuel conditioner method. Of course, if the engine is to be stored for an undetermined amount of time (more than 1 or 2 seasons), removing and completely draining the carburetors is probably the best option.

2. Stabilize the engine's fuel supply and fog the motor using a high quality fuel stabilizer and storage fogging oil. At the same time, take this opportunity to thoroughly flush the engine cooling system as well:

a. Add an appropriate amount of fuel stabilizer to the fuel tank and top off to minimize the formation of moisture through condensation in the fuel tank.

■ Now back in the day when these motors were produced Johnson/Evinrude normally recommended fogging the motor manually and either draining the fuel system completely or treating it with fuel stabilizer. These days they recommend using a fogging and stabilizing brew which is introduced to the fuel system and motor by running it off a special portable tank. We mention it here (even though most people don't bother with it even on modern outboards) because the company obviously has a reason for suggesting this is the best way. Perhaps it is overkill. We've done it for some winters on our own motors, but have used the straight fuel stabilizer and manual fogging method for other winters. We can't tell any difference, but it could show up in long term reliability tests. Lastly, the Johnson/Evinrude recommendations usually leave you with about 4 or more gallons of this mix and nothing to do with it (unless you're going to winterize 4-5 more boats), so you might want to reduce the amounts of the fluids proportionately.

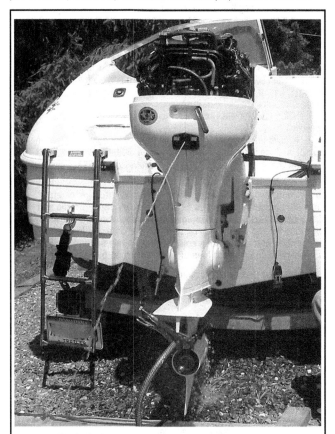

Fig. 181 Flush the motor. . .

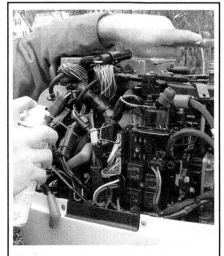

Fig. 182 . . .and on most models it's a good idea to manually fog it through the spark plug holes

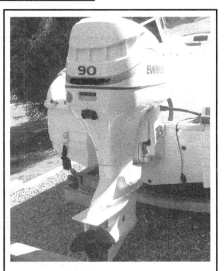

Fig. 183 Always store with the engine vertical so the cooling system drains

Fig. 184 If possible, block up the trailer during storage

b. Next, prepare a fuel storage mixture as directed. Use a portable 6.0 gal. (23L) gas tank to mix:

- 5.0 gal. (19L) of gas
- 2.0 qt. (1.9L) of Johnson/Evinrude Storage Fogging Oil
- For carbureted motors also add 1 pt. (0.47L) of Johnson/Evinrude 2-stroke engine oil.
- Add 2.5 oz. (74ml) of Johnson/Evinrude 2+4 Fuel Conditioner or equivalent storage fluids.

c. Connect this tank to the engine in order to provide a treated fuel mixture to the engine for storage.

d. Attach a flushing attachment as a cooling water/flushing source. For details, please refer to the information on Flushing the Cooling System.

e. Start and run the engine at about 1500 rpm for approximately 5 minutes. This will ensure the entire fuel supply system contains the appropriate storage mixtures.

f. If you wish to spray fogging oil down the throats of the carburetors while the motor is running (and we recommend it), follow the instructions provided on the can of fogging oil and spray the oil into the mouth of each carburetor.

■ **If the motor is equipped with a flushing plug, remove the plug to help ensure all water can drain. Reinstall the plug after the motor has fully drained.**

g. Stop the engine and remove the flushing source, keeping the outboard perfectly vertical. Allow the cooling system to drain completely, especially if the outboard might be exposed to freezing temperatures during storage.

✳✳ WARNING

NEVER keep the outboard tilted when storing in below-freezing temperatures as water could remain trapped in the cooling system. Any water left in cooling passages (or even captured in the prop housing through the exhaust due to rain) might freeze and could cause severe engine damage by cracking the powerhead or lower unit.

h. Finish the fogging procedure by removing all of the spark plugs, and placing the engine is the fully-tilted position then spraying a generous amount of fogging oil directly into each of the cylinders.

■ **On models equipped with power steering, allow the engine to sit in the fully-tilted position for at LEAST 5 minutes in order to completely drain the oil cooler. But, the engine must be returned to a fully-vertical position for long term storage.**

i. Turn the crankshaft slowly (by hand) in a **clockwise** direction a few complete turns to evenly distribute the fogging oil throughout the cylinders.

j. Spray a small amount of additional fogging lubricant into each of the cylinders, then reinstall and tighten the Spark Plugs (as detailed in this section).

k. To prevent accidental starting, leave the spark plug wires tagged and disconnected. To prevent potential damage to the ignition system, make sure the motor is not cranked with the wires disconnected. The best thing to do is to ground the spark plug wires to the powerhead. But, alternately, you could secure a reminder note to the ignition switch (a wire tie and a note in a plastic bag has infinitely better chance of lasting than a piece of paper and some tape. May we suggest the following text for the note "LISTEN DUMMY, DON'T CRANK THE MOTOR, THE SPARK PLUG WIRES ARE DISCONNECTED."

3. Drain and refill the engine gearcase while the oil is still warm (for details, refer to the Lower Unit Oil procedures in this section). Take the opportunity to inspect for problems now, as storage time should allow you the opportunity to replace damaged or defective seals. More importantly, remove the old, contaminated gear oil now and place the motor into storage with fresh oil to help prevent internal corrosion.

4. On models equipped with portable fuel tanks, disconnect and relocate them to a safe, well-ventilated, storage area, away from the motor. Drain any fuel lines that remain attached to the tank.

■ **On VRO motors (and there shouldn't be any here), DO NOT disconnect the oil tank lines. Top off the tanks and leave the lines connected to help protect the system from moisture.**

5. Remove the battery or batteries from the boat and store in a cool dry place. If possible, place the battery on a smart charger or Battery Tender®, otherwise, trickle charge the battery once a month to maintain proper charge.

✳✳ WARNING

Remember that the electrolyte in a discharged battery has a much lower freezing point and is more likely to freeze (cracking/destroying the battery case) when stored for long periods in areas exposed to freezing temperatures. Although keeping the battery charged offers one level or protection against freezing; the other is to store the battery in a heated or protected storage area.

6. On models equipped with a boat-mounted fuel filter or filter/water canister, clean or replace the boat mounted fuel filter at this time. The engine mounted fuel filters should be left intact, so the sealed system remains filled with treated fuel during the storage period.

7. On motors with external oil tanks, if possible, leave the oil supply line connected to the motor. This is the best way to seal moisture out of the system. If the line must be disconnected for any reason (such as to remove the motor or oil tank from the boat), seal the line by sliding a snug fitting cap over the end. Most motors equipped with remote oil tanks are equipped with a cap, mounted somewhere on the engine, such as on the fuel line, near the fuel pump. Top off the oil tank to displace moisture-laden air and help prevent contamination of the oil in storage.

8. Perform a complete lubrication service following the procedures in this section.

9. Remove the propeller and check thoroughly for damage. Clean the propeller shaft and apply a protective coating of grease.

10. Check the motor for loose, broken or missing fasteners. Tighten fasteners and, again, use the storage time to make any necessary repairs.

■ **If the motor is to be removed from the boat for storage, carefully examine all mounting fasteners as well as the steering, throttle and shift systems. Replace any damaged or missing components. Also, keep close track of the fasteners, during installation NEVER substitute the mounting hardware from a smaller motor since a mounting failure during service could cause loss of control (or loss of the motor).**

11. Inspect and repair all electrical wiring and connections at this time. Make sure nothing was damaged during the season's use. Repair any loose connectors or any wires with broken, cracked or otherwise damaged insulation.

12. Clean all components under the engine cover and apply a corrosion preventative spray.

13. Too many people forget the boat and trailer, don't be one of them.

a. Take the opportunity to touch-up any damaged paint on the motor cases or trailer (if you're using a painted trailer).

b. Coat the boat and all outside painted surfaces of the motor with a fresh coating of wax then cover it with a breathable cover

c. If possible place the trailer on stands or blocks so the wheels are supported off the ground.

d. Check the air pressure in the trailer tires. If it hasn't been done in a while, remove the wheels to clean and repack the wheel bearings.

14. Sleep well, since you know that your baby will be ready for you come next season.

Re-Commissioning

REMOVAL FROM STORAGE

◆ **See Figures 185 and 186**

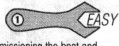

The amount of service required when re-commissioning the boat and motor after storage depends on the length of non-use, the thoroughness of the storage procedures and the storage conditions.

At minimum, a thorough spring or pre-season tune-up and a full lubrication service is essential to getting the most out of your engine. If the engine has been properly winterized, it is usually no problem to get it in top running condition again in the springtime. If the engine has just been put in the garage and forgotten for the winter, then it is doubly important to perform a complete tune-up before putting the engine back into service. If you have ever been stranded on the water because your engine has died and you had to suffer the embarrassment of having to be towed back to the marina you know how it can be a miserable experience. Now is the time to prevent that from occurring.

Take the opportunity to perform any annual maintenance procedures that were not conducted immediately prior to placing the motor into storage. If the motor was stored for more than one off-season, pay special attention to inspection procedures, especially those regarding hoses and fittings. Check the engine gear oil for excessive moisture contamination. The same goes for oil tanks on oil injected motors. If necessary, change the lower unit or drain and refill the injection tank oil to be certain no bad or contaminated fluids are used.

■ Although not absolutely necessary, it is a good idea to ensure optimum cooling system operation by replacing the water pump impeller at this time. In the old days, seasonal replacement was a regular thing. To be honest, the impellers and pumps are usually made of better materials now and can easily last a couple of seasons, but it is cheap insurance.

Other items that require attention include:
Install the battery (or batteries) if so equipped.
Inspect all wiring and electrical connections. Rodents have a knack for feasting on wiring harness insulation over the winter. If any signs of rodent life are found, check the wiring carefully for damage, do not start the motor until damaged wiring has been fixed or replaced.
On models with a remote oil tank, if the line was disconnected, remove the cover and reconnect the line, then prime the system to ensure proper operation once the motor is started.
If not done when placing the motor into storage clean and/or replace the boat fuel filters at this time. Also, clean or replace the engine mounted filters (which should have been neglected during winterization so the fuel system would remain sealed with treated fuel.
If the fuel tank was emptied, or if it must be emptied because the fuel is stale fill the tank with fresh fuel. Keep in mind that even fuel that was treated with stabilizer will eventually become stale, especially if the tank is stored for more than one off-season. Pump the primer bulb and check for fuel leakage or flooding at the carburetor or vapor separator tank.
Attach a flush device or place the outboard in a test tank and start the engine. Run the engine at idle speed and warm it to normal operating temperature. Check for proper operation of the cooling, electrical and warning systems.

✳✳ CAUTION

Before putting the boat in the water, take time to verify the drain plug is installed. Countless number of spring boating excursions have had a very sad beginning because the boat was eased into the water only to have the boat begin to fill with it.

CLEARING A SUBMERGED MOTOR

The good news is that motors of this size are rarely lost overboard (unless there is serious neglect of the mounting fasteners or unless there is a catastrophic failure to the boat, transom or motor). It is rare enough that Johnson/Evinrude does not even discuss the possibility in literature for many of the motors covered in this guide. On the other hand, accidents do occur, and if you're reading this for some other reason than morbid curiosity then you've obviously got a situation with which you must deal. Should a large outboard become submerged, it is usually possible to salvage, service and enjoy the motor again.

In order to prevent severe damage, be sure to recover an engine that is dropped overboard or otherwise completely submerged as soon as possible. It is really best to recover it immediately. But, keep in mind that once a submerged motor is recovered exposure to the atmosphere will allow corrosion to begin etching highly polished bearing surfaces of the crankshaft, connecting rods and bearings. For this reason, not only do you have to recover it right away, but you should service it right away too. Make sure the motor is serviced within 3 hours of initial submersion.

OK, maybe now you're saying "3 hours, it will take me that long to get it to a shop or to my own garage." Well, if the engine cannot be serviced immediately (or sufficiently serviced so it can be started), re-submerge it in a tank of fresh water to minimize exposure to the atmosphere and slow the corrosion process. Even if you do this, do not delay any more than absolutely necessary, service the engine as soon as possible. This is especially important if the engine was submerged in salt, brackish or polluted water as even submersion in fresh water will not preserve the engine indefinitely. Service the engine, at the **MOST** within a few days of protective submersion.

Fig. 185 Start and test run the motor

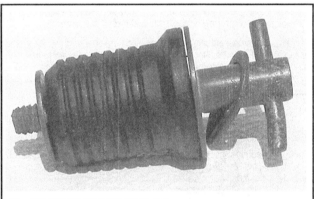
Fig. 186 DON'T FORGET THE PLUG

✳✳ WARNING

Keep in mind that even fresh water will cause etching on the highly polished bearing surfaces of the crankshaft, connecting rods and bearings. We simply cannot over-emphasize the need to purge the motor of moisture once submersion has occurred.

After the engine is recovered, vigorously wash all debris from the engine using pressurized freshwater.

■ If the engine was submerged while still running, there is a good chance of internal damage (such as a bent connecting rod). Under these circumstances, don't start the motor, follow the beginning of this procedure to try turning it over slowly by hand, feeling for mechanical problems. If necessary, refer to Powerhead for complete disassembly and repair instructions.

✳✳ WARNING

NEVER try to start a recovered motor until at least the first few steps (the ones dealing with draining the motor and checking to see it if is hydro-locked or damaged) are performed. Keep in mind that attempting to start a hydro-locked motor could cause major damage to the powerhead, including bending or breaking a connecting rod.

If the motor was submerged for any length of time it should be thoroughly disassembled and cleaned. Of course, this depends on whether or not water intruded into the motor itself. To help determine this check the lower unit oil for signs of contamination. Also, be sure to remove the spark plugs and visually check for signs of moisture.

The extent of cleaning and disassembly that must take place depends also on the type of water in which the engine was submerged. Engines totally submerged, for even a short length of time, in salt, brackish or polluted water will require more thorough servicing than ones submerged in fresh water for the same length of time. But, as the total length of submerged time or time before service increases, even engines submerged in fresh water will require more attention. Complete powerhead disassembly and inspection is required when sand, silt or other gritty material is found inside the engine cover.

Many engine components suffer the corrosive effects of submersion in salt, brackish or polluted water. The symptoms may not occur for some time after the event. Salt crystals will form in areas of the engine and promote significant corrosion.

Electrical components should be dried and cleaned or replaced, as necessary. If the motor was submerged in salt water, the wire harness and connections are usually affected in a shorter amount of time. Since it is difficult (or nearly impossible) to remove the salt crystals from the wiring connectors, it is best to replace the wire harness and clean all electrical component connections. The starter motor, relays and switches on the engine usually fail if not thoroughly cleaned or replaced.

To ensure a thorough cleaning and inspection:

1. Remove the engine cover and wash all material from the engine using pressurized freshwater. If sand, silt or gritty material is present inside the engine cover, completely disassemble and inspect the powerhead.

2. Tag and disconnect the spark plugs leads. Be sure to grasp the spark plug cap and not the wire, then twist the cap while pulling upward to free it from the plug. Remove the spark plugs. For more details, refer to the Spark Plug procedure in this section.

3. Disconnect the fuel supply line from the engine, then drain and clean all fuel lines. Depending on the circumstances surrounding the submersion, inspect the fuel tank for contamination and drain, if necessary.

4. On oil injected models, drain and clean the oil supply system. Be sure to drain and clean the VRO oil reservoir. Purge any potentially contaminated oil from the supply lines. Properly prime the oil system before attempting to start and run the motor.

✳✳ WARNING

When attempting to turn the flywheel for the first time after the submersion, be sure to turn it SLOWLY, feeling for sticking or binding that could indicate internal damage from hydro-lock. This is a concern, especially if the engine was cranked before the spark plug(s) were removed to drain water or if the engine was submerged while still running.

5. Support the engine horizontally with the spark plug port(s) facing downward, allowing water, if present, to drain. Force any remaining the water out by slowly rotating the flywheel by hand about 20 times or until there are no signs of water. If there signs of water are present, spray some fogging oil into the spark plug ports before turning the flywheel. This will help dislodge moisture and lubricate the cylinder walls.

6. Drain the carburetors. The best method to thoroughly drain/clean the carburetor is to remove and disassemble it. For details refer to the Carburetor procedures under Fuel System.

7. Support the engine in the normal upright position. Check the engine lower unit oil for contamination. Refer to the procedures for Lower Unit Oil in this section. The lower unit is sealed and, if the seals are in good condition, should have survived the submersion without contamination. But, if contamination is found, look for possible leaks in the seals, then drain the lower unit and make the necessary repairs before refilling it. For more details, refer to the section on Lower Units.

8. Remove all external electrical components for disassembly and cleaning. Spray all connectors with electrical contact cleaner and then apply a small amount of dielectric grease prior to reconnection to help prevent corrosion. On electric start models, remove, disassemble and clean the starter components. For details on the electrical system components, refer to the Ignition & Electrical section.

9. Reassemble the motor and mount the engine or place it in a test tank. Start and run the engine for 1/2 hour using a break-in fuel/oil mixture. If the engine won't start, remove the spark plugs again and check for signs of moisture on the tips. If necessary, use compressed air to clean moisture from the electrodes or replace the plugs.

10. Stop the engine and recheck the lower unit oil.

11. Perform all other lubrication services.

12. Try not to let it get away from you (or anyone else) again!

SPECIFICATIONS

TWO-STROKE MOTOR FUEL:OIL RATIO CHART

Desired Fuel:Oil Ratio	Amount of oil needed when mixed with:				
	3 G (11.4 L) of Gas	6 G (22.7 L) of Gas	18 G (68.1 L) of Gas	30 G (114 L) of Gas	45 G (171 L) of Gas
100:1 (1% oil)	4 fl. oz. (118 mL)	8 fl. oz. (236 mL)	24 fl. oz. (708 mL)	40 fl. oz. (1180 mL)	60 fl. oz. (1770 mL)
50:1 (2% oil)	8 fl. oz. (236 mL)	16 fl. oz. (473 mL)	48 fl. oz. (1419 mL)	80 fl. oz. (2360 mL)	120 fl. oz. (3.54 L)
25:1 (4% oil)	16 fl. oz. (473 mL)	32 fl. oz. (946 mL)	96 fl. oz. (2838 mL)	160 fl. oz. (4.73 L)	240 fl. oz. (7.1 L)

NOTE: Fuel:Oil ratios listed here are for calcuation purposes. Refer to the fuel:oil recommendations for your engine before mixing. Remember that a pre-mix system designed to produce a 50:1 ratio will produce a 25:1 ratio if a 50:1 ratio is already in the fuel tank feeding the motor.

Tune-Up Specifications
1.5-50 Hp, 1-2 Cylinder

Year	Model (Hp)	No. of Cyl	Model Johnson	Model Evinrude	Displace cu. in. (cc)	Spark Plug Make	Spark Plug Type	Spark Plug Gap In. (mm)	WOT MAX (rpm)	Shift Removal Note	Primary Pick-Up Location ①	Primary Pick-Up Adjustment ①
1958	3	2	JW-14	3026-3028	5.28 (85.5)	Champion	J6J	0.030 (0.8)	4000	No Rod	C	1
	5.5	2	CD-15	5516, 5517	8.84 (145)	Champion	J6J	0.030 (0.8)	4000	②③	C	1
	7.5	2	AD-12	7524, 7524	12.4 (203)	Champion	J6J	0.030 (0.8)	4000	②③	C	1
	10	2	QD-19	10016, 10017	16.6 (281)	Champion	J6J	0.030 (0.8)	4000	④	C	1
	18	2	FD, FDE-12	15024, 15025	22 (361)	Champion	J4J	0.030 (0.8)	4500	④	C	1
	35	2	RD, RDE, RDS-19C	25034, 25936, 35514	40.5 (664)	Champion	J4J	0.030 (0.8)	4500	④	E	5
1959	3	2	JW-15	3030-3032	5.28 (85.5)	Champion	J6J	0.030 (0.8)	4000	No Rod	C	1
	5.5	2	CD-16	5518, 5519	8.84 (145)	Champion	J6J	0.030 (0.8)	4000	②③	C	1
	10	2	QD-20	10018, 10019	16.6 (281)	Champion	J4J	0.030 (0.8)	4000	④	C	1
	18	2	FD-13	15028, 15029	22 (361)	Champion	J4J	0.030 (0.8)	4500	④	C	1
	35	2	RD, RDS-20	35012, 35516	40.5 (664)	Champion	J4J	0.030 (0.8)	4500	④	E	5
1960	3	2	JW-16	3034-3036	5.28 (85.5)	Champion	J6J	0.030 (0.8)	4000	No Rod	C	1
	5.5	2	CD-17	5520, 5521	8.84 (145)	Champion	J6J	0.030 (0.8)	4000	②③	C	1
	10	2	QD-21	10020, 10021	16.6 (281)	Champion	J4J	0.030 (0.8)	4500	④	C	1
	18	2	FD-14	15032, 15033	22 (361)	Champion	J4J	0.030 (0.8)	4500	④	C	1
	40	2	RD, RDS-22	35018-35520	43.9 (719)	Champion	J4J	0.030 (0.8)	4500	④	E	5
1961	3	2	JW-17	3038-3040	5.28 (85.5)	Champion	J6J	0.030 (0.8)	4000	No Rod	C	1
	5.5	2	CD-18	5522, 5523	8.84 (145)	Champion	J6J	0.030 (0.8)	4000	②③	C	1
	10	2	QD-22	10022, 10023	16.6 (281)	Champion	J4J	0.030 (0.8)	4500	④	C	1
	18	2	FD-15	15034, 15035	22 (361)	Champion	J4J	0.030 (0.8)	4500	④	C	1
	40	2	RD, RDS-23	35022-35524	43.9 (719)	Champion	J4J	0.030 (0.8)	4500	④	E	5
1962	3	2	JW-17	3042-3044	5.28 (85.5)	Champion	J6J	0.030 (0.8)	4000	No Rod	C	1
	5.5	2	CD-19	5524, 5525	8.84 (145)	Champion	J6J	0.030 (0.8)	4000	②③	C	1
	10	2	QD-23	10024, 15025	16.6 (281)	Champion	J4J	0.030 (0.8)	4500	④	C	1
	18	2	FD-16	15036, 15037	22 (361)	Champion	J4J	0.030 (0.8)	4500	④	C	1
	28	2	RX-10	28202, 28203	35.7 (585)	Champion	J4J	0.030 (0.8)	4500	④	E	5
	40	2	RD, RDS-23	35022-35524	43.9 (719)	Champion	J4J	0.030 (0.8)	4500	④	E	5
1963	3	2	JW-18	3302-3312	5.28 (85.5)	Champion	J6J	0.030 (0.8)	4000	No Rod	C	1
	5.5	2	CD-20	5302, 5303	8.84 (145)	Champion	J6J	0.030 (0.8)	4000	②③	C	1
	10	2	QD-24	10302, 10303	16.6 (281)	Champion	J4J	0.030 (0.8)	4500	④	C	1
	18	2	FD-17	18302, 18303	22 (361)	Champion	J4J	0.030 (0.8)	4500	④	C	1
	28	2	RX-11	28302, 28303	35.7 (585)	Champion	J4J	0.030 (0.8)	4500	④	E	5
	40	2	RD, RDS, RK-25	40302-352, 40362-372	43.9 (719)	Champion	J4J	0.030 (0.8)	4500	④	E	3

Tune-Up Specifications
1.5-50 Hp, 1-2 Cylinder

Year	Model (Hp)	No. of Cyl	Model Johnson	Model Evinrude	Displace cu. in. (cc)	Spark Plug Make	Spark Plug Type	Spark Plug Gap In. (mm)	WOT MAX (rpm)	Shift Removal Note	Primary Pick-Up Location ①	Primary Pick-Up Adjustment ①
1964	3	2	JW, JH-19	3402-12, 3432	5.28 (85.5)	Champion	J6J	0.030 (0.8)	4000	No Rod	C	1
	5.5	2	CD-21	5402, 5403	8.84 (145)	Champion	J6J	0.030 (0.8)	4000	② ③	C	1
	9.5	2	MQ-10	9422, 9423	15.2 (249)	Champion	J4J	0.030 (0.8)	4500	⑤	C	2
	18	2	FD-18	18402, 18403	22 (361)	Champion	J4J	0.030 (0.8)	4500	④	C	1
	28	2	RX-12	28402, 28403	35.7 (585)	Champion	J4J	0.030 (0.8)	4500	④	E	5
	40	2	RD, RDS, RK-26	40402-352, 40462-472	43.9 (719)	Champion	J4J	0.030 (0.8)	4500	④	E	3
1965	3	2	JW, JH-20B	3502-12, 3532	5.28 (85.5)	Champion	J6J	0.030 (0.8)	4000	No Rod	C	1
	5	2	LD-10S	5502, 5503	8.84 (145)	Champion	J6J	0.030 (0.8)	4000	No Rod, ②	C	1
	6	2	CD-22M	6502, 6503	8.84 (145)	Champion	J6J	0.030 (0.8)	4500	② ⑤	C	1
	9.5	2	MQ-11C	9522, 9523	15.2 (249)	Champion	J4J	0.030 (0.8)	4500	⑤	C	2
	18	2	FD-19DL	18502, 18503	22 (361)	Champion	J4J	0.030 (0.8)	4500	④	C	1
	33	2	RX, RXW-13BE	33502-552	40.5 (664)	Champion	J4J	0.030 (0.8)	4500	④	E	5
	40	2	RD, RDS, RK, AM-26	40502-552, 40562-572	43.9 (719)	Champion	J4J	0.030 (0.8)	4500	④	E	3
1966	3	2	JW, JWL, JH, JHL-21	3602-603, 3612-633	5.28 (85.5)	Champion	J6J	0.030 (0.8)	4000	No Rod	C	1
	5	2	LD, LDL-11	5602, 5603	8.84 (145)	Champion	J6J	0.030 (0.8)	4000	No Rod, ②	C	1
	6	2	CDL, CD-23	6602, 6603	8.84 (145)	Champion	J6J	0.030 (0.8)	4500	② ⑤	C	1
	9.5	2	MQ, MQ-12	9622, 9623	15.2 (249)	Champion	J4J	0.030 (0.8)	4500	⑤	C	2
	18	2	—	18602, 18603	22 (361)	Champion	J4J	0.030 (0.8)	4500	④	C	1
	20	2	FD, FDL-20	—	22 (361)	Champion	J4J	0.030 (0.8)	4500	④	C	1
	33	2	RX, RXL, RXE, RXEL	33602, 33603, 33652, 33653	40.5 (664)	Champion	J4J	0.030 (0.8)	4500	④	E	5
	40	2	RD, RDL-28	40602-652, 40662-672	43.9 (719)	Champion	J4J	0.030 (0.8)	4500	④	E	3
1967	3	2	JW, JWF, JH, JHF, JHL-22	3712-3716, 3732-3733, 3706-3707	5.28 (85.5)	Champion	J6J	0.030 (0.8)	4000	No Rod	B	1
	5	2	LD, LDL-12	5702, 5703	8.84 (145)	Champion	J6J	0.030 (0.8)	4000	No Rod, ②	B	1
	6	2	CDL, CD-24	6702, 6703	8.84 (145)	Champion	J6J	0.030 (0.8)	4500	② ⑤	B	1
	9.5	2	MQ, MQ-13	9722, 9723	15.2 (249)	Champion	J4J	0.030 (0.8)	4500	⑤	C	2
	18	2	—	18702, 18703	22 (361)	Champion	J4J	0.030 (0.8)	4500	④	D	1
	20	2	FD, FDL-21	—	22 (361)	Champion	J4J	0.030 (0.8)	4500	④	D	1

Tune-Up Specifications
1.5-50 Hp, 1-2 Cylinder

Year	Model (Hp)	No. of Cyl	Model Johnson	Model Evinrude	Displace cu. in. (cc)	Spark Plug Make	Type	Gap In. (mm)	WOT MAX (rpm)	Shift Removal Note	Primary Pick-Up Location ①	Primary Pick-Up Adjustment ①
1967 (cont'd)	33	2	RX, RXL, RXE, RXEL-15	33702, 33703, 33752, 33753	40.5 (664)	Champion	J4J	0.030 (0.8)	4500	④	E	5
	40	2	RD, RDL, RDS, RDSL, RK, RKL-29	40702, 40703, 40752, 40753, 40772, 40773	43.9 (719)	Champion	J4J	0.030 (0.8)	4500	④	E	3
1968	1.5	1	SC-10	1802	2.64 (43.3)	Champion	J6J	0.030 (0.8)	4000	No Rod	A	1
	3	2	JH, JHL JHF, JW, JWF, JWF/C-23	3802, 3803, 3806, 3807, 3832, 3822	5.28 (85.5)	Champion	J6J	0.030 (0.8)	4000	No Rod	B	1
	5	2	LD, LDL-13	5802, 5803	8.84 (145)	Champion	J6J	0.030 (0.8)	4000	No Rod, ②	B	1
	6	2	CDL, CD-25	6802, 6803	8.84 (145)	Champion	J6J	0.030 (0.8)	4500	② ⑤	B	1
	9.5	2	MQ, MQ-14	9822, 9823	15.2 (249)	Champion	J4J	0.030 (0.8)	4500	⑤	C	2
	18	2	-	18802, 18803	22 (361)	Champion	J4J	0.030 (0.8)	4500	④	D	1
	20	2	FD, FDL-22	-	22 (361)	Champion	J4J	0.030 (0.8)	4500	④	D	1
	33	2	RX, RXL, RXE, RXEL-16	33802, 33803, 33852, 33853	40.5 (664)	Champion	J4J	0.030 (0.8)	4500	④	E	5
	40	2	RD, RDL, RDS, RDL-30	40802, 40803, 40852, 40853, 40872, 40873	43.9 (719)	Champion	J4J	0.030 (0.8)	4500	④	E	3
1969	1.5	1	IR-69	1902	2.64 (43.3)	Champion	J6J	0.030 (0.8)	4000	No Rod	A	1
	4	2	4R, 4W, 4WF-69	4902, 4906, 4936	5.28 (85.5)	Champion	J6J	0.030 (0.8)	4500	No Rod	B	1
	6	2	6RL, 6RL-69	6902, 6903	8.84 (145)	Champion	J6J	0.030 (0.8)	4500	② ⑤	B	1
	9.5	2	9R, 9RL-69	9922, 9923	15.2 (249)	Champion	J4J	0.030 (0.8)	4500	⑤	C	2
	18	2	-	18902, 18903	22 (361)	Champion	J4J	0.030 (0.8)	4500	④	D	3
	20	2	20R, 20RL-69	-	22 (361)	Champion	J4J	0.030 (0.8)	4500	④	D	3
	25	2	25R, 25RL-69	25902, 25903	22 (361)	Champion	J4J	0.030 (0.8)	5500	④	D	3
	33	2	33R, 33RL, 33E, 33EL-69	33902, 33903, 33952, 33953	40.5 (664)	Champion	J4J	0.030 (0.8)	4500	④	E	5

Tune-Up Specifications
1.5-50 Hp, 1-2 Cylinder

Year	Model (Hp)	No. of Cyl	Model Johnson	Evinrude	Displace cu. in. (cc)	Spark Plug Make	Type	Gap In. (mm)	WOT MAX (rpm)	Shift Removal Note	Primary Pick-Up Location ①	Primary Pick-Up Adjustment ①
1967 (cont'd)	33	2	RX, RXL, RXE, RXEL-15	33702, 33703, 33752, 33753	40.5 (664)	Champion	J4J	0.030 (0.8)	4500	④	E	5
	40	2	RD, RDL, RDS, RDSL, RK, RKL-29	40702, 40703, 40752, 40753, 40772, 40773	43.9 (719)	Champion	J4J	0.030 (0.8)	4500	④	E	3
1968	1.5	1	SC-10	1802	2.64 (43.3)	Champion	J6J	0.030 (0.8)	4000	No Rod	A	1
	3	2	JH, JHL JHF, JW, JWF, JWF/C-23	3802, 3803, 3806, 3807, 3832, 3822	5.28 (85.5)	Champion	J6J	0.030 (0.8)	4000	No Rod	B	1
	5	2	LD, LDL-13	5802, 5803	8.84 (145)	Champion	J6J	0.030 (0.8)	4000	No Rod, ②	B	1
	6	2	CDL, CD-25	6802, 6803	8.84 (145)	Champion	J6J	0.030 (0.8)	4500	② ⑤	B	1
	9.5	2	MQ, MQ-14	9822, 9823	15.2 (249)	Champion	J4J	0.030 (0.8)	4500	⑤	C	2
	18	2	-	18802, 18803	22 (361)	Champion	J4J	0.030 (0.8)	4500	④	D	1
	20	2	FD, FDL-22	-	22 (361)	Champion	J4J	0.030 (0.8)	4500	④	D	1
	33	2	RX, RXL, RXE, RXEL-16	33802, 33803, 33852, 33853	40.5 (664)	Champion	J4J	0.030 (0.8)	4500	④	E	5
	40	2	RD, RDL, RDS, RDL-30	40802, 40803, 40852, 40853, 40872, 40873	43.9 (719)	Champion	J4J	0.030 (0.8)	4500	④	E	3
1969	1.5	1	IR-69	1902	2.64 (43.3)	Champion	J6J	0.030 (0.8)	4000	No Rod	A	1
	4	2	4R, 4W, 4WF-69	4902, 4906, 4936	5.28 (85.5)	Champion	J6J	0.030 (0.8)	4500	No Rod	B	1
	6	2	6RL, 6RL-69	6902, 6903	8.84 (145)	Champion	J6J	0.030 (0.8)	4500	② ⑤	B	1
	9.5	2	9R, 9RL-69	9922, 9923	15.2 (249)	Champion	J4J	0.030 (0.8)	4500	⑤	C	2
	18	2	-	18902, 18903	22 (361)	Champion	J4J	0.030 (0.8)	4500	④	D	3
	20	2	20R, 20RL-69	-	22 (361)	Champion	J4J	0.030 (0.8)	4500	④	D	3
	25	2	25R, 25RL-69	25902, 25903	22 (361)	Champion	J4J	0.030 (0.8)	5500	④	D	3
	33	2	33R, 33RL, 33E 33EL-69	33902, 33903, 33952, 33953	40.5 (664)	Champion	J4J	0.030 (0.8)	4500	④	E	5
	40	2	40R, 40RL, 40E, 40EL, 40ES, 40ESL-69	40902, 40903, 40952, 40953, 40972, 40973	43.9 (719)	Champion	J4J	0.030 (0.8)	4500	④	E	3
1970	1.5	1	IR-70	1002	2.64 (43.3)	Champion	J6J	0.030 (0.8)	4000	No Rod	A	1
	4	2	4R, 4W-70	4006, 4036	5.28 (85.5)	Champion	J6J	0.030 (0.8)	4500	No Rod	B	1
	6	2	6RL, 6RL-70	6002, 6003	8.84 (145)	Champion	J6J	0.030 (0.8)	4500	② ⑤	B	1
	9.5	2	9R, 9RL-70	9022, 9023	15.2 (249)	Champion	J4J	0.030 (0.8)	4500	⑤	C	2
	18	2	-	18002, 18003	22 (361)	Champion	J4J	0.030 (0.8)	4500	④	D	3
	20	2	20R, 20RL-70	-	22 (361)	Champion	J4J	0.030 (0.8)	4500	④	D	3
	25	2	25R, 25RL-70	25002, 25003	22 (361)	Champion	J4J	0.030 (0.8)	5500	④	D	3
	33	2	33R, 33RL, 33E 33EL-70	33002, 33003, 33052, 33053	40.5 (664)	Champion	J4J	0.030 (0.8)	4500	④	E	5
	40	2	40R, 40RL, 40E, 40EL, 40ES, 40ESL-70	40002, 40003, 40052, 40053, 40072, 40073	43.9 (719)	Champion	J4J	0.030 (0.8)	4500	④	E	3
1971	2	1	2R-71	2102	2.64 (43.3)	Champion	J6J	0.030 (0.8)	4000	-	A	1
	4	2	4R, 4W-71	4106, 4136	5.28 (85.5)	Champion	J6J	0.030 (0.8)	4500	-	B	1
	6	2	6RL, 6RL-71	6102, 6103	8.84 (145)	Champion	J6J	0.030 (0.8)	4500	-	B	1
	9.5	2	9R, 9RL-71	9122, 9123	15.2 (249)	Champion	J4J	0.030 (0.8)	4500	-	C	2
	18	2	-	18102, 18103	22 (361)	Champion	J4J	0.030 (0.8)	4500	-	D	3
	20	2	20R, 20RL-71	-	22 (361)	Champion	J4J	0.030 (0.8)	4500	-	D	3
	25	2	25R, 25RL-71	25102, 25103	22 (361)	Champion	J4J	0.030 (0.8)	5500	-	D	3
	40	2	40R, 40RL, 40E, 40EL-71	40102, 40103, 40152, 40153	43.9 (719)	Champion	J4J	0.030 (0.8)	4500	-	D	5
	50	2	50ES, 50ESL-71	50172, 50173	41.5 (680)	Champion	L77J4	0.040 (01.0)	5500	- ⑥ ⑦	C	4
1972	2	1	2R-72	2202	2.64 (43.3)	Champion	J6J	0.030 (0.8)	4000	-	B	1
	4	2	4R, 4W-72	4206, 4236	5.28 (85.5)	Champion	J6J	0.030 (0.8)	4500	-	B	1
	6	2	6RL, 6RL-72	6202, 6203	8.84 (145)	Champion	J6J	0.030 (0.8)	4500	-	B	1
	9.5	2	9R, 9RL-72	9222, 9223	15.2 (249)	Champion	J4J	0.030 (0.8)	4500	-	C	2
	18	2	-	18202, 18203	22 (361)	Champion	J4J	0.030 (0.8)	4500	-	C	3
	20	2	20R, 20RL-71	-	22 (361)	Champion	J4J	0.030 (0.8)	4500	-	C	3
	25	2	25R, 25RL, 25E, 25EL-72	25202, 25203, 25252, 25253	22 (361)	Champion	J4J	0.030 (0.8)	5500	-	C	3
	40	2	40R, 40RL, 40E, 40EL-72	40202, 40203, 40252, 40253	43.9 (719)	Champion	J4J	0.030 (0.8)	4500	-	D	3
	50	2	50R, 50RL, 50ES, 50ESL-72	50272, 50273, 50202, 50203	41.5 (680)	Champion	L77J4	0.040 (1.0)	5500	- ⑥ ⑦	C	4

① Refer to end of chart for description of codes in this column. Full details provided in the Timing & Syncronization section
② Pin in upper end of driveshaft
③ Remove powerhead
④ Remove window
⑤ Drop lower unit
⑥ Timing at WOT: 19°
⑦ Models using UL77V - not adjustable

Primary Pick-Up Location   **Primary Pick-Up Adjustment**

A: Port side of mark   **1:** 1968-70: loosen 2 screws under armature plate and move primary pick-up in or out to meet follower
B: Starboard side of mark   1971-72: Loosen 2 screws under armatuer plate and move plate in or out to make coontact with roller
C: Center of mark   **2:** Loosen center screw on throttle lever and move lever in or out to match line on cam
D: Between the marks   **3:** Loosen clamp on throttle shaft and move roller to meet armature cam
E: Mark at pointer   **4:** On the starboard side, loosen the throttle arm screw and move the roller in or out to align with the mark on the cam
   5: Loosen eccentric lock screw on throttle shaft and turn eccentric to move roller to meet armature cam

Tune-Up Specifications
50-125 Hp, 3-4 Cylinder

Year	Model (Hp)	No. of Cyl	Model Johnson	Model Evinrude	Displace cu. in. (cc)	Spark Plug Make	Spark Plug Type	Gap In. (mm)	WOT MAX (rpm)	Point or Sensor Gap (In.)	Timing	Primary Pick-Up Location	Ignition Type	Carb Type
1958	50	V4	V4, V4S-10	50012, 50014	70.7 (1159)	Champion	J4J	0.030 (0.8)	4000	0.020	-	Port Side	①	②
1959	50	V4	V4, V4S-11	50016, 50518	70.7 (1159)	Champion	J4J	0.030 (0.8)	4000	0.020	-	Port Side	①	②
1960	75	V4	V4S-12	50522	89.5 (1467)	Champion	J4J	0.030 (0.8)	4500	0.020	-	Port Side	①	②
1961	75	V4	V4S, V4A-13	50524, 50926	89.5 (1467)	Champion	J4J	0.030 (0.8)	4500	0.020	-	Port Side	① or ③	④
1962	75	V4	V4S, V4A-14	50528, 50930	89.5 (1467)	Champion	J4J	0.030 (0.8)	4500	0.020	-	Port Side	① or ③	④
1963	75	V4	V4S, V4A-15	75352, 75382, 75392	89.5 (1467)	Champion	J4J	0.030 (0.8)	4500	0.020	-	Port Side	① or ③	④
1964	60	V4	VX-10S	60432-52	70.7 (1159)	Champion	J4J	0.030 (0.8)	4500	0.020	-	Port Side	①	②
	75	V4	V4S, V4A-16	75432-52, 75482-92	89.5 (1467)	Champion	J4J	0.030 (0.8)	4500	0.020	-	Port Side	① or ③	④
	90	V4	V4M, V4ML-10S	90482-92	89.5 (1467)	Champion	J4J	0.030 (0.8)	4500	0.020	-	Top	③	⑤
1965	60	V4	VX, VXH-11S	60532, 60552	70.7 (1159)	Champion	J4J	0.030 (0.8)	4500	0.020	-	Port Side	①	②
	75	V4	V4S, V4H, V4A-17S	75532-52, 75582-92	89.5 (1467)	Champion	J4J	0.030 (0.8)	4500	0.020	-	Port Side	① or ③	④
	90	V4	V4M, V4ML-11C	90582-92	89.5 (1467)	Champion	J4J	0.030 (0.8)	4500	0.020	-	Top	③	⑤
1966	60	V4	VX, VXH-12	60652-53, 60632-33	70.7 (1159)	Champion	J4J	0.030 (0.8)	4500	0.020	-	Port Side	①	④
	80	V4	V4S, V4SA-18	80652-53, 80692-93	89.5 (1467)	Champion	J4J	0.030 (0.8)	4500	0.020	-	Port Side	① or ③	④
	100	V4	V4ML-12	100693	89.5 (1467)	Champion	J4J	0.030 (0.8)	4500	0.020	-	Top	③	⑤
1967	60	V4	VX, VXO, VXH, VXHL-13	60752-53, 60732-33	70.7 (1159)	Champion	J4J	0.030 (0.8)	4500	0.020	-	Port Side	③	④
	80	V4	V4S, V4SL, V4A, V4AL-19	80752-53, 80792-93	89.5 (1467)	Champion	J4J	0.030 (0.8)	4500	0.020	-	Port Side	③	④
	100	V4	V4TL-13	100783	89.5 (1467)	Champion	V40FFX	0.030 (0.8)	5000	0.028	4500 Mark	Top	⑥	⑤
1968	55	3	TR, TRL-10	55872-73	49.7 (814)	Champion	L76V	⑫	5000	0.010	4500 Mark	Stbd Side	⑦	⑧
	65	V4	VX, VXH, VXL, VXHL-14	60852-53, 60832-33	70.7 (1159)	Champion	L76V	⑫	5000	0.010	4500 Mark	Port Side	⑦	④
	85	V4	V4S, V4SL, V4A, V4AL-20	80852-53, 80892-93	89.5 (1467)	Champion	L76V	⑫	5000	0.010	4500 Mark	Port Side	⑦	④
	100	V4	V4TL-14	100882-83	89.5 (1467)	Champion	V40FFX	⑫	5000	0.028	4500 Mark	Top	⑥	⑤
1969	55	3	55ES, 55ESL-69	55972-73	49.7 (814)	Champion	L76V	⑫	5000	0.010	4500 Mark	Stbd Side	⑦	⑧
	85	V4	85ESL-69	85993	92.6 (1517)	Champion	L76V	⑫	5000	0.010	4500 Mark	Stbd Side	⑦	⑨
	115	V4	115ESL-69	115983	96.1 (1575)	Champion	L76V	⑫	5000	0.028	4500 Mark	Stbd Side	⑥	⑨
1970	60	3	60ES, 60ESL-70	60072-73	49.7 (814)	Champion	L76V	⑫	5000	0.010	4500, 22°	Stbd Side	⑦	⑧
	85	V4	85ESL-70	85093	92.6 (1517)	Champion	L76V	⑫	5000	0.010	4500 Mark	Stbd Side	⑦	⑨
	115	V4	115ESL-70	115083	96.1 (1575)	Champion	L76V	⑫	5000	0.028	4500 Mark	Stbd Side	⑥	⑨
1971	60	3	60ES, 60ESL-71	60172-73	49.7 (814)	Champion	L76V	⑫	5000	0.010	4500, 22°	Stbd Side	⑦	⑧
	85	V4	85ESL-71	85193	92.6 (1517)	Champion	L76V	⑫	5000	0.010	4500, 28°	Stbd Side	⑦	⑨
	100	V4	100ESL-71	100193	92.6 (1517)	Champion	L76V	⑫	5000	0.010	4500, 28°	Stbd Side	⑦	⑨
	125	V4	125ESL-71	125183	99.6 (1632)	Champion	L76V	⑫	5000	0.028	4500, 26°	Stbd Side	⑥	⑨
1972	65	3	65ES, 65ESL-72	65272-73	49.7 (814)	Champion	L76V	⑫	5000	0.010	4500, 26°	Stbd Side	⑩	⑧
	85	V4	85ESL-72	85293	92.6 (1517)	Champion	L76V	⑫	5000	0.010	4500, 28°	Stbd Side	⑦	⑪
	100	V4	100ESL-72	100293	92.6 (1517)	Champion	L76V	⑫	5000	0.028	4500, 28°	Stbd Side	⑥	⑪
	125	V4	125ESL-72	125283	99.6 (1632)	Champion	L76V	⑫	5000	0.028	4500, 26°	Stbd Side	⑥	⑪

① Distributor Magneto Ignition
② Downdraft, 2 bbl with high and low speed needle valves
③ Distributor Battery Ignition
④ Downdraft, 2 bbl with fixed high orifice and low speed needle valve
⑤ Frontdraft, 4 bbl with fixed high speed orifice and low speed needle valve
⑥ Capacitor Discharge (CD) Ignition with sensor
⑦ Capacitor Discharge (CD) Ignition with breaker points
⑧ Frontdraft, 1 bbl with fixed high speed orifice and low speed needle valve. Three carbs per engine
⑨ Frontdraft, 2 bbl with low speed needle valve and fixed high speed orifice. Two carbs per engine
⑩ Capacitor Discharge (CD) Ignition with flywheel ignition
⑪ Frontdraft, 2 bbl with fixed low speed and high speed orifices. Two carbs per engine
⑫ Not adjustable

Capacities

Year	Model (Hp)	No. of Cyl	Displace cu. in. (cc)	Gear Oil Oz. (ml)	Fuel:Oil Ratio
1958	3	2	5.28 (85.5)	2.9 (68)	50:1
	5.5	2	8.84 (145)	8.5 (200)	50:1
	7.5	2	12.4 (203)	8.5 (200)	50:1
	10	2	16.6 (281)	10 (236)	50:1
	18	2	22 (361)	8.3 (196)	50:1
	35	2	40.5 (664)	13.9 (328)	50:1
	50	V4	70.7 (1159)	34.8 (820)	50:1
1959	3	2	5.28 (85.5)	2.9 (68)	50:1
	5.5	2	8.84 (145)	8.5 (200)	50:1
	10	2	16.6 (281)	10 (236)	50:1
	18	2	22 (361)	8.3 (196)	50:1
	35	2	40.5 (664)	13.9 (328)	50:1
	50	V4	70.7 (1159)	34.8 (820)	50:1
1960	3	2	5.28 (85.5)	2.9 (68)	50:1
	5.5	2	8.84 (145)	8.5 (200)	50:1
	10	2	16.6 (281)	10 (236)	50:1
	18	2	22 (361)	8.3 (196)	50:1
	40	2	43.9 (719)	15 (354)	50:1
	75	V4	89.5 (1467)	19.5 (460)	50:1
1961	3	2	5.28 (85.5)	2.9 (68)	50:1
	5.5	2	8.84 (145)	8.5 (200)	50:1
	10	2	16.6 (281)	10 (236)	50:1
	18	2	22 (361)	8.3 (196)	50:1
	40	2	43.9 (719)	15 (354)	50:1
	75	V4	89.5 (1467)	19.5 (460)	50:1
1962	3	2	5.28 (85.5)	2.9 (68)	50:1
	5.5	2	8.84 (145)	8.5 (200)	50:1
	10	2	16.6 (281)	10 (236)	50:1
	18	2	22 (361)	8.3 (196)	50:1
	28	2	35.7 (585)	13.9 (328)	50:1
	40	2	43.9 (719)	15 (354)	50:1
	75	V4	89.5 (1467)	19.5 (460) ③	50:1
1963	3	2	5.28 (85.5)	2.9 (68)	50:1
	5.5	2	8.84 (145)	8.5 (200)	50:1
	10	2	16.6 (281)	10 (236)	50:1
	18	2	22 (361)	8.3 (196)	50:1
	28	2	35.7 (585)	13.9 (328)	50:1
	40	2	43.9 (719)	15 (354)	50:1
	75	V4	89.5 (1467)	19.5 (460) ③	50:1
1964	3	2	5.28 (85.5)	2.9 (68)	50:1
	5.5	2	8.84 (145)	8.5 (200)	50:1
	9.5	2	15.2 (249)	9.7 (229)	50:1

Capacities

Year	Model (Hp)	No. of Cyl	Displace cu. in. (cc)	Gear Oil Oz. (ml)	Fuel:Oil Ratio
1964 (cont'd)	18	2	22 (361)	8.3 (196)	50:1
	28	2	35.7 (585)	13.9 (328)	50:1
	40	2	43.9 (719)	15 (354)	50:1
	60	V4	70.7 (1159)	19.5 (460) ④	50:1
	75	V4	89.5 (1467)	19.5 (460) ③④	50:1
	90	V4	89.5 (1467)	17.4 (410)	50:1
1965	3	2	5.28 (85.5)	2.9 (68)	50:1
	5	2	8.84 (145)	2.9 (68)	50:1
	6	2	8.84 (145)	8.5 (200)	50:1
	9.5	2	15.2 (249)	9.7 (229)	50:1
	18	2	22 (361)	8.3 (196)	50:1
	33	2	40.5 (664)	13.9 (328)	50:1
	40	2	43.9 (719)	15 (354)	50:1
	60	V4	70.7 (1159)	19.5 (460) ④	50:1
	75	V4	89.5 (1467)	19.5 (460) ③④	50:1
	90	V4	89.5 (1467)	17.4 (410)	50:1
1966	3	2	5.28 (85.5)	2.9 (68)	50:1
	5	2	8.84 (145)	2.9 (68)	50:1
	6	2	8.84 (145)	8.5 (200)	50:1
	9.5	2	15.2 (249)	9.7 (229)	50:1
	18	2	22 (361)	8.3 (196)	50:1
	20	2	22 (361)	8.3 (196)	50:1
	33	2	40.5 (664)	13.9 (328)	50:1
	40	2	43.9 (719)	15 (354)	50:1
	60	V4	70.7 (1159)	19.5 (460) ④	50:1
	80	V4	89.5 (1467)	19.5 (460) ③	50:1
	100	V4	89.5 (1467)	32.2 (759)	50:1
1967	3	2	5.28 (85.5)	2.9 (68)	50:1
	5	2	8.84 (145)	2.9 (68)	50:1
	6	2	8.84 (145)	8.5 (200)	50:1
	9.5	2	15.2 (249)	9.7 (229)	50:1
	18	2	22 (361)	8.3 (196)	50:1
	20	2	22 (361)	8.3 (196)	50:1
	33	2	40.5 (664)	13.9 (328)	50:1
	40	2	43.9 (719)	15 (354)	50:1
	60	V4	70.7 (1159)	19.5 (460) ④	50:1
	80	V4	89.5 (1467)	19.5 (460) ③	50:1
	100	V4	89.5 (1467)	37.2 (877)	50:1
1968	1.5	1	2.64 (43.3)	0.75 (18)	50:1
	3	2	5.28 (85.5)	2.9 (68)	50:1

Capacities

Year	Model (Hp)	No. of Cyl	Displace cu. in. (cc)	Gear Oil Oz. (ml)	Fuel:Oil Ratio
1968 (cont'd)	5	2	8.84 (145)	2.9 (68)	50:1
	6	2	8.84 (145)	8.5 (200)	50:1
	9.5	2	15.2 (249)	9.7 (229)	50:1
	18	2	22 (361)	8.3 (196)	50:1
	20	2	22 (361)	8.3 (196)	50:1
	33	2	40.5 (664)	13.9 (328)	50:1
	40	2	43.9 (719)	15 (354)	50:1
	55	3	49.7 (814)	25.3 (596)	50:1
	65	V4	70.7 (1159)	19.5 (460) ④	50:1
	85	V4	89.5 (1467)	19.5 (460) ③	50:1
	100	V4	89.5 (1467)	37.2 (877)	50:1
1969	1.5	1	2.64 (43.3)	0.75 (18)	50:1
	4	2	5.28 (85.5)	1.3 (31) ①	50:1
	6	2	8.84 (145)	8.5 (200)	50:1
	9.5	2	15.2 (249)	9.7 (229)	50:1
	18	2	22 (361)	8.3 (196)	50:1
	20	2	22 (361)	8.3 (196)	50:1
	25	2	22 (361)	8.3 (196)	50:1
	33	2	40.5 (664)	13.9 (328)	50:1
	40	2	43.9 (719)	13.9 (328) ②	50:1
	55	3	49.7 (814)	25.3 (596)	50:1
	85	V4	92.6 (1517)	27.9 (658)	50:1
	115	V4	96.1 (1575)	26.9 (634)	50:1
1970	1.5	1	2.64 (43.3)	0.75 (18)	50:1
	4	2	5.28 (85.5)	1.3 (31) ①	50:1
	6	2	8.84 (145)	8.5 (200)	50:1
	9.5	2	15.2 (249)	9.7 (229)	50:1
	18	2	22 (361)	8.3 (196)	50:1
	20	2	22 (361)	8.3 (196)	50:1
	25	2	22 (361)	8.3 (196)	50:1
	33	2	40.5 (664)	13.9 (328)	50:1
	40	2	43.9 (719)	13.9 (328) ②	50:1
	60	3	49.7 (814)	25.3 (596)	50:1
	85	V4	92.6 (1517)	27.9 (658)	50:1
	115	V4	96.1 (1575)	26.9 (634)	50:1
1971	2	1	2.64 (43.3)	1.3 (31)	50:1
	4	2	5.28 (85.5)	1.3 (31) ①	50:1
	6	2	8.84 (145)	8.5 (200)	50:1
	9.5	2	15.2 (249)	9.7 (229)	50:1
	18	2	22 (361)	8.3 (196)	50:1
	20	2	22 (361)	8.3 (196)	50:1
	25	2	22 (361)	8.3 (196)	50:1
	40	2	43.9 (719)	13.9 (328)	50:1
	50	2	41.5 (680)	25.3 (596)	50:1
	60	3	49.7 (814)	25.3 (596)	50:1
	85	V4	92.6 (1517)	27.9 (658)	50:1
	100	V4	92.6 (1517)	26.9 (634)	50:1
	125	V4	99.6 (1632)	26.9 (634)	50:1
1972	2	1	2.64 (43.3)	1.3 (31)	50:1
	4	2	5.28 (85.5)	1.3 (31) ①	50:1
	6	2	8.84 (145)	8.5 (200)	50:1
	9.5	2	15.2 (249)	9.7 (229)	50:1
	18	2	22 (361)	8.3 (196)	50:1
	20	2	22 (361)	8.3 (196)	50:1
	25	2	22 (361)	8.3 (196)	50:1
	40	2	43.9 (719)	13.9 (328)	50:1
	50	2	41.5 (680)	25.3 (596)	50:1
	65	3	49.7 (814)	19.5 (460)	50:1
	85	V4	92.6 (1517)	27.9 (658)	50:1
	100	V4	92.6 (1517)	26.9 (634)	50:1
	125	V4	99.6 (1632)	26.9 (634)	50:1

NOTE: Fuel/Oil ratio based on normal operating conditions, some severe or high performance applications may need higher ratio, refer to Fuel Recommendations in the Maintenance section

① 4.0W: 3.4 (80)   ③ Electric: 17.4 (410)
② 40E: 15 (354)   ④ HD: 34.8 (820)

Maintenance Intervals Chart

Component	Each Use	Monthly or As Needed	First 20-Hour Check	Every 12mths/100hrs	Off Season
Anode(s)	I	—	—	—	—
Battery condition and connections (if equipped)*	I (condition/connections)	I (charge/fluid level)	—	—	—
Breaker points and condenser (Magneto only)	—	—	—	R	R
Boat hull*	I	—	—	—	—
Bolts and nuts (all accessible fasteners)*	I	—	—	—	—
Case finish (wash and wax)	C (salt / brackish / polluted water)	C	—	C	C
Cylinder head bolts	—	—	T	T	—
Electrical wiring and connectors*	I	—	—	—	—
Emergency stop switch, clip &/or lanyard*	I	I	—	—	—
Engine mounting clamps/bolts	I	I	—	—	—
Flush cooling system	if in salt / brackish / polluted water	—	—	—	P
Fuel filter (clean or replace, as applicable)	P	P	P	P	P
Fuel hose and system components*	I	—	—	—	—
Gear oil	I (for signs of leakage)	I (level and condition)	R	R	R
Lubrication points	I	①	①	①	①
Oil system hose and components*	I	—	—	—	—
Pistons (Decarbon)	—	—	—	②	—
Power steering belt, fluid, filter (if equipped)	I (quick-check of belt/fluid)	I (check fluid level monthly)	—	R (500 hours)	—
Power trim and tilt (if equipped)	—	I	L	L	L
Propeller (or rotor on RescuePro motors)	I	—	—	—	—
Propeller (or rotor) shaft and nut	—	—	I, L / T	I, L / T	I, L / T
Remote control*	I	—	—	—	—
Spark plugs	—	R (as needed)	—	—	—
Steering cable*	I	L (as needed)	L	L	L
Steering friction	I	A (as needed)	—	—	—
Tune-up	—	A (as needed)	—	I (annually)	Perform pre-season tune-up
Valve clearance	—	—	—	I (annually)	Perform with pre-season tune-up
Water pump intake grate and indicator	I	—	—	—	—

A–Adjust
C–Clean
I–Inspect and Clean, Adjust, Lubricate or Replace, as necessary
L–Lubricate
R–Replace
T–Tighten

* Denotes possible safety item (although, all maintenance inspections/service can be considered safety related when it means not being stranded on the water should a component fail.)
① Varies with use, generally every 30 days when used in salt, brackish or polluted water and every 60 days when used in fresh water (refer to Lubrication Chart for more details)
② Every 50 hours is OMC Carbon Guard additive is NOT used consistently with fuel.

Lubrication Chart

Component	Applicable Models	Recommended Lubricant	Minimum Frequency	
			Fresh Water	Salt, Polluted or Brackish Water
Choke linkage	2-15 Hp motors	OMC Triple-Guard or equivalent marine grease	every 60 days ①	every 30 days ①
Electric starter motor pinion	Electric start models only	OMC Start pinion lube	every 60 days ①	every 30 days ①
Engine cover latches	737cc and larger 2-strokes	OMC Triple-Guard or equivalent marine grease	every 60 days ①	every 30 days ①
Engine mount clamp screws	1 and 2-Cylinder motors	OMC Triple-Guard or equivalent marine grease	every 60 days ①	every 30 days ①
Shift, carb and/or throttle shafts, cables and/or linkage	All	OMC Triple-Guard or equivalent marine grease	every 60 days ①	every 30 days ①
Steering (remote arm/ tiller shaft pivot and friction screw)	All	OMC Triple-Guard or equivalent marine grease	every 60 days ①	every 30 days ①
Swivel bracket	All	OMC Triple-Guard or equivalent marine grease	every 60 days ①	every 30 days ①
Tilt assembly (bracket, tube, pin and/or tilt lever shaft)	All	OMC Triple-Guard or equivalent marine grease	every 60 days ①	every 30 days ①

① Lubrication points should be checked weekly or with each use, whichever is LESS frequent. Based upon individual motor/use needs frequency of actual lubrication should occur at recommended intervals or more often during season. Perform all lubrication procedures immediately prior to extended motor storage.

SPARK PLUG DIAGNOSIS

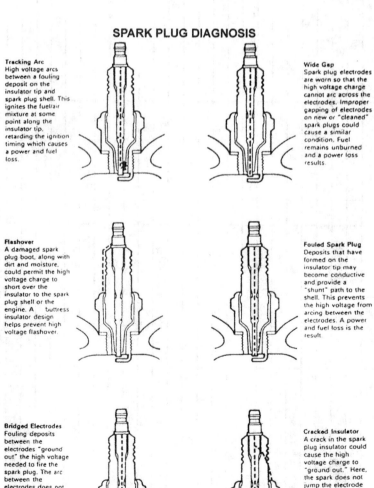

Tracking Arc
High voltage arcs between a fouling deposit on the insulator tip and spark plug shell. This ignites the fuel/air mixture at some point along the insulator tip, retarding the ignition timing which causes a power and fuel loss.

Wide Gap
Spark plug electrodes are worn so that the high voltage charge cannot arc across the electrodes. Improper gapping of electrodes on new or "cleaned" spark plugs could cause a similar condition. Fuel remains unburned and a power loss results.

Flashover
A damaged spark plug boot, along with dirt and moisture, could permit the high voltage charge to short over the insulator to the spark plug shell or the engine. A buttress insulator design helps prevent high voltage flashover.

Fouled Spark Plug
Deposits that have formed on the insulator tip may become conductive and provide a "shunt" path to the shell. This prevents the high voltage from arcing between the electrodes. A power and fuel loss is the result.

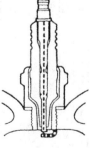

Bridged Electrodes
Fouling deposits between the electrodes "ground out" the high voltage needed to fire the spark plug. The arc between the electrodes does not occur and the fuel air mixture is not ignited. This causes a power loss and exhausting of raw fuel.

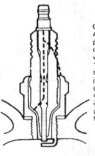

Cracked Insulator
A crack in the spark plug insulator could cause the high voltage charge to "ground out." Here, the spark does not jump the electrode gap and the fuel air mixture is not ignited. This causes a power loss and raw fuel is exhausted.

3

FUEL SYSTEM

FUEL SYSTEM BASICS

✳✳ CAUTION

If equipped, disconnect the negative battery cable ANYTIME work is performed on the engine, especially when working on the fuel system. This will help prevent the possibility of sparks during service (from accidentally grounding a hot lead or powered component). Sparks could ignite vapors or exposed fuel. Disconnecting the cable on electric start motors will also help prevent the possibility fuel spillage if an attempt is made to crank the engine while the fuel system is open.

✳✳ CAUTION

Fuel leaking from a loose, damaged or incorrectly installed hose or fitting may cause a fire or an explosion. ALWAYS pressurize the fuel system and run the motor while inspecting for leaks after servicing any component of the fuel system.

The carburetion, and the ignition principles of engine operation must be understood in order to troubleshoot and repair an outboard motor's fuel system or to perform a proper tune-up on carbureted motors.

If you have any doubts concerning your understanding of engine operation, it would be best to study The Basic Operating Principles of an engine as detailed under Troubleshooting in the General Information, Safety & Tools section, before tackling any work on the fuel system.

The fuel systems used on engines covered by this manual include multiple carburetors. For the most part, the carbureted motors covered here utilize fuel-enrichment for quicker cold starting in the form of an electric primer solenoid (though many models are equipped with a manual primer).

Fuel

GENERAL

◆ See Figure 1

Fuel recommendations have become more complex as the chemistry of modern gasoline changes. The major driving force behind many of the changes in gasoline chemistry was the search for additives to replace lead as an octane booster and lubricant, these additives are governed by the types of emissions they produce in the combustion process. Also, the replacement additives do not always provide the same level of combustion stability, making a fuel's octane rating less meaningful.

In the 1950's, 1960's and 1970's, leaded fuel was common. The lead served two functions. First, it served as an octane booster (combustion stabilizer) and second, in 4-stroke engines, it served as a valve seat lubricant. For 2-stroke engines, the primary benefit of lead was to serve as a combustion stabilizer. Lead served very well for this purpose, even in high heat applications.

For decades now, all lead has been removed from the refining process. This means that the benefit of lead as an octane booster has been eliminated. Several substitute octane boosters have been introduced in the place of lead. While many are adequate in automobile engines, most do not perform nearly as well as lead did, even though the octane rating of the fuel is the same.

OCTANE RATING

◆ See Figure 1

A fuel's octane rating is a measurement of how stable the fuel is when heat is introduced. Octane rating is a major consideration when deciding whether a fuel is suitable for a particular application. For example, in an engine, we want the fuel to ignite when the spark plug fires and not before, even under high pressure and temperatures. Once the fuel is ignited, it must burn slowly and smoothly, even though heat and pressure are building up while the burn occurs. The unburned fuel should be ignited by the traveling flame front, not by some other source of ignition, such as carbon deposits or the heat from the expanding gasses. A fuel's octane rating is known as a measurement of the fuel's anti-knock properties (ability to burn without exploding). Essentially, the octane rating is a measure of a fuel's stability.

Usually a fuel with a higher octane rating can be subjected to a more severe combustion environment before spontaneous or abnormal

combustion occurs. To understand how two gasoline samples can be different, even though they have the same octane rating, we need to know how octane rating is determined.

The American Society of Testing and Materials (ASTM) has developed a universal method of determining the octane rating of a fuel sample. The octane rating you see on the pump at a gasoline station is known as the pump octane number. Look at the small print on the pump. The rating has a formula. The rating is determined by the R+M/2 method. This number is the average of the research octane reading and the motor octane rating.

• The Research Octane Rating is a measure of a fuel's anti-knock properties under a light load or part throttle conditions. During this test, combustion heat is easily dissipated.

• The Motor Octane Rating is a measure of a fuel's anti-knock properties under a heavy load or full throttle conditions, when heat buildup is at maximum.

In general, 2-stroke engines tend to respond more to the motor octane rating than the research octane rating, because a 2-stroke engine has a power stroke (with heat buildup) every revolution. Therefore, in a 2-stroke outboard motor, the motor octane rating of the fuel is one of the best indications of how it will perform.

VAPOR PRESSURE

Fuel vapor pressure is a measure of how easily a fuel sample evaporates. Many additives used in gasoline contain aromatics. Aromatics are light hydrocarbons distilled off the top of a crude oil sample. They are effective at increasing the research octane of a fuel sample but can cause vapor lock (bubbles in the fuel line) on a very hot day. If you have an inconsistent running engine and you suspect vapor lock, use a piece of clear fuel line to look for bubbles, indicating that the fuel is vaporizing.

One negative side effect of aromatics is that they create additional combustion products such as carbon and varnish. If your engine requires high-octane fuel to prevent detonation, de-carbon the engine more frequently with an internal engine cleaner to prevent ring sticking due to excessive varnish buildup.

ALCOHOL-BLENDED FUELS

When the Environmental Protection Agency mandated a phase-out of the leaded fuels in January of 1986; fuel suppliers needed an additive to improve the octane rating of their fuels. Although there are multiple methods currently employed, the addition of alcohol to gasoline seems to be favored because of its favorable results and low cost. Two types of alcohol are used in fuel today as octane boosters, methanol (wood alcohol) or ethanol (grain alcohol).

When used as a fuel additive, alcohol tends to raise the research octane of the fuel, so these additives will have limited benefit in an outboard motor.

Fig. 1 Damaged piston, possibly caused by; using too-low an octane fuel; using fuel that had "soured" or by insufficient oil

There are, however, some special considerations due to the effects of alcohol in fuel.

• Since alcohol contains oxygen, it replaces gasoline without oxygen content and tends to cause the air/fuel mixture to become leaner.

• On older outboards, the leaching affect of alcohol will, in time, cause fuel lines and plastic components to become brittle to the point of cracking. Unless replaced, these cracked lines could leak fuel, increasing the potential for hazardous situations.

• When alcohol blended fuels become contaminated with water, the water combines with the alcohol then settles to the bottom of the tank. This leaves the gasoline (and the oil for 2-stroke models using pre-mix) on a top layer.

■ **Modern outboard fuel lines and plastic fuel system components have been specially formulated to resist alcohol-leaching effects.**

HIGH ALTITUDE OPERATION

At elevated altitudes there is less oxygen in the atmosphere than at sea level. Less oxygen means lower combustion efficiency and less power output. As a general rule, power output is reduced three percent for every thousand feet above sea level.

On carbureted engines, re-jetting for high altitude does not restore lost power, it simply corrects the air-fuel ratio for the reduced air density and makes the most of the remaining available power. The most important thing to remember when re-jetting for high altitude is to reverse the jetting when return to sea level. If the jetting is left lean when you return to sea level conditions, the correct air/fuel ratio will not be achieved (the motor will run very lean) and possible powerhead damage may occur.

Fuel System Service Cautions

There is no way around it. Working with gasoline can provide for many different safety hazards and requires that extra caution be used during all steps of service. To protect yourself and others, you must take all necessary precautions against igniting the fuel or vapors (which will cause a fire at best or an explosion at worst).

✳✳ CAUTION

Take extreme care when working with the fuel system. NEVER smoke (it's bad for you anyhow, but smoking during fuel system service could kill you much faster!) or allow flames or sparks in the work area. Flames or sparks can ignite fuel, especially vapors, resulting in a fire at best or an explosion at worst.

For starters, disconnect the negative battery cable **EVERY** time a fuel system hose or fitting is going to be disconnected. It takes only one moment of forgetfulness for someone to crank the motor, possibly causing a dangerous spray of fuel from the opening.

Gasoline contains harmful additives and is quickly absorbed by exposed skin. As an additional precaution, always wear gloves and some form of eye protection (regular glasses help, but only safety glasses offer any significant protection for your eyes).

■ **Throughout this service, pay attention to ensure that all components, hoses and fittings are installed them in the correct location and orientation to prevent the possibility of leakage. Matchmark components before they are removed as necessary.**

Because of the dangerous conditions that result from working with gasoline and fuel vapors always take extra care and be sure to follow these guidelines for safety:

• Keep a Coast Guard-approved fire extinguisher handy when working.

• Allow the engine to cool completely before opening a fuel fitting. Don't all gasoline to drip on a hot engine.

• The first thing you must do after removing the engine cover is to check for the presence of gasoline fumes. If strong fumes are present, look for leaking or damage hoses, fittings or other fuel system components and repair.

• Do not repair the motor or any fuel system component near any sources of ignition, including sparks, open flames, or anyone smoking.

• Clean up spilled gasoline right away using clean rags. Keep all fuel soaked rags in a metal container until they can be properly disposed of or cleaned. NEVER leave solvent, gasoline or oil soaked rags in the hull.

• Don't use electric powered tools in the hull or near the boat during fuel system service or after service, until the system is pressurized and checked for leaks.

• Fuel leaking from a loose, damaged or incorrectly installed hose or fitting may cause a fire or an explosion. ALWAYS pressurize the fuel system and run the motor while inspecting for leaks after servicing any component of the fuel system.

Fuel System Pressurization

When it comes to safety and outboards, the condition of the fuel system is of the utmost importance. The system must be checked for signs of damage or leakage with every use and checked, especially carefully when portions of the system have been opened for service.

The best method to check the fuel system is to visually inspect the lines, hoses and fittings once the system has been properly pressurized.

PRESSURIZING THE FUEL SYSTEM - CHECKING FOR LEAKS

✳✳ CAUTION

Fuel leaking from a loose, damaged or incorrectly installed hose or fitting may cause a fire or an explosion. ALWAYS pressurize the fuel system and run the motor while inspecting for leaks after servicing any component of the fuel system.

Carbureted engines are equipped only with a low-pressure fuel system, making pressure release before service a non-issue. But, even a low-pressure fuel system should be checked following repairs to make sure that no leaks are present. Only by checking a fuel system under normal operating pressures can you be sure of the system's integrity.

Carbureted engines should be equipped with a fuel primer bulb mounted inline between the fuel tank and engine. The bulb can be used to pressurize that portion of the fuel system. Squeeze the bulb until it and the fuel lines feel firm with gasoline. At this point check all fittings between the tank and motor for signs of leakage and correct, as necessary.

Once fuel reaches the engine it is the job of the fuel pump to distribute it to the carburetors. On pre-mix 2-stroke motors, the fuel is pumped directly from the pump to the carburetor

No matter what system you are inspecting, start and run the motor with the engine top case removed, then check each of the system hoses, fittings and gasket-sealed components to be sure there is no leakage after service.

Fuel System Troubleshooting

The following paragraphs provide an orderly sequence of tests to pinpoint problems in the system. If an engine has not been for some time and fuel has remained in the carburetor, it is possible that varnish may have formed. Such a condition could be the cause of hard starting or complete failure of the engine to operate.

FUEL PROBLEMS

General

◆ **See Figures 2, 3 and 4**

Many times fuel system troubles are caused by a plugged fuel filter, a defective fuel pump, or by a leak in the line from the fuel tank to the fuel pump. Aged fuel left in the carburetor and the formation of varnish could cause the needle to stick in its seat and prevent fuel flow into the bowl. A defective choke may also cause problems. WOULD YOU BELIEVE, a majority of starting troubles, which are traced to the fuel system, are the result of an empty fuel tank or aged fuel.

Fuel will begin to sour in three to four months and will cause engine starting problems. Therefore, leaving the motor sitting idle with fuel in the carburetor, lines, or tank, during the off-season, usually results in very serious problems. A fuel additive such as Sta-Bil or Johnson/Evinrude 2+4 Fuel Conditioner may be used to prevent gum from forming during storage or prolonged idle periods.

For many years there has been the widespread belief that simply disconnecting the fuel line at the engine or at the tank, and then running the engine until it stops is the proper procedure before storing the engine for any length of time. Right? WRONG!

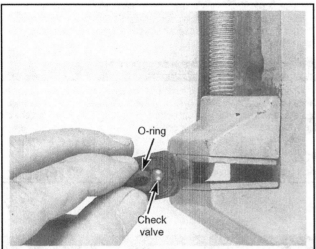

Fig. 2 Fuel connector with the O-ring visible. These O-rings have a short life and must be replaced regularly

Fig. 3 A good look at a fouled spark plug - choke problems can make this happen quicker than you think

Fig. 4 A female line connector ready to be popped into position

First, it is NOT possible to remove all fuel in the carburetor by operating the engine until it stops. Considerable fuel is trapped in the float chamber, other passages, and in the line leading to the carburetor. The ONLY guaranteed method of removing ALL fuel, is to take the time to remove the carburetor and drain the fuel. On all engines using high-speed orifice carburetors, the high-speed orifice plug can be removed to drain fuel from the carburetor.

Secondly, if the engine is operated with the fuel supply disconnect at the "quick-disconnect" until it stops, the fuel and oil mixture inside the engine is removed, leaving bearings, pistons, rings, and other parts without any protective lubricant.

Checking For Stale/Contaminated Fuel

◆ See Figures 5, 6 and 7

Outboard motors often sit weeks at a time making them the perfect candidate for fuel problems. Gasoline has a short life, as combustibles begin evaporating almost immediately. Even when stored properly, fuel starts to deteriorate within a few months, leaving behind a stale fuel mixture that can cause hard-starting, poor engine performance and even lead to possible engine damage.

Furthermore, as gasoline evaporates it leaves behind gum deposits that can clog filters, lines and small passages. Carburetors, due to their tiny passages and naturally vented designs are the most susceptible components.

As mentioned under Alcohol-Blended Fuels, many modern fuels contain alcohol, which is hydroscopic (meaning it absorbs water). And, over time, fuel stored in a partially filled tank or a tank that is vented to the atmosphere will absorb water. The water/alcohol settles to the bottom of the tank, promoting rust (in metal tanks) and leaving a non-combustible mixture at the bottom of a tank that could leave a boater stranded.

One of the first steps to fuel system troubleshooting is to make sure the fuel source is not at fault for engine performance problems. Check the fuel if the engine will not start and there is no ignition problem. Stale or contaminated fuels will often exhibit an unusual or even unpleasant unusual odor.

■ **The best method of disposing stale fuel is through a local waste pickup service, automotive repair facility or marine dealership. But, this can be a hassle. If fuel is not too stale or too badly contaminated, it may be mixed with greater amounts of fresh fuel and used to power lawn/yard equipment or even an automobile (if non-pre-mix fuel is greatly diluted so as to prevent misfiring, unstable idle or damage to the automotive engine). But we feel that it is much less of a risk to have a lawn mower stop running because of the fuel problem than it is to have your boat motor quit or refuse to start.**

Carburetors are normally equipped with a float bowl drain screw that can be used to drain fuel from the carburetor for long-term storage or for inspection. For some motors, it may be easier to drain a fuel sample from the hoses leading to or from the low-pressure fuel filter or fuel pump. Removal and installation instructions for the fuel filters are provided in the Maintenance Section, while fuel pump procedures are found in this section. To check for stale or contaminated fuel:

1. Disconnect the negative battery cable for safety. Secure it or place tape over the end so that it cannot accidentally contact the terminal and complete the circuit.

✷✷ CAUTION

Throughout this procedure, clean up any spilled fuel to prevent a fire hazard.

2. Remove the float bowl drain screw (and orifice plug, if equipped), then allow a small amount of fuel to drain into a glass container.

■ **If there is no fuel present in the carburetor, disconnect the inlet line from the fuel pump and use the fuel primer bulb to obtain a sample.**

3. If necessary, obtain a sample from the low pressure circuit by disconnecting the fuel supply hose from the pump or low pressure fuel filter (as desired), then squeezing the fuel primer bulb to obtain a small sample of fuel. Place the sample in a clear glass container and reconnect the hose.

■ **If a sample cannot be obtained from the fuel filter or pump supply hose, there is a problem with the fuel tank-to-motor fuel circuit. Check the tank, primer bulb, fuel hose, fuel pump, fitting or inlet needle on carbureted models.**

Fig. 5 Carburetor float bowls are normally equipped with a drain screw. . .

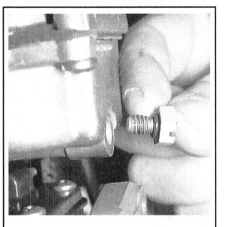

Fig. 6 . . .to drain the carburetor, remove the drain screw

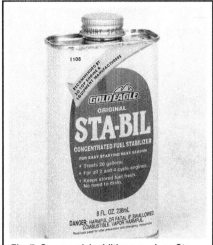

Fig. 7 Commercial additives, such as Sta-bil, may be used to help prevent "souring"

4. Check the appearance and odor of the fuel. An unusual smell, signs of visible debris or a cloudy appearance (or even the obvious presence of water) points to a fuel that should be replaced.

5. If contaminated fuel is found, drain the fuel system and dispose of the fuel in a responsible manner, then clean the entire fuel system.

■ **If debris is found in the fuel system, clean and/or replace all fuel filters.**

6. When finished, reconnect the negative battery cable, then properly pressurize the fuel system and check for leaks.

CHOKE PROBLEMS

◆ **See Figures 8 and 9**

When the engine is hot, the fuel system can cause starting problems. After a hot engine is shut down, the temperature inside the fuel bowl may rise to 200°F and cause the fuel to actually boil. All carburetors are vented to allow this pressure to escape to the atmosphere; however, some of the fuel may percolate over the high-speed nozzle.

If the choke should stick in the open position, the engine will be hard to start. If the choke should stick in the closed position, the engine will flood making it VERY difficult to start.

In order for this raw fuel to vaporize enough to burn, considerable air must be added to lean out the mixture. Therefore, the only remedy is to remove the spark plugs; ground the leads; crank the engine through about 10 revolutions; clean the plugs; install the plugs again; and start the engine.

If the needle valve and seat assembly is leaking, an excessive amount of fuel may enter the intake manifold in the following manner: after the engine is shut down, the pressure left in the fuel line will force fuel past the leaking needle valve. This extra fuel will raise the level in the fuel bowl and cause fuel to overflow into the intake manifold.

A continuous overflow of fuel into the intake manifold may be due to a sticking inlet needle or to a defective float which would cause an extra high level of fuel in the bowl and overflow into the intake manifold.

Procedures to troubleshoot the primer choke system are given later in this section.

FUEL PUMP TESTING

Testing System with Squeeze Bulb

◆ **See Figures 10 thru 13**

② MODERATE

✳✳ WARNING

Gasoline will be flowing in the engine area during this test. Therefore, guard against fire by grounding the high-tension wire to prevent it from sparking.

Fig. 8 Fouled spark plug, possibly caused by over-choking or a malfunctioning enrichment circuit

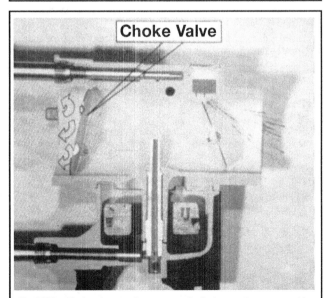

Choke Valve

Fig. 9 The choke plays an important role during engine start and in controlling the amount of air entering the carburetor under various load conditions

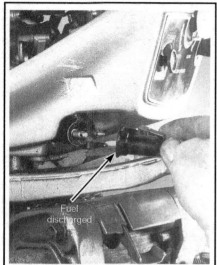

Fig. 10 The fuel line quick-disconnect fitting at the engine separated in preparation of a fuel test

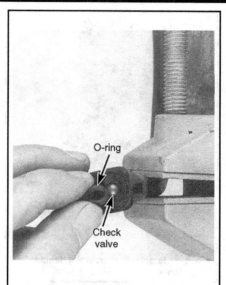

Fig. 11 The O-ring must be replaced on a regular basis

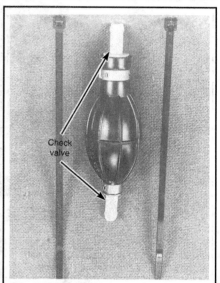

Fig. 12 A good look at a standard squeeze bulb replacement kit

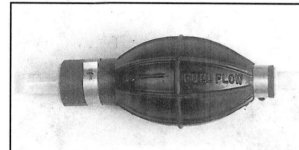

Fig. 13 Always make sure that the directional arrow is pointing in the direction of fuel flow

An adequate safety method is to ground the (or each) spark plug lead. Disconnect the fuel line at the quick-disconnect at the engine. Place a suitable container over the end of the fuel line to catch the fuel discharged. Insert a small screwdriver into the end of the line to open the check valve, and then squeeze the primer bulb and observe if there is satisfactory fuel flow from the line.

If there is no fuel discharged from the line, the check valve in the squeeze bulb may be defective, or there may be a break or obstruction in the fuel line.

If there is a good fuel flow, remove the fuel lines at the carburetors and connect the quick-disconnect at the engine. Crank the engine. If the fuel pump is operating properly, a healthy stream of fuel should pulse out of the line.

Continue cranking the engine and catching the fuel for about 15 pulses to determine if the amount of fuel decreases with each pulse or maintains a constant amount. A decrease in the discharge indicates a restriction in the line. If the fuel line is plugged, the fuel stream may stop. If there is fuel in the fuel tank but no fuel flows out the fuel line while the engine is being cranked, the problem may be in one of several areas:

• Plugged fuel line from the fuel pump to the carburetor.

• Defective O-ring in fuel line connector into the fuel tank.

• Defective O-ring in fuel line connector into the engine.

• Defective fuel pump.

• The line from the fuel tank to the fuel pump may be plugged; the line may be leaking air; or the squeeze bulb may be defective.

• Defective fuel tank.

If the engine does not start even though there is adequate fuel flow from the fuel line, the fuel inlet needle valve and the seat may be gummed together and prevent adequate fuel flow.

FUEL LINE TESTING

◆ See Figures 10 thru 13

On most factory installations covered in this service, the fuel line is provided with quick-disconnect fittings at the tank and at the engine, as produced by the manufacturer. Owners may install a built-in tank with a permanent-type fuel line to the engine. If there is reason to believe the problem is at the quick-disconnects, the hose ends can be replaced as an assembly, or new O-rings may be installed. A supply of new O-rings should be carried on board for use in isolated areas where a marine store is not available.

For a small additional expense, the entire fuel line can be replaced eliminating this entire area as a problem source for many future seasons.

The primer squeeze bulb can be replaced in a short time. A squeeze bulb assembly kit, complete with the check valves installed, may be obtained from the local Johnson/Evinrude dealer. The replacement kit will normally include two tie straps to secure the bulb properly in the line.

An arrow is clearly visible on the squeeze bulb to indicate the direction of fuel flow. The squeeze bulb MUST be installed correctly in the line because the check valves in each end of the bulb will allow fuel to flow in ONLY one direction. Therefore, if the squeeze bulb should be installed backwards, in a moment of haste to get the job done, fuel will not reach the carburetor.

Boats equipped with larger horsepower engines (or even twin motors) usually have built-in fuel tanks. The fuel system is provided with an anti-siphon valve to prevent fuel from being siphoned from the tank in the event the fuel line is broken or disconnected. The valve is mounted on the tank top. It should be removed periodically and checked to verify an adequate fuel flow under normal conditions and no fuel flow when the valve is closed.

ROUGH ENGINE IDLE

◆ See Figures 8 and 14

If an engine does not idle smoothly, the most reasonable approach to the problem is to perform a tune-up to eliminate such areas as: defective ignition parts; faulty spark plugs; and synchronization out of adjustment.

Other problems that can prevent an engine from running smoothly include: an air leak in the intake manifold; uneven compression between the cylinders; and sticky or broken reeds.

Of course any problem in the carburetor affecting the air/fuel mixture will also prevent the engine from operating smoothly at idle speed. These problems usually include: too high a fuel level in the bowl; a heavy float; leaking needle valve and seat; defective automatic choke; or an improper orifice installed.

"Sour" fuel (fuel left in a tank without a preservative additive) will cause an engine to run rough and idle with great difficulty.

If the fuel/oil mixture is too strong on the oil side, the engine will run rough. The only solution to this problem is to drain the fuel and fill the tank with a FRESH ACCURATE mixture.

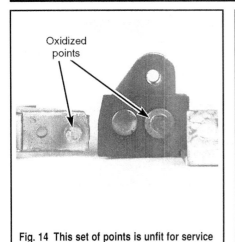

Fig. 14 **This set of points is unfit for service**

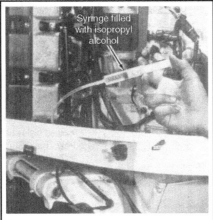

Fig. 15 **Using a syringe to make the test**

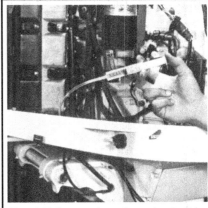

Fig. 16 **Pull out on the syringe when testing the check valve**

DRAIN & PRIMER TESTING

◆ **See Figures 15 and 16**

On certain models a check valve may be installed in the cylinder drain. If this valve is not operating correctly, the engine will operate roughly, will fail to idle properly, and will not achieve maximum rpm at full throttle.

A syringe, a 1/8 in. ID piece of clear plastic hose, a short piece of 3/32 in. OD brass or copper tubing, and some Isopropyl Alcohol are all that is required to make these simple tests. This equipment may also be used to test idle and air bleed passages in carburetors.

Install the plastic tubing onto the syringe and fill them both with alcohol. Remove the primer hose from the nipple on the bypass cover. Install the end of the syringe hose to the fitting. Now, push on the syringe plunger slowly. If bubbles or liquid moves forward, the orifice is open or clean. If the liquid does not move, the orifice is plugged.

1. Disconnect the hose from the intake bypass nipple.
2. Install the syringe tubing to the drain hose. Now, push in slightly on the syringe plunger. If the fluid fails to move, the check valve is closed and operating properly. If fluid flows in the hose, the check valve is stuck open and should be replaced.
3. Pull out on the syringe plunger. Some dark fluid should be pulled through the hose. This action further verifies the check valve is operating properly.
4. If the valve fails either test, it MUST be replaced.
5. After all orifices and check valves have been tested, reconnect the primer hoses.

EXCESSIVE FUEL CONSUMPTION

◆ **See Figures 17 and 18**

Excessive fuel consumption can result from one of three conditions, or a combination of all three:
- Inefficient engine operation.
- Damaged condition of the hull, including excessive marine growth.
- Poor boating habits of the operator.

If the fuel consumption suddenly increases over what could be considered normal, then the cause can probably be attributed to the engine or boat and not the operator.

Marine growth on the hull can have a very marked effect on boat performance. This is why sail boats always try to have a haul-out as close to race time as possible. While you are checking the bottom take note of the propeller condition. A bent blade or other damage will definitely cause poor boat performance.

If the hull and propeller are in good shape, then check the fuel system for possible leaks. Check the line between the fuel pump and the carburetor while the engine is running and the line between the fuel tank and the pump when the engine is not running. A leak between the tank and the pump many times will not appear when the engine is operating, because the suction created by the pump drawing fuel will not allow the fuel to leak. Once the engine is turned off and the suction no longer exists, fuel may begin to leak.

If a minor tune-up has been performed and the spark plugs, ignition parts, and synchronization are properly adjusted, then the problem most likely is in the carburetor, indicating an overhaul is in order. Check for leaks at the

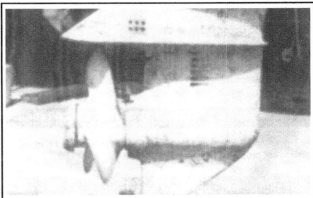

Fig. 17 **Marine growth on the lower unit will create drag and seriously hamper boat performance**

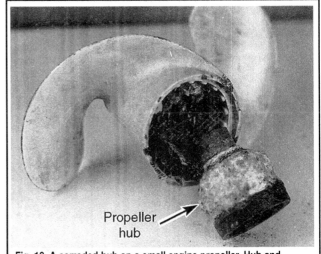

Fig. 18 **A corroded hub on a small engine propeller. Hub and propeller damage will also cause poor performance**

needle valve and seat. Use extra care when making any adjustments affecting the fuel consumption, such as the float level or automatic choke.

ENGINE SURGE

If the engine operates as if the load on the boat is being constantly increased and decreased, even though an attempt is being made to hold a constant engine speed, the problem can most likely be attributed to the fuel pump. The only corrective solution is to replace the pump.

FUEL SYSTEM TROUBLESHOOTING - BY SYMPTOM

Powerhead Fails to Start
- Poor quality or old "sour" fuel.
- Fuel supply restricted to the fuel pump and carburetors.
- Fuel primer valve leaking, flooding the powerhead.
- Out of fuel.

Powerhead Fails to Idle
- Fuel supply restricted to the powerhead.
- Fuel primer valve is leaking.
- Fuel pump is sucking air.
- Throttle linkage misadjusted.
- Carburetors gummed or dirty.
- Recirculation hoses and fittings plugged or disconnected.
- Intake manifold air leaks.
- Reed valves bent or broken.

Powerhead Fails to Idle Slowly
- Fuel supply anti-siphon valve or filter restricted.
- Fuel primer valve is leaking.
- Fuel pump malfunction.
- Throttle linkage misadjusted.
- Internal carburetor leakage.

Powerhead Coughs/Spits at Idle
- Fuel system leaks air.
- Throttle linkage misadjusted.
- Carburetor leaking air.
- Recirculation fittings plugged or hoses misrouted.
- Intake manifold leaking air.

Powerhead Operates Rich at Idle
- Fuel primer valve leaks.
- Carburetor malfunction.
- Recirculation system hoses loose.

Powerhead Stalls During Acceleration
- Restricted fuel supply.
- Fuel pump malfunction.
- Throttle linkage misadjusted.
- Carburetor malfunction.
- Intake manifold leaking air.
- Reed plates bent or broken.

Powerhead Surges at High RPM
- Restricted fuel supply.
- Fuel pump malfunction.
- Carburetors gummed or dirty.
- Air silencer/baffle missing.

Runs Rich at High Speed
- Fuel primer valve leaks.
- Internal carburetor leakage.

Powerhead Fails to Obtain WOT RPM
- Fuel supply restricted to the powerhead.
- Fuel pump malfunction.
- Carburetors gummed or dirty.
- Reed valves bent or broken.

Fuel Spits From Carburetor(s)
- Reed valves bent, broken.
- Manifold seal is leaking air.

FUEL TANK & LINES - PRESSURE TYPE (1958-59)

■ The procedures outlined in this section cover service of original equipment produced by Johnson/Evinrude.

Pressurized Fuel System

◆ See Figures 19, 20 and 21

A pressure-type fuel tank was used with the early model engines (1958) not equipped with a fuel pump. Two hoses were connected between the tank and the engine. One hose served as the fuel transfer line and the other supplied pressurized air to the tank. For the system to operate, the fuel fill cap on the tank must be completely closed, making the tank air-tight.

The system is primed by operating the primer pump located on top of the fuel tank. The primer pump is operated until the carburetor is full of fuel. The engine is then started. Once the engine is operating, even at idle speed, pressurized air is fed from the intake manifold through a check valve and the hose to the fuel tank. As pressure is increased inside the fuel tank, the fuel is forced through the fuel hose to the carburetor.

Each tank should build and retain the following pressures:
- Model CD: 2-4 1/2 lbs.
- Model AD: 2-5 lbs.
- Model QD: 2-5 lbs.
- Model FD: 2-5 1/2 lbs.
- Model RD: 2-4 1/2 lbs.

When the engine is shut down and the fuel hose is disconnected from the engine, the fuel tank cap should be opened slightly to allow the air pressure in the tank to escape. It is not a good practice to allow the tank to remain pressurized.

Any engine equipped with this pressurized fuel system can be updated by installing a fuel pump. The cost of replacing the pressurized-type fuel tank actually exceeds the cost involved in making the conversion. The necessary work involves drilling a hole in the bypass cover to mount the pump and providing the pump with vacuum from one of the cylinders. Installation of new hoses to the carburetor is not a difficult or expensive task.

The accompanying illustration clearly shows the 2 hoses connected to the fuel tank. The hose on the left transfers fuel to the carburetor and the hose on the right is the air pressure line. To assist in identifying the fuel line, the fuel hose has a rib extending its full length from the tank to the carburetor. The primer pump is also clearly shown in the illustration.

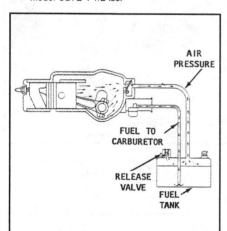

Fig. 19 Functional diagram of an old style pressure type fuel system

Fig. 20 A good look at an old style pressure tank

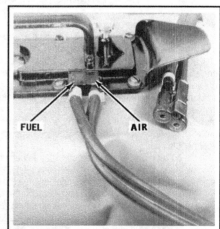

Fig. 21 Fuel and air pressure lines attached to pressure fuel tank at the double connector. The fuel line will be ribbed

DISASSEMBLY & ASSEMBLY

◆ See Figures 22 thru 44

** WARNING

TAKE EXTREME CARE during work with the fuel tank, because highly flammable fumes are present and the danger of fire or explosion is present. Demand and observe NO SMOKING or open flame in the work area. Clean the tank outdoors.

1. Check to be sure the knob on the pump shaft is able to rotate on the shaft. The knob is secured to the shaft with a cotter pin. If the knob does not turn during operation, the diaphragm in the pump will be damaged or ruptured.

■ The following description of pump operation may be helpful in explaining how damage to the diaphragm may be caused by failure of the button to turn.

When a person operates the pump, the action is not in a straight down direction. The natural tendency is to turn the thumb or hand while operating the plunger. The shaft is connected directly to the diaphragm. Therefore, if the knob fails to turn, the shaft will be torn away from the diaphragm.

2. Check to be sure the proper hose is connected to the correct fitting at the carburetor. The hose on the left in the accompanying illustration is the air pressure line and the hose on the right is the fuel supply line. Notice the rib extending the full length of the fuel hose.

3. Remove the fuel tank cap. Disconnect the retaining chain from inside of the fuel cap. The chain may hang free inside the fuel tank.

Fig. 22 A look at the pump knob. . .

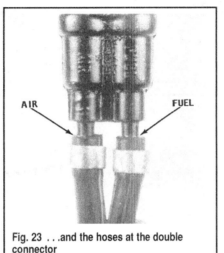

Fig. 23 . . .and the hoses at the double connector

Fig. 24 Remove the fuel tank cap. . .

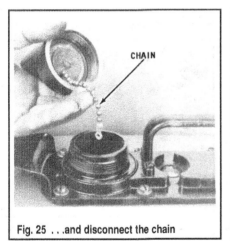

Fig. 25 . . .and disconnect the chain

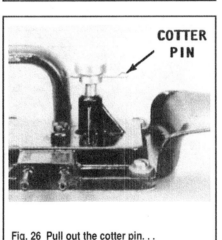

Fig. 26 Pull out the cotter pin. . .

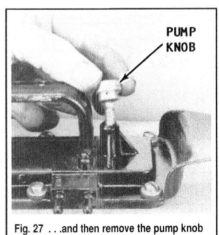

Fig. 27 . . .and then remove the pump knob

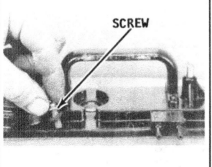

Fig. 28 Remove the retaining screws. . .

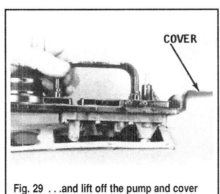

Fig. 29 . . .and lift off the pump and cover

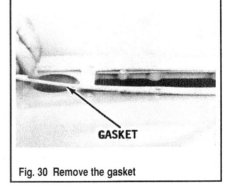

Fig. 30 Remove the gasket

Fig. 31 Secure the pump housing in a vise before removing the pick-up

If the chain cannot be disconnected from the cap by removing the retaining screw, then remove the cotter pin from inside the tank. The chain and cap can then be removed as an assembly.

4. Remove the cotter pin from the push-knob on the fuel pump plunger and then remove the push knob from the pump plunger. If the knob is "frozen" and cannot be removed easily, hold the plunger with a pair of pliers and work the knob free with a second pair of pliers.

5. Remove the retaining screws from around the pump housing. Take care to save the washer on each screw for use during installation.

6. Lift the pump and cover from the fuel tank, as shown.

7. Remove and discard the gasket from the fuel tank. Clean any old or residual gasket material from the fuel tank surface.

8. Secure the pump housing in a vise as shown in the illustration. This position will provide freedom to work for the pump and pick-up assembly. Remove the nut and separate the pick-up from the pump body. The pick-up and screen are sold only as an assembly.

9. Test the operation of the check valve by attempting to suck air up through the tube. The attempt should be successful. Attempt to blow air down through the tube - the attempt should fail. The check valve should allow air to pass up through the tube but prevent the movement of air in the opposite direction. If the check valve fails the tests, clean the assembly in carburetor cleaner and then wash it thoroughly with soap and water to

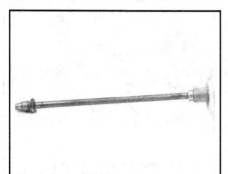

Fig. 32 Test the check valve by sucking air though the tube

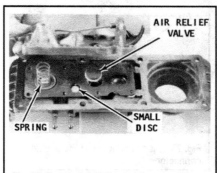

Fig. 33 Be careful of the small parts when removing the pump body

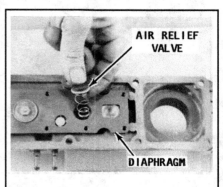

Fig. 34 Lift out the air relief valve. . .

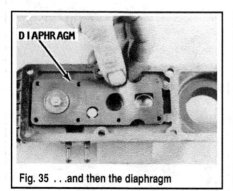

Fig. 35 . . .and then the diaphragm

Fig. 36 Remove the spring. . .

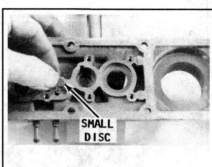

Fig. 37 . . .and small disc

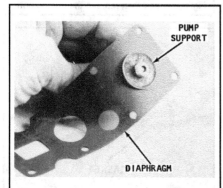

Fig. 38 Now remove the nut and washer from the diaphragm

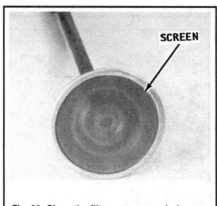

Fig. 39 Clean the filter screen regularly

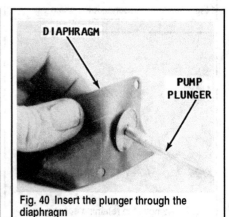

Fig. 40 Insert the plunger through the diaphragm

prevent acid burn to the lips and mouth when the test is repeated. Perform the test a second time. If the unit fails to pass the second test, the entire pick-up assembly MUST be replaced.

10. Remove the 8 screws from the pump body and lift the body free of the pump. Take care not to damage or lose the 2 springs, 2 washers, disc and the diaphragm.

■ Observe the relationship of parts: the spring on top of the diaphragm; the rubber washer and spring to the right, and the small disc in the center, as shown in the accompanying illustration.

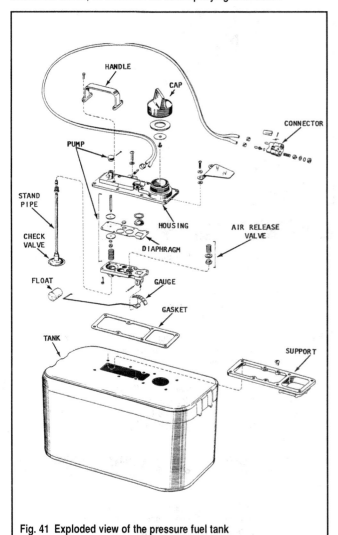

Fig. 41 Exploded view of the pressure fuel tank

11. Remove the spring from the top of the diaphragm. Lift the rubber washer and spring from the right side of the diaphragm.

12. Lift the diaphragm free of the pump body with the pump plunger passing out through the housing.

13. Remove the spring from the housing located under the diaphragm.

14. Remove the small disc from the center of the housing.

15. Remove the small nut and washer from the old diaphragm.

To assemble:

16. Wash the interior of the fuel tank with solvent. Agitate the solvent violently while rapidly changing position of the tank to remove any foreign material in the tank. After the tank has been cleaned with the solvent, rinse it thoroughly with clear water and then dry and remove any moisture from the inside.

17. Slide the large concave washer onto the pump plunger. Insert the pump plunger through the diaphragm.

18. Slide a flat washer onto the plunger on the opposite side of the diaphragm from the concave washer. Thread the nut onto the plunger and tighten it just snugly. To prevent possible damage to the diaphragm, DO NOT over-tighten the nut.

19. Insert the small disc washer into the center hole of the housing.

20. Insert the glass sight gauge in the housing and check to be sure it is properly seated.

21. Install the spring over the boss on the plunger.

22. Work the pump plunger down through the spring and the housing with the diaphragm seated on the housing. Install the relief spring and washer into the housing and hole of the diaphragm.

23. Install the spring over the top of the pump washer. Check to be sure the housing is level to allow the spring to remain in position when the pump housing is installed.

24. Position the pump housing down over the spring and relief valve.

25. Start to thread the 8 screws securing the pump housing to the pump body, and then tighten them evenly and alternately.

26. Place the pick-up check valve assembly in position and then start the compression nut *by hand* to prevent cross-threading. Tighten the nut securely with the proper size wrench.

27. Position a NEW gasket in place around the fuel pump opening on the tank.

28. Lower the complete assembly into the fuel tank. Align the retaining screw holes with the matching holes in the gasket and tank.

29. Secure the pump in place with the attaching screws. Use NEW tiny gaskets on each screw to prevent fuel linkage through the screw holes.

30. Slide the push button onto the pump plunger.

31. Secure the push button in place with a NEW cotter pin. After the pin is in place and the ends have been bent back in the usual manner, clip off the ends of the pin to prevent scratching a finger or hand when the pump is operated. Check to be sure the bottom will rotate in a full circle, without binding.

32. Install the fuel cap and chain. If the chain and cap were removed, use a NEW cotter pin through the flange in the tank to secure the chain in place.

33. Install the fuel cap. If only the cap was removed, attach the chain to the cap with the retaining screw.

34. After the work is completed, add a quantity of fuel/oil mixture to the tank and test the system for proper operation and no leaks.

35. Insert a small screwdriver into the end of the fuel connector at the end of the line to open the check valve. Operate the fuel tank pump and be prepared to catch fuel being discharged from the end of the hose.

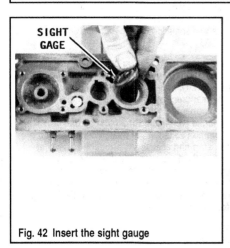

Fig. 42 Insert the sight gauge

Fig. 43 Tighten the housing screws alternately and evenly

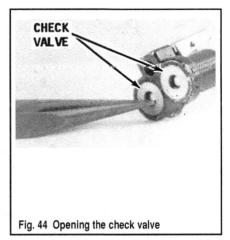

Fig. 44 Opening the check valve

Fuel Line Service

◆ **See Figures 45 and 46**

The only service work to be performed on the fuel lines is replacement of the O-rings in the fuel line connectors, and replacement of the hoses. New O-rings and fuel lines are available at the local OMC dealer.

1. Use 2 ice picks or similar tool, and push down the center plunger of the connector and work the O-ring out of the hole. Repeat the procedure to remove the O-ring from the other check valve.

2. Apply just a drop of oil into the hole of the connector. Apply a thin coating of oil to the surface of the O-ring. Pinch the O-ring together and work it into the hole while simultaneously using a punch to depress the plunger inside the connector.

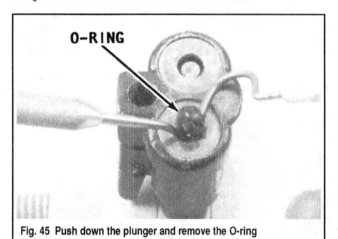

Fig. 45 Push down the plunger and remove the O-ring

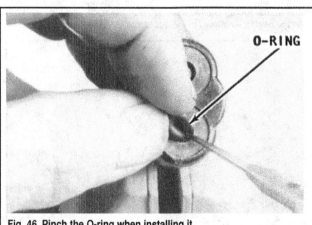

Fig. 46 Pinch the O-ring when installing it

FUEL TANK & LINES - NON-PRESSURE TYPE (1959 & LATER)

General Information

◆ **See Figure 47**

✳✳ CAUTION

If equipped, disconnect the negative battery cable ANYTIME work is performed on the engine, especially when working on the fuel system. This will help prevent the possibility of sparks during service (from accidentally grounding a hot lead or powered component). Sparks could ignite vapors or exposed fuel. Disconnecting the cable on electric start motors will also help prevent the possibility fuel spillage if an attempt is made to crank the engine while the fuel system is open.

✳✳ CAUTION

Fuel leaking from a loose, damaged or incorrectly installed hose or fitting may cause a fire or an explosion. ALWAYS pressurize the fuel system and run the motor while inspecting for leaks after servicing any component of the fuel system.

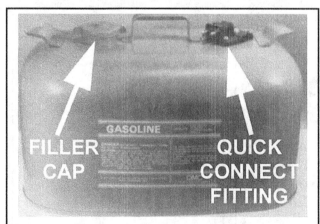

Fig. 47 Although some larger 2-cylinder motors (and most 3 cyl/V4) are not rigged with portable tanks, they are useful to keep around as shop or test tanks

If a problem is suspected in the fuel supply, tank and/or lines, by far the easiest test to eliminate these components as possible culprits is to substitute a known good fuel supply. This is known as running a motor on a test tank (as opposed to running a motor IN a test tank, which is an entirely different concept). If possible, buy or borrow a portable tank, fill it with fresh gasoline (or gas and oil for pre-mix models) and connect it to the motor.

■ When using a test fuel tank, make sure the inside diameter of the fuel hose and fuel fittings is the same size as the lines which are normally rigged to the motor.

■ A 6-gallon shop tank is recommended for the most modern OEM winterization procedure (for details, see the information on Winterization in the Maintenance & Tune-Up section). So, it's probably not a bad idea to obtain a shop or test tank if you need to diagnose fuel system problems.

Fuel Tank

◆ **See Figure 48**

There are 2 different types of fuel tanks that might be used along with these Johnson/Evinrude motors. Some commercial or specially rigged boats might choose to use portable fuel tanks. But, some larger 2-cylinder motors, and almost all 3 cylinder and V4 motors, will use a permanently mounted, integral (boat mounted) fuel tank. In both cases, a tank that is not mounted to the engine itself (as occurs with some tiny, portable outboards) is commonly called a remote tank.

■ Although many Johnson/Evinrude dealers rig boats using Johnson/Evinrude fuel tanks, there are many other tank manufacturers and tank designs may vary greatly. Your outboard might be equipped with a tank from the engine manufacturer or more likely, the boat manufacturer. Although components used, as well as the techniques for cleaning and repairing tanks are similar for almost all fuel tanks, be sure to use caution and common sense. If the design varies from the instructions included here, stop and assess the situation instead of following the instructions blindly. If we reference 2 or 4 screws for something and the component is still tight after removing that many, look for another or for another means of securing the component, don't force it. Refer to a reputable marine repair shop or marine dealership when parts are needed for aftermarket fuel tanks.

Whether or not your boat is equipped with a boat mounted, built-in tank depends mostly on the boat builder and partially on the initial engine installer.

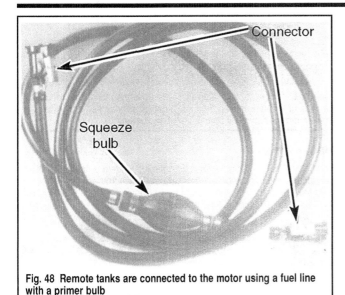

Fig. 48 Remote tanks are connected to the motor using a fuel line with a primer bulb

Boat mounted tanks can be hard to access (sometimes even a little hard to find if parts of the deck must be removed. When dealing with boat mounted tanks, look for access panels (as most manufacturers are smart or kind enough to install them for tough to reach tanks). At the very least, all manufacturers must provide access to fuel line fittings and, usually, the fuel level sender assembly.

No matter what type of tank is used, all must be equipped with a vent (either a manual vent or an automatic one-way check valve) that allows air in (but should prevent vapors from escaping). An inoperable vent (one that is blocked in some fashion) would allow the formation of a vacuum that could prevent the fuel pump from drawing fuel from the tank. A blocked vent could cause fuel starvation problems. Whenever filling the tank, check to make sure air does not rush into the tank each time the cap is loosened (which could be an early warning sign of a blocked vent).

If fuel delivery problems are encountered, first try running the motor with the fuel tank cap removed to ensure that no vacuum lock will occur in the tank or lines due to vent problems. If the motor runs properly with the cap removed but stall, hesitates or misses with the cap installed, you know the problem is with the tank vent system.

SERVICE

Portable Fuel Tanks

◆ See Figures 49 thru 53

Modern fuel tanks are vented to prevent vapor lock of the fuel supply system, but are normally vented by a one-way valve to prevent pollution through the evaporation of vapors. A squeeze bulb is used to prime the system until the powerhead is operating. Once the engine starts, the fuel pump, mounted on the powerhead pulls fuel from the tank and feeds the carburetors. The pick-up unit in the tank is usually sold as a complete unit, but without the gauge and float.

② MODERATE

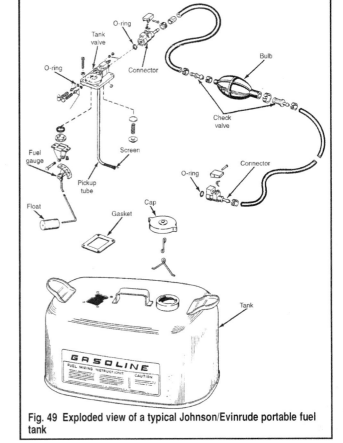

Fig. 49 Exploded view of a typical Johnson/Evinrude portable fuel tank

To disassemble and inspect or replace tank components, proceed as follows:

1. For safety, remove the filler cap and drain the tank into a suitable container.

2. Disconnect the fuel supply line from the tank fitting.

3. To replace the pick-up unit, first remove the screws (normally 4) securing the unit in the tank. Next, lift the pick-up unit up out of the tank.

4. Remove the Phillips screws (usually 2) securing the fuel gauge to the bottom of the pick-up unit and set the gauge aside for installation onto the new pick-up unit.

■ If the pick-up unit is not being replaced, clean and check the screen for damage. It is possible to bend a new piece of screen material around the pick-up and solder it in place without purchasing a complete new unit.

5. If equipped with a level gauge assembly, check for smooth, non-binding movement of the float arm and replace if binding is found. Check the float itself for physical damage or saturation and replace, if found.

6. Check the fuel tank for dirt or moisture contamination. If any is found use a small amount of gasoline or solvent to clean the tank. Pour the solvent

Fig. 50 To service the tank, disconnect the fuel line from the quick-connect fitting. . .

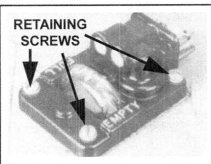

Fig. 51 . . . then remove the screws holding the pick-up and float assembly to the tank

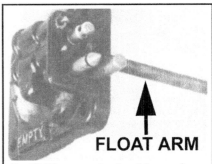

Fig. 52 When removing the pick-up and float, be careful not to damage the float arm

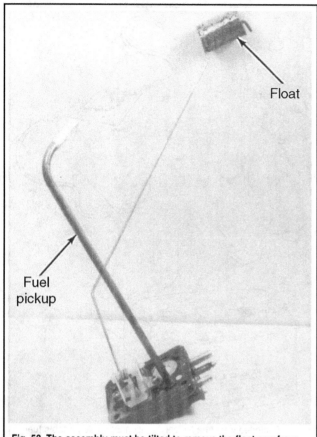

Fig. 53 The assembly must be tilted to remove the float arm from the tank

in and slosh it around to loosen and wash away deposits, then pour out the solvent and recheck. Allow the tank to air dry, or help it along with the use of an air hose from a compressor.

✳✳ CAUTION

Use extreme care when working with solvents or fuel. Remember that both are even more dangerous when their vapors are concentrated in a small area. No source of ignition from flames to sparks can be allowed in the workplace for even an instant.

To Install:

7. Attach the fuel gauge to the new pick-up unit and secure it in place with the Phillips screws.
8. Clean the old gasket material from fuel tank and, if being used, the old pickup unit. Position a new gasket/seal, then work the float arm down through the fuel tank opening, and at the same time the fuel pick-up tube into the tank. It will probably be necessary to exert a little force on the float arm in order to feed it all into the hole. The fuel pick-up arm should spring into place once it is through the hole.
9. Secure the pick-up and float unit in place with the attaching screws.
10. If removed, connect the fuel tank, then pressurize the fuel system and check for leaks.

Boat Mounted Fuel Tanks

The other type of remote fuel tank usually used on many larger models is a boat mounted, built-in tank. Depending on the boat manufacturer, built-in tanks may vary greatly in actual shape/design and access. All should be of a one-way vented to prevent a vacuum lock, but capped to prevent evaporation design.

Most boat manufacturers are kind enough to incorporate some means of access to the tank should fuel lines, fuel pick-up or floats require servicing. But, the means of access will vary greatly from boat-to-boat. Some might

contain simple access panels, while others might require the removal of one or more minor or even major components for access. If you encounter difficulty, seek the advice of a local dealer for that boat builder. The dealer or his/her techs should be able to set you in the right direction.

✳✳ CAUTION

Observe all fuel system cautions, especially when working in recessed portions of a hull. Fuel vapors tend to gather in enclosed areas causing an even more dangerous possibility of explosion.

Anti-Siphon Valve

On many boats with built-in fuel tanks, the fuel system is provided with an anti-siphon valve to prevent fuel from being siphoned from the tank in the event the fuel line is broken or disconnected. The valve is mounted on top of the tank. It should be removed periodically and checked to verify adequate fuel flow under normal conditions and no fuel flow when the valve is closed.

Fuel Lines & Fittings

◆ See Figures 48

In order for an engine to run properly it must receive an uninterrupted and unrestricted flow of fuel. This cannot occur if improper fuel lines are used or if any of the lines/fittings are damaged. Too small a fuel line could cause hesitation or missing at higher engine rpm. Worn or damaged lines or fittings could cause similar problems (also including stalling, poor/rough idle) as air might be drawn into the system instead of fuel. Similarly, a clogged fuel line, fuel filter or dirty fuel pickup or vacuum lock (from a clogged tank vent as mentioned under Fuel Tank) could cause these symptoms by starving the motor for fuel.

If fuel delivery problems are suspected, check the tank first to make sure it is properly vented, then turn your attention to the fuel lines. First check the lines and valves for obvious signs of leakage and then check for collapsed hoses that could cause restrictions.

■ **If there is a restriction between the primer bulb and the fuel tank, vacuum from the fuel pump may cause the primer bulb to collapse. Watch for this sign when troubleshooting fuel delivery problems.**

✳✳ CAUTION

Only use the proper fuel lines containing suitable Coast Guard ratings on a boat. Failure to do so may cause an extremely dangerous condition should fuel lines fail during adverse operating conditions.

TESTING

Fuel Line Quick Check

◆ See Figure 48

Stalling, hesitation, rough idle, misses at high rpm are all possible results of problems with the fuel lines. A quick visual check of the lines for leaks, kinked or collapsed lengths or other obvious damage may uncover the problem. If no obvious cause is found, the problem may be due to a restriction in the line or a problem with the fuel pump.

If a fuel delivery problem due to a restriction or lack of proper fuel flow is suspected, operate the engine while attempting to duplicate the miss or hesitation. While the condition is present, squeeze the primer bulb rapidly to manually pump fuel from the tank to (and through) the fuel pump to the carburetors. If the engine then runs properly while under these conditions, suspect a problem with a clogged restricted fuel line, a clogged fuel filter or a problem with the fuel pump.

Checking Fuel Flow at Motor

◆ See Figures 10, 11 and 48

To perform a more thorough check of the fuel lines and isolate or eliminate the possibility of a restriction, proceed as follows:

1. For safety, disconnect the spark plug leads, then ground each of them to the powerhead to prevent sparks and to protect the ignition system.

2. Disconnect the fuel line from the engine. Place a suitable container over the end of the fuel line to catch the fuel discharged. If using a quick-connector, insert a small screwdriver into the end of the line to hold the valve open.

3. Squeeze the primer bulb and observe if there is satisfactory fuel flow from the line. If there is no fuel discharged from the line, the check valve in the squeeze bulb may be defective, or there may be a break or obstruction in the fuel line.

4. If there is a good fuel flow, reconnect the tank-to-motor fuel supply line and disconnect the fuel line from the carburetors, directing that line into a suitable container. Crank the powerhead. If the fuel pump is operating properly, a healthy stream of fuel should pulse out of the line. If sufficient fuel does not pulse from the line, compare flow at either side of the inline fuel filter (if equipped) or check the fuel pump.

5. Continue cranking the powerhead and catching the fuel for about 15 pulses to determine if the amount of fuel decreases with each pulse or maintains a constant amount. A decrease in the discharge indicates a restriction in the line. If the fuel line is plugged, the fuel stream may stop. If there is fuel in the fuel tank but no fuel flows out the fuel line while the powerhead is being cranked, the problem may be in one of several areas:

a. Plugged fuel line from the fuel pump to the carburetors.

b. Defective O-ring or seal in fuel line connector into the fuel tank.

c. Defective O-ring or seal in fuel line connector into the engine.

d. Defective fuel pump.

e. The line from the fuel tank to the fuel pump may be plugged; the line may be leaking air; or the squeeze bulb may be defective.

f. Defective fuel tank.

6. If the engine does not start even though there is adequate fuel flow from the fuel line, the fuel inlet needle valve and the seat may be gummed together and prevent adequate fuel flow into the float bowl.

Checking the Primer Bulb

◆ See Figures 10, 11 and 48

The way most outboards are rigged, fuel will evaporate from the system during periods of non-use. Also, anytime quick-connect fittings on portable tanks are removed, there is a chance that small amounts of fuel will escape and some air will make it into the fuel lines. For this reason, outboards are normally rigged with some method of priming the fuel system through a hand-operated pump (primer bulb).

When squeezed, the bulb forces fuel from inside the bulb, through the one-way check valve toward the motor filling the carburetor float bowls with the fuel necessary to start the motor. When the bulb is released, the one-way check valve on the opposite end (tank side of the bulb) opens under vacuum to draw fuel from the tank and refill the bulb.

When using the bulb, squeeze it gently as repetitive or forceful pumping may flood the carburetor. The bulb is operating normally if a few squeezes will cause it to become firm, meaning the float bowl/tank is full, and the float valve is closed. If the bulb collapses and does not regain its shape, the bulb must be replaced.

For the bulb to operate properly, both check valves must operate properly and the fuel lines from the check valves back to the tank or forward to the motor must be in good condition (properly sealed). To check the bulb and check valves use hand operated vacuum/pressure pump (available from most marine or automotive parts stores):

1. Remove the fuel hose from the tank and the motor. If equipped, remove the clamps for the quick-connect valves at the ends of the hose and carefully remove the quick-connect valve from the motor side of the fuel line.

■ **Most quick-connect valves are secured to the fuel supply hose using disposable plastic ties that must be cut and discarded for removal. If equipped, spring-type or threaded metal clamps may be reused, but be sure they are in good condition first. Do not over-tighten threaded clamps and crack the valve or cut the hose.**

2. Place the end of the line into the filler opening of the fuel tank. Gently pump the primer bulb to empty the hose into the fuel tank.

■ **Be careful when removing the quick-connect valve from the fuel line as fuel will likely still be present in the hose and will escape (drain or splash) if the valve is jerked from the line. Also, make sure the primer bulb is empty of fuel before proceeding.**

3. If equipped, remove the quick-connect valve from the tank side of the fuel line, draining any residual fuel into the tank.

■ **For proper orientation during testing or installation, the primer bulb is marked with an arrow that faces the engine side check valve.**

4. Securely connect the pressure pump to the hose on the tank side of the primer bulb. Using the pump, slowly apply pressure while listening for air escaping from the end of the hose that connects to the motor. If air escapes, both one-way check valves on the tank side and motor side of the prime bulb are opening.

5. If air escapes prior to the motor end of the hose, hold the bulb, check valve and hose connections under water (in a small bucket or tank). Apply additional air pressure using the pump and watch for escaping bubbles to determine what component or fitting is at fault. Repair the fitting or replace the defective hose/bulb component.

6. If no air escapes, attempt to draw a vacuum form the tank side of the primer bulb. The pump should draw and hold a vacuum without collapsing the primer bulb, indicating that the tank side check valve remained closed.

7. Securely connect the pressure pump to the hose on the motor side of the primer bulb. Using the pump, slowly apply pressure while listening for air escaping from the end of the hose that connects to the motor. This time, the check valve on the tank side of the primer bulb should remain closed, preventing air from escaping or from pressurizing the bulb. If the bulb pressurizes, the motor side check valve is allowing pressure back into the bulb, but the tank side valve is operating properly.

8. Replace the bulb and/or check valves if they operate improperly.

SERVICE

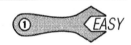

◆ See Figures 11, 12, and 13

Whenever work is performed on the fuel system, check all hoses for wear or damage. Replace hoses that are soft and spongy or ones that are hard and brittle. Fuel hoses should be smooth and free of surface cracks, and they should definitely not have split ends (there's a bad hair joke in there, but we won't sink that low). Do not cut the split ends of a hose and attempt to reuse it, whatever caused the split (most likely time and deterioration) will cause the new end to follow soon. Fuel hoses are safety items, don't scrimp on them, instead, replace them when necessary. If one hose is too old, check the rest, as they are likely also in need of replacement.

■ **When replacing fuel lines, make sure the inside diameter of the fuel hose and fuel fittings is the same size as the ones you are replacing. Also, be certain to use only marine fuel line the meets or exceeds United States Coast Guard (USCG) A1 or B1 guidelines.**

When replacing fuel lines only use Johnson/Evinrude replacement hoses or other marine fuel supply lines that meet United States Coast Guard (USCG) requirements A1 or B1 for marine applications. All lines must be of the same inner diameter as the original to prevent leakage and maintain the proper seal that is necessary for fuel system operation.

■ **Using a smaller fuel hose than specified could cause fuel starvation problems leading to misfiring, hesitation, rough idling and possibly even engine damage.**

The USCG ratings for fuel supply lines have to do with whether or not the lines have been testing regarding length of time it might take for them to succumb to flame (burn through) in an emergency situation. A line is "A" rated if it passes specific requirements regarding burn-through times, while "B" rated lines are not tested in this fashion. The A1 and B1 lines (normally recommended on Johnson/Evinrude applications) are capable of containing liquid fuel at all times. The A2 and B2 rated lines are designed to contain fuel vapor, but not liquid.

✳✳ CAUTION

To help prevent the possibility of significant personal injury or death, do not substitute "B" rated lines when "A" rated lines are required. Similarly, *do not* use "A2" or "B2" lines when "A1" or "A2" lines are specified.

Various styles of fuel line clamps may be found on these motors. Many applications will simply secure lines with plastic wire ties or special plastic locking clamps. Although some of the plastic locking clamps may be released and reconnected, it is usually a good idea to replace them. Obviously wire ties are cut for removal, which requires that they be replaced.

Some applications use metal spring-type clamps that contain tabs which are squeezed, releasing pressure and allowing the clamp to slid up the hose and over the end of the fitting so the hose can be pulled from the fitting. Threaded metal clamps are nice since they are very secure and can be reused, but do not over-tighten threaded clamps as they will start to cut into the hose and they can even damage some fittings underneath the hose. Metal clamps should be replaced anytime they've lost tension (spring type clamps), are corroded, bent or otherwise damaged.

■ **The best way to ensure proper fuel fitting connection is to use the same size and style clamp that was originally installed (unless of course the "original" clamp never worked correctly, but in those cases, someone probably replaced it with the wrong type before you ever saw it).**

To avoid leaks, replace all displaced or disturbed gaskets, O-rings or seals whenever a fuel system component is removed.

On most installations with portable tanks, the fuel line is provided with quick-disconnect fittings at the tank and at the powerhead. If there is reason to believe the problem is at the quick-disconnects, the hose ends can be replaced as an assembly, or new O-rings may be installed. A supply of new O-rings should be carried on board for use in isolated areas where a marine store is not available (like dockside, or worse, should you need one while on the water). For a small additional expense, the entire fuel line can be replaced and eliminate this entire area as a problem source for many future seasons. (If the fuel line is replaced, keep the old one around as a spare, just in case).

CARBURETED FUEL SYSTEMS

General Information

✳✳ CAUTION

If equipped, disconnect the negative battery cable ANYTIME work is performed on the engine, especially when working on the fuel system. This will help prevent the possibility of sparks during service (from accidentally grounding a hot lead or powered component). Sparks could ignite vapors or exposed fuel. Disconnecting the cable on electric start motors will also help prevent the possibility fuel spillage if an attempt is made to crank the engine while the fuel system is open.

✳✳ CAUTION

Fuel leaking from a loose, damaged or incorrectly installed hose or fitting may cause a fire or an explosion. ALWAYS pressurize the fuel system and run the motor while inspecting for leaks after servicing any component of the fuel system.

The carburetion and the ignition principles of engine operation must be understood in order to troubleshoot and repair an outboard motor's fuel system or to perform a proper tune-up on carbureted motors.

Certain newer units may have been equipped with a manual primer system.

Description & Operation

THE ROLE OF A CARBURETOR

◆ **See Figures 54 thru 57**

The carburetor is merely a metering device for mixing fuel and air in the proper proportions for efficient engine operation. At idle speed, an outboard engine requires a mixture of about 8 parts air to 1 part fuel. At high speed or under heavy load, the mixture may change to as much as 12 parts air to 1 part fuel.

Carburetors are wonderful devices that succeed in precise air/fuel mixture ratios based on tiny passages, needle jets or orifices and the variable vacuum that occurs as engine rpm and operating conditions vary.

Because of the tiny passages and small moving parts in a carburetor (and the need for them to work precisely to achieve exact air/fuel mixture ratios) it is important to retain fuel system integrity. Introduction of water (that might

If a quick-connect O-ring must be replaced, use two small punches, picks or similar tools, one to push down the check valve of the connector and the other to work the O-ring out of the hole. Apply just a drop of oil into the hole of the connector. Apply a thin coating of oil to the surface of the O-ring. Pinch the O-ring together and work it into the hole while simultaneously using a punch to depress the check valve inside the connector.

The primer squeeze bulb can be replaced in a short time. A squeeze bulb assembly kit, complete with the check valves installed, may be obtained from your local Johnson/Evinrude dealer. The replacement kit will usually include two tie straps to secure the bulb properly in the line.

An arrow is clearly visible on the squeeze bulb to indicate the direction of fuel flow. The squeeze bulb must be installed correctly in the line because the check valves in each end of the bulb will allow fuel to flow in only one direction. Therefore, if the squeeze bulb should be installed backwards, in a moment of haste to get the job done, fuel will not reach the carburetors.

To replace the bulb, first unsnap the clamps on the hose at each end of the bulb. Next, pull the hose out of the check valves at each end of the bulb. New clamps are included with a new squeeze bulb.

If the fuel line has been exposed to considerable sunlight, it may have become hardened, causing difficulty in working it over the check valve. To remedy this situation, simply immerse the ends of the hose in boiling water for a few minutes to soften the rubber. The hose will then slip onto the check valve without further problems. After the lines on both sides have been installed, snap the clamps in place to secure the line. Check a second time to be sure the arrow is pointing in the fuel flow direction, towards the powerhead.

lead to corrosion), debris (that could clog passages) or even the presence of unstabilized fuel that could evaporate over time can cause big problems for a carburetor. Keep in mind that when fuel evaporates it leaves behind a gummy deposit that can clog those tiny passages, preventing the carburetor (and therefore preventing the engine) from operating properly.

FLOAT SYSTEMS

◆ **See Figure 58**

Ever lift the tank lid off the back of your toilet? Pretty simple stuff once you realize what's going on in there. A supply line keeps the tank full until a valve opens allowing all or some of the liquid in the tank to be drawn out through a passage. The dropping level in the tank causes a float to change position, and, as it lowers in the tank it opens a valve allowing more pressurized liquid back into the tank to raise levels again. OK, we were talking about a toilet right, well yes and no, we're also talking about the float bowl on a carburetor. The carburetor uses a more precise level control, uses vacuum to draw out fuel from the bowl through a metered passage and, most importantly, stores gasoline instead of water, but otherwise, they basically work in the same way.

A small chamber in the carburetor serves as a fuel reservoir. A float valve admits fuel into the reservoir to replace the fuel consumed by the engine.

Fuel level in each chamber is extremely critical and must be maintained accurately. Accuracy is obtained through proper adjustment of the float. This adjustment will provide a balanced metering of fuel to each cylinder at all speeds.

Following the fuel through its course, from the fuel tank to the combustion chamber of the cylinder, will provide an appreciation of exactly what is taking place. In order to start the engine, the fuel must be moved from the tank to the carburetor by a squeeze bulb installed in the fuel line.

The fuel systems for engines covered here are equipped with a manually-operated squeeze bulb in the line to transfer fuel from the tank to the engine until the engine starts.

After the engine starts, the fuel is pulled by the fuel pump and pushed to the carburetor. All systems have some type of filter installed somewhere in the line between the tank and the carburetor.

At the carburetor, the fuel passes through the inlet passage to the needle and seat, and then into the float chamber (reservoir). A float in the chamber rides up and down on the surface of the fuel. After fuel enters the chamber and the level rises to a predetermined point, a tang on the float closes the inlet needle and the fuel flow entering the chamber is cutoff. When fuel leaves the chamber as the engine operates, the fuel level drops and the float tang allows the inlet needle to move off its seat and fuel once again enters

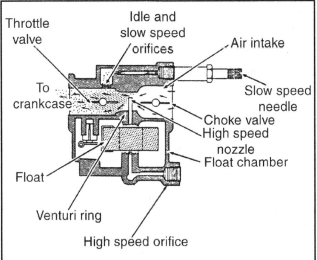

Fig. 54 Fuel flow through the venturi, showing principal and related parts controlling intake and outflow (carburetor with manual choke circuit shown)

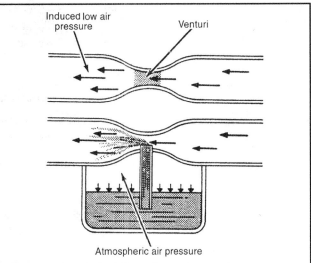

Fig. 55 Air flow principal of a modern carburetor, demonstrates how the low pressure induced behind the venturi draws fuel through the high speed nozzle

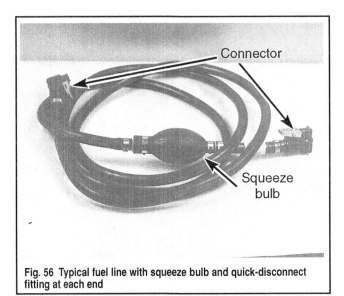

Fig. 56 Typical fuel line with squeeze bulb and quick-disconnect fitting at each end

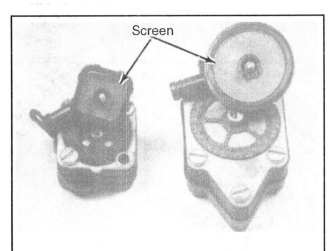

Fig. 57 Two different types of fuel pumps with their covers removed to show their filter screens. These pumps cannot be rebuilt

Fig. 58 A properly adjusted float, notice how the surface of the float is parallel to the surface of the carburetor

the chamber. In this manner a constant reservoir of fuel is maintained in the chamber to satisfy the demands of the engine at all speeds.

A fuel chamber vent hole is located near the top of the carburetor body to permit atmospheric pressure to act against the fuel in each chamber. This pressure assures an adequate fuel supply to the various operating systems of the engine.

AIR/FUEL MIXTURE

◆ See Figures 54 and 55

A suction effect is created each time the piston moves upward in the cylinder. This suction draws air through the throat of the carburetor. A restriction in the throat, called a venturi, controls air velocity and has the effect of reducing air pressure at this point.

The difference in air pressures, at the throat and in the fuel chamber, causes the fuel to be pushed out metering jets extending down into the fuel chamber. When the fuel leaves the jets, it mixes with the air passing through the venturi. This fuel/air mixture should then be in the proper proportion for burning in the cylinders for maximum engine performance (provided that the venturi is of proper size and is not damaged or gummed).

In order to obtain the proper air/fuel mixture for all engine speeds, different high- and low-speed needle valves are installed. On late-model engines, the high-speed needle valve was replaced with a high-speed orifice.

There is no adjustment with the orifice type (except potentially to replace the orifice with one of a different size). These needle valves are used to compensate for changing atmospheric conditions. Only 15% to 20% of the engines covered in this service have an adjustable high- and low-speed needle valve.

Engine operation at sea level compared with performance at high altitudes is quite noticeable. A throttle valve controls the volume of air/fuel mixture drawn into the engine. A cold engine requires a richer fuel mixture to start and during the brief period it is warming to normal operating temperature. On most early-model motors, a choke valve is placed ahead of the metering jets and venturi to provide the extra amount of air required for start and while the engine is cold.

When this choke valve is closed, a very rich fuel mixture is drawn into the engine.

■ Late model motors *may* use an electric or manual primer system to add fuel instead of taking away air, thus enriching the air/fuel mixture for cold start.

The throat of the carburetor is usually referred to as the "barrel." Carburetors installed on engines covered in this service all have a single metering jet with a single throttle and choke plate. Single barrel carburetors are fed by one float and chamber.

Fig. 59 The fuel filter is replaceable on some pumps

CARBURETORS - 1 & 2 CYLINDER MOTORS

This section provides complete detailed procedures for removal, disassembly, cleaning and inspecting, assembling including bench adjustments, installation, and operating adjustments for the four Johnson/Evinrude carburetors installed on engines covered in this service.

There are 4 general types of carburetors, we've designated as Type I, II, III and IV for reference throughout this guide. However, there are some minor variations within each carburetor Type. These variations generally occur by model and year and are listed under Carburetor Installation in this section. Where necessary, differences will be pointed out in procedures.

These carburetors are equipped with either a manual choke, or an electric choke with a manual back-up. The manual back-up permits the operator to operate the choke in the event the battery is dead. The following section lists the types of carburetors, the horsepower and model engines equipped with each.

■ Late-models may be equipped with a manual primer instead of a choke. Otherwise, the carburetors are the same as the early model counterparts.

FUEL SYSTEM

◆ See Figures 59 and 60

The fuel system includes the fuel tank, fuel pump, fuel filters, carburetor, fuel lines with a squeeze bulb, and the associated parts to connect it all together. Regular maintenance of the fuel system to obtain maximum performance is limited to changing the fuel filter at regular intervals and using FRESH fuel. Even with the high price of fuel, removing gasoline that has been standing unused over a long period of time is still the easiest and least expensive preventive maintenance possible.

■ The best method of disposing stale fuel is through a local waste pickup service, automotive repair facility or marine dealership. But, this can be a hassle. If fuel is not too stale or too badly contaminated, it may be mixed with greater amounts of fresh fuel and used to power lawn/yard equipment or even an automobile (if it is NOT pre-mix and if greatly diluted so as to prevent misfiring, unstable idle or damage to the automotive engine). But we feel that it is much less of a risk to have a lawn mower stop running because of the fuel problem than it is to have your boat motor quit or refuse to start.

If a sudden increase in gas consumption is noticed, or if the engine does not perform properly, a carburetor overhaul, including boil-out, or attention to the fuel pump, may be required. All engines covered in this service have a non-serviceable throw away fuel pump.

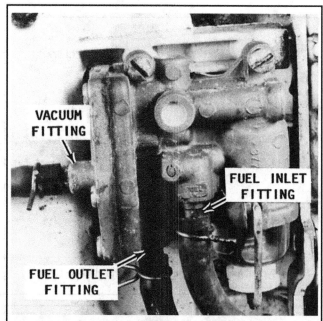

Fig. 60 Some models will utilize a pump similar to this one

Carburetor Applications

For information on fuels, tanks and lines please refer to the section for General Information on Carbureted Fuel Systems, as well as the section on Fuel Tanks & Lines.

The most important fuel system maintenance that a boat owner can perform is to provide and to stabilize fuel supplies before allowing the system to sit idle for any length of time more than a few weeks. The next most important item is to provide the system with fresh gasoline if the system has stood idle for any length of time, especially if it was without fuel system stabilizer during that time.

If a sudden increase in gas consumption is noticed, or if the engine does not perform properly, a carburetor overhaul, including cleaning or replacement of the fuel pump may be required.

TYPE I

Single barrel, front draft, with adjustable low-speed and high-speed needle valves.

- 1.5 hp: 1968-20
- 2 hp: 1971-72
- 3 hp: 1958-62
- 4 hp: 1969-72
- 5 hp: 1965-68
- 5.5 hp: 1959-63
- 10 hp: 1962-63
- 18 hp: 1958-63
- 28 hp: 1962-64

Single barrel, front draft, with adjustable low-speed and high-speed needle valve. Filter bowl attached to carburetor

- 5.5 hp: 1958
- 7.5 hp: 1958
- 10 hp: 1958-61

Single barrel, front draft, with adjustable low-speed needle valve. High-speed is fixed orifice in float bowl.

- 3 hp: 1963-68
- 5.5 hp: 1964
- 6 hp: 1965-72
- 18 hp: 1964-68
- 20 hp: 1966-68

Single barrel, front draft, with adjustable low-speed needle valve with special packing. High-speed is fixed orifice in float bowl.

- 18 hp: 1969-70
- 20 hp: 1969-70
- 25 hp: 1969-70

TYPE II

Single barrel, front draft, with adjustable low-speed needle valve with special packing. High-speed is fixed orifice in the float bowl

- 18 & 20 hp: 1971-72

- 25 hp: 1971-72

Single barrel, front draft, with adjustable low-speed and high-speed needle valves.

- 33 hp: 1965-70
- 35 hp: 1958-59
- 40 hp: 1960, 1971

Single barrel, front draft with adjustable low-speed needle valve. High-speed is fixed orifice in the float bowl.

- 40 hp: 1961-70, 1972

TYPE III

Single barrel, down draft, with adjustable low-speed needle valve. High-speed is fixed orifice in float bowl.

- 9.5 hp: 1964-72

TYPE IV

Single barrel, front draft, with adjustable low-speed needle valve and special packing. High-speed is fixed orifice in the float bowl

- 50 hp: 1971-72

Type I Carburetor

◆ See Figures 61 and 62

There are 4 slightly different versions of the Type I carburetor used on various years and models as follows:
Single barrel, front draft, with adjustable low-speed and high-speed needle valves.

- 1.5 hp: 1968-20
- 2 hp: 1971-72

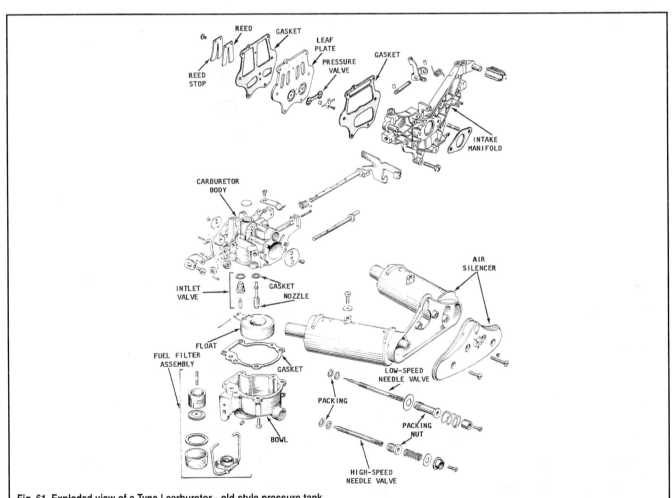

Fig. 61 Exploded view of a Type I carburetor - old style pressure tank

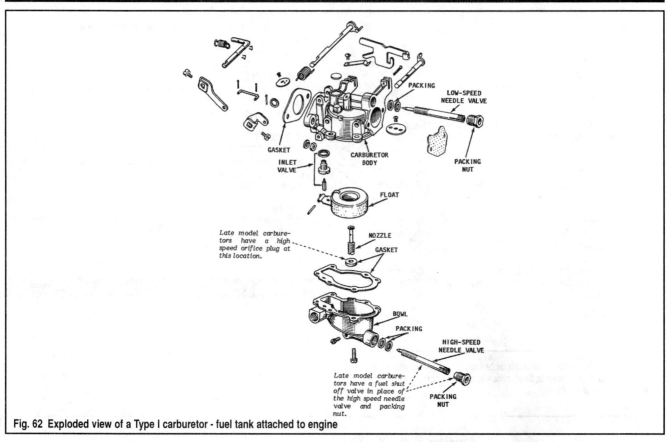

Fig. 62 Exploded view of a Type I carburetor - fuel tank attached to engine

- 3 hp: 1958-62
- 4 hp: 1969-72
- 5 hp: 1965-68
- 5.5 hp: 1959-63
- 10 hp: 1962-63
- 18 hp: 1958-63
- 28 hp: 1962-64

Single barrel, front draft, with adjustable low-speed and high-speed needle valve. Filter bowl attached to carburetor

- 5.5 hp: 1958
- 7.5 hp: 1958
- 10 hp: 1958-61

Single barrel, front draft, with adjustable low-speed needle valve. High-speed is fixed orifice in float bowl.

- 3 hp: 1963-68
- 5.5 hp: 1964
- 6 hp: 1965-72
- 18 hp: 1964-68
- 20 hp: 1966-68

Single barrel, front draft, with adjustable low-speed needle valve with special packing. High-speed is fixed orifice in float bowl.

- 18 hp: 1969-70
- 20 hp: 1969-70
- 25 hp: 1969-70

REMOVAL, INSTALLATION & ADJUSTMENT

◆ **See Figures 61 thru 73**

1. On engines using an electric starter motor, remove the battery leads from the battery terminals.
2. Disconnect the fuel line from the engine or from the fuel tank at the quick-disconnect fitting.
3. Remove the hood assembly from the engine.
4. Remove the choke and throttle linkage to the carburetor.
5. Remove the fuel line from the carburetor. This may be accomplished by either one of 2 methods:

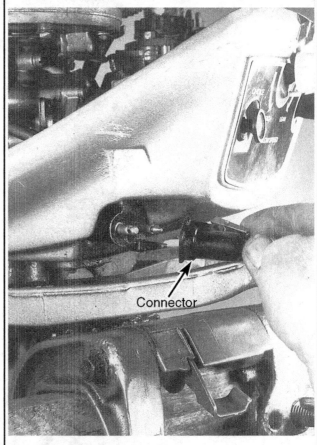

Fig. 63 Disconnect the quick-disconnect

a. On engines equipped with a self-contained fuel tank, close the fuel shut-off valve, located just below the tank. Disconnect both ends, and then remove the copper fuel line between the shut-off valve and the carburetor.

b. On engines utilizing a separate fuel tank, remove the tie-strap or clamp securing the rubber hose connecting the fuel connector to the carburetor. Remove the rubber hose from the carburetor.

6. Remove the nuts securing the carburetor to the crankcase. The carburetor may have to be moved slightly forward as the nuts are loosened in order to obtain clearance for the nuts to clear the studs. After the nuts are clear, lift the carburetor from the intake manifold.

7. Remove and DISCARD the gasket from the intake manifold or the carburetor, if the gasket stuck to the carburetor when it was removed. A new gasket is included in a carburetor repair kit.

To Install:

8. Check to be sure the surface of the intake manifold is clean and free of any old gasket material. Place a NEW gasket in position on the intake manifold.

9. Connect the fuel line to the carburetor; or, on engines equipped with a self- contained fuel tank, replace the line between the shut-off valve and the carburetor.

10. Slide the carburetor over the intake manifold mounting studs, and then secure it in place with the 2 nuts. Tighten the nuts ALTERNATELY to avoid warping the carburetor body.

11. Connect the manual choke rod and the front cover onto the carburetor. Thread the knobs onto the low- and high-speed needle valves.

12. Check the synchronization of the fuel and ignition systems according to the procedures outlined in the Maintenance & Tune-Up section. Mount the engine in a body of water. Connect the fuel line to a fuel source. Prime the engine. If the engine is equipped with a self-contained fuel tank, open the fuel shut-off valve and allow the carburetor to fill with fuel.

⁂ CAUTION

Water must circulate through the lower unit to the engine any time the engine is run to prevent damage to the water pump in the lower unit. Just 5 seconds without water will damage the water pump.

13. Start the engine and allow it to warm to operating temperature. Shift the engine into Forward gear. Adjust the low-speed idle by turning the low-speed needle valve CLOCKWISE until the engine begins to misfire or the rpm drops noticeably. From this point, rotate the needle valve COUNTERCLOCKWISE until the engine is operating at the highest rpm.

14. Advance the throttle to the wide open position (WOT). Adjust the high-speed by rotating the high-speed needle valve CLOCKWISE until the number of rpm begins to drop, then rotate the high-speed valve COUNTERCLOCKWISE until the highest rpm is reached.

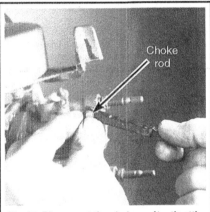

Fig. 64 Disconnect the choke and/or throttle linkage

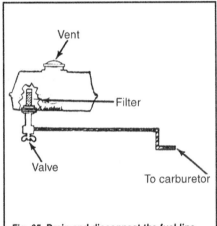

Fig. 65 Drain and disconnect the fuel line

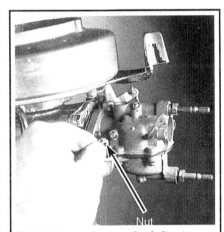

Fig. 66 Remove the mounting bolt nuts

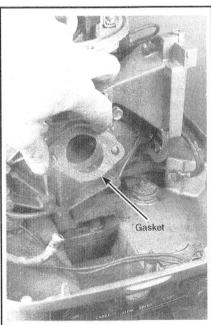

Fig. 67 Make sure you remove the old gasket

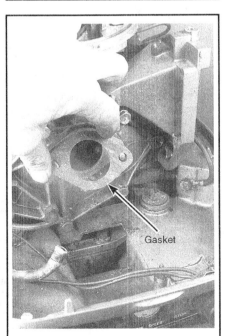

Fig. 68 Position a new gasket on the manifold. . .

Fig. 69 . . .connect the fuel line. . .

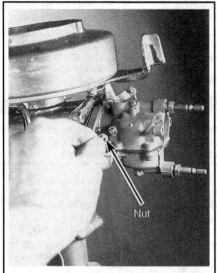

Fig. 70 . . .and then install the carburetor

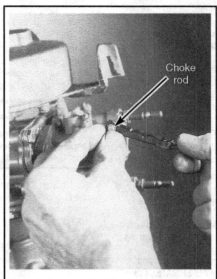

Fig. 71 Connect all of the linkage. . .

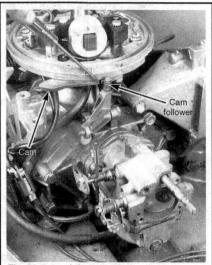

Fig. 72 . . .and then check the synchronization

Fig. 73 The idle stop is almost always on the port side

15. Return the throttle to idle speed. Adjust the idle speed a second time as described earlier. Again, advance the throttle to the WOT position and check the high-speed adjustment. Return the engine to idle speed. If the engine coughs and operates as if the fuel is too lean, but the idle and high-speed adjustments have been correctly made, then re-check the synchronization between the fuel and ignition systems.

16. Now, shut off the fuel supply and allow the engine to run until it first begins to misfire from lack of fuel. Retard the spark and shut the engine down. Tighten the sleeve nut securely to prevent the needle valves from changing position through engine vibration while it is operating, but still allow the needle valves to be adjusted by hand using the knob on the end of the valve.

17. The idle stop is located on the port side of the engine, on the outside of the cowling. Adjust the nylon screw inward or outward to obtain the desired idle speed.

DISASSEMBLY & ASSEMBLY

◆ See Figures 61 thru 100

On early powerheads equipped with this type carburetor, observe how both the low-speed needle valve (the top one), and the high-speed needle

valve (the bottom one), are secured in the carburetor body with a packing sleeve. Use a 7/16 in. box-end wrench and remove each sleeve.

After the sleeves are removed, turn each needle valve counterclockwise until they are free of the carburetor body. If the carburetor being serviced does not have a high-speed needle valve, then it is equipped with a high-speed orifice and covered with a plug in approximately the same location in the carburetor as the high-speed needle valve. Therefore, if there is a plug, instead of the high-speed needle valve, remove the plug.

On later model powerheads (but unlikely any covered here), the high speed orifice is located on the side of the nozzle well as shown in the illustration on the following page.

1. Use a small screwdriver, and remove the packing from the needle valve cavities in the carburetor body.

2. If installed, remove the high-speed orifice, using the proper size screwdriver.

3. Turn the carburetor upside down. If the unit uses a filter bowl assembly (early models), loosen the hinge and swing it aside - remove the glass bowl. Back off the knurled washer and then remove the filter and gasket from inside the bowl.

4. Remove the 5 attaching screws, and then lift the bowl free of the carburetor.

5. Remove the hinge pin and lift the float from the carburetor bowl cavity. As the float is lifted from the carburetor, observe the small spring attached to the needle. The needle will come out with the float assembly. Remove the needle seat from the carburetor.

■ Further disassembly of the carburetor is not necessary. The nozzle in the center of the carburetor does NOT have to be removed in order to properly clean the carburetor. However, the nozzle can be removed with a screwdriver if it is damaged and needs to be replaced.

To assemble:

✱✱ CAUTION

NEVER dip rubber parts, plastic parts, or nylon parts, in carburetor cleaner. These parts should be cleaned ONLY in solvent, and then blown dry with compressed air.

Place all metal parts in a screen-type tray and dip them in carburetor cleaner until they appear completely clean, then wash with solvent or clean water, and blow dry with compressed air.

■ Interestingly, in 1983 Johnson/Evinrude states that carburetor parts should NOT be submerged in carburetor cleaner, as has been the practice since carburetors were invented. Their approved procedure is to place the parts in a shallow tray and then spray them with an aerosol carburetor cleaner. Take your pick, but we've provided conventional procedures here.

6. A syringe, short section of clear plastic hose, and Isopropyl Alcohol should be used to clear passages and jets

7. Blow out all passages in the castings with compressed air. Check all parts and passages to be sure they are not clogged or contain any deposits. NEVER use a piece of wire or any type of pointed instrument to clean drilled passages or calibrated holes in a carburetor.

8. Move the throttle shaft back and forth to check for wear. If the shaft appears to be too loose, replace the complete throttle body because individual replacement parts are NOT normally available.

9. If any part of the float is damaged, the unit must be replaced. Check the float arm needle contacting surface and replace the float if this surface has a groove worn in it.

10. Inspect the tapered section of the idle adjusting needles and replace any that have developed a groove.

11. If a high-speed orifice is installed on the carburetor being serviced, check the orifice for cleanliness. The orifice has a stamped number. This number represents a drill size. Check the orifice with the shank of the proper size drill to verify the proper orifice is used. The correct size orifice for the engine and carburetor being serviced may be obtained from the local Johnson/Evinrude dealer.

Most of the parts that should be replaced during a carburetor overhaul are included in overhaul kits available from your local marine dealer. One of

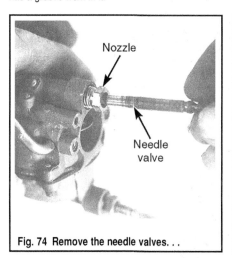

Fig. 74 Remove the needle valves. . .

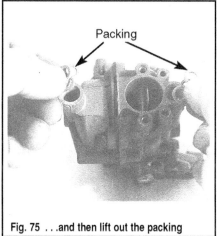

Fig. 75 . . .and then lift out the packing

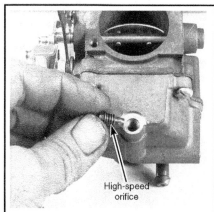

Fig. 76 Remove the high speed orifice if equipped

Fig. 77 On certain early models, remove the float bowl. . .

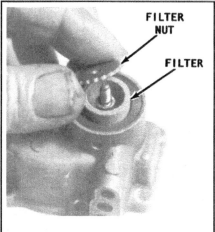

Fig. 78 . . .and lift out the gasket and filter

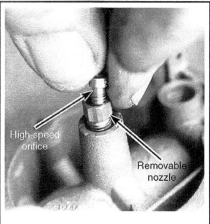

Fig. 79 Many later models have a removable high speed nozzle and orifice

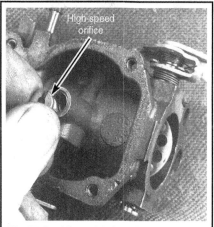

Fig. 80 On this model, the orifice is removable but the nozzle is not

Fig. 81 Lift the float bowl free of the carburetor. . .

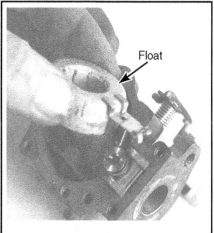

Fig. 82 . . .and then remove the float

these kits will contain a matched fuel inlet needle and seat. This combination should be replaced each time the carburetor is disassembled as a precaution against leakage.

12. Install a NEW inlet nozzle seat with a NEW gasket into the carburetor. Slide a NEW gasket down over the inlet nozzle onto the surface of the carburetor.

13. Attach a NEW inlet needle onto the spring included in the carburetor repair kit, and slip the spring over the edge of the float, as shown. Apply a drop of oil to the inlet needle and then lower the needle into the seat. Install a NEW hinge pin as included in the kit.

14. Hold the carburetor in a horizontal position, as shown, and observe the attitude of the float. The float must be level (parallel) with the surface of the carburetor. If the float is not level, use a pair of needle-nose pliers and *carefully* bend the float tab *squarely* until the float is in the correct position (level with the carburetor). Check to be sure the float is square with the carburetor cavity (one side is not further away than the other).

15. **Measuring Float Drop** Allow the float to drop under its own weight, and then measure the distance between the base of the carburetor body and the lowest edge of the float. Dimension should be as listed.
- 1.25-4.5 hp: 1-1/8 - 1 1/2 in. (28-33mm)
- 5-8 hp: 1 - 1-3/8 in. (25-35mm)
- 9.9-15 hp: 1-1/8 - 1 1/2 in. (28-33mm)

16. Place the bowl gasket in position on the carburetor, and then position the bowl on top of the gasket. Secure the bowl in place with the 5 screws.

17. If you have a unit that used a filter in the float bowl, slide the filter over and down the center post in the bowl. Follow it with the gasket and tighten the nut. Position the bowl and secure it with the bail.

18. Insert 3 new packing washers into the high-speed and low-speed cavities. Place a white plastic washer into the high and low-speed cavities. If a high-speed orifice is used, install the orifice with the proper size screwdriver until the orifice just BARELY seats. Install a new gasket on the orifice plug, and then the plug.

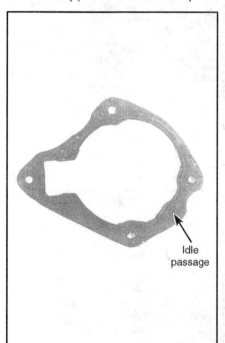

Fig. 83 Make sure the float bowl gasket is positioned correctly

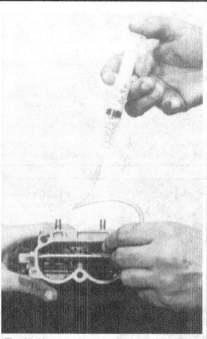

Fig. 84 Use a syringe to clean the idle air bleed system

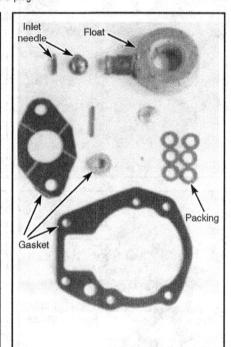

Fig. 85 Here's a standard carburetor rebuild kit

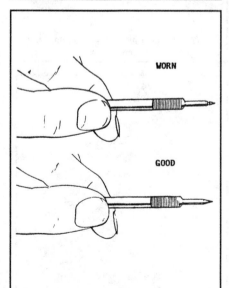

Fig. 86 The top idle adjustment screw is unfit for service, while the bottom one is new

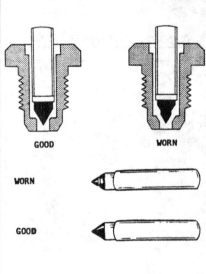

Fig. 87 Cross section of the needle and seat arrangements showing a worn and new needle

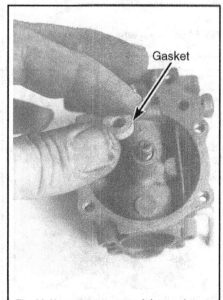

Fig. 88 You must use a new inlet nozzle seat and gasket

19. Start the packing nuts into the carburetor, but ONLY far enough to allow the needle valves to be installed.

■ **Carburetors not equipped with a high speed needle valve have a high speed orifice covered with a plug in approximately the same location as valve.**

20. If the carburetor has a high speed needle valve, install the valve into the bottom cavity by rotating it CLOCKWISE. Allow the needle valve to seat *lightly*, then back it out (COUNTERCLOCKWISE) 3/4 turn. Now, tighten the sleeve nut until it is just difficult to turn the valve by hand.

21. Install the low-speed needle valve in to the upper cavity in the same manner as the high-speed needle valve in the previous step. After the valve is seated *lightly*, back it out (COUNTERCLOCKWISE) 1 1/2 turns. Now, tighten the packing nut until there is drag on the needle valve, but the valve may still be rotated by hand; with just a little difficulty.

22. If the carburetor has a high speed orifice, proceed as follows:

a. Install the high-speed orifice, if one is used, into the float bowl. *Always* take time to use the proper size screwdriver to prevent damaging the orifice. Tighten the orifice only until it just seats.

b. Install the high-speed orifice plug using a NEW gasket. Tighten the plug securely.

Fig. 89 Install the float...

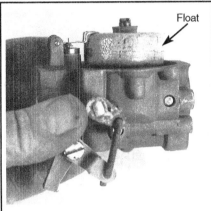

Fig. 90 ...and then make sure it is level with the surface of the carburetor

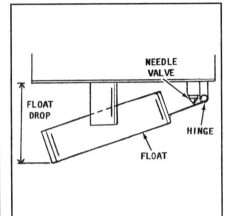

Fig. 91 Measuring float drop on a Type I carburetor

Fig. 92 Install the bowl after adjusting the float drop

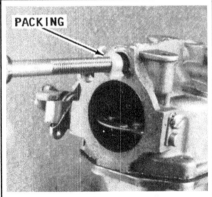

Fig. 93 A new type packing may be used for some low speed needles - if you have this, there will be conventional packing nuts

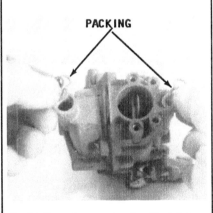

Fig. 94 If not, then go ahead and install the packing into the cavities

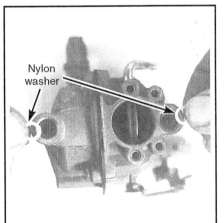

Fig. 95 Don't forget the nylon washers...

Fig. 96 ...or the packing nuts

Fig. 97 Install the high speed needle valve

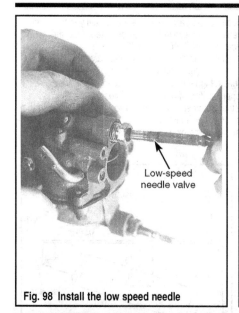

Fig. 98 Install the low speed needle

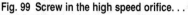

Fig. 99 Screw in the high speed orifice. . .

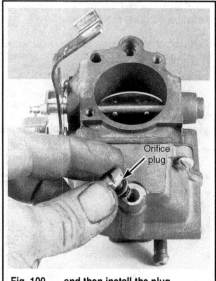

Fig. 100 . . .and then install the plug

Type II Carburetor

◆ **See Figures 101 thru 105**

There are 3 slightly different versions of the Type II carburetor used on various years and models as follows:

Single barrel, front draft, with adjustable low-speed needle valve with special packing. High-speed is fixed orifice in the float bowl
- 18 & 20 hp: 1971-72
- 25 hp: 1971-72

Single barrel, front draft, with adjustable low-speed and high-speed needle valves.
- 33 hp: 1965-70

- 35 hp: 1958-59
- 40 hp: 1960, 1971

Single barrel, front draft with adjustable low-speed needle valve. High-speed is fixed orifice in the float bowl.
- 40 hp: 1961-70, 1972

REMOVAL, INSTALLATION & ADJUSTMENT

◆ **See Figures 101 thru 105**

A number of different choke systems are used on the Type II carburetors covered here. On later models, one system utilizes a manual choke; the other

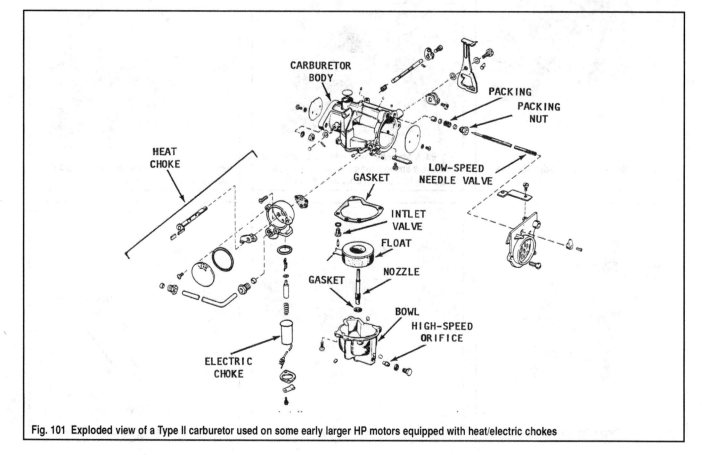

Fig. 101 Exploded view of a Type II carburetor used on some early larger HP motors equipped with heat/electric chokes

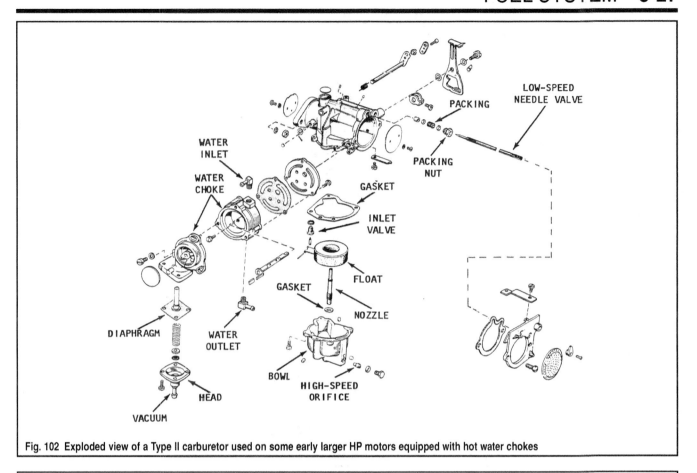

Fig. 102 Exploded view of a Type II carburetor used on some early larger HP motors equipped with hot water chokes

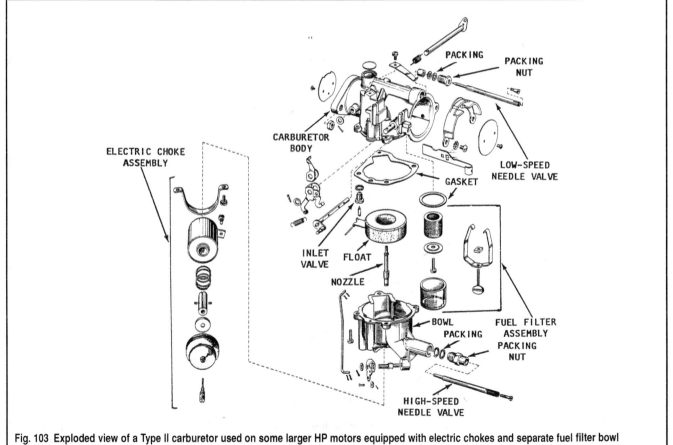

Fig. 103 Exploded view of a Type II carburetor used on some larger HP motors equipped with electric chokes and separate fuel filter bowl

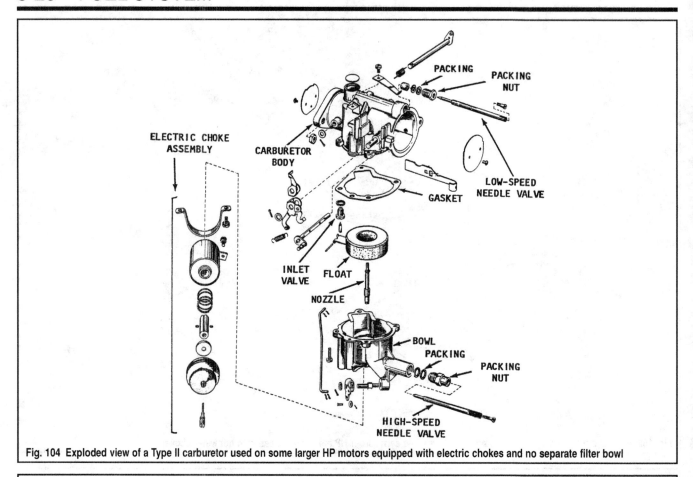

Fig. 104 Exploded view of a Type II carburetor used on some larger HP motors equipped with electric chokes and no separate filter bowl

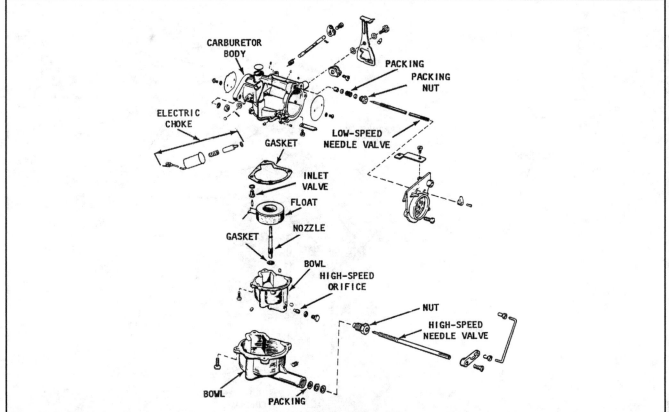

Fig. 105 Exploded view of a later model Type II carburetor with the electric choke mounted on the side instead of the bottom (note the illustration shows 2 float bowls. The upper bowl has a fixed high-speed orifice, while the lower bowl uses an adjustable high-speed needle valve)

system has an electric choke. Earlier models used a heat/electric, manual, electric or a water choke and these systems will be covered separately after the carburetors. None of these choke systems should not be confused with the manual or electric primers also covered later in this section.

The later electric choke system is equipped with a manual override to permit operation of the choke without electrical power (a dead battery). Detailed instructions are outlined in this section for removal of the electric choke mounted on the bottom of the carburetor.

Many larger powerheads covered here (18-40 hp) are equipped with electric starter motors and a generator system. In most cases, these items must be shifted out of the way to provide clearance for the carburetor. Only attaching hardware need be removed in order to shift the item to provide carburetor clearance. Electrical connections may be left intact.

The electric choke is mounted on the bottom side of the carburetor on later models.

1. Disconnect the electrical wire from the choke solenoid.
2. Disconnect the fuel line from the carburetor.
3. Remove the 2 nuts securing the carburetor to the intake manifold.
4. Lift the choke from the engine (if equipped). Scribe a mark on the side of the bracket to ensure the choke solenoid will be installed in its original position.
5. Remove the cotter key, washers, and pin from the choke plunger (if equipped).
6. Remove the 2 screws securing the clamp to the carburetor.
7. Remove the choke solenoid from the carburetor. The choke solenoid *cannot* be serviced. The boot should be in good condition. It may be removed by sliding it off the solenoid. Observe the 2 washers and spring under the boot. Take care not to lose these three items. Clean the solenoid, and then slide the boot back in place. The solenoid is then ready to be installed.
8. Remove the cotter pins from each end of the rod extending from the upper section to the lower section of the carburetor, and remove the rod. This rod works the choke in the upper body of the carburetor when the electric choke is activated. Further disassembly of the all electric choke is not necessary.

To Install:

All late model electric choke Installation
9. Install the rod extending from the upper choke assembly to the lower hinge of the choke. Secure the rod in place with a cotter pin at each end.
10. Insert the choke assembly into the cavity of the float bowl with the electrical connector facing DOWNWARD to permit installation of the electrical wires after the carburetor is installed. Now, bring the clamp over the top of the choke to secure the choke assembly in place.
11. Install the pin through the end of the shaft of the choke solenoid and into the lever attached to the carburetor. Install the washer and cotter pin.
12. Clean the surface of the intake manifold thoroughly. Check to be sure all old gasket material has been removed.

All carburetor installation
13. Place a NEW gasket over the studs and into place on the manifold.
14. Slide the carburetor onto the intake manifold studs and secure it in place with the attaching nuts. Tighten the nuts EVENLY and ALTERNATELY.
15. Connect the fuel line to the carburetor.
16. If equipped, attach the electrical wires to the choke solenoid.
17. Install the starter motor and generator if the engine being serviced is equipped with these two units. Now, proceed make the carburetor adjustments under a load condition, as outlined in the following paragraphs.

■ Under all conditions, the ignition and fuel system *must* be synchronized before the fine adjustments to the carburetor are made. See the information on Timing & Synchronization in the Maintenance & Tune-Up section. After the synchronization has been completed, proceed with the following work.

18. Mount the engine in a test tank or body of water. If this is not possible, connect a flush attachment and garden hose to the lower unit. ONLY the low-speed adjustment may be made using the flush attachment. If the engine is operated above idle speed with no load on the propeller, the engine could RUNAWAY resulting in serious damage or destruction of the unit.

✳✳ CAUTION

Water must circulate through the lower unit to the engine any time the engine is run to prevent damage to the water pump in the lower unit. Just 5 seconds without water will damage the water pump.

19. Start the engine and allow it to warm to operating temperature. Adjust the low-speed idle by turning the low-speed needle valve CLOCKWISE until the engine begins to misfire or the rpm drops noticeably. From this point, rotate the needle valve COUNTERCLOCKWISE until the engine is operating at the highest rpm.
20. If the engine is equipped with an adjustable high-speed needle valve, shift the engine into Forward gear, and then advance the throttle to the wide open position (WOT).

✳✳ WARNING

NEVER attempt to make this adjustment with a flush attachment and garden hose attached to the lower unit.

21. Adjust the high-speed by rotating the high-speed needle valve CLOCKWISE until the number of rpm begins to drop, then rotate the high speed needle valve COUNTERCLOCKWISE until the highest rpm is reached.
22. Return the throttle to idle speed. Adjust the idle speed a second time as described earlier.
23. Again, advance the throttle to the WOT position and check the high-speed adjustment.
24. Return the engine to idle speed. If the engine coughs and operates as if the fuel is too lean, but the idle and high-speed adjustments have been correctly made, then re-check the synchronization between the fuel and ignition systems.
25. Now, shut off the fuel supply and allow the engine to run until it first begins to misfire from lack of fuel. Retard the spark and shut the engine down.
26. Tighten the sleeve nut securely to prevent the needle valves from changing position through engine vibration while it is operating, but still allow the needle valves to be adjusted by hand using the knob on the end of the valve.
27. The idle stop is located on the port side of the engine, on the outside of the cowling. Adjust the nylon screw inward or outward to obtain the desired idle speed.

DISASSEMBLY & ASSEMBLY

◆ **See Figures 86, 87 and 101 thru 130**

The following procedures outline service procedures for the carburetor with an electric choke system installed. If you have a much older model with other kinds of chokes, please skip the choke steps and refer to the specific choke section after this carburetor section.

1. Remove the knobs from the low-speed needle valve by holding the knob firmly and at the same time backing out the retaining screw from the center of the knob. Once the screw is removed, the knob may be pulled free of the needle valve. If the carburetor being serviced has a front shield installed, remove the top screw.
2. Remove the 2 screws on the front side of the carburetor cover. Remove the cover from the front of the carburetor.
3. Remove the small screen and gasket.
4. Loosen the packing nut and rotate the low-speed needle valve counterclockwise to remove it from the carburetor. After the needle valve has been removed, back-out the packing nut.
5. The low-speed needle valve has a removable sleeve. To remove this sleeve, use an OLD needle valve as follows: First screw the old valve into the sleeve. Next, clamp the needle valve in a vise. Now, tap on the carburetor and remove the packing and sleeve. The packing is installed in front of the sleeve.
6. Remove the 7/16 in. nut from the bowl at the bottom of the carburetor. Using the proper size screwdriver, remove the high-speed orifice from the bowl.

■ The following steps are to be performed only if the carburetor being serviced has an adjustable high-speed needle valve.

7. Loosen the packing nut securing the high-speed needle valve to the carburetor bowl. Turn the high-speed needle valve COUNTERCLOCKWISE until it is free and then remove it from the carburetor bowl. Remove the carburetor flange nut from the bowl.

8. With a small screwdriver, reach into the bowl cavity and remove the packing glands from the carburetor.

9. Remove the 4 screws securing the bowl to the carburetor, and then remove the bowl.

10. If your unit is equipped with a fuel filter bowl, turn the locking retainer disc counterclockwise and then flip the bail free of the bowl. Remove the bowl, center screw, washer, filter and gasket from the carb.

11. Remove the bowl gasket from the carburetor.

12. Remove the gasket from the high-speed nozzle.

13. Remove the high-speed nozzle from the carburetor using the proper size screwdriver to prevent possible damage to the nozzle. If difficulty is experienced in removing the high-speed nozzle, leave it in place. When the carburetor is immersed in the carburetor cleaner the high-speed nozzle will be cleaned suitable for further service.

14. Work the hinge pin free of the float and then remove the float.

15. Remove the inlet needle from the seat. Remove the seat and gasket from the carburetor.

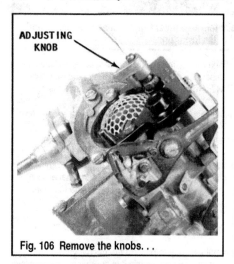

Fig. 106 Remove the knobs. . .

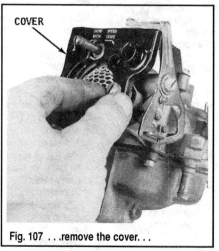

Fig. 107 . . .remove the cover. . .

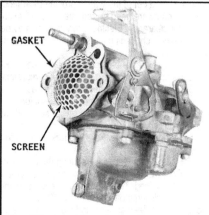

Fig. 108 . . .and then the small screen and gasket

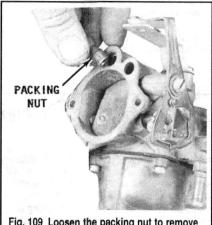
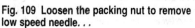

Fig. 109 Loosen the packing nut to remove low speed needle. . .

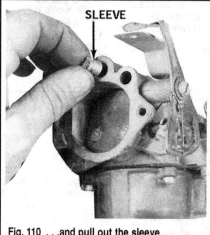

Fig. 110 . . .and pull out the sleeve

Fig. 111 Remove the high speed orifice

Fig. 112 Loosen this packing nut to remove the high speed needle. . .

Fig. 113 . . .and then the packing glands

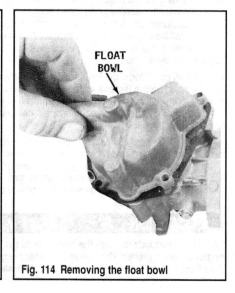

Fig. 114 Removing the float bowl

To assemble:

✳✳ CAUTION

NEVER dip rubber parts, plastic parts, or nylon parts, in carburetor cleaner. These parts should be cleaned ONLY in solvent, and then blown dry with compressed air.

Place all metal parts in a screen-type tray and dip them in carburetor cleaner until they appear completely clean, then wash with solvent or clean water, and blow dry with compressed air.

■ Interestingly, in 1983, Johnson/Evinrude states that carburetor parts should NOT be submerged in carburetor cleaner, as has been the practice since carburetors were invented. Their approved procedure is to place the parts in a shallow tray and then spray them with an aerosol carburetor cleaner. We will give standard procedures, but you be the judge.

16. A syringe, short section of clear plastic hose, and Isopropyl Alcohol should be used to clear passages and jets

17. Blow out all passages in the castings with compressed air. Check all parts and passages to be sure they are not clogged or contain any deposits.

Fig. 115 On models with a glass filter bowl, flip over the retaining bail. . .

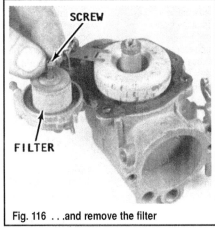

Fig. 116 . . .and remove the filter

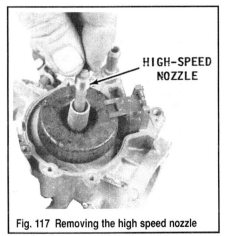

Fig. 117 Removing the high speed nozzle

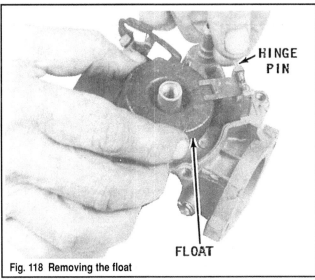

Fig. 118 Removing the float

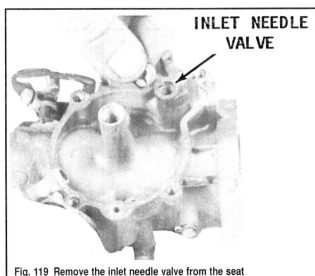

Fig. 119 Remove the inlet needle valve from the seat

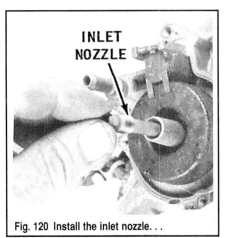

Fig. 120 Install the inlet nozzle. . .

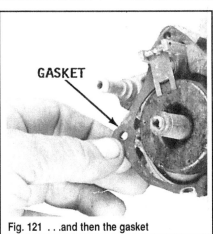

Fig. 121 . . .and then the gasket

Fig. 122 The float must be level

NEVER use a piece of wire or any type of pointed instrument to clean drilled passages or calibrated holes in a carburetor.

18. Move the throttle shaft back and forth to check for wear. If the shaft appears to be too loose, replace the complete throttle body because individual replacement parts are NOT normally available.

19. If any part of the float is damaged, the unit must be replaced. Check the float arm needle contacting surface and replace the float if this surface has a groove worn in it.

20. Inspect the tapered section of the idle adjusting needles and replace any that have developed a groove.

21. If a high-speed orifice is installed on the carburetor being serviced, check the orifice for cleanliness. The orifice has a stamped number. This number represents a drill size. Check the orifice with the shank of the proper size drill to verify the proper orifice is used. The correct size orifice for the engine and carburetor being serviced may be obtained from the local Johnson/Evinrude dealer.

Most of the parts that should be replaced during a carburetor overhaul are included in overhaul kits available from your local marine dealer. One of these kits will contain a matched fuel inlet needle and seat. This combination should be replaced each time the carburetor is disassembled as a precaution against leakage.

Purchase of a carburetor repair kit is almost a must when servicing this unit. All parts required for a complete rebuild, including the necessary gaskets, are contained in the kit. Most repair kits contain more parts and gaskets than are needed because the kit may be used to service a wide range of carburetor models.

22. Install a NEW inlet seat and gasket into the carburetor base. Make sure you use the proper size screwdriver as a precaution against damaging the inside surface of the seat. Apply a drop of oil into the seat to prevent the needle from sticking when it is installed. Insert the inlet needle into the seat.

23. Install a NEW float and hinge pin from the repair kit so the tang of the float faces DOWNWARD toward the inside of the carburetor.

24. If the inlet nozzle was removed, install the nozzle into the center of the carburetor and secure it tightly with the proper size screwdriver. Be careful not to burr the edges of the nozzle. The end of the nozzle seats in the bowl of the carburetor and burrs on the nozzle will result in fuel leakage, or the bowl will not fit properly onto the carburetor. Install a NEW gasket over the nozzle. Force the gasket down onto the carburetor stem.

25. Position a NEW gasket on the carburetor base. *Never* attempt to install a used gasket at this location. As the gasket is used, the low-speed hole has a tendency to shrink slightly and prevent sufficient fuel from passing through. Hence, the need for installation of a new gasket each time the carburetor is serviced.

26. Hold the carburetor in a horizontal position, as shown, and observe the attitude of the float - the float must be level (parallel) with the surface of the carburetor. If the float is not level, use a pair of needle-nose pliers and *carefully* bend the float tab SQUARELY until the float is in the correct position (level with the carburetor). Check to be sure the float is square with the carburetor cavity (one side is not further away than the other).

27. **Measuring Float Drop:** Allow the float to drop under its own weight. Measure the distance between the base of the carburetor body and the lowest edge of the float. Acceptable distance:

- 20, 25 & 30 hp: 1-1/8 - 1-1/2 in. (28-33mm).
- 40 hp: 1-1/8 - 1-5/8 in. (28-41mm).

28. Lower the bowl onto the carburetor and secure it with the 4 screws.

29. Install the bushing into the low-speed needle valve opening.

30. Insert the 2 fiber washers and the 1 nylon washer into the low-speed cavity.

31. Start the low-speed needle valve packing nut into the opening. DO NOT tighten the nut at this time.

32. Thread the low-speed needle valve into the low-speed opening. Continue threading it into the opening until it barely seats. After the needle valve seats, back it out 1 1/2 turns COUNTERCLOCKWISE.

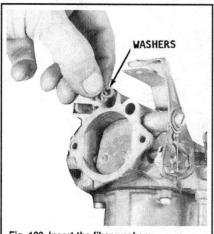

Fig. 123 Insert the fiber washers. . .

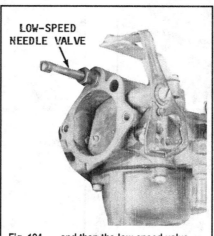

Fig. 124 . . .and then the low speed valve

Fig. 125 This is the new style packing

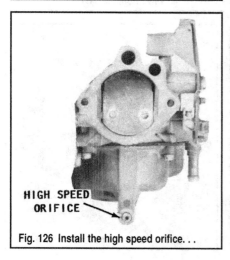

Fig. 126 Install the high speed orifice. . .

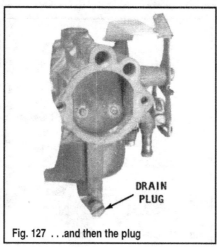

Fig. 127 . . .and then the plug

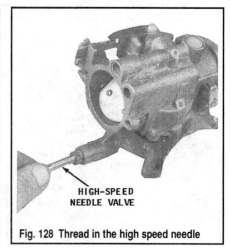

Fig. 128 Thread in the high speed needle

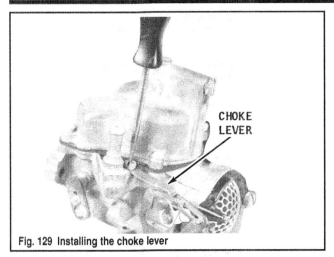

Fig. 129 Installing the choke lever

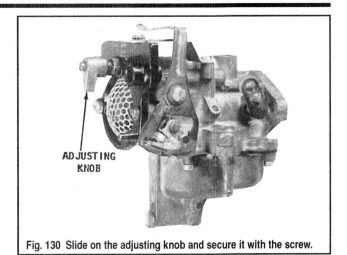

Fig. 130 Slide on the adjusting knob and secure it with the screw.

33. Now, tighten the packing nut until there is drag on the needle valve, but the valve may still be rotated by hand with just a little difficulty.

34. Install the high-speed orifice into the carburetor bowl using the proper size screwdriver.

35. Install a NEW drain plug gasket, and then thread the drain plug into the carburetor bowl cavity.

■ **The following step is to be performed if the carburetor being serviced has an adjustable high-speed needle valve.**

36. Install the 3 packing washers into the carburetor bowl.

37. Start the high-speed needle valve packing nut into the high-speed opening. DO NOT tighten the nut at this time.

38. Thread the high-speed needle valve into the high-speed opening. Continue to thread it into place until it BARELY seats. After the needle valve seats, back it out 3/4 turn COUNTERCLOCKWISE. Now, tighten the packing nut until there is drag on the needle valve, but the valve may still be rotated by hand, but with just a little difficulty.

39. Install a NEW gasket and screen onto the front of the carburetor.

40. If a front cover is used on the unit being serviced, install the front cover and secure it in place with the two attaching screws.

41. Install the spring steel lever into the indent of the manual choke (if equipped) on the bottom side of the carburetor. This lever holds the choke in the neutral position and prevents loose movement of the handle.

42. Install the low-speed needle valve (adjusting) knob by sliding it onto the valve stem, and then secure it in place with the screw.

Type III Carburetor

◆ See Figure 131

This carburetor is installed on the 1964-72 9.5 hp engine. The unit has a single barrel, down draft, with an adjustable low-speed needle valve and fixed high-speed orifice.

The only changes that have been made to this particular model carburetor over the years were a redesign of the idle needle valves. On early model engines, the reeds set directly below the carburetor. On later models, the reeds were removed from this location and installed between the engine block and the intake manifold.

The early model idle needle valves extended out of the carburetor with an O-ring, spring, and E-clip. The needle valve was controlled from the front of the engine by means of a flexible cable. On this model carburetor, the cable and control knob must be removed before the carburetor is removed from the engine.

Later model carburetors are equipped with a long packing nut and linkage. An adjustable knob located on the front of the engine controls movement of the valve. When servicing the late model carburetors, the low-speed needle valve adjustment knob and linkage must be removed before the carburetor is removed from the engine.

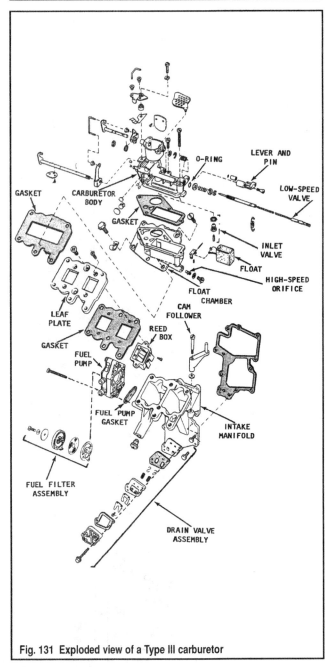

Fig. 131 Exploded view of a Type III carburetor

REMOVAL & INSTALLATION

◆ See Figures 131 thru 140

1. Remove the choke rod that extends over the top of the carburetor by snapping the choke rod out of the nylon snap.

2. Disconnect the fuel line from the carburetor.

3. If the carburetor has a packing nut with the needle through the nut and a nylon adjustment knob, remove the knob or remove the linkage from the knob.

4. If the carburetor has a flexible line to the front of the engine, remove the knob on the front of the engine and then turn the flexible cable COUNTERCLOCKWISE until the needle valve is removed from the carburetor.

■ **As the needle valve is being removed, take care to retain the washer, O-ring, and spring installed between the E-ring on the valve and the carburetor. The washer and spring will be used again. The O-ring may be discarded.**

5. Remove the 5 screws securing the carburetor to the intake manifold. Notice that four of the screws have slots and one is a countersunk screw.

6. Lift the carburetor from the engine.

To Install:

■ **If the carburetor being serviced has packing nuts and washers for the needle valves, perform first step. If the carburetor has the flexible line extending from the valve to the front of the engine, the needle valve will be installed *after* the carburetor is in place on the engine.**

7. Install the packing nut washers into the needle valve openings. Thread the packing nut into the opening but DO NOT tighten them at this time. Thread the low-speed needle valve into place until it just *barely* seats. From this position, back it out (COUNTERCLOCKWISE) 1 1/2 turns as a preliminary rough adjustment.

8. Check the surface of the intake manifold to be sure it has been thoroughly cleaned and is free of any old gasket material. Place a NEW gasket in position on the manifold.

9. Set the carburetor into place on the manifold and secure it with the 5 attaching screws. Note that one of the screws is a countersunk type. This screw MUST be installed into the countersunk hole.

10. If the carburetor being serviced has the flexible low-speed needle valve arrangement, check to be sure the snapring is in place and then install the spring, washer and NEW O-ring onto the needle. Apply just a drop of oil onto the O-ring to ease installation of the needle valve. Thread the low-speed needle valve into the carburetor until it just BARELY seats. From this position, back it out (COUNTERCLOCKWISE) 1 1/2 turns as a preliminary rough adjustment.

11. Install the choke rod by snapping it into place in the nylon retainer.

12. Connect the fuel line to the carburetor.

■ **It is best to synchronize the fuel and ignition systems at this time. For details, please refer to the Timing & Synchronization procedures in the Maintenance & Tune-Up section. After the synchronization has been completed, proceed with the following work.**

13. Mount the engine in a test tank or body of water. If this is not possible, connect a flush attachment and garden hose to the lower unit.

✳✳ WARNING

NEVER operate the engine above idle speed using the flush attachment. If the engine is operated above idle speed with no load on the propeller, the engine could RUNAWAY resulting in serious damage or destruction of the unit.

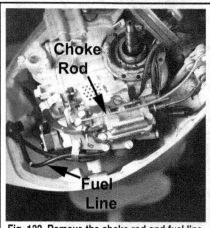

Fig. 132 Remove the choke rod and fuel line

Fig. 133 Remove the packing nut

Fig. 134 Remove the flexible cable

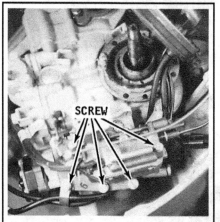

Fig. 135 Remove the mounting screws

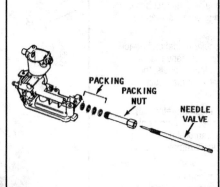

Fig. 136 Install the packing nut washers into the needle valve openings, then thread but do NOT tighten the packing nut

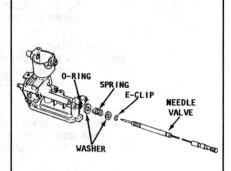

Fig. 137 Models with a flexible low-speed needle use a spring, washer and C-clip arrangement

Fig. 138 Mount the engine in a test tank

Fig. 139 Idle adjustment linkage

Fig. 140 Flexible idle cable set-up

✳✳ WARNING

Water must circulate through the lower unit to the engine any time the engine is run to prevent damage to the water pump in the lower unit. Just five seconds without water will damage the water pump.

14. Start the engine and allow it to warm to operating temperature.

15. Adjust the low-speed idle by turning the low-speed needle valve CLOCKWISE until the engine begins to misfire or the rpm drops noticeably. From this point, rotate the needle valve COUNTERCLOCKWISE until the engine is operating at the highest rpm. If the engine coughs and operates as if the fuel is too lean, but the idle and high-speed adjustments have been correctly made, then re-check the synchronization between the fuel and ignition systems

DISASSEMBLY & ASSEMBLY

◆ See Figures 131 and 141 thru 153   ③ DIFFICULT

1. Unscrew the 4 screws and then remove the float bowl.

2. Remove and DISCARD the float bowl gasket.

3. Remove the hinge pin and then lift the float assembly from the carburetor body.

4. Reach inside the inlet seat and remove the inlet needle. Remove the inlet needle seat.

5. Remove the drain plug from the bottom of the float bowl. Use the proper size screwdriver and remove the high-speed orifice from the float bowl.

6. Loosen the low-speed needle valve packing nut by turning it COUNTERCLOCKWISE and remove the low-speed needle valve. Remove the packing nut and washers.

To assemble:

✳✳ CAUTION

NEVER dip rubber parts, plastic parts, or nylon parts, in carburetor cleaner. These parts should be cleaned ONLY in solvent, and then blown dry with compressed air.

Place all metal parts in a screen-type tray and dip them in carburetor cleaner until they appear completely clean, then wash with solvent or clean water, and blow dry with compressed air.

■ Interestingly, in 1983 Johnson/Evinrude states that carburetor parts should NOT be submerged in carburetor cleaner, as has been the practice since carburetors were invented. Their approved procedure is to place the parts in a shallow tray and then spray them with an aerosol carburetor cleaner. You be the judge.

7. Blow out all passages in the castings with compressed air. Check all parts and passages to be sure they are not clogged or contain any deposits. NEVER use a piece of wire or any type of pointed instrument to clean drilled passages or calibrated holes in a carburetor.

8. Move the throttle shaft back and forth to check for wear. If the shaft appears to be too loose, replace the complete throttle body because individual replacement parts are not normally available.

Fig. 141 Remove the float bowl. . .

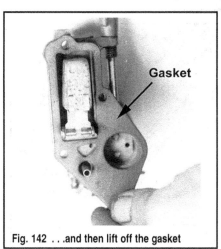

Fig. 142 . . .and then lift off the gasket

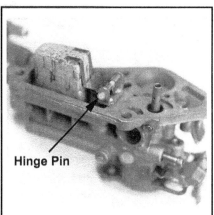

Fig. 143 Remove the hinge pin and lift out the float

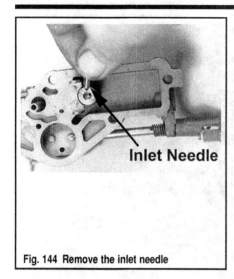

Fig. 144 Remove the inlet needle

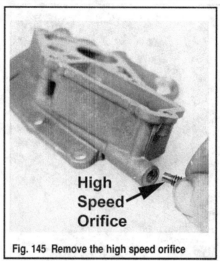

Fig. 145 Remove the high speed orifice

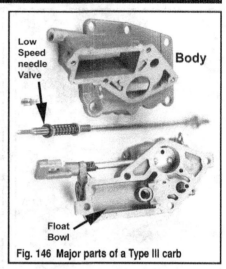

Fig. 146 Major parts of a Type III carb

9. Inspect the main body, air horn, and venturi cluster gasket surfaces for cracks and burrs which might cause a leak.

10. If any part of the float is damaged, the unit must be replaced. Check the float for deterioration. Check to be sure the float spring has not been stretched. If any part of the float is damaged, the unit must be replaced. Check the float arm needle contacting surface and replace the float if this surface has a groove worn in it.

11. Inspect the tapered section of the idle adjusting needles and replace any that have developed a groove.

12. If a high-speed orifice is installed on the carburetor being serviced, check the orifice for cleanliness. The orifice has a stamped number. This number represents a drill size. Check the orifice with the shank of the proper size drill to verify the proper orifice is used. The correct size orifice for the

engine and carburetor being serviced may be obtained from the local Johnson/Evinrude dealer.

Most of the parts that should be replaced during a carburetor overhaul are included in overhaul kits available from your local marine dealer. One of these kits will contain a matched fuel inlet needle and seat. This combination should be replaced each time the carburetor is disassembled as a precaution against leakage.

13. Install the high-speed orifice into the float body using the proper size screwdriver to prevent burring the edges of the orifice. Any damage to the orifice will result fuel leakage and poor engine performance.

14. Install the drain plug with a NEW gasket and tighten it securely with a 7/16 in. wrench.

15. Install the inlet needle seat and gasket, being sure to use the proper size screwdriver to install the seat. If the inside diameter of the seat is damaged the needle valve will leak fuel causing a flooding condition in the carburetor.

16. Apply a drop of oil into the seat, and then insert the inlet needle into seat.

17. Position a NEW float over the needle, and then slide a NEW hinge pin into place.

18. Hold the carburetor in a vertical position (up-and-down), and observe the float. The float MUST be parallel (align evenly) with the surface of the carburetor body. If the float is not parallel, *carefully* bend the float ever so slightly, as shown, until the correct positioning is obtained.

19. Slide the float bowl gasket over the float and nozzle into position on the carburetor base.

20. Lower the float bowl down over the top of the carburetor body and secure it in place with the 4 attaching screws.

21. If your unit uses packing nuts and washers for the needle valves, move to the next step. If the carb uses the flexible line from the valve to the front of the engine, the valve should be installed *after* the carb has been installed on the engine.

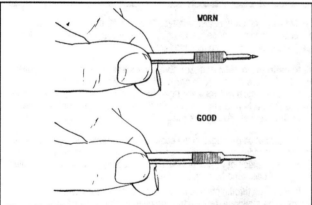

Fig. 147 Comparing a worn and new carburetor adjustment screw. The upper one is unfit for service

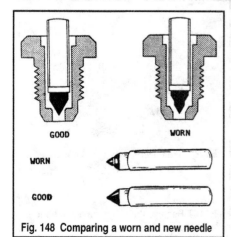

Fig. 148 Comparing a worn and new needle

Fig. 149 A good look at the carburetor overhaul kit

Fig. 150 Install the high speed orifice. . .

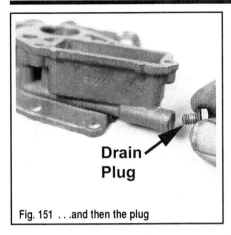

Fig. 151 . . .and then the plug

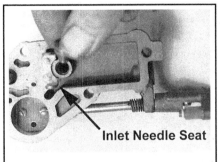
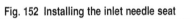

Fig. 152 Installing the inlet needle seat

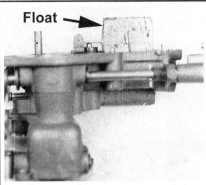

Fig. 153 The float must be parallel with the carburetor surface

22. Position the packing nut washers into the needle valve openings and then thread the nut in without tightening them fully yet.

23. Thread in the low speed valve until it just begins to seat itself. Now, back it out (counterclockwise) 1 1/2 turns for the preliminary adjustment.

24. Confirm that the intake manifold mounting surface is clean and free of any older gasket residue. Position a new gasket and then install the carburetor and tighten the 5 mounting screws; making certain that the countersunk screw goes in the correct hole.

25. If you unit uses the flexible hose mentioned previously, make sure the snap-ring is in position and then install the spring, washer and a NEW O-ring onto the needle. Use a drop of oil on the O-ring to help with installation.

26. Thread in the low speed valve until it just begins to seat itself. Now, back it out (counterclockwise) 1 1/2 turns for the preliminary adjustment.

27. Install the choke rod by snapping it into position in the nylon retainer.

28. Connect the fuel line to the carburetor.

Type IV Carburetor

◆ See Figure 154

The 1971-72 50 hp (2 cyl) motor uses a single barrel, front draft carburetor, with an adjustable low-speed needle valve and special packing. High-speed is fixed orifice in the float bowl

Two carburetors are installed on each powerhead, one for each cylinder.

All of these carburetors utilize an electric choke mounted on the side of the engine. On most models, the electric choke does not have to be removed in order to remove and service the carburetor.

Fig. 154 Starboard side view showing the throttle and choke linkage

REMOVAL, INSTALLATION & ADJUSTMENT

◆ See Figures 154 thru 160

 MODERATE

1. Disconnect the battery cables at the battery as a precaution against an accidental spark igniting the fuel or fuel fumes present during the service work.

2. Remove the cowling.

3. Disconnect the fuel line from the junction on the port side of the engine. This is the line from the fuel pump to the carburetors.

4. Remove the low-speed adjustment knobs at the bottom of the air silencer and remove the silencer cover.

5. Disconnect the hose at the bottom of the air silencer. This hose is connected to a fitting at the bottom of the crankcase, and then the air silencer.

6. Disconnect the choke and throttle linkage between the two carburetors on the port side of the engine. The linkage will snap out of a retainer on the carburetor cams. The nylon retainers MUST be removed before the carburetor is immersed in any type of cleaning solution. Disconnect the choke wire from the electric choke, by first removing the O-ring and coil spring.

7. Remove the 4 nuts securing the carburetor to the intake manifold. Identify the top carburetor as an aid to installation. The carburetors MUST be installed in their original positions. The fuel connections and the choke arrangement are different for each carburetor. Both carburetors can now be removed as an assembly.

■ The carburetor parts should be kept with the individual unit to ensure all items are installed back in their original positions.

To Install:

8. Connect the fuel line between the top and bottom carburetor.

9. Clean the mating surface of the intake manifold. Check to be sure all old gasket material has been removed.

10. Slide a NEW gasket down over the studs into place on the manifold.

11. Install both carburetors at the same time onto the studs and secure them in place with the nuts. Tighten the nuts ALTERNATELY and EVENLY.

12. Connect the fuel line between the fuel pump and the carburetors.

13. On the starboard side of the engine, as applicable: Install the 4 choke and throttle retainers, 2 for each carburetor. Connect the choke and throttle linkage between the 2 carburetors. Connect the choke coil solenoid wire onto the linkage stud, and then slide the O-ring and coil spring onto the stud to secure the wire in place. Adjust the choke butterflies by loosening the screw between the top and bottom linkage. Make the adjustment to close BOTH choke butterflies, then tighten the screw.

14. Adjust the throttle butterflies in the same manner. Loosen the screw between the top and bottom linkage. Make the adjustment to close BOTH throttle butterflies, and then tighten the screw.

15. Position a NEW air silencer gasket in place on the front of the engine. Connect the hose from the crankcase to the back of the air silencer.

16. Coat the threads of the air silencer retaining screws with Loctite and then install the air silencer. Install the air silencer cover and at the same time slide the plastic adjusting knob onto the low-speed needle for each carburetor. DO NOT push the knobs all the way home onto the needle at this time. You may seriously consider NOT to install and use the piece of linkage connecting the low-speed needle valve of each carburetor.

Fig. 155 One method of disconnecting the fuel line is to remove the fuel pump cover

Fig. 156 Remove the air silencer cover. . .

Fig. 157 . . .and then the silencer itself

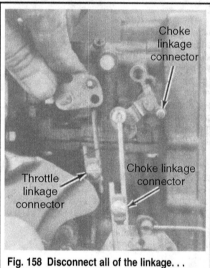

Fig. 158 Disconnect all of the linkage. . .

Fig. 159 . . .and then remove the carburetor

Fig. 160 Install the choke linkage.

■ This linkage permits adjusting both carburetors simultaneously while using only one knob. Sounds great! But when the one carburetor is adjusted, the other is also changed. A great many professional mechanics have discovered the linkage is not required and it is far more efficient to adjust each carburetor individually.

17. It is best to synchronize the fuel and ignition systems at this time. For details, please refer to the Timing & Synchronization procedures found in the Maintenance & Tune-Up section.

✳✳ CAUTION

Water must circulate through the lower unit to the engine any time the engine is run to prevent damage to the water pump in the lower unit. Just five seconds without water will damage the water pump.

18. Mount the engine in a test tank or body of water. If this is not possible, connect a flush attachment and garden hose to the lower unit.

✳✳ WARNING

NEVER operate the engine above idle speed using the flush attachment. If the engine is operated above idle speed with no load on the propeller, the engine could RUNAWAY resulting in serious damage or destruction of the unit.

19. Start the engine and allow it to warm to operating temperature. Pop the two rubber caps, one for each carburetor, out of the front cover of the air silencer.

20. Adjust the low-speed idle by turning the low-speed needle valve CLOCKWISE until the engine begins to misfire or the rpm drops noticeably. From this point, rotate the needle valve COUNTERCLOCKWISE until the engine is operating at the highest rpm. If the engine coughs and operates as if the fuel is too lean, but the idle and high-speed adjustments have been correctly made, then re-check the synchronization between the fuel and ignition systems.

■ On engines with the flexible low-speed extension to the front of the engine, the retainer maintains tension on the needle and adjustment will not be lost because of vibration during operation.

21. After the idle has been adjusted, push the idle knob retainers onto the needle until it engages the spline on the low-speed needle.

22. Repeat the procedure for the 2nd carburetor.

23. Replace the rubber caps into the front cover of the air silencer.

DISASSEMBLY & ASSEMBLY

◆ See Figures 91, 154 and 161 thru 171

Only one carburetor will be rebuilt in the following procedures. Service of the second unit is to be performed in the same manner.

■ **The carburetor parts should be kept with the individual unit to ensure all items are installed back in their original positions.**

1. Disconnect the fuel hoses between the 2 carburetors. This is accomplished by simply cutting the tie strap or working the clip securing the hose to the carburetor fitting.

2. Remove the drain plug or the high-speed orifice plug from the bottom of the carburetor bowl.

3. Remove the orifice from the float bowl, using the proper size screwdriver.

4. Remove the low-speed needle valve, if used. After removing the needle valve, observe and remove the retainer that accepts the needle valve. Good shop practice is to replace the retainer, because the retainer is actually the component holding the needle valve in adjustment. If the retainer has become worn, it is not possible to hold the needle valve in an accurate adjustment.

■ The low-speed needle valve has a bearing deep inside the carburetor. Removal of this bearing is no small task. A special tool is not available. However, a paper clip with a hook on the end, can be used to reach inside the bore and remove the bearing. Actually this bearing is made of plastic and the needle valve rotates inside. The bearing centers the needle valve in the carburetor.

5. Remove the low-speed needle valve bearing using a paper clip as described in the previous paragraph.

6. If the carburetor being serviced does not have the low-speed needle valve, remove the screw plug and washer from the top of the carburetor. Remove the orifice using the proper size screwdriver.

7. If the carburetor being serviced has the intermediate orifice, remove the plug and gasket from the starboard side of the carburetor. Remove the intermediate orifice using the proper size screwdriver.

8. Turn the carburetor upside-down and remove the screws securing the float bowl to the carburetor body. Lift the bowl free of the carburetor.

9. Remove and DISCARD the bowl gasket and the small gasket around the high-speed nozzle. The high-speed nozzle is NOT removable.

10. Remove the float hinge pin, and then remove the float.

11. Remove the inlet needle and seat from the carburetor body

Fig. 161 Remove the drain or high speed orifice plug. . .

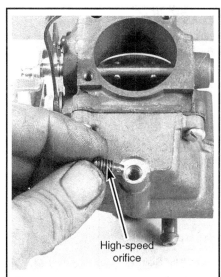
Fig. 162 . . .and then remove the high speed orifice

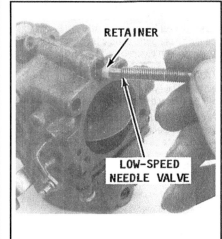
Fig. 163 Remove the low speed needle valve. . .

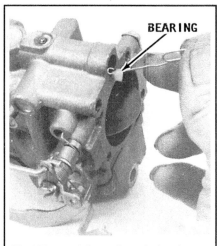
Fig. 164 . . .and then pull out the bearing with a paper clip

Fig. 165 Remove the fuel bowl. . .

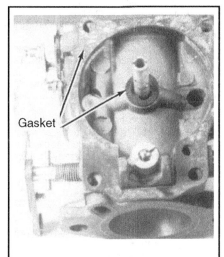
Fig. 166 . . .and then remove the gaskets

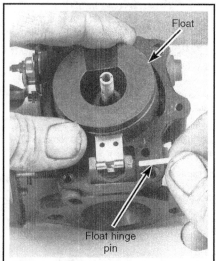

Fig. 167 Remove the hinge pin and lift out the float

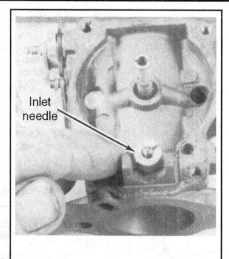

Fig. 168 Remove the inlet needle and seat

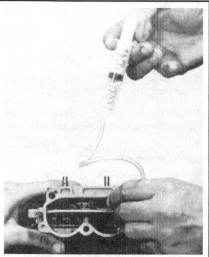

Fig. 169 Use a syringe to clean the idle air bleed system

To assemble:

✱✱ CAUTION

NEVER dip rubber parts, plastic parts, or nylon parts, in carburetor cleaner. These parts should be cleaned ONLY in solvent, and then blown dry with compressed air.

Place all metal parts in a screen-type tray and dip them in carburetor cleaner until they appear completely clean, then wash with solvent or clean water, and blow dry with compressed air.

■ Interestingly, in 1983 Johnson/Evinrude states that carburetor parts should NOT be submerged in carburetor cleaner, as has been the practice since carburetors were invented. Their approved procedure is to place the parts in a shallow tray and then spray them with an aerosol carburetor cleaner. You be the judge.

12. A syringe, short section of clear plastic hose, and Isopropyl Alcohol should be used to clear passages and jets

13. Blow out all passages in the castings with compressed air. Check all parts and passages to be sure they are not clogged or contain any deposits. NEVER use a piece of wire or any type of pointed instrument to clean drilled passages or calibrated holes in a carburetor.

14. Move the throttle shaft back and forth to check for wear. If the shaft appears to be too loose, replace the complete throttle body because individual replacement parts are NOT normally available.

15. If any part of the float or spring is damaged, the unit must be replaced. Check the float arm needle contacting surface and replace the float if this surface has a groove worn in it.

16. Inspect the tapered section of the idle adjusting needles and replace any that have developed a groove.

17. If a high-speed orifice is installed on the carburetor being serviced, check the orifice for cleanliness. The orifice has a stamped number. This number represents a drill size. Check the orifice with the shank of the proper size drill to verify the proper orifice is used. The correct size orifice for the engine and carburetor being serviced may be obtained from the local Johnson/Evinrude dealer.

Most of the parts that should be replaced during a carburetor overhaul are included in overhaul kits available from your local marine dealer. One of these kits will contain a matched fuel inlet needle and seat. This combination should be replaced each time the carburetor is disassembled as a precaution against leakage.

18. Install the inlet seat using the proper size screwdriver. Inject just a drop of light weight oil into the seat. Thread the inlet needle valve into the seat, and tighten it until it barely seats. DO NOT over-tighten or the needle valve may be damaged.

■ Two different type gaskets are used on this carburetor. One MUST be installed before the float, because it is a one-piece gasket with the nozzle gasket incorporated. The other type consists of two individual gaskets, one for the float and a separate one for the nozzle.

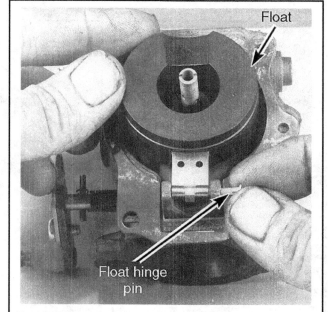

Fig. 170 Position the float and slide in the hinge pin

Fig. 171 The float must be parallel with the surface of the carburetor

19. Position NEW gaskets in place on the carburetor body and on the nozzle.

20. Lower the float down over the nozzle, and then slide the hinge pin into place.

21. Hold the carburetor body in a horizontal position and check to be sure the float is parallel to the carburetor surface, as should be. If the float is not parallel,

CAREFULLY bend the float tang until the float is in a parallel position and both sides are equal distance from the carburetor surface.

22. **Measuring Float Drop:** Allow the float to drop under its own weight. Measure the distance between the base of the carburetor body and the lowest edge of the float, as indicated in the accompanying illustration. The measurement should be 1 1/8 - 1 5/8 in. (28-41mm).

23. Place the float bowl in position, and then secure it in place with the attaching screws.

24. Install the low-speed retainer in the carburetor body.

25. Slide the nylon bearing onto the end of the low-speed needle valve, and then thread the valve into the retainer. Continue threading the valve into the retainer until it seats LIGHTLY, and then back it out COUNTERCLOCKWISE 5/8 turn.

CARBURETORS - 3 CYLINDER & V4 MOTORS

General Information

✳✳ CAUTION

If equipped, disconnect the negative battery cable ANYTIME work is performed on the engine, especially when working on the fuel system. This will help prevent the possibility of sparks during service (from accidentally grounding a hot lead or powered component). Sparks could ignite vapors or exposed fuel. Disconnecting the cable on electric start motors will also help prevent the possibility fuel spillage if an attempt is made to crank the engine while the fuel system is open.

✳✳ CAUTION

Fuel leaking from a loose, damaged or incorrectly installed hose or fitting may cause a fire or an explosion. ALWAYS pressurize the fuel system and run the motor while inspecting for leaks after servicing any component of the fuel system.

This section provides complete detailed procedures for: removal; disassembly; cleaning and inspecting; assembling including bench adjustments; installation; and operating adjustments for the 6 carburetors installed on powerheads covered in this service.

These carburetors are equipped with a manual choke, an electric choke, or the "primer choke system" (unlikely though). The manual back-up permits the operator to operate the choke in the event the battery is dead. The Carburetor Applications section lists the types of carburetors installed on the various horsepower and model years for engines covered in this service.

Carburetor Applications

■ These applications should be viewed as a guide only - you may find that your engine has a different unit than specified. Please review all pictures from each type if in doubt.

TYPE I

Downdraft, double barrel with high speed and low speed needle valves.
- 50 hp: 1958-59
- 60 hp: 1964-65
- 75 hp: 1960

Downdraft, double barrel with fixed high speed orifice and low speed needle valve.
- 60 hp: 1966-67
- 65 hp: 1968
- 75 hp: 1961-65
- 80 hp: 1966-67
- 85 hp: 1968

✳✳ WARNING

Take time to use the proper size screwdriver to install an orifice. If the orifice is damaged and the edge of the opening burred, because the wrong size screwdriver was used, the flow of fuel will be restricted and the orifice will not function properly. Damage to the orifice will also make it very difficult to remove during the next carburetor service.

26. If an orifice is used instead of the needle valve, install the low-speed orifice into its recess, using the proper size screwdriver to prevent damaging the orifice. Tighten the orifice until it seats *lightly*. Thread the plug, with a NEW gasket, into place and tighten it snugly.

27. If the carburetor being serviced has the intermediate orifice, install the orifice, using the proper size screwdriver to prevent damaging the orifice. Tighten the orifice until it seats *lightly*. Thread the plug, with a NEW gasket, in place and tighten it snugly.

28. Install the high-speed orifice into the carburetor bowl, using the proper size screwdriver to prevent damaging the orifice.

29. Thread the plug, with a new gasket, into place and tighten it snugly.

30. Repeat these steps for the second carburetor.

TYPE II

Front draft, 4 barrel with fixed high speed orifice and adjustable low speed needle valves.
- 90 hp: 1964-65
- 100 hp: 1966-68

TYPE III

Three carburetors per powerhead. Front draft, single barrel with fixed high speed orifice and low speed needle valve.
- 55 hp: 1968-69
- 60 hp: 1970-71
- 65 hp: 1972

TYPE IV

Two or three carburetors per powerhead, front draft, double barrel with low-speed orifice and fixed high-speed orifice.
- 85 hp: 1969-71
- 100 hp: 1971
- 115 hp: 1969-70
- 125 hp: 1971

Same as Type previous except with fixed low speed and high speed orifices.
- 85 hp: 1972
- 100 hp: 1972
- 125 hp: 1972

For information on fuels, tanks and lines please refer to the section for General Information on Carbureted Fuel Systems, as well as the section on Fuel Tanks & Lines.

The most important fuel system maintenance that a boat owner can perform is to provide and to stabilize fuel supplies before allowing the system to sit idle for any length of time more than a few weeks. The next most important item is to provide the system with fresh gasoline if the system has stood idle for any length of time, especially if it was without fuel system stabilizer during that time.

If a sudden increase in gas consumption is noticed, or if the engine does not perform properly, a carburetor overhaul, including cleaning or replacement of the fuel pump may be required.

Type I Carburetors

◆ See Figures 172 thru 175

Type IA
Downdraft, double barrel with high speed and low speed needle valves.
- 50 hp: 1958-59
- 60 hp: 1964-65
- 75 hp: 1960

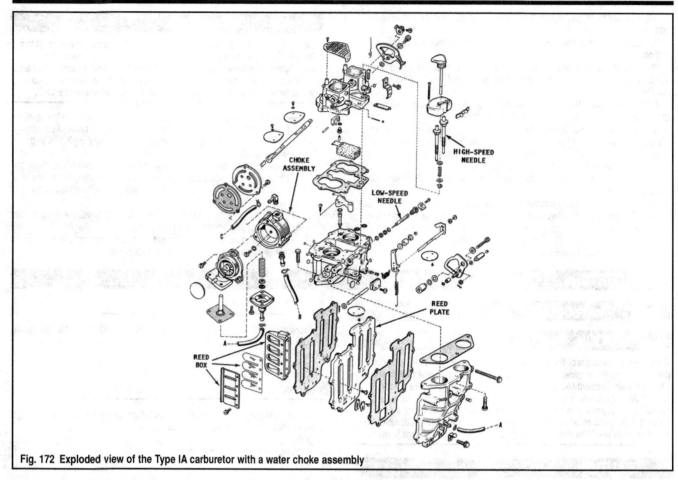

Fig. 172 Exploded view of the Type IA carburetor with a water choke assembly

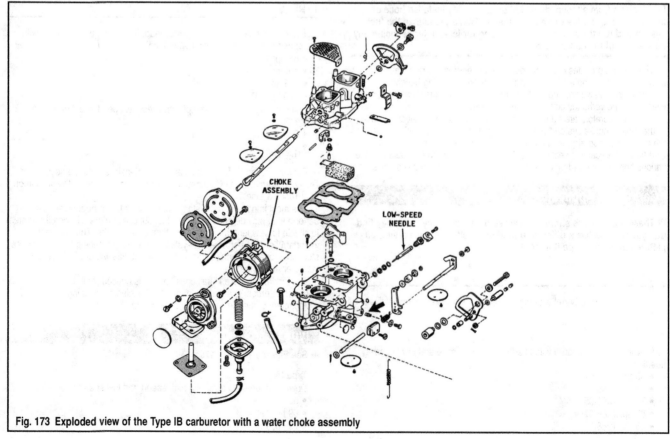

Fig. 173 Exploded view of the Type IB carburetor with a water choke assembly

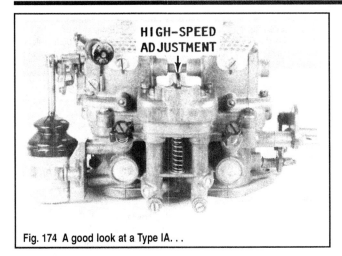

Fig. 174 A good look at a Type IA. . .

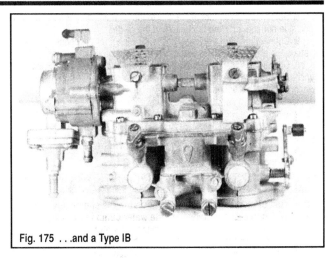

Fig. 175 . . .and a Type IB

Type IB

Downdraft, double barrel with fixed high speed orifice and low speed needle valve.

- 60 hp: 1966-67
- 65 hp: 1968
- 75 hp: 1961-65
- 80 hp: 1966-67
- 85 hp: 1968

The Type I carburetor has a double throat, commonly referred to as a "double barrel". High speed operation of large horsepower engines requires a great deal of air. In order to retain the efficiency and fuel economy of small horsepower engines with relatively small carburetor throats, a double barrel is used instead of one large single barrel. A single barrel would be wasteful of fuel and unresponsive at low speed operation. Air velocities become too low for proper air/fuel mixing and vaporization. To solve this problem, carburetors with two or more throats are used.

When facing the front of the carburetor, the left barrel crosses and feeds the No. 2 and No. 4 cylinders, port bank. The right barrel feeds the No.1 and No.3 cylinders, starboard bank.

■ **Since both the Type I carburetors are identical except for the high-speed adjustable needle valve and the venturis on the 1958-65 model engines. The following service procedures will cover the Type IB, with a fixed high-speed orifice. Any differences, due to the high-speed needle valve used on a few of the early model engines, will be clearly indicated.**

These carburetors are equipped with either a manual choke, an electric choke, a heat choke, water choke, or an override choke. The manual back-

up permits the operator to operate the choke in the event the battery is dead. See the appropriate Choke System Service section.

REMOVAL & INSTALLATION

◆ **See Figures 176 thru 182**

MODERATE

1. Disconnect the cables at the battery terminals. Remove the hood.
2. Disconnect the fuel line at the engine and at the fuel tank; by separating the quick-disconnect.
3. Disconnect the advance throttle linkage between the tower shaft and the carburetor. This is accomplished either by removing the 2 sets crews and then the rod; or by removing the screw on the cam. If this screw is removed, be sure to thread a nut onto the screw to prevent losing the parts. Remove the return spring.
4. Remove either the choke wire to the choke, or the water lines, if a water choke is used. If a heat choke is used, disconnect the heat tube.
5. Remove the 3 bolts securing the carburetor to the manifold. One bolt is located on each side of the carburetor and the third comes up through the manifold from the bottom.
6. Lift the carburetor free of the manifold and remove the fuel line from the back side. Remember that this fuel line MUST he connected before the carburetor is placed in position on the manifold.

To Install:

7. Position a NEW gasket in place on the intake manifold and then install the carburetor with the inlet hose towards the rear of the carburetor.

Fig. 176 Disconnect the fuel lines and battery cables

Fig. 177 Disconnect the tower shaft at the throttle link and then remove the spring

Fig. 178 Remove the choke lines

8. Secure the carburetor in place with the 3 retaining screws, one on each side and the 3rd from the bottom in the center.

9. Connect the choke linkage, hoses, or wiring, for the choke system used on the engine being serviced.

10. Connect the advance arm from the tower shaft to the carburetor. To synchronize the carburetor with the ignition system, see Timing & Synchronization section in Maintenance & Tune-Up.

11. On carburetors with high speed adjustment, mount the engine in a test tank or body of water. If this is not possible, connect a flush attachment and garden hose to the lower unit. NEVER operate the engine above idle speed using the flush attachment. If the engine is operated above idle speed with no load on the propeller, the engine could RUNAWAY resulting in serious damage or destruction of the unit.

✳✳ CAUTION

Water must circulate through the lower unit to the engine any time the engine is run to prevent damage to the water pump in the lower unit. Just 5 seconds without water will damage the water pump.

12. Start the engine and allow it to warm to operating temperature at about 500 rpm.

13. Back out the low speed adjustment needle COUNTERCLOCKWISE until the engine begins to run rough. From this point, rotate the adjustment needle inward, CLOCKWISE, very slowly until engine performance smoothes out.

Repeat this procedure for the other needle.

14. On carburetors with high speed adjustment, the engine MUST be mounted in an adequate size test tank or body of water, preferably with a test wheel attached to the propeller shaft. Push upward on the knob in the center between the 2 high speed needle valves and then turn it 180°. This will take the knob out of engagement with the 2 needle valves.

Fig. 179 Installing the E-clip

15. Start the engine and increase engine rpm to full advance throttle.

16. Rotate the high speed needle CLOCKWISE until engine rpm begins to drop (Illustration shows carburetor not mounted). From this point, rotate the adjustment needle COUNTERCLOCKWISE until the highest rpm speed is attained.

17. Repeat the procedure for the other high speed needle valve. After the adjustments have been made, lift the center rod between the 2 needle valves, and rotate the knob 180°, then allow the knob to return to its normal position in mesh with the 2 high speed needle valves.

DISASSEMBLY

◆ See Figures 172, 173 and 183 thru 199

1. Remove the carburetor as detailed previously.

2. Remove all screws from the top of the carburetor. Notice the one screw on the port side that passes through a small elongated clip. This clip secures the choke cam in place. Remember under which screw, this clip is secured.

3. On carburetors with a high speed adjustment, after all the screws on top of the carburetor have been removed: notice that 2 screws pass through the cap covering the high speed adjustment needle valve.

   a. Remove the E-clip from the small rod between the 2 high speed needle valves.

   b. Pull the rod, with the knob attached, free. Remove the cap.

■ **TAKE CARE not to lose the washers and spring in the carburetor recess from which the rod was removed.**

4. On carburetors with the high speed adjustment, both high-speed needles must be removed *before* the carburetor cover is removed. Simply rotate the needles COUNTERCLOCKWISE until they are free.

5. Lift the top of the carburetor from the body. Turn the top upside down. Notice that the top contains the float along with the needle and seat.

6. Remove and discard the hinge pin from the float. A carburetor repair kit contains a new float and hinge pin.

7. Remove the needle valve from the seat, and then remove the seat from the carburetor. It is not necessary to be careful with the seat because a NEW seat should be installed in the kit during a carburetor overhaul.

8. Remove and discard the gasket.

■ **Two gaskets are included in the replacement kit. One is used with carburetors having the fixed high speed orifice, and the other gasket is used with the carburetors with the adjustable high speed needle. Therefore, always take note of the gasket removed, to ensure the NEW gasket installed is the proper one for the carburetor being serviced.**

9. Remove and DISCARD the small O-ring from the side of the throttle advance cam roller, then remove the roller.

Fig. 180 Connect the inlet line before positioning the unit on the manifold

Fig. 181 Turn the high speed adjustment knob to disengage the needle valves. . .

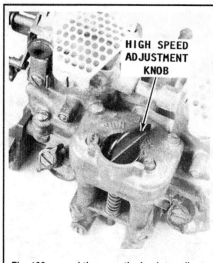

Fig. 182 . . .and then use the knob to adjust the engine speed

10. Remove the low speed adjusting knob by removing the screw in the center of the knob, and then pull the knob free.

11. Loosen the packing nut. Rotate the low speed adjusting needle COUNTERCLOCKWISE until it is free of the carburetor body.

12. Remove the low speed packing nut and the packing from the carburetor.

13. On carburetors with high speed adjustment, remove the low speed adjustment needle by pushing on the knob and then, using a screwdriver, rotate the needle COUNTERCLOCKWISE until it is free of the carburetor body. Remove and DISCARD the O-ring.

14. On the carburetor body, remove the 2 screws securing the venturis in place, and then remove both venturis.

15. Remove the O-ring from around the high speed tube. Use the proper size screwdriver and rotate the high speed tube COUNTERCLOCKWISE until it is free of the carburetor body. Remove the other high-speed tube in a similar manner.

16. On carburetors with high speed adjustment, remove the 2 screws securing the venturis in place, and then remove both venturis. The high-speed tube is part of the venturi and will come free with the venturi.

17. Remove the two 7/16 In. drain orifice plugs. These plugs are located on the bottom side of the carburetor in front.

18. Use the proper size screwdriver or special tool (# 317002), and remove the high-speed orifices.

■ On some carburetors, even though a high speed needle valve is used, a high speed orifice is also used. Check to determine if the high speed needle valve opening has an orifice.

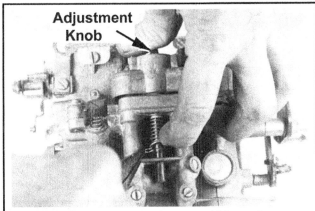
Fig. 184 Removing the E-clip on models with high speed adjustment. . .

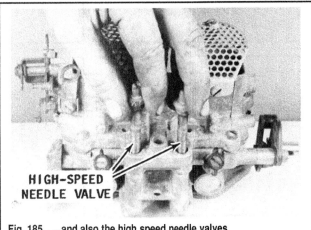
Fig. 185 . . .and also the high speed needle valves

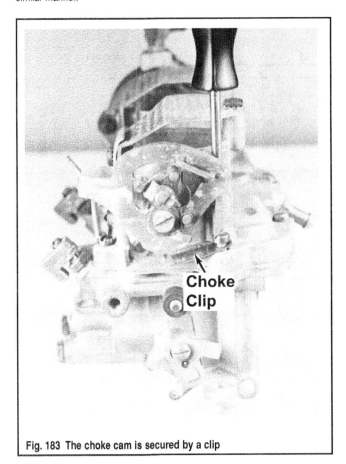
Fig. 183 The choke cam is secured by a clip

Fig. 186 Lift the top off of the body

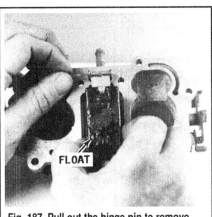
Fig. 187 Pull out the hinge pin to remove the float. . .

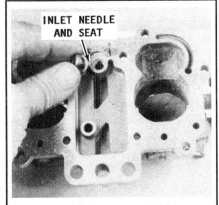
Fig. 188 . . .and then remove the inlet needle and seat

Fig. 189 Remove the gasket. . .

Fig. 190 . . .and then the O-ring

Fig. 191 Remove the low speed knob. . .

Fig. 192 . . .then the needle. . .

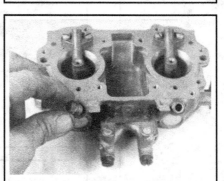

Fig. 193 . . .and finally the packing nut

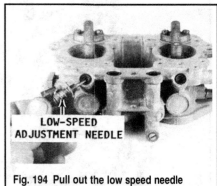

Fig. 194 Pull out the low speed needle

Fig. 195 Remove the mounting screws and lift out the 2 venturis. . .

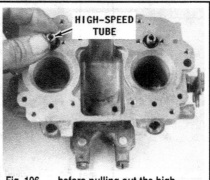

Fig. 196 . . .before pulling out the high speed tube and O-ring

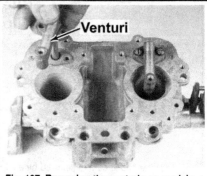

Fig. 197 Removing the venturis on models with high speed adjustment

Fig. 198 Remove the drain orifice plugs. .

Fig. 199 . . .and then the 2 high speed orifices

CLEANING & INSPECTION

◆ **See Figures 172, 173, 200 and 201**

NEVER dip rubber parts, plastic parts, or nylon parts, in carburetor cleaner. These parts should be cleaned ONLY in solvent and then blown dry with compressed air.

Place all metal parts in a screen-type tray and dip them in carburetor cleaner until they appear completely clean; then wash with solvent, or clean water, and blow dry with compressed air.

Blow out all passages in the castings with compressed air. Check all parts and passages to be sure they are not clogged or contain any deposits. NEVER use a piece of wire or any type of pointed instrument to clean drilled passages or calibrated holes in a carburetor.

Move the throttle shaft back-and-forth to check for wear. If the shaft appears to be too loose, replace the complete throttle body because individual replacement parts are NOT available.

Inspect the main body, airhorn and venturi cluster gasket surfaces for cracks and burrs which might cause a leak. Check the float for deterioration. Check to be sure the float spring has not been stretched. If any part of the float is damaged, the unit must be replaced. Check the float arm needle contacting surface and replace the float if this surface has a groove worn in it.

Inspect the tapered section of the idle adjusting needles and replace any that have developed a groove.

Check the orifices for cleanliness. The orifice has a stamped number. This number represents a drill size. Check the orifice with the shank of the proper size drill to verify the proper orifice is used. The local dealer will be able to provide the correct size orifice for the engine and carburetor being serviced.

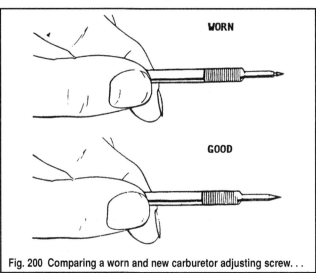

Fig. 200 Comparing a worn and new carburetor adjusting screw. . .

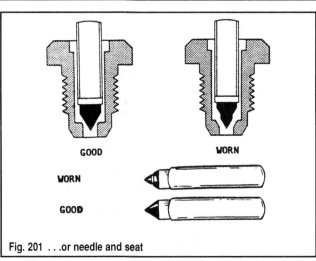

Fig. 201 . . .or needle and seat

Most of the parts that should be replaced during a carburetor overhaul are included in overhaul kits available from your local marine retailer. One of these kits will contain a matched fuel inlet needle and seat. This combination should be replaced each time the carburetor is disassembled as a precaution against leakage.

ASSEMBLY

◆ **See Figures 172, 173 and 202 thru 207**

■ **Please refer to the Disassembly procedures for more illustrations to aid in assembly.**

1. Install the high speed orifices using the special tool (#317002), or a proper size screwdriver.

2. After the orifices are installed, thread the orifice plugs into place. Use NEW gaskets when installing the orifice plugs.

3. Lubricate the high speed tube O-rings with light-weight oil. Slide the O-ring onto the tube and then thread the tube CLOCKWISE into the carburetor body. Tighten the tube.

Install the other high speed needle tube in the same manner.

4. Place the 2 venturis over the high speed tubes and start the retaining screws. Bring the screws up just SNUG, but DO NOT tighten them at this time.

5. On carburetors with high speed adjustment, place a NEW gasket in position on the carburetor base and then install a venturi into the carburetor throat. Install the venturis with the retaining screws, but DO NOT tighten the screws at this time.

6. Measure from each side of the venturi to center the venturi in the throat. A special locating tool (#379242), is available for use on the Type I carburetor with fixed high-speed orifices. Tighten the venturi at this time. Install the second venturi in the same manner.

7. Slide the packing onto the low-speed needle and then thread the needle into the carburetor body just enough to hold it in place. DO NOT tighten it even snugly at this time.

8. Thread the packing nut into the carburetor body finger-tight. Push the knob onto the adjusting needle and then secure it in place with the screw. Rotate the low-speed adjusting needle CLOCKWISE until it just barely seats, then back it out 1 1/2 turns as a preliminary adjustment. Tighten the packing nut to hold this preliminary adjustment.

9. On carburetors with high speed adjustment, place a new O-ring into the carburetor opening for the low speed valve - squirt just a drop of oil onto the O-ring. Push in on the knob and, at the same time, thread the low speed needle valve into the carburetor body using the proper size screwdriver. Tighten the needle valve until it just BARELY seats, then back it out 1 1/2 turns as a preliminary adjustment. Install the 2nd low speed needle valve in the same manner.

10. Place the cam roller onto the throttle shaft, and then install a new O-ring onto the side of the roller.

11. Place a new gasket in position on the carburetor body.

12. On carburetors with high speed adjustment, insert a new O-ring into each opening for the high speed needle valve. Squirt a drop of oil onto each O-ring.

13. Place a new gasket in position on the carburetor body - the gasket *must* lie over the top of the O-rings.

14. Install the seat into the carburetor cover with a new gasket, taking care not to damage the seat. Insert a drop of oil into the seat and then install the inlet needle.

15. Install a new float into the holder, followed by a new hinge pin.

16. Hold the carburetor cover upside down, and observe the float position. The float should be parallel with the cover surface. If the float is not positioned properly, CAREFULLY bend the float arm at a point close to the float on both sides. This is accomplished by holding the arm as close to the float as possible, and then moving the float *slightly* in the desired direction.

17. Lower the cover down over the carburetor with the float properly positioned in the body. Secure the cover with the screws. REMEMBER the small tang that must be installed under one of the screws on the port side of the carburetor. Tighten the screws alternately and securely.

18. On carburetors with high speed adjustment, lubricate the high speed needle valve O-rings with light-weight oil. Slide the O-ring onto the needle, and then thread the needle CLOCKWISE into the carburetor body. Tighten the needle valve just *snugly*, then back it out COUNTERCLOCKWISE 3/4 of

Fig. 202 Positioning the venturis on models with fixed high speed orifices

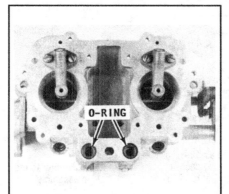

Fig. 203 On models with high speed adjustment, insert new O-rings. . .

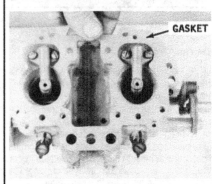

Fig. 204 . . .and then position the new gasket

Fig. 205 Always check float positioning carefully

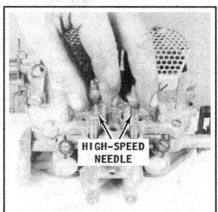

Fig. 206 On models with high speed adjustment, insert new high speed needles. . .

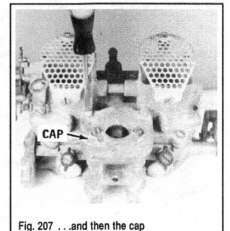

Fig. 207 . . .and then the cap

a turn as a preliminary adjustment. Check to be sure the O-ring is seated properly in the recess of the carburetor body.

Install and adjust the other high-speed needle valve in the same manner.

19. Install the cap over the high speed adjustment knobs and secure them in place with the screws.

20. Work the center high speed adjustment knob through the housing, washers, and spring, and into the carburetor body. Lift up on the bottom washer and install the retaining E-clip.

■ See the Choke System Service sections to service the choke system installed on the engine being serviced.

21. Install the carburetor.

Type II Carburetor

◆ See Figures 208 and 209

Front draft, 4 barrel with fixed high speed orifice and adjustable low speed needle valves.
- 90 hp: 1964-65
- 100 hp: 1966-68

The Type II carburetor is a front draft unit with 4 throats (barrels) utilizing fixed high speed orifices and adjustable low speed needle valves. The carburetor might actually be described as 4 individual carburetors incorporated into one unit. Four fuel delivery items are incorporated to provide a separate "barrel", inlet needle and seat, float, high speed orifice and low speed adjustment; for each of the 4 cylinders.

As a result of its design, construction and performance demands, the carburetor is considered a very complex piece of equipment. Therefore, when servicing the Type II carburetor or when making any adjustments, the reader must take extra care to follow each step carefully and in sequence. DO NOT anticipate what will be done next, skip any steps thinking they are not necessary, or perform extra tasks because of experience with other carburetor units.

A heat-type choke system is used on the Type II carburetor. Heat is drawn from the exhaust baffle just behind the exhaust plate at the rear of the powerhead, through a pipe to the carburetor. This heat warms a bi-metal spring, which expands as its temperature rises. The choke is activated through the expansion and contraction of the bi-metal spring.

See Choke System Service section service procedures.

REMOVAL & INSTALLATION

MODERATE

◆ See Figures 208 and 210 thru 216

1. Tag and disconnect the cables at the battery terminals. Remove the hood. Disconnect the fuel line at the engine and at the fuel tank - by separating the quick-disconnects. Disconnect the fuel hose between the fuel pump and the carburetor, at the carburetor.

2. On the starboard side of the engine, observe the tube extending from the exhaust manifold at the rear of the engine to the automatic choke. Loosen the packing gland nut on each end of the tube and then slide the tube back out of the way to permit removal of the carburetor.

3. On the port side of the engine, observe the rod extending from the tower shaft to a nylon yoke. Remove the cotter key and pin from the yoke and then remove the yoke from the carburetor.

4. Remove the cotter pin from the bottom side of the throttle cam assembly. Be very careful not to lose the 2 washers and spring as the key is removed. Lift the throttle cam assembly free of the carburetor. These items must be removed BEFORE the bolts are removed securing the carburetor to the intake manifold.

5. Working at the front of the carburetor, remove the hose from the lower end of the shield. Remove the shield attaching screws and then the shield.

6. Remove the 10 carburetor attaching bolts. Three bolts are located on the starboard side. Loosen these bolts, but you will not be able to remove them at first. One bolt is long and extends through the carburetor with a nut. This bolt has the ground cable attached. Another bolt is located on the bottom of the carburetor. This one is very difficult to remove and to install. The only advice is PATIENCE. Continue removing and loosening the bolts while working the carburetor free of the manifold. Do not force the carburetor free. If all bolts are removed or loosened properly, the carburetor will come free without difficulty.

To Install:

7. Remove the 3 bolts that were loosened during removal of the carburetor from the intake manifold. Position a NEW gasket in place on the manifold.

8. Start the 3 bolts just removed in the previous step. Slide the carburetor into place on the intake manifold and secure it in place with the mounting bolts. Tighten the bolts ALTERNATELY and EVENLY. Do not forget the ground strap secured by one of the carburetor mounting bolts.

9. Install the shield with the attaching hardware. Connect the hose at the lower end of the shield.

10. Install and connect the fuel line between the carburetor and the fuel pump. Connect the override choke wire connector.

11. Slide the throttle cam assembly down into the carburetor retainer; and at the same time, work up the washer, spring and other washer, securing them with the cotter pin. As an aid to accomplishing this task, use a screwdriver between the flywheel guard and the top of the tower shaft to hold it while the washers and spring are worked upward. As the tower shaft is lowered into place, make sure the roller on the end of the throttle cam rides on the throttle cam.

12. Install the yoke that is attached to the rod extending from the tower shaft to the throttle cam assembly. Attach the clevis on the end of the yoke and insert the pin to hold it in place, then the washer, and finally the cotter pin.

13. Perform all necessary Timing & Synchronization procedures as detailed in the Maintenance & Tune-Up section.

14. Mount the engine in a test tank or body of water. If this is not possible, connect a flush attachment and garden hose to the lower unit. NEVER operate the engine above idle speed using the flush attachment. If the engine is operated above idle speed with no load on the propeller, the engine could RUNAWAY resulting in serious damage or destruction of the unit.

✳✳ CAUTION

Water must circulate through the lower unit to the engine any time the engine is run to prevent damage to the water pump in the lower unit. Just 5 seconds without water will damage the water pump.

15. Start the engine and allow it to warm to operating temperature at about 500 rpm.

Back out each low speed adjustment needle COUNTERCLOCKWISE until the engine begins to run rough. From this point, rotate each adjustment needle inward, CLOCKWISE, very slowly until engine performance smoothes out. If the engine should "POP" while the adjustment is being made, the adjustment is already too lean.

16. After all needles have been properly adjusted, snap the knobs over the low-speed needles and gang bar.

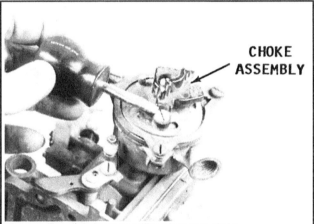

Fig. 209 Type II carburetors utilize a heat-type choke

Fig. 210 Disconnect the fuel lines. . .

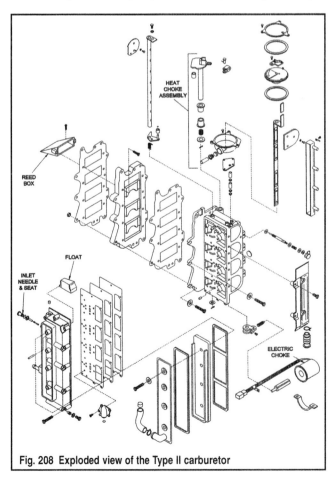

Fig. 208 Exploded view of the Type II carburetor

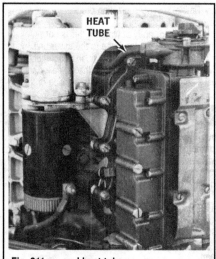

Fig. 211 . . .and heat tubes

Fig. 212 Disconnect the yoke at the carb. . .

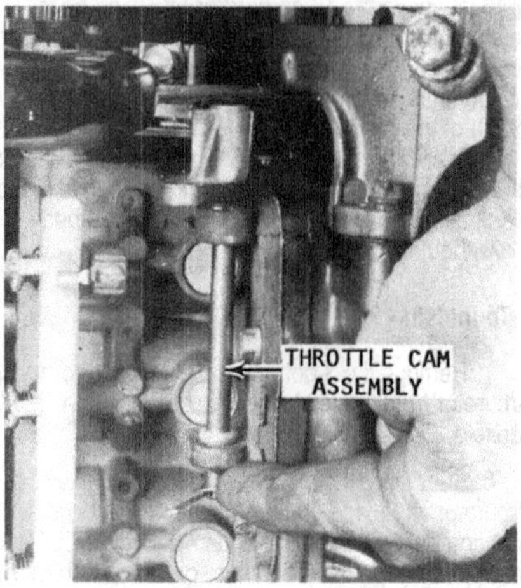

Fig. 213 . . .and then remove the throttle cam assembly

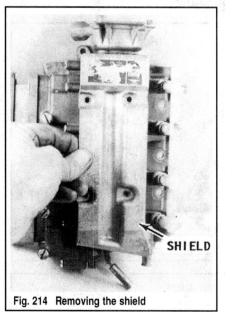

Fig. 214 Removing the shield

Fig. 215 There are 10 mounting bolts in all

Fig. 216 Remove the 3 bolts still on from removal before installing the new gasket

DISASSEMBLY

OEM ② MODERATE

◆ See Figures 208 and 217 thru 232

1. Remove the carburetor.

2. Remove the choke wire clamp screw and then shift the clamp away from the carburetor body. Leave the clamp attached to the wire.

3. Remove the 2 screws and bracket holding the override choke from the bottom side of the carburetor body. Remove the plunger, spring, and small rubber washer from the choke. Do not clean these items in solvent.

4. Observe what is termed the "gang bar" on the front side of the carburetor body. Remove the plastic knobs from each needle valve.

5. Remove the screws securing the gang bar and lift it free.

6. Remove the 4 screws securing the cover plate over the float chamber and then remove the plate.

7. Position the carburetor on the bench on its side, with the float chamber facing towards you. Remove all screws securing the float chamber to the carburetor body. STOP! DO NOT lift the float chamber from the carburetor body at this time.

8. Hold the float chamber in place and turn the carburetor body 180° until the float chamber is facing downward towards the bench surface. Lift the carburetor body upward and away from the float chamber.

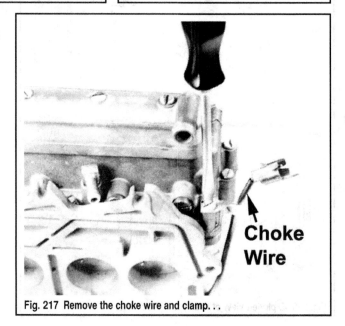

Fig. 217 Remove the choke wire and clamp. . .

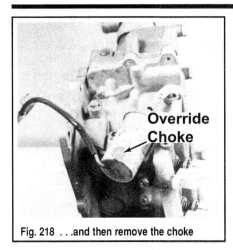

Fig. 218 . . .and then remove the choke

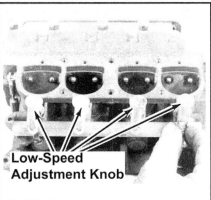

Fig. 219 Remove the low speed adjustment knobs. .

Fig. 220 . . .and then lift off the gang bar

Fig. 221 Remove the cover plate. . .

Fig. 222 . . .loosen the float chamber bolts. .

Fig. 223 . . .and then remove the chamber *exactly* as detailed in the instructions

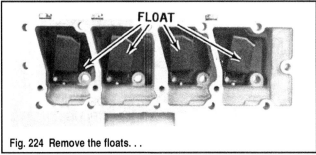

Fig. 224 Remove the floats. . .

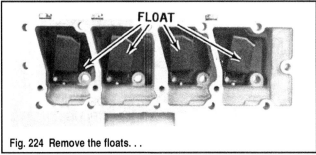

Fig. 225 . . .and then the float seats and needles

9. With the float chamber on the bench facing upward, notice the 4 floats and 4 needles and seats. Using a pair of needle nose pliers, reach in and remove the floats.

10. From the outside of the carburetor, use a screwdriver and remove the float seats, needle seats and then the needles.

11. Notice how a spring, small pin, and needle are incorporated into each needle valve.

These parts will be supplied in a new carburetor repair kit. If, for some reason, a repair kit has not been purchased - take care not to lose any of these parts.

12. Remove the 4 drain plugs from the carburetor body.

13. Obtain the special screwdriver (#317002) and remove the 4 high speed orifices.

14. Remove the low speed needle valves by rotating them COUNTERCLOCKWISE until each one is free of the carburetor body. Be careful not to lose the washers and spring from each needle. These parts are NOT included in a repair kit.

15. Use a narrow blade screwdriver and pop the O-ring out of each of the 4 low speed needle openings.

16. On the float chamber side of the carburetor, remove the 4 high speed tube assemblies.

17. Remove the gasket, inner metal plate and second gasket from the carburetor body.

■ Because of the time involved in realigning throttle valves during assembly, the throttle shaft should ONLY be removed if it is damaged, if the return spring requires replacement, or if the throttle shaft is bent.

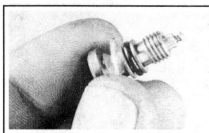
Fig. 226 Don't lose the needle hardware if you didn't buy a rebuild kit!

Fig. 227 Remove the drain plugs. . .

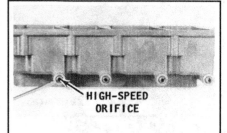
Fig. 228 . . .and then the high speed orifices

Fig. 229 Remove the low speed needles. . .

Fig. 230 . . .and then pop out the O-rings

Fig. 231 Remove the high speed tube assemblies

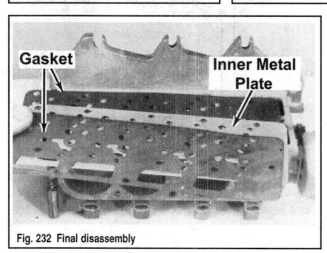

Fig. 232 Final disassembly

CLEANING & INSPECTION

■ Please refer to the cleaning and inspection procedures detailed in the Type IA/IB Carburetor section for all steps and illustrations.

ASSEMBLY

◆ See Figures 208 and 233 thru 239

1. Check to be sure the inner plate is clean. Position a new gasket to the side of the carburetor, then the inner plate, followed by another new gasket.

2. Install the 4 high speed tube assemblies. Do not tighten the plate mounting screws until all 4 tubes are in place. Tighten the inner plate mounting screws alternately and evenly.

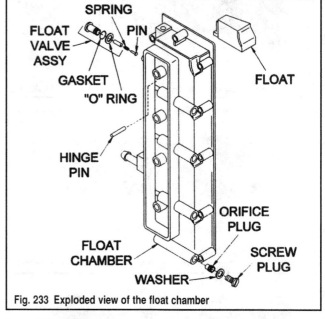
Fig. 233 Exploded view of the float chamber

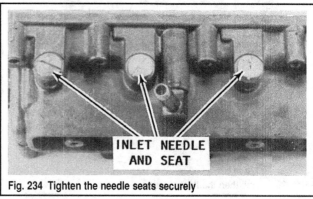
Fig. 234 Tighten the needle seats securely

Low speed needle valve installation

3. Lubricate each O-ring with a drop of oil and then work each ring into place in the 4 low speed needle openings.

4. The E-clip on the end of each needle valve should not have been removed. However, if it was, snap it into place.

5. Install the washer, spring and 2nd washer onto each of the 4 low speed needle valves.

6. Install the 4 low-speed needle valves into the carburetor body. Seat each needle GENTLY and then back it out 3/4 turn, as a preliminary adjustment.

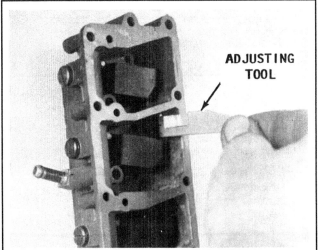

Fig. 235 Use the special tool when adjusting the floats. . .

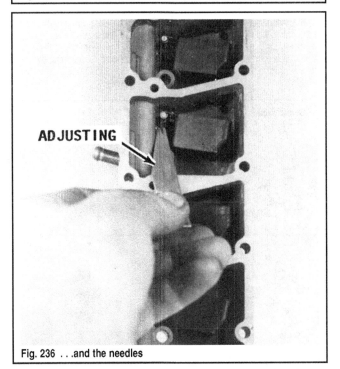

Fig. 236 . . .and the needles

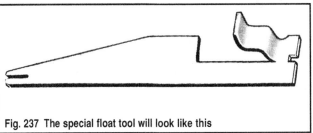

Fig. 237 The special float tool will look like this

High speed orifice installation

7. Install the 4 high speed orifices into the float chamber, using the special tool (#317002).

8. Install the 4 orifice drain plugs using NEW gaskets.

Float chamber

9. Assemble all 4 needle seats and have them ready for installation. This is accomplished as follows:

a. Set the seat on the bench with the screwdriver slot facing downward.

b. Install the fiber washer down over the seat.

c. Place a new O-ring in place in the seat groove.

d. Insert the needle into the seat, then the small spring and finally the pin into the center of the spring.

■ Once the assembled needles are installed, the float chamber cannot be rotated or the parts will be dislodged. Therefore, proceed with caution and patience.

10. Position the carburetor body with the float chamber opening facing towards you. Insert each needle and seat into place from the bottom side of the float chamber. Thread each needle into the float chamber finger-tight - it is not important for them to be real tight at this time.

11. After all 4 needles have been threaded into place, take each float, one at-a-time, and work the float down the pin with the tab on the float set over the pin in the needle. The float arm will hold the needle in the seat.

12. Lay the float chamber down on the carburetor with the opening facing upward. Tighten the 4 needle seats securely at this time.

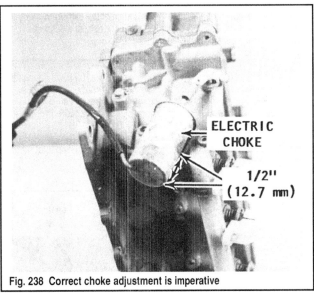

Fig. 238 Correct choke adjustment is imperative

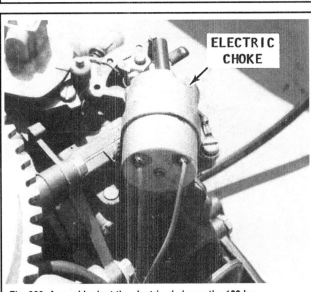

Fig. 239 A good look at the electric choke on the 100 hp

Float adjustment

■ Experience has proven that it is almost impossible to adjust the floats properly without the special float adjustment tool (#38054-6). This tool is not expensive, but it is essential to ensure proper carburetor performance following the service work.

13. Turn the float body allowing the float to hang down and so the tangs of each float are resting up against the needles. In this position, the 4 floats will have weight against the inlet needle valves. Use the adjusting tool and insert it above the float at the top of the float chamber, as shown.

14. If a needle requires adjustment, move the tang inward or outward using the adjusting tool.

■ It is MOST important that all 4 floats be adjusted properly and equally to ensure proper carburetor performance. If one float is just a wee bit out of adjustment, the cylinder served by that float and needle valve will not have the proper fuel/air mixture.

After the four floats are properly adjusted, set the float chamber aside with the floats facing upward.

15. Hold the assembled float chamber in one hand, and then lower the carburetor body down over the float chamber. Once the two have been brought together, CAREFULLY rotate the two 180° and set it on the bench. DO NOT lift the float chamber from the carburetor body.

16. Start all of the retaining screws around the perimeter of the float chamber.

17. Install and secure the drain cover using a new gasket. Check to be sure the outlet nipple extends through the bottom of the carburetor body.

18. Install the choke and choke plunger into the bottom side of the carburetor body. Connect the linkage.

   Override choke adjustment

19. Measure the distance from the back of the choke to the base of the carburetor body - this measurement should be 1/2 in. (12.7mm). Loosen the 2 screws and slide the override choke inward or outward until the proper dimension is obtained. Tighten the 2 screws to hold the adjustment.

20. Move the choke wire towards the carburetor and secure the wire clamp with the attaching screw.

21. Install the "gang bar" over the top of the 4 needles. Secure the bar in place with the retaining clip. DO NOT install the nylon adjusting knob at this time.

22. Install the carburetor.

Type III Carburetor

◆ See Figure 240

Three carburetors per powerhead. Front draft, single barrel with fixed high speed orifice and adjustable low speed needle valve.
- 55 hp: 1968-69
- 60 hp: 1970-71
- 65 hp: 1972

Three carburetors are installed on each engine, one for each cylinder.

This carburetor is the single barrel, front draft. It has a adjustable low-speed orifice and a fixed high-speed orifice in the float bowl.

All of the Type I carburetors utilize an electric choke mounted on the side of the engine. On most models, the electric choke does not have to be removed in order to remove and service the carburetor.

REMOVAL & INSTALLATION

◆ See Figures 240 thru 244

1. Disconnect the battery cables at the battery as a precaution against an accidental spark igniting the fuel or fuel fumes present during the service work.

2. Disconnect the quick-disconnect on the fuel line to the engine.

3. Remove the hood.

4. Disconnect the fuel line from the junction on the port side of the engine. This is the line from the fuel pump to the carburetors.

5. Remove the low-speed adjustment lever at the bottom of the air silencer.

6. Remove the air silencer cover, and then the air silencer.

7. Disconnect the hose at the bottom of the air silencer (this hose is connected to a fitting at the bottom of the crankcase).

8. Disconnect the choke and throttle linkage between the 3 carburetors on the starboard side of the engine. The linkage will snap out of a retainer on the carburetor cams. The nylon retainers *must* be removed before the carburetor is immersed in any type of cleaning solution. Disconnect the choke wire from the electric choke, by first removing the O-ring and coil spring.

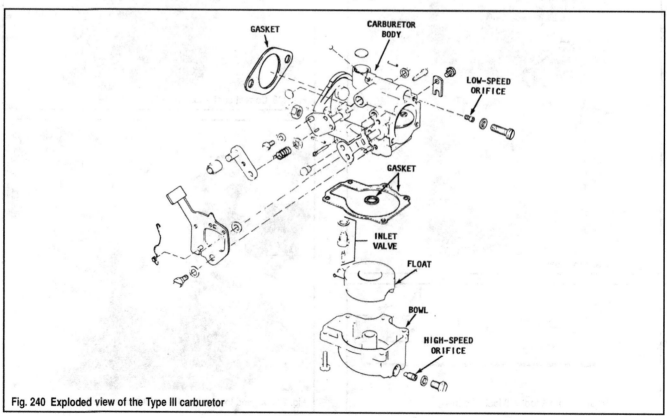

Fig. 240 Exploded view of the Type III carburetor

9. Remove the 6 nuts securing the carburetor to the intake manifold. Identify all of the carburetors as an aid to installation. The carburetors *must* be installed in their original positions. The fuel connections and the choke arrangement are different for each carburetor. The carburetors can now be removed as an assembly.

To Install:

10. Connect the fuel line between the carburetors.

11. Clean the mating surface of the intake manifold. Check to be sure all old gasket material has been removed. Slide a NEW gasket down over the studs into place on the manifold. Install the 3 carburetors at the same time onto the studs and secure them in place with the nuts. Tighten the nuts ALTERNATELY and EVENLY.
Connect the fuel line between the fuel pump and the carburetors.

12. Position a NEW air silencer gasket in place on the front of the engine.

13. Connect the hose from the crankcase to the back of the air silencer. Coat the threads of the air silencer retaining screws with Loctite and then install the air silencer.

14. Install the air silencer cover and at the same time slide the plastic adjusting knob onto the low-speed needle for each carburetor. DO NOT push the knobs all the way home onto the needle at this time. You may seriously consider NOT installing and using the piece of linkage connecting the low-speed needle valve of each carburetor.

15. On the starboard side of the engine, install the choke and throttle retainers.
Connect the choke and throttle linkage between the carburetors. Connect the choke coil solenoid wire onto the linkage stud, and then slide the O-ring

and coil spring onto the stud to secure the wire in place. Adjust the choke butterflies by loosening the screw between the top and bottom carburetor linkage. The center carburetor is the base unit. Adjustments are made from the center carburetor to the top and bottom carburetors. Make the adjustment to close the upper and lower carburetor choke butterflies when the center butterfly is closed. After the adjustment is satisfactory, tighten the screw.

16. Adjust the throttle butterflies in the same manner as the choke butterflies. Loosen the screw between the top and center and the bottom and center carburetor linkage. Make the adjustment to close upper and lower throttle butterflies when the center butterfly is closed. After the adjustment is correct, tighten the screw.

■ This linkage permits adjustment of the three carburetors simultaneously, while using only one knob. Sounds great! But when the one carburetor is adjusted, the others are also changed. A great many professional mechanics have discovered that the linkage is not required. They feel it is far more efficient to adjust each carburetor individually.

■ It is best to synchronize the fuel and ignition systems at this time. See the information under Timing & Synchronization in the Maintenance & Tune-Up section. After the synchronization has been completed, proceed with the following work.

✳✳ CAUTION

NEVER operate the engine above idle speed using the flush attachment. If the engine is operated above idle speed with no load on the propeller, the engine could RUNAWAY resulting in serious damage or destruction of the unit.

17. Mount the engine in a test tank or body of water. If this is not possible, connect a flush attachment and garden hose to the lower unit.

✳✳ CAUTION

Water must circulate through the lower unit to the engine any time the engine is run to prevent damage to the water pump in the lower unit. Just 5 seconds without water will damage the water pump.

18. Start the engine and allow it to warm to operating temperature.

19. Pop the 3 rubber caps, one for each carburetor, out of the front cover of the air silencer. Adjust the low-speed idle by turning the low-speed needle valve CLOCKWISE until the engine begins to misfire or the rpm drops noticeably. From this point, rotate the needle valve COUNTERCLOCKWISE until the engine is operating at the highest rpm. If the engine coughs and operates as if the fuel is too lean, but the idle adjustments have been correctly made, then re-check the synchronization between the fuel and ignition systems.

20. After the idle has been adjusted, push the idle knob retainers onto the needle until it engages the spline on the low-speed needle.

21. Repeat the adjustment portion of this procedure for the other 2 carburetors. Replace the rubber caps into the front cover of the air silencer.

Fuel Outlet Hose

Fig. 241 Disconnect the fuel lines. . .

Air Silencer Cover

Fig. 242 . . .and remove the silencer cover

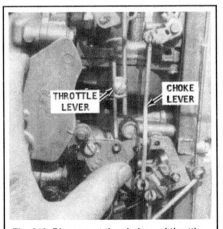

THROTTLE LEVER CHOKE LEVER

Fig. 243 Disconnect the choke and throttle linkage. . .

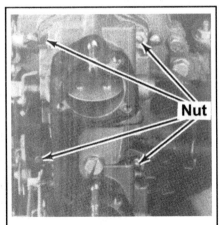

Nut

Fig. 244 . . .and remove the carburetor from the manifold

DISASSEMBLY & ASSEMBLY

◆ See Figures 240 and 245 thru 256

■ Only one carburetor will be rebuilt in the following procedures. Service of the other 2 units is to be performed in the same manner. The carburetor parts should be kept with the individual unit to ensure all items are installed back in their original positions.

1. Disconnect the fuel hoses between the 3 carburetors. This is accomplished by simply cutting the tie strap or working the clip securing the hose to the carburetor fitting.

2. Remove the drain plug or the high speed orifice plug from the bottom of the carburetor bowl.

3. Remove the orifice from the float bowl, using the proper size screwdriver or special OMC tool (#317002).

4. Remove the low speed needle valve. After removing the needle valve, observe and remove the retainer that accepts the needle valve. Good shop practice is to replace the retainer because the retainer is actually the component holding the needle valve in adjustment. If the retainer has become worn, it is not possible to hold the needle valve in an accurate adjustment.

■ The low speed needle valve has a bearing deep inside the carburetor. Removal of this bearing is no small task. A special tool is not available. However, a paper clip with a hook on the end can be used to reach inside the bore and remove the bearing. Actually this bearing is made of plastic and the needle valve rotates inside. The bearing centers the needle valve in the carburetor.

5. Remove the low speed needle valve bearing using a paper clip as described in the previous Note.

6. Turn the carburetor upside-down and remove the screw securing the float bowl to the carburetor body. Lift the bowl free of the carburetor.

7. Remove and discard the bowl gasket and the small gasket around the high speed nozzle. The high speed nozzle is NOT removable.

8. Remove the float hinge pin, and then remove the float.

9. Remove the inlet needle and seat from the carburetor body.

To assemble:

10. Install the inlet seat using the proper size screwdriver. Squeeze just a drop of lightweight oil into the seat. Thread the inlet needle into the seat and tighten it until it barely seats. DO NOT over-tighten or the needle valve may be damaged.

■ Two different type gaskets are used on this carburetor. One MUST be installed before the float, because it is a 1-piece gasket with the nozzle gasket incorporated. The other type consists of 2 individual gaskets, one for the float and a separate one for the nozzle.

11. Position NEW gasket(s) in place on the carburetor body and on the nozzle.

12. Lower the float down over the nozzle, and then slide the hinge pin into place.

13. Hold the carburetor body in a horizontal position and check to be sure the float is parallel to the carburetor surface. If the float is not parallel, *carefully* bend the float tang until the float is in a parallel position and both sides are equal distance from the carburetor surface.

14. Place the float bowl in position and then secure it in place with the attaching screws.

15. Install the low speed retainer in the carburetor body.

16. Slide the nylon bearing onto the end of the low speed needle valve and then thread the valve into the retainer. Continue threading the valve into the retainer until it seats *lightly*, and then back it out COUNTERCLOCKWISE 5/8 of a turn.

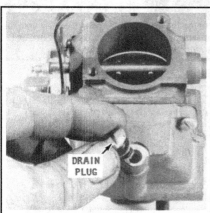

Fig. 245 Remove the drain plug or the high speed orifice plug. . .

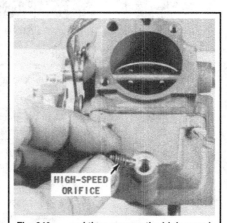

Fig. 246 . . .and then remove the high speed orifice

Fig. 247 Remove the low speed needle valve and retainer. . .

Fig. 248 . . .and then pull out the bearing with a paper clip

Fig. 249 Remove the float bowl. . .

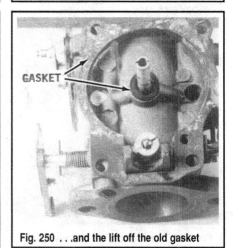

Fig. 250 . . .and the lift off the old gasket

Fig. 251 Older original float bowl gaskets were one piece, make sure that you replace them with the newer style two piece gasket set

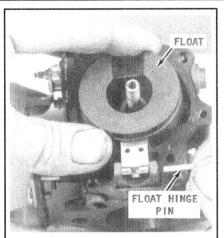

Fig. 252 Pull out the hinge pin and lift off the float

Fig. 253 Don't forget to remove the inlet needle and seat

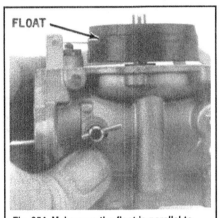

Fig. 254 Make sure the float is parallel to the carb surface

Fig. 255 On early versions, insert the retainer. . .

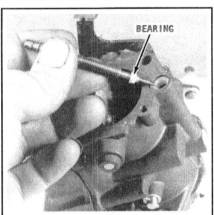

Fig. 256 . . .and then install the low speed needle with the bearing

■ Take time to use the proper size screwdriver, or special tool (#317002), to install an orifice. If the orifice is damaged and the edge of the opening burred because the wrong size screwdriver was used, the flow of fuel will be restricted and the orifice will not function properly. Damage to the orifice will also make it very difficult to remove during the next carburetor service.

17. Install the high speed orifice into the carburetor bowl, using the proper size screwdriver to prevent damaging the orifice.
18. Thread the plug, with a new gasket, into place and tighten it snugly.
19. Repeat the previous procedures for the 2nd or 3rd carbs if necessary.

CLEANING & INSPECTION

② MODERATE

■ Please refer to Cleaning & Inspection in the Type IA/IB Carburetor section for detailed procedures and illustrations.

Type IV Carburetors

◆ See Figures 257 and 258

Type IV A
Two or three carburetors per powerhead, front draft, double barrel with low-speed orifice and fixed high-speed orifice.
- 85 hp: 1969-71
- 100 hp: 1971
- 115 hp: 1969-70
- 125 hp: 1971

Type IV B
Same as Type IVA except with fixed low speed and high speed orifices.

- 85 hp: 1972
- 100 hp: 1972
- 125 hp: 1972

Both the "A" and "B" Type IV carburetors are front draft, with two throats (double barrel). The Type IV A has a low-speed adjustable needle valve and fixed high-speed orifice. The Type IV B has fixed low-speed and high-speed orifices. Both models are covered in this section.

Two of the Type IV carburetors are installed on the model engines listed in the heading. This arrangement provides a separate "barrel" for each cylinder. The float in each carburetor serves 2 cylinders.

REMOVAL & INSTALLATION

◆ See Figures 257 thru 262

② MODERATE

1. Disconnect the battery cables at the battery as a precaution against an accidental spark igniting the fuel or fuel fumes present during the service work.
2. Disconnect the quick-disconnect on the fuel line to the engine.
3. Remove the hood.
4. Remove the retaining screws from the front of the air silencer. Remove the drain hose from the bottom of the air silencer.
5. Remove the adjusting arms to the low-speed needle valves, if used. Remove the 4 screws securing the retainer base to the carburetors. Lift the retainer base upward and out of the way.
6. Disconnect the choke linkage. Slip the retaining ring and the choke solenoid spring back from the upper choke arm. Remove the screws securing the choke yoke to the air silencer. Unclamp the wires to the choke solenoid. Set the choke solenoid aside.
7. Remove the fuel pump retaining screws. Move the fuel pump aside to allow the carburetors to be removed. It is not necessary to disconnect the

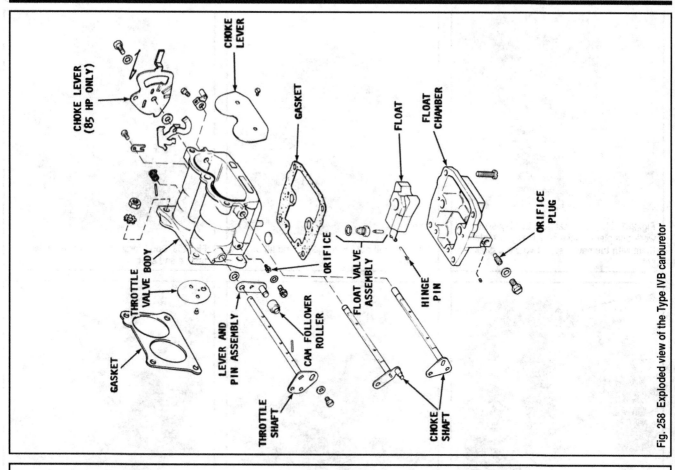

Fig. 258 Exploded view of the Type IVB carburetor

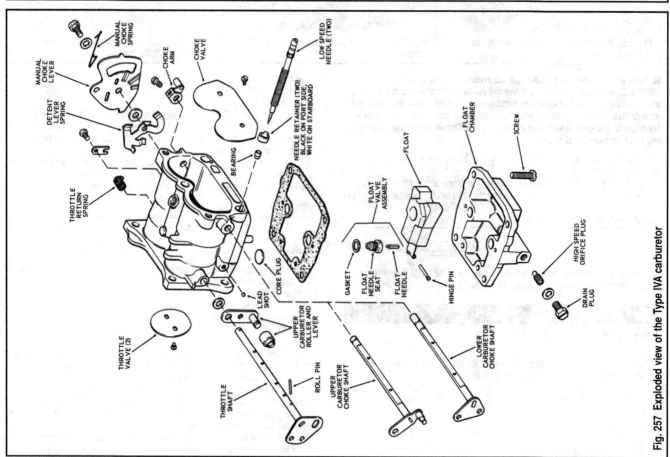

Fig. 257 Exploded view of the Type IVA carburetor

lines to and from the fuel pump. Disconnect the fuel line hoses between the 2 carburetors.

8. On the starboard side of the engine, notice the arms between the carburetors for throttle and choke action. Loosen the linkage fastener; slide the fastener upward and the linkage arms will then come free.

9. Remove the carburetor retaining nuts and lock washers. Lift the carburetors free of the intake manifold.

To Install:

10. Install and connect the fuel hoses between the carburetors.

11. Place a NEW gasket in position on the intake manifold. Install the lower carburetor first, and secure it in place with the 4 attaching screws. Install the upper carburetor on the V4 engines or the middle and upper carburetors on the V6 engines.

12. Connect the fuel supply hose to the carburetors.

13. Install a NEW gasket to the face of the carburetor, and then the air silencer base. Install the choke to the air silencer base. Connect the linkage to the carburetor choke arm. Secure the linkage in place with the small O-ring. Sometimes an E-clip is used.

14. Connect the linkage between the carburetors on the starboard side of the engine. Make sure the throttle butterflies are fully closed when the connections are made. If the carburetor has the "primer choke system" the throttle butterflies are not used. Tighten the retainer securely. Connect the choke linkage between the carburetors. Make sure the choke butterflies are wide open when the connections are made. Tighten the retainer securely.

15. If equipped, check to be sure the choke butterfly in each carburetor seats fully closed. Check the throttle butterflies to be sure both are fully

closed. Now, move the linkage and check to be sure they both open fully and to the same degree. If an adjustment is required, the retainer on the linkage can be moved.

✳✳ CAUTION

NEVER operate the engine above idle speed using the flush attachment. If the engine is operated above idle speed with no load on the propeller, the engine could RUNAWAY resulting in serious damage or destruction of the unit.

16. Mount the engine in a test tank or body of water, then properly synch the carburetors as detailed under Timing & Synchronization in the Maintenance & Tune-Up section.

✳✳ CAUTION

Water must circulate through the lower unit to the engine any time the engine is run to prevent damage to the water pump in the lower unit. Just five seconds without water will damage the water pump.

17. Start the engine and allow it to warm to operating temperature. Adjust the low-speed idle by turning the low-speed needle valve CLOCKWISE until the engine begins to misfire or the rpm drops noticeably. From this point, rotate the needle valve COUNTERCLOCKWISE until the engine is operating at the highest rpm. If the engine coughs and operates as if the fuel is too lean, but the idle adjustments have been correctly made, then re-check the synchronization between the fuel and ignition systems.

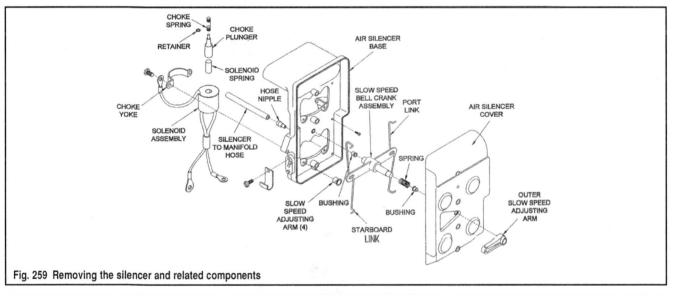

Fig. 259 Removing the silencer and related components

Fig. 260 Remove the fuel pump and lines

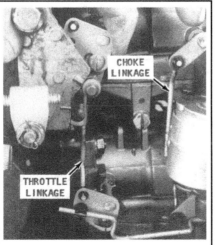

Fig. 261 Disconnect the choke and throttle linkages

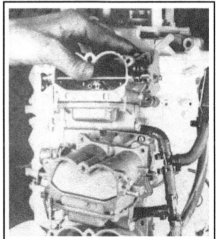

Fig. 262 Remove the mounting bolts and lift off the carbs

DISASSEMBLY & ASSEMBLY

◆ See Figures 257, 258 and 263 thru 272

■ Only one carburetor will be rebuilt in the following procedures. Service of the other unit(s) is to be performed in the same manner. The basic carburetor is the unit with the orifices. Differences for units with the low-speed adjustments will be noted. The carburetor parts should be kept with the individual unit to ensure all items are installed back in their original positions.

1. Lay the carburetors on the bench on the throat openings. Disconnect and remove the hoses between the carburetors.

2. Turn the carburetor body upside down, and remove the 4 float chamber screws.

3. Lift the bowl assembly free of the carburetor body. Remove and discard the gasket.

4. Remove the hinge pin and then the float.

5. Remove the needle from the seat, then the seat, and finally the gasket.

6. Remove the 2 low speed orifice screws. These screws are located just above the high speed orifices on both sides of the carburetor body. Use the proper size screwdriver, or special tool (OMC #317002), and remove the 2 low speed orifices.

7. If the unit being serviced has the adjustable low speed needle valves, remove the low speed needle valves. After removing the needle valve(s), observe and remove the retainer that accepts the needle valve. Good shop practice is to replace the retainer, because the retainer is actually the component holding the needle valve in adjustment.

If the retainer has become worn, it is not possible to hold the needle valve in an accurate adjustment.

8. Observe the low speed needle valve bearing deep inside the carburetor. Removal of this bearing is no small task. A special tool is not available. However, a paper clip with a hook on the end can be used to reach inside the bore and remove the bearing. Actually this bearing is made of plastic and the needle valve rotates inside. The bearing centers the needle valve in the carburetor.

9. Identify the orifices (or needle valves) to ensure each will be installed into the opening from which it was removed.

10. Remove the 2 drain plugs, one on each side, from the carburetor body.

11. Use the proper size screwdriver, or special tool (OMC #317002), and remove the 2 high-speed orifices.

To assemble:

Replace all gaskets and O-rings. Never attempt to use the original parts a second time.

12. Install the inlet seat using the proper size screwdriver. Squeeze a drop of lightweight oil into the seat. Thread the inlet needle valve into the seat, and tighten it until it barely seats. DO NOT over-tighten or the needle valve may be damaged.

13. Position NEW gaskets in place on the carburetor body and on the nozzle.

14. Lower the float down over the nozzle, and then slide the hinge pin into place. Hold the carburetor body in such a manner that it allows the float to hang free. The float should be parallel with the carburetor body. If it is not, bend the tab as close to the float as possible until the float is parallel with the body. Take care to bend the tab on both sides of the float and to bend it squarely.

■ Take time to use the proper size screwdriver, or special tool (OMC #317002), to install an orifice. If the orifice is damaged and the edge of the opening burred, because the wrong size screwdriver was used, the flow of fuel will be restricted and the orifice will not function properly. Damage to the orifice will also make removal very difficult during the next carburetor service.

Fig. 263 Disconnect the fuel lines at each carburetor

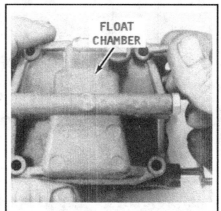

Fig. 264 Remove the float chamber screws. . .

Fig. 265 . . .and then lift off the bowl and gasket

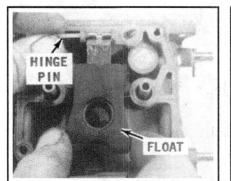

Fig. 266 Slide out the hinge pin and lift out the float

Fig. 267 Remove the 2 low speed screws

Fig. 268 Remove the 2 low speed orifices

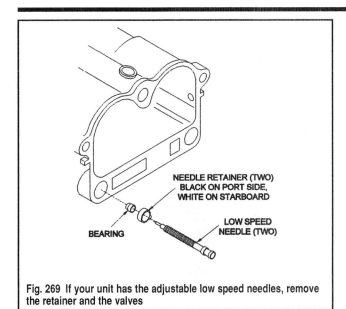

Fig. 269 If your unit has the adjustable low speed needles, remove the retainer and the valves

15. Use special tool (OMC #317002) and install the high speed orifices, one on each side of the carburetor body.

16. Install the drain plugs with NEW gaskets.

Models with low speed orifices

17. Install the low speed orifices into the carburetor body just above the high speed orifices.

18. Install the screws, with NEW washers behind the orifices.

Models with low speed needle valves

19. Install the low speed retainer in the carburetor body. Slide the nylon bearing onto the end of the low speed needle valve, and then thread the valve into the retainer. Continue threading the valve into the retainer until it seats *lightly*, and then back it out counterclockwise 5/8 of a turn.

20. Install the float chamber and secure it with the 4 screws. Tighten the screws alternately and evenly.

CLEANING & INSPECTION

■ Please refer to Cleaning & Inspection in the Type IA/IB Carburetor section for detailed procedures and illustrations.

Fig. 270 Use a paper clip to pull out the bearing

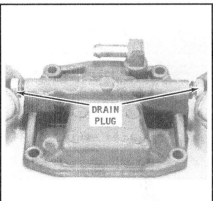

Fig. 271 Remove the 2 drain plugs. . .

Fig. 272 . . .and then thread out the high speed orifices

FUEL SYSTEM COMPONENTS

Heat/Electric Choke

REMOVAL & INSTALLATION

◆ See Figures 273 thru 283

■ The images shown are of a carburetor off a 2 cyl motor, but are representative of all heat/electric choke systems.

On the 90 and 100 hp engines the heat choke is mounted on top of the carburetor and the electric portion is at the bottom of the carburetor.

On all other engines covered here, the heat choke is mounted on the side of the carburetor and the electric portion is a part of the heat choke.

■ The carburetor does NOT have to be removed in order to service the choke system. Step No.1 need only be performed if the carburetor is to be removed.

1. Observe the compression nut under the plate securing the heat tube to the exhaust manifold. Loosen the nut and slide it back on the heat tube. At the other end of the heat tube is another compression nut securing the tube in the choke chamber of the carburetor. Loosen the compression nut and slide it back on the tube. Remove the heat tube from the engine.

2. Observe the marks on the cover plate and on the base - notice the cover plate has only one mark, but the base has several marks. Identify the relationship of the cover plate and the marks on the base to ensure the cover

plate will be installed back into its original position. Remove the 3 screws securing the cover plate to the choke base.

3. Lift the cover and gasket from the choke base. Note that the cover has a spring on the inside; this spring performs a very important role in the choke system. As the heat from the exhaust manifold moves through the tube and around the spring, the spring expands and releases its pressure on the choke. Also notice how the spring is attached to the choke lever because the spring is attached to a plunger which actuates the electrical part of the choke system.

4. Remove the 2 screws securing the electric choke solenoid to the carburetor. Remove the solenoid and take care to retain the spring inside the solenoid. Remove the choke solenoid gasket.

5. Using a pair of needle-nose pliers, reach inside the choke base and remove the spring from the choke lever. Lower the spring and plunger from the carburetor. Further disassembly of the combination heat/electrical choke is not necessary.

If the choke is to be assembled without rebuilding the carburetor, proceed with the following steps.

If the carburetor is to be rebuilt at this time, proceed to the appropriate section, depending on the unit being serviced.

To Install:

6. Install the choke plunger and spring by hooking the spring into the choke lever.

7. Insert a NEW gasket into the cavity of the choke base and then insert the spring into the bore of the choke solenoid.

Fig. 273 Loosen the compression nuts to remove the heat tube

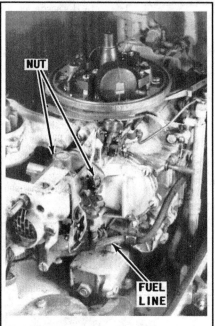

Fig. 274 Disconnect the fuel line and remove the carburetor

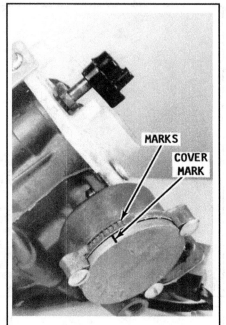

Fig. 275 Matchmark the cover plate to the base before removing it. . .

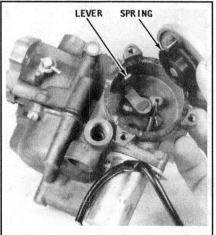

Fig. 276 . . .and the lift off the cover, paying attention to the spring/lever orientation

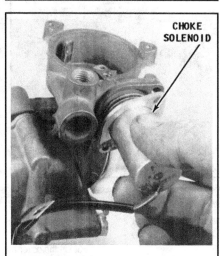

Fig. 277 Remove the choke solenoid. . .

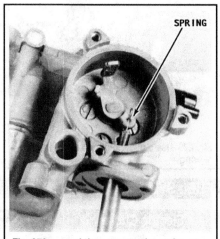

Fig. 278 . . .and then remove the spring at the choke lever

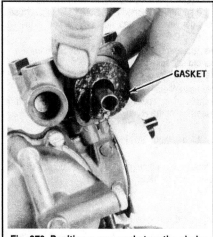

Fig. 279 Position a new gasket on the choke base cavity. . .

Fig. 280 . . .and then insert the spring into the solenoid bore before installing the unit

Fig. 281 Don't forget the heat shield after the tube has been installed

8. Work the choke solenoid up the plunger and secure the plunger with the 2 retaining screws.

9. Position a NEW gasket onto the surface of the carburetor.

10. Install the choke cover onto the carburetor. Notice how the spring has a clip and the choke lever has a protrusion? The clip *must* fit around the lever and protrusion.

11. Align the mark on the cover plate with the matching mark on the choke base and then tighten the cover plate in this position with the 3 attaching screws.

12. Connect the heat tube to the engine and secure it in place with the compression nuts at each end.

Electric Choke

REMOVAL & INSTALLATION

◆ See Figures 282 thru 286    MODERATE

The electric choke is usually mounted on the bottom of the carburetor for 1-2 cylinder motors, or the starboard side of the carburetor got the 3 cyl/V4 motors.

■ **Although not strictly necessary, we recommend removing the carburetor before working on the choke. Take a look at the set-up on your particular engine and make your own decision.**

1. Disconnect the electrical wire from the choke solenoid.

2. Disconnect the fuel line at the carburetor, plug it, and position it out of the way.

3. Remove the 2 mounting bolts and carefully lift the carb/choke assembly from the manifold.

4. Scribe a mark on the side of the bracket and over the solenoid body and clamp.

5. Remove the cotter pin, washers and pin from the choke plunger and pin.

6. Remove the 2 screws securing the choke clamp to the carburetor body. Remove the choke solenoid from the carburetor.

■ **The choke solenoid CANNOT be serviced.**

7. If the boot is unfit for further service, it may be removed by sliding it off the solenoid. Note the 2 washers and spring under the boot. Take care not to lose these items.

8. Clean the solenoid, and then slide the boot back in place.

9. If necessary, straighten and remove the cotter pins at each end of the choke rod and then remove the rod.

To Install:

10. Install the choke rod as it was removed - although not imperative, we recommend using new pins.

11. Position the choke assembly into the choke bracket on the carburetor body. Install the retainer clamp around the choke and start the 2 attaching screws on each side.

12. Shift the choke assembly until the bottom of the choke is even with the bottom of the bracket on the carburetor. Rotate the choke assembly until the choke wires are accessible. Make sure that all of your match marks line up and then tighten the screws securely.

13. Install the pin through the end of the shaft of the choke solenoid and into the lever attached to the carburetor. Install the washer and cotter pin. Again, we'd use new cotter pins if it were us doing the work!

14. Position the assembly onto the manifold and tighten the mounting bolts securely (and alternately). Unplug and reconnect the fuel line.

15. Connect the choke wires to the choke.

Water Choke

DESCRIPTION

An understanding of exactly how the water choke functions is essential to ensure that the service procedures are properly performed. Hot water from the engine is allowed to circulate inside a chamber of the choke. As the hot water heats the spring, the metal expands and the spring relieves its tension on the choke. Once the tension is released, the choke opens.

The hot circulating water is routed to the port side of the engine and is discharged into the exhaust chamber. An additional feature is included in the

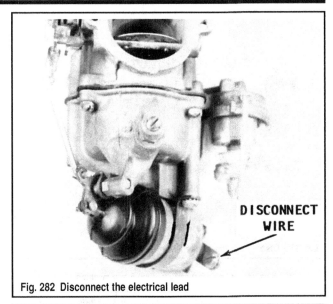
Fig. 282 Disconnect the electrical lead

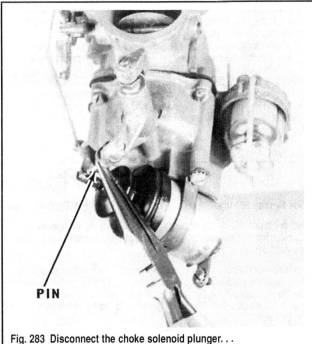
Fig. 283 Disconnect the choke solenoid plunger. . .

Fig. 284 . . .and then remove the solenoid

automatic choke assembly. A spring-loaded diaphragm and plunger assembly, activated by intake manifold pressure, closes the choke the instant the engine is shut down, whether the engine is hot or cold. The plunger will release its hold on the choke only when the engine is again started. At low manifold pressure, with the engine operating, the spring-loaded plunger is released to open the choke and the thermal (water) system is in control. This system operates independently of the water choke system.

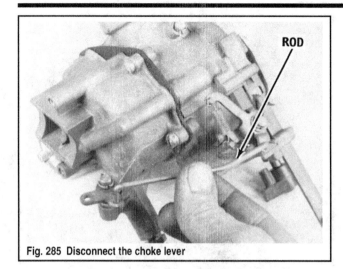

Fig. 285 Disconnect the choke lever

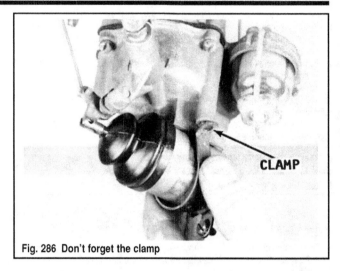

Fig. 286 Don't forget the clamp

A check valve and orifice are installed in the water system to prevent flooding of the diaphragm when the manifold pressure fluctuates during slow speed operation. A puncture, or other damage, to the diaphragm will cause the choke to be held in the closed position at all times, when the vacuum system is in the automatic position. However, the choke can still be opened by using the manual lever.

During installation of the carburetor, considerable attention must be exercised to install the proper hose to the correct fitting. If the hoses are crossed, water will be discharged into the intake manifold and the engine, resulting in a series of horrendous problems. This water will then be released from the cylinder through the spark plug openings when the spark plugs are removed and the engine is cranked in an effort to start it. *Therefore*, take time to identify the hoses to ensure each is correctly connected to the proper carburetor fitting.

REMOVAL & INSTALLATION

◆ See Figures 287 thru 296

■ Although not absolutely necessary, we recommend removing the carburetor for this procedure. Take a look at your particular engine and make your decision.

1. Remove the hoses connected to the choke, 1 from the top, and the other 2 from the bottom. The hose connected to the bottom of the choke and closest to the carburetor is the water outlet line. The other hose on the bottom is the vacuum line. The hose connected to the top of the choke is the water inlet line.

2. Disconnect the fuel line, plug it and move it out of the way. Loosen the 2 mounting bolts and then remove the carb/choke assembly.

3. Make sure there is a mark on the choke cover and a matching mark on the choke base. These marks must be aligned in the same position when the choke is assembled to the carburetor. Don't forget to do this!

4. Remove the 3 screws securing the choke cover to the base. Lift the cover from the base and take notice of the spring on the inside of the cover. This spring relieves pressure on the choke as heated hot water from the engine circulates through the chamber.

5. After the cover has been removed, it is not necessary to remove the choke base from the carburetor. Remove the 4 screws securing the choke head, and then remove the head.

6. Remember that there is a spring hidden by the diaphragm. Remove the diaphragm and spring from the choke head.

7. Remove the small valve and support located under the spring.

■ Further disassembly of the water choke valve is not necessary. If the choke is to be assembled without rebuilding the carburetor, proceed with the following steps. If the carburetor is to be rebuilt at this time, proceed to the appropriate section.

To Install:

8. Position the check valve support into the vacuum opening of the choke head.

9. Position the check valve into the head on top of the support.

10. Position the spring onto the top of the support.

11. Lower the stem down through the spring and diaphragm and onto the base.

Fig. 287 A good look at the carb with a water choke

Fig. 288 Matchmark the choke cover to the base. . .

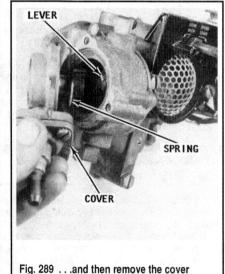

Fig. 289 . . .and then remove the cover

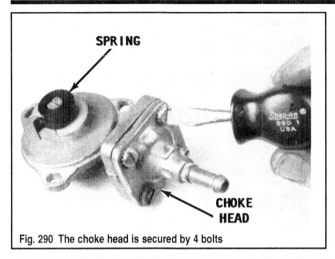

Fig. 290 The choke head is secured by 4 bolts

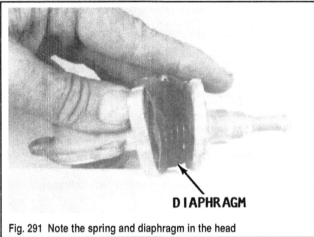

Fig. 291 Note the spring and diaphragm in the head

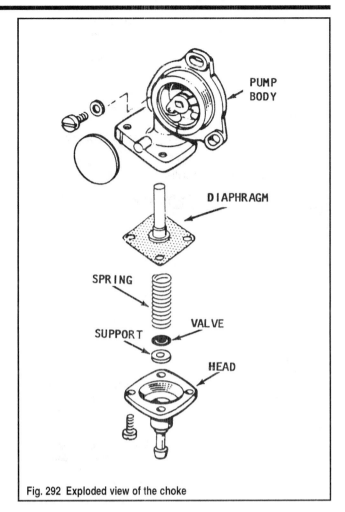

Fig. 292 Exploded view of the choke

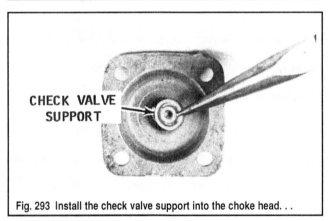

Fig. 293 Install the check valve support into the choke head. . .

12. Install the head and secure it in place with the attaching screws. Tighten the screws ALTERNATELY as a precaution against warping the cover.

13. Take the cover assembly and install a new gasket onto the base of the carburetor/choke. Work the spring clip over the protrusion of the choke lever and lower it into the cavity.

14. Align the mark on the cover with the mark on the body, and then secure the cover in this position with the 3 attaching screws.

15. If removed, install the assembly onto the manifold and tighten the nuts/bolts. Connect the fuel line.

16. Connect the vacuum and water hoses to the carburetor paying careful attention to connect the proper hose to the correct fitting according to the identification given to the hoses during disassembly. If identification was not made during disassembly, follow the hose to the other end to ensure proper connection to the carburetor.

   a. Connect the water hose to the top fitting of the water choke.

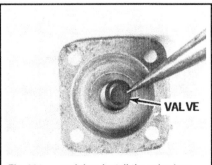

Fig. 294 . . .and then install the valve into the vacuum opening

Fig. 295 Install the spring onto the top of the support. . .

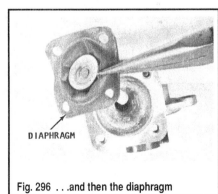

Fig. 296 . . .and then the diaphragm

b. Connect the outlet water hose to the fitting closest to the carburetor. The other end of this line is connected to a fitting on the exhaust chamber on the port side of the engine.

c. Connect the vacuum line to the vacuum diaphragm fitting. The other end of this line is connected to the intake manifold.

■ **If the hoses are not connected correctly, water will be injected into the cylinder when the engine is cranked, resulting in a series of horrendous problems.**

Install the starter motor and generator if the engine being serviced is equipped with these two units.

17. Mount the engine in a test tank or on the boat in a body of water. Make any carburetor adjustments under a load condition.

Fuel Pump

◆ **See Figures 297 and 298**

A considerable number of fuel pump designs and sizes have been installed on the Johnson/Evinrude engines covered here. Only one can truly be rebuilt and detailed procedures with illustrations are given in this section.

This fuel pump has 3 nipples providing the means of connecting fuel and vacuum lines. The vacuum line is connected to one nipple. The other end of this hose is connected to the vacuum side of the engine. The inlet hose (from the fuel tank) is connected to the second nipple. The outlet hose (to the carburetor) is connected to the third nipple.

Minor changes have been incorporated into the fuel pump over the years. These changes will be identified in the text.

All other fuel pumps must be replaced as a unit; although they can be overhauled (sort of). However, the pump cover can be removed, the filter screen cleaned or replaced, and a NEW cover gasket installed.

The accompanying illustrations show only a couple of these fuel pumps, including the unit that can be rebuilt.

■ **Have a shop towel and a suitable container handy when testing or servicing a fuel pump as fuel will likely spill from hoses disconnected during these procedures. To ensure correct assembly and hose routing, mark the orientation of the fuel pump and hoses before removal.**

TROUBLESHOOTING

If the spark plug of the cylinder to which the vacuum line is connected becomes wet fouled, the cause may very well be a ruptured fuel pump diaphragm. This reasoning is sound for both types of fuel pumps. Both types of pumps have an inlet hose from the fuel source, an outlet hose to the carburetors, and a vacuum hose from the powerhead.

A good test for the pump is to disconnect the vacuum line from the engine, operate the squeeze bulb in the fuel line until it is firm, and then to carefully observe the end of the vacuum hose to detect any fuel leakage. The smallest amount of fuel from the hose indicates a damaged diaphragm.

The pump must be rebuilt to restore satisfactory service of the pump. If the pump is a non-rebuildable type, the unit must be replaced.

TESTING

General

◆ **See Figure 299**

The problem most often seen with diaphragm-displacement fuel pumps (the non-rebuildable kind) is fuel starvation, hesitation or missing due to inadequate fuel pressure/delivery. In extreme cases, this might lead to a no start condition as all but total failure of the pump prevents fuel from reaching the carburetor(s). More likely, pump failures are not total, and the motor will start and run fine at idle, only to miss, hesitate or stall at speed when pump performance falls short of the greater demand for fuel at high rpm.

Before replacing a suspect fuel pump, be absolutely certain the problem is the pump and NOT with fuel tank, lines or filter. A plugged tank vent could create vacuum in the tank that will overpower the pump's ability to create vacuum and draw fuel through the lines. An obstructed line or fuel filter could also keep fuel from reaching the pump. Any of these conditions could partially restrict fuel flow, allowing the pump to deliver fuel, but at a lower pressure/rate. A pump delivery or pressure test under these circumstances would give a low reading that might be mistaken for a faulty pump. Before testing the fuel pump, refer to the testing procedures found under Fuel Lines & Fitting to ensure there are no problems with the tank, lines or filter.

If inadequate fuel delivery is suspected and no problems are found with the tank, lines or filters, a conduct a quick-check to see how the pump affects performance. Use the primer bulb to supplement fuel pump. This is done by operating the motor under load and otherwise under normal operating conditions to recreate the problem. Once the motor begins to hesitate, stumble or stall, pump the primer bulb quickly and repeatedly while listening for motor response. Pumping the bulb by hand like this will force fuel through the lines to the carburetor, regardless of the fuel pump's ability to deliver fuel. If the engine performance problem goes away while pumping the bulb, and returns when you stop, there is a good chance you've isolated the fuel pump as the culprit. Perform a pressure test to be certain, then repair or replace the pump assembly.

■ **Keep in mind that low vacuum supply from the crankcase or insufficient vacuum at the pump itself due to bad seals can also be the culprit for poor fuel delivery.**

✳✳ WARNING

Never run a motor without cooling water. Use a test tank, a flush/test device or launch the craft. Also, never run a motor at speed without load, so for tests running over idle speed, make sure the motor is either in a test tank with a test wheel or on a launched craft with the normal propeller installed.

Pump Pressure Test

◆ **See Figure 299**

By far the most accurate way to test the fuel pump is using a low-pressure fuel gauge while running the engine at various speeds, under load.

Fig. 297 Many models will use a pump that is not serviceable...

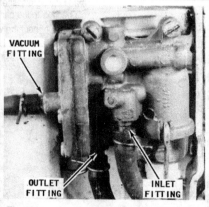

Fig. 298 ...but if yours looks like this, it is rebuildable

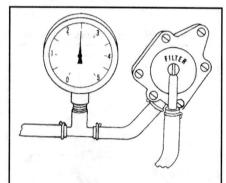

Fig. 299 Use a fuel pressure gauge connected in the pump outlet line to check operating pressure - typical Johnson/Evinrude pump shown

To prevent the possibility of severe engine damage from over-speed, the test must be conducted under load, either in a test tank (with a proper test propeller) or mounted on the boat with a suitable propeller.

1. Test the Fuel Lines & Fittings as detailed in this section to be sure there are no vacuum/fuel leaks and no restrictions that could give a false low reading.

2. Make sure the fuel filter(s) is(are) clean and serviceable.

3. Start and run the engine in forward gear, at idle, until normal operating temperature is reached. Then shut the motor down to prepare for the test.

4. Remove the fuel tank cap to make sure there is no pressure in the tank (the fuel tank vent must also be clear to ensure there is no vacuum). Check the tank location, for best results, make sure the tank is not mounted any more than 30 in. (76mm) below the fuel pump mounting point. On portable tanks, reposition them, as necessary to ensure accurate readings.

■ **The fuel outlet line from the fuel pump may be disconnected at either the pump or the carburetor whichever provides easier access. If you disconnect it from the pump itself you might have to provide a length of fuel line (depending on whether or not the gauge contains a length of line to connect to the pump fitting).**

5. Disconnect the fuel output hose from the carburetor or fuel pump, as desired.

6. Connect a fuel pressure gauge inline between the pump and the carburetor(s).

7. Run the engine at or around each of the following speeds and observe the pressure on the gauge. At idle the gauge must read at least 1 psi (7 kPa). Pressure should increased with engine speed. Typically these pumps will read in the 4-5 psi (27-34 kPA) at or near wide open throttle.

8. If the pump fails to produce sufficient pressure for proper operation, and other causes such as fuel line or filter restrictions have been eliminated, repair or replace the pump.

Pump Leak Test

Pressurize the fuel system using the primer bulb. Squeeze repeatedly, but slowly, until the bulb is firm, then check the pump body and connections for leaks. Remove the pump from the powerhead, leaving the fuel lines connected. Observe the vacuum port at the rear of the pump (where is connects to the port on the powerhead). The leakage of any fuel at this point indicates a damaged diaphragm.

Repair or replace any pump that shows signs of leakage.

REMOVAL, DISASSEMBLY & INSTALLATION

Old Style Rebuildable Type

◆ **See Figures 300 thru 314**

1. Identify each hose and its location, then disconnect the vacuum hose and 2 fuel hoses from the fuel pump.

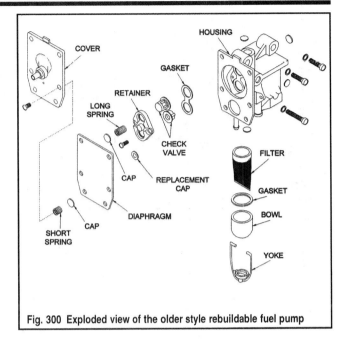
Fig. 300 Exploded view of the older style rebuildable fuel pump

2. Remove the 3 attaching screws securing the pump to the engine. Two screws are visible on top of the pump and the third is hidden behind the fuel inlet nipple.

3. Mount the fuel pump in a vise and loosen the plastic screw securing the filter bowl to the pump.

4. Swing the bail/hinge down and lift the fuel bowl free of the pump. Remove and discard the bowl gasket and filter.

5. Shift the pump position in the vise and remove the 6 screws securing the pump cover. Remove the cover. Take care not to lose the disc washer and spring from the top of the diaphragm.

6. Remove the disc washer, spring, and diaphragm. Save the small disc washer because it will be used again.

7. Some model pumps may have a small disc washer and long spring installed under the diaphragm. Other models may have a large nylon washer and spring. Discard the small disc washer because it has been replaced with the larger nylon washer and is included in the new pump repair kit.

8. Remove the 2 screws from the check valve retainer. Reach into the pump and remove the check valve retainer.

9. Note how one check valve is facing downward and the other valve is facing upward. Also notice the groove in the fuel pump body. A small boss on the retainer fits into the groove as a prevention against installing the retainer incorrectly. Remove the 2 check valves and the check valve gasket.

Further disassembly of the pump is not necessary.

To assemble and install:

Wash the fuel pump body thoroughly and then blow it dry with compressed air.

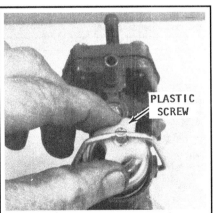
Fig. 301 Mount the pump in a vise and remove the screw

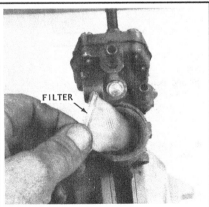
Fig. 302 Lift off the fuel bowl and remove the filter

Fig. 303 Remove the cover screws and cover. . .

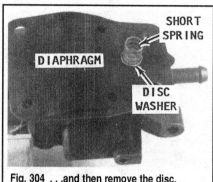

Fig. 304 . . .and then remove the disc, spring and diaphragm

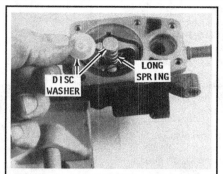

Fig. 305 Some models will have a slightly different set-up

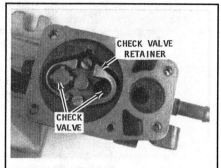

Fig. 306 Loosen the screws and remove the check valve retainer. . .

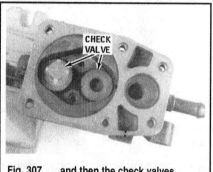

Fig. 307 . . .and then the check valves

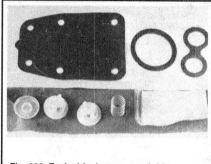

Fig. 308 Typical fuel pump repair kit

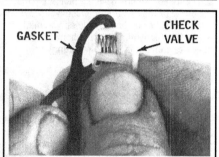

Fig. 309 Insert the check valve through the gasket. . .

All internal parts necessary to rebuild the pump, including diaphragm, check valves, gaskets, etc., will be included in the pump repair kit. At one time, these kits were available from the local OMC dealer at modest cost. However, OMC has discontinued packaging the parts in kit form. Therefore, unless the dealer still has one of the old kits in stock, the fuel pump parts must be purchased individually at your Bombardier dealer.

10. Insert one of the new check valves through the check valve gasket.

11. Install the gasket and check valve into the pump body with the valve facing *downward*.

12. Position the other check valve on top of the gasket facing *upward*.

13. Slide the retainer down over the check valves with the boss on the retainer in the groove of the pump body. Observe the large and small hole in the retainer. The large hole *must* be positioned over the check valve facing UPWARD.

14. Secure the retainer in place with the 2 attaching screws.

15. Install the new long spring over the boss of the retainer and then place the nylon disc washer on top of the spring that was provided in the kit. NEVER use the small disc on top of the long spring.

■ **The following steps may only be properly accomplished by exercising patience and a little time.**

16. Mount the pump in a vise in a vertical position. Lay the diaphragm over the top of the nylon disc washer and onto the pump body. Notice how the spring holds the disc up and partially lifts the diaphragm from the pump. This is a normal condition.

17. Insert the small disc and the short spring into the cavity. This spring and disc help to cushion the vacuum impulses from the engine.

18. Ease the fuel pump cover down over the diaphragm and then thread the 6 cover attaching screws into the pump body. As each screw is started, pull on the edge of the diaphragm to align the screw holes in the diaphragm with the matching holes in the pump body. Tighten the attaching screws securely.

19. Slide the filter element into the fuel pump. The end of the filter element must slide over the indexing peg in the bottom of the pump. Force the element onto the peg until it is fully seated.

20. Place a NEW gasket into position in the fuel pump.

21. Place the fuel pump bowl over the filter element and into position on the gasket.

22. Swing the hinge up and over the pump bowl, and then secure it in place by tightening the plastic thumb screw.

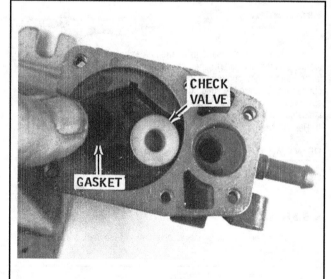

Fig. 310 . . .and then position in the pump body

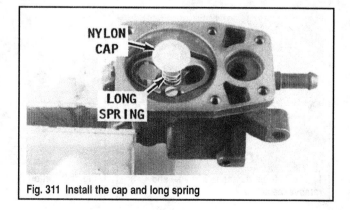

Fig. 311 Install the cap and long spring

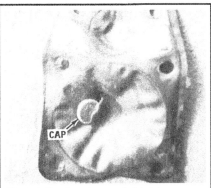

Fig. 312 Here's a damaged old style diaphragm with the small cap that was installed on early engines. The small cap has been replaced with a larger one

Fig. 313 Position the new gasket. . .

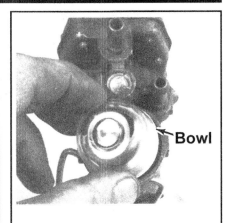

Fig. 314 . . .and then install the pump bowl

23. Mount the fuel pump onto the engine and secure it with the 2 screws on top and 1 behind the fuel inlet nipple.

24. CAREFULLY connect the vacuum hose and the 2 fuel hoses to their proper nipples, as identified during disassembly. If identification was not made, follow the hoses as described. Connect the vacuum hose (from the engine) to the nipple on the pump cover. Connect the inlet hose (from the fuel tank) to the inside nipple (the one closest to the engine). Connect the outlet hose (to the carburetor) to the remaining nipple.

REMOVAL & INSTALLATION

Diaphragm-Displacement Type

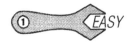

◆ See Figures 315 and 316

1. For safety, either disconnect the negative battery cable (if so equipped) and/or disconnect the spark plug lead(s) and ground them to the powerhead.

2. Locate the fuel pump on the powerhead and determine if it will be easier to remove the lower engine covers. On some models equipped with split (2-piece) lower covers, it is easier to access the pump if the lower engine covers are removed. For details, refer to the Engine Cover procedure in the Engine Maintenance section.

■ On many models, fuel hoses are retained by a plastic wire tie (which must be cut to remove the hose). Use a pair or dikes or cutters to carefully remove the wire tie. Be sure not to cut, nick or otherwise damage the fuel hose or it will have to be replaced. On other models, threaded or spring-type clamps are used on the connections. Threaded clamps are unscrewed to loosen them, while spring-type clamps are released by squeezing the protruding ends of the clamp arms together.

3. Place a small drain basin or a shop rag under the fuel line fittings (to catch escaping fuel), then tag and disconnect the fuel hoses from the pump. Make sure you plug them!

■ The fuel pumps used on these motors are equipped with 2 or 3 sets of bolts visible on the surface of the pump. On most models, a round inlet cover is mounted to the center of the pump with a single bolt. Then, of the remaining bolts, 2 are usually used to secure the pump to the powerhead and the balance of the bolts are used to secure the halves of the body together around the diaphragm.

4. Loosen the pump mounting bolts (the bolts that thread not just through the inlet cover, but all the way through the body of the pump and into the powerhead). If in doubt as to which bolts secure the pump, look at the back of the pump (as can be seen at the pump-to-powerhead seam line) to determine which bolts continue through the pump assembly and into the powerhead. These are the only bolts that should be loosened for pump removal.

■ On the 5 sided pumps used by most Johnson/Evinrude 2-stroke models, there are 2 mounting bolts at the bottom of the pump assembly.

5. If necessary, remove the cover screw or screws and disassemble the fuel pump for inspection or overhaul, as applicable. For details, refer to the Overhaul procedure in this section.

6. Clean the mating surface of the pump and powerhead of any remaining gasket material, dirt, or debris. Be careful not to damage the surface as that could lead to vacuum leaks.

To Install:

7. Apply a coating of Johnson/Evinrude nut lock to the fuel pump retaining screws.

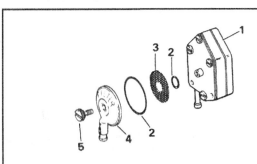

① FUEL PUMP BODY
② O-RING or GASKET
③ FILTER ELEMENT
④ INLET COVER
⑤ SCREW

Fig. 315 Typical diaphragm-displacement fuel pump used on some Johnson/Evinrude outboards

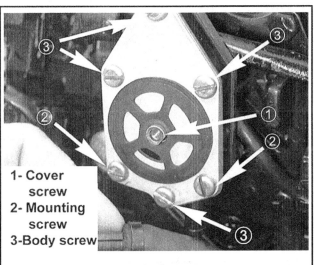

1- Cover screw
2- Mounting screw
3- Body screw

Fig. 316 Typical Johnson/Evinrude fuel pump screw identification

8. Position a new gasket and install the pump to the powerhead using the retaining screws. Tighten the screws to 24-36 inch lbs. (2.8-4.0 Nm).

9. Connect the fuel lines as noted during removal and secure using the clamp or new wire ties, as applicable.

10. Gently squeeze the primer bulb while checking for fuel leakage. Correct any fuel leaks before returning the engine to service.

11. Connect the negative battery cable and/or spark plug lead(s).

OVERHAUL

Diaphragm-Displacement Type

◆ See Figures 315 and 317

Most of the displacement-diaphragm fuel pumps used on Johnson/Evinrude outboards may be disassembled for overhaul, but they're not really rebuildable like the older ones. These pumps are of a fairly simple design with relatively few moving parts. Check with your local parts supplier to make sure that an overhaul kit containing the necessary parts are available for your model. In most cases, the parts are limited to the diaphragm(s), gasket(s) and a fuel inlet screen (if equipped).

If overhaul is required due to damage from contamination or debris (as opposed to simple deterioration) disassemble and clean the rest of the fuel supply system prior to the fuel pump. Failure to replace filters and clean or replace the lines and fuel tank could result in damage to the overhauled pump after it is placed back into service.

All diaphragms and seals should be replaced during assembly, regardless of their condition. Check for fuel leakage after completing the repair and verify proper operating pressures before returning the motor to service.

✳✳ WARNING

No sealant should be used on fuel pump components unless otherwise specifically directed. If small amounts of a dried sealant were to break free and travel through the fuel supply system it could easily clog passages (especially the small, metered orifices and needle valves of the carburetor).

1. Remove the fuel pump from the powerhead as detailed in this section.

2. Matchmark the fuel pump cover, housing and base to ensure proper assembly.

■ **To ease inspection and assembly, lay out each piece of the fuel pump as it is removed. In this way, keep track of each component's orientation in relation to the entire assembly.**

3. Remove the center cover screw, then remove the fuel inlet cover, filter element (screen) and gasket or O-rings (as applicable).

4. Remove the pump housing-to-base screws (usually 4 flat-head screws) and carefully separate the housing from the base, removing the diaphragm(s) from the center.

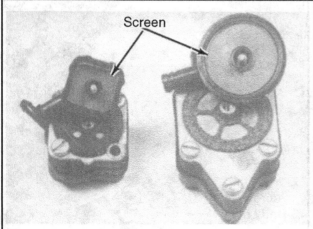

Fig. 317 Many of the pumps used on these motors are equipped with a filter screen under the cover

5. Clean the metallic components thoroughly using solvent and carefully remove all traces of gasket material.

6. Inspect the diaphragm closely for cracks or tears.

■ **It is advisable to replace the diaphragm ANYTIME the fuel pump is disassembled to ensure reliability and proper performance.**

7. Inspect the fuel pump body for cracks. Check gasket surfaces for nicks, scratches, or irregularities. Inspect the mating surfaces of the fuel cover, body and base using a straight edge to ensure that they are not warped from heat or other damage Replace warped or damaged components.

To assemble:

8. Assemble the components of the fuel pump housing, base and diaphragm noting the following:

a. Use new gaskets. Make sure each gasket and the components it seals are aligned properly.

b. Align the matchmarks made during disassembly to ensure proper component mounting.

9. Apply a coating of Johnson/Evinrude Nut Lock, or an equivalent thread locking compound to the pump housing and base screws, then install and tighten them to 24-36 inch lbs. (2.8-4.0 Nm).

10. If equipped, install the fuel pump cover and filter screen using a new gasket or O-ring(s), then tighten the center screw securely.

11. Install the fuel pump and then check for leaks and for proper operation.

Manual Fuel Primer

GENERAL INFORMATION

◆ See Figure 318

Some late-model motors are equipped (or may have been retro-fitted) with a manual fuel primer system to aid with cold starts. Although we doubt any engines covered here will utilize this, we have provided coverage just in case. The basic design of the manual primer is that of a small, hand-operated plunger-type pump. The primer works by drawing fuel into the pump housing through a fuel line with a one-way check valve when the shaft is withdrawn. The fuel is then forced out, toward the motor, through a second one-way check valve when the shaft is pushed back inward.

The primer performs the same function of a choke (aiding cold starting by making sure the engine receives a richer fuel mixture), but by opposite means. Whereas a choke reduces the amount of air provided to the combustion chamber (thus increasing the fuel portion of the air/fuel ratio), a primer works on the fuel side of the ratio by manually increasing the amount of fuel. The extra fuel provided by the manual primer enriches the air/fuel mixture for cold start purposes only. Use of the primer on an engine that is at or near operating temperature can flood the motor preventing starting.

TESTING

General

◆ See Figure 318

An inoperable manual primer will cause hard start or possibly even a no start condition during attempts to start a cold motor. The colder the ambient temperature, the more trouble an inoperable primer will cause. A primer with internal leakage (allowing fuel to bypass the air/fuel metering system) will cause rich running conditions that could include hesitation, stumbling, rough running, especially at idle and lead to spark plug fouling.

Function Test

If the motor is operable, but trouble is suspected with the primer system, perform a function test with the engine running. Although this test can be conducted on a flush-fitting, engine speed will reach 2000 rpm and it is much safer to conduct the test in a test tank or with the boat launched.

1. Start and run the engine until it reaches normal operating temperature.

2. Once the engine warms, set the throttle so it runs at 2000 rpm.

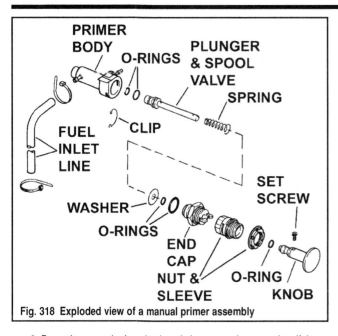

Fig. 318 Exploded view of a manual primer assembly

3. Pump the manual primer knob and observe engine operation. If the primer is operating correctly, the engine should run rich and speed should drop to about 1000 rpm.

4. If the primer seems ineffective, stop the engine, then remove the primer hose from its fitting(s). Check each fitting for clogs using a syringe filled with isopropyl and a clear vinyl 1/8 in. inner diameter hose. Attach the hose to the fitting being checked and press lightly on the syringe. Fluid will move through the fitting unless it is clogged.

5. If any clogs are found, use a thin pick to carefully clean the fitting. Johnson/Evinrude makes a cleaning tool for this purpose (#326623).

6. Be sure to check the primer hose T-fitting for clogs as well.

7. If no clogs are found, perform the Primer Check procedure to see if the problem lies within the primer assembly itself.

Primer Check

If you suspect the manual primer system is not functioning correctly (and no clogs were found in the lines or fittings), check the primer as follows:

1. Remove the fuel line from the primer fitting at the carburetor.

2. Place the end of the fuel line just removed into a suitable container. Squeeze the fuel tank primer bulb to make sure the carburetor bowls are full of fuel.

3. Operate the primer choke lever twice. If fuel squirts from the disconnected fuel line into the container, the manual primer system is functioning correctly. If not, a kinked or restricted fuel line may be the problem, or if no kinks/clogs are found, the primer is at fault. Check the primer nipple to ensure the nipple is free of obstructions.

■ **The most probable cause of a malfunctioning primer system is internal leakage past the O-rings. Therefore if the primer itself is still suspected, proceed to Servicing the Manual Primer.**

REMOVAL & DISASSEMBLY

◆ See Figures 318, 319 and 320

1. Disconnect and plug the inlet and outlet fuel lines to prevent loss of fuel and contamination. Remove the primer assembly from the engine.

2. Carefully pull or pry the retaining clip from the primer body housing. Pull out the end cap, plunger, and spool valve assembly. Slide the end cap from the plunger. Remove and discard the O-ring around the end cap.

■ **Observe the small O-rings, there are usually 2 on the spool valve and 1 around the plunger shaft. These O-rings are made from a special material and must be replaced with a genuine Johnson/Evinrude replacement part. Just matching O-ring sizes will not work!**

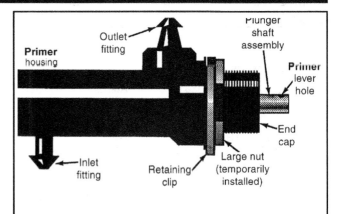

Fig. 319 The primer choke valve removed from the powerhead. The large nut is temporarily installed onto the threads of the end cap for safe keeping

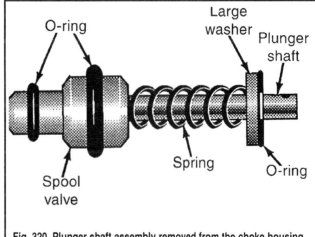

Fig. 320 Plunger shaft assembly removed from the choke housing with major parts identified

3. Remove and discard the O-rings.

4. Remove the large washer and spring from the plunger shaft.

CLEANING & INSPECTION

◆ See Figures 318, 319 and 320

1. Inspect the grooves of the spool valve and the shaft of the plunger for any scratches or burrs. Polish away any imperfections using crocus cloth. If a smooth finish cannot be obtained without removing excessive material, replace the spool valve and plunger assembly.

2. Check the plunger where the cross hole meets the inside hole. The slightest burrs around the cross hole will cause rapid O-ring wear. Remove any burrs and polish using crocus cloth.

3. Inspect the condition of the plunger spring, replace as required.

4. Test each of the one-way valves (there is 1 at each fuel fitting) by blowing through them in turn. Each valve is functioning correctly if it allows air to pass one direction, but not in the other direction. If a valve allows air to be drawn both in and out, the valve is defective. Individual valves are not serviceable. The primer body must be replaced.

■ **The valves can also be checked using a syringe filled with isopropyl alcohol and a length of tube. Squeeze the syringe lightly to force alcohol through the hose, although it is permissible for a drop or two to pass through the wrong direction of a check valve, a steady stream indicates the valve has failed and must be replaced.**

ASSEMBLY & INSTALLATION

◆ See Figures 318, 319 and 320

1. Install the 2 new O-rings around the spool valve. Slide the spring, followed by the large washer and the 3rd O-ring, over the plunger.

2. Install a new O-ring over the end cap and place the end cap over the plunger end.

3. If desired, bench test the assembly before installation, as follows:

   a. Connect a 5 in. (127mm) long piece of hose to the large nipple on the primer assembly, then place the other end of the hose in a container of alcohol.

   b. Connect a length of hose to the small primer nipple and place the other end in a small container (preferable a graduated cylinder or measuring cup).

   c. Hold the primer horizontally (the same way it would be installed on the motor) and pump the plunger 10 times. The primer should deliver approximately 10cc of alcohol to the graduated cylinder total as a result of the 10 strokes.

   d. If the pump does not deliver sufficient volume, disassemble it again and check for torn, missing or dislodged O-rings.

4. Insert the assembly into the primer housing and install the retaining clip to secure everything together.

5. Slide the assembled primer into the opening in the lower cowling and thread the large nut over the protruding threads. Tighten the nut securely.

6. Install the fuel lines to the appropriate fittings and snap the choke lever into the vertical hole in the plunger.

7. Pressurize the fuel system using the fuel tank supply line primer bulb and thoroughly inspect for leaks.

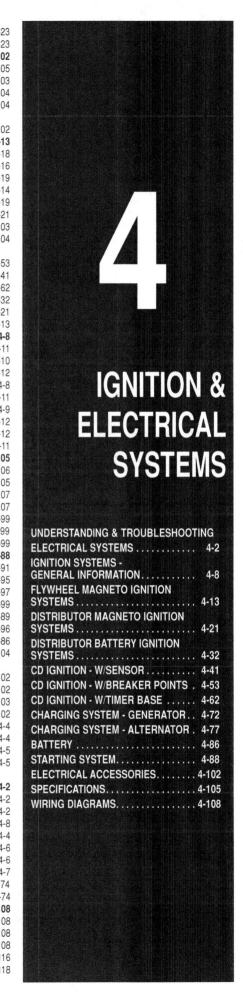

4

IGNITION & ELECTRICAL SYSTEMS

UNDERSTANDING & TROUBLESHOOTING ELECTRICAL SYSTEMS

Basic Electrical Theory

WHERE TO START

◆ See Figure 1

For any 12-volt, negative ground, electrical system to operate, the electricity must travel in a complete circuit. This simply means that current (power) from the positive terminal (+) of the battery must eventually return to the negative terminal (-) of the battery. Along the way, this current will travel through wires, fuses, switches and components. If, for any reason, the flow of current through the circuit is interrupted, the component fed by that circuit would cease to function properly.

Perhaps the easiest way to visualize a circuit is to think of connecting a light bulb (with 2 wires attached to it) to the battery - 1 wire attached to the negative (-) terminal of the battery and the other wire to the positive (+) terminal. With the 2 wires touching the battery terminals, the circuit would be complete and the light bulb would illuminate. Electricity would follow a path from the battery to the bulb and back to the battery. It's easy to see that with wires of sufficient length, our light bulb could be mounted nearly anywhere on the boat. Further, 1 wire could be fitted with a switch inline so that the light could be turned on and off without having to physically remove the wire(s) from the battery.

The normal marine circuit differs from this simple example in two ways. First, instead of having a return wire from each bulb to the battery, the current travels through a single ground wire that handles all the grounds for a specific circuit. Secondly, most marine circuits contain multiple components that receive power from a single circuit. This lessens the overall amount of wire needed to power components.

HOW DOES ELECTRICITY WORK: THE WATER ANALOGY

Electricity is the flow of electrons - the sub-atomic particles that constitute the outer shell of an atom. Electrons spin in an orbit around the center core of an atom. The center core is comprised of protons (positive charge) and neutrons (neutral charge). Electrons have a negative charge and balance out the positive charge of the protons. When an outside force causes the number of electrons to unbalance the charge of the protons, the electrons will split off the atom and look for another atom to balance out. If this imbalance is kept up, electrons will continue to move and an electrical flow will exist.

Many people find electrical theory easier to understand when using an analogy with water. In a comparison with water flowing through a pipe, the electrons would be the water and the wire is the pipe.

The flow of electricity can be measured much like the flow of water through a pipe. The unit of measurement used is amperes, frequently abbreviated as amp (a). You can compare amperage to the volume of water flowing through a pipe (for water that would mean a measurement of mass usually measured in units delivered over a set amount of time such as gallons or liters per minute). When connected to a circuit, an ammeter will measure the actual amount of current flowing through the circuit. When relatively few electrons flow through a circuit, the amperage is low. When many electrons flow, the amperage is high.

Water pressure is measured in units such as pounds per square inch (psi). The electrical pressure is measured in units called volt (v). When a voltmeter is connected to a circuit, it is measuring the electrical pressure.

The actual flow of electricity depends not only on voltage and amperage, but also on the resistance of the circuit. The higher the resistance, the higher the force necessary to push the current through the circuit. The standard unit for measuring resistance is an ohm (omega). Resistance in a circuit varies depending on the amount and type of components used in the circuit. The main factors that determine resistance are:

• Material - some materials have more resistance than others. Those with high resistance are said to be insulators. Rubber materials (or rubber-like plastics) are some of the most common insulators used, as they have a very high resistance to electricity. Very low resistance materials are said to be conductors. Copper wire is among the best conductors. Silver is actually a superior conductor to copper and is used in some relay contacts, but its high cost prohibits its use as common wiring. Most marine wiring is made of copper.

• Size - the larger the wire size being used, the less resistance the wire will have (just as a large diameter pipe will allow small amounts of water to

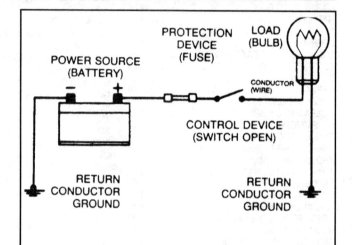

Fig. 1 This example illustrates a simple circuit. When the switch is closed, power from the positive (+) battery terminal flows through the fuse and the switch, and then to the light bulb. The electricity illuminates the bulb and the circuit is completed through the ground wire back to the negative (-) battery terminal.

just trickle through). This is why components that use large amounts of electricity usually have large wires supplying current to them.

• Length - for a given thickness of wire, the longer the wire, the greater the resistance. The shorter the wire, the less the resistance. When determining the proper wire for a circuit, both size and length must be considered to design a circuit that can handle the current needs of the component.

• Temperature - with many materials, the higher the temperature, the greater the resistance (positive temperature coefficient). Some materials exhibit the opposite trait of lower resistance with higher temperatures (these are said to have a negative temperature coefficient). These principles are used in many engine control sensors (especially those found on modern fuel injection systems such as those likely found in your car or truck).

OHM'S LAW

There is a direct relationship between current, voltage and resistance. The relationship between current, voltage and resistance can be summed up by a statement known as Ohm's law.

Voltage (E) is equal to amperage (I) times resistance (R): $E = I \times R$
Other forms of the formula are $R = E/I$ and $I = E/R$

In each of these formulas, E is the voltage in volts, I is the current in amps and R is the resistance in ohms. The basic point to remember is that if the voltage of a circuit remains the same, as the resistance of that circuit goes up, the amount of current that flows in the circuit will go down.

The amount of work that electricity can perform is expressed as power. The unit of power is the watt (w). The relationship between power, voltage and current is expressed as:

Power (W) is equal to amperage (I) times voltage (E): $W = I \times E$

This is only true for direct current (DC) circuits; the alternating current formula is a tad different, but since the electrical circuits in most vessels are DC type, we need not get into AC circuit theory.

Electrical Components

POWER SOURCE

◆ See Figure 2

Typically, power is supplied to a vessel by two devices: The battery and the stator (or battery charge coil). The stator supplies electrical current anytime the engine is running in order to recharge the battery and in order to operate electrical devices of the vessel. The battery supplies electrical power during starting or during periods when the current demand of the vessel's electrical system exceeds stator output capacity (which includes times when the motor is shut off and stator output is zero).

The Battery

In most modern vessels, the battery is a lead/acid electrochemical device consisting of six 2-volt subsections (cells) connected in series, so that the unit is capable of producing approximately 12 volts of electrical pressure. Each subsection consists of a series of positive and negative plates held a short distance apart in a solution of sulfuric acid and water.

The two types of plates in each battery cell are of dissimilar metals. This sets up a chemical reaction, and it is this reaction which produces current flow from the battery when its positive and negative terminals are connected to an electrical load. Power removed from the battery in use is replaced by current from the stator and restores the battery to its original chemical state.

The Stator

Alternators and generators are devices that consist of coils of wires wound together making big electromagnets. The coil is normally referred to as a stator or battery charge coil. Either, one group of coils spins within another set (or a set of permanently charged magnets, usually attached to the flywheel, are spun around a set of coils) and the interaction of the magnetic fields generates an electrical current. This current is then drawn off the coils and fed into the vessel's electrical system.

■ **Some vessels utilize a generator instead of an alternator. Although the terms are often misused and interchanged, the main difference is that an alternator supplies alternating current that is changed to direct current for use on the vessel, while a generator produces direct current. Alternators tend to be more efficient and that is why they are used on almost all modern engines.**

GROUND

Two types of grounds are used in marine electric circuits. Direct ground components are grounded to the electrically conductive metal through their mounting points. All other components use some sort of ground wire that leads back to the battery. The electrical current runs through the ground wire and returns to the battery through the ground or negative (-) cable; if you look, you'll see that the battery ground cable connects between the battery and a heavy gauge ground wire.

■ **A large percentage of electrical problems can be traced to bad grounds.**

If you refer back to the basic explanation of a circuit, you'll see that the ground portion of the circuit is just as important as the power feed. The wires delivering power to a component can have perfectly good, clean connections, but the circuit would fail to operate if there was a damaged ground connection. Since many components ground through their mounting or through wires that are connected to an engine surface, contamination from dirt or corrosion can raise resistance in a circuit to a point where it cannot operate.

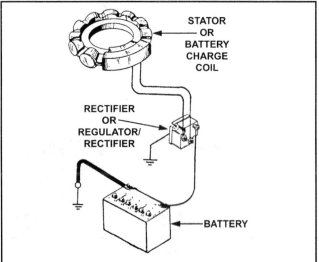

Fig. 2 Functional diagram of a typical charging circuit showing the relationship between the stator (battery charge coil), rectifier (or regulator/rectifier) and battery

PROTECTIVE DEVICES

◆ **See Figure 3**

Problems can occur in the electrical system that will cause large surges of current to pass through the electrical system of your vessel. These problems can be the fault of the charging circuit, but more likely would be a problem with the operating electrical components that causes an excessively high load. An unusually high load can occur in a circuit from problems such as a seized electric motor (like a damaged starter) or the excessive resistance caused by a bad ground (from loose or damaged wires or connections). A short to ground that bypasses the load and allows the battery to quickly discharge through a wire can also cause current surges.

If this surge of current were to reach the load in the circuit, the surge could burn it out or severely damage it. It can also overload the wiring, causing the harness to get hot and melt the insulation. To prevent this, fuses, circuit breakers and/or fusible links are connected into the supply wires of the electrical system. These items are nothing more than a built-in weak spot in the system. When an abnormal amount of current flows through the system, these protective devices work as follows to protect the circuit:

• Fuse - when an excessive electrical current passes through a fuse, the fuse blows (the conductor melts) and opens the circuit, preventing current flow.

• Circuit Breaker - a circuit breaker is basically a self-repairing fuse. It will open the circuit in the same fashion as a fuse, but when the surge subsides, the circuit breaker can be reset and does not need replacement. Most circuit breakers on marine engine applications are self-resetting, but some that operate accessories (such as on larger vessels with a circuit breaker panel) must be reset manually (just like the circuit breaker panels in most homes).

• Fusible Link - a fusible link (fuse link or main link) is a short length of special, high temperature insulated wire that acts as a fuse. When an excessive electrical current passes through a fusible link, the thin gauge wire inside the link melts, creating an intentional open to protect the circuit. To repair the circuit, the link must be replaced. Some newer type fusible links are housed in plug-in modules, which are simply replaced like a fuse, while older type fusible links must be cut and spliced if they melt. Since this link is very early in the electrical path, it's the first place to look if nothing on the vessel works, yet the battery seems to be charged and is otherwise properly connected.

✳✳ CAUTION

Always replace fuses, circuit breakers and fusible links with identically rated components. Under no circumstances should a component of higher or lower amperage rating be substituted. A lower rated component will disable the circuit sooner than necessary (possibly during normal operation), while a higher rated component can allow dangerous amounts of current that could damage the circuit or component (or even melt insulation causing sparks or a fire).

Fig. 3 Fuses and circuit breakers may be found in a central location, or mounted to individual holders in the wiring harness

SWITCHES & RELAYS

◆ **See Figure 4**

Switches are used in electrical circuits to control the passage of current. The most common use is to open and close circuits between the battery and the various electric devices in the system. Switches are rated according to the amount of amperage they can handle. If a sufficient amperage rated switch is not used in a circuit, the switch could overload and cause damage.

Some electrical components that require a large amount of current to operate use a special switch called a relay. Since these circuits carry a large amount of current, the thickness of the wire in the circuit is also greater. If this large wire were connected from the load to the control switch, the switch would have to carry the high amperage load and the space needed for wiring in the vessel would be twice as big to accommodate the increased size of the wiring harness. A relay is used to prevent these problems.

Think of relays as essentially "remote controlled switches." They allow a smaller current to throw the switch that operates higher amperages devices. Relays are composed of a coil and a set of contacts. When current is passed through the coil, a magnetic field is formed that causes the contacts to move together, closing the circuit. Most relays are normally open, preventing current from passing through the main circuit until power is applied to the coil. But, relays can take various electrical forms depending on the job for which they are intended. Some common circuits that may use relays are horns, lights, starters, electric fuel pumps and other potentially high draw circuits.

LOAD

Every electrical circuit must include a load (something to use the electricity coming from the source). Without this load, the battery would attempt to deliver its entire power supply from one pole to another. This is called a short circuit. All this electricity would take a short cut to ground and cause a great amount of damage to other components in the circuit (including the battery) by developing a tremendous amount of heat. This condition could develop sufficient heat to melt the insulation on all the surrounding wires and reduce a multiple wire cable to a lump of plastic and copper. A short can allow sparks that could ignite fuel vapors or other combustible materials in the vessel, causing an extremely hazardous condition.

WIRING & HARNESSES

The average vessel contains miles of wiring, with hundreds of individual connections. To protect the many wires from damage and to keep them from becoming a confusing tangle, they are organized into bundles, enclosed in plastic or taped together and called wiring harnesses. Different harnesses serve different parts of the vessel. Individual wires are color coded to help trace them through a harness where sections are hidden from view.

Marine wiring or circuit conductors can be either single strand wire, multi-strand wire, or, printed circuitry. Single strand wire has a solid metal core and is usually used inside such components as stator coil windings, motors, relays and other devices. Multi-strand wire has a core made of many small strands of wire twisted together into a single conductor. Most of the wiring in a marine electrical system is made up of multi-strand wire, either as a single conductor or grouped together in a harness. All wiring is color coded on the insulator, either as a solid color or as a colored wire with an identification stripe. A printed circuit is a thin film of copper or other conductor that is printed on an insulator backing. Occasionally, a printed circuit is sandwiched between 2 sheets of plastic for more protection and flexibility. A complete printed circuit, consisting of conductors, insulating material and connectors is called a printed circuit board. Printed circuitry is used in place of individual wires or harnesses in places where space is limited, such as behind 1-piece instrument clusters.

Since marine electrical systems are very sensitive to changes in resistance, the selection of properly sized wires is critical when systems are repaired. A loose or corroded connection or a replacement wire that is too small for the circuit will add extra resistance and an additional voltage drop to the circuit.

The wire gauge number is an expression of the cross-section area of the conductor. Vessels from countries that use the metric system will typically describe the wire size as its cross-sectional area in square millimeters. In this method, the larger the wire, the greater the number. Another common system for expressing wire size is the American Wire Gauge (AWG) system. As gauge number increases, area decreases and the wire becomes smaller.

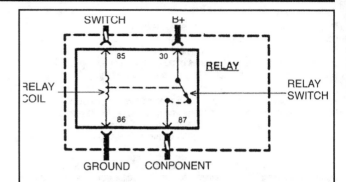

Fig. 4 Relays are composed of a coil and a switch. These two components are linked together so that when one is operated it actuates the other. The large wires in the circuit are connected from the battery to one side of the relay switch (B+) and from the opposite side of the relay switch to the load (component). Smaller wires are connected from the relay coil to the control switch for the circuit and from the opposite side of the relay coil to ground

Using the AWG system, an 18 gauge wire is smaller than a 4 gauge wire. A wire with a higher gauge number will carry less current than a wire with a lower gauge number. Gauge wire size refers to the size of the strands of the conductor, not the size of the complete wire with insulator. It is possible, therefore, to have 2 wires of the same gauge with different diameters because one may have thicker insulation than the other.

It is essential to understand how a circuit works before trying to figure out why it doesn't. An electrical schematic shows the electrical current paths when a circuit is operating properly. Schematics break the entire electrical system down into individual circuits. In most schematics no attempt is made to represent wiring and components as they physically appear on the vessel; switches and other components are shown as simply as possible. But, this is usually **not** the case on Evinrude/Johnson schematics and some of the wiring diagrams provided here. So, when using a Evinrude/Johnson schematic if the component in question is represented by something more than a small square or rectangle with a label, it is likely a true representation of the component. On most schematics, the face views of harness connectors show the cavity or terminal locations in all multi-pin connectors to help locate test points.

Test Equipment

TROUBLESHOOTING WITH TEST EQUIPMENT

Pinpointing the exact cause of trouble in an electrical circuit is usually accomplished by the use of special test equipment, but the equipment does not always have to be expensive. The following sections describe different types of commonly used test equipment and briefly explains how to use them in diagnosis. In addition to the information covered below, be sure to read and understand the tool manufacturer's instruction manual (provided with most tools) before attempting any test procedures.

JUMPER WIRES

◆ **See Figure 5**

✳✳ CAUTION

Never use jumper wires made from a thinner gauge wire than the circuit being tested. If the jumper wire is of too small a gauge, it may overheat and possibly melt. Never use jumpers to bypass high resistance loads in a circuit. Bypassing resistances, in effect, creates a short circuit. This may, in turn, cause damage and fire. Jumper wires should only be used to bypass lengths of wire or to simulate switches.

Jumper wires are simple, yet extremely valuable, pieces of test equipment. They are basically test wires that are used to bypass sections of a circuit. Although jumper wires can be purchased, they are usually fabricated from lengths of standard marine wire and whatever type of connector (alligator clip, spade connector or pin connector) that is required for the particular application being tested. In cramped, hard-to-reach areas, it is advisable to have insulated boots over the jumper wire terminals in order to prevent accidental grounding. It is also advisable to include a standard marine fuse in any jumper wire. This is commonly referred to as a fused jumper. By inserting an in-line fuse holder between a set of test leads, a fused jumper wire is created for bypassing open circuits. Use a 5-amp fuse to provide protection against voltage spikes.

Jumper wires are used primarily to locate open electrical circuits, on either the ground (-) side of the circuit or on the power (+) side. If an electrical component fails to operate, connect the jumper wire between the component and a good ground. If the component operates only with the jumper installed, the ground circuit is open. If the ground circuit is good, but the component does not operate, the circuit between the power feed and component may be open. By moving the jumper wire successively back from the component toward the power source, you can isolate the area of the circuit where the open is located. When the component stops functioning, or the power is cut off, the open is in the segment of wire between the jumper and the point previously tested.

You can sometimes connect the jumper wire directly from the battery to the hot terminal of the component, but first make sure the component uses a full 12 volts in operation. Some electrical components are designed to operate on smaller voltages like 4 or 5 volts, and running 12 volts directly to these components can damage or destroy them.

TEST LIGHTS

◆ See Figure 6

The test light is used to check circuits and components while electrical current is flowing through them. It is used for voltage and ground tests. To use a 12-volt test light, connect the ground clip to a good ground and probe connectors the pick where you are wondering if voltage is present. The test light will illuminate when voltage is detected. This does not necessarily mean that 12 volts (or any particular amount of voltage) is present; it only means that some voltage is present. It is advisable before using the test light to touch its ground clip and probe across the battery posts or terminals to make sure the light is operating properly and to note how brightly the light glows when 12 volts is present.

✳✳ WARNING

Do not use a test light to probe electronic ignition, spark plug or coil wires, as the circuit is much, much higher than 12 volts. Also, never use a pick-type test light to probe wiring on electronically controlled systems unless specifically instructed to do so. Whenever possible, avoid piercing insulation with the test light pick, as you are inviting shorts or corrosion and excessive resistance. But, any wire insulation that is pierced by necessity, must be sealed with silicone and taped after testing.

Like the jumper wire, the 12-volt test light is used to isolate opens in circuits. But, whereas the jumper wire is used to bypass the open to operate the load, the 12-volt test light is used to locate the presence or lack of voltage in a circuit. If the test light illuminates, there is power up to that point in the circuit; if the test light does not illuminate, there is an open circuit (no power). Move the test light in successive steps back toward the power

source until the light in the handle illuminates. The open is between the probe and the point that was previously probed.

The self-powered test light is similar in design to the 12-volt test light, but contains a 1.5 volt penlight battery in the handle. It is most often used in place of a multi-meter to check for open or short circuits when power is isolated from the circuit (thereby performing a continuity test).

The battery in a self-powered test light does not provide much current. A weak battery may not provide enough power to illuminate the test light even when a complete circuit is made (especially if there is high resistance in the circuit). Always make sure that the test battery is strong. To check the battery, briefly touch the ground clip to the probe; if the light glows brightly, the battery is strong enough for testing.

■ A self-powered test light should not be used on any electronically controlled system or component. Even the small amount of electricity transmitted by the test light is enough to damage many electronic components.

MULTI-METERS

◆ See Figure 7

Multi-meters are extremely useful for troubleshooting electrical problems. They can be purchased in either analog or digital form and have a price range to suit nearly any budget. A multi-meter is a voltmeter, ammeter and ohmmeter (along with other features) combined into one instrument. It is often used when testing solid state circuits because of its high input impedance (usually 10 mega-ohms or more). A high-quality digital multi-meter or Digital Volt Ohm Meter (DVOM) helps to ensure the most accurate test results and, although not absolutely necessary for electronic components such as EFI systems and charging systems, is highly recommended. A brief description of the main test functions of a multi-meter follows:

• Voltmeter - the voltmeter is used to measure voltage at any point in a circuit, or to measure the voltage drop across any part of a circuit. Voltmeters usually have various scales and a selector switch to allow metering and display of different voltage ranges. The voltmeter has a positive and a negative lead. To avoid damage to the meter, connect the negative lead to the negative (-) side of the circuit (to ground or nearest the ground side of the circuit) and connect the positive lead to the positive (+) side of the circuit (to the power source or the nearest power source). This is mostly a concern on analog meters, as DVOMs are not normally adversely affected (as they are usually designed to take readings even with reverse polarity and display accordingly). Note that the negative voltmeter lead will always be black and that the positive voltmeter will always be some color other than black (usually red).

• Ohmmeter - the ohmmeter is designed to read resistance (measured in ohms) in a circuit or component. Most ohmmeters will have a selector switch which permits the measurement of different ranges of resistance (usually the selector switch allows the multiplication of the meter reading by 10, 100, 1,000 and 10,000). Most modern ohmmeters (especially DVOMs) are auto-ranging which means the meter itself will determine which scale to use. Since ohmmeters are powered by an internal battery, the ohmmeter can be used like a self-powered test light. When the ohmmeter is connected, current from the ohmmeter flows through the circuit or component being tested. Since the ohmmeter's internal resistance and voltage are known

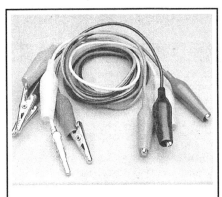

Fig. 5 Jumper wires are simple, but valuable pieces of test equipment

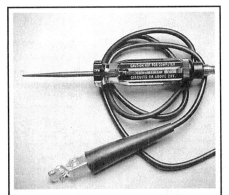

Fig. 6 A 12-volt test light is used to detect the presence of voltage in a circuit

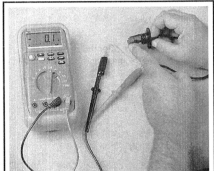

Fig. 7 Multi-meters are probably the most versatile and handy tools for diagnosing faulty electrical components or circuits

values, the amount of current flow through the meter depends on the resistance of the circuit or component being tested. The ohmmeter can also be used to perform a continuity test for suspected open circuits. When using the meter for continuity checks, do not be concerned with the actual resistance readings. Zero resistance, or any ohm reading, indicates continuity in the circuit. Infinite resistance indicates an opening in the circuit. A high resistance reading where there should be none indicates a problem in the circuit. Checks for short circuits are made in the same manner as checks for open circuits, except that the circuit must be isolated from both power and normal ground. Infinite resistance indicates no continuity, while zero resistance indicates a dead short.

✳✳ WARNING

Never use an ohmmeter to check the resistance of a component or wire while there is voltage applied to the circuit. Voltage in the circuit can damage or destroy the meter.

* **Ammeter** - an ammeter measures the amount of current flowing through a circuit in units called amperes or amps. At normal operating voltage, most circuits have a characteristic amount of amperes, called current draw that can be measured using an ammeter. By referring to a specified current draw rating, then measuring the amperes and comparing the two values, you can determine what is happening within the circuit to aid in diagnosis. An open circuit, for example, will not allow any current to flow, so the ammeter reading will be zero. A damaged component or circuit will have an increased current draw, so the reading will be high. The ammeter is always connected in series with the tested circuit. All of the current that normally flows through the circuit must also flow through the ammeter; if there is any other path for the current to follow, the ammeter reading will not be accurate. The ammeter itself has very little resistance to current flow and, therefore, will not affect the circuit, but it will measure current draw only when the circuit is closed and electricity is flowing. Excessive current draw can blow fuses and drain the battery, while a reduced current draw can cause motors to run slowly, lights to dim and other components to not operate properly.

Troubleshooting Electrical Systems

When diagnosing a specific problem, organized troubleshooting is a must. The complexity of a modern marine vessel demands that you approach any problem in a logical, organized manner. There are certain troubleshooting techniques, however, which are standard:

* **Establish when the problem occurs**. Does the problem appear only under certain conditions? Were there any noises, odors or other unusual symptoms? Isolate the problem area. To do this, make some simple tests and observations, then eliminate the systems that are working properly. Check for obvious problems, such as broken wires and loose or dirty connections. Always check the obvious before assuming something complicated is the cause.

* **Test for problems systematically to determine the cause once the problem area is isolated**. Are all the components functioning properly? Is there power going to electrical switches and motors? Performing careful, systematic checks will often turn up most causes on the first inspection, without wasting time checking components that have little or no relationship to the problem.

* **Test all repairs after the work is done to make sure that the problem is fixed**. Some causes can be traced to more than one component, so a careful verification of repair work is important in order to pick up additional malfunctions that may cause a problem to reappear or a different problem to arise. A blown fuse, for example, is a simple problem that may require more than another fuse to repair. If you don't look for a problem that caused a fuse to blow, a shorted wire (for example) may go undetected and cause the new fuse to blow right away (if the short is still present) or during subsequent operation (as soon as the short returns if it is intermittent).

Experience shows that most problems tend to be the result of a fairly simple and obvious cause, such as loose or corroded connectors, bad grounds or damaged wire insulation that causes a short. This makes careful visual inspection of components during testing essential to quick and accurate troubleshooting.

Electrical Testing

VOLTAGE

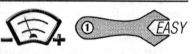

◆ **See Figure 8**

This test determines the voltage available from the battery and should be the first step in any electrical troubleshooting procedure after visual inspection. Many electrical problems, especially on electronically controlled systems, can be caused by a low state of charge in the battery. Many circuits cannot function correctly if the battery voltage drops below normal operating levels.

Loose or corroded battery cable terminals can cause poor contact that will prevent proper charging and full battery current flow.

1. Set the voltmeter selector switch to the 20V position.
2. Connect the meter negative lead to the battery's negative (-) post or terminal and the positive lead to the battery's positive (+) post or terminal.
3. Turn the ignition switch **ON** to provide a small load.
4. A well charged battery should register over 12 volts. If the meter reads below 11.5 volts, the battery power may be insufficient to operate the electrical system properly. Check and charge or replace the battery as detailed under Engine Maintenance before further tests are conducted on the electrical system.

VOLTAGE DROP

◆ **See Figure 9**

When current flows through a load, the voltage beyond the load drops. This voltage drop is due to the resistance created by the load and also by small resistances created by corrosion at the connectors (or by damaged insulation on the wires). Since all voltage drops are cumulative, the maximum allowable voltage drop under load is critical, especially if there is more than one load in the circuit.

1. Set the voltmeter selector switch to the 20 volts position.
2. Connect the multi-meter negative lead to a good ground.
3. Operate the circuit and check the voltage prior to the first component (load).
4. There should be little or no voltage drop in the circuit prior to the first component. If a voltage drop exists, the wire or connectors in the circuit are suspect.
5. While operating the first component in the circuit, probe the ground side of the component with the positive meter lead and observe the voltage readings. A small voltage drop should be noticed. This voltage drop is caused by the resistance of the component.
6. Repeat the test for each component (load) down the circuit.
7. If an excessively large voltage drop is noticed, the preceding component, wire or connector is suspect.

RESISTANCE

◆ **See Figure 10**

✳✳ WARNING

Never use an ohmmeter with power applied to the circuit. The ohmmeter is designed to operate on its own power supply. The normal 12-volt electrical system voltage will damage or destroy many meters!

1. Isolate the circuit from the vessel's power source.
2. Ensure that the ignition key is **OFF** when disconnecting any components or the battery.
3. Where necessary, also isolate at least one side of the circuit to be checked, in order to avoid reading parallel resistances. Parallel circuit resistances will always give a lower reading than the actual resistance of either of the branches.
4. Connect the meter leads to both sides of the circuit (wire or component) and read the actual measured ohms on the meter scale. Make

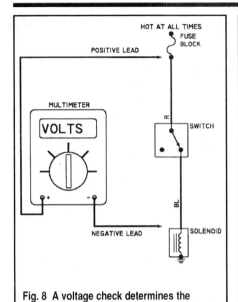

Fig. 8 A voltage check determines the amount of battery voltage available and, as such, should be the first step in any troubleshooting procedure

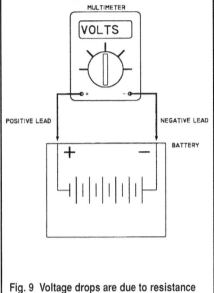

Fig. 9 Voltage drops are due to resistance in the circuit, from the load or from problems with the wiring

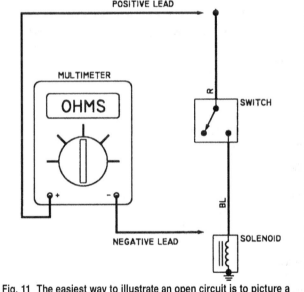

Fig. 10 Resistance tests must be conducted on portions of the circuit, isolated from battery power

sure the selector switch is set to the proper ohm scale for the circuit being tested, to avoid misreading the ohmmeter test value.

■ **The resistance reading of most electrical components will vary with temperature. Unless otherwise noted, specifications given are for testing under ambient conditions of 68°F (20°C). If the component is tested at higher or lower temperatures, expect the readings to vary slightly. When testing engine control sensors or coil windings with smaller resistance specifications (less than 1000 ohms) it is best to use a high quality DVOM and be especially careful of your test results. Whenever possible, double-check your results against a known good part before purchasing the replacement. If necessary, bring the old part to the marine parts dealer and have them compare the readings to prevent possibly replacing a good component.**

OPEN CIRCUITS

◆ See Figure 11

This test already assumes the existence of an open in the circuit and it is used to help locate position of the open.
  1. Isolate the circuit from power and ground.
  2. Connect the self-powered test light or ohmmeter ground clip to the ground side of the circuit and probe sections of the circuit sequentially.
  3. If the light is out or there is infinite resistance, the open is between the probe and the circuit ground.
  4. If the light is on or the meter shows continuity, the open is between the probe and the end of the circuit toward the power source.

SHORT CIRCUITS

◆ See Figure 12

■ **Never use a self-powered test light to perform checks for opens or shorts when power is applied to the circuit under test. The test light can be damaged by outside power.**

  1. Isolate the circuit from power and ground.
  2. Connect the self-powered test light or ohmmeter ground clip to a good ground and probe any easy-to-reach point in the circuit.
  3. If the light comes on or there is continuity, there is a short somewhere in the circuit.

Fig. 11 The easiest way to illustrate an open circuit is to picture a circuit in which the switch is turned OFF (creating an opening in the circuit) that prevents power from reaching the load

  4. To isolate the short, probe a test point at either end of the isolated circuit (the light should be on or the meter should indicate continuity).
  5. Leave the test light probe engaged and sequentially open connectors or switches, remove parts, etc. until the light goes out or continuity is broken.
  6. When the light goes out, the short is between the last two circuit components that were opened.

Wire & Connector Repair

Almost anyone can replace damaged wires, as long as the proper tools and parts are available. Wire and terminals are available to fit almost any need. Even the specialized weatherproof, molded and hard shell connectors used on Evinrude/Johnson engines are usually available for purchase individually.

Be sure the ends of all the wires are fitted with the proper terminal hardware and connectors. Wrapping a wire around a stud is not a permanent solution and will only cause trouble later. Replace wires one at a time to avoid confusion. Always route wires in the same manner of the manufacturer.

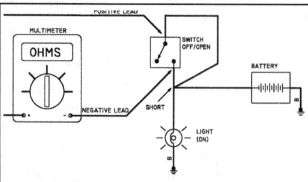

Fig. 12 In this illustration a load (the light) is powered when it should not be (since the switch should be creating an open condition), but a short to power (battery) is powering the circuit. Shorts like this can be caused by chaffed wires with worn or broken insulation

When replacing connections, make absolutely certain that the connectors are certified for marine use. Automotive wire connectors may not meet United States Coast Guard (USCG) specifications.

■ **If connector repair is necessary, only attempt it if you have the proper tools. Weatherproof and hard shell connectors require special tools to release the pins inside the connector. Attempting to repair these connectors with conventional hand tools will damage them. See a Evinrude/Johnson dealer about the proper connector terminal tools available from the manufacturer for these engines.**

IGNITION SYSTEMS - GENERAL INFORMATION

Introduction & Identifying Ignition Systems

◆ **See Figures 13 thru 18**

The less an outboard engine is operated, the more care it needs. Allowing an outboard engine to remain idle will do more harm than if it is used regularly. To maintain the engine in top shape and always ready for efficient operation at any time, the engine should be operating every 3-4 weeks throughout the year.

The carburetion and ignition principles of 2-cycle engine operation *must* be understood in order to perform a proper tune-up on an outboard motor or to diagnose problems with the ignition system.

If you have any doubts concerning your understanding of 2-cycle engine operation, it would be best to study the information on Basic Operating Principles, under Troubleshooting in the General Information, Safety & Tools section before tackling any work on the ignition system.

Six ignition systems are used on the Johnson/Evinrude engines covered here: the flywheel magneto system, the distributor-type magneto system driven by a belt from the flywheel, a battery distributor-type system also driven by a flywheel belt, and 3 types of capacitor discharge (CD) magneto systems (w/sensor, w/breaker points and flywheel magneto).

The ignition system's main purpose is to provide the spark necessary for engine combustion, and to do so at the proper time. It does so by converting a low voltage power source (such as the low voltage alternating current produced by the stator's charge coil) into a high voltage DC current. This is accomplished in the primary circuit of the ignition coil. Power is then conducted from the primary circuit, through the ignition coil's secondary circuit (spark plug wires) to the spark plugs.

This section provides information for troubleshooting and repairing the various ignition systems found on these Johnson/Evinrude motors.

The first few topics of this section apply to ALL engines and ignition systems covered here. This includes topics like Spark Plug Evaluation, Polarity Check, Wiring Harness and a starting point for Ignition System Troubleshooting. Later sections descriptions, troubleshooting and repair instructions that differ on the various systems used on these models.

■ **Remember that Timing & Synchronization procedures are covered in the Maintenance & Tune-Up section.**

Electrical System Precautions

• Wear safety glasses when working on or near the battery.

• Don't wear a watch with a metal band when servicing the battery or starter. Serious burns can result if the band completes the circuit between the positive battery terminal (or a hot wire) and ground.

• Be absolutely sure of the polarity of a booster battery before making connections. Remember that even momentary connection of a booster battery with the polarity reversed will damage charging system diodes. Connect the cables positive-to-positive, and negative (of the good battery)-to-a good ground on the engine (away from the battery to prevent the possibility of an explosion if hydrogen vapors are present from the electrolyte in the discharged battery). Connect positive cables first (starting with the discharged battery), and then make the last connection to ground on the body of the booster vessel so that arcing cannot ignite hydrogen gas that may have accumulated near the battery. • Disconnect both vessel battery cables before attempting to charge a battery.

• Never ground the alternator or generator output or battery terminal. Be cautious when using metal tools around a battery to avoid creating a short circuit between the terminals.

• When installing a battery, make sure that the positive and negative cables are not reversed.

• Always disconnect the battery (negative cable first) when charging.

• Never smoke or expose an open flame around the battery. Hydrogen gas is released from battery electrolyte during use and accumulates near the battery. Hydrogen gas is **highly** explosive.

IDENTIFICATION

Flywheel Magneto Ignition Systems
The following motors are normally equipped with a flywheel magneto ignition system:
• All 1 and 2 cylinder engines except the 1971-72 50 hp 2 cylinder.

Distributor Magneto Ignition Systems
The distributor magneto system is normally found on the following motors:
• 1958-59 50 hp
• 1964-66 60 hp
• 1960-65 75 hp
• 1966 80 hp

Distributor Battery Ignition Systems
The distributor battery ignition system is normally found on the following motors:
• 1961-65 75 hp
• 1966-67 80 hp
• 1964-65 90 hp
• 1966 100 hp

Capacitor Discharge (CD) Ignition System - With Sensor
The capacitor discharge (CD) ignition system with a sensor is normally found on the following motors:
• 1967-68 and 1972 100 hp
• 1969-70 115 hp
• 1971-72 125 hp

Capacitor Discharge (CD) Ignition System - With Breaker Points
The capacitor discharge (CD) ignition system with breaker points is normally found on the following motors:
• 1968-69 55 hp
• 1970-71 60 hp
• 1968 65 hp
• 1968-72 85 hp
• 1971 100 hp

Capacitor Discharge (CD) Ignition System - With Timer Base
The following motors are normally equipped with a capacitor discharge CD flywheel ignition system which utilizes a timer base:
• 50 hp, 1971-72
• 65 hp, 1972

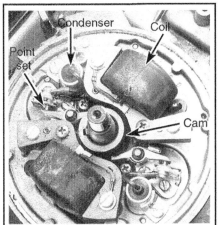

Fig. 13 A good view of the magneto. A single cylinder engine will only have 1 set of points

Fig. 14 A V4 engine with a belt-driven distributor magneto system

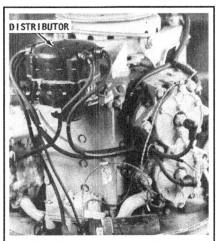

Fig. 15 A V4 engine with a belt-driven distributor battery system

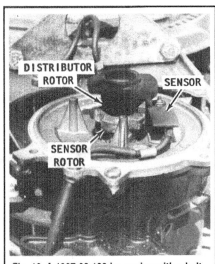

Fig. 16 A 1967-68 100 hp engine with a belt-driven distributor using the CD system with sensor

Fig. 17 A V4 engine showing components of the CD system after the flywheel, stator, distributor cap and rotor have been removed

Fig. 18 A nice shot of the CD system using 2 sets of points system after the flywheel, stator, distributor cap and rotor have been removed

Spark Plug Evaluation

◆ See Figures 19 thru 22

Removal: Tag and disconnect the spark plug wires by pulling and twisting on only the molded cap. NEVER pull on the wire or the connection inside the cap may become separated or the boot damaged. Remove the spark plugs and keep them in order. Take care not to tilt the socket as you remove the plug or the insulator may be cracked.

Examine: Line the plugs in order of removal and carefully examine them to determine the firing conditions in each cylinder. If the side electrode is bent down onto the center electrode, the piston is traveling too far upward in the cylinder and striking the spark plug. Such damage indicates the wrist pin or the rod bearing is worn excessively (an engine overhaul is required to correct the condition). To verify the cause of the problem, crank the engine by hand. As the piston moves to the full up position, push on the piston crown with a screwdriver inserted through the spark plug hole and at the same time rock the flywheel back and forth. If any play in the piston is detected, the engine must be rebuilt.

Correct Color: A proper firing plug should be dry and powdery. Hard deposits inside the shell indicate too much oil is being mixed with the fuel. The most important evidence is the light gray color of the porcelain, which is an indication this plug has been running at the correct temperature. This means the plug is one with the correct heat range and also that the air-fuel mixture is correct.

Rich Mixture: A black, sooty condition on both the spark plug shell and the porcelain is caused by an excessively rich air-fuel mixture, both at low and high speeds. The rich mixture lowers the combustion temperature so the spark plug does not run hot enough to burn off the deposits.

Deposits formed only on the shell are an indication the low-speed air-fuel mixture is too rich. At high speeds with the correct mixture, the temperature in the combustion chamber is high enough to burn off the deposits on the insulator.

Too Cool: A dark insulator, with very few deposits, indicates the plug is running too cool. This condition can be caused by low compression or by using a spark plug of an incorrect heat range. If this condition shows on only 1 plug it is most usually caused by low compression in that cylinder. If all of the plugs have this appearance, then it is probably due to the plugs having a too-low heat range.

Fouled: A fouled spark plug may be caused by the wet oily deposits on the insulator shorting the high-tension current to ground inside the shell. The condition may also be caused by ignition problems which prevent a high-tension pulse being delivered to the spark plug.

Carbon Deposits: Heavy carbon-like deposits are an indication of excessive oil in the fuel. This condition may be the result of poor oil grade, (automotive-type instead of a marine-type); improper oil-fuel mixture in the fuel tank; or by worn piston rings.

Overheating: A dead white or gray insulator, which is generally blistered, is an indication of overheating and pre-ignition. The electrode gap wear rate will be more than normal and in the case of pre-ignition, will actually cause the electrodes to melt. Overheating and pre-ignition are usually caused by

Fig. 19 This plug is foul from operating with an over-rich condition, possibly an improper carburetor adjustment

Fig. 20 Damaged spark plugs. Notice the broken electrode on the left plug. The broken part must be found and removed before returning the engine to service

Fig. 21 This plug has been operating in a too-cool condition because its heat range was too low for the engine

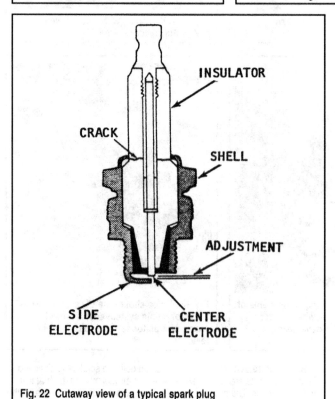

Fig. 22 Cutaway view of a typical spark plug

improper point gap adjustment; detonation from using too-low an octane rating fuel; an excessively lean air-fuel mixture; or problems in the cooling system.

Electrode Wear: Electrode wear results in a wide gap and if the electrode becomes carbonized it will form a high-resistance path for the spark to jump across. Such a condition will cause the engine to misfire during acceleration. If all plugs are in this condition, it can cause an increase in fuel consumption and very poor performance during high-speed operation. The solution is to replace the spark plugs with a rating in the proper heat range and gapped to specification.

Red rust-colored deposits on the entire firing end of a spark plug can be caused by water in the cylinder combustion chamber. This can be the first evidence of water entering the cylinders through the exhaust manifold because of scale accumulation. This condition *must* be corrected at the first opportunity.

Coil Polarity Check

◆ See Figures 23, 24 and 25

Coil polarity is extremely important for proper battery ignition system operation. If a coil is connected with reverse polarity, the spark plugs may demand 30-40 percent more voltage to fire. Under such demand conditions, in a very short time the coil would be unable to supply enough voltage to fire the plugs. Any 1 of the following 3 methods may be used to quickly determine coil polarity.

1. The polarity of the coil can be checked using an ordinary D.C. voltmeter. Connect the positive lead to a good ground. With the engine running, momentarily touch the negative lead to a spark plug terminal. The needle should swing upscale. If the needle swings downstage, the polarity is reversed.

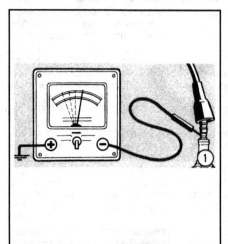

Fig. 23 Checking coil polarity using a voltmeter

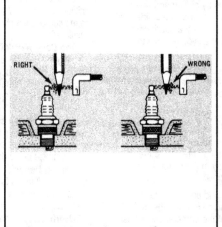

Fig. 24 Checking coil polarity using a pencil

Fig. 25 If the ground electrode is dished, it may mean the polarity has been reversed

2. If a voltmeter is not available, a pencil may be used in the following manner: Disconnect a spark plug wire and hold the metal connector at the end of the cable about 1/4 in. from the spark plug terminal. Now, insert an ordinary pencil tip between the terminal and the connector. Crank the engine with the ignition switch in the **ON** position. If the spark feathers on the plug side and has a slight orange tinge, the polarity is correct. If the spark feathers on the cable connector side, the polarity is reversed.

3. The firing end of a used spark plug can give a clue to coil polarity. If the ground electrode is "dished", it may mean polarity is reversed.

Ignition Wiring Harness

◆ See Figures 26, 27 and 28

These next 2 paragraphs may well be the most important words in this section. Misuse of the wiring harness is one of the most single common causes of electrical problems with outboard power plants.

Fig. 26 This coil was destroyed when 12 volts was connected to the key switch

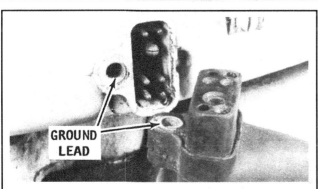

Fig. 27 Certain early engines had a harness plug that must be checked regularly

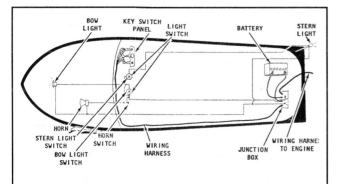

Fig. 28 Functional diagram to illustrate correct hook-up of accessories through a junction box.

A wiring harness is used between the key switch and the engine. This harness seldom contains wire of sufficient size to allow connecting accessories. Therefore, anytime a new accessory is installed, NEW wiring should be used between the battery and the accessory.

A separate fuse panel *must* be installed on the dash. To connect the fuse panel, use 1 red and 1 black No. 10 gauge wire from the battery. If a small amount of 12-volt current should be accidentally attached to the magneto system, the coil may be damaged or destroyed. Such a mistake in wiring can easily happen if the source for the 12-volt accessory is taken from the key switch. Therefore, again let it be said, NEVER connect accessories through the key switch.

The wiring harness installed on 1958-59 50 hp (3 cyl) and 1960 75 hp engines was connected to the side of the engine through an electrical plug using male and female connectors. This connector has been a contributing factor to a host of problems in the ignition system due to its susceptibility to corrosion. Whenever troubleshooting work is done on these particular engines, always disconnect the plug and check it carefully for any sign of corrosion.

Key Switch

A magneto key switch in the *reverse* of any other type key switch. When the key is moved to the **OFF** position, the circuit is closed between the magneto and ground. In some cases the points are grounded as well.

An automotive style key switch must never be used or else the circuit would be opened or closed in the reverse than is necessary.

Breaker Points

◆ See Figures 29 and 30

The breaker points in an outboard motor are an extremely important part of the ignition system. A set of points may appear to be in good condition, but they may be the source of hard starting, misfiring, or poor engine performance. The rules and knowledge gained from association with 4-cycle engines does not necessarily apply to a 2-cycle engine.

The points should be replaced every 100 hours of operation or at least once a year. Remember. . . the *less* an outboard engine is operated, the *more* care it needs. Allowing an outboard engine to remain idle will do more harm than if it is used regularly.

A breaker point set consists of 2 points. One is attached to a stationary bracket and does not move. The other point is attached to a moveable mount. A spring is used to keep the points in contact with each other, except when they are separated by the action of the cam. Both points are constructed with a steel base and a tungsten cap fused to the base.

To properly diagnose magneto (spark) problems, the theory of electricity flow must be understood. The flow of electricity through a wire may be compared with the flow of water through a pipe. Consider the voltage in the wire as the water pressure in the pipe and the amperes as the volume of water. Now, if the water pipe is broken, the water does not reach the end of the pipe. In a similar manner if the wire is broken the flow of electricity is broken. If the pipe springs a leak, the amount of water reaching the end of the pipe is reduced. Same with the wire - if the installation is defective or the wire becomes grounded, the amount of electricity (amperes) reaching the end of the wire is reduced.

Check the wiring carefully. Inspect the points closely and adjust them accurately. The point setting for *most* engines covered here using magneto

Fig. 29 These points are worn and corroded

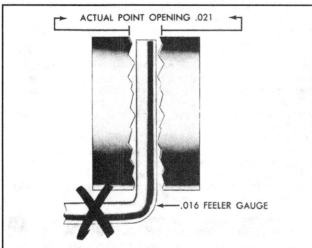

Fig. 30 This illustration shows how a 0.016 in. feeler gauge can be inserted between a badly worn set of points and the actual opening is 0.021 in.

and battery ignition systems is usually 0.020 in. (5.1mm) - check the Tune-Up chart.

Condenser

◆ **See Figures 31 and 32**

A condenser is only used on units equipped with magneto or battery ignition systems. One condenser is used for both sets of points.

In simple terms, a condenser is composed of 2 sheets of tin or aluminum foil laid one on top of the other, but separated by a sheet of insulating material such as waxed paper, etc. The sheets are rolled into a cylinder to conserve space and then inserted into a metal case for protection and to permit easy assembly.

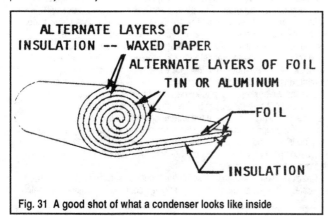

Fig. 31 A good shot of what a condenser looks like inside

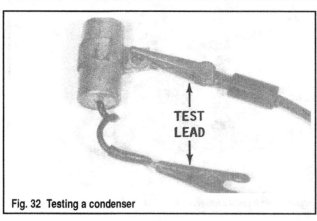

Fig. 32 Testing a condenser

The purpose of the condenser is to absorb or store the secondary current built-up in the primary winding at the instant the breaker points are separated. By absorbing or storing this current, the condenser prevents excessive arcing and the useful life of the breaker points is extended. The condenser also gives added force to the charge produced in the secondary winding as the condenser discharges.

Modern condensers seldom cause problems, therefore, it is not necessary to install a new one each time the points are replaced. However, if the points show evidence of arcing, the condenser may be at fault and should be replaced. A faulty condenser may not be detected without the use of special test equipment. The modest cost of a new condenser justifies its purchase and installation to eliminate this item as a source of trouble.

Vacuum Cut-Out Switch

◆ **See Figures 33, 34 and 35**

On some 40 hp engine models, a cut-out vacuum switch is installed. This switch is connected to 1 of the cylinders in the ignition system. The switch is actuated by vacuum from the cylinder. When a high vacuum pull is exerted against the switch, during engine operation in gear without the lower unit in the water, the switch is closed and the engine is shut down.

This feature is a safeguard against the engine "running away" while operating with a no-load condition on the propeller. A 2-cycle engine will continue to increase rpm under a no-load condition and attempts to shut it down will fail, resulting in serious damage or destruction of the unit.

The vacuum switch also serves as a safety feature when the boat is operating in the water. If the propeller is released from the shaft, because of an accident (striking an underwater object, whatever), the engine would then be operating under a no-load condition. The vacuum switch will shut down the engine and prevent extensive damage, resulting from a "runaway" condition.

This cut-out switch arrangement was installed on the 1971-72 40 hp engines. Therefore, if spark is not present at the spark plug, disconnect the wires from the vacuum switch and again test for spark at the spark plug. If spark is present with the vacuum switch disconnected, the switch is defective and must be replaced.

Ignition System Troubleshooting

Always attempt to proceed with the troubleshooting in an orderly manner. The shotgun approach will only result in wasted time, incorrect diagnosis, replacement of unnecessary parts, and frustration.

Begin the ignition system troubleshooting with the spark plugs and continue through the system until the source of trouble is located.

The test equipment listed in this section is a MUST when troubleshooting this system. Stating it another way - there is simply no way to properly and accurately test the complete system or individual components without the special items listed.

PRELIMINARY TEST

The first area to check on an ignition system is the system ground. Also check the battery charge and to verify the battery cables are connected properly for correct polarity. If the battery cables are connected backwards, testing will indicate NO spark.

If the battery is below a full charge, it will not be possible to obtain full cranking speed during the tests.

COMPRESSION CHECK

■ **For complete details on checking the compression, please refer to Compression Check in the Maintenance & Tune-up section.**

SPARK PLUGS

Check the plug wire(s) to be sure they are properly connected. Check the entire length of the wires from the plugs to the coils. If the wire is to be removed from the spark plug, ALWAYS use a pulling and twisting motion as a precaution against damaging the connection.

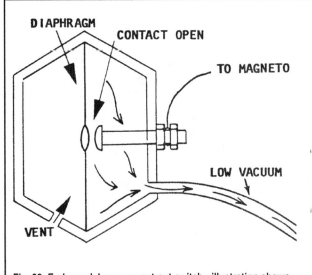

Fig. 33 Early-model vacuum cut-out switch - illustration shows diaphragm positioning during normal operation (no grounding)

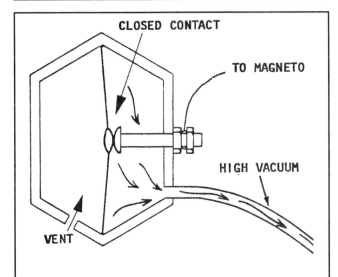

Fig. 34 Early-model vacuum cut-out switch - illustration shows diaphragm positioning when grounded (such as from abnormally high manifold suction which might occur on rapid deceleration)

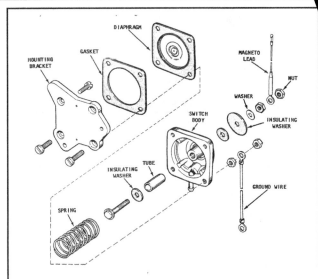

Fig. 35 Exploded view of a "runaway" cut-out switch as used on most late-model

Attempt to remove the spark plugs by hand. This is a rough test to determine if the plug is tightened properly. You should not be able to remove the plug without using the proper socket size tool. Remove the spark plugs and keep them in order on twins. Examine each plug and evaluate its condition as described earlier and also detailed in the Maintenance & Tune-Up section.

If the spark plugs have been removed and the problem cannot be determined, but the plug appears to be in satisfactory condition, electrodes, etc., then reinstall the plugs in the spark plug openings.

A conclusive spark plug test should always be performed with the spark plugs installed. A plug may indicate satisfactory spark when it is removed and tested, but under a compression condition may fail. An example would be the possibility of a person being able to jump a given distance on the ground, but if a strong wind is blowing, his distance may be reduced by half. The same is true with the spark plug. Under good compression in the cylinder, the spark may be too weak to ignite the fuel properly.

Therefore, to test the spark plug under compression, install and tighten the plug in the engine. Another reason for testing for spark with the plugs installed is to duplicate actual operating conditions regarding flywheel speed. If the flywheel is rotated with the pull cord with the plugs removed, the flywheel will rotate much faster because of the no-compression condition in the cylinder, giving the FALSE indication of satisfactory spark.

A spark tester capable of testing for spark while cranking and also while the engine is operating can be purchased from your local marine dealer or automotive parts house. An inexpensive tester will give the same information as a more costly unit.

FLYWHEEL MAGNETO IGNITION SYSTEMS

Identification

The following motors are normally equipped with a flywheel magneto ignition System:
- All 1 and 2 cylinder engines except the 1971-72 50 hp 2 cylinder.

Description & Operation

◆ See Figure 36

A battery installed to crank the engine DOES NOT mean the engine is equipped with a battery-type ignition system. A magneto system uses the battery only to crank the engine. Once the engine is running, the battery has absolutely no affect on engine operation.

Therefore, if the battery is low and fails to crank the engine properly for starting, the engine may be cranked manually, started, and operated. Under these conditions, the key switch must be turned to the ON position or the engine will not start by hand cranking.

A magneto system is a self-contained unit, that does not require assistance from an outside source for starting or continued operation. Therefore, as previously mentioned, if the battery is dead the engine may be cranked manually and the engine started.

The flywheel. The ignition coil, condenser and breaker points are mounted on the armature plate.

As the pole pieces of the magnet pass over the heels of the coil, a magnetic field is built up about the coil, causing a current to flow through the primary winding.

Now, at the proper time, the breaker points are separated by action of a cam, and the primary circuit is broken. When the circuit is broken, the flow of primary current stops and causes the magnetic field about the coil to break down instantly. At this precise moment, an electrical current of extremely high voltage is induced in the fine secondary windings of the coil. This high voltage is conducted to the spark plug where it jumps the gap between the points of the plug to ignite the compressed charge of air-fuel mixture in the cylinder.

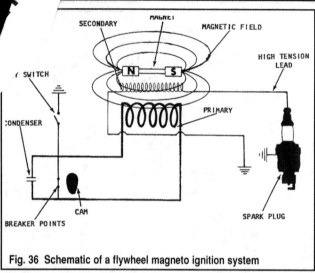

Fig. 36 Schematic of a flywheel magneto ignition system

Flywheel

REMOVAL & INSTALLATION

◆ See Figures 37 thru 46  MODERATE

Magnetos installed on outboard engines will usually operate over extremely long periods of time without requiring adjustment or repair. However, if ignition system problems are encountered and the usual corrective actions such as replacement of spark plugs do not correct the problem, the magneto output should be checked to determine if the unit is functioning properly.

Magneto overhaul procedures may differ slightly on various outboard models, but the following general basic instructions will apply to all Johnson/Evinrude high speed flywheel-type magnetos.

1. Remove the hood or enough of the engine cover to expose the flywheel.

2. Disconnect the battery connections from the battery terminals, if a battery is used to crank the engine. If a hand starter is installed, remove the attaching hardware from the legs of the starter assembly and lift the starter free.

3. On hand started models, a round ratchet plate is attached to the flywheel to allow the hand starter to engage in the ratchet and thus turn the flywheel. This plate must be removed before the flywheel nut is removed.

4. Remove the nut securing the flywheel to the crankshaft. It may be necessary to use some type of flywheel strap to prevent the flywheel from turning as the nut is loosened.

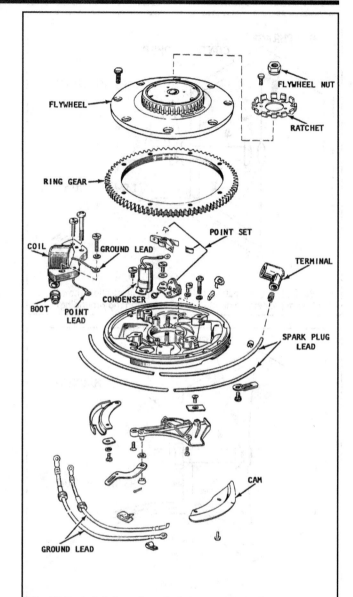

Fig. 37 Exploded view of a typical magneto system (note, only 1 coil and set of points are depicted)

Fig. 38 Remove the ratchet plate on hand start models. . .

Fig. 39 . . .and then remove the flywheel nut

Fig. 40 You may find a washer under the flywheel nut on some engines

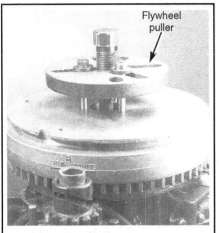

Fig. 41 Install a puller to remove the flywheel

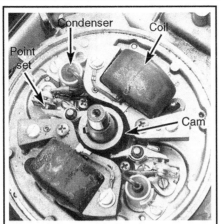

Fig. 42 Take a good look at the magneto and associated components before performing any work on it

Fig. 43 Some engines will have a plate installed that covers the inspection hole, and it will be clearly marked with regard to which side is up

Fig. 44 Move the flywheel into position

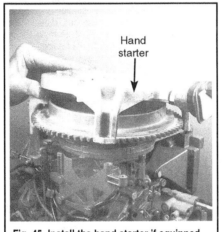

Fig. 45 Install the hand starter if equipped

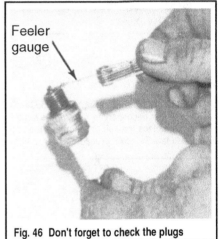

Fig. 46 Don't forget to check the plugs

5. Install the proper flywheel puller using the same screw holes in the flywheel that are used to secure the ratchet plate removed in Step 2. NEVER attempt to use a puller which pulls on the outside edge of the flywheel or the flywheel may be damaged. After the puller is installed, tighten the center screw onto the end of the crankshaft. Continue tightening the screw until the fly wheel is released from the crankshaft. Remove the flywheel. DO NOT strike the puller center bolt with a hammer in an attempt to dislodge the flywheel. Such action could seriously damage the lower seal and/or lower bearing.

6. Stop, and carefully observe the magneto and associated wiring layout. Study how the magneto is assembled. Take time to make notes on the wire routing. Observe how the heels of the laminated core, with the coil attached, is flush with the boss on the armature plate. These items must be replaced in their proper positions. You may even want to take a digital picture of the engine with the flywheel removed: one from the top, and a couple from the sides showing the wiring and arrangement of parts.

To Install:

7. Check to be sure the flywheel magnets are free of any metal parts.

8. Place the key in the crankshaft key way. Check to be sure the inside taper of the flywheel and the taper on the crankshaft are clean of dirt or oil, to prevent the flywheel from "walking" on the crankshaft while the engine is operating. Slide the flywheel down over the crankshaft with the key way in the flywheel aligned with the key on the crankshaft.

9. Rotate the flywheel clockwise and check to be sure the flywheel does not contact any part of the magneto or the wiring.

10. Place the ratchet for the starter on top of the flywheel and install the three 7/16 in. screws. On some models, a plate retainer covers these screws.

11. Thread the flywheel nut onto the crankshaft and tighten it to the value given in the Torque Specifications chart.

12. After the ratchet and flywheel nut have been installed, install the hand starter over the flywheel, if one is used. Check to be sure the ratchet engages the flywheel properly.

13. Set the gap on each spark plug at 0.030 in. (or as detailed in the Tune-Up Specifications chart).

14. Install the spark plugs and tighten them to 210-246 inch lbs.

15. Connect the battery leads to the battery terminals, if a battery is used with a starter motor to crank the powerhead.

16. Perform the Timing & Synchronization procedures in the Maintenance & Tune-Up section.

CLEANING & INSPECTION

◆ **See Figures 37, 47 and 48**

1. Inspect the flywheel for cracks or other damage, especially around the inside of the center hub. Check to be sure metal parts have not become attached to the magnets. Verify each magnet has good magnetism by using a screwdriver or other tool.

2. Thoroughly clean the inside taper of the flywheel and the taper on the crankshaft to prevent the flywheel from "walking" on the crankshaft while the engine is running.

3. Check the top seal around the crankshaft to be sure no oil has been leaking onto the armature plate. If there is *any* evidence the seal has been leaking, it must be replaced, as outlined earlier in this section.

4. Test the armature plate to verify it is not loose. Attempt to lift each side of the plate. There should be little or no evidence of movement.

5. Clean the surface of the armature plate where the points and condenser attach. Install a new condenser into the recess and secure it with the hold-down screw.

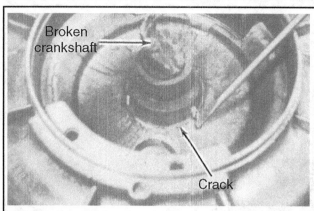

Fig. 47 A broken crankshaft and cracked flywheel damaged when the engine was operated at high rpm without a flush attachment connected to the lower unit

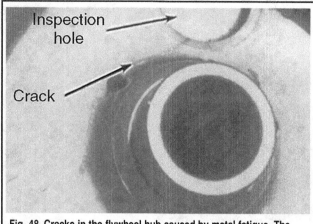

Fig. 48 Cracks in the flywheel hub caused by metal fatigue. The inspection hole is not incorporated on later models

Breaker Points/Condenser

SERVICE

MODERATE

◆ See Figures 30, 37, 42, and 49 thru 57

■ The armature plate does not have to be removed to service the magneto. If it is necessary to remove the plate for other service work, such as to replace the coil or to replace the top seal, refer to the Armature Plate Removal procedure later in this section.

For simplicity and clarity, the following procedures and accompanying illustrations cover a 1-cylinder ignition system. If a larger engine is being serviced, repeat the procedures for each coil and breaker point assembly.

1. Remove the screw attaching the wires from the coil and condenser to 1 set of points. On engines equipped with a key switch, kill button, or run-away switch, a ground wire is also connected to this screw.

2. Using a pair of needle-nose pliers remove the wire clip from the post protruding through the center of the points.

3. Again, with the needle-nose pliers, remove the flat retainer holding the set of points together.

4. Lift the movable side of the points free of the other half of the set.

5. Remove the hold-down screw securing the non-movable half of the point set to the armature plate.

Fig. 49 Remove the screw attaching the wires to the points

Fig. 50 Remove the wire clip . . .

Fig. 51 . . .and then the flat retainer

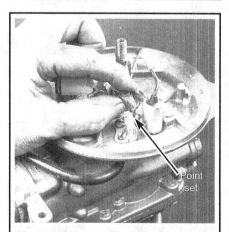

Fig. 52 Lift the moveable side of the points. . .

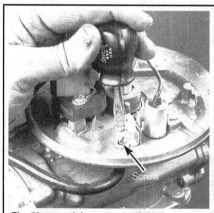

Fig. 53 . . .and then remove the screw on the non-movable side of the points

Fig. 54 Remove the condenser hold down scre

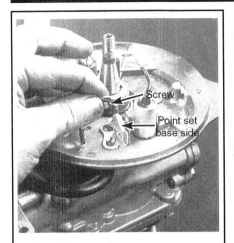

Fig. 55 Slide the base side of the points over the anchor pin. Make sure you use a new wavy washer

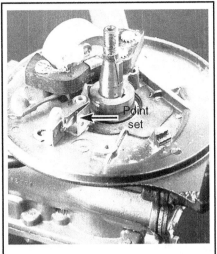

Fig. 56 Check that the points look like this

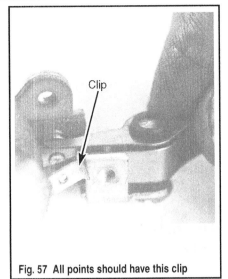

Fig. 57 All points should have this clip

6. Remove the hold down screw securing the condenser to the armature plate. Observe how the condenser sets into a recess in the armature plate.

7. Repeat the procedure for the other set of points.

To Install:

■ All engines covered in this section use the same set of points (#580148). The points MUST be assembled as they are installed. One side of each point set has the base and is non-movable. The other side of the set has a movable arm. A small wire clip and a flat retainer are included in each point set package.

Hold the base side of the points and the flat retainer. Notice how the base has a bar at right angle to the points. Observe the hole in the bar. Observe the flat retainer. Notice that one side has a slight indentation. When the points are installed, this indentation will slip into the hole in the base bar.

8. Install the condensers and secure them in place with their hold-down screws.

9. Hold the base side of the points and slide it down over the anchor pin onto the armature plate. Install the wavy washer and hold-down screw to secure the point base to the armature plate. Tighten the hold-down screw securely.

10. Hold the movable arm and slide the points down over post, and at the same time, hold back on the points and work the spring arm to the inside of the post of the base points. Continue to work the points on down into the base.

11. Observe the points. The points should be together and the spring part of the movable arm on the inside of the flat post.

12. Install the flat retainer onto the flat bar of the base points. Check to be sure the flat spring from the other side of the points is on the inside of the retainer. Push the retainer inward until the indentation slips into the hole in the base. The retainer must be horizontal with the armature plate.

13. Install the wire clip into the groove of the post.

14. Repeat for the 2nd set of points (if equipped).

■ As the coil, condenser, and kill switch wire are being attached to the point set, take the following precautions and adjustments:

• The wire between the coil and the points should be tucked back under the coil and as far away from the crankshaft as possible.

• The condenser wire leaving the top of the condenser and connected to the point set should be bent downward to prevent the flywheel from making contact with the wire. A countless number of installations have been made only to have the flywheel rub against the condenser wire and cause failure of the ignition system.

• Check to be sure all wires connected to the point set are bent downward toward the armature plate. The wires MUST NOT touch the plate. If any of the wires make contact with the armature plate, the ignition system will be grounded and the engine will fail to start.

• Connect the wire leads to the set of points, with the attaching screw.

• Repeat these for the 2nd set of points.

ADJUSTMENT

◆ See Figures 30, 37, 42, 58 and 59

■ The point spring tension is predetermined at the factory and does not require adjustment. Once the point set is properly installed, all should be well. In most cases, breaker contact and alignment will not be necessary. If a slight alignment adjustment should be required, *carefully* bend the insulated part of the point set.

1. Install the flywheel nut onto the end of the crankshaft.

2. Now, turn the crankshaft clockwise and at the same time observe the cam on the crankshaft.

3. Continue turning the crankshaft until the rubbing block of the point set is at the high point of the cam.

4. At this position, use a wire gauge or feeler gauge and set the points at 0.020 in. for all models covered in this section. A wire gauge will always give a more accurate adjustment than a feeler gauge.

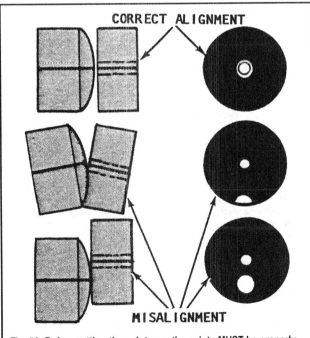

Fig. 58 Before setting the point gap, the points MUST be properly aligned

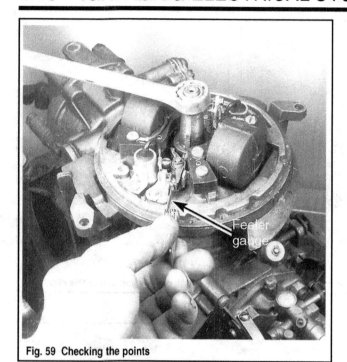

Fig. 59 Checking the points

5. Work the gauge between the points and, at the same time, turn the eccentric on the armature plate until the proper adjustment (0.020 in.) is obtained.

6. Rotate the crankshaft a complete revolution and again check the gap adjustment.

7. After the crankshaft has been turned and the points are on the high point of the cam, check to be sure the hold-down screw is tight against the base. There is enough clearance to allow the eccentric on the base points to turn. If the hold-down screw is tightened *after* the point adjustment has been made, it is very likely the adjustment will be changed.

8. Follow the same procedure and adjust the other set of points.

9. Remove the nut from the crankshaft.

Armature Plate

REMOVAL & INSTALLATION

◆ See Figures 37, 60 and 61

■ It is not necessary to remove the armature plate unless the top seal or the coil is to be replaced.

1. Disconnect the advance arm connecting the armature plate with the power shaft on the side of the engine.

2. Next, remove the wires connecting the underside of the armature plate with the kill switch. If a kill switch is not installed, these wires are connected to the wiring harness plug. The wires of most units have a quick-disconnect fitting. Remove the wires from the vacuum (runaway) switch, if one is installed.

3. Observe the 4 screws in a square pattern through the armature plate. Two of these screws pass through the laminated core and the armature plate into the powerhead retainer. The other two pass just through the plate. Loosen these 4 screws.

4. After the screws are loose, lift the armature plate up the crankshaft and clear of the engine. If any oil is present on top of the armature plate, or on the points, the top seal *must* be replaced.

To Install:

5. Slide the armature plate down over the crankshaft and onto the engine.

6. Align the screw holes in the armature plate with the holes in the powerhead retainer.

7. After the armature plate is in place, install and tighten the 2 screws securing the armature plate to the retainer.

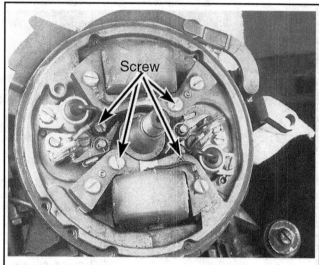

Fig. 60 Remove the 4 armature plate mounting screws. . .

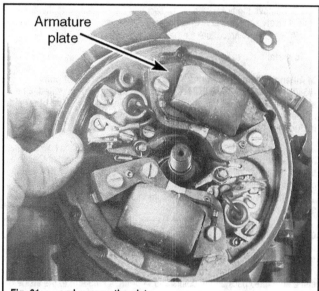

Fig. 61 . . .and remove the plate

8. Now, take up on the 3 screws through the laminated core closest to the crankshaft. Tighten the screws securely.

9. Attach the advance arm from the magneto to the tower shaft arm.

CLEANING & INSPECTION

◆ See Figure 37

1. Inspect the flywheel for cracks or other damage, especially around the inside of the center hub. Check to be sure metal parts have not become attached to the magnets. Verify each magnet has good magnetism by using a screwdriver or other tool.

2. Thoroughly clean the inside taper of the flywheel and the taper on the crankshaft to prevent the flywheel from "walking" on the crankshaft while the engine is running.

3. Check the top seal around the crankshaft to be sure no oil has been leaking onto the armature plate. If there is *any* evidence the seal has been leaking, it MUST be replaced, as outlined earlier in this section.

4. Test the armature plate to verify it is not loose. Attempt to lift each side of the plate. There should be little or no evidence of movement.

5. Clean the surface of the armature plate where the points and condenser attach. Install a new condenser into the recess and secure it with the hold-down screw.

Top Seal

REPLACEMENT

 DIFFICULT

◆ **See Figures 37, 62, 63 and 64**

Replacement of the top seal on a Johnson/Evinrude engine is not a difficult task, with the proper tools: a seal remover and seal installer. NEVER attempt to remove the seal with screwdrivers, punch, pick, or other similar tool. Such action will most likely damage the collars in the powerhead.

1. Obtain a Seal Remover tool (#387780). A 1 1/8 in. open end wrench is needed to hold the remover portion of the tool, while a 3/8 in. open end wrench is used on the top bolt.

2. To remove the seal, first, work the point cam up and free of the driveshaft. Next, remove the Woodruff key from the crankshaft. A pair of side-cutters is a handy tool for this job. Grasp the Woodruff key with the side-cutters and use the leverage of the pliers against the crankshaft to remove the key.

3. Work the special tool into the seal. Observe how the special tool is tapered and has threads. Continue working and turning the tool until it has a firm grip on the inside of the seal. Now, tighten the center screw of the puller against the end of the crankshaft and the seal will begin to lift from the collars. Continue turning this center screw until the seal can be raised manually from the crankshaft.

4. To install the new seal: Coat the inside diameter of the seal with a thin layer of oil. Apply Johnson/Evinrude sealer to the outside diameter of the seal. Slide the seal down the crankshaft and start it into the recess of the powerhead. Use the special tool and work the seal completely into place in the recess,

5. Install the Woodruff key into the crankshaft. On some models, a pin was used to locate the cam for the points. If the pin was used, install it at this time. Observe the difference to the sides of the cam. On almost all cams, the word **TOP** is stamped on one side. Also, on some cams, the groove does not go all the way through. Therefore, it is very difficult to install the cam incorrectly, with the wrong side up.

6. Slide the cam down the crankshaft with the word **TOP** facing upward. Continue working the cam down the crankshaft until it is in place over the Woodruff key or pin.

If the coil is not to be removed, proceed directly to Wick Replacement. To remove the coil, perform the procedures in the following section.

CLEANING & INSPECTION

 EASY

◆ **See Figure 37**

1. Inspect the flywheel for cracks or other damage, especially around the inside of the center hub. Check to be sure metal parts have not become attached to the magnets. Verify each magnet has good magnetism by using a screwdriver or other tool.

2. Thoroughly clean the inside taper of the flywheel and the taper on the crankshaft to prevent the flywheel from "walking" on the crankshaft while the engine is running.

3. Check the top seal around the crankshaft to be sure no oil has been leaking onto the armature plate. If there is *any* evidence the seal has been leaking, it MUST be replaced, as outlined earlier in this section.

4. Test the armature plate to verify it is not loose. Attempt to lift each side of the plate. There should be little or no evidence of movement.

5. Clean the surface of the armature plate where the points and condenser attach. Install a new condenser into the recess and secure it with the hold-down screw.

Coil

REMOVAL & INSTALLATION

 MODERATE

◆ **See Figures 37 and 65 thru 75**

The armature plate must be removed as described earlier in this section. Notice how the coil has a laminated core. The coil cannot be separated, that is, the laminations from the core.

1. Turn the armature plate over and notice how the high-tension leads are installed on the plate in a recess. Take a picture of make a note because the routing of the wires is misleading. . .the wire to the No. 1 spark plug is NOT connected to the No. 1 coil as might be expected.

2. Remove the 3 screws attaching the coils to the armature plate.

3. Hold the armature plate and separate the coils from the plate. As the coil is separated from the plate, observe the high-tension lead to the spark plug inside the coil. Work the small boot, if used, and the high-tension lead from the coil.

To Install:

4. To install a new coil, first turn the armature plate over, and loosen the spark plug lead wires, and push them through the armature plate. Now, work the leads into the coil.

Fig. 62 Remove the point cam

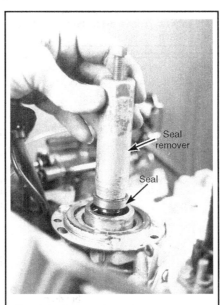

Fig. 63 Pulling out the seal

Fig. 64 Driving in the new seal

5. After the leads are into the coil, work the small boot up onto the coil. Apply a coating of rubber seal material underneath the boot, if a boot is used.

6. Start the 3 screws through the laminated core into the armature plate, but DO NOT tighten them. If the engine being serviced has a 2nd coil, install the other coil in the same manner.

7. Check to be sure the spark plug (high-tension) leads are properly positioned in the coil and are securely attached to the bottom side of the armature plate.

8. To adjust the coils, a special ring tool is required that fits down over the armature plate. This tool will properly locate the coil in relation to the flywheel. Install this special tool over the armature plate. Push outward on the coil and secure the 2 outer screws.

9. If a special ring tool is not available, and in an emergency, hold a straight edge against the boss on the armature plate and bring the heel of the laminated core out square against the edge of the boss on the armature plate. The ground wire for the coil should be attached under the head of the top screw passing through the laminated core.

■ **Special Tip** Make sure the ground wire for the coil is not broken and has not been damaged during service. If the ground wire is broken or damaged you will likely experience a no spark condition. Pay close attention not to damage this or any of the other wires during service.

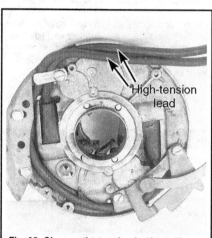

Fig. 65 Observe the tension leads routing

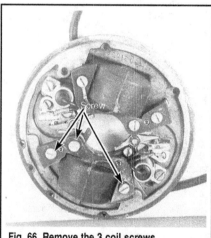

Fig. 66 Remove the 3 coil screws

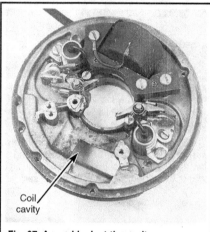

Fig. 67 A good look at the cavity

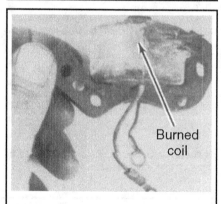

Fig. 68 A coil burned by arcing where the high tension lead enters the coil

Fig. 69 The side blew out of this coil when 12 volts was connected to the circuit at the key switch

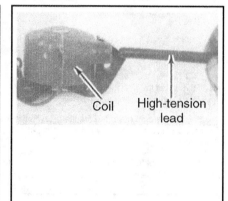

Fig. 70 Work the lead into the coil. . .

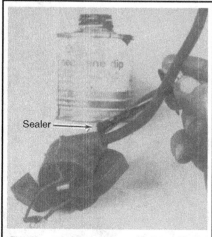

Fig. 71 . . .and then slide on the boot. Make sure that you use rubber sealer

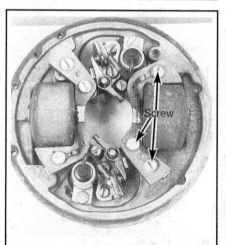

Fig. 72 Install the 3 screws, but do not tighten them

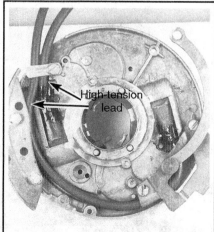

Fig. 73 Make sure the plug leads are positioned properly

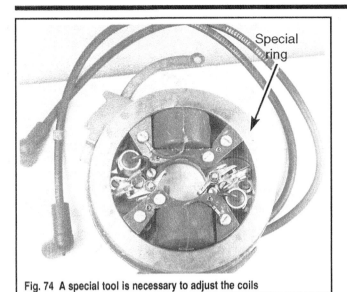

Fig. 74 A special tool is necessary to adjust the coils

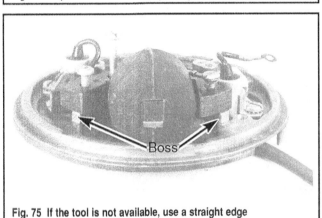

Fig. 75 If the tool is not available, use a straight edge

Wick

REPLACEMENT

◆ See Figures 37 and 76

② *MODERATE*

The wick, mounted in a bracket under the coil, can be replaced without removing the armature plate. The wick should be replaced each and every time the breaker points are replaced.

1. To replace the wick, simply loosen all 3 coil retaining screws and remove the 1 screw through the wick holder.

2. Lift the coil slightly and remove the wick and wick holder.

3. Slide the new wick into the holder; install the holder and wick under the coil; and secure it in place with the retaining screw. Adjust the coil as described earlier in this section, then tighten the 3 screws.

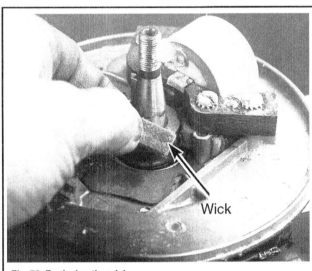

Fig. 76 Replacing the wick

DISTRIBUTOR MAGNETO IGNITION SYSTEMS

Identification

The distributor magneto system is normally found on the following motors:

- 1958-59 50 hp
- 1964-66 60 hp
- 1960-65 75 hp
- 1966 80 hp

Description & Operation

◆ See Figure 77

A battery installed to crank the engine DOES NOT mean the engine is equipped with a battery-type ignition system. A magneto system uses the battery only to crank the engine. Once the engine is running, the battery has absolutely no affect on engine operation. Therefore, if the battery is low and fails to crank the engine properly for starting, the engine may be cranked manually, started, and operated. Under these conditions, the key switch must be turned to the **ON** position or the engine will not start by hand cranking.

A magneto system is a self-contained unit. The unit does not require assistance from an outside source for starting or continued operation. Therefore, as previously mentioned, if the battery is dead, the engine may be cranked manually and the engine started.

A magneto ignition system can easily be distinguished from a battery system. The high-tension spark plug leads on the magneto system are routed out the bottom of the distributor - distributor cap is on the bottom. On a battery ignition system, the distributor cap is on top.

THE MAGNETO

The magneto assembly installed on V4 engines is a self-contained unit. The magneto is driven at crankshaft speed through a belt around a sprocket pulley attached to the flywheel and a similar pulley keyed to the distributor rotor shaft. Both pulleys are of equal diameter, therefore the rotor shaft turns at the same speed as the crankshaft. The pulley also acts as a cover for the breaker assembly housing.

BREAKER ASSEMBLY

◆ See Figure 77

The breaker assembly consists of 2 sets of (#580290). Each package includes one complete set of points, a wick, retainer and clip. The 2-lobed breaker cam, which is keyed to the rotor shaft, and the condenser are sold separately. A tune-up kit includes 2 sets of points, a condenser and a rotor.

Now, with the engine running at 4,000 rpm, 16,000 power impulses occur every minute. Therefore, 16,000 spark impulses are required to fire the compressed charges in each of the 4 cylinders.

Two sets of breaker points in parallel and spaced at 90° are actuated by the two-lobed breaker cam. This arrangement equalizes the individual breaker point "load". Each set of breaker points shares half the load by operating, or breaking, 8,000 times per minute or twice the engine running speed to fire just 2 cylinders. The other set of points fires the other 2 cylinders.

A check of the schematic wiring diagram in this section will assist in understanding how this is accomplished. Looking at it another way, a single set of breaker points operating on a four-lobed cam would be required to interrupt the primary circuit 16,000 times per minute, if the engine were running at 4,000 rpm, to fire the 4 cylinders.

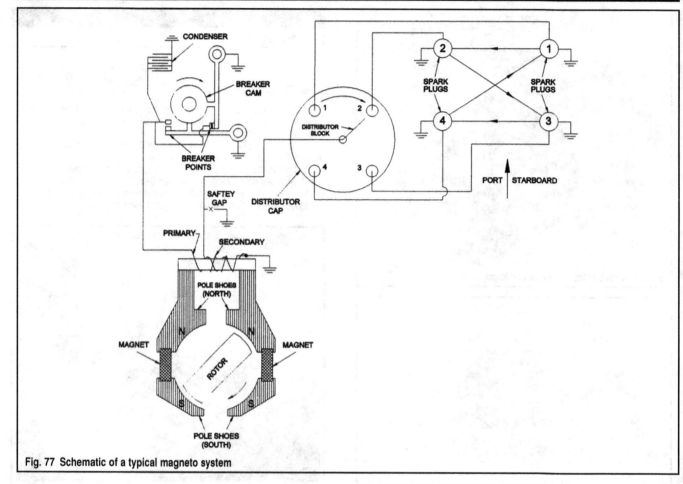

Fig. 77 Schematic of a typical magneto system

With the double breaker installation spaced at 90°, the contour of the 2-lobed breaker cam has been calibrated to permit both sets of points to remain open simultaneously for short intervals, but never closed simultaneously. The breaker points thus alternately make and break the primary circuit to share the load.

The breaker base and point assemblies may appear to be the same as those installed in the flywheel magneto system but they are NOT interchangeable. The difference is in the mounting base. The diameter of the breaker point face has been increased to 3/16 in. (4.76mm) to prolong its active life. Therefore, when purchasing a new assembly, take time at the marine dealer to ensure the proper set is obtained for the engine being serviced.

SPARK PLUG TERMINAL

◆ See Figures 78 and 79

A spark plug terminal kit, consisting of the terminal and a rubber boot, may be purchased at modest cost. To install, pierce the high-tension lead in the center with the terminal prong as shown. If the lead is not pierced in the center, a good contact may not be accomplished. Moisten the inside of the boot with just a drop of saliva. Work the boot over the terminal and high-tension lead until the terminal is in the center of the boot opening for the spark plug.

CONDENSER

◆ See Figure 31

The condenser (#580256) is of conventional design and construction, shunted or bridged across the breaker points. The condenser prevents excessive arcing between the operating breaker points and therefore contributes significantly to prolonging the active life of the point set. Of more importance, however, the condenser makes possible the surge of high voltage intensity required to ignite the compressed fuel vapor charge in each cylinder.

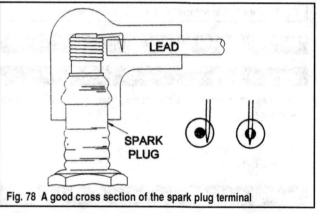

Fig. 78 A good cross section of the spark plug terminal

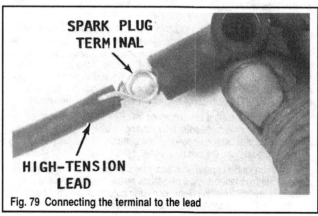

Fig. 79 Connecting the terminal to the lead

IGNITION COIL

◆ See Figure 80

An ignition coil, of basic design and construction, is mounted inside the lower half of the distributor. The coil consists of primary and secondary windings around an electric steel laminated core which in assembly "bridges" the magnet pole shoes. The distributor cap covers the coil and the rotor.

DISTRIBUTOR ROTOR

◆ See Figure 81

A distributor rotor is attached to and turns with the rotor shaft to properly distribute spark impulses while the engine is running.

DISTRIBUTOR CAP

◆ See Figure 82

A conventional distributor cap is used. The cap is mounted on the lower end of the distributor and covers the coil and the rotor. Each opening has a number to indicate the proper lead to be installed. The high-tension leads are

Fig. 80 A good shot of the ignition coil

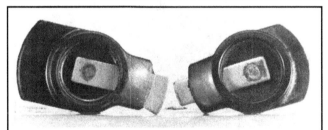

Fig. 81 Although they appear similar, the rotor used on the magneto system (left) and battery ignition (right), are different

Fig. 82 The distributor cap on these systems is mounted on the bottom of the distributor

identified by number with a tag. The high-tension spark plug leads are threaded into the cap. Whenever a high-tension lead is disconnected from the cap, it must be rotated COUNTERCLOCKWISE until it is free.

Troubleshooting

Always attempt to proceed with the troubleshooting in an orderly manner. The shotgun approach will only result in wasted time, incorrect diagnosis, replacement of unnecessary parts, and frustration.

Begin the ignition system troubleshooting with the spark plugs and continue through the system until the source of trouble is located.

On a V4 engine, the cylinders are numbered 1 and 3 in the starboard bank, 2 and 4 in the port bank.

Remember, a magneto system is a self-contained unit. Therefore, if the engine has a key switch and wire harness, remove them from the engine and then make a test for spark. A black "Kill" wire between the wire harness and the magneto can be disconnected to eliminate the wiring in the key switch. If this black "Kill" wire is disconnected, there is no convenient way to shut the engine down, should it start. In such a case the magneto would have to be grounded in order to shut the engine down.

If a good spark is obtained with these 2 items disconnected, but no spark is available at the plug when they are connected, then the trouble is in the harness or the key switch. If a test is made for spark at the plug with the harness and switch connected, check to be sure the key switch is turned to the **ON** position.

SPARK PLUGS

■ Please refer to the Spark Plug procedures in the Ignition System - General Information section found previously in this section.

COMPRESSION

■ For complete details on checking your engine's compression, please refer to the appropriate procedures in the Maintenance & Tune-Up section.

Service

Magneto ignition overhaul procedures may differ slightly on various outboard models, but the following general basic instruction will apply to all V4 magneto engines.

The breaker points should be carefully inspected before proceeding with more involved troubleshooting. On outboard engines, a very small amount of pitting, dirt, oil, or oxidation will rob the engine of power and should be corrected. If in doubt, replace the point set.

When replacing a point set or performing other distributor service work, it is always best to remove the distributor from the engine. Realistically, removal of the distributor is necessary to do any work properly.

REMOVAL & DISASSEMBLY

◆ See Figures 81 and 83 thru 99

1. Disconnect the battery leads from the battery terminals.
2. Remove the high-tension leads from the spark plugs. Use a pulling and twisting motion as a precaution against damaging the connections. Release the high-tension leads to the distributor by opening the retainers on each head. These retainers are split on the side and by using a pair of pliers they may be bent slightly to facilitate removal. Take care not to lose the rubber bushings from inside the retainers. Notice how each bushing has a groove on the inside to allow it to seat properly in the retainer.
3. Remove and examine each spark plug as an aid in determining how that particular cylinder has been operating. Check the electrodes carefully. A bent electrode is an indication of a faulty wrist pin in that piston assembly. If a compression check has not been made, as described previously, perform a

check on each cylinder at this time. Correcting conditions in the ignition system will NOT restore the engine to satisfactory performance if the compression is weak in one cylinder or uneven between the four.

4. Remove the 3 screws securing the plate to the top of the distributor pulley. Notice how the plate has a beveled edge facing upward? The plate must be installed with the bevel facing upward. If the bevel faces downward the distributor belt will be cut. Squeeze the belt together as tightly as possible with one hand and at the same time, remove the nut on top of the distributor pulley with the other hand and a wrench. After the nut has been removed, slip the timing belt free of the pulley. Disconnect the "kill" wire on the starboard side of the magneto.

■ **Do not worry about removing the timing belt or other items. No problem! Detailed timing procedures are presented later in this section.**

5. Remove the 2 screws from the distributor advance arm. This is the arm that connects the distributor to the tower shaft.

6. Slip the "kill" wire free of the switch. If this switch is a part of the system, it is installed on the port side of the magneto.

7. Remove the 3 bolts securing the magneto bracket to the block. These bolts are located just aft and to the port side of the flywheel. Lift the distributor free of the engine.

8. Clean the exterior of the magneto using a wire brush, solvent and compressed air, as required. NEVER submerge the magneto in solvent.

9. Mount the magneto on a test block and slowly turn the pulley by hand. As the pulley is turned, check for any feeling of binding or rubbing. If a definite binding or rubbing is felt, no further testing is necessary. The magneto must be disassembled because the bearings are badly worn and

Fig. 83 Remove the pulley nut. . .

Fig. 84 . . .and then disconnect the advance arm and kill switch lead

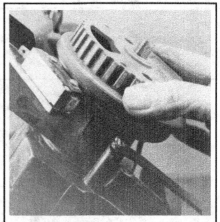
Fig. 85 Check the magneto for binding. . .

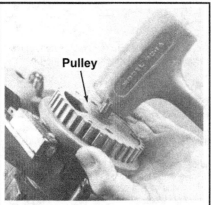
Fig. 86 . . .and then remove the pulley

Fig. 87 Remove the distributor cap. . .

Fig. 88 . . .and then the rotor

Fig. 89 Remove the cam. . .

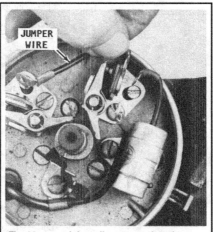
Fig. 90 . . .and then disconnect the wire

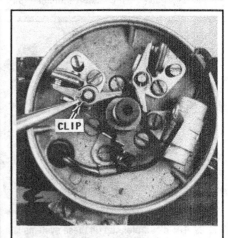
Fig. 91 Remove the retaining clips. . .

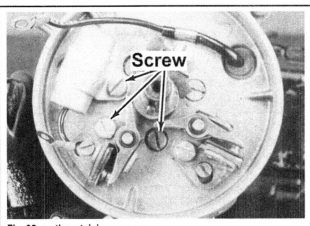

Fig. 92 ...the retaining screws...

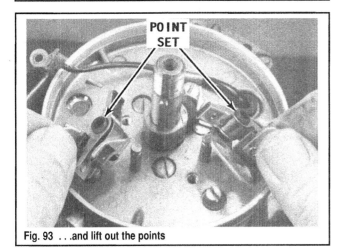

Fig. 93 ...and lift out the points

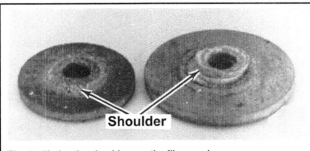

Fig. 94 Notice the shoulders on the fiber washers

must be replaced. Do not confuse the "pull" due to the magnetic field during rotation with binding or rubbing.

10. Hold the distributor a couple inches (5.0 cm) above the surface of the bench. Now, hold the distributor pulley firmly with one hand and at the same time tap the center rotor shaft with a soft headed mallet. This action will "pull" the pulley from the shaft. If pulley fails to come free, a flywheel puller will have to be installed. Use the 3 cover plate screws with the puller to remove the pulley from the distributor shaft.

11. Remove the 4 screws securing the distributor cap. Lift the cap and gasket free of the magneto assembly. Pull the rotor free of the distributor shaft.

■ Two different type rotors are used on Johnson/Evinrude engines. Both rotors appear very similar and both will fit on the distributor shaft. *However*, the rotor used with the magneto system has a wider metal surface on the end of the rotor than the one used on a battery ignition system. If the wrong rotor is installed, the engine will never run satisfactorily at full power. When purchasing a new rotor, be definite at the dealer parts counter as to which unit the rotor is to be installed.

12. If new bearings are to be installed in the distributor base, remove the cam, then the key from the distributor shaft. A single key is used for the pulley and the cam. If the only work to be done is to replace the points, it is NOT necessary to remove the cam and key.

■ Take a few moments and notice the wire routing. You may elect to follow the practice of many professional mechanics and take a Polaroid-type picture (ok, who's got a Polaroid anymore...use your digital) of the wire connections as an aid during the installation process. Notice how one wire extends from the condenser to the set of points closest to the condenser, then a jumper wire connects that set of points with the other set. Observe the "Kill" wire extending from the second set of points through the distributor plate to the kill switch.

13. Disconnect the wire from the point sets and from the condenser. Notice how the wires are attached sideways and are tucked neatly downward to prevent them from rubbing on the belt pulley.

14. Using a pair of needle-nose pliers, remove the wire clips from both posts protruding through the center of the point sets.

15. Remove the screws securing the point sets and the condenser. Notice that the point set screws have wavy washers and the condenser screw has a lockwasher. DO NOT remove the adjusting screws through the point sets. Be sure to save the jumper wire between the 2 sets of points.

16. Slide the point sets up and free of the posts.

■ If the only work to be performed is to replace the point sets and/or the condenser, further disassembly is not necessary. Proceed directly to the following Cleaning & Inspection section.

17. Remove the small nut from the side of the coil housing. Save the fiber washer.

18. Pull the bolt through the housing and save the fiber washer on the bolt inside the housing. These fiber washers prevent the bolt from grounding to the housing. Notice how each of the fiber washers has a shoulder on the inside. The washers are installed with the shoulders fitting into the hole. These shoulders keep the screw in the center of the hole through the housing and prevent it from "shorting out" against the housing.

Fig. 95 Remove the nut on the side of the coil housing...

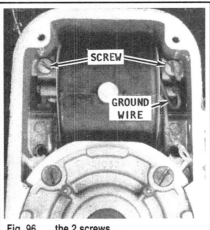

Fig. 96 ...the 2 screws...

Fig. 97 ...the retaining clip...

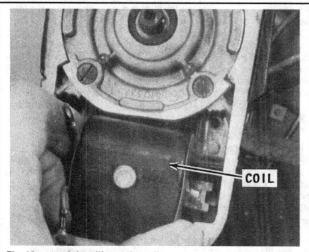

Fig. 98 . . .and then lift out the coil

19. Remove the screw from each end of the coil. Observe that 1 of the screws secures a ground wire from the coil.

20. Secure a good grip with a pair of needle-nose pliers onto 1 of the coil retainers and then pull the retainer free. Remove the second retainer in a similar manner.

21. Pull on the coil and remove it. The coil may appear to be stuck, but it is only the force of the magnets acting against the coil core.

CLEANING & INSPECTION

◆ **See Figures 29, 81 and 99 thru 104**

Inspect all parts for wear or damage. Check the timing belt for cracks, cuts or other damage. Test the coil, condenser, rotor, end caps, high-tension leads and breaker assemblies.

Check the coil closely to determine if there has been any leakage to the distributor housing. If in doubt as to the integrity of the coil, have it checked at a shop or replace it with a new coil.

Perform a resistance test on the high-tension wires. The high-tension leads may be removed. DO NOT attempt to pull the high-tension leads from

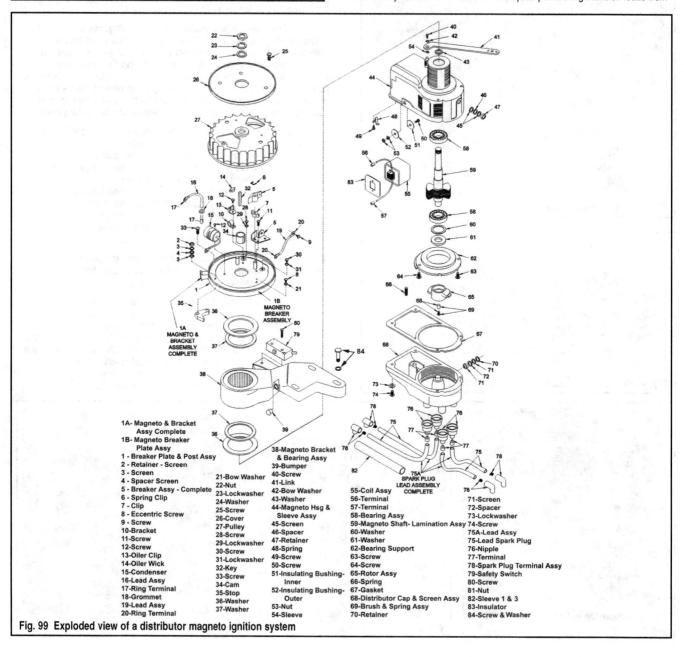

1A- Magneto & Bracket Assy Complete
1B- Magneto Breaker Plate Assy
1 - Breaker Plate & Post Assy
2 - Retainer - Screen
3 - Screen
4 - Spacer Screen
5 - Breaker Assy - Complete
6 - Spring Clip
7 - Clip
8 - Eccentric Screw
9 - Screw
10-Bracket
11-Screw
12-Screw
13-Oiler Clip
14-Oiler Wick
15-Condenser
16-Lead Assy
17-Ring Terminal
18-Grommet
19-Lead Assy
20-Ring Terminal

21-Bow Washer
22-Nut
23-Lockwasher
24-Washer
25-Screw
26-Cover
27-Pulley
28-Screw
29-Lockwasher
30-Screw
31-Lockwasher
32-Key
33-Screw
34-Cam
35-Stop
36-Washer
37-Washer

38-Magneto Bracket & Bearing Assy
39-Bumper
40-Screw
41-Link
42-Bow Washer
43-Washer
44-Magneto Hsg & Sleeve Assy
45-Screen
46-Spacer
47-Retainer
48-Spring
49-Screw
50-Screw
51-Insulating Bushing-Inner
52-Insulating Bushing-Outer
53-Nut
54-Sleeve

55-Coil Assy
56-Terminal
57-Terminal
58-Bearing Assy
59-Magneto Shaft- Lamination Assy
60-Washer
61-Washer
62-Bearing Support
63-Screw
64-Screw
65-Rotor Assy
66-Spring
67-Gasket
68-Distributor Cap & Screen Assy
69-Brush & Spring Assy
70-Retainer

71-Screen
72-Spacer
73-Lockwasher
74-Screw
75A-Lead Assy
75-Lead Spark Plug
76-Nipple
77-Terminal
78-Spark Plug Terminal Assy
79-Safety Switch
80-Screw
81-Nut
82-Sleeve 1 & 3
83-Insulator
84-Screw & Washer

Fig. 99 Exploded view of a distributor magneto ignition system

Fig. 100 Always make sure the springs are in position and in good condition

Fig. 101 The plug lead numbers are embossed into the cap

Fig. 102 A good look at the coil. . .

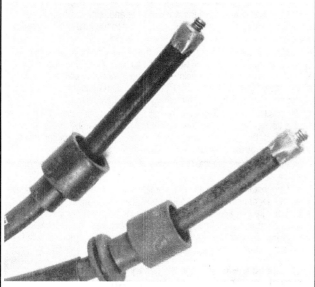

Fig. 103 . . .and the end of the plug leads

Fig. 104 Check the belt for cracks and other damage

the distributor cap. These wires are screwed into the cap. First pull the rubber cap back onto the high-tension lead, and then unscrew the wire in a COUNTERCLOCKWISE direction until the wire is free.

Leakage Paths

The high-voltage surge of the secondary circuit may establish a path to ground by a different route than across the spark plug gap. Once such a path is established, the spark will most likely continue to jump across the ground.

A surface leakage path can usually be detected because of the burning affect the high-voltage spark has on the plastic insulating material. The condition causing the high-voltage spark to stray from its intended circuit must be corrected. Any repairs of the unit should be performed very carefully,

and should include discarding any insulating parts with evidence of high-voltage flashover.

Corrosion

One cause of complete magneto failure is oxidation inside the unit. Such oxidation is the result of continued high-voltage arcing within the housing. Interior corrosion is easily detected by the green discoloration of the copper and brass parts. A brownish deposit is usually found throughout the unit, and sometimes evidence of moisture condensation may be found. Oxidation may be eliminated, if it is detected in time, by removing the cause. Three common causes of oxidation inside a distributor magneto are: a spark gap across a loose connection in the high-voltage circuit; carbon paths inside the magneto; and broken or sticking brush leads.

Usually an oxidized distributor magneto can be cleaned and returned to satisfactory service. Examine the cam for wear. The cam surface must be smooth and free of rough spots or any indication of corrosion.

INSTALLATION & ASSEMBLY

Ignition Coil

◆ See Figures 95 thru 99

■ Please refer to the Removal section for additional illustrations.

1. Lower the coil into the housing with the high-tension center post of the coil facing upward. Push the coil towards the rotor. The magnetic pull of the pulley magnets acting on the coil core will hold the coil in the proper position. Work the primary and ground wires up alongside the coil to permit installation of the coil retainers.

2. Slip the 2 coil retainers down alongside the coil, and then align the retainer screw holes with the holes in the housing.

3. Install the 2 screws through the coil retainers into the housing. *Remember*, one of these screws secures the ground wire. Tighten the screws securely against the retainers.

4. Place the screw through the primary wire, and then slip the fiber washer onto the screw so the washer shoulder, will enter the hole in the housing first. Insert the screw through the fiber washer and housing. Hold the screw in place and slip the other fiber washer onto the screw with the shoulder of the washer indexed into the hole in the housing. This arrangement, properly installed, will keep the screw in the center of the hole and prevent it from "shorting out" against the housing. Slide the metal washer onto the screw, then thread the nut on and tighten it securely. Check to be sure the primary wire is tucked down out of the way.

Distributor

◆ See Figures 87 and 88

■ Please refer to the Removal section for additional illustrations.

1. Place the rotor onto the distributor shaft and snap it down into place. Check to be sure the proper rotor with the wide tip is being installed, as described during the disassembly procedures.

2. Thread the spark plug high-tension leads into the cap (if they were removed). The cylinder numbers are embossed on the cap at each hole. The correct lead must be threaded into the proper opening in the cap or the engine will not operate. Thread the lead CLOCKWISE into the cap. Slide the rubber caps into place on the distributor cap. Check to be extra sure each wire is threaded into the proper hole in the cap. The cap has numbers

stamped into the material near each hole and the high-tension leads have a tag with an identifying number.

3. Place a NEW gasket onto the housing. Install the distributor cap and secure it in place with the four attaching screws. Tighten the screws alternately to prevent warping the cap or putting it under a strain which could develop into a crack at a later date.

Cam Wick

◆ See Figures 105 and 106

The cam wick should be replaced each time a point set is installed or after every 100 hours of engine operation. The cam wick is especially lubricated and requires no further lubrication. Any additional lubrication would shorten breaker point life.

1. Install the cam wick.

2. Install the condenser into the recess in the breaker plate and secure it in place with the screw and lockwasher.

■ **All V4 engines covered in this section use the same set of points (#580390) and the same condenser (#580256). The points must be assembled as they are installed, one side of each point set has the base and is non-moveable. The other side of the set has a moveable arm. A small wire clip and a flat retainer are included in each point set package. NEVER touch the breaker point contact surfaces with your fingers. Such action will deposit an oily residue on the contact surfaces which will reduce the amount of current conducted through the points and will lead to premature pitting.**

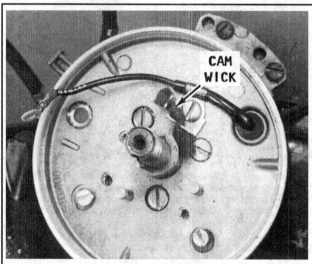

Fig. 105 Install the cam wick. . .

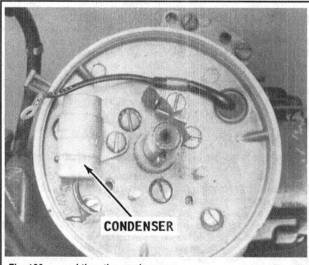

Fig. 106 . . .and then the condenser

Point Set

◆ See Figures 107 thru 112

■ **All V4 engines covered in this Section use the same set of points (#580390) and the same condenser (#580256). The points must be assembled as they are installed. One side of each point set has the base and is non-moveable. The other side of the set has a moveable arm. A small wire clip and a flat retainer are included in each point set package. NEVER touch the breaker point contact surfaces with your fingers. Such action will deposit an oily residue on the contact surfaces which will reduce the amount of current conducted through the points and will lead to premature pitting.**

1. Hold the base side of the points and the flat retainer. Notice how the base has a bar at right angle to the points. Observe the hole in the bar and the flat retainer; notice that one side has a slight indentation. When the points are installed, this indentation will slip into the hole in the base bar.

2. Hold the base side of the points and slide it down over the anchor pin onto the breaker plate. Install the wavy washer and hold-down screw to secure the point base to the breaker plate. Tighten the hold-down screw securely.

3. Hold the moveable arm and slide the points down over the post, and at the same time, hold back on the points and work the spring arm to the inside of the post of the base points. Continue to work the points on down into the base. Observe the points. The points should be together and the spring part of the moveable arm on the inside of the flat post.

4. Install the flat retainer onto the flat bar of the base points. Check to be sure the flat spring from the other side of the points is on the inside of the retainer. Push the retainer inward until the indentation slips into the hole in the base. The retainer *must* be horizontal with the breaker plate.

5. Install the wire clip into the groove of the post. Repeat for the second set of points.

6. Connect the condenser wire, primary wire, and one end of the jumper wire to the set of points closest to the condenser. Connect the other end of the jumper wire to the second set of points. Connect the Kill switch wire to the second set of points. After all connections have been made, check to be sure all wires are neatly tucked down and safe from being rubbed by the pulley as it rotates.

■ **Point spring tension is set at the factory and does not require adjustment after the point set has been properly installed. In most cases, the contact points do not require alignment. Should the occasion arise and the contact points need alignment, carefully bend only the insulated part of the breaker set to achieve satisfactory alignment.**

Breaker Point Adjustment

◆ See Figures 113 thru 117

Two methods are available for setting the breaker points. Follow either procedure to adjust the points.

■ **Check to be sure the hold down screw with the wavy washer is tight. This screw must be tight *before* the adjustment is made. If the screw is tightened after the adjustment, in most cases, the adjustment will be disturbed as the screw is rotated.**

First Method

1. Use a feeler gauge and adjust the points when the cam is at the "highest" point. This is an effective procedure, especially when it is necessary to make an adjustment away from the shop. With a feeler gauge, adjust the points at 0.020 in. (0.52mm). A wire gauge will permit a more accurate setting than the blade-type.

Second Method

2. An alternate and much more accurate method of setting the points is accomplished in the shop and requires the use of special equipment. First, modify an old magneto pulley by cutting a 90° section out just to the side of the timing mark. Cut out the section to the hub.

3. Insert a small piece of cardboard between the set of points that is not being adjusted. Remove the screw from the condenser to the base plate, and then insulate the condenser from the housing by inserting a piece of paper between the condenser and the base plate.

4. Next, install the modified pulley onto the distributor shaft. Rotate the distributor shaft until the timing mark on the pulley is aligned with the first mark on the side of the magneto base. Now, one set of points is visible through the cutout section.

■ **Check to be sure the hold down screw with the wavy washer is tight. This screw must be tight BEFORE the adjustment is made. If the screw is tightened after the adjustment, in most cases the adjustment will be disturbed as the screw is rotated.**

5. Connect an ohmmeter to one side of the point set to be adjusted and the other side of the meter to a good ground on the distributor housing. With the mark on the pulley aligned with the mark on the distributor base, adjust the points until they just close and the light comes on or the meter indicates continuity. From this position, keep an eye on the needle or the light and SLOWLY turn the point adjusting screw in the opposite direction until the light or the meter indicates a broken circuit. This is it! The first set of points is properly adjusted.

6. Remove the piece of cardboard from the second set of points and insert it between the points of the set just adjusted. Rotate the distributor shaft until the second set of points is visible through the cut out section and the mark on the pulley is aligned with the mark on the distributor base. Connect the ohmmeter and repeat the procedure until the second set of

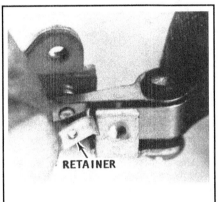

Fig. 107 Positioning the retainer

Fig. 108 Installing the point set base. . .

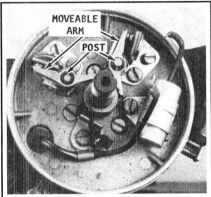

Fig. 109 . . .and then the points

Fig. 110 Press in the retainer. . .

Fig. 111 . . .and then the retaining clip

Fig. 112 Connect the leads

Fig. 113 Using a feeler gauge in Method 1

Fig. 114 In Method 2, modify an old magneto pulley. . .

Fig. 115 . . .and then insert some cardboard

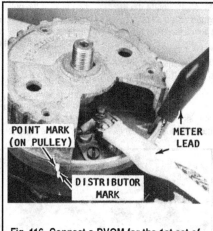

Fig. 116 Connect a DVOM for the 1st set of points. . .

Fig. 117 . . .and then the other

points is properly adjusted. Remove the test equipment and the piece of cardboard from underneath the condenser and from between the point set. Secure the condenser to the breaker plate.

Distributor Installation

◆ See Figures 118 and 119

1. Place the distributor in position on the engine and secure it with the 3 attaching bolts. Bring the bolts up just snug, at this time, to allow adjustment of the distributor belt.

2. Slide the engine magneto pulley onto the distributor shaft with the keyway indexed over the key. Install a flat washer, then a lockwasher onto the shaft, and then thread on the pulley nut. Tighten the pulley nut.

3. Connect the primary and key switch wires to the starboard side of the distributor housing. Tighten the nut securely.

4. Connect the kill switch wire to the switch mounted on the distributor, if one is used.

DISTRIBUTOR BELT REPLACEMENT

◆ See Figures 120, 121 and 122

■ If the distributor belt is worn, frayed, or requires replacement for any reason, it should be replaced as soon as possible.

■ Usually, a distributor belt breaks because the engine is out of time. Now, ask yourself the question: "What caused the engine to jump out of time?" Possibly a broken flywheel key or the key in the distributor shaft has sheared. Should either of these conditions happen, the engine will backfire breaking the belt. Therefore, the necessity of verifying the engine is properly timed with the No.1 cylinder at TOC when a new belt is installed.

If an engine is out of time, there must be a good reason.

1. Check to determine if the Woodruff key is broken or damaged. The Woodruff key is the most essential item in the timing operation. To check the key, rotate the flywheel CLOCKWISE until the No.1 cylinder is at top dead center (TOC). In the accompanying illustration, the head has been removed for photographic clarity to show the piston at TOC. Insert a small screwdriver into the No.1 spark plug opening. The piston should be felt as being at top dead center (TDC).

2. Observe the flywheel timing mark. The mark on the flywheel must be aligned with the embossed mark on the water cover jacket. If the marks are not aligned and the No.1 cylinder is verified as being at TOC, the Woodruff key is most likely broken. Remove the flywheel using a proper flywheel puller and install a new Woodruff key.

3. If the belt requires replacement, simply cut it free. No loss. To install a new belt, first remove the hand starter. Remove the 3 screws securing the plate to the top of the distributor pulley. Notice how the plate has a beveled edge facing upward. The plate must be installed with the bevel facing upward. If the bevel faces downward the distributor belt will be cut. Work the belt around the flywheel and onto the pulley under the flywheel.

Two tasks must now be accomplished. The engine must be in time and the distributor must be adjusted in time with the engine.

Fig. 118 Secure the distributor with the 3 bolts

Fig. 119 Connect the kill switch

Fig. 120 Bump the engine around so the No.1 piston is at TDC. . .

Fig. 121 . . .make sure the timing marks are in alignment. . .

Fig. 122 . . .and install the belt

The flywheel will first be set to bring the cylinders to the proper position and then the distributor will be timed to provide the necessary current to ignite the fuel/air mixture in the proper cylinder at precisely the proper instant.

Either the flywheel or the flywheel protector will have definite marks as an aid to timing the engine. The flywheel mark may be just above or below the ring gear, or in some cases; there may be a red mark between the teeth on the ring gear. A matching mark may be visible on the port side of the hand starter bracket.

On all engines a water cover jacket is installed on top of the powerhead. A mark is embossed on this cover.

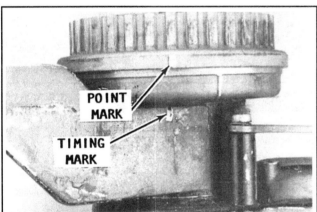

Fig. 123 Align the mark on the pulley with the cut-out mark on the base plate

CHECKS & ADJUSTMENT

Flywheel Alignment

◆ See Figure 121

Rotate the flywheel CLOCKWISE until the mark on the flywheel is aligned with the embossed mark on the water cover jacket. The No.1 (top, starboard bank) is now in the firing position and the engine block is considered timed.

Distributor to Engine Timing

◆ See Figures 123 thru 126

Timing the distributor to the engine is not an easy task. Therefore, patience and attention to detail is absolutely necessary to complete the task for maximum engine performance.

■ DO NOT use the marks on the distributor base that were used to adjust the points. These marks are of no concern when timing the distributor with the engine.

1. Rotate the distributor pulley until the boss mark on the pulley is aligned with cut-out mark on the distributor base plate. This cutout is at approximately the 9 o'clock position of the distributor base plate when viewed from the top of the engine. The cut-out is just a bit hard to see, but it is underneath the bracket to the powerhead.

2. If the distributor has a starter cut-out switch mounted on the side of the distributor, bring the pulley mark to the center of the switch plunger.

3. Hold the pulley in this position and slip the belt over the pulley and into place.

4. Pull back on the distributor and tighten the three mounting bolts.

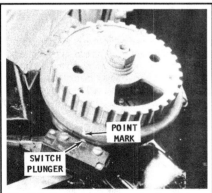

Fig. 124 If equipped with a cut-out switch, line up the pulley mark with the center of the switch plunger

Fig. 125 Install the belt while holding the pulley. . .

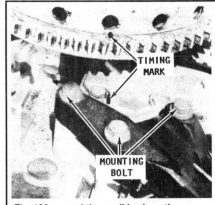

Fig. 126 . . .and then pull back on the distributor while tightening the bolts

Belt Check

◆ See Figures 127 and 128

1. The distance between 2 sides of the belt should be 4 in. (10.16 cm). The belt tension is correct when the 2 halves of the belt can be brought together with finger pressure to 3 5/8 in. (9.21 cm) to 3 11/16 in. (9.37 cm).

2. If the adjustment is not correct, keep the pulley from turning; loosen the 3 distributor mounting bolts slightly, make the adjustment, and then tighten the mounting bolts securely. Again, check the tension.

3. Place the distributor cover on top of the distributor with the flanges of the plate facing UPWARD. Secure the cover in place with the 3 screws.

4. Install the spark plugs into the openings and tighten them appropriately. Connect the high tension leads to the spark plugs.

5. Insert the spark plug leads back into the holder on the head with the rubber bushings in place. Check to be sure the No.1 and No. 3 wires are on the starboard side and No. 2 and No. 4 wires are on the port side.

Fig. 127 Make sure that the belt deflection is within specifications. . .

6. Mount the engine in a large test tank or body of water and check the completed work.

■ **The engine must be mounted in a large test tank or body of water to adjust the timing. NEVER attempt to make this adjustment with a flush attachment connected to the lower unit or in a small confined test tank. The no-load condition on the propeller would cause the engine to 'runaway', resulting in serious damage or destruction of the unit.**

✳✳ CAUTION

Water must circulate through the lower unit to the engine any time the engine is run to prevent damage to the water pump in the lower unit. Just 5 seconds without water will damage the water pump.

7. Perform the timing and synchronization procedures as detailed in the Maintenance & Tune-Up section.

Fig. 128 . . .and then secure the cover

DISTRIBUTOR BATTERY IGNITION SYSTEMS

Identification

The distributor battery system is normally found on the following motors:
- 1961-65 75 hp
- 1966-67 80 hp
- 1964-65 90 hp
- 1966 100 hp

Description & Operation

◆ See Figure 129

The battery ignition system used on all these engines has an alternator included as a necessary component.

It would be well worth the time spent to read and understand the information presented at the beginning of this section, covering the condenser, points, wiring harness, key switch, and other general aspects of an ignition system BEFORE studying this battery distributor ignition system. The principle of induction for sparking purposes is alike in both the magneto and battery systems except that in a battery ignition system, the source of primary current is the storage battery while in the magneto system the magneto generates its own primary current to make it a self contained system. Standard established maintenance procedures for both systems of ignition are basically alike.

The low-voltage current of the ignition system is carried by the primary circuits. Parts of each primary circuit include the ignition switch, and a double set of contact points, coils, condensers, and resistor wire.

The secondary circuit carries the high voltage surges from the ignition coil which result in a high voltage spark between the electrodes of each spark plug. Each secondary circuit includes the secondary winding of each coil,

coil-to-distributor high tension lead, distributor rotor and cap, ignition cables and the spark plugs.

ELECTRICAL FLOW

The following paragraphs describe, in detail, the electrical flow through the primary and secondary circuits for 1 bank of 2 cylinders. The flow is identical for the other bank.

When one set of contact points is closed and the ignition switch is on, current from the battery, or from the alternator, flows through the primary winding of the coil, through the contact points to ground. The current flowing through the primary winding of the coil creates a magnetic field around the coil windings and energy is stored in the coil.

Now, when the contact points are opened by rotation of the distributor cam, the primary circuit is broken. The current attempts to surge across the gap as the points begin to open, but the condenser absorbs the current. In so doing, the condenser creates a sharp break in the current flow and a rapid collapse of the magnetic field in the coil. This sudden change in the strength of the magnetic field causes a voltage to be induced in each turn of the secondary windings in the coil.

The ratio of secondary windings to the primary windings in the coil increases the voltage to about 20,000 volts. This high voltage travels through a cable to the center of the distributor cap, through the rotor to an adjacent distributor cap contact point, and then on through one of the ignition wires to a spark plug.

When the high voltage surge reaches the spark plug it jumps the gap between the insulated center electrode and the grounded side electrode. This high voltage jump across the electrodes produces the energy required to ignite the compressed air/fuel mixture in the cylinder.

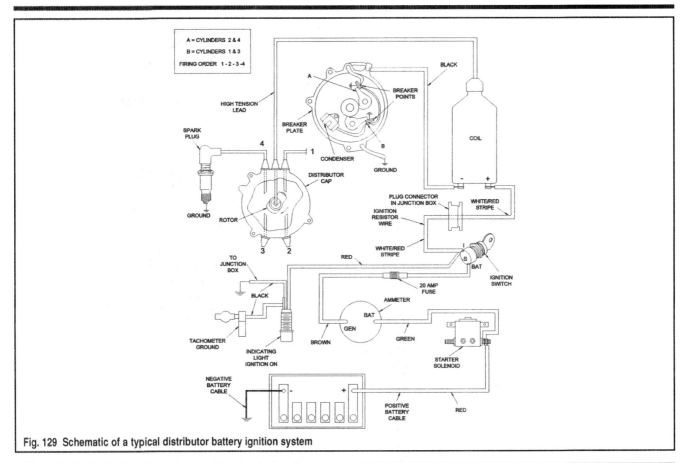

A = CYLINDERS 2 & 4
B = CYLINDERS 1 & 3
FIRING ORDER 1 - 2 - 3 -4

Fig. 129 Schematic of a typical distributor battery ignition system

The entire electrical build-up, break down, and transfer of voltage is repeated as each lobe of the distributor cam passes the rubbing block on the contact breaker arm, causing the contact points to open and close. When the engine is operating at a high rpm rate, the number of times this sequence of action takes place is staggering. All battery ignition systems use the same set of points (#580290) and the same condenser (#580256).

BREAKER ASSEMBLY

◆ See Figure 130

The breaker assembly consists of 2 sets of points (#580290). Each package includes 2 sets of primary breaker points connected in parallel and spaced at 90°; a two-lobed breaker cam which is keyed to the rotor shaft; and a condenser. The lobes of the cam are spaced 180° apart.

With the engine running at 4,000 rpm, 16,000 power impulses occur every minute. Therefore, 16,000 spark impulses are required to fire the compressed charges in each of the 4 cylinders.

Two sets of breaker points in parallel and spaced at 90° are actuated by the 2-lobed breaker cam. This arrangement equalizes the individual breaker point "load". Each set of breaker points share half the load by operating or breaking 8,000 times per minute or twice the engine running speed to fire just 2 cylinders. The other set of points fires the other 2 cylinders.

A check of the schematic wiring diagram in this section will assist in understanding how this is accomplished. Looking at it another way, a single set of breaker points operating on a 4-lobed cam would be required to interrupt the primary circuit 16,000 times per minute, if the engine were running at 4,000 rpm, to fire the 4 cylinders.

With the double breaker installation spaced at 90°, the contour of the 2-lobed breaker cam has been calibrated to permit both sets of points to remain open simultaneously for short intervals, but never closed simultaneously. The breaker points thus alternately make and break the primary circuit to share the load.

The breaker base and point assemblies may appear to be the same as those installed in the flywheel magneto system but they are NOT interchangeable. The difference is in the mounting base. The diameter of the breaker point face has been increased to 3/16 in. (4.76 mm) to prolong its active life.

POINT SET

Fig. 130 A nice shot of the breaker points

CONDENSER

The condenser (#580256) is of conventional design and construction, shunted or bridged across the breaker points. The condenser prevents excessive arcing between the operating breaker points and therefore contributes significantly to prolonging the active life of the point set. Of more importance, however, the condenser makes possible the surge of high voltage intensity required to ignite the compressed fuel vapor charge in each cylinder.

RESISTOR WIRE

◆ **See Figure 131**

As mentioned several times, this system utilizes a resistor wire. This is the white lead with the red tracer running from the positive post of the ignition coil to the wiring harness of the motor assembly. On some later model units, this resistor wire is in the wiring harness from the engine to the key switch. This wire is a definite length and must NEVER be shortened.

The following description of current flow is valid for each resistor wire.

Beginning at the key switch, current flows to the resistor wire and then to the positive side of the coil. When the resistor wire is cold, its resistance is approximately 1.0 ohm. As the temperature of the resistor wire rises, the resistance of the wire increases in a definite proportion.

While the engine is operating at idle or slow speed, the cam on the distributor shaft revolves at a relatively slow rate. Therefore, the breaker points remain closed for a slightly longer period of time. Because the points remain closed longer, more current is allowed to flow. This current flow heats the resistor wire and increases its resistance to cut down on current flow. This current reduction minimizes burning of the contact points.

During high rpm engine operation, the reduced current flow allows the resistor wire to cool enough to reduce resistance, thus increasing the current flow and effectiveness of the ignition system for high-speed performance.

The voltage drops about 25% during engine cranking due to the heavy current demands of the starter. These demands reduce the voltage available for the ignition system.

■ **If the coil appears to have a very short life and requires replacement at short intervals, the cause is almost always with the resistor wire. Someone has removed it; the wire has become disconnected, or is defective and not able to perform its job.**

When replacing a coil, make every attempt to use a marine replacement. If an automobile type coil is installed, it should be the type requiring an outside resistor. If the coil has a built-in resistor and the marine factory resistor wire is used, obviously too much resistance would be in the system.

In brief, if an automobile coil with an internal resistor is used, the resistor wire on the outboard must be disconnected. If the coil uses an outside resistor, no problem, proceed as if it were a marine coil.

Troubleshooting

Always attempt to proceed with the troubleshooting in an orderly manner. The shotgun approach will only result in wasted time, incorrect diagnosis, replacement of unnecessary parts, and frustration.

Begin the ignition system troubleshooting with the spark plugs and continue through the system until the source of trouble is located.

Fig. 131 Resistor wire location

SPARK PLUGS

■ For complete details on troubleshooting the plugs, please refer to the appropriate procedures in the Ignition Systems - General Information section.

COMPRESSION

■ For complete details on checking the compression, please refer to the appropriate procedures in the Maintenance & Tune-up section.

Service

It would be well worth the time spent to read and understand the information presented at the beginning of this chapter, covering the condenser, points, wiring harness, key switch, and other general aspects of an ignition system BEFORE working on this ignition system.

This section contains detailed instructions to service the battery ignition system. The procedures include the necessary steps to service the distributor including instructions to replace, adjust, and synchronize the double set of contact points. Replacement of the distributor drive belt is also covered.

REMOVAL & DISASSEMBLY

◆ **See Figures 132 thru 137**

1. Disconnect the battery leads from the battery terminals.

2. Remove the spark plug wires. Use a pulling and twisting motion as a precaution against damaging the connections.

3. Release the high tension leads to the distributor by opening the retainers on each head. These retainers are split on the side and by using a pair of pliers they may be bent slightly to facilitate removal. Take care not to lose the rubber bushings from inside the retainers. Notice how each bushing has a groove on the inside to allow it to seat properly in the retainer.

4. Remove and examine each spark plug as an aid in determining how each cylinder has been operating. Check the electrodes carefully. A bent electrode is an indication of a faulty wrist pin in that piston assembly. If a compression check has not been made, perform a check on each cylinder at this time. Correcting conditions in the ignition system will NOT restore the engine to satisfactory performance if the compression is weak in 1 cylinder or uneven between the four.

5. Remove the 3 screws securing the distributor cap to the distributor housing, and then lift the cap free.

■ **After the cap is removed, take time to notice the 4 posts extending up from the breaker point base. The purpose of these posts is to allow the high voltage current in the distributor that normally ignites the fuel/air mixture in the cylinder, to jump to the post if the spark plug wire should become disconnected from the spark plug while the engine is operating. Providing a path for this released current prevents damage to expensive parts in the ignition system, such as the coil.**

■ **Two different type rotors are used on Johnson/Evinrude engines. Both rotors appear very similar and both will fit on the distributor shaft. *However*, the rotor used with the magneto system has a wider metal surface on the end of the rotor than the one used on a battery ignition system. If the wrong rotor is installed, the engine will never run satisfactorily at full power. Observing the illustration will be helpful in distinguishing between the 2 rotors. When purchasing a new rotor, be definite at the dealer parts counter as to which unit the rotor is to be installed.**

6. Remove the rotor from the distributor shaft.

7. Disconnect the wire from the point sets and from the condenser. Notice how the wires are attached sideways and are tucked neatly downward to prevent them from rubbing on the belt pulley.

8. Using a pair of needle nose pliers, remove the wire clips from both posts protruding through the center of the point sets.

9. Remove the screws securing the point sets and the condenser. Notice that the point set screws have wavy washers and the condenser screw has a lockwasher.

10. Remove the point sets and condenser. DO NOT remove the adjusting screws through the point sets. Be sure to save the jumper wire between the 2 sets of points. Slide the point sets up and free of the posts.

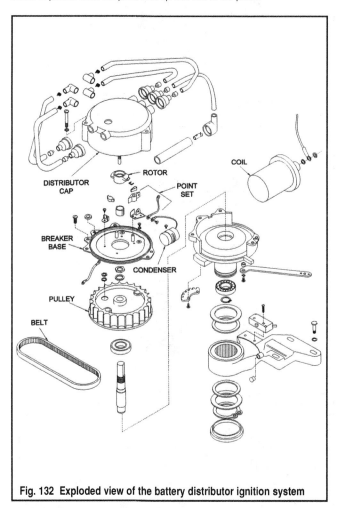

Fig. 132 Exploded view of the battery distributor ignition system

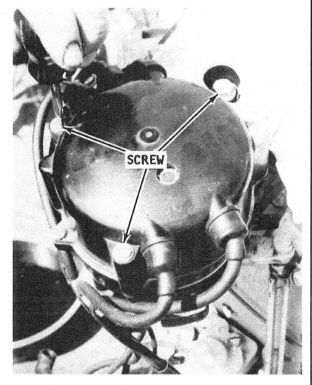

Fig. 133 Remove the distributor cap mounting screws

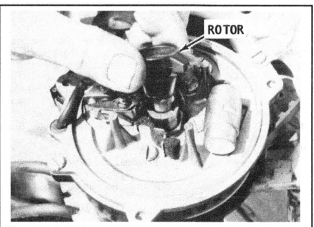

Fig. 134 Lift off the rotor. . .

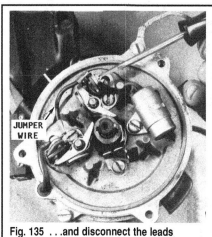

Fig. 135 . . .and disconnect the leads

Fig. 136 Remove the wire clips. . .

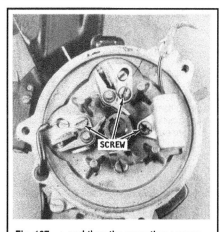

Fig. 137 . . .and then the mounting screws

CLEANING & INSPECTION

◆ **See Figures 29, 78, 79, 132, 138 and 139**

1. Inspect all parts for wear or damage.

2. Check the timing belt for cracks, cuts, or other damage. To replace the belt, see the appropriate procedures later in this section.

3. Test the coil, condenser, rotor, end caps, high-tension leads and breaker assemblies.

4. Check the coil closely to determine if there has been any leakage to the coil mounting bracket. If in doubt as to the integrity of the coil have it checked at a shop or replace it with a new coil.

5. Make a resistance test on the high tension wires. The high tension leads may be removed. DO NOT attempt to pull the high tension leads from the distributor cap.

These wires are screwed into the cap. First pull the rubber cap back onto the high tension lead, and then unscrew the wire in a counterclockwise direction until the wire is free.

6. Check the center area of the distributor cap to be sure the brush and the spring are in good condition and seated properly in the cap recess. In some cases the brush and spring are missing (they were not installed during the last service work).

Leakage Paths

The high-voltage surge of the secondary circuit may establish a path to ground by a different route than across the spark plug gap. Once such a path is established, the spark will most likely continue to jump across the ground.

A surface leakage path can usually be detected because of the burning effect the high voltage spark has on the plastic insulating material. The condition causing the high voltage spark to stray from its intended circuit must be corrected. Any repairs of the unit should be performed very carefully, and should include discarding any insulating parts with evidence of high voltage flashover.

Corrosion

One cause of complete failure is oxidation inside the unit. Such oxidation is the result of continued high-voltage arcing within the housing. Interior corrosion is easily detected by the green discoloration of the copper and brass parts. A brownish deposit is usually found throughout the unit and sometimes evidence of moisture condensation may be found. Oxidation may be eliminated, if it is detected in time by removing the cause. Three common causes of oxidation inside a distributor are: a spark gap across a loose connection in the high-voltage circuit; carbon paths inside the distributor cap; and, broken or sticking brush leads.

Overhaul

Usually an oxidized distributor can be cleaned and returned to satisfactory service.

Check the cam to be sure it is smooth without evidence of roughness or corrosion.

If the cam is not in excellent condition it should be replaced.

INSTALLATION & ASSEMBLY

Breaker Cam

◆ **See Figures 132 and 140**

An oil-impregnated breaker actuating cam with double lobes is used on all V4 battery ignition systems. Both the lobes on the cam are identical. The cam has 2 keyways machined on the inside to time the cam to the distributor shaft and to dampen cam vibration. A Woodruff key is used for timing to the shaft and a detent spring is used to dampen vibration. Since the 2 cam keyways are exactly 180° apart and since each keyway is in exact relationship to the same degree of cam angle with either lobe, it is permissible to install either keyway over the Woodruff key - providing the side of the cam marked **TOP** is facing upward.

1. Install the cam (if it was removed) to the distributor shaft with the cam aligned with the Woodruff key and the detent spring. When both are aligned, push the cam against the distributor shaft nut.

2. The cam wick should be replaced each time a point set is installed or after every 100 hours of engine operation. A new cam wick is included in the new point set package.

Fig. 138 The small spring and carbon cylinder must be in place before installation of the cap

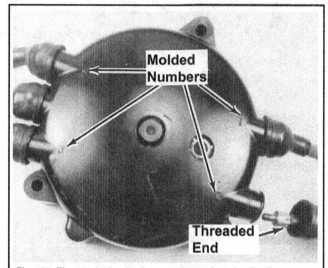

Fig. 139 The spark plug leads are embossed on the distributor cap

Fig. 140 It's always a good idea to replace the wick

The wick is pre-lubricated and requires no further lubrication. Any additional lubrication would shorten breaker point life. Install the cam wick. Install the condenser into the recess in the breaker plate and secure it in place with the screw and lockwasher.

Breaker Point Installation

◆ See Figures 29, 81 and 141 thru 145

All engines covered in this section use the same set of points (#580390) and the same condenser (#580256). The points *must* be assembled as they are installed. One side of each point set has the base and is non-moveable. The other side of the set has a moveable arm. A small wire clip and a flat retainer are included in each point set package. NEVER touch the breaker point contact surfaces with your fingers. Such action will deposit an oily residue on the contact surfaces which will reduce the amount of current conducted through the points and will lead to premature pitting.

■ **The position of the short connector lead attached between the two sets of points is of extreme importance during breaker point installation. This lead must bend around the breaker post, as shown in the accompanying illustration. DO NOT allow the lead to route over the top of the breaker post. Such routing will cause the No. 4 cylinder to misfire.**

1. Hold the base side of the points and the flat retainer. Notice how the base has a bar at right angle to the points. Observe the hole in the bar. Observe the flat retainer. Notice that one side has a slight indentation. When the points are installed, this indentation will slip into the hole in the base bar. Hold the base side of the points and slide it down over the anchor pin onto the breaker plate. Install the wavy washer and hold-down screw to secure the point base to the breaker plate. Tighten the hold down screw securely.

2. Hold the moveable arm and slide the points down over the post. At the same time, hold back on the points and work the spring arm to the inside of the post of the base points. Continue to work the points on down into the base. Observe the points - the points should be together and the spring part of the moveable arm should be on the inside of the flat post.

3. Install the flat retainer onto the flat bar of the base points. Check to be sure the flat spring from the other side of the points is on the inside of the retainer. Push the retainer inward until the indentation slips into the hole in the base. The retainer must be horizontal with the breaker plate.

4. Install the wire clip into the groove of the post. Repeat steps for the second set of points.

5. Connect the condenser wire and one end of the jumper wire to the set of points closest to the condenser. Connect the other end of the jumper wire and the primary wire to the second set of points. After all connections have been made, check to be sure all wires are neatly tucked down and safe from being rubbed by the rotor as it rotates.

■ **Point spring tension is set at the factory and does not require adjustment after the point set has been properly installed. In most cases, the contact points do not require alignment. Should the occasion arise and the contact points need alignment,**

6. CAREFULLY bend only the insulated part of the breaker set to achieve satisfactory alignment.

Fig. 141 Install the base side of the points onto the breaker plate...

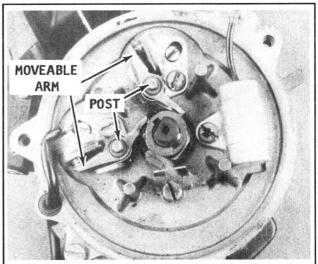

Fig. 142 ...and then slide on the points

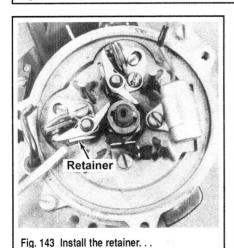

Fig. 143 Install the retainer...

Fig. 144 ...and then the clip

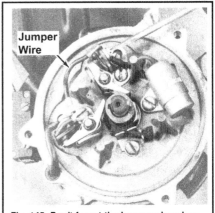

Fig. 145 Don't forget the jumper wire when reconnecting the leads

Breaker Point Adjustment

◆ **See Figures 146 thru 150**

Two methods are available for setting the breaker points. Follow either procedure to adjust the points.

■ **When adjusting ignition breaker points, ALWAYS rotate the distributor pulley in the normal operating direction by turning the engine flywheel CLOCKWISE (when viewed from above the engine). If the flywheel should be turned in the wrong direction, the water pump in the lower unit would be damaged.**

■ **Check to be sure the hold down screw with the wavy washer is tight. This screw must be tight BEFORE the adjustment is made. If the screw is tightened after the adjustment, in most cases the adjustment will be disturbed as the screw is rotated. The belt has been removed for the next 2 steps only for photographic clarity. The belt does NOT need to be removed to make the adjustment.**

First Method

1. Rotate the flywheel CLOCKWISE until the point set heel is on the highest part of the cam, and then use a feeler gauge and adjust the points. This is an effective procedure, especially when it is necessary to make an adjustment away from the shop.

2. With a feeler gauge, adjust the points at 0.020 in. (0.52mm). A wire gauge will permit a more accurate setting than the blade-type. Repeat the procedure for the second set of points.

3. Now, rotate the flywheel a couple complete turns, and then check the point gap setting again.

Second Method

A much more accurate method than using either of the feeler gauges in the first method, is to use a continuity meter and the 3 timing marks. Observe the 2 raised marks on the distributor housing 90° apart and the timing mark on the distributor pulley.

4. Disconnect the primary lead on the coil and remove the condenser hold down screw.

5. DO NOT disconnect the lead from the point set. Lay the condenser on a piece of cardboard. Slide a piece of cardboard between the points of the set identified as **B** in the accompanying illustration. The purpose of the cardboard is to prevent meter current flow to ground through the points.

6. Connect an ohmmeter to one side of the point set, identified as **A** and the other side of the meter to a good ground on the engine. Rotate the flywheel CLOCKWISE until the distributor pulley timing mark is aligned with the mark identified as **A** on the casting of the distributor base. Adjust the points until they just close and the light comes on or the meter indicates continuity.

From this position, keep an eye on the needle or the light and SLOWLY turn the point adjusting screw in the opposite direction until the light or the meter indicates a broken circuit. This is it! The first set of points is properly adjusted.

7. Remove the piece of cardboard from point set **B** and insert it between the points of the set **A**. To adjust the second point set, first rotate the flywheel CLOCKWISE until the timing mark on the distributor pulley is aligned with the mark **B**. Connect the ohmmeter to the second set of points **B** and repeat the procedure until the second set of points is properly adjusted.

8. Remove the test equipment and piece of cardboard from the point set. Install the condenser. Secure the condenser in the recess with the hold down screw and lockwasher. Connect the primary lead to the coil.

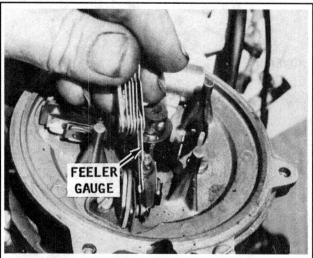

Fig. 146 Using a feeler gauge in Method I

Fig. 147 In Method II, position the cardboard and paper as detailed...

Fig. 148 ...and connect the meter leads

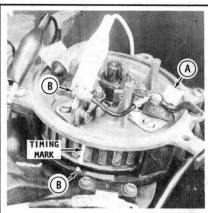

Fig. 149 Next move the cardboard and test again

Fig. 150 Final set-up

Final Assembly/Installation

◆ See Figures 81, 133, 134 and 139

Two different type rotors are used on Johnson/Evinrude engines.

Both rotors appear very similar and both will fit on the distributor shaft. *However*, the rotor used with the magneto system has a wider metal surface on the end of the rotor than the one used on a battery ignition system. If the wrong rotor is installed, the engine will never run satisfactorily at full power. The accompanying illustrations will be helpful in distinguishing between the 2 rotors. When purchasing a new rotor, be definite at the dealer parts counter as to which unit the rotor is to be installed.

1. Slide the distributor rotor into place on the distributor shaft.

2. Screw the spark plug leads into the cap (if they were removed).

3. Slide the rubber caps into place on the distributor cap. Check to be extra sure each wire is threaded into the proper hole in the cap. The cap has numbers stamped into the material near each hole and the high tension leads have a tag with an identifying number.

4. Install the distributor cap onto the distributor housing with the brush and brush spring properly centered in the cap. Secure the cap in place with the three retaining screws.

BELT REPLACEMENT & TIMING

If an engine is out of time, there must be a good reason. First, check to determine if the flywheel Woodruff key is broken or damaged. This key is the most essential item in the timing operation. To check the key, rotate the flywheel CLOCKWISE until the No.1 cylinder is at top dead center (TDC). Observe the flywheel timing mark. The mark on the flywheel must be aligned with the arrow on the flywheel ring gear guard. If the marks are not aligned and the No.1 cylinder is verified as being at TDC, the Woodruff key is most likely broken. Of course, the flywheel must be removed to replace the Woodruff key.

Second, the Woodruff key on the distributor driveshaft may be broken. Replacement of this key is covered in this section.

Third, the timing belt may be stripped or damaged with a few teeth missing on the inside of the belt.

Removal & Installation

◆ See Figures 151 thru 163

If the belt is being replaced because it is broken, OR if the belt is worn and unfit for further service, the following steps are not necessary to remove the belt. If the belt is broken, naturally it does not have to be removed. If it is worn, simply cut it free.

However, if the belt is being removed in order to accomplish other task, all steps MUST be performed in the order given.

1. Remove the high tension lead from the center of the coil by first pulling the rubber boot back, and then pulling the lead free of the coil.

2. Remove the coil from the flywheel guard. Disconnect the alternator leads from the flywheel guard where the diodes are located (if diodes are used). Remove the flywheel guard from the engine.

3. Remove the distributor cap retaining screws and then lift the cap free of the distributor.

4. Slide the rotor up and free of the distributor shaft. Remove the 2 screws securing the breaker plate to the distributor housing. CAREFULLY lift the plate from the housing. Work the heels of the point set free of the cam. DO NOT remove the breaker point cam from the distributor shaft.

5. Loosen the 3 distributor bracket bolts and push the distributor towards the crankshaft to relieve tension on the belt. Bring one bracket bolt up snug, and then lift the drive belt free of the distributor pulley.

6. Remove the flywheel cover.

7. Position a flywheel holder wrench over the flywheel ring gear to prevent the flywheel from turning.

■ NEVER attempt to hold the flywheel from turning by placing a screwdriver or other object between the flywheel ring teeth and the Bendix assembly of the starter motor. Such practice will surely cause damage and misalignment of the starter motor pinion gear or to the engine block. An expensive mistake just to save the trouble of securing the flywheel properly to remove the flywheel nut.

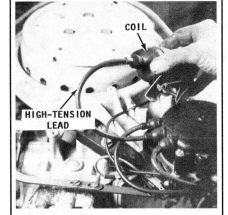

Fig. 151 Remove the coil

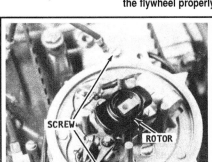

Fig. 152 Lift off the rotor and then remove the breaker plate

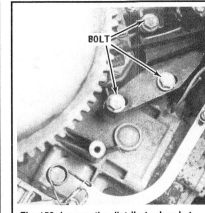

Fig. 153 Loosen the distributor bracket bolts to relieve belt tension

Fig. 154 Remove the flywheel cover. . .

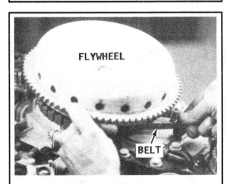

Fig. 155 . . .and then the flywheel

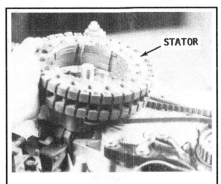

Fig. 156 Remove the stator. . .

8. Hold firm on the flywheel wrench and remove the flywheel nut.

9. Obtain an approved OMC flywheel puller that uses 3 puller screws. Attach the puller with the flat side facing upward. Install the puller screws into the holes normally used for the cover plate screws.

10. Take up on the puller center screw and at the same time have an assistant lift up on the flywheel rim and strike a sharp blow or two against the center screw with a medium-sized hammer. This combination of efforts should unseat the flywheel from the crankshaft taper. If this action fails on the first try, repeat the procedure until the flywheel is unseated.

11. Slide the drive belt free of the flywheel. Lift the flywheel from the crankshaft, taking care not to snag the stator windings with the flywheel as it is removed.

12. Uncouple the stator electrical connector on the starboard side of the engine.

Remove the stator hold down bolts and lift the stator free of the engine.

13. Lift the distributor timing belt off of the engine. These procedures cover removal and installation of the belt in order to accomplish other work. However, NOW is the time to check the condition of the belt and replace it with a new one if there is any sign of wear.

To Install:

14. Lay the belt down around the crankshaft and into the recess of the upper bearing cap. The bearing cap has a slight flange on the upper edge. The stator rests on this flange. Two slots are provided at the back side of the bearing cap to permit the belt to run through.

15. Lower the stator down the crankshaft with the holes in the stator plate indexed over the locating pins. Coat the retaining bolts with Loctite, or equivalent material, and then install and tighten the bolts.

■ Loctite, or a similar material, MUST be used on the stator retaining bolts and the bolts tightened securely to prevent engine vibration from causing them to back out and catch on the flywheel while the engine is running.

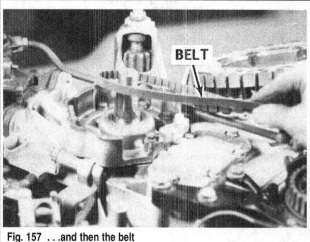

Fig. 157 . . .and then the belt

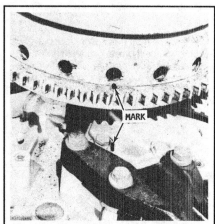

Fig. 158 Line up the red mark on the flywheel with the mark on the water jacket cover

Fig. 159 Align the pulley notch with the mark on the distributor housing. . .

Fig. 160 . . .and then work the belt over the pulley

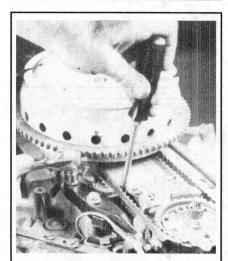

Fig. 161 Loosen the distributor bracket screw

Fig. 162 Adjust the belt tension to specifications

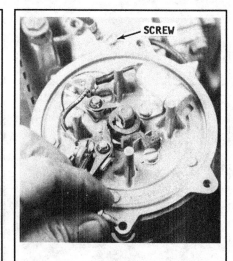

Fig. 163 Install the breaker plate

■ A very tight fit is achieved between the flywheel and the crankshaft taper after the specified torque value tension is applied to the flywheel nut. Surprising as it may sound, it is this tight fit that actually drives the flywheel during engine operation and *not* the Woodruff key. The key merely serves to time the flywheel to the crankshaft so the ignition timing marks can be used properly. For this reason the crankshaft taper and the inside flywheel taper must be absolutely dean before the flywheel is installed. There can be no trace of grease and/or oil. Both tapers must be dry.

16. Check to be sure the Woodruff key is properly positioned in the crankshaft. Lower the flywheel down the crankshaft and at the same time, work the belt towards the outside perimeter of the upper bearing cap. Continue to lower the flywheel and work the belt into the pulley on the underside of the flywheel. Install and tighten the flywheel nut to 70-85 ft. lbs (94.9-115.2 Nm) using an accurate torque wrench and a flywheel holder wrench.

17. Install the flywheel cover plate onto the flywheel and secure it in place with the 3 bolts.

■ After the proper torque value has been achieved, approximately one crankshaft thread will protrude above the top surface of the flywheel nut. If one thread appears, the flywheel keyway is properly indexed over the Woodruff key. If no threads are protruding above the flywheel nut, remove the flywheel and align the flywheel keyway with the Woodruff key. DO NOT attempt to operate the engine unless the flywheel is properly installed.

18. Rotate the flywheel CLOCKWISE until the red mark on the flywheel rim is aligned with the mark on the aft port side of the water jacket cover, identified as **BELT TIMING**.

✳✳ CAUTION

Take care when working with the timing belt loose to prevent it from becoming caught between the flywheel and the bearing head flanges. A new belt can easily be cut or damaged during the installation and timing work.

19. Rotate the distributor pulley until the notch in the pulley is aligned with the mark on the distributor bracket.

20. Work the timing belt onto the distributor pulley. Hold the distributor pulley timing mark aligned with the bracket mark.

21. Loosen the distributor bracket screw and slide the distributor away from the crankshaft until both the distributor pulley and the flywheel pulley are engaged with the distributor belt. Take up on one bolt until it is snug.

22. Adjust the distributor until the belt tension will allow 5/16 to 3/8 in. (7.94-9.53mm) deflection with slight finger pressure. After the belt has been adjusted properly, tighten the 3 distributor bracket bolts securely.

23. Place the breaker plate assembly into position in the distributor housing. Take care to work the point arms out and around the cam. Install the 2 hold down screws.

24. Slide the rotor onto the distributor shaft.

25. Install the distributor cap and secure it in place with the 3 retaining screws.

26. Install the flywheel ring gear guard and secure it in place with the attaching screws.

27. Install the coil and the coil holding bracket. Connect the high tension lead into the center of the coil.

Timing Check

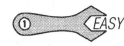

1. Rotate the flywheel CLOCKWISE until the red mark on the flywheel is aligned with the arrow on the flywheel ring gear guard marked **BELT TIMING**.

2. The notch in the distributor pulley must be in exact alignment with the mark on the distributor bracket.

3. Mount the engine in a large test tank or body of water and check the completed work.

4. Synchronize the distributor and carburetor as detailed in the Maintenance & Tune-Up section.

The engine must be mounted in a large test tank or body of water to adjust the timing. NEVER attempt to make this adjustment with a flush attachment connected to the lower unit or in a small confined test tank. The no-load condition on the propeller would cause the engine to 'runaway' resulting in serious damage or destruction of the unit.

✳✳ CAUTION

Water must circulate through the lower unit to the engine any time the engine is run to prevent damage to the water pump in the lower unit. Just 5 seconds without water will damage the water pump.

CAPACITOR DISCHARGE (CD) IGNITION SYSTEM - WITH SENSOR

Identification

The Type III capacitor discharge (CD) system is normally found on the following motors:
- 1967-68 and 1972 100 hp
- 1969-70 115 hp
- 1971-72 125 hp

Description & Operation

It would be worth the time to read and understand the information presented in the General before starting troubleshooting or service work on the ignition system.

This section covers the ignition system for the horsepower engines and model years listed in the heading. The Description & Operation section applies to all engines. however, some of the troubleshooting procedures differ somewhat for each horsepower. Therefore, each troubleshooting section will clearly indicate exactly what horsepower and model years the procedures cover.

This system consists of 3 major components: A pulse transformer, a sensor and an electronic pack. The pulse transformer replaces the conventional ignition coil. On the 1967 100 hp engine, the sensor is installed inside the distributor and replaces the breaker points and condenser. On the other models, the sensor, rotor and distributor cap are installed under the flywheel. The electronic power pack is completely sealed in epoxy and attached to a bracket on the lower rear motor cover. Special surface gap spark plugs are used with this system.

A transistorized converter steps up the 12 volt DC battery current to 300 volts DC, and then stores it in an energy storage capacitor located within the electronic pack.

A triggering circuit senses the trigger wheel position and doses an electronic switch. This transfers the 300 volt energy from the storage capacitor to the pulse transformer where it is boosted again, this time to 25,000 volts DC. This high voltage is transmitted to the spark plugs through the high tension leads. For this reason, NEVER hold a high tension lead with your hand while the engine is being cranked or running. Such action could result in severe electrical shock.

Now, when the storage capacitor has discharged its energy at the spark plug gap, the electronic switch opens and the converter recharges the storage capacitor. Service on this type of system is limited to detailed troubleshooting, and then replacing the faulty component.

✳✳ CAUTION

This system generates approximately 20,000 volts which is fed to the spark plugs. Therefore, perform each step of the troubleshooting procedures exactly as presented as a precaution against personal injury.

The following safety precautions should ALWAYS be observed:
- DO NOT attempt to remove any of the potting in the back of the power pack. Repair of the power pack is impossible.
- DO NOT attempt to remove the high tension leads from the ignition pulse transformer.
- DO NOT open or dose any plug-in connectors, or attempt to connect or disconnect any electrical leads while the engine is being cranked or is running.
- DO NOT set the timing advance any further than as specified.
- DO NOT hold a high tension lead with your hand while the engine is being cranked or is running. Remember, the system can develop

approximately 20,000 volts which will result in a severe shock if the high tension lead is held. ALWAYS use a pair of approved insulated pliers to hold the leads.

- DO NOT attempt any tests except those listed in this troubleshooting section.
- DO NOT connect an electric tachometer to the system unless it is a type which has been approved for such use.
- DO NOT connect this system to any voltage source other than given in this troubleshooting section.

Troubleshooting - General

The following troubleshooting procedures cover the 3 most common causes requiring service: the engine fails to start, the engine misses badly, or the unit lacks proper power - performs poorly.

Always attempt to proceed with the troubleshooting in an orderly manner. The shotgun approach will only result in wasted time, incorrect diagnosis, replacement of unnecessary parts and frustration.

Begin the ignition system troubleshooting with the spark plugs and continue through the system until the source of trouble is located.

BATTERY CONDITION

The battery is one of the most important items in this or any CD ignition system. The battery should be rated at least 70 amps with good clean terminals and tight cable connections. Double check the battery hook-up. The red cable should be connected to the positive battery terminal and the black cable connected to the negative terminal. The other end of the positive cable (red) should be connected to the starter solenoid terminal. The other end of the negative cable (black) should be connected to a good ground on the engine. If a battery should accidentally be connected to a CD system backwards (wrong polarity), the diodes in the charging system will be damaged and the engine will fail to start. The cranking system will function but the ignition system will NOT operate.

Before spending time troubleshooting the ignition system, check the complete battery circuit to be sure all areas are in satisfactory condition: electrolyte level; battery charge; clean terminals; tight connections at the terminals; and battery leads free of any frayed areas, cracks or other damage.

POWER PACK GROUNDING

◆ **See Figure 164**

The power pack on the original engines listed in the heading of this section was grounded through the power pack case and mounting bracket to the power head. This method of grounding was not as satisfactory as expected. Therefore, a service bulletin was issued from the factory to install a No. 12 size flexible ground wire (OMC #376349) from one of the power pack mounting screws to a bolt on the powerhead.

On the engines listed, if the engine continues to run after the ignition key switch has been turned to the **OFF** position, or if the engine is turned off and the boat appears to suddenly move forward, the shift diodes may be at fault. See the Lower Unit section for complete testing and service of the shift diodes.

SPARK PLUGS

■ **For complete information on testing the spark plugs, please refer to the Spark Plug procedures in the Ignition Systems - General Information section.**

COMPRESSION

■ **For complete information on checking your engine's compression, please refer to the appropriate procedures in the Maintenance & Tune-Up section.**

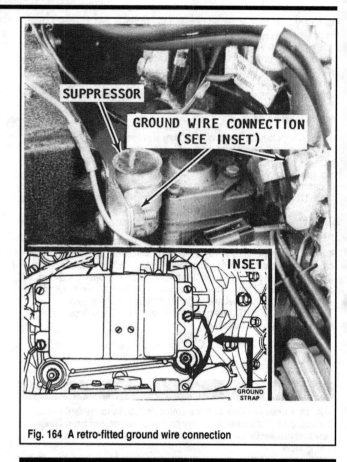

Fig. 164 A retro-fitted ground wire connection

Troubleshooting - 1967-68 100 Hp Engines

◆ **See Figures 165 and 166**

Before starting the following troubleshooting procedures for this ignition system, take the time to read and understand the system description at the beginning of this section and perform the General Troubleshooting procedures that apply to all models found in the Ignition Systems - General Information section.

The following test numbers are matched to test numbers on the accompanying wire diagram as an assist in performing the work.

Power Pack

The 100 hp engines have a power pack and voltage suppressor installed as original equipment. Now, if the power pack is to be replaced or if the pack was replaced sometime in the past, the suppressor must be removed and discarded.

In other words, the engine will NOT perform properly (in some cases it will not run at all) if the voltage suppressor is not removed when the new power pack is installed. The early model power pack can easily be identified by the horizontal cooling ribs visible on the front of the pack.

Spark Plugs

Any time a CD ignition system is serviced or troubleshooting procedures performed, the spark plugs must be removed and the high tension leads grounded. Grounding the leads provides a path for the electrical energy to follow when the engine is being cranked during tests. This simple procedure will prevent a cylinder from firing during a test and will also prevent any fuel vapors around the power head from being ignited during a test.

Model Difference

The 1967 100 hp model is equipped with a belt driven distributor. The 1968 model is equipped with the same components, but they are all installed under the flywheel.

If the engine being serviced has the old power pack, perform all of the following steps. If the engine being serviced has the new power pack installed, perform Test No.1, then skip to Test No.4 and carry on with the remainder of tests.

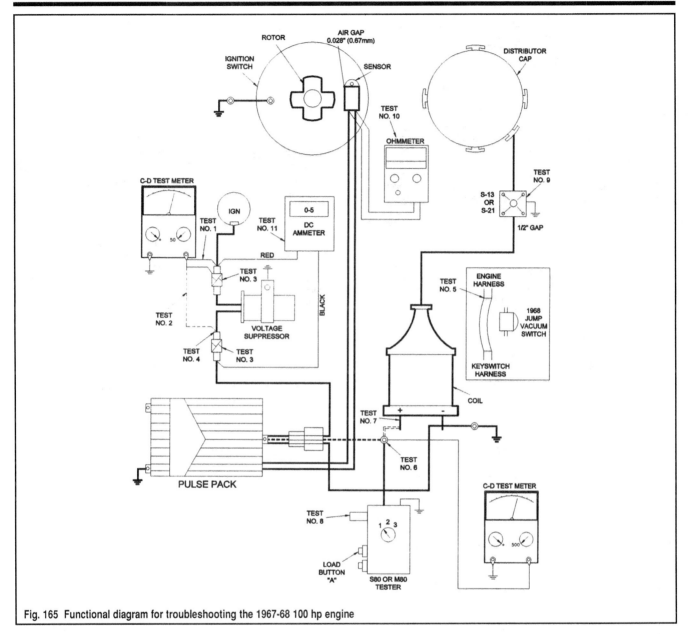

Fig. 165 Functional diagram for troubleshooting the 1967-68 100 hp engine

Fig. 166 These models utilize a voltage suppressor along with the power pack

TEST NO. 1 - VOLTAGE TEST TO SUPPRESSOR

◆ See Figure 165

■ If the engine being serviced has the old power pack, perform all of the following steps. If the engine being serviced has the new power pack installed, perform Test No.1, then skip to Test No.4 and carry on with the remainder of tests.

Turn the ignition switch to the **ON** position. Connect a voltmeter into the circuit at the connector just before the voltage suppressor. The meter should indicate 12 volts. If voltage is not present, the ignition switch is faulty, or the battery is not up to a full charge.

TEST NO. 2 - VOLTAGE SUPPRESSOR CHECK

◆ See Figure 165

If voltage was present in Test No.1, connect the voltmeter into the circuit after the voltage suppressor. If voltage is not present, the voltage suppressor must be checked. If voltage is present, proceed directly to Test No.4.

TEST NO. 3 - SUPPRESSOR OHM CHECK

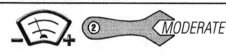

◆ **See Figure 165**

1. Disconnect both leads, one to the key and the other to the power pack. Connect the ohmmeter to the 2 leads. The meter should indicate 0.5 ohms resistance on the low scale. If 0.5 ohms is not indicated, replace the suppressor. If 0.5 ohms is indicated, proceed with the testing.

2. Carefully check the lead that was connected to the power pack. Turn the ignition switch to the **OFF** position. Disconnect the lead coming out of the suppressor to the power pack.

3. Connect the ohmmeter to the red lead and to a ground on the voltage suppressor. Check the meter reading. Reverse the ohmmeter connections. The ohmmeter should give a reading when the leads arc connected one way but NOT the other way. If there is no reading in either direction, or there is a reading in both directions, the voltage suppressor is definitely defective and must be replaced. If the voltage suppressor passes the test proceed with the testing.

TEST NO. 4 - VOLTAGE TEST

◆ **See Figure 165**

1. Connect at voltmeter probe into the red lead to the power pack and the other probe to a good ground on the engine.

2. Crank the engine and observe the reading. The voltmeter should indicate not less than 9.5 volts. If the reading is less than 9.5 volts, check the condition of the battery.

3. One other possibility is a defective starter motor drawing an excessive amount of current from the battery. If the starter motor is defective, conduct a "draw test" on the motor as described elsewhere in this section.

TEST NO. 5 - VACUUM SAFETY SWITCH TEST (1968 ONLY)

◆ **See Figure 165**

This model unit has a vacuum switch mounted on the starboard side of the engine. As the engine is cranked, this switch doses and allows current to pass through from the key switch to the voltage suppressor and the power pack.

The safety vacuum switch does not close and allow current to pass to the power pack until the engine is cranked and vacuum pulls the switch in permitting current to flow.

This safety switch also prevents a hot engine from starting if the key switch is turned to the **ON** position immediately after a shutdown. Such a circumstance could result in a 'runaway' engine or boat.

1. Connect a voltmeter to the output side of the vacuum switch and to a good ground on the engine.

2. Crank the engine and observe the voltmeter reading. The meter should indicate not less than 9.5 volts. If the reading is less than 9.5 volts, connect the voltmeter to the input side of the switch and again observe the reading. If the reading is 9.5 volts or more, the vacuum switch is defective and must be replaced.

TEST NO. 6 - POWER PACK OUTPUT

◆ **See Figure 165**

1. Set the voltmeter to the 500 volt scale. If the voltmeter is not available, a spark tester may be used for this test. The tester will flash indicating power pack output.

2. Disconnect the green wire from the pulse transformer.

3. Connect the red test lead to the green wire running to the power pack. Connect the black test lead to ground.

4. Crank the engine and observe the meter reading. The meter should indicate 250 volts or more. If the meter falls short, the indication is an intermittent short in the power pack. A satisfactory meter reading indicates a good power pack. The problem is most likely in the pulse transformer.

TEST NO. 7 - INTERMITTENT MISS, ENGINE RUNNING

◆ **See Figure 165**

1. Connect the green wire back to the pulse transformer.

2. Connect the red test lead to the pulse transformer terminal where the green wire is connected.

3. Leave the black lead connected to ground.

4. Start the engine and allow it to warm to operating temperature.

✳✳ CAUTION

Water must circulate through the lower unit to the engine any time the engine is run to prevent damage to the water pump in the lower unit. Just 5 seconds without water will damage the water pump.

5. Advance the throttle until it begins to misfire and observe the meter. The meter reading should not waver, but should increase to 300 volts. If the meter needle does waver, an intermittent short is indicated in the power pack. If the voltmeter is satisfactory, a safe assumption is that the power pack is suitable for service. If there is no meter reading while the engine was cranked, or if the engine fails to start, proceed with the testing in the next paragraph.

6. On 1967 models, remove the distributor cap and rotate the distributor until the flat edge of the rotor is parallel with the sensor.

7. On 1968 models, connect a jumper wire to bypass the vacuum switch. If the flywheel, stator, distributor cap and rotor have not been removed, remove them at this time. Disconnect the 2 blue wires from the sensor.

8. On all models, connect the voltmeter to the positive terminal of the pulse transformer and to a good ground. On the 1967 model only, insert a feeler gauge between the rotor and the sensor. The voltmeter should indicate 250 volts or more. On the 1968 model, momentarily ground the blue wire to the power pack. The voltmeter should indicate 2.50 volts or more. On both models, if the voltmeter reading is satisfactory, the power pack is acceptable for service. If the voltmeter reading is not satisfactory proceed with the testing.

TEST NO. 8 - POWER PACK TEST

◆ **See Figure 165**

1. Disconnect the green wire from the pulse transformer.

2. Connect the blue wire from a neon tester (Stevens #S-80 or M-80) to the green wire disconnected from the pulse transformer.

3. Connect the other test lead to a good ground.

4. Crank the engine and observe the light. If the light is strong and bright, the power pack is acceptable for service and the problem is in the pulse transformer. If the light is weak or fails to come on, either the power pack or the sensor is defective. To check the sensor, proceed with Test No. 9.

TEST NO. 9 - PULSE TRANSFORMER OUTPUT & SENSOR CHECK (1967)

◆ **See Figure 165**

1. Connect one side of a spark tester to the high tension lead coming out of the pulse transformer and the other side of the tester to a good ground. If a spark tester is not available, use a pair of insulated pliers and hold the high tension lead about 1/2 in. (12.7 mm) from the engine block.

2. Remove the distributor cap and rotate the flywheel until the rotor is parallel with the sensor.

3. Slide a feeler gauge between the rotor and the sensor. A spark should be observed at the spark tester or between the high tension lead and the engine block. If a spark is not observed, remove the blue wires from the sensor and momentarily ground them. Each time the blue wires are grounded a spark should be observed at the tester or between the high tension lead and the engine block.

If no spark is observed the pulse transformer is defective and must be replaced.

TEST NO. 10 - SENSOR CIRCUIT

◆ **See Figure 165**

1. On 1967 models, remove the distributor cap and connect an ohmmeter to the terminals of the sensor.

2. Set the ohmmeter to the low scale. The meter should indicate 4.0-6.0 ohms. If the reading is not within the required range, the sensor is defective and must be replaced. The sensor on this model may be checked visually. Remove the sensor and check it thoroughly for cracks that would allow the unit to short out to ground.

3. On 1968 models, if the flywheel, stator, distributor cap and rotor have not been removed, remove them at this time.

4. Disconnect the blue wires from the sensor and momentarily ground them to the engine block. Each time they are grounded the spark tester should indicate a strong spark. If no spark is observed, the pulse transformer is defective and must be replaced.

5. Still on 1968 models, disconnect the 2 blue wires at the sensor.

6. Connect an ohmmeter to each lead from the sensor. Set the ohmmeter to the low scale. The ohmmeter should indicate 4.0-6.0 ohms. If the reading is not within the required range, the sensor is defective and must be replaced.

7. Set the ohmmeter to the high ohm scale. Connect one meter lead to the sensor lead and the other lead to a good ground on the engine. Advance and retract the throttle to shift the anti-reverse spring to and from ground. The ohmmeter should read at least 100,000 ohms. With the meter still connected, work the wires back-and-forth for indication of a ground condition.

8. On all models now, when the sensor is replaced, the air gap between the rotor and the sensor must be adjusted to 0.028 in. (7.1mm).

■ **On 1968 models, the blue wires from the power pack are routed onto an arm on the sensor base plate. The wires are then routed into a groove on the underside of the base plate and held in place with a retainer and screw. Where the wires are held by the retainer is a common source of problems - the wires may be grounded. The base plate should be removed and the wires checked for a grounded condition.**

TEST NO. 11 - AMPERAGE INPUT TO POWER PACK CHECK

◆ **See Figure 165**

The engine MUST be mounted in a large test tank or body of water to check the amperage input to the power pack. NEVER attempt to make this check with a flush attachment connected to the lower unit or in a small confined test tank. The no-load condition on the propeller would cause the engine to 'runaway', resulting in serious damage or destruction of the unit.

✳✳ CAUTION

Water must circulate through the lower unit to the engine any time the engine is run to prevent damage to the water pump in the lower unit. Just 5 seconds without water will damage the water pump.

1. Connect a low reading DC ammeter (0-5 amps) in series between the key switch and the power pack.

2. Start the engine and allow it to warm to operating temperature at 500 rpm.

3. With the engine operating at idle speed, the meter should indicate 0.4-0.6 amps. At high rpm, the meter should hold steady and not exceed 2.75 amps. If the meter reading is not within these limits, the power pack is defective (drawing excessive amperage) and must be replaced.

If the power pack is to be replaced, the voltage suppressor will not be used. Connect the lead from the key switch directly to the power pack bypassing the voltage suppressor.

In other words, if the power pack is to be replaced, the voltage suppressor is not used with the new pack and can be removed and discarded. An old power pack has ribs on the front side while the newer unit is smooth.

Restore the system to its original condition by removing all test equipment and making the necessary connections.

Troubleshooting - 1969-70 115 Hp Models

◆ **See Figure 167**

Before starting the following troubleshooting procedures for this ignition system, take the time to read and understand the system description at the beginning of the General Troubleshooting procedures that apply to all models.

The following troubleshooting procedures cover the 3 most common causes requiring service: the engine fails to start, the engine misses badly, or the unit lacks proper power - performs poorly.

Always attempt to proceed with the troubleshooting in an orderly manner. The shotgun approach will only result in wasted time, incorrect diagnosis, replacement of unnecessary parts, and frustration.

Begin the ignition system troubleshooting with the spark plugs and continue through the system until the source of trouble is located.

If the safety switch fails to close and allow current to pass to the power pack, several sources may be at fault: the safety switch circuit, rectifier, shift diodes, stator, tachometer (if one is installed), or the flywheel magnets. The flywheel magnets usually do not cause problems unless they become corroded, or foreign particles become attached to them.

■ **Before proceeding with the following tests, the spark plugs should be removed and the high tension leads connected to a spark tester or grounded. This procedure will prevent a cylinder from firing accidentally during a test. Grounding the leads will provide a path for the current generated during a test and drastically reduce the chances of a spark igniting fuel or fuel vapors around the powerhead.**

TEST NO. 1 - KEY SWITCH OUTPUT

◆ **See Figure 167**

1. Attach the black lead of a low-voltage voltmeter to a good ground on the engine and the red lead to either No. 8 or No. 9 posts on the engine terminal board.

2. Turn the ignition key to the **ON** position. The meter should indicate 12 volts at either terminal. If the meter fails to register the required voltage, either the key switch is defective or the battery circuit is defective.

Voltage Check

3. Disconnect the red voltmeter test lead from the engine terminal board and connect it to No. 7 post on the terminal board.

4. Connect the other meter lead to a good ground on the engine.

5. Crank the engine. The meter should indicate no less than 9.5 volts. If the meter resisters the required voltage, the safety switch is satisfactory. If the meter indicated less than 9.5 volts, one of 3 areas could be at fault: the battery circuit is not up to full service, the safety switch is defective, or the starter circuit is drawing an excessive amount of voltage.

Ignition Safety Switch

This safety switch is installed between the key switch and the power pack. The switch does not close and allow current to pass to the power pack until the alternator is rotating and creating current. This safety switch also prevents a hot engine from accidentally starting if the key switch is turned to the **ON** position immediately after a shutdown. Such a circumstance could result in a RUNAWAY engine or boat.

TEST NO. 2 - SAFETY SWITCH LEAKAGE TEST

◆ **See Figure 167**

1. Remove the lead from No. 7 post at the engine terminal board.

2. Connect one lead of a test light to the wire leading to the safety switch. Connect the other lead to a good ground on the engine.

3. Turn the ignition switch to the **ON** position. The test light should come on (glow).

4. Disconnect the wires from No. 1, 2, 3 and 4 posts at the terminal board. If the light continues to glow after the wires have been disconnected, the safety switch is defective and must be replaced. If the light goes out when the wires are disconnected at the terminal board, the shift diodes and the rectifier must be checked.

Procedures to check the shift diodes are presented in the Lower Unit section.

TEST NO. 3 - FIRST POWER PACK OUTPUT CHECK

◆ **See Figure 167**

1. Connect the black meter lead of a high voltage voltmeter to a good ground on the engine.

2. Connect the red meter lead to No. 5 post at the engine terminal board. Crank the engine and observe the meter. If the meter indicates at least 250 volts, the power pack is satisfactory for continued service. If less than 250 volts is registered when the engine is cranked, proceed to Test No. 4.

TEST NO. 4 - SECOND POWER PACK OUTPUT CHECK

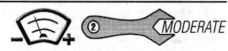

◆ **See Figure 167**

1. Disconnect the blue wire from the No. 5 post at the terminal board.

2. Connect a Stevens S-80 or M-80 tester blue lead to the blue power pack lead and the black test lead to a good ground on the engine.

3. Connect one lead to the blue lead to the power pack, and the other lead to engine ground.

4. Crank the engine. The test light should flicker. If the light does flicker, the power pack is satisfactory. If the test light failed to flicker, disconnect the test light and connect the blue wire back to the No. 5 post at the terminal board. Proceed to Test No.5.

TEST NO. 5 - THIRD POWER PACK OUTPUT CHECK

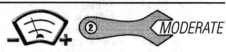

◆ **See Figure 167**

1. Disconnect the plug on the wires between the sensor and the power pack. Connect 2 jumper wires to the wires to the power pack.

2. With a small wire, bridge between No. 7 and No. 8 posts on the terminal board. Connect the blue wire from a Stevens S-80 or M-80 test light to the No. 5 post on the terminal board. Connect this test light to the No. 5 post and a good ground on the engine.

3. Turn the key switch to the **ON** position. Now, momentarily ground the wires from the sensor. Each time the wires are grounded, the neon light should flicker. If the light flickers, the power pack is satisfactory for further service. If the light failed to flicker, the power pack is defective and must be replaced.

4. Turn the key switch to the **OFF**. Disconnect the jumper wire from the No. 7 and No. 8 posts, also the tester lead from the No. 5 post.

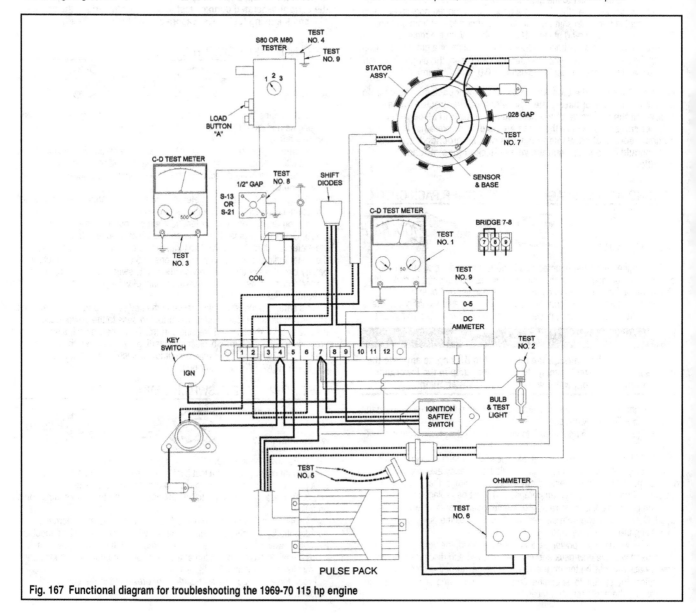

Fig. 167 Functional diagram for troubleshooting the 1969-70 115 hp engine

TEST NO. 6 - SENSOR OUTPUT CHECK

◆ **See Figure 167**

First Sensor Check

1. While the plug is still disconnected per Test No. 5, disconnect the 2 jumper leads out of the plug on the side to the power pack.

2. Connect the 2 leads from an ohmmeter to the 2 leads in the connector that run to the sensor under the flywheel. If the ohmmeter indicates other than 4.0-6.0 ohms, the sensor installed under the flywheel is defective and must be replaced. If the required resistance is indicated, proceed to Test No. 7.

Second Sensor Check - for Ground

3. Connect one ohmmeter lead to the sensor lead at the plug and the other lead to a good ground on the engine. Open and close the throttle to shift anti-reverse spring to and from ground. The meter should indicate at least 100,000 ohms.

4. With the meter still connected, work the wires back-and-forth to indicate ground. If a ground is discovered, the sensor or the sensor wires must be replaced. If the sensor is found to be satisfactory, proceed to Test No. 7. Disconnect the ohmmeter and connect the sensor plug together.

■ **The blue wires from the power pack are routed onto an arm on the sensor base plate. The wires are then routed into a groove on the underside of the base plate and held in place with a retainer and screw. Where the wires are held by the retainer is a common source of problems - the wires may be grounded. The base plate should be removed and the wires checked for a grounded condition.**

TEST NO. 7 - OTHER IGNITION ITEMS CHECK

◆ **See Figure 167**

Remove the flywheel and sensor.

Check the rotor and distributor cap for cracks, carbon trails or corrosion. Inspect the sensor for cracks across the sensor eye.

If all appears to be in satisfactory condition, install and adjust the sensor to 0.028 in. (0.71mm). Install the rotor, cap, stator and flywheel.

TEST NO. 8 - COIL OUTPUT

◆ **See Figure 167**

With all units in place and secure, with all test equipment disconnected and with all original wires connected back in their proper positions; disconnect the pulse transformer wire from the distributor cap.

This high-tension lead is threaded into the cap, therefore, rotate it COUNTERCLOCKWISE until it is free.

As the wire is turned, count the number of revolutions as an aid to connecting it later. Before threading it back into the cap, the wire must be twisted the required number of revolutions in the opposite direction. Then, when it is properly installed, the wire will not have any twists, but follow its natural direction.

Use a pair of insulated pliers and hold the high tension lead approximately 1/2 in. (12.7mm) from ground to the engine. Crank the engine. If the spark is weak or no spark is observed, the coil is defective and must be replaced.

TEST NO. 9 - INTERMITTENT MISS CHECK

◆ **See Figure 167**

1. Connect the black lead of the neon tester to a good ground on the powerhead.

2. Connect the other meter lead to the No. 5 post of the engine terminal board.

3. Connect the ammeter to the purple wire from the No. 7 post on the terminal board, and to the purple wire leading to the power pack. This places the ammeter in series with the power pack.

4. Start and operate the engine.

---

5. Observe both the neon tester light and the ammeter. If the light flashes or the ammeter needle wavers, replace the power pack.

6. Restore the system to its original condition by removing all test equipment and making the necessary connections.

Troubleshooting - 1972 100 Hp & 1971-72 125 Hp Models

◆ **See Figure 168**

Before starting the following troubleshooting procedures for this ignition system, take the time to read and understand the system description at the beginning of the General Troubleshooting procedures that apply to all models.

The following troubleshooting procedures cover the 3 most common causes of requiring service: the engine fails to start, the engine misses badly, or the unit lacks proper power - performs poorly.

Always attempt to proceed with the troubleshooting in an orderly manner. The shotgun approach will only result in wasted time, incorrect diagnosis, replacement of unnecessary parts and frustration.

Ground the spark plug leads to the engine to render the ignition system inoperative while performing the compression check.

■ **The spark plug leads are grounded as a safety measure and to prolong coil life. By grounding the leads the current has a path to follow without creating a spark. Any amount of fuel in or on the engine, even the fumes from under the cowling, creates a dangerous fire hazard. By eliminating the spark, the chances of igniting any fuel or fuel vapors is drastically reduced.**

TEST NO. 1 - BASIC COMPONENTS AT REST

◆ **See Figure 168**

1. Connect the black lead of the low-voltage voltmeter to a good ground on the engine.

2. Connect the red lead to either the No. 8 or No. 9 post at the engine terminal board.

3. Turn the ignition key to the **ON** position and observe the meter reading. If the reading is less than 12 volts the problem is either a blown fuse in the fuse holder; a defective key switch, connection, or wires leading from the key switch; a poor connection at the positive terminal on the starter solenoid; or, a battery that is not up to full charge.

TEST NO. 2 - BASIC COMPONENTS, ENGINE CRANKING

◆ **See Figure 168**

With the voltmeter still connected to the No. 8 or No.9 post at the engine terminal board, per Test No. 1, crank the engine and observe the meter reading. If the reading is less than 9.5 volts, the battery may not be up to a full charge, or the starter motor is drawing an excessive amount of voltage. Check out the battery circuit, perform the test a second time, and then check out the starter motor circuit.

Disconnect the test equipment.

TEST NO. 3 - FIRST POWER PACK OUTPUT CHECK

◆ **See Figure 168**

1. Connect the back lead of a high-voltage voltmeter to a good ground on the engine.

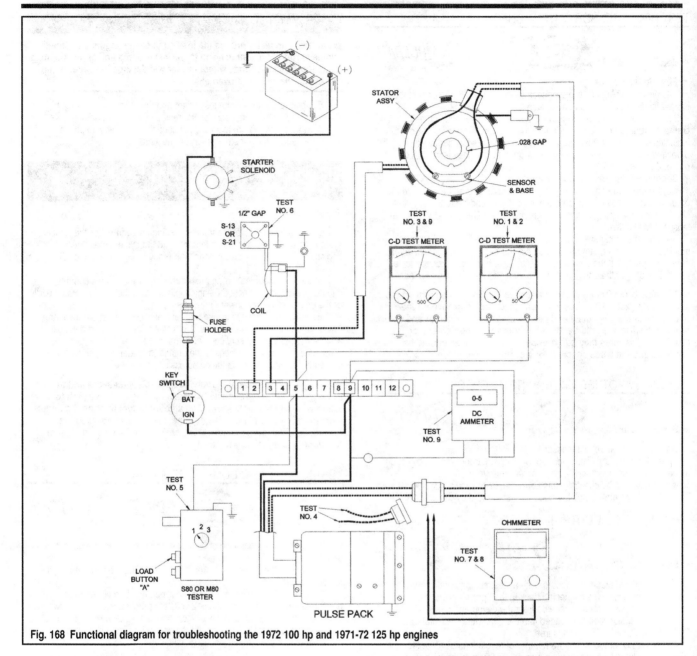

Fig. 168 Functional diagram for troubleshooting the 1972 100 hp and 1971-72 125 hp engines

2. Connect the red lead to the No. 5 post on the engine terminal board.

3. Crank the engine and observe the meter reading. If the reading is less than 250 volts, proceed to Test No. 4. If the meter reading is satisfactory, the power pack is delivering sufficient voltage to the coil and the unit is acceptable for further service.

TEST NO. 4 - SECOND POWER PACK OUTPUT CHECK

◆ See Figure 168

1. Turn the ignition switch to the **ON** position.

2. With the high-voltage voltmeter still connected to the No. 5 post at the engine terminal board per Test No. 3, disconnect the sensor plug at the power pack.

3. Use 2 jumper wires and insert them into the plug half with the wires to the power pack. Momentarily ground one of the wires and at the same time observe the meter reading. Each time the wire is grounded, the meter should indicate at least 250 volts. If the meter fails to register the required voltage, the problem is in the sensor or in the coil.

4. Proceed to Test No. 5.

TEST NO. 5 - THIRD POWER PACK OUTPUT CHECK

◆ See Figure 168

1. Disconnect the voltmeter from the No. 5 post at the engine terminal board. Disconnect the blue wire from the No. 5 post.

2. Turn the ignition switch to the **ON** position.

3. Connect the blue wire, from either a Stevens S-80 or M-80 tester, to the blue wire that was just removed from the No. 5 post.

4. Connect the other test lead to a good ground on the engine. With the jumper wires still connected per Test No. 4, momentarily touch one of the jumper wires to ground. Each time the wire is grounded, the tester light should flash indicating the power pack is sending current to the coil. If the light fails to flash, the problem is in the coil or in the sensor. Proceed to Test No .6.

5. Disconnect the jumper leads from the sensor plug and connect the plug together.

6. Connect the blue wire back to the No. 5 post on the terminal board.

TEST NO. 6 - COIL OUTPUT

◆ See Figure 168

1. With all units in place and secure, all test equipment disconnected, and with all original wires connected back in their proper positions; disconnect the high tension lead from the distributor cap. This high-tension lead is threaded into the cap; therefore, rotate it COUNTERCLOCKWISE until it is free. As the wire is turned, count the number of revolutions as an aid to connecting it later. Before threading it back into the cap, the wire must be twisted the required number of revolutions in the opposite direction. Then, when it is properly installed, the wire will not have any twists, but follow its natural direction.

2. Use a pair of insulated pliers and hold the high tension lead approximately 1/2 in. (12.7mm) from ground on the engine.

3. Crank the engine. If the spark is weak or no spark is observed, the coil is defective and must be replaced.

TEST NO. 7 - SENSOR CIRCUIT

◆ See Figure 168

1. Disconnect the plug in the lines leading to the power pack.

2. Set the ohmmeter to the low-scale. Now, insert the 2 meter test probes into the plug on the sensor half (wires leading to the sensor under the flywheel). The meter should indicate 4.0-6.0 ohms. If the reading is low, the sensor is defective and must be replaced. If the reading is satisfactory, proceed with the next Test.

TEST NO. 8 - SENSOR CHECK FOR GROUNDING

◆ See Figure 168

1. Set the ohmmeter to the high scale.

2. Connect one meter lead to a good ground on the engine.

3. Connect the other meter lead into the sensor plug at either terminal. Open and close the throttle to shift the anti-reverse spring to-and-from ground. The meter should indicate 100,000 ohms; then zero. With the meter still connected, work the wires back-and-forth to indicate ground. If a ground is indicated, either the wires or the sensor is grounded. Correct the condition by replacing the wires and/or the sensor. Restore the system to its original condition.

TEST NO. 9 - INTERMITTENT MISS

◆ See Figure 168

1. Set the high-voltage voltmeter to the 500 volt scale.

2. Connect 1 meter lead to a good ground on the engine.

3. Connect the other meter lead to the No. 5 post at the engine terminal board.

4. Connect 1 lead of the low-amp ammeter to the purple wire of the power pack and the other lead to the No. 8 or No. 9 post of the terminal board.

5. Start the engine and operate it at the rpm at which the miss occurs.

✳✳ CAUTION

Water must circulate through the lower unit to the: engine any time the engine is run to prevent damage to the water pump and an overheating condition. Just 5 seconds without water will damage the water pump and cause the engine to overheat.

6. Observe both meters. The needles should hold steady. If either meter fluctuates (wavers) the power pack is defective and must be replaced. If the ammeter reading is greater than 1.8 amps, the power pack is defective and must be replaced. A NEW rotor must be installed if the power pack is replaced. The rotor is sold with the new style power pack. DO NOT attempt to use an old rotor with the new style power pack.

7. Restore the system to its original condition by removing all test equipment and making the necessary connections.

Service - 1967 100 Hp Models

ROTOR REPLACEMENT

◆ See Figures 169 and 170

1. Remove the 3 attaching screws, and then lift the cap free of the distributor.

2. Two rotors are installed on this model engine. The top rotor conducts current to the distributor cap. The lower metal rotor performs the function of the trigger and works with the sensor. Grasp the top rotor and pull it free of the distributor shaft. If the lower rotor must be replaced, use 2 screwdrivers, one on the opposite side of the rotor, and work the rotor upward and free of the distributor shaft.

3. Push the new lower rotor down the shaft and into place.

4. Push the top rotor down into place on the distributor shaft.

5. Place the distributor cap into position and secure it with the 3 attaching screws.

SENSOR REPLACEMENT

◆ See Figure 171

1. Remove the 3 attaching screws, and then lift the cap free of the distributor.

2. Remove the hold down screw; disconnect the 2 blue wires at the sensor, then remove the sensor.

3. Install the new sensor to the base plate.

4. Rotate the flywheel until the lower metal rotor is horizontal with the sensor; adjust the sensor to 0.028 in. (7.1mm).

5. Connect the blue wires to the sensor.

6. Install and secure the distributor cap with the 3e attaching screws.

Fig. 169 Remove the distributor cap. . .

Fig. 170 . . .and then the upper and lower rotors

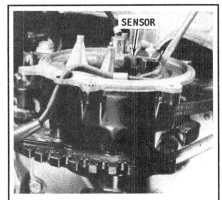

Fig. 171 Removing the sensor

POWER PACK REPLACEMENT

The distributor cap must be removed and the sensor wires disconnected from the sensor. These wires are an integral part of the power pack.

1. Disconnect the positive (green) wire running to the pulse transformer and the ground wire.

2. Remove the 3 attaching bolts, separate the 2 quick disconnects and remove the power pack.

3. Install the new power pack, secure it in place with the 3 attaching screws, and connect the wire to the sensor. The green wire is positive and the other wire is the ground wire.

4. Install the distributor cap.

■ **If a new power pack is being installed on the 1967-68 models, the voltage suppressor is not used with the system. Remove and discard the suppressor.**

Service - All Other Models

VACUUM SWITCH SERVICE - 1968 100 HP MODELS

◆ **See Figure 172**

The vacuum switch should remain open when the ignition key is turned to the **ON** position. The switch should close after the key is turned to the **START** position and the starter motor begins to crank the engine.

To test the vacuum switch, see Test No. 5 in the Troubleshooting - 1967-68 100 Hp Models section.

Disassemble the switch components. Wash the parts in solvent and blow them dry with compressed air. Inspect the valve and seat for any condition that might prevent the valve from seating properly.

Inspect the contacts to be sure they are clean and not corroded. Inspect the diaphragm and the diaphragm contact. Check the diaphragm fit over the switch housing. Any leakage in this area would prevent the contacts from closing. Assemble the check valve and test the action by alternately blowing and drawing air through the valve. The disc valve should close under pressure and open under suction.

Replace any defective parts.

DISASSEMBLY

◆ **See Figures 173 thru 181**

Flywheel

1. Use a holding tool to prevent the flywheel from rotating and remove the flywheel nut. Use the proper flywheel puller and "pull" the flywheel free of the crankshaft. After the flywheel has been removed, notice how the wiring is routed and tucked to prevent chafing or other damage during engine operation. In particular notice how the sensor wire and stator wires are routed and held down by clamps to protect them from the flywheel during engine operation.

Fig. 172 Vacuum switch

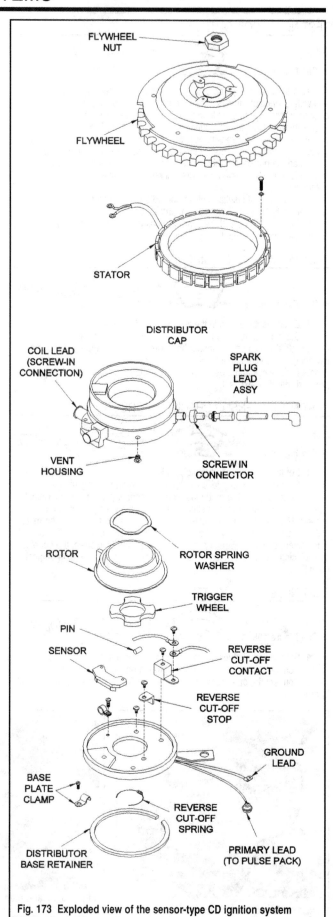

Fig. 173 Exploded view of the sensor-type CD ignition system

Stator

2. Note how the stator surrounds the distributor cap. Remove the Phillips screws securing the stator and then lift the stator free of the distributor cap.

3. Lift the distributor cap free of the powerhead.

4. Lift the rotor free of the crankshaft. Observe the keyway in the rotor. Turn the rotor over and remove the wavy washer. This washer keeps the rotor from vibrating during engine operation.

5. The sensor and the sensor rotor are now visible. Notice how the sensor rotor is also keyed to the crankshaft. Use 2 small prybars under the rotor and work the rotor upward and free of the crankshaft - carefully.

6. The sensor is held in place with 2 attaching screws. The wires from the sensor are held in place on the distributor base by a retainer and a small screw between the 2 wires. These sensor wires are a potential source of a ground; therefore, work with care not to damage the wires in any way. Remove the wire retainer screw.

7. Check the condition of each wire carefully. Remove the sensor, remove the wire retainers, disconnect the wires at the power pack, and then remove the sensor.

Distributor Base

8. Remove the 4 retaining screws and the retainer clips. Lift the distributor base free and check the anti-reverse spring underneath the base.

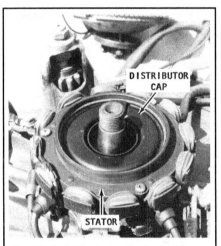
Fig. 174 A nice look after the flywheel has been removed

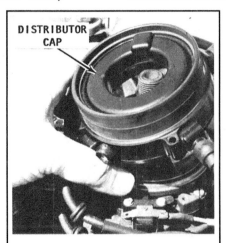
Fig. 175 Lift out the distributor cap. . .

Fig. 176 . . .and then the rotor (upper) and washer

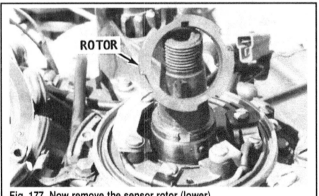
Fig. 177 Now remove the sensor rotor (lower). . .

Fig. 178 . . .and the sensor

Fig. 179 Remove the distributor base

Fig. 180 Pop out the anti-reverse spring. . .

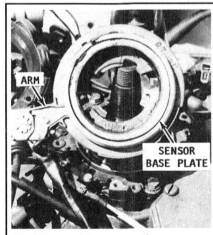
Fig. 181 . . .and then remove the sensor base plate

9. The function of the anti-reverse spring is to shut the engine down immediately if it should start with the crankshaft turning in the wrong direction. This spring should be replaced any time the distributor base has been removed.

10. The blue wires from the power pack are routed onto an arm on the sensor base plate. The wires are then routed into a groove on the underside of the base plate and held in place with a retainer and screw. Where the wires are held by the retainer is a common source of problems - the wires may be grounded. The base plate should be removed and the wires checked for a grounded condition.

CLEANING & INSPECTION

◆ **See Figure 173**

Check and clean the distributor cap and rotor with a good greaseless cleaning solvent, Tempo Instant Spray Alternator/Generator Cleaner No. 20-013, or equivalent. Denatured alcohol or trichloroethylene may also be used. Wipe the parts dry with a clean cloth.

Inspect the cap and rotor to be sure they are free of any cracks which could cause a voltage leak. If the distributor cap is to be replaced, the 5 high tension leads (4 to the spark plugs and 1 to the coil) must be unscrewed from the cap. Simply slide the small rubber caps back, and then rotate each wire COUNTERCLOCKWISE until the wire is free. The high tension leads all have numbers and the distributor cap has matching numbers. Install the correct numbered lead into the proper numbered hole.

Disconnect the high tension lead from the spark plug one-at-a-time as it is installed into the distributor cap to prevent any twisting condition as the lead is threaded into the cap. Connect the wire to the spark plug after the other end is secure in the distributor cap.

Check the electrodes for a burned condition.

Inspect the sensor for cracks across the sensor eye.

Check the plastic ring on the underneath side of the distributor base. This ring is the wearing surface for the distributor base as it rotates during engine operation. The ring is a replaceable item.

ASSEMBLY

◆ **See Figures 173 and 182 thru 187**

Distributor Base

1. Install a NEW anti-reverse spring. After the spring is in place, lubricate it with Shell EP No. 2 Grease. Lubricate the nylon plastic ring with light-weight oil, and then install it to the underside of the distributor base with the FLAT side facing UPWARD.

2. Lower the distributor base down into the top bearing cap, paying careful attention to be sure the anti-reverse spring slides into the recess in the underside of the base plate. Work the nylon plastic ring down into the recess of the bearing head.

3. Replace the 4 clips and screws securing the base plate to the bearing head. Rotate the base to be sure it moves freely without any evidence of binding. Install the sensor rotor so the cut-out in the rotor indexes over the pin in the crankshaft. Work the rotor down onto the crankshaft shoulder.

Fig. 182 Install the NEW anti-reverse spring...

Fig. 183 ...and make sure that it slides into the recess on the bottom of the distributor base when its installed

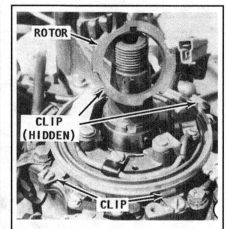

Fig. 184 Don't forget the 4 clips securing the base

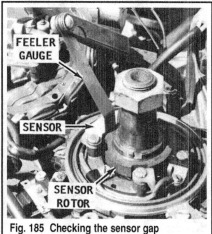

Fig. 185 Checking the sensor gap

Fig. 186 Slide the rotor down over the shaft

Fig. 187 Make sure that you use Loctite (or similar) on the stator mounting screws

Sensor

4. Install the sensor to the base plate. Rotate the crankshaft until the flat side of the sensor rotor is horizontal with the sensor. Adjust the sensor to 0.028 in. (7.1mm).

5. Install the upper rotor with the small knob on the rotor indexed into the small recess on the crankshaft.

6. Slide the wavy washer down the crankshaft and into place inside the distributor rotor.

7. Install the distributor cap over the rotor and seat it with the small protrusion indexed into the cutout of the bearing head. This position prevents the cap from rotating and maintains engine timing.

Stator

8. Slide the stator over the distributor cap. Coat the attaching screws with Loctite, or similar holding compound. Secure the cap with the screws and tighten the screws to 40-60 inch lbs. (4.52-6.78 Nm).

Final

9. Check the flywheel magnets to be sure they are absolutely free of any foreign particles. Install the flywheel and flywheel nut. Tighten the nut to the torque value given in the appropriate Specifications chart.

CAPACITOR DISCHARGE (CD) IGNITION SYSTEM - WITH BREAKER POINTS

Identification

This capacitor discharge (CD) system is normally found on the following motors:

- 1968-69 55 hp
- 1970-71 60 hp
- 1968 65 hp
- 1968-72 85 hp
- 1971 100 hp

Description & Operation

This system contains 3 major units: the ignition coil, 2 individual sets of points and an amplifier - the amplifier is a throwaway-type. If troubleshooting isolates the problem to the amplifier, it is simply removed, discarded, and a new unit installed.

This section is divided into 3 parts covering the following horsepower engines and model years:

First part:
- 1968-69 55 hp, 3-cyl.
- 1968 65 hp, V4
- 1968 85 hp, V4

Second part:
- 1969-72 85 hp, V4
- 1971 100 hp, V4

Third part:
- 1970-71 60 hp, 3-cyl.

The amplifier contains a transistorized converter, which steps up the 12 volts from the battery to approximately 350 volts. This voltage is stored in an energy capacitor. The breaker points function in the same manner as a switch. Current passes or does not pass through. The points are actuated through a cam on the crankshaft.

The points trigger the amplifier which releases the voltage stored in the capacitor to the coil. The coil again boosts the voltage, this time to approximately 20,000 volts, to the spark plugs. For this reason, NEVER hold a high tension lead with your hand while the engine is being cranked or running. Such action could result in severe electrical shock.

When the storage capacitor has discharged its energy at the spark plug gap, the electronic switch opens and the converter recharges the storage capacitor.

After the storage capacitor has discharged its energy at the spark plug gap, the electronic switch opens and the converter recharges the storage capacitor.

Service on this type of system is limited to detailed troubleshooting, and then replacing the faulty component.

The wiring on these engines is connected directly to the individual units by means of connectors. The wiring for the other CD systems in this section is connected through an engine terminal board located on the aft side of the engine.

Normally, three conditions require troubleshooting: the engine fails to start; fails to run smoothly at all rpm (misses in certain ranges); or lacks power to perform properly.

■ **This system generates approximately 20,000 volts which is fed to the spark plugs. Therefore, perform each step of the troubleshooting procedures exactly as presented as a precaution against personal injury.**

The following safety precautions should always be observed:
- DO NOT attempt to remove any of the potting in the back of the power pack. Repair of the power pack is impossible
- DO NOT attempt to remove the high tension leads from the ignition coil

- DO NOT open or close any plug-in connectors, or attempt to connect or disconnect any electrical leads while the engine is being cranked or is running
- DO NOT set the timing advanced any further than as specified
- DO NOT hold a high tension lead with your hand while the engine is being cranked or is running. Remember, the system can develop approximately 20,000 volts which will result in a severe shock if the high tension lead is held. ALWAYS use a pair of approved insulated pliers to hold the leads
- DO NOT attempt any tests except those listed in this troubleshooting section.
- DO NOT connect an electric tachometer to the system unless it is a type which has been approved for such use
- DO NOT connect this system to any voltage source other than given in this troubleshooting section.

Perform the following troubleshooting procedures in the sequence presented and always make certain that you have already performed the general troubleshooting procedures found in the Ignition Systems - General section:
- DO NOT skip a step or add checks of your own
- DO NOT anticipate what is to be done next, just keep in step with the procedures and the work will move smoothly and orderly. Only in this manner will the faulty component be discovered in the shortest time and without frustration.

Replacement of a faulty component is covered in the sections following the Troubleshooting procedures.

The following test equipment is a must when troubleshooting this system. Stating it another way. . .there is no way to properly, and accurately, test the complete system or individual components without most of the special items listed.

- Voltmeter (0-50 volts)
- Voltmeter (to 500 volts)
- Ohmmeter
- DC Ammeter (0-5 amps)
- Timing Light
- Stevens 5-80 or M-80 neon test light
- Spark tester

Always attempt to proceed with the troubleshooting in an orderly manner. The shotgun approach will only result in wasted time, incorrect diagnosis, replacement of unnecessary parts, and frustration.

Begin the ignition system troubleshooting with the spark plugs and continue through the system until the source of trouble is located.

Troubleshooting - 1968-69 55 Hp & 1968 65/85 Hp Engines

◆ **See Figures 188 and 189**

Before starting the following troubleshooting procedures for this ignition system, take time to read and understand the system description and the General Troubleshooting procedures found in the Ignition Systems - General section.

The following test numbers are matched to test numbers on the accompanying wire diagram as an assist in performing the work.

1968-69 55 hp

A vacuum switch is installed to allow current flow to continue to the lower unit after the ignition switch has been turned to the OFF position. This

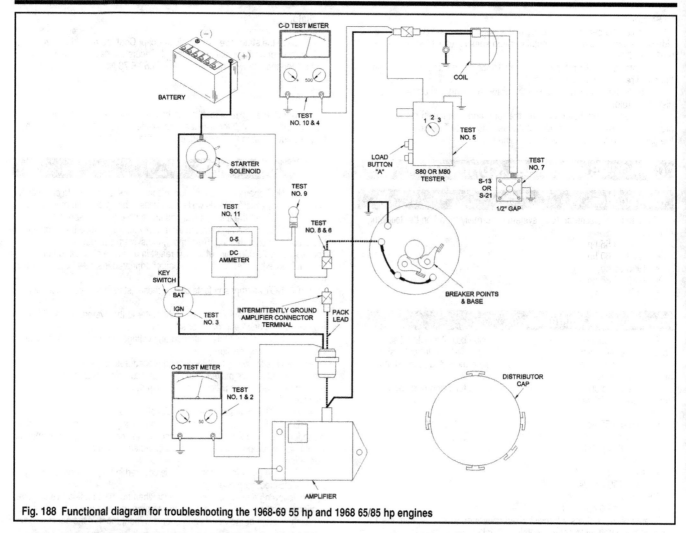

Fig. 188 Functional diagram for troubleshooting the 1968-69 55 hp and 1968 65/85 hp engines

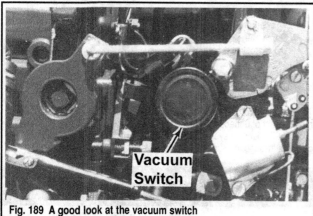

Fig. 189 A good look at the vacuum switch

arrangement prevents the shift diodes in the lower unit from immediately moving the shift mechanism into forward gear. Therefore, if the boat suddenly seems to move forward, after the key switch is turned to the **OFF** position, the shift diodes may be at fault. For complete troubleshooting and service procedures of the shift diodes, see the Lower Unit section.

■ Before making any test on a CD ignition system the spark plugs should be removed and the high tension leads grounded. This procedure will prevent the engine from firing accidentally during a test and grounding the spark plug leads will provide a path for the high voltage generated in the system. This path will prevent any fuel or fuel vapors around the powerhead from igniting from a spark.

An assistant should be available for most tests involving turning the key switch. In case of a short, the assistant can quickly turn the key off before an expensive part is damaged.

TEST NO. 1 - IGNITION SWITCH VOLTAGE OUTPUT

◆ See Figure 188

1. Connect 1 lead from the low voltage voltmeter to a good ground on the engine.
2. Probe with the other lead into the red wire of the harness to the key switch.
3. Turn the ignition switch to the **ON** position. The meter should indicate 12 volts. If the meter fails to indicate 12 volts, the battery is not up to a full charge or the key switch is defective. If the meter indicates 12 volts, proceed to Test No. 2.

TEST NO. 2 - IGNITION SWITCH VOLTAGE DROP

◆ See Figure 188

With the voltmeter probe still inserted, as in Test No. 1, crank the engine and observe the voltmeter. The voltage should have dropped to approximately 9.5 volts. If the voltage dropped to less than 9.5 volts, the battery circuit may need attention, or the starter may be drawing too many volts.

TEST NO. 3 - BATTERY VOLTAGE OUTPUT

◆ See Figure 188

Remove the test probe inserted in Test No. 1 and Test No. 2. Use the voltmeter and check the battery voltage and the voltage at the key switch. If no voltage is indicated at the key switch, use the voltmeter and check the voltage at the starter solenoid - the meter should indicate 12 volts.

TEST NO. 4 - AMPLIFIER OUTPUT TO THE COIL

◆ See Figure 188

1. Set the high voltage voltmeter to the 500 volt scale.
2. Insert 1 lead of the voltmeter into the lead going to the coil from the amplifier.
3. Connect the other meter lead to a good ground on the engine.
4. Crank the engine and observe the meter reading. The meter should indicate at least 250 volts. If the meter fails to indicate any voltage at this point, proceed with the next Test.

TEST NO. 5 - SECOND AMPLIFIER OUTPUT CHECK

◆ See Figure 188

1. Disconnect the connector to the coil and connect one lead of a Stevens 5-80 or M-80 tester to the lead to the amplifier; and the other lead to a good ground on the powerhead.
2. Depress load button **A** on the tester, and at the same time, crank the engine. The light should flash. If the light flashes, the amplifier is good and the coil should be checked, per Test No. 7.

TEST NO. 6 - THIRD AMPLIFIER OUTPUT CHECK

◆ See Figure 188

1. With the tester still connected to the coil as in Test No. 5, disconnect the wire connector between the amplifier and the point set.
2. Turn the ignition switch to the **ON** position.
3. Momentarily make contact with the wire from the amplifier to a good ground on the engine. Each time the wire is grounded, the tester light should flash. If the light fails to flash, the amplifier is defective and must be replaced.
4. Restore the system to its original condition by removing all test equipment and making the necessary connections.

TEST NO. 7 - COIL OUTPUT

◆ See Figure 188

1. First, the coil wire must be removed from the distributor cap. Rotate the lead COUNTERCLOCKWISE and count the number of turns until the lead is free. During installation, the lead must be twisted COUNTERCLOCKWISE the correct number of turns *before* it is threaded into the cap. Only in this manner will the lead be at rest without a twist after it is properly installed.
2. Connect 1 lead of a spark tester to the lead coming from the coil. If the spark tester is not available, use a pair of insulated pliers and hold the high tension lead approximately 1/2 in. (12.7mm) from ground on the engine.
3. Crank the engine. If the spark is weak or no spark is observed, the coil is defective and must be replaced. If the spark is strong, thread the coil wire back into the distributor cap.

TEST NO. 8 - POINT SET CONDITION & ADJUSTMENT

◆ See Figure 188

1. First, check to be sure the ignition switch is in the **Off** position. At the amplifier: disconnect the lead from the amplifier to the breaker point set under the flywheel.
2. Connect 1 lead from a continuity light to the positive terminal of the starter solenoid and the other lead into the lead going to the point set.
3. Advance the distributor to the full advance position. Manually rotate the flywheel CLOCKWISE. The light should glow until each set of points opens. This will occur as the flywheel timing mark passes the pointer on the lifting ring. If either set of points fails to break at the proper time, the point set is shorted or they are out of adjustment. Adjust the point sets as outlined later in this section. Perform the test again. If the points still fail the test, they are defective and must be replaced.

TEST NO. 9 - POINT SET CHECK

◆ See Figures 188 and 190

55/60 hp Engines Only

The 2 point sets are connected in a delayed parallel circuit. Both sets affect each cylinder.

Another timing mark must be added to the flywheel in order to check each point set. Measure exactly 1 3/8 in. (34.9mm) COUNTERCLOCKWISE around the flywheel perimeter from the existing mark.

All Models

1. Disconnect the wire to the amplifier from the point set, and connect 1 lead of a continuity light to the wire to the point set.
2. Connect the other test lead to the positive terminal of the starter solenoid.
3. With the key in the **OFF** position and the distributor fully advanced, manually rotate the flywheel CLOCKWISE until the existing mark on the flywheel is aligned with the boss on the ring gear guard. The light should go out indicating one set of points has opened. Continue to rotate the flywheel CLOCKWISE (the light should come on again) until the new added mark on the flywheel is aligned with the boss on the flywheel ring gear guard. When the mark is aligned with the boss, the light should go out again, indicating the second set of points has opened.
4. If the timing is not accurate, as described in these 2 checks, the flywheel must be removed and the points adjusted.

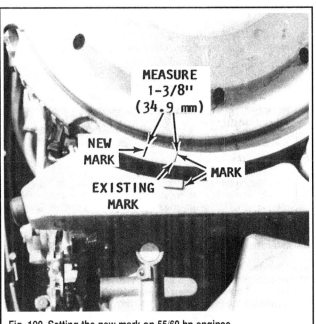

Fig. 190 Setting the new mark on 55/60 hp engines

TEST NO. 10 - INTERMITTENT MISS

◆ See Figure 188

1. Set the voltmeter to the 500 volt scale.

2. Connect 1 meter lead to a good ground on the engine.

3. Insert a probe into the connector lead to the blue wire going into the coil. Start the engine and operate it at the rpm at which the miss occurs.

■ The engine must be mounted in a body of water with a test wheel to make the following check. NEVER attempt to run an engine at above idle speed with a flush attachment connected or in a confined test tank. Such practice, with a no-load condition on the propeller, would cause the engine to 'run-away', causing serious damage or destruction of the unit. The test cannot be performed with the boat moving through the water.

✳✳ CAUTION

Water must circulate through the lower unit to the engine any time the engine is run to prevent damage to the water pump and an overheating condition. Just 5 seconds without water will damage the water pump and cause the engine to overheat.

4. With the engine operating at the rpm at which the miss occurs, observe the voltmeter. The meter should indicate at least 250 volts and hold steady. If the needle fluctuates (wavers), or fails to register 250 volts, check the complete battery circuit. If the battery is up to full charge and the circuit, including all connections and wiring are in good condition, test the coil output as outlined earlier in this section.

5. If the voltage check was satisfactory and the engine still misses within a certain rpm range, perform the next test. Disconnect the voltmeter and check to be sure all connections are in their original positions.

TEST NO. 11 - AMPLIFIER CHECK

◆ See Figure 188

1. Connect the 0-5 amp ammeter in series between the key and the amplifier.

2. Start and operate the engine at idle speed, 500 rpm. The ammeter should indicate less than 1.0 amps.

3. Increase engine speed to the range where the miss occurs. The ammeter should hold steady. If the needle wavers, check the battery wire to the amplifier. If all wires and connections are in good condition, replace the amplifier.

4. Restore the system to its original condition by removing all test equipment and making the necessary connections.

Troubleshooting - 1969-72 85 Hp & 1971 100 Hp Engines

◆ See Figure 191

Before starting the following troubleshooting procedures for this ignition system, take time to read and understand the system description at the beginning and the General Troubleshooting procedures (Ignition Systems - General) that apply to all models.

The following test numbers are matched to test numbers on the accompanying wire diagram as an assist in performing the work.

■ Before making any test on a CD ignition system the spark plugs should be removed and the high-tension leads grounded. This procedure will prevent the engine from firing accidentally during a test, and grounding the spark plug leads will provide a path for the high voltage generated in the system. This path will prevent a spark from igniting any fuel or fuel vapors in the area of the power head.

TEST NO. 1 - IGNITION SWITCH CHECK

◆ See Figure 191

1. Connect 1 lead of the low voltage voltmeter to a good ground on the engine.

2. Turn the ignition switch to the **ON** position, and then momentarily make contact with the other meter lead to the No. 8 or No. 9 post at the engine terminal board. The meter should indicate 12 volts. If voltage is present, the current between the key switch and the terminal board is satisfactory. If voltage is not present, the battery is low; the key switch is defective; or the wiring to-and-from the terminal board requires attention.

TEST NO. 2 - BATTERY CIRCUIT CHECKS

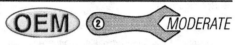

◆ See Figure 191

If voltage was not present in Test No.1, test the voltage at the ignition switch, then between the switch and the starter solenoid, and finally between the starter solenoid and the battery.

TEST NO. 3 - VOLTAGE DRAW TEST

◆ See Figure 191

1. With 1 lead of the voltmeter still connected to a good ground on the engine, connect the other lead to either the No. 8 or No. 9 post of the terminal board.

2. Now, crank the engine and observe the voltmeter. The meter should indicate at least 9.5 volts. If the voltage reading is low, either the battery circuit requires service, or the starter is drawing excessive voltage.

TEST NO. 4 - AMPLIFIER OUTPUT

◆ See Figure 191

1. Connect the high scale voltmeter to ground and to the No. 5 post at the terminal board.

2. Set the meter to the 500 volt scale.

3. Crank the engine and observe the meter reading. The meter should indicate at least 250 volts. If the required amount of voltage is not present, proceed with the Test No. 5.

TEST NO. 5 - FIRST AMPLIFIER CHECK

◆ See Figure 191

1. If the engine will not run and there was no indication of voltage while the engine was being cranked during Test No. 4, disconnect the coil primary wire (the blue wire) from the No. 5 post at the terminal board.

2. Connect one lead of a Stevens S-80 or M-80 test light to the blue wire leading to the amplifier and the other lead to a good ground on the powerhead.

3. Again crank the engine. If the light flashes, the coil is defective and must be replaced.

TEST NO. 6 - SECOND AMPLIFIER CHECK

◆ See Figure 191

1. If the light failed to flash during Test No. 5, disconnect the quick-disconnect between the amplifier and the point set.

2. Turn the ignition switch to the **ON** position.

3. With the tester still connected to the blue wire to the amplifier and ground, momentarily make contact to ground with the wire from the amplifier.

Each time the wire makes contact with ground, the tester light should flash on. If the light flashes during this test, check the point set to be sure they are properly adjusted.

4. Adjust the points as required. If the light fails to flash, the amplifier is defective and must be replaced.

5. Restore the system to its original condition by removing all test equipment and making the necessary connections.

TEST NO. 7 - COIL OUTPUT

◆ **See Figure 191**

1. Disconnect the coil high tension lead from the distributor cap. The lead is threaded into the cap and must be rotated COUNTERCLOCKWISE until it is free. As the lead is turned, count the revolutions and make a note as an aid during installation. The lead must be twisted COUNTERCLOCKWISE the correct number of turns before it is threaded into the distributor cap to ensure it is at rest and not twisted after installation.

2. Use a pair of insulated pliers and hold the lead about 1/2 in. (12.7mm) from a good engine ground.

3. Crank the engine. If the spark is weak or no spark is observed, the coil is defective and must be replaced. If the spark is strong, thread the coil wire back into the distributor cap.

TEST NO. 8 - POINT SET CONDITION

◆ **See Figure 191**

1. Disconnect the wire from the amplifier to the points at the quick-disconnect.

2. Connect 1 lead of a continuity light to the lead to the point set.

3. Connect the other continuity light lead to the positive terminal of the starter solenoid.

4. With the ignition key in the **OFF** position, advance the distributor to full advance.

5. Remove the spark plugs as an aid to rotating the flywheel by hand. Manually rotate the flywheel CLOCKWISE. Observe the timing mark on the flywheel guard and the 2 timing marks on the flywheel. As the flywheel is rotated and each mark passes the mark on the flywheel guard, the light should go out indicating the points are open.

6. If the light fails to go out, the flywheel must be removed, and the point set adjustment carefully checked.

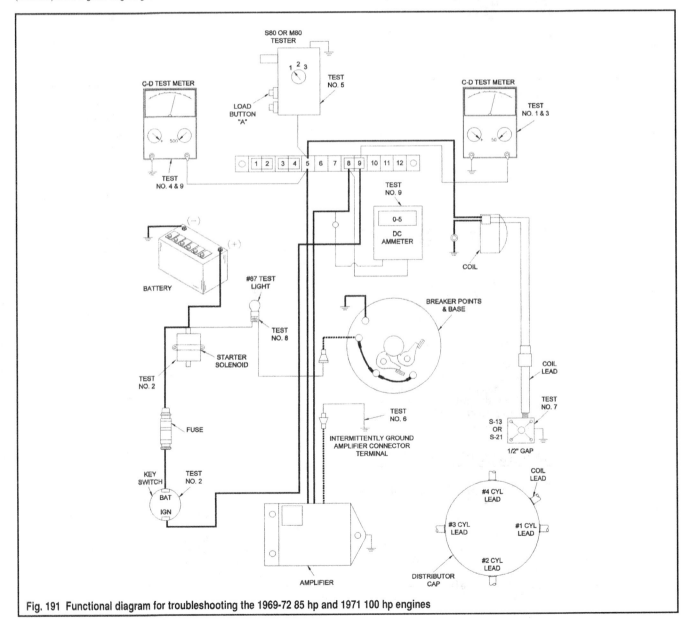

Fig. 191 Functional diagram for troubleshooting the 1969-72 85 hp and 1971 100 hp engines

TEST NO. 9 - INTERMITTENT MISS

◆ **See Figure 191**

1. Restore the system to its original condition by removing all test equipment and making the necessary connections.

2. Connect the black lead of a high voltage voltmeter to a good ground on the engine. Set the meter to the 500 volt scale.

3. Connect the other lead to the blue lead at the No. 5 post at the terminal board.

4. Connect a low amp ammeter (0-5 amps) in series between the amplifier and the No. 8 or No. 9 post of the terminal board.

■ **The engine must be mounted in a body of water with a test wheel to make the following check. NEVER attempt to run an engine at above idle speed with a flush attachment connected or in a confined test tank. Such practice, with a no-load condition on the propeller, would cause the engine to 'runaway', causing serious damage or destruction of the unit. The test cannot be performed with the boat moving through the water.**

✳✳ CAUTION

Water must circulate through the lower unit to the engine any time the engine is run to prevent damage to the water pump and an overheating condition. Just 5 seconds without water will damage the water pump and cause the engine to overheat.

5. Start the engine and allow it to warm to operating temperature at 500 rpm.

6. Increase the throttle until the engine is operating at the rpm at which the miss occurs and observe the voltmeter and the ammeter. The voltmeter should indicate at least 250 volts. If either needle fluctuates (wavers), check the complete battery circuit. If the battery is up to full charge and the circuit, including all connections and wiring, are in good condition, the amplifier is defective and must be replaced.

7. Restore the system to its original condition by removing all test equipment and making the necessary connections.

Troubleshooting - 1970-71 60 Hp Engines

◆ **See Figure 192**

Before starting the following troubleshooting procedures for this ignition system, take time to read and understand the system description at the beginning and the General Troubleshooting procedures (Ignition systems - General) that apply to all models.

The following test numbers are matched to test numbers on the accompanying wire diagram as an assist in performing the work.

■ **Before making any test on a CD ignition system, the spark plugs should be removed and the high tension leads grounded. This procedure will prevent the engine from firing accidentally during a test. Grounding the spark plug leads will provide a path for the high voltage generated in the system. This path will prevent a spark from igniting any fuel or fuel vapors in the area of the powerhead.**

TEST NO. 1 - IGNITION SWITCH & BATTERY CIRCUIT

◆ **See Figure 192**

1. Connect 1 lead of the 0-50 voltmeter to a good ground on the engine.

2. Turn the ignition switch to the **ON** position, and then momentarily make contact with the other meter lead to the No. 8 or No. 9 post at the engine terminal board on the aft side of the engine.

3. The meter should indicate 12 volts. If voltage is not present, or the voltage is low, the battery is low; the key switch is defective; or the wiring to-and-from the terminal board requires attention.

TEST NO. 2 - VOLTAGE INPUT TO AMPLIFIER, SYSTEM AT REST

◆ **See Figure 192**

1. Connect the black lead of a low voltage (0-50) meter to a good ground on the engine.

2. Connect the other voltmeter lead to the No. 6 post of the terminal board.

3. Turn the ignition key to the **ON** position.

4. The voltmeter should indicate 12 volts. If the meter fails to register 12 volts, the fuse connected in series with the starter solenoid is blown; the connection at the starter solenoid is not clean and tight; or the battery circuit requires attention.

TEST NO. 3 - VOLTAGE INPUT TO AMPLIFIER, ENGINE BEING CRANKED

◆ **See Figure 192**

With the test equipment still connected as for Test No. 2, crank the engine with the starter motor. The voltmeter should indicate at least 9.5 volts. If the meter reading is less than 9.5 volts, the battery circuit may need attention or, the starter circuit is drawing excessive voltage.

TEST NO. 4 - FIRST AMPLIFIER OUTPUT CHECK

◆ **See Figure 192**

1. Connect the black lead of a high voltage voltmeter to a good ground on the engine. Set the meter to the 500 volt scale.

2. Connect the other voltmeter lead to the No. 9 post of the terminal board.

3. Crank the engine and the voltmeter should indicate at least 250 volts. If the meter reading is not satisfactory, or the meter fails to indicate any voltage, proceed to the next test.

TEST NO. 5 - SECOND AMPLIFIER OUTPUT CHECK

◆ **See Figure 192**

1. Leave the test equipment connected as for Test No. 4.

2. Disconnect the amplifier lead going to the point set. With the ignition key still in the **ON** position, momentarily make contact with the amplifier lead to engine ground.

3. Each time the amplifier lead makes contact with engine ground, the meter should indicate at least 250 volts. If the voltage reading is not satisfactory, perform the next test.

4. Connect the amplifier lead going to the point set.

TEST NO. 6 - THIRD AMPLIFIER OUTPUT CHECK

◆ **See Figure 192**

1. Disconnect the high voltage voltmeter.

2. Disconnect the blue lead from the No. 9 post at the terminal board.

3. Connect one lead of a Stevens S-80 or M-80 tester to the blue lead going to the amplifier.

4. Connect the other lead to a good ground on the powerhead.

5. Turn the ignition key to the **ON** position. Momentarily make contact with the amplifier point lead to engine ground.

6. Each time contact is made to ground, the test light should flash on. If the light does flash, the amplifier is satisfactory for further service and the coil is defective. If the light did NOT flash, the amplifier is defective. If the coil is suspected, proceed to the next test.

TEST NO. 7 - COIL OUTPUT

◆ See Figure 192

1. Disconnect all test equipment connected thus far, and then connect all wiring back to their original positions.

2. Disconnect the coil high tension lead from the distributor cap. The lead is threaded into the cap and must be rotated COUNTERCLOCKWISE until it is free. As the lead is turned, count the revolutions and make a note as an aid during installation. The lead must be twisted COUNTERCLOCKWISE the correct number of turns before it is threaded into the distributor cap to ensure it is at rest and not twisted after installation.

3. Use a pair of insulated pliers and hold the lead about 1/2 in. (12.7mm) from a good engine ground.

4. Crank the engine. If the spark is weak or no spark is observed, the coil is defective and must be replaced. If the spark appears to be good, replace the coil wire into the distributor cap.

TEST NO. 8 - POINT SET CONDITION

◆ See Figure 192

1. Insert 1 lead of a test light into the connector wire to the point set under the flywheel.

2. Connect the other lead of the test light to the positive terminal of the starter solenoid.

3. Turn the ignition switch to the **OFF** position.

4. Move the distributor to the full advance position. Manually rotate the flywheel CLOCKWISE until the timing mark on the flywheel is aligned with the timing mark on the flywheel guard. The test light should go out indicating the point set is open.

5. Continue to rotate the flywheel CLOCKWISE 1 3/8 in. (34.9mm), and the light comes on. If the light failed to come on when the flywheel passed the measurement given, the points must he replaced or adjusted.

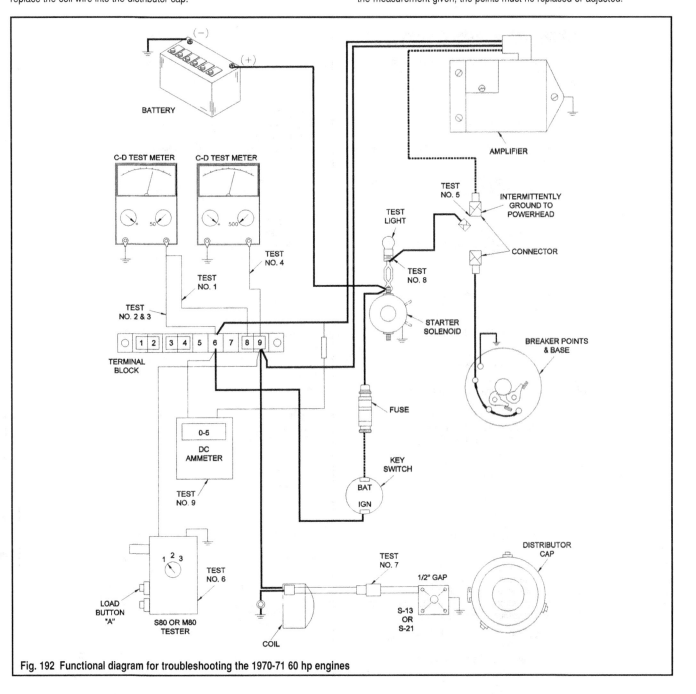

Fig. 192 Functional diagram for troubleshooting the 1970-71 60 hp engines

TEST NO. 9 - INTERMITTENT MISS

◆ **See Figure 192**

1. Connect the black lead of a high volt voltmeter to a good engine ground.
2. Connect the other voltmeter lead to the No. 9 post of the terminal board. Set the voltmeter to the 500 volt scale.
3. Connect a DC ammeter (0-5 amp range) in series with the amplifier purple wire and the No. 6 post.
4. Start the engine and increase the rpm until the miss occurs.

■ **The engine must be mounted in a body of water with a test wheel to make the following check. NEVER attempt to run an engine at above idle speed with a flush attachment connected or in a confined test tank. Such practice, with a no-load condition on the propeller, would cause the engine to 'runaway', causing serious damage or destruction of the unit. The test cannot be performed with the boat moving through the water.**

✳✳ CAUTION

Water must circulate through the lower unit to the engine any time the engine is run to prevent damage to the water pump and an overheating condition. Just 5 seconds without water will damage the water pump and cause the engine to overheat.

5. If either needle fluctuates (wavers), the amplifier could be breaking down and must be replaced.
6. Restore the system to its original condition by removing all test equipment and making the necessary connections.

Service

DISASSEMBLY

◆ **See Figures 193 thru 202**

The following procedures outline the necessary steps required to replace or adjust a component in a wide range of CD ignition systems. If the troubleshooting procedures indicate a part is defective and must be replaced, only the procedures necessary for that particular part need be performed. Therefore, proceed directly to the paragraph heading for the part to be removed and installed.

Flywheel

1. Hold the flywheel with a holding fixture and remove the flywheel nut from the crankshaft. Obtain an appropriate flywheel remover and "pull" the flywheel from the crankshaft. Notice how the stator fits over the top of the distributor cap.

Other Components

2. Separate the disconnect for the stator leads. Remove the 3 stator attaching screws and then lift the stator free of the distributor cap.
3. Lift the distributor clear of the engine.
4. After the distributor has been removed, notice the wavy washer inside the rotor. Remove the rotor and the wavy washer from the distributor.
5. The 2 sets of points are now exposed. Notice how these points are different from the normal set of outboard motor points. The adjusting screw head is knurled. The grooves forming the knurls on the screw head help to hold the proper adjustment. When a point adjustment is made, the screw will make a definite "clicking" sound as the grooves move past a slight protrusion in the breaker point base plate.

■ **The breaker point base does not normally have to be removed. However, the breaker point leads are routed under the base and secured in place with a retainer. This wire has been known to break and "ground" the point set. If such a condition is suspected, the distributor base must be removed and the wire replaced or repaired.**

6. To remove the distributor base plate, first, remove the 4 retaining clips located on the plate perimeter.
7. Lift the plate clear of the distributor. Turn the plate over and inspect the wires.

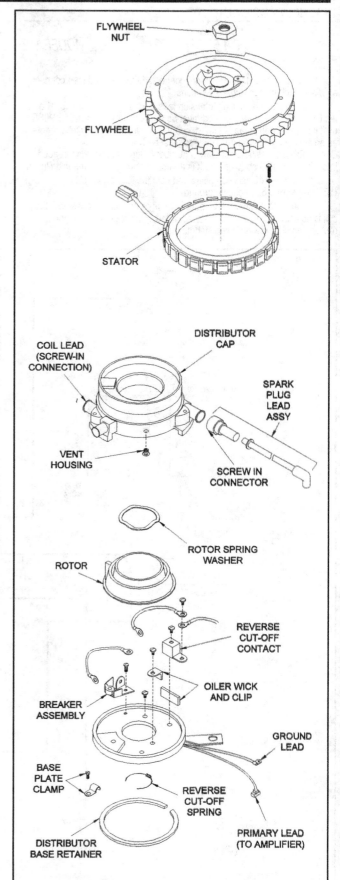

Fig. 193 Exploded view of the CD ignition system with breaker points

■ Notice the wire-type spring on top of the breaker base plate that fits around the crankshaft. This spring performs the function of grounding the ignition system if the engine should happen to backfire and start operating in a COUNTERCLOCKWISE direction.

8. During service of the point set, the anti-reverse spring should be replaced. After the new spring is installed, use a very small amount of OMC Type D lubricant on the crankshaft where the spring rides to reduce friction and prolong spring life. Take note how the point set leads are routed to a terminal block, and then under the base plate.

Point Set Replacement

Seldom does the point set require replacement. The explanation is simple: the set is not subjected to high voltage charges, as in the conventional system. The set merely acts similar to a light switch to open and close the circuit, in this case to ground the circuit and prevent current flow.

9. Remove the 2 attaching screws through the set. Disconnect the point set lead from the terminal block. Remove the point set.

CLEANING & INSPECTION

◆ See Figure 193

Inspect all parts for wear or damage. Test the coil, rotor, end caps, high-tension leads and breaker assemblies.

Service of the point set is limited to cleaning them with a calling card pulled between the points and using acetone, or similar material, to remove any small amount of corrosion that may have started to form.

Check the coil closely to determine if there has been any leakage to the powerhead. If in doubt as to the integrity of the coil, have it checked at a shop or replace it with a new coil.

Make a resistance test on the high tension wires. The high tension leads may be removed. DO NOT attempt to pull the high tension leads from the distributor cap. These wires are screwed into the cap. First pull the rubber

Fig. 194 Install the puller and remove the flywheel

Fig. 195 Remove the mounting screws and lift off the stator

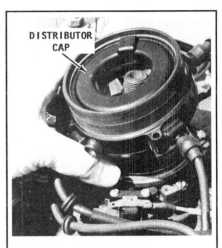

Fig. 196 Remove the distributor. . .

Fig. 197 . . .and then the rotor and washer

Fig. 198 A good look at the points and base

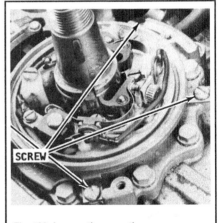

Fig. 199 Loosen the mounting screws. . .

Fig. 200 . . .and remove the base

Fig. 201 Remove the anti-reverse spring

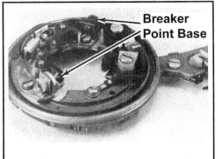

Fig. 202 Removing the points from the base

cap back up onto the high tension lead, and then unscrew the wire In a COUNTERCLOCKWISE direction until the wire is free.

Leakage Paths

The high voltage surge of the secondary circuit may establish a path to ground by a different route than across the spark plug gap. Once such a path is established, the spark will most likely continue to jump across the ground.

A surface leakage path can usually be detected because of the burning effect the high voltage spark has on the plastic insulating material. The condition causing the high voltage spark to stray from its intended circuit must be corrected. Any repairs of the unit should be performed very carefully, and should include discarding any insulating parts with evidence of high voltage flashover.

ASSEMBLY

◆ **See Figures 193, 203, 204 and 205**

Base Plate and Points

1. To install the distributor base plate, first notice the nylon ring on the underneath side. This nylon ring must be worked down into the recess in the upper bearing head as the base plate is installed. Lower the plate down onto the power head, and then work the plate to enable the nylon retainer ring to seat in the upper bearing head recess.

2. Secure the plate in place with the 3 clips on the perimeter. Place the new set of points into place on the distributor base plate, and then secure

them with the 2 attaching screws. Make sure you carefully route the wires along the edge of the base plate, and then connect them to the terminal block.

3. Thread the flywheel nut onto the crankshaft, and then using a wrench on the nut; rotate the crankshaft until the point heel is on the high point of the cam. Adjust the points by rotating the adjusting screw to achieve a clearance of 0.010 in. (0.254mm) for a used set of points and to 0.012 in. (0.305mm) for a new point set.

4. Repeat the procedure for the second set of points.

5. Remove the flywheel nut.

Distributor Parts

6. Slide the distributor rotor down the crankshaft with the tab on the rotor indexed into the recess on the crankshaft.

7. Slide the wavy washer onto the crankshaft and seat it in the rotor.

8. Slide the distributor cap down the crankshaft and into place on the bearing head.

Stator

9. Install the stator over the top of the distributor cap.

10. Dip the stator attaching screws in Loctite (or similar substance). Secure the stator and distributor cap with the screws. Tighten the screws to 48-60 inch lbs. (5.42-6.78 Nm).

11. Connect the quick-disconnect for the stator leads.

Flywheel

12. Check the inside tapered surface of the flywheel to be sure it is clean and DRY. Lower the flywheel onto the crankshaft with the cut-out in the flywheel indexed over the Woodruff key. Thread the flywheel nut onto the crankshaft, and then using a holding fixture on the flywheel; tighten the nut to the correct value.

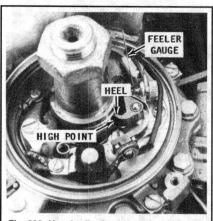

Fig. 203 Use the flywheel nut as an aid while adjusting the points

Fig. 204 Make sure you use Loctite (or similar) when installing the stator bolts

Fig. 205 Installing the flywheel and nut

CAPACITOR DISCHARGE (CD) IGNITION SYSTEM - WITH TIMER BASE

Identification

The following motors are normally equipped with a Type I CD Flywheel Ignition System which utilizes a timer base:
- 50 hp, 1971-72
- 65 hp, 1972

Description & Operation

◆ **See Figures 206 thru 209**

A battery installed to crank the engine DOES NOT mean the engine is equipped with a battery-type ignition system. A magneto system uses the battery only to crank the engine. Once the engine is running, the battery has absolutely no affect on engine operation.

Therefore, if the battery is low and fails to crank the engine properly for starting, the engine may be cranked manually, started, and operated. Under these conditions, the key switch must be turned to the **ON** position or the engine will not start by hand cranking.

A CD magneto system is a self-contained unit that does not require assistance from an outside source for starting or continued operation. Therefore, as previously mentioned, if the battery is dead the engine may be cranked manually and the engine started.

This capacitor discharge (CD) magneto ignition system consists of the flywheel and ring gear assembly; timer base and sensor assembly installed under the flywheel; a power pack installed on the starboard side of the powerhead; and 2 (50 hp 2 cyl) or 3 (65 hp 3 cyl) ignition coils mounted at the rear of the powerhead. On some models, an alternator stator and charge coils assembly is installed directly under the flywheel. The spark plugs might be considered a part of the ignition system.

Repair of these components is not possible. Therefore, if troubleshooting indicates a part unfit for further service, the entire assembly must be removed and replaced in order to restore the outboard to satisfactory performance. As an example the coil and coil wire leading to the spark plug is one assembly. If the coil or wire is found to be faulty the coil and wire must be replaced as an assembly.

Before performing maintenance work on the system, it would be well to take time to read and understand the introduction information presented about Understanding & Troubleshooting Electrical Systems as well as the general information about Ignition Systems, Spark Plugs, Ignition System Troubleshooting, at the beginning of this section and the Theory of Operation in the following paragraphs.

Theory Of Operation

This system generates approximately 30,000 volts which is fed to the spark plugs without the use of a point set or an outside voltage source.

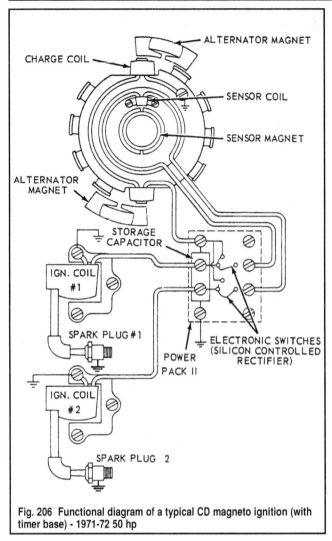

Fig. 206 Functional diagram of a typical CD magneto ignition (with timer base) - 1971-72 50 hp

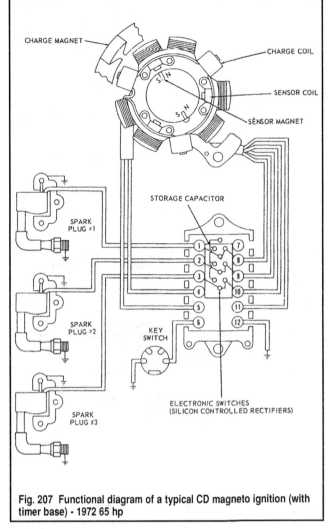

Fig. 207 Functional diagram of a typical CD magneto ignition (with timer base) - 1972 65 hp

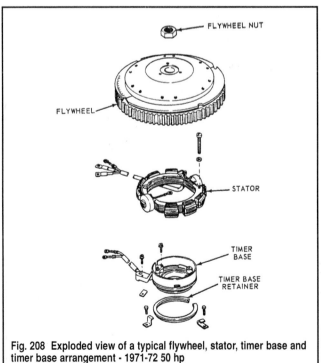

Fig. 208 Exploded view of a typical flywheel, stator, timer base and timer base arrangement - 1971-72 50 hp

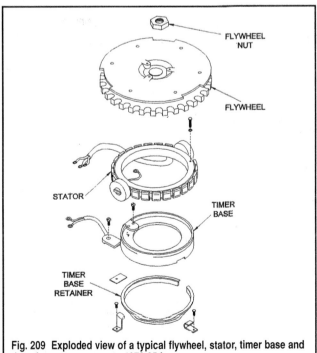

Fig. 209 Exploded view of a typical flywheel, stator, timer base and timer base arrangement - 1972 65 hp

To understand how high voltage current is generated and reaches a spark plug, imagine the flywheel turning very slowly. As the flywheel rotates, flywheel magnets induce current in the alternator stator and also generate about 300 volts AC in the charge coils. Therefore, no external voltage source is required. The 300 volts AC is converted to DC in the power pack, and is stored in the power pack capacitor.

Sensor magnets are a part of the flywheel hub. Gaps exist between the sensor magnets. As one gap passes the sensor coil, voltage is generated. This small voltage generated in the sensor coil activates 1 of 2 electronic switches in the power pack. The switch discharges the 300 volts stored in the capacitor into one of the ignition coils. The ignition coil steps the voltage up to approximately 30,000 volts. This high voltage is fed to the spark plug igniting the fuel/air mixture in the cylinder.

Now, as the flywheel continues to rotate, the next sensor magnet gap on the flywheel hub, which is opposite in polarity from the first, generates a reverse polarity voltage in the sensor coil.

This voltage activates the second electronic switch in the power pack, and discharges the capacitor into the other ignition coil. The voltage is stepped up to approximately 30,000 volts and fed to the next spark plug. The cycle is repeated as the flywheel continues to rotate.

Troubleshooting - 1971-72 50 Hp

GENERAL

◆ **See Figures 206 thru 209**

Always attempt to proceed with the troubleshooting in an orderly manner. The shotgun approach will only result in wasted time, incorrect diagnosis, replacement of unnecessary parts, and frustration.

Begin the ignition system troubleshooting with the spark plugs and a compression check and then continue through the system until the source of trouble is located.

The following test equipment is a must when troubleshooting this system. Stating it another way. . .there is no way to properly and accurately test the complete system or individual components without the special items listed.
- Continuity Meter
- Ohmmeter
- Timing Light
- S-80 or M-80 neon test light.
- Neon spark tester.

✳✳ WARNING

This system generates approximately 30,000 volts which is fed to the spark plugs. Therefore, perform each step of the troubleshooting procedures exactly as presented as a precaution against personal injury.

The following safety precautions should always be observed:
- DO NOT attempt to remove any of the potting in the back of the power pack. Repair of the power pack is impossible.
- DO NOT attempt to remove the high tension leads from the ignition coil.
- DO NOT open or close any plug-in connectors, or attempt to connect or disconnect any electrical leads while the engine is being cranked or is running.
- DO NOT set the timing advanced any further than as specified.
- DO NOT hold a high tension lead with your hand while the engine is being cranked or is running. Remember, the system can develop approximately 30,000 volts which will result in a severe shock if the high tension lead is held. ALWAYS use a pair of approved insulated pliers to hold the leads.
- DO NOT attempt any tests except those listed in this troubleshooting section.
- DO NOT connect an electric tachometer to the system unless it is a type which has been approved for such use.
- DO NOT connect this system to any voltage source other than given in this troubleshooting section.

Each cylinder has its own ignition system in a flywheel-type ignition system. This means if a strong spark is observed on any 1 cylinder and not at another, only the weak system is at fault. However, it is always a good idea to check and service all systems while the flywheel is removed.

Preliminary Test

The first area to check on a CD flywheel magneto ignition system is the system ground. Connect 1 lead of a continuity meter to the No. 4 Power Pack terminal and the other lead to a good ground. The meter should indicate continuity.

Also check the battery charge. If the battery is below a full charge it will not be possible to obtain full cranking speed during the tests.

TEST NO. 1 - IGNITION COIL OUTPUT CHECK

◆ **See Figures 210 and 211**

1. Use a spark tester and check for spark at each cylinder. If a spark tester is not available, use a pair of insulated pliers and hold the plug wire about 1/4 in. from the engine.

2. Turn the flywheel with a pull starter or electrical starter and check for spark. A strong spark over a wide gap must be observed when testing in this manner, because under compression a strong spark is necessary in order to ignite the air/fuel mixture in the cylinder. This means it is possible to think you have a strong spark, when in reality the spark will be too weak when the plug is installed.
- If there is no spark, or if the spark is weak, from 1 coil, proceed directly to Test No. 2, Trigger Coil Input Check.
- If there is no spark or the spark is weak from both coils, proceed directly to Test No. 5, Charge Coil Output Check.
- If the spark is strong and steady, across the 1/4 in. gap indicating it is satisfactory, then the problem is in the spark plugs, the fuel system, or the compression is weak.

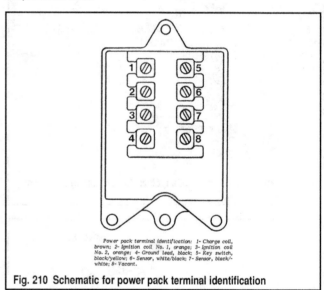

Power pack terminal identification: 1- Charge coil, brown; 2- Ignition coil No. 1, orange; 3- Ignition coil No. 2, orange; 4- Ground lead, black; 5- Key switch, black/yellow; 6- Sensor, white/black; 7- Sensor, black/white; 8- Vacant.

Fig. 210 Schematic for power pack terminal identification

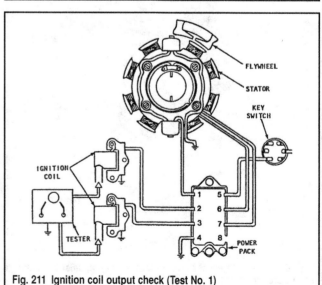

Fig. 211 Ignition coil output check (Test No. 1)

TEST NO. 2 - TRIGGER COIL INPUT CHECK

◆ **See Figures 210 and 212**

1. Remove the power pack cover, and then disconnect the sensor leads from the No. 6 and No. 7 terminals.

2. Obtain a neon tester, No. S-80 or M-80. If one of these testers is not available, a 1 1/2 volt battery with a short lead soldered to each terminal may be used. If neither of these items is available, an ohmmeter may be used.

3. Connect the black lead of the neon tester to the No. 6 terminal and the blue lead to the No. 7 terminal. Set the neon light selector to the No. 3 position.

4. Crank the engine with the electric starter motor, and at the same time depress the **B** load button rapidly. Stop; reverse the lead at the No. 6 and No. 7 terminals.

5. Repeat the cranking and depressing procedure. Observe the spark across the tester. If both coils fired at the same time, during this test, the power pack is at fault. Only 1 coil should fire when the button is depressed.

■ **If the neon tester is not available, an assistant is required. Have the assistant make contact with the 2 leads from the small 1 1/2 volt battery to the No. 6 and No. 7 terminals while the engine is being cranked. Reverse the leads at the No. 6 and No. 7 terminals and repeat the test. One coil should fire when the leads are connected one way and the other coil fire when the leads are reversed.**

• If adequate spark was observed from both coils in the previous tests, the problem is in the sensor. To test the sensor assembly, see Test No. 3 and Test No. 4.

• If no spark was observed from either coil during the previous tests, the charge coil must be checked as outlined in Test No. 5.

• If spark was observed on only 1 coil, the power pack must be checked, see Test No. 6.

TEST NO. 3 - SENSOR COIL LOW OHM CHECK

◆ **See Figures 210 and 213**

1. Disconnect the sensor leads from the power pack terminals No. 6 and No. 7.

2. Connect the ohmmeter leads to the sensor coil leads (the white lead with the black stripes and the black lead with white stripes).

3. Use the low ohm scale and observe the reading. The meter should indicate 15.0 +/-5.0 ohms (10-20 ohms) at room temperature (70° F). If the ohmmeter reading is not satisfactory, proceed as follows:

  a. Turn the ohmmeter to the High Ohm scale. The meter should indicate zero ohms - no reading.

  b. If the test fails, the sensor coil and timer base must be replaced as a unit.

  c. If the test is successful disconnect the ohmmeter and connect the lead at the No. 6 and No. 7 Power Pack terminals. If the test is not successful, proceed with the next test, Sensor Coil High Ohm Test.

TEST NO. 4 - SENSOR COIL HIGH OHM CHECK

◆ **See Figures 210 and 214**

1. Set the ohmmeter to the High Ohm scale.

2. With the sensor leads still disconnected from the power pack, as in Test No. 3, connect the red ohmmeter test lead to either one of the sensor leads.

3. Connect the black ohmmeter lead to a good ground. The meter should indicate infinity. Any resistance indicates a short to ground.

4. If the test fails, the sensor coil and timer base must be replaced as a unit.

5. If the test is successful disconnect the ohmmeter and connect the leads at the No. 6 and No. 7 power pack terminals.

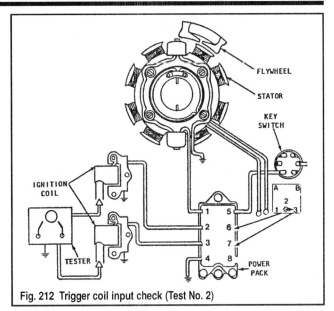

Fig. 212 Trigger coil input check (Test No. 2)

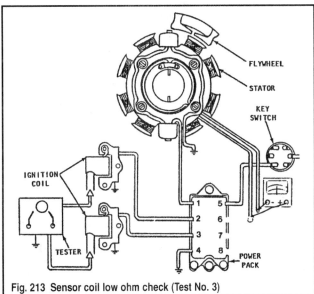

Fig. 213 Sensor coil low ohm check (Test No. 3)

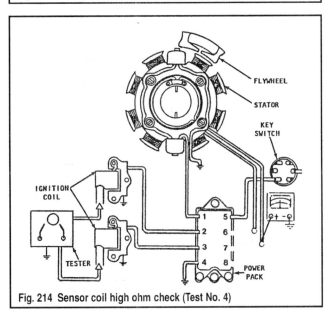

Fig. 214 Sensor coil high ohm check (Test No. 4)

TEST NO. 5 - CHARGE COIL OUTPUT CHECK

◆ **See Figures 210 and 215**

1. Connect the black lead from the S-80 or M-80 neon tester to the power pack No. 1 terminal.

2. Connect the tester blue lead to the No. 4 power pack terminal, or to a good ground on the engine. (The No. 4 terminal is ground.)

3. Turn the neon tester to position No. 2. If a neon test light is being used, connect 1 lead to the power pack No. 1 terminal and the other lead to the No. 4 power pack terminal or a good ground on the engine.

4. Crank the engine and at the same time, depress load button "B" on the tester. Observe the light. If the light glows steadily, the charge coils are good.

5. Check the power pack output. Disconnect the coil lead at terminal No. 1 on the power pack.

6. Connect the black tester lead to the wire just removed from the power pack terminal.

7. Crank the engine. If the neon light glows intermittently or does not glow at all, the charge coil and stator assembly must be replaced.

■ **If a neon tester is not available, obtain an ohmmeter. Disconnect the lead at the No. 1 power pack terminal. Connect 1 ohmmeter lead to the lead just disconnected from the terminal and the other ohmmeter lead to a good ground. Use the high ohmmeter scale and observe the reading. The meter should indicate 750 +/- 75 ohms (675-825 ohms) at room temperature (70°F). If the proper reading cannot be obtained, the charge coil and stator must be replaced as an assembly.**

TEST NO. 6 - POWER PACK OUTPUT CHECK

◆ **See Figures 210 and 216**

1. Check all connections at the power pack to be sure they are clean and secure.

2. Disconnect the lead from the No. 2 and No. 3 terminals.

3. If the S-80 or M-80 neon tester is used, connect the black lead to the No. 2 power pack terminal and the blue lead to a good ground or to the No. 4 terminal.

4. If a neon test light is used, connect 1 lead to the No. 2 terminal and the other lead to a good ground or the No. 4 terminal.

5. Move the neon tester switch to the No. 1 position, and then depress load button "A" and crank the engine with the electric starter motor. Observe the light. If the test light is used, crank the engine with the electric starter motor and observe the light.

6. Now, move the lead from the No. 2 terminal to the No. 3 terminal. Crank the engine again and observe the light.

• If the light is strong and steady on both outputs, replace the faulty coil or coils. The No. 2 terminal connection is the top coil and the No. 3 terminal is the bottom coil.

• If no light is visible on both outputs, check the key switch as outlined in Key -Switch Check, Test No. 7.

• If a steady light was observed on 1 output but no light on the other, replace the power pack.

• If the light was very dim or intermittent on one or both outputs during the tests, the power pack must be replaced.

7. Connect the wires disconnected at the start of the tests.

TEST NO. 7 - KEY SWITCH CHECK

◆ **See Figures 210 and 217**

A magneto key switch operates in the *reverse* of any other type key switch. When the key is moved to the **OFF** position, the circuit is CLOSED between the magneto and ground.

In some cases, when the key is turned to the **OFF** position the points are grounded. For this reason, an automotive-type switch must NEVER be used, because the circuit would be opened and closed in reverse, and if 12-volts should reach the coil, the coil will be destroyed.

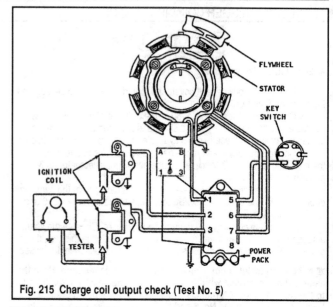

Fig. 215 Charge coil output check (Test No. 5)

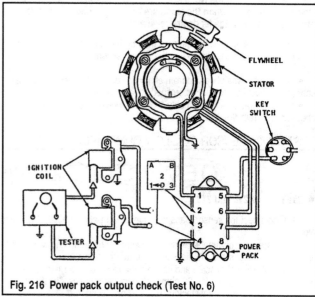

Fig. 216 Power pack output check (Test No. 6)

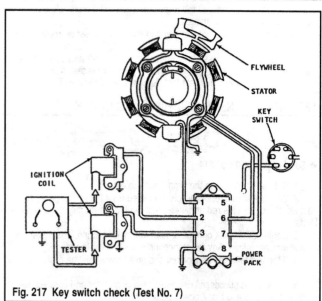

Fig. 217 Key switch check (Test No. 7)

1. Connect the spark tester to the high tension leads.

2. Disconnect the lead at the No. 5 power pack terminal. This is the wire from the key.

3. Crank the engine with the key and observe the spark.

• If there is no indication of spark on either coil or on only one, replace the power pack.

• If spark is indicated on both coils, the problem is most likely in the lead from the key.

• If the key switch leads appear to be in good condition, replace the key switch.

Troubleshooting - 1972 65 Hp

SPECIAL EQUIPMENT & ADVICE

There is no way on this green earth to properly or accurately check the system without the following test equipment.

• Continuity Meter
• Ohmmeter
• Timing Light
• Stevens S-80 or M-80 neon test light
• Neon spark tester

SAFETY WORDS

✳✳ WARNING

This system generates approximately 30,000 volts which is fed to the spark plugs. Therefore, perform each step of the troubleshooting procedures exactly as presented as a precaution against personal injury.

The following safety precautions should always be observed:

DO NOT attempt to remove any of the potting in the back of the power pack. Repair of the Power Pack is impossible.

DO NOT attempt to remove the high tension leads from the ignition coil.

DO NOT open or close any plug-in connectors, or attempt to connect or disconnect any electrical leads while the engine is being cranked or is running.

DO NOT set the timing advanced any further than as specified.

DO NOT hold a high tension lead with your hand while the engine is being cranked or is running. Remember, the system can develop approximately 30,000 volts which will result in a severe shock if the high tension lead is held. ALWAYS use a pair of approved insulated pliers to hold the leads.

DO NOT attempt any tests except those listed in this troubleshooting section.

DO NOT connect an electric tachometer to the system unless it is a type which has been approved for such use.

DO NOT connect this system to any voltage source other than given in this troubleshooting section.

Each cylinder has its own ignition system in a flywheel-type ignition system. This means if a strong spark is observed on any one cylinder and not at another, only the weak system is at fault. However, it is always a good idea to check and service all systems while the flywheel is removed.

Before starting the troubleshooting work, it would be well worth the time and effort to read and understand the information presented earlier in this section for Description and Operation, as well as to conduct the general troubleshooting procedures listed earlier. Considerable time, money, and possible frustration can be saved by having a thorough knowledge of the system and appreciating why the tests are conducted in a particular sequence.

TEST NO. 1 - IGNITION COIL OUTPUT CHECK

◆ See Figure 218

1. Use a standard spark tester and check for spark at each cylinder. If a spark tester is not available, use a pair of insulated pliers and hold the plug wire about 1/2 in. (12.7mm) from the engine.

2. Turn the flywheel with an electrical starter and check for spark. A strong spark over a wide gap must be observed when testing in this manner, because under compression a strong spark is necessary in order to ignite the air/fuel mixture in the cylinder. This means it is possible to think you have

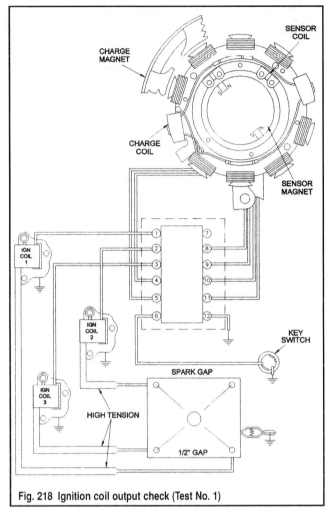

Fig. 218 Ignition coil output check (Test No. 1)

a strong spark, when in reality the spark will be too weak when the plug is installed.

3. If there is no spark, or if the spark is weak, from one coil, proceed directly to Test No. 2, Trigger Coil Input Check. If there is no spark or the spark is weak from the coils, proceed directly to Test No. 3, Charge Coil Output Check. If the spark is strong and steady, across the 1/2 in. (12.7mm) gap, or the spark tester indicates the spark is satisfactory, then the problem is in the spark plugs, the fuel system, or the compression is weak.

TEST NO. 2 - TRIGGER COIL INPUT CHECK

◆ See Figures 219 and 220

1. Remove the power pack cover.

■ **QUICK WORD: Always identify leads prior to disconnecting to ensure they will be connected to the proper terminal after the test. Sometimes the leads are the same color. Therefore, without a tag or other identification, the lead may not be connected properly.**

2. Disconnect the sensor leads from the No. 8, 9, 10 and 11 terminals at the power pack Disconnect the ignition coil lead from the power pack at No. 2 and No. 3 terminals. Obtain a neon tester (#S-80 or M-80). Connect the black lead of the neon tester to the No. 8 terminal and the blue lead to the No. 11 terminal. Set the neon light selector to the No. 3 position.

3. Crank the engine with the electric starter motor, and at the same time momentarily depress (tap) the **B** load button rapidly.

a. If adequate spark was observed from the coil in the previous tests, the problem is in the sensor. To test the sensor assembly, see Test No. 6.

b. If no spark was observed from the coil during the previous tests, the charge coil must be checked as outlined in Test No. 3.

c. Check the second coil: Disconnect the ignition coil lead from the No. 1 terminal on the power pack. Connect the No. 2 coil wire to the power pack and repeat the above test.

d. Check the third coil: Disconnect the lead from the No. 2 terminal and connect the lead from the No. 3 coil to the power pack to the No. 3 terminal. Repeat the above test.

e. If spark was not observed on all three coils, the Power Pack must be checked, see Test No. 4.

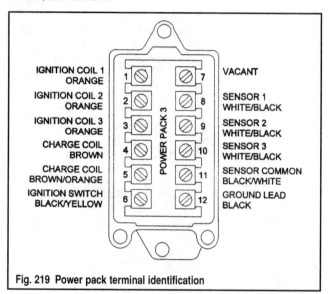

Fig. 219 Power pack terminal identification

TEST NO. 3 - CHARGE COIL OUTPUT CHECK

◆ **See Figures 219 and 221**

1. Identify the leads, then disconnect at the No. 4 and No. 5 terminals. Connect the tester lead across the charge coil leads to the timer base. Turn the neon tester to position No. 2.

2. Crank the engine and at the same time, depress load button **B** on the tester. Observe the light. If the light glows steadily, the charge coils are good.

3. Check the power pack output. If the light comes on intermittently, check for a broken or shorted wire. If the wires are in good shape, replace the charge coil. If the light fails to come on, replace the charge coil and stator assembly.

TEST NO. 4 - POWER PACK OUTPUT CHECK

◆ **See Figure 222**

1. Check all connections at the power pack to be sure they are clean and secure.

2. Disconnect the lead from the No. 1, 2, and 3 terminals. If the neon tester is used, connect the black lead to the No. 1 power pack terminal and the blue lead to a good ground.

3. Move the neon tester switch to the No. 1 position, and then depress load button **A** and crank the engine with the electric starter motor. Observe the light.

4. Now, move the black lead from the No. 1 terminal to the No. 2 terminal. Crank the engine and repeat the above test.

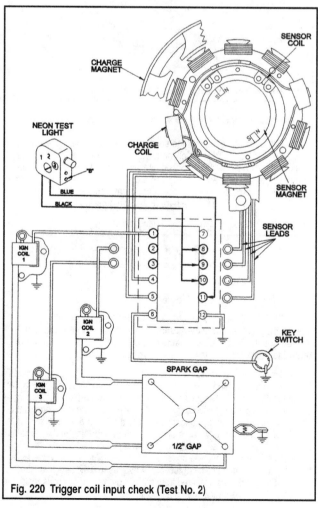

Fig. 220 Trigger coil input check (Test No. 2)

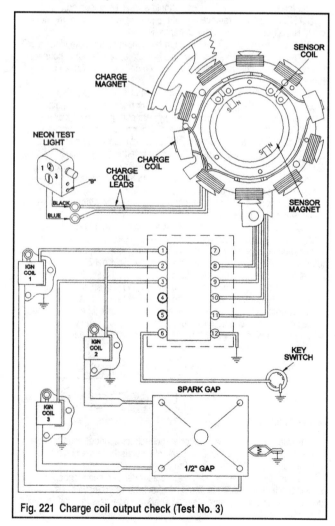

Fig. 221 Charge coil output check (Test No. 3)

5. Move the black lead from the No. 2 terminal to the No. 3 terminal. Crank the engine and repeat the above test.

   a. If the light is strong and steady on all three outputs, replace the faulty coil or coils. The No. 1 terminal connection is the top coil, the No. 2 terminal Is the center coil, and the No. 3 terminal Is the bottom coil.

   b. If no light is visible on any of the three outputs, check the key switch as outlined in Test No. 5.

   c. If a steady light was observed on one or two output, but no light on the other, replace the power pack.

   d. If the light was very dim or intermittent on any of the three outputs during the tests, the power pack must be replaced.

6. Connect the wires disconnected at the start of the tests.

TEST NO. 5 - KEY SWITCH CHECK

◆ **See Figure 223**

A CD magneto key switch operates in the reverse of any other type key switch. When the key is moved to the **OFF** position, the circuit is CLOSED between the CD magneto and ground. In some cases, when the key is turned to the **OFF** position the points are grounded. For this reason, an automotive-type switch should NEVER be used, because the circuit would be opened and closed in reverse, and if 12-volts should reach the coil, the coil will be destroyed.

1. Connect the spark tester to the high tension leads. Disconnect the lead at the No. 6 power pack terminal. This is the wire from the key.

2. Crank the engine with the key and observe the spark.

   a. If there is no indication of spark on either coil or on only one, replace the power pack.

   b. If spark is indicated on all three coils, the problem is most likely in the lead from the key.

   c. If the key switch leads appear to be in good condition, replace the key switch.

TEST NO. 6 - SENSOR COIL LOW AND HIGH OHM CHECKS

◆ **See Figure 224**

1. Identify the leads with a tag or other means, and then disconnect them at the No. 8, 9, 10 and 11 terminals.

2. Disconnect the sensor leads from the power pack terminals No. 1, 2 and 3. Connect the ohmmeter leads to the sensor coil leads alternately (the white lead with the black stripes and the black lead with white stripes).

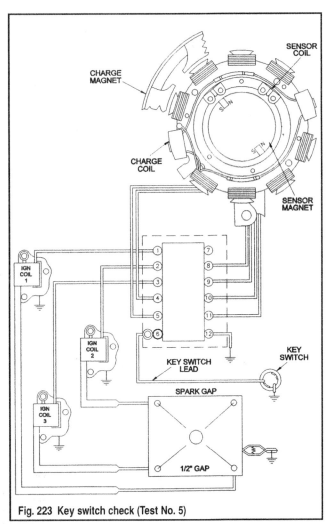

Fig. 223 Key switch check (Test No. 5)

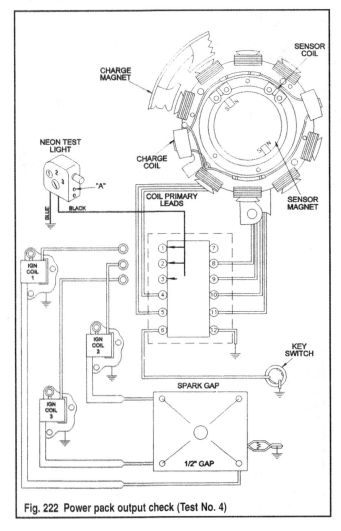

Fig. 222 Power pack output check (Test No. 4)

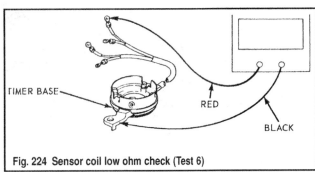

Fig. 224 Sensor coil low ohm check (Test 6)

3. Use the low ohm scale and observe the reading. The meter should indicate 8.5 -1.0 ohms at room temperature (70° F). If the ohmmeter reading is not satisfactory, proceed with the Sensor Coil High Ohm Check

Sensor Coil High Ohm Check:

4. Move the ohmmeter to the High Ohm scale.

5. With the sensor leads still disconnected from the power pack, as in Low Ohm Test, connect the red ohmmeter test lead to either one of the sensor leads. Connect the black ohmmeter lead to a good ground. The meter should indicate infinity. A ZERO reading indicates a short to ground.

a. If the test fails, the sensor coil and timer base must be replaced as a unit.

b. If the test is successful disconnect the ohmmeter and connect the leads at the No. 8, 9, 10 and 11 power pack terminals. Connect the leads at the No. 1, 2 and 3 terminals.

Stator & Charge Coil

REPLACEMENT

◆ **See Figures 208, 209 and 225 thru 229**

CD magneto systems installed on outboard engines will usually operate over extremely long periods of time without requiring adjustment or repair. However, if ignition system problems are encountered, and the usual corrective actions, such as replacement of spark plugs, does not correct the problem, the magneto output should be checked to determine if the unit is functioning properly.

CD magneto overhaul procedures may differ slightly on various outboard models but the following general basic instructions will apply to all Johnson/Evinrude high speed flywheel-type CD magnetos.

1. Hold the flywheel with a proper tool and remove the flywheel nut. Obtain special tool (#378103) or an equivalent puller.

NEVER use a puller that exerts a force on the rim or ring gear of the flywheel.

2. Remove the flywheel.

3. Observe closely how the stator and trigger base is secured to the powerhead. You may want to take a couple of digital pictures as an aid during the assembling and installation work.

4. Disconnect the wires at the terminal block and power pack. The leads from the charge coil are connected to the No. 1 terminal of the power pack and the leads from the stator are connected to the No. 1 and No. 2 terminals of the terminal board.

5. Remove the 4 screws securing the stator to the powerhead. Lift the stator and charge coil assembly free of the powerhead. The stator cannot be repaired; therefore, if troubleshooting and testing indicates the stator or the charge coils are unfit for service, the complete unit must be replaced as an assembly.

To Install:

6. Slide the stator assembly down the crankshaft and into place on the powerhead.

7. Apply just a drop of Loctite to the threads of the attaching screws, and then install and tighten them ALTERNATELY and EVENLY until secure.

8. Check the crankshaft and flywheel tapers for any traces of oil, burrs, nicks, or other damage. Clean the tapered surfaces with solvent, and then blow them dry with compressed air. These two surfaces must be absolutely dry.

9. Slide the flywheel onto the crankshaft with the slot in the flywheel indexed over the Woodruff key in the crankshaft.

10. Thread the flywheel nut onto the crankshaft. Hold the flywheel from rotating with a proper tool and tighten the nut to the specification given in the Torque Specifications chart.

Fig. 225 Remove the flywheel. . .

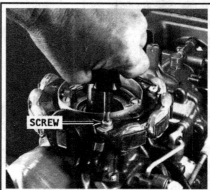

Fig. 226 . . .and then the stator

Fig. 227 A good shot of the stator assembly

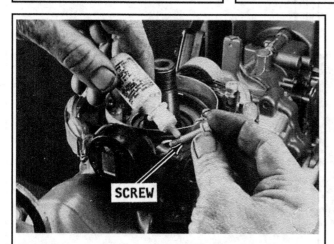

Fig. 228 Always use a drop of Loctite on the mounting screws

Fig. 229 Tighten the flywheel nut

Timer Base & Sensor Assembly

REPLACEMENT

 MODERATE

◆ See Figures 208, 209 and 230 thru 233

CD magneto systems installed on outboard engines will usually operate over extremely long periods of time without requiring adjustment or repair. However, if ignition system problems are encountered, and the usual corrective actions, such as replacement of spark plugs, does not correct the problem, the magneto output should be checked to determine if the unit is functioning properly.

CD magneto overhaul procedures may differ slightly on various outboard models but the following general basic instructions will apply to all Johnson/Evinrude high speed flywheel-type CD magnetos.

1. Remove the stator and charge coil assembly as outlined elsewhere in this section.

2. Remove the 4 retaining clips and screws. The clips engage in a Delrin ring which fits around the timer base.

3. Lift the timer base and Delrin ring free of the power-head.

■ **A brass bushing is an integral part of the timer base. This bushing has a very close tolerance with the upper bearing and seal assembly. The bushing rotates as the spark is advanced or retarded. After the assembly has been removed, make a careful check for dirt, chips, or damage which might prevent the timer base from rotating freely.**

To Install:

4. Coat the upper bearing and seal assembly with Johnson/Evinrude Sea-Lube (Trade Mark) or equivalent.

5. Apply a coating of light-weight oil to the Delrin ring.

6. Slip the ring into place on the timer base.

Fig. 230 Removing the retaining clips

7. Position the timer base assembly into position on the power-head and secure it with the 4 retaining clips and screws.

8. If the Woodruff key on the crankshaft was removed, insert it into the key way with the outer edge of the key parallel to the center line of the crankshaft.

9. Install the stator assembly and flywheel.

Power Pack

REPLACEMENT

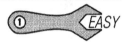 EASY

◆ See Figures 208, 209 and 234

CD magneto systems installed on outboard engines will usually operate over extremely long periods of time without requiring adjustment or repair. However, if ignition system problems are encountered, and the usual corrective actions, such as replacement of spark plugs, does not correct the problem, the magneto output should be checked to determine if the unit is functioning properly.

CD magneto overhaul procedures may differ slightly on various outboard models but the following general basic instructions will apply to all Johnson/Evinrude high speed flywheel-type CD magnetos.

1. Disconnect the battery from the engine.

2. Tag and disconnect all leads at the power pack terminal board.

3. Remove the attaching hardware and lift the power pack free of the powerhead.

To Install:

4. Place the new power pack in place on the powerhead. Secure it with the attaching hardware.

5. Connect the electrical leads to the terminal board following the color code designations given on the power pack cover or following the tags made during removal.

6. Install and secure the cover in place.

Timing

CHECK & ADJUSTMENT

 OEM MODERATE

◆ See Figures 235 and 236

■ **Under normal operating conditions, the timing should not change. If the spark advance stop screw has been moved, or if the power pack assembly has been replaced, the timing should be checked.**

The timing cannot be properly or accurately adjusted without a timing light.

✳✳ CAUTION

NEVER attempt to make this adjustment with a flush attachment connected to the lower unit. The no-load condition on the propeller would cause the engine to 'runaway', resulting in serious damage or destruction of the unit.

Fig. 231 Check the brass bushing. . .

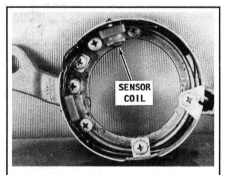

Fig. 232 . . .and the other side of the base

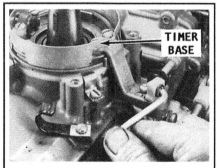

Fig. 233 Installing the base

Fig. 234 Removing the power pack

Fig. 235 Performing the preliminary adjustment

Fig. 236 A good shot of the timing marks and pointer

The engine *must* be mounted in a test tank or body of water to adjust the timing.

✳✳ CAUTION

Water must circulate through the lower unit to the engine any time the engine is run to prevent damage to the water pump in the lower unit. Just 5 seconds without water will damage the water pump.

1. As a preliminary adjustment, loosen the lock-nut on the spark advance screw and rotate the screw inward or outward until the exposed portion of the screw outside the bracket is 1/2 in. (12.7 mm).

2. Connect the timing light to the No. 1 cylinder.

3. Start the engine and warm it to normal operating temperature.

4. With the unit in neutral or in gear, set engine speed to a minimum of 3500 rpm, full spark advance as listed in the Tune-Up Specifications chart.

5. If necessary, advance or retard the spark to obtain the proper degree reading, as follows:

   a. Shut down the engine.

   b. Loosen the lock-nut and move the advance stop adjustment screw to obtain the proper setting. One full turn of the screw will result in approximately 1° of adjustment.

   c. Again start the engine and check the degree reading. Tighten the lock-nut securely after the final adjustment.

6. Perform the Timing & Synchronization procedures in the Maintenance & Tune-Up section.

CHARGING SYSTEM - GENERATOR

Description

■ Almost all 1 and 2 cylinder engines covered here utilize a generator (late model 50 hp motor use an alternator) and many 3 cylinder and V4 motors do as well. Simply, if your engine has a battery ignition system or a CD system, it is using an alternator; if not, it's got a generator.

A generator is installed as an accessory on engines equipped with a distributor magneto ignition system. The generator has 2 terminals on the lower end. One terminal is larger than the other and the wires connected have different size connectors to ensure the proper wire is connected to the correct terminal.

If several electrical accessories are used and the engine is operating at idle speed, or below 1500 rpm for extended periods of time, the battery will not receive adequate current to remain in serviceable condition.

The rated capacity of the generator is 10 amps. Therefore, the electrical accessory load should not exceed 10 amps or current will be drawn from the battery at a greater rate than the generator is able to produce. Such a negative draw on the battery will result in a run-down condition and failure of the battery to provide the required current to the starter for cranking the engine.

To calculate the amperage draw of an accessory the following simple formula may be used: Amps equals watts divided by volts. Amps = Watts/Volts

The volts will always be 12. Accessories will usually be given in watts. If the obsolete measurement of candlepower is used, then one candle power is equal to approximately one watt. Example: A boat has running lights requiring 8 watts; auxiliary lights use 10 watts; and a radio rated at 30 watts.

- Amps = 48 Watts/12 volts = 4 Amps.

In this case, if all the lights are on and the radio is being used, the total draw on the battery would be 4 amps. If the engine is running at 1500 rpm or higher, and the generator circuit is performing properly by charging the battery with 10 amps, then a net positive gain of 6 amps is being received by the battery.

The battery can be externally charged or the engine can be equipped with a generator to charge the battery while the engine is operating.

A voltage regulator, mounted in a junction box on the rear of the engine, is connected between the generator and the battery to prevent overcharging of the battery while the engine is operating. The junction box also houses a fuse to protect the charging circuit.

The generator circuit requires at least 1500 rpm engine speed to effectively charge the battery. At this speed, the ampere meter on the dash will indicate a positive charge to the battery.

If the boat has a twin engine installation, the usual practice is to use only one battery for cranking both units. With such a twin installation, only one engine generator should be used to charge and maintain the battery at its full amperage rating.

Most mechanics have discovered if both generators of a twin installation are connected to charge the battery, one seems to "fight" the other. Instead of having an improved system, this type of hook-up causes many serious electrical problems that are unexplainable.

Troubleshooting

GENERAL

◆ See Figures 237 and 238

One of 3 areas, identified as a, b, and c, below, may cause problems in the generating circuit. Any one of the problems will result in failure of the system to provide sufficient current to maintain the battery at a satisfactory charge. Remember, the generator will only produce approximately 10 amps of current. Most ammeters have a 20 amp scale. Therefore, it is only necessary for the scale to register in the 10 amp area, while the engine is operating above 1500 rpm, to indicate satisfactory performance

a. The 4 amp or 20 amp fuse in the junction box may have burned/blown, opening the circuit. If the fuse requires replacement, a check should be made immediately, to determine why the fuse burned protecting the circuit.

b. The voltage regulator may be defective. If the regulator has failed, a thorough check of the circuit to determine the cause. Simply replacing the regulator usually will not solve the problem and the new regulator may be damaged when the engine and generator are operating.

c. The generator may be defective and fail to produce the current necessary to maintain the battery. The problem may simply be worn brushes. Replacement with new brushes may solve the problem. However, if the brushes are in good condition, and testing reveals the generator must be replaced, the conservative mechanic will install a new voltage regulator at the same time.

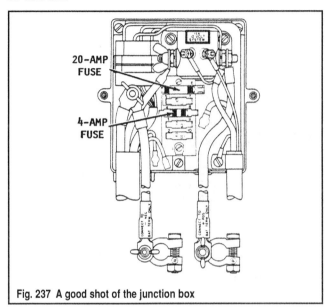

Fig. 237 A good shot of the junction box

■ The engine must be operated, in gear, at speeds in excess of 1500 rpm to test the generator circuit. Therefore, the engine must be mounted in a body of water to prevent a 'runaway' condition and serious damage to internal parts, or destruction of the unit. NEVER attempt to operate the engine above idle speed with a flush attachment connected to the lower unit or with the engine mounted in a small test tank, such as a fifty-gallon drum.

✳✳ CAUTION

Water must circulate through the lower unit to the engine any time the engine is run to prevent damage to the water pump in the lower unit. Just 5 seconds without water will damage the water pump.

1. Check to be sure all electrical connections in the circuit are secure and free of corrosion. Double check the battery connections and terminals. If the terminals are badly corroded, there is no way on this green earth for the current produced by the generator to reach the battery cells. A special wire brush can be purchased at very modest cost to clean the inside of the wire connectors. A common wire brush may be used to clean the battery terminals. Baking soda and water is a good cleaning agent for the battery surface.

2. Check the wiring in the circuit for broken insulation, or an actual break in the line. Disconnect the Positive electrical lead from the battery as a precaution against an accidental short causing damage to the voltage regulator. Remove the junction box cover. Check the condition of the 4 amp or 20 amp fuse with continuity light, or install a new fuse and check the charging circuit again with the engine operating. The 4 amp and 20 amp fuses can easily be "popped" out of their retainers in the panel of the junction box base and replaced after the cover has been removed.

3. Ground the field of the generator very QUICKLY and only MOMENTARILY with a jumper wire while the engine is operating at approximately 2000 rpm. By momentarily grounding the field, the voltage regulator is actually bypassed and the generator will "run wild". If the amp meter registers a high reading while the generator field is grounded, the circuit has a broken wire, or the voltage regulator is defective. If the ammeter reading does not change while the generator field is grounded, then the indication is a faulty generator.

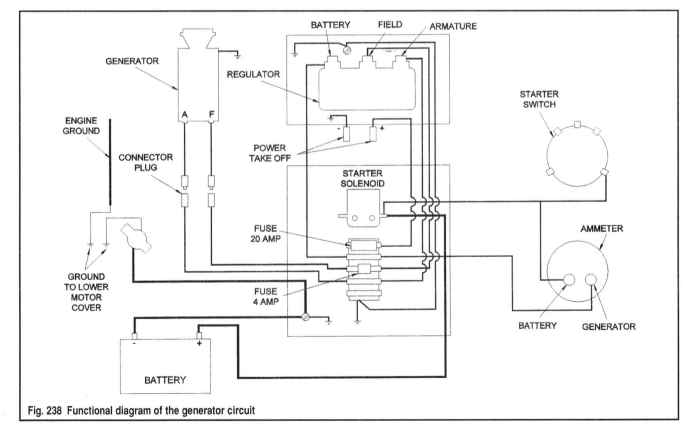

Fig. 238 Functional diagram of the generator circuit

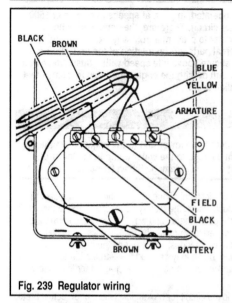

Fig. 239 Regulator wiring

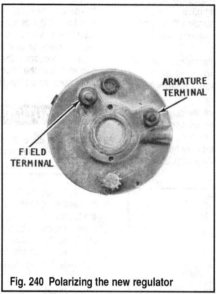

Fig. 240 Polarizing the new regulator

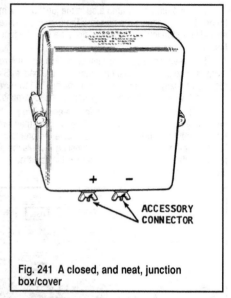

Fig. 241 A closed, and neat, junction box/cover

Voltage Regulator

REMOVAL & INSTALLATION

◆ See Figures 237 and 239, 240 and 241

1. Disconnect the positive lead from the battery terminal to prevent an accidental short causing damage in the circuit.

2. Loosen the 2 wing nuts on both sides of the junction box cover. These wing nuts are "captive" with the cover and cannot be completely removed, only released from the junction box (this arrangement prevents loss of the wing nuts).

3. Tag and disconnect the wires between the generator and the terminal board in the junction box at the board, if the regulator is to be replaced.

4. Place the junction box cover on its back and remove the 5 leads to the voltage regulator. Notice how the leads are color-coded, as an assist in connecting them correctly during installation.

5. Remove the attaching hardware securing the voltage regulator to the cover. Remove the regulator.

To Install:

6. Place the new regulator in position in the junction box cover and secure it with the attaching hardware.

7. Connect the 4 color-coded wires to the regulator: yellow, from the generator armature; blue, from the generator field; brown, from the battery; and the black is the ground wire. A second brown wire is connected to the bottom of the junction box to allow additional electrical accessories to be connected. This second wire would not have been disconnected to remove the regulator.

■ When a new voltage regulator is installed, the generator must be "polarized" *before* the cover is installed.

8. "Polarize" the new regulator by first connecting the positive lead to the battery, and then using a small jumper wire to make a MOMENTARY connection between the battery terminal and the armature terminal of the regulator. The generator is now properly "polarized" with the new regulator for service.

9. Now, disconnect the positive battery lead again, before installing the junction box cover. The few moments involved in disconnecting and connecting the positive lead at the battery is well spent. This small task will prevent any possible short from causing damage to the circuit when working with the wires.

10. Install the junction box cover to the box base. As the cover is moved into place, work the wires alongside the regulator. Secure the cover in place with the 2 captive wing nuts.

Generator

REMOVAL & DISASSEMBLY

◆ See Figures 242 thru 249

1. Disconnect the positive lead at the battery terminal.
2. Remove the hood from the engine.
3. Remove the retaining bolts and lift off the hand starter (if so equipped).
4. Prevent the shaft from turning by engaging an open-end wrench with the flats on the generator shaft underneath the pulley. Now, while continuing to hold the wrench on the shaft, remove the nut from the top of the generator pulley.

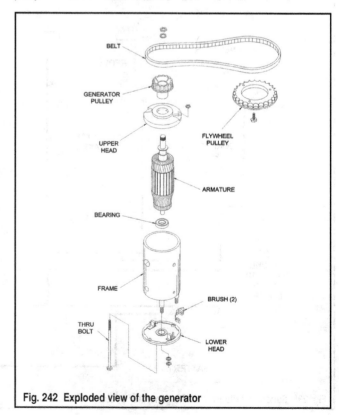

Fig. 242 Exploded view of the generator

Fig. 243 Removing the pulley nut on 1-2 cyl motors. . .

Fig. 244 . . .and then the belt

Fig. 245 Removing the pulley nut on 3 cyl and V4 motors. . .

Fig. 246 . . .and the pulley/belt

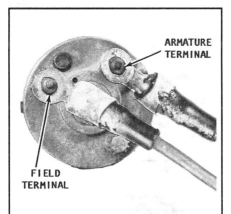

Fig. 247 Disconnect the wires at the bottom of the unit

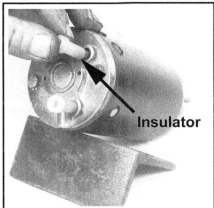

Fig. 248 Don't forget to remove the 2 insulators

Fig. 249 Remove the end cap

5. Remove the cover from the inside of the generator pulley (usually just 1-2 cyl motors).

6. Remove the generator belt from the pulley. Use a small prybar or other similar tool and pry the pulley up and free of the generator shaft.

7. Hold the generator, and at the same time, remove the nuts from the top of the generator support bracket, and the generator is free.

8. Remove the 2 wires on the bottom side of the generator. Notice how one generator stud is smaller than the other and the electrical connectors are different sizes to match the studs.

9. Remove the 2 nuts from the wire terminals at the bottom of generator. Work out the 2 white insulators from around the studs.

10. Remove the 2 thru-bolts.

11. Remove the end cap from the generator. After the cap has been removed, take notice of the small dowel in the end of the generator. This dowel ensures the cap will be installed correctly when the dowel is indexed in a matching hole in the cap. Pull the armature and upper cap out of the frame.

ARMATURE TESTING

Testing for a Short

◆ See Figure 250

1. Position the armature on a growler, then hold a hacksaw blade over the armature core.

2. Turn the growler switch to the **ON** position. Slowly rotate the armature. If the hacksaw blade vibrates, the armature or commutator has a short.

3. Clean the grooves between the commutator bars on the armature.

4. Perform the test again. If the hacksaw blade still vibrates during the test; the armature has a short and must be replaced.

Testing for a Ground

◆ See Figure 251

Obtain a test lamp or continuity meter. Make contact with 1 probe lead on the armature core and the other probe lead on the commutator bar. If the lamp lights or the meter indicates continuity, the armature is grounded and MUST be replaced.

Checking The Commutator Bar

◆ See Figure 252

Check between or check bar-to-bar as shown in the accompanying illustration. The test light should light, or the meter should indicate continuity. If the commutator fails the test, the armature MUST be replaced.

Field Coil Test For Ground

◆ See Figure 253

1. Check to be sure the free end of the field wire is not grounded to the frame and the field insulation is not broken.
2. Using a test lamp or ohmmeter, make contact with 1 probe lead to the ground of the generator frame.
3. Make contact with the other lead to the field terminal. If the lamp lights or the ohmmeter indicates continuity, the field coils are grounded. If the location of the ground in the field coils cannot be determined, or repaired, the coils MUST be replaced.

Armature Terminal Test For Ground

◆ See Figure 254

1. Check to be sure the loose end of the armature terminal lead of the generator is NOT grounded to the frame.
2. Using a test lamp or ohmmeter, make contact with 1 probe lead to the armature terminal of the generator.
3. Make contact with the other probe lead to a good ground on the generator frame. If the test lamp lights or the ohmmeter indicates continuity,

the positive terminal insulation through the generator frame is broken down and MUST be replaced.

Positive Brush Test For Ground

◆ See Figure 255

1. Using a test lamp or ohmmeter, make contact with 1 probe lead to the positive or insulated brush holder.
2. Make contact with the other probe lead to a good ground on the generator frame. If the lamp lights or the ohmmeter indicates continuity, the brush holder is grounded due to defective insulation at the frame.

Field Test

◆ See Figure 256

1. Using a test lamp or ohmmeter, make contact with 1 probe lead to the armature stud.
2. Make contact with the other probe lead to the armature brush. The lamp should light or the ohmmeter should indicate continuity. If this test is not successful, check for a poor connection between the stud and the brush.

CLEANING & INSPECTION

◆ See Figures 242 and 257

Check the ball bearing at the end of the commutator bar. Verify that the bearing turns free with no sign of "rough spots" or binding. Hold the armature in one hand and turn the upper cap on the shaft with the other hand. The

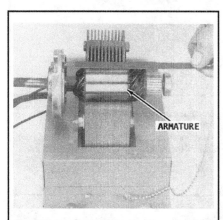

Fig. 250 Testing the armature for a short with a growler

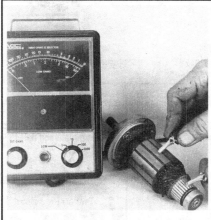

Fig. 251 Testing the armature for a ground

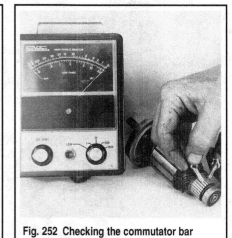

Fig. 252 Checking the commutator bar

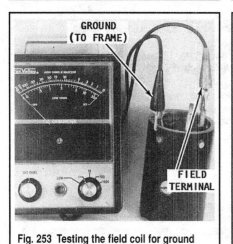

Fig. 253 Testing the field coil for ground

Fig. 254 Testing the armature terminal for a ground

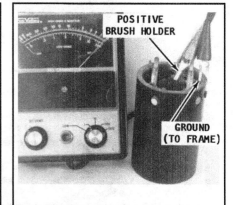

Fig. 255 Testing the positive brush for a ground

cap and shaft should turn freely with no sign of binding. If either of these tests are not successful, the bearing MUST be replaced.

Check the amount of brush wear. If the brush is worn more than 50% of its original size, or to within 1/4 in. (6.35mm) of the base, it should be replaced. Replacement of the brushes is a simple task. First, remove the brush retaining screw, and then remove the old brush and install a new brush. Secure the new brush in place with the retaining screw.

If the armature commutator requires turning, it should be turned in a lathe to ensure accuracy. The local generator shop can perform this task, usually for a very reasonable fee. If the turning is accomplished by other than generator shop personnel, the following words are necessary. After the turning, an undercut should be made. The insulation between the

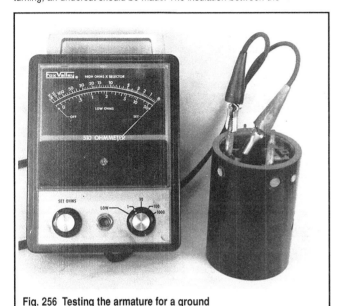

Fig. 256 Testing the armature for a ground

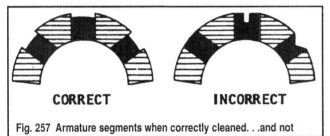

CORRECT   **INCORRECT**

Fig. 257 Armature segments when correctly cleaned. . .and not

commutator bars should be 1 3/4 in. (4.44 cm). This undercut must be the full width of the insulation and flat at the bottom. A triangular groove is NOT satisfactory. After the undercut work is completed the slot should be thoroughly cleaned to remove any foreign material, dirt, or copper dust. Sand the commutator *lightly* with "00" sandpaper to remove any slight burrs left from the undercutting. After all work has been completed, test the unit again, on the growler.

ASSEMBLY & INSTALLATION

◆ **See Figures 242 and 258 thru 261**

1. Slide the armature into the frame and align the top armature cap with the dowel in the frame. Proper alignment is achieved when the dowel in the frame indexes into a matching hole in the cap. As the armature is moved into place, pull back on the brushes, and work them around the commutator bar.

2. Install the end cap down over the studs of the field and armature. Check to be sure the dowel in the frame has indexed with the hole in the cap.

3. Install the 2 thru-bolts and secure the complete assembly with the nuts.

4. Place the 2 bushings over the terminal studs of the armature and field. Secure the bushings in place with the washers and proper nuts (one terminal is larger than the other).

Testing by Rotating the Armature

Performing this test will also "polarize" the new or rebuilt generator. If this test is not performed, the new or rebuilt generator must still be polarized following installation.

Polarization at that time is accomplished by first connecting the battery to the system in the normal manner, and then connecting a jumper lead to the **BAT** terminal of the voltage regulator. Next, MOMENTARILY make contact with the other end of the jumper lead to the **GEN** terminal of the regulator. The generator is now "polarized" for service.

Return to the bench for testing after rebuilding the generator.

■ **The armature will turn rapidly during this test. Therefore, the generator MUST be well secured before making the test to prevent personal injury or damage to the generator.**

5. Connect a jumper wire between the field terminal and a good ground on the case.

6. Connect a second jumper wire between the positive battery terminal and the armature stud. Momentarily make contact with the negative lead from the battery to any good ground on the generator. The generator should rotate rapidly. If the generator fails to rotate, the generator must be disassembled again and the service work carefully checked. Sorry about that, but some phase of the rebuild task was not performed properly.

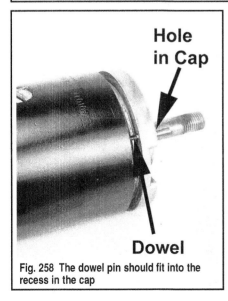

Hole in Cap

Dowel

Fig. 258 The dowel pin should fit into the recess in the cap

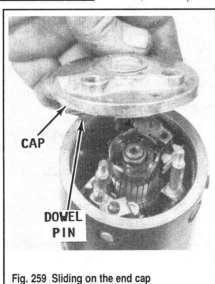

CAP

DOWEL PIN

Fig. 259 Sliding on the end cap

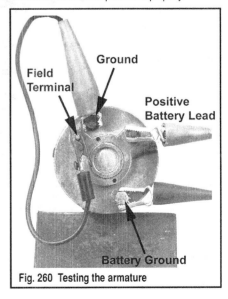

Field Terminal   **Ground**

Positive Battery Lead

Battery Ground

Fig. 260 Testing the armature

Fig. 261 Adjusting the tension on the pulley/belt

7. Install the holding bracket to the generator, or if the bracket remained on the engine, install the generator into the bracket and secure it in place with the attaching hardware, but DO NOT tighten the nuts on the generator thru-bolts at this time.

8. Check to be sure the Woodruff key is in place in the generator shaft. Install the generator into the bracket and at the same time slide the pulley onto the armature shaft with the slot in the pulley indexed with the Woodruff key.

9. Secure the generator in place with the attaching hardware, but DO NOT tighten the nuts on the generator thru-bolts at this time.

10. Install the drive belt around the pulley on the bottom side of the flywheel and onto the generator pulley. Install the lockwasher and nut to secure the generator pulley in place. Hold the generator shaft from turning with an open-wrench on the flats of the shaft underneath the pulley.

11. Adjust tension on the generator pulley by pulling the generator away from the engine, and then tightening the thru-bolt nuts securing the generator in the bracket. The pulley is properly adjusted when it may be depressed approximately 1/4 in. (6.35mm) at a point mid-way between the two pulleys.

12. Install the hand starter, if one is used.

CHARGING SYSTEM - ALTERNATOR

Description

An alternator system is installed on engines equipped with a battery or CD ignition system. The system replaces the current drained from the battery to start and operate the engine.

On engines equipped with battery ignition systems, a 15-20-amp alternator is installed. On engines with the CD ignition system, the alternator is rated at approximately 2-3 amps above the requirements for operating the engine. Therefore, any accessories added to the engine or boat would require a higher output alternator. This higher output is usually accomplished through the installation of an alternator kit available from the local Bombardier dealer.

The alternator consists of a stator mounted underneath the flywheel, a rotor attached to the inside diameter of the flywheel, and a regulator on the larger output alternators. A set of positive and negative diodes is installed in the system to change the alternating current to direct (DC) current. The necessary wiring ties the system together.

Under normal operating conditions, very few problems are encountered with the alternator system. The most prevalent problem is connecting the battery backwards. Such action will damage the diodes in the system. The red cable from the starter motor solenoid terminal must be connected to the positive battery terminal. The black cable from engine ground must be connected to the negative battery terminal. Another problem area is the use of a charging system with the battery disconnected. This practice will damage the diodes or the voltage regulator.

Operation

◆ See Figure 262

Immediately after engine start, the ammeter, if installed, on the dash may indicate a full charging rate equal to the capacity of the alternator. As engine operation continues, the rate of charge will fall off, depending on the number of electrical accessories in use. Any high demand on the battery will result in an increase in the charging rate until the battery approaches a full or nearly full state of charge. The alternator system contains two distinct circuits, the field circuit and the charging circuit.

FIELD CIRCUIT

When the ignition switch is turned passed the **ON** position to the **START** position, the field circuit is closed to cause a current flow from the positive pole of the battery directly to and through the field coils; on through the transistor: then through a ground return to the negative pole of the battery to complete the circuit. The ignition switch is spring loaded for starting.

Once the engine starts, the key will return to the **ON** position and remain there while the engine is operating. An indicator light bridges the field circuit and glows Green when the key is set to the **ON** position. This light will continue to glow until the switch is turned to the **OFF** position when the engine is shut down.

Bear in mind, battery current continues flowing through the field circuit as long as the ignition switch is in the **ON** position. This is true regardless of whether the engine is operating or not. Therefore, if the switch should be accidentally turned to or left in the **ON** position, the battery will discharge itself through the field and ignition circuits and continue doing so until the battery is completely discharged or, until the ignition switch is turned to the **OFF** position.

CHARGING CIRCUIT

When the field coil is energized, as just described, the stator core becomes magnetized. The upper row of pole segments assumes a North polarity and the lower row of segments a South polarity.

A magnetic field is built-up around the poles, shifting through the surrounding atmosphere, from north to south poles. Now, the irregularly shaped rotor attached to and revolving with the flywheel around the stator poles, passes through intermittent areas of variable magnetic field density. It does so in such a manner as to cause a current surge traveling in one direction to be induced in every other stator coil. At the next instant, a surge traveling in the opposite direction is induced in the oppositely wound intervening coils as it enters and passes through the adjacent field, creating alternating (AC) current.

The stator includes 36 stator coils. Therefore, the current surges alternately 36 times per EACH revolution of the flywheel, which amounts to 36,000 such reversals per each 1000 motor rpm. At 4500 engine rpm, the rate is 162,000 alternating surges. This amount of current cannot be employed to charge the battery but must be first rectified or changed to direct (DC) current. This function is accomplished by the diodes installed in series with the charging circuit. Two diodes are of positive polarity and two are negative. The diodes permit voltage to pass in one direction and prevent passage in the opposite direction.

On engines equipped with a higher output alternator, a transistorized voltage regulator is installed in series with the battery field circuit. This regulator confines alternator voltage rise to within predetermined limits by automatically regulating intensity of the stator field.

Troubleshooting

GENERAL - READ PLEASE

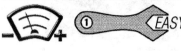

◆ See Figures 263 and 264

The following alternator troubleshooting numbered procedures are divided into 2 groups. The first covers an alternator installed on an engine equipped with a battery-type ignition system and a 20 amp alternator. The second set of procedures covers engines with CD ignition.

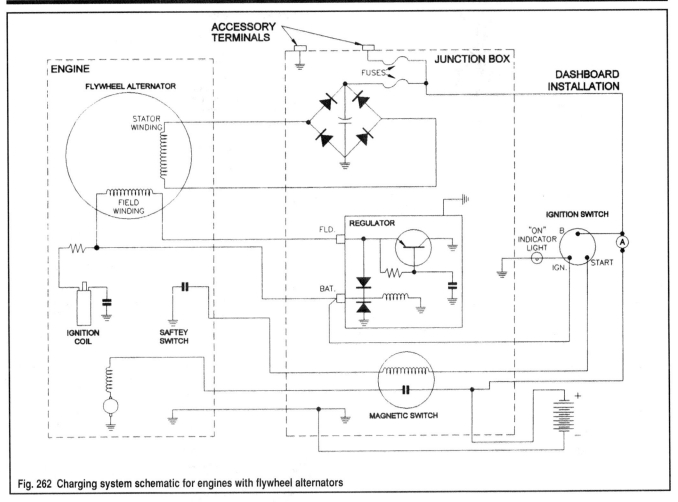

Fig. 262 Charging system schematic for engines with flywheel alternators

■ Before assuming the alternator may be at fault, check all wiring and connections associated with the alternator circuit. Frayed wires, loose connections, corroded terminals, or a defective battery, will cause problems in the alternator circuit. The battery must be FULLY charged, the terminals clean, the cable connections tight, and the battery polarity properly connected before any troubleshooting work is commenced on the alternator circuit. Proper polarity means the correct cables are connected to the positive and negative battery terminals. One end of the positive (red) cable should be connected to the positive battery terminal and the other end to the starter motor solenoid terminal. One end of the negative (black) cable should be connected to the negative battery terminal and the other end to a good ground on the engine.

Fuse Check

Check the heavy-duty 60 amp fuse in the junction box. On early model engines, this fuse is located in the junction box on the bottom side of the bar to which the diodes are attached.

On later model engines, this fuse is located underneath the small black cover of the diodes installed on the flywheel guard cover.

The fuse does not appear as a normal fuse. The fuse is square with a piece of metal on the back side and another on the front. The metal pieces are separated by a heavy insulating material. A narrow piece of metal connects the front and back metal strips. It is this small connecting piece of metal that will burn through and render the fuse defective.

All Alternator Systems

Check to be sure all electrical accessories are turned OFF, such as radio, lights, bait pump, blower, etc.

Turn the ignition switch to the **ON** position several times and observe the needle of the ammeter on the dash and the indicator light (if one is installed). A slight amount of needle movement to the negative side should be observed each time the switch is moved to the **ON** position, and the light should come on.

Fig. 263 On early models, the fuse is located here. . .

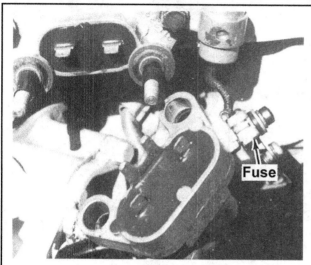

Fig. 264 . . .while on later models it can be found here

If needle movement is to the positive side, the battery or some other component is connected in reverse polarity. A considerable amount of needle movement to the negative side indicates a short some place in the electrical or ignition systems. Such a short would prevent the alternator from overcoming the drain on the battery.

If there is no needle movement, check all wiring and terminal connections to be sure they are in good condition.

ALTERNATORS WITH BATTERY IGNITION - 1961-67 3 CYL./V4 MODELS

Voltage Regulator Check

◆ See Figure 265

1. Mount the engine in an adequate size test tank or in a body of water.
2. Remove the junction box cover at the rear of the boat. Start the engine and allow it to warm to operating temperature at 500 rpm.

✳✳ CAUTION

Water must circulate through the lower unit to the engine any time the engine is run to prevent damage to the water pump in the lower unit. Just 5 seconds without water will damage the water pump.

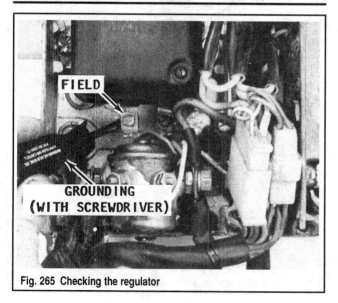

Fig. 265 Checking the regulator

3. Increase engine speed to approximately 1500 rpm. While the engine is running, use a screwdriver and ground the field terminal at the voltage regulator. Grounding the field terminal effectively removes the voltage regulator from the alternator circuit.
4. Observe the ammeter on the dash. A high positive reading indicates a defective voltage regulator. A low, or negative, reading indicates problems in the stator, wiring, or diodes. Unless you have considerable experience with voltage regulator repair work, simply replace the regulator.
5. The voltage regulator is secured on rubber mounts and therefore is allowed to move to compensate for engine vibration. Therefore, take care when installing the regulator and the junction box cover to allow slack in the wires to permit the regulator to move.
6. If the ammeter on the dash does not move while the engine is running and the field terminal is grounded, shut the engine down and proceed with the testing.

Diode Check

◆ See Figures 266 and 267

A diode is simply an electrical check valve, allowing current to flow in one direction but preventing it from flowing in the opposite direction. Two positive diodes and two negative diodes are installed on all alternator circuits.

1. Disconnect the battery from the system. On early model engines, disconnect the 4 diode leads from the junction box, 2 positive and 2 negative.
2. On later models, remove the cover on the flywheel guard to expose the diodes.

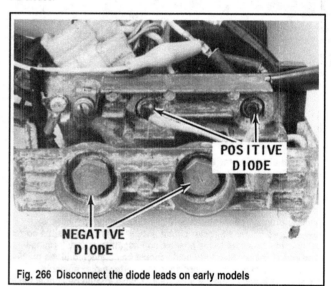

Fig. 266 Disconnect the diode leads on early models

Fig. 267 Using the ommeter on later models

3. Connect 1 lead of an ohmmeter to the ground terminal of the junction box and the other lead to the terminal of the diode to be tested. Observe the ohmmeter.

4. Reverse the leads of the ohmmeter. Again observe the ohmmeter. Continuity should be indicated with the leads connected one way and no continuity when connected in reverse. If continuity is indicated in both directions, the diode is defective and must be replaced.

If continuity is not indicated in one direction, the diode is defective.

5. Check the other 3 diodes in the same manner.

Stator Field Winding Checks

◆ See Figures 268 and 269

■ In the illustration, the stator has been removed for photography sake, it is not necessary to remove during testing.

1. Disconnect the quick-disconnect plug on the starboard side of the engine just behind the starter motor. This plug connects the wires from the stator to the wires to engine wiring harness. This connector will contain a red and blue wire from the stator.

2. Obtain an ohmmeter and set the reading for the high scale. Connect one lead of the meter to a good ground on the engine and the other lead to either the blue or the red wire going to the stator. The meter should indicate

NO continuity (an infinite reading). If continuity is indicated (less than 5000 ohms), the stator is grounded and should be replaced.

3. Set the ohmmeter to the low scale. Connect the meter leads to the red and blue stator field leads. The meter reading should be from 1.69-3.14 ohms. A too high reading indicates an open (circuit) winding. A too low reading indicates a shorted field winding. In either case, the stator is defective and must be replaced.

Stator Winding Tests

◆ See Figures 270 and 271

1. Connect the 2 halves of the quick-disconnect that were separated in the last test.

2. Disconnect the yellow leads to the stator. Set the ohmmeter selector to the high ohm scale. Leave 1 meter lead connected to ground and connect the other lead to either yellow stator winding leads from the stator. Observe the ohmmeter.

3. Connect the meter lead to the other yellow wire and again observe the meter. The meter should indicate NO continuity (an infinite reading) during either test. If continuity is indicated (less than 5000 ohms) during either test, the stator is defective and must be replaced.

4. Set the meter selector to the low ohm scale. Connect the meter leads to the yellow leads to the stator. Observe the meter reading. If less than 1 ohm is indicated, the stator is satisfactory. A reading higher than 1 ohm indicates a broken (open) winding and the stator must be replaced.

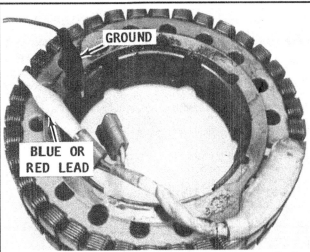

Fig. 268 Testing the stator field winding should show no continuity here. . .

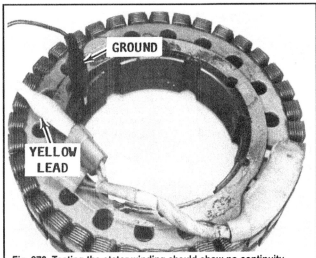

Fig. 270 Testing the stator winding should show no continuity here. . .

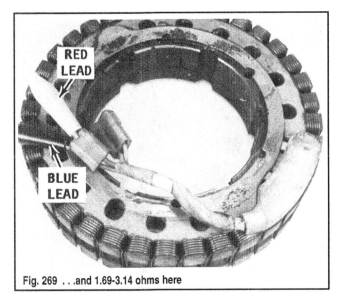

Fig. 269 . . .and 1.69-3.14 ohms here

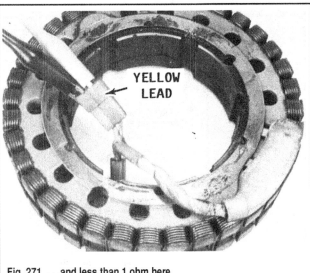

Fig. 271 . . .and less than 1 ohm here

ALTERNATORS WITH CD IGNITION - 1971-72 50 HP 2 CYL. MODELS

Failure in the alternator charging circuit will usually become evident when the battery reaches a run-down condition. To determine why the charging system has failed to maintain the battery in satisfactory condition, first, check the condition of the battery and all electrical connections, especially at the battery terminals. Many times, this visual inspection will reveal the problem area.

Verify the battery cables have been *properly* connected. If the battery has not been connected properly the diodes in the circuit will be blown instantly. Determine the number of accessories connected their draw, and the capability of the system to maintain the battery in a fully charged condition.

✳✳ WARNING

Before conducting any troubleshooting tests or actual service work, the battery cables should be disconnected at the battery terminals. Battery current is NOT required for any of the following troubleshooting tests. This safety measure will prevent accidental current from reaching a component resulting in possible damage to the part or personal injury.

TESTING DIODES USE HIGH-OHM SCALE

◆ **See Figures 272 and 273**

1. Check the diodes by connecting 1 test lead of an ohmmeter to one of the yellow wires at the diode and the other test lead to a good ground.
2. Now reverse the connections. The ohmmeter should indicate current flow in only one direction.
3. Perform the test on the other yellow lead. If current flow is indicated in both directions, the diode is defective and MUST be replaced.
4. Connect 1 ohmmeter test lead to the yellow diode lead and the other test lead to the red lead of the rectifier. Observe the ohmmeter reading.
5. Now, reverse the test leads and again observe the ohmmeter. Current should be indicated in only one direction.
6. Next, connect the test lead to the other diode yellow lead and perform the same 2 tests.
7. If current flow is indicated in both directions when the test lead is connected to either of the yellow leads, the diode is defective and MUST be replaced.

TESTING THE STATOR USE LOW-OHM SCALE

◆ **See Figures 274 and 275**

Very seldom does the stator cause problems in the charging circuit. However, if other checks have been performed and the stator is suspected as the problem area, first make a careful visual check of the stator for physical damage. If the visual inspection fails to indicate the problem, proceed as follows to test the stator:

1. Disconnect both battery cables. Disconnect the Yellow, Yellow/Blue and the Yellow/Grey stator leads from the terminal block.
2. Set the meter to the Rx 1000 scale. Make contact with the Black meter lead to a good engine ground.
3. Make contact with the Red meter lead to the Yellow/Blue stator lead and then to the Yellow lead, and finally to the Yellow/Grey lead. In each case, the meter should register NO continuity. If any other reading is registered, the stator is grounded and MUST be replaced.
4. Set the ohmmeter to the R x 1 scale. Make contact with the Black meter lead to the Yellow/Blue stator lead. Make contact with the Red meter lead to the Yellow stator lead. The meter should register between 1.0-2.0 ohms.
5. Keep the black meter lead in place and move the Red meter lead to contact the Yellow/Grey stator lead. The meter should register the same reading as in the previous test.

ALTERNATORS WITH CD IGNITION - 1967 100 HP & ALL 1968-72 3CYL./V4 MODELS

Most engines covered in this section were manufactured and distributed with 9 amp alternators as standard factory equipment. The one exception is the 1972 65 hp which had a 6 amp alternator. However, due to the addition of accessories, these units may have a 15 amp alternator installed. If the higher rated alternator is used, a voltage regulator must be installed.

The following troubleshooting procedures give detailed steps for checking the voltage regulator, alternator output, the rectifier and the stator. Before assuming the alternator may be at fault, check all wiring and connections associated with the alternator circuit. Frayed wires, loose connections, corroded terminals, or a defective battery will cause problems in the alternator circuit.

The battery must be FULLY charged, the terminals clean, the cable connections tight and the battery polarity properly connected, before any

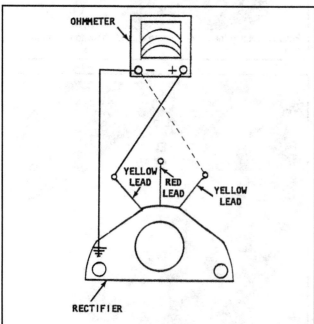

Fig. 272 First check the diodes by connecting between a ground and each of the yellow leads

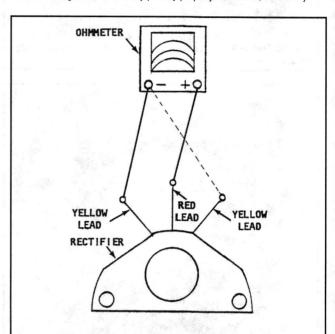

Fig. 273 Second, check between the red and each of the yellow leads for the rectifier

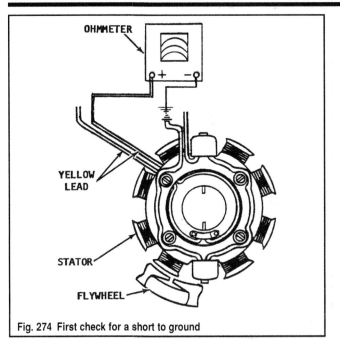

Fig. 274 First check for a short to ground

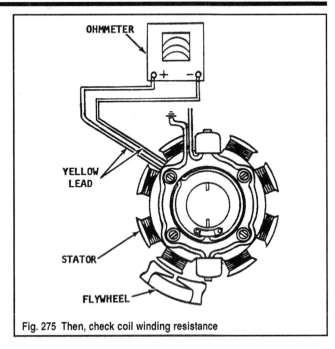

Fig. 275 Then, check coil winding resistance

troubleshooting work is commenced on the alternator circuit. Proper polarity means the correct cables are connected to the positive and negative battery terminals. One end of the positive (red) cable should be connected to the positive battery terminal and the other end to the starter motor solenoid terminal. One end of the negative (black) cable should be connected to the negative battery terminal and the other end to a good ground on the engine.

Voltage Regulator Check

◆ See Figure 276    MODERATE

A voltage regulator is only installed with a 15 amp alternator following engine purchase as an accessory. It is not a factory item nor is it original equipment. The voltage regulator installed with a CD ignition system is a solid state type. This means, if the regulator is proven to be defective, it must be discarded; it cannot be repaired. Therefore, if the following checks indicate the voltage regulator is defective, the only remedy to restore the circuit to satisfactory performance is to replace the regulator.

If the regulator is defective, it may allow too much voltage to pass through, or it may prevent sufficient voltage from passing to meet the engine and accessory demands to keep the battery fully charged.

Continually adding water to the battery at an unreasonable frequency could be an indication the voltage regulator is allowing excessive voltage to pass through to the battery. If the battery cannot be maintained at a satisfactory charge, a defective regulator could be one of several items to blame.

1. Mount the engine in an adequate size test tank or in a body of water.

2. If an ammeter is not installed on the dash, disconnect the red wire on the positive side of the starter solenoid. Connect 1 ammeter lead to this red wire and the other meter lead to the positive terminal of the starter solenoid. The meter is now in series with the alternator circuit. Start the engine and allow it to warm to operating temperature at 500 rpm.

✳✳ CAUTION

Water must circulate through the lower unit to the engine any time the engine is run to prevent damage to the water pump in the lower unit. Just 5 seconds without water will damage the water pump.

3. Increase engine speed to approximately 1500 rpm. Observe the ammeter on the dash or the meter connected into the circuit. The ammeter should indicate maximum output for the alternator rating, and then gradually fall back.

4. If the ammeter fails to indicate the required current, shut down the engine and disconnect the voltage regulator from the circuit, either at the quick-disconnect at the rear of the engine, or from the terminal board.

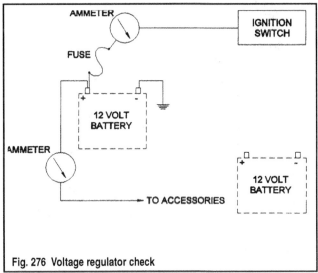

Fig. 276 Voltage regulator check

5. Again, start and operate the engine at approximately 1500 rpm and observe the ammeter. If the meter indicates adequate current passing through, the voltage regulator is defective and must be replaced. If the ammeter still fails to indicate adequate current, proceed with testing other components in the system.

Alternator Output Check

 MODERATE

◆ See Figure 277

1. Connect 1 lead of an AC voltmeter to a good ground on the engine. Start the engine.

✳✳ CAUTION

Water must circulate through the lower unit to the engine any time the engine is run to prevent damage to the water pump in the lower unit. Just 5 seconds without water will damage the water pump.

2. Momentarily make contact with the other meter lead to first one of the yellow leads from the stator and then to the other yellow lead. This momentary contact will be made to the yellow leads inside the quick-

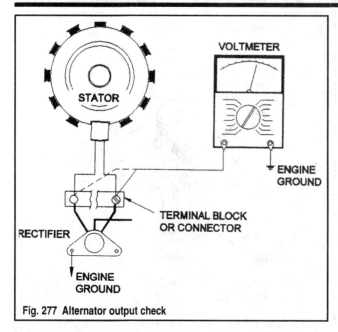

Fig. 277 Alternator output check

disconnect fitting, or at the terminal board, depending on the model engine being serviced. The meter should indicate 12 volts when the meter contact is made to either yellow lead. If the meter fails to indicate l2 volts, the rectifier or the stator may be defective. Continue with the testing.

Rectifier Checks

◆ See Figures 278 and 279

Four checks must be performed on the rectifier to determine if it is acceptable for further service. Two tests are performed for the negative diodes and 2 for the positive diodes.

1. Disconnect both the positive and negative cables from the battery terminals.

2. Obtain an ohmmeter and set the selector to the high ohm scale. Disconnect the rectifier leads at the terminal board (if a terminal board is not used, disconnect the rectifier leads at the connector).

3. Connect 1 ohmmeter lead to a good ground on the rectifier and the other lead to 1 of the yellow leads. Observe the meter reading.

4. Reverse the connections and again observe the meter. The meter should indicate continuity in one direction but not in the other. Therefore, if a reading is obtained in both directions, or if a reading is not obtained in one direction, the diode in the rectifier is defective and the rectifier must be replaced.

5. Repeat the test for the other yellow wire. If the test fails, the other diode is defective and the rectifier must be replaced.

6. Connect 1 ohmmeter lead to 1 of the yellow leads from the rectifier and the other lead to the red or purple wire. Observe the meter reading.

7. Reverse the connections and again observe the meter. The meter should indicate continuity in only 1 direction. If the test fails, the diode in the rectifier is defective and the rectifier must be replaced.

8. Leave 1 meter lead connected to the red wire and connect the other meter lead to the other yellow wire. Observe the meter reading. If continuity is observed in both directions or not in either direction, the diode in the rectifier is defective and the rectifier must be replaced.

Stator Check

◆ See Figures 280 and 281

1. Check to be sure both cables are disconnected from the battery.

2. Separate the quick-disconnect from the stator leads.

3. Obtain an ohmmeter and set the selector to the low ohm scale. Connect the meter leads to the 2 yellow leads going to the stator. The meter should indicate as follows: 9 amp alternator: 0.75 + 0.2 ohms or 15 amp alternator: 0.4 + 0.1 ohm.

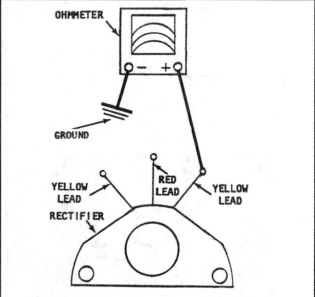

Fig. 278 When checking the rectifier, connect the meter to ground and the yellow lead. . .

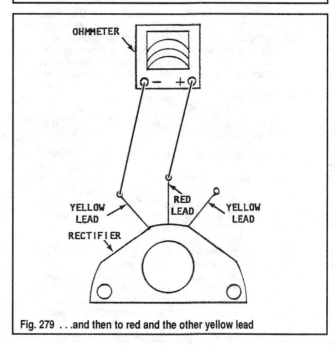

Fig. 279 . . .and then to red and the other yellow lead

■ The 15 amp alternator installation has a voltage regulator; the 9 amp unit does not. Therefore, the alternator output rating can quickly be determined.

4. If the reading is not as indicated, the stator is defective and must be replaced.

5. Set the ohmmeter selector to the high ohm scale. Connect 1 meter lead to a good ground on the engine and the other lead to 1 of the yellow wires going to the stator. The meter should indicate no continuity. If continuity is indicated, the stator is defective and must be replaced.

Clipper Circuit Check

◆ See Figures 282 thru 286

The clipper circuit protects the CD ignition system against an intermittent battery current, or a sudden surge of current from the alternator. The clipper

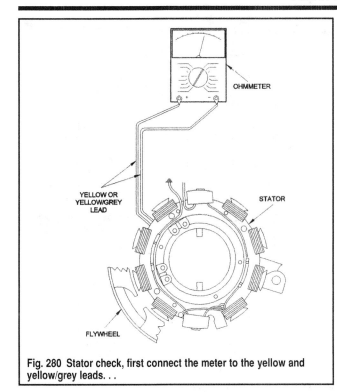

Fig. 280 Stator check, first connect the meter to the yellow and yellow/grey leads. . .

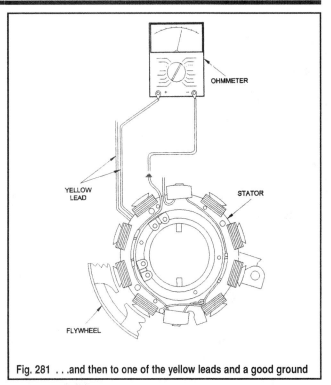

Fig. 281 . . .and then to one of the yellow leads and a good ground

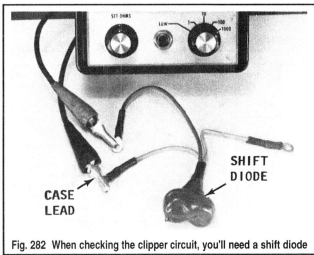

Fig. 282 When checking the clipper circuit, you'll need a shift diode

circuit must be checked if the battery suddenly went dead, the CD amplifier became inoperative, or if the battery current has fluctuated or been interrupted.

1. To check the clipper circuit properly, a shifting diode (#383840) and an ohmmeter are required. The diode is used to determine which meter lead is the ohmmeter case lead.

2. Connect the purple and green lead from the diode to 1 of the ohmmeter leads. Connect the yellow lead to the other meter lead. If the meter indicates zero or has a very low reading, the test lead connected to the yellow diode lead is the case lead. If the meter has a high reading, the meter lead connected to the purple and green diode leads is the case lead. Identify the case lead with a tag or piece of tape.

3. Disconnect the clipper circuit from the engine. The tests will be made on the work bench. Early model clipper circuits had 3 wires - 2 yellow and 1 purple wire. Later circuits had a 4th wire, black, for a ground wire. The following tests are identical except for grounding.

4. Momentarily ground the purple wire to the case to discharge any current in the clipper circuit.

5. Connect the case test lead to the ground wire of the clipper circuit or directly to the case. Connect the other test lead to 1 of the yellow wires.

The meter should indicate at least 300 ohms. If the reading is lower than 300 ohms, the clipper circuit must be replaced.

6. Momentarily ground the purple wire to the case to discharge any current in the clipper circuit. Connect the meter case wire to a ground wire of the clipper circuit. Connect the other meter lead to the yellow wire with the gray stripe. If the meter indicates less than 300 ohms, the clipper circuit must be replaced.

7. Momentarily ground the purple wire to the case to discharge any current in the clipper circuit. Connect the meter case lead to the ground wire of the clipper circuit and the other meter lead to the purple wire. If the meter indicates more than 300 ohms, proceed to the next test. If the reading is less than 300 ohms, or if the needle moves toward zero and then returns to infinite ohms, the clipper circuit is defective and must be replaced.

8. Momentarily ground the purple wire to the case to discharge any current in the clipper circuit. Connect the meter case wire to the purple wire and the other lead to the ground wire of the clipper circuit. If the needle moves towards zero and returns to infinite ohms, the clipper circuit is satisfactory. If the needle fails to move or moves to zero and remains at zero, the clipper circuit is defective and must be replaced.

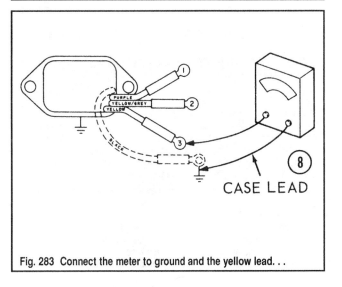

Fig. 283 Connect the meter to ground and the yellow lead. . .

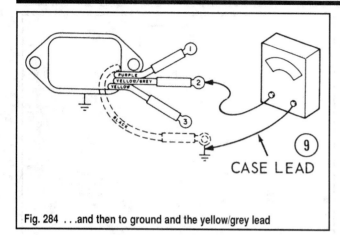

Fig. 284 . . .and then to ground and the yellow/grey lead

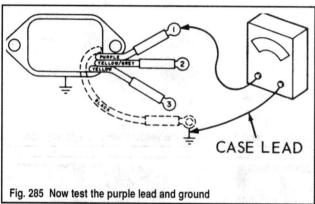

Fig. 285 Now test the purple lead and ground

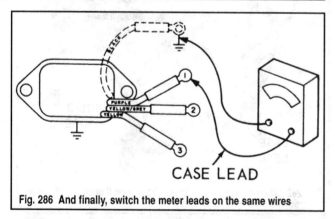

Fig. 286 And finally, switch the meter leads on the same wires

Stator

REPLACEMENT

◆ See Figures 226 and 227

Because the stator is located under the flywheel, the proper puller must be obtained; the flywheel removed; the 2 and sometimes 3 yellow wires disconnected at the junction box; and then the stator assembly removed.

The new stator is installed and secured with the attaching hardware; the wires connected to the junction box; and the flywheel installed and secured.

Use Loctite on threads of the stator attaching screws to prevent them from vibrating loose and being struck by the flywheel.

Diode

REPLACEMENT

◆ See Figure 287

The diode assembly is a sealed unit secured to the terminal block with 2 attaching screws. Take note of the wire color coding and the terminals to which each is connected as an aid during installation of the new diode.

1. Disconnect the 2 yellow and 1 red wire from the diode.
2. Remove the 2 attaching screws, and then the diode assembly.
3. Secure the new diode assembly in place with the 2 attaching screws, and then connect the 2 yellow and 1 red wire to the same assembly terminals from which they were disconnected.

Fig. 287 Removing the diode assembly

BATTERY

Marine Batteries

◆ See Figure 288

The battery is one of the most important parts of the electrical system. In addition to providing electrical power to start the engine, it also provides power for operation of the running lights, radio, and electrical accessories.

Because of its job and the consequences (failure to perform in an emergency), the best advice is to purchase a well-known brand, with an extended warranty period, from a reputable dealer.

The usual warranty covers a pro-rated replacement policy, which means the purchaser would be entitled to a consideration for the time left on the warranty period if the battery should prove defective before its time.

Do not consider a battery of less Cold Cranking Amperage (CCA) or Amp Hour (AH) rating than the battery that was originally installed for your motor. In fact, due to the increased resistance that will occur in circuits over time (from things like corrosion or internal wire strands that wear and break inside

the insulation), it is advisable to buy a replacement battery with higher capacity than the original (but do not go overboard, pun intended).

Because marine batteries are required to perform under much more rigorous conditions than automotive batteries, they are constructed differently than those used in automobiles or trucks. Therefore, a marine battery should always be the No. 1 unit for the boat and other types of batteries used only in an emergency (or possibly as a second battery).

Marine batteries have a much heavier exterior case to withstand the violent pounding and shocks imposed on it as the boat moves through rough water and in extremely tight turns. The plates are thicker and each plate is securely anchored within the battery case to ensure extended life. The caps are spill proof to prevent acid from spilling into the bilge when the boat heels to one side in a tight turn, or is moving through rough water. Because of these features, the marine battery will recover from a low charge condition and give satisfactory service over a much longer period of time than any type intended for automotive use.

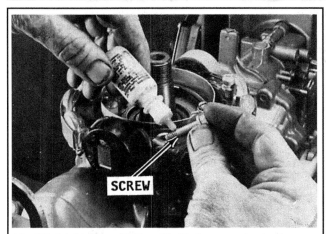

Fig. 228 A fully charged battery, filled to the proper level with electrolyte, is the heart of the ignition and electrical systems. Engine cranking and efficient performance of electrical items depend on a full rated battery

✳✳ WARNING

Never use a maintenance-free battery with an outboard engine that is not voltage regulated. The charging system will continue to charge as long as the engine is running and it is possible that the electrolyte could boil out rendering the battery useless.

Battery Construction

◆ See Figure 289

A battery consists of a number of positive and negative plates immersed in a solution of diluted sulfuric acid. The plates contain dissimilar active materials and are kept apart by separators. The plates are grouped into elements. Plate straps on top of each element connect all of the positive plates and all of the negative plates into groups.

The battery is divided into cells holding a number of the elements apart from the others. The entire arrangement is contained within a hard plastic case. The top is a 1-piece cover and contains the filler caps for each cell. The terminal posts protrude through the top where the battery connections for the boat are made. Each of the cells is connected to its neighbor in a positive-to-negative manner with a heavy strap called the cell connector.

Battery Ratings

◆ See Figure 290

Three different methods are used to measure and indicate battery electrical capacity:
- Amp/hour (AH) rating
- Cold Cranking Amp (CCA) performance
- Reserve capacity

The AH rating of a battery refers to the battery's ability to provide a set amount of amps for a given amount of time under test conditions at a constant temperature. Therefore, if the battery is capable of supplying 4 amps of current for 20 consecutive hours, the battery is rated as an 80 amp/hour battery. The amp/hour rating is useful for some service operations, such as slow charging or battery testing.

CCA performance is measured by cooling a fully charged battery to 0°F (-17°C) and then testing it for 30 seconds to determine the maximum current flow. In this manner the cold cranking amp rating is the number of amps available to be drawn from the battery before the voltage drops below 7.2 volts.

The illustration depicts the amount of power in watts available from a battery at different temperatures and the amount of power in watts required of the engine at the same temperature. It becomes quite obvious - the colder the climate, the more necessary for the battery to be fully charged.

Reserve capacity of a battery is considered the length of time, in minutes, at 80°F (27°C), a 25 amp current can be maintained before the voltage drops below 10.5 volts. This test is intended to provide an approximation of how

long the engine, including electrical accessories, could operate satisfactorily if the stator assembly or lighting coil did not produce sufficient current. A typical rating is 100 minutes.

If possible, the new battery should have a power rating equal to or higher than the unit it is replacing.

Battery Location

Every battery installed in a boat must be secured in a well protected, ventilated area. If the battery area lacks adequate ventilation, hydrogen gas, which is given off during charging may gather in a concentrated quantity, and could cause a hazardous condition as it is very explosive.

Battery Service

Details regarding cleaning the battery, checking fluid level and testing it and maintaining a proper charge while the battery is in storage can be found under Batteries in the Boat Maintenance section.

Jumper Cables

■ If booster batteries are used for starting an engine the jumper cables must be connected correctly and in the proper sequence to prevent damage to either battery, or to the alternator diodes. And, perhaps most importantly, to prevent the possibility of sparks which could ignite the explosive gases given off by a discharged battery.

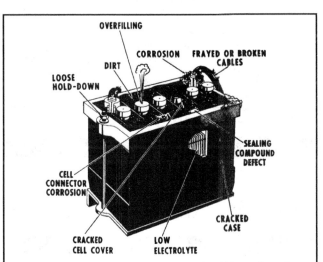

Fig. 289 A visual inspection of the battery should be made each time the boat is used. Such a quick check may reveal a potential problem in its early stages. A dead battery in a busy waterway or far from assistance could have serious consequences

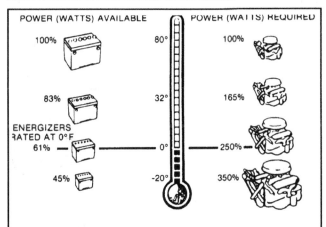

Fig. 290 Comparison of battery efficiency and engine demands at various temperatures

Always connect a cable from the positive terminal of the dead battery to the positive terminal of the good battery *first*. Next, connect one end of the other cable to the negative terminal of the good battery and the other end to the engine for a good ground. By making the ground connection on the engine, if there is an arc when you make the connection it will not be near the battery. An arc near the battery could cause an explosion, destroying the battery and causing serious personal injury.

Disconnect the battery ground cable before replacing an alternator or before connecting any type of meter to the alternator.

If it is necessary to use a fast-charger on a dead battery, *Always* disconnect one of the boat cables from the battery first, to prevent burning out the diodes in the rectifier.

NEVER use a fast-charger as a booster to start the engine because the voltage regulator may be damaged.

Storage

If the boat is to be laid up for the winter or for more than a few weeks, special attention must be given to the battery to prevent complete discharge or possible damage to the terminals and wiring.

Before putting the boat in storage, disconnect and remove the batteries. Clean them thoroughly of any dirt or corrosion, and then charge them to full specific gravity reading. After they are fully charged, store them in a clean cool dry place where they will not be damaged or knocked over.

✳✳ WARNING

NEVER store the battery with anything on top of it or cover the battery in such a manner as to prevent air from circulating around the filler-caps.

All batteries, both new and old, will discharge during periods of storage, more so if they are hot than if they remain cool. Therefore, the electrolyte level and the specific gravity should be checked at regular intervals. A drop in the specific gravity reading is cause to charge them back to a full reading.

In cold climates, care should be exercised in selecting the battery storage area. A fully-charged battery will freeze at about 60° below zero. A discharged battery, almost dead, will have ice forming at about 19° above zero.

Dual Battery Installation

Three methods are available for utilizing a dual-battery hook-up.

1. A high-capacity switch can be used to connect the 2 batteries. The accompanying illustration details the connections for installation of such a switch. This type of switch installation has the advantage of being simple, inexpensive, and easy to mount and hookup. However, if the switch is accidentally left in the closed position, it will cause the convenience loads to run down both batteries and the advantage of the dual installation is lost. The switch may be closed intentionally to take advantage of the extra capacity of the 2 batteries, or it may be temporarily closed to help start the engine under adverse conditions.

2. A relay can be connected into the ignition circuit to enable both batteries to be automatically put in parallel for charging or to isolate them for ignition use during engine cranking and start. By connecting the relay coil to the ignition terminal of the ignition-starting switch, the relay will close during the start to aid the starting battery. If the second battery is allowed to run down, this arrangement can be a disadvantage since it will draw a load from the starting battery while cranking the engine. One way to avoid such a condition is to connect the relay coil to the ignition switch accessory terminal. When connected in this manner, while the engine is being cranked, the relay is open. But when the engine is running with the ignition switch in the normal position, the relay is closed, and the second battery is being charged at the same time as the starting battery.

3. A heavy duty switch installed as close to the batteries as possible can be connected between them. If such an arrangement is used, it must meet the standards of the American Boat and Yacht Council, Inc. or the Fire Protection Standard for Motor Craft, N.F.P.A. No. 302.

Battery Selector Switch

Two differently designed selector switches are available. One design makes contact on the new side before breaking contact on the old; the other design breaks contact with the old before making contact with the new side.

✳✳ CAUTION

DO NOT use the type which breaks contact with the old side before making contact with the new. This type switch will result in failure of the rectifier/regulator.

NEVER switch to the OFF position with the powerhead operating. Such action will result in failure of the rectifier/regulator.

When purchasing a battery selector switch, INSIST on the type which makes contact on the new side *before* breaking with the old side. Make the clerk prove it.

STARTING SYSTEM

Description & Operation

◆ See Figures 291 thru 297

In the old days (when everyone walked barefoot to school, uphill, both ways), all outboards were started by pulling on a rope wrapped around the flywheel. As time passed and owners were reluctant to use muscle power (or came up short especially as larger and larger outboards were built), it was necessary to replace the rope starter with some form of power cranking system. Today, only the smaller outboards are normally equipped with a starting rope, but many smaller and almost all large engines, may also be equipped with an electric starter system.

The system used to replace the rope starter is an electric starter motor coupled with a mechanical gear mesh between the starter motor and the powerhead flywheel, similar to the method used to crank an automobile engine.

As the name implies, the sole purpose of the starter circuit is to control operation of the starter motor causing it to crank the powerhead until the engine catches and runs. The circuit usually includes a solenoid or magnetic switch that connects the motor to the battery when the circuit is actuated and disconnects the motor from the battery when the circuit is deactivated. The operator controls the solenoid switch with a key switch or starter button, depending on the model.

A cut-out switch is installed in the system to prevent starting the engine if the throttle is advanced too far, beyond idle speed. When the throttle is advanced, the starter solenoid is not grounded and the starter motor will not rotate. The cut-out switch is usually installed inside the shift box.

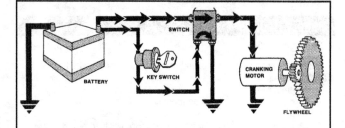

Fig. 291 **A typical starting system converts electrical energy into mechanical energy to turn the engine. The basic components are: Battery, to provide electricity to operate the starter; Ignition switch, to control the energizing of the starter relay or solenoid; Starter relay or solenoid switch, to make and break the circuit between the battery and starter; Starter, to convert electrical energy into mechanical energy to rotate the engine; Starter drive gear, to transmit the starter rotation to the engine flywheel**

Delco-Remy, Autolite and Prestolite starter motors are used on the engines covered here. Any one of the 3 may be installed on the engine. The early model starters (especially the Delco-Remy) were shorter and just a bit less powerful than the later models. If a replacement unit must be purchased, anyone of the 3 may be obtained and installed. The recommendation is to spend a few dollars more for the longer, more powerful unit. One more word - if a long starter motor is replacing the short model, an

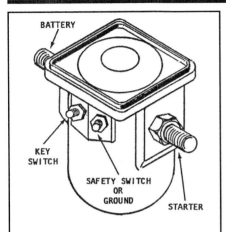

Fig. 292 Typical starter solenoid (note the terminal for the ground wire, this particular unit is NOT grounded through the mounting bracket)

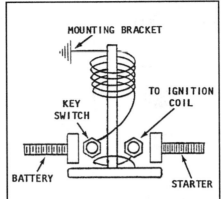

Fig. 293 Functional diagram of a "slave-type" starter solenoid used on 4-cycle engines (this CANNOT be used on 2-cycle engines)

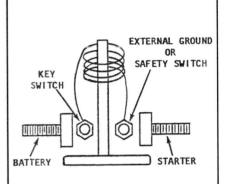

Fig. 294 Functional diagram of a typical 2-cycle outboard starter motor solenoid. Notice the right hand small terminal is connected to ground or the safely switch, NOT through the mounting bracket

Fig. 295 Typical mounting of a standard starter motor on a 2 cyl engine

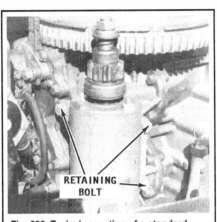

Fig. 296 Typical mounting of a standard starter motor on a 3 cyl engine

Fig. 297 Typical mounting of a standard starter motor on a V4 engine

additional bracket MUST be bought and installed on the lower end of the starter.

Marine starter motors are very similar in construction and operation to the units used in the automotive industry. Some marine starter motors use the inertia-type drive assembly. This type assembly is mounted on an armature shaft with external spiral splines which mate with the internal splines of the drive assembly. Until about 1967, marine starter motors had a splined gear that meshed with an intermediate gear which in turn meshed with the Bendix mechanism.

The starter motor is a series-wound electric motor which draws a heavy current from the battery. It is designed to be used only for short periods of time to crank the engine for starting. To prevent overheating the motor, cranking should not be continued for more than 30 seconds without allowing the motor to cool for at least 3 minutes. Actually, this time can be spent in making preliminary checks to determine why the engine fails to start.

Most starter motors operate in much the same manner and the service work involved in restoring a defective unit to service is almost identical. Therefore, the information in this section is grouped together for the major components of the starter under separate headings. Differences, where they occur, between the various manufacturers, are clearly indicated.

Theory of Operation

With one type of Bendix drive, power is transmitted from the starter motor to the engine flywheel directly through the Bendix drive. This drive has a pinion gear mounted on screw threads. When the motor is operated, the pinion gear moves up to mesh with the teeth on the flywheel ring gear.

On another type of drive gear, the pinion gear and shaft are mounted above a powerhead bracket and the starter motor is mounted below. The starter motor has a small gear installed on the armature shaft which meshes with a much larger gear on the lower end of the drive gear assembly. This difference in gear sizes gives the arrangement a much greater mechanical

advantage (power) than the other model. The pinion gear of the drive gear assembly then moves upward and the drive gear meshes with the teeth of the flywheel ring gear.

When the engine starts, the pinion gear is driven faster than the shaft, and as a result, it screws out of mesh with the flywheel. A rubber cushion is built into the Bendix drive to absorb the shock when the pinion meshes with the flywheel ring gear. The parts of the drive MUST be properly assembled for efficient operation. If the drive is removed for cleaning, take care to assemble the parts as shown in the accompanying illustration. If the screw shaft assembly is reversed, it will strike the splines and the rubber cushion will not absorb the shock.

The sound of the motor during cranking is a good indication of whether the starter motor is operating properly or not. Naturally, temperature conditions will affect the speed at which the starter motor is able to crank the engine. The speed of cranking a cold engine will be much slower than when cranking a warm engine. An experienced operator will learn to recognize the favorable sounds of the cranking engine under various conditions.

Troubleshooting

FAULT SYMPTOMS

If the starter spins, but fails to crank the engine, the cause is often a corroded or gummy Bendix drive. The drive should be removed, cleaned, and given an inspection.

If the starter motor cranks the engine too slowly, the following are a list of possible causes and the corrective actions that may be taken:

- Battery charge is low. Charge the battery to full capacity.
- High resistance connections at the battery, solenoid, or motor. Clean and tighten all connections.

- Undersize battery cables. Replace cables with sufficient size.
- Battery cables too long. Relocate the battery to shorten the run to the starter solenoid.

MAINTENANCE

◆ **See Figures 298**

Two more areas may cause the engine to turn over slowly even though the starter motor circuit is in excellent condition a tight or "frozen" engine and/or, water in the lower unit causing the bearings to tighten up.

The starter motor does not require periodic maintenance or lubrication *except* just a drop of light-weight oil on the starter shaft to ease movement of the Bendix drive. If the motor fails to perform properly, the checks outlined in the previous paragraph should be performed.

The frequency of starts governs how often the motor should be removed and reconditioned. The manufacturer recommends removal and reconditioning every 1000 hours.

Naturally, the motor will have to be removed if the corrective actions outlined, does not restore the motor to satisfactory operation.

TESTING

◆ **See Figure 299**

■ The starter solenoid is actually nothing more than a switch between the battery and the starter motor. Several types of solenoids are used and many appear similar. NEVER attempt to use an automotive-type solenoid in a marine installation. Such practice will lead to more problems than can be imagined. An automotive-type solenoid has a completely different internal wiring circuit. If such a solenoid is connected into the starter system, and the system is activated, current will be directed to ground. The wires will be burned and the cutout switch will be burned and rendered useless. Therefore, when installing replacement parts in the starter or other circuits on a marine installation, always take time to obtain parts from a marine outlet to ensure proper service and to prevent damage to other expensive components.

✳✳ WARNING

Before making any test of the cranking system, disconnect the spark plug leads at the spark plugs to prevent the engine from possibly starting during the test and causing personal injury.

The tests in this section are to be performed according to the faulty condition described. The numbers referenced in the steps are correlated with numbers on the accompanying circuit diagram, to identify exactly where the connection or test is to be made.

Starter Motor Turns Slowly

◆ **See Figure 299**

As noted earlier, this is one starting system fault which has a number of common causes. Check and remedy each, as necessary:
- Battery charge is low. Charge the battery to full capacity.
- High resistance connections at the battery, solenoid, or motor. Clean and tighten all connections.
- Undersize battery cables. Replace cables with sufficient size.
- Battery cables too long. Relocate the battery to shorten the run to the starter solenoid.
- Defective starter motor. Perform an amp draw test. Connect an amp draw gauge on the cable leading to the starter motor terminal No. 5 (in the accompanying illustration). Turn the key to the **START** position and attempt to crank the engine. If the gauge indicates an excessive amperage draw the starter motor MUST be replaced or rebuilt.

Fig. 298 Lube the pinion shaft around the spring

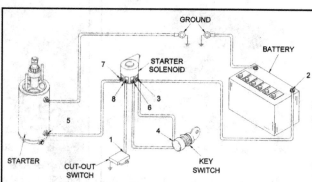

Fig. 299 Use the numbers in this diagram to help identify starter circuit testing locations. Certain 2 cyl engines may have the No. 6 and No.8 terminals reversed on the solenoid - in either situation, the No. 6 wire is from the key switch and the No. 8 wire is from the cut-out switch

Starter Motor Fails To Rotate - Voltage Check

◆ **See Figure 299**

If the starter motor fails to rotate completely, check for voltage in the circuit as follows:

1. Check the voltage at No. 2 (in the accompanying illustration), the battery and ground.

2. If satisfactory voltage is indicated at the battery, check the voltage at No. 3 (in the accompanying illustration), the positive side of the starter solenoid. Weak or no voltage at this point indicates corroded battery terminals, poor connection at the solenoid, or defective wiring between the battery and the solenoid.

3. Test the voltage at No. 4 (in the accompanying illustration), the key. A full 12-volt reading should be registered at the key. Weak or no voltage at the key indicates a poor connection at the solenoid, or a broken wire between the starter solenoid and the key.

4. If satisfactory voltage is indicated during the proceeding steps, connect a voltmeter at No. 5 (in the accompanying illustration) and ground, and then turn the key switch to the **START** position. If 12-volts are registered at No. 5 and the starter still fails to operate, the starter is defective and requires service. If voltage is NOT present at No. 5, proceed to the next section, Testing Starter Solenoid.

Testing Starter Solenoid

◆ See Figure 299

1. Remove the heavy starter cable at No. 5 (in the accompanying illustration), at the starter. This cable MUST be disconnected prior to performing this test to prevent the starter motor from turning and cranking the engine. Connect a voltmeter to No. 6 (the starter solenoid in the accompanying illustration), and ground. Turn the key to the **START** position. The meter should indicate 12-volts. If voltage is not present at No. 6, the key switch is defective, or the wire is broken between the key switch and the starter solenoid.

2. If voltage is present at No. 6, connect a voltmeter at No. 3 and to No. 7 (in the accompanying illustration). Connect one end of a jumper wire to No. 2, the positive terminal of the battery and MOMENTARILY make contact with the other end at No. 6, the starter solenoid. If voltage is indicated through the starter solenoid, the solenoid is satisfactory and the problem has been corrected while making the tests. Sometimes, when working with electrical circuits, corrective action has been taken almost accidentally, a bad connection has been made good, etc. If the solenoid test failed, it does not necessarily mean the solenoid is defective. The solenoid may not be properly grounded through the cutout switch. Therefore, the cutout switch may be defective and should be checked as outlined later in this section.

3. With the voltmeter still connected at No. 3 and No. 7, connect one end of a jumper wire at No. 8 (in the accompanying illustration), the starter solenoid, and the other lead to a good ground. Connect a second jumper wire at No. 2, the positive terminal of the battery, to No. 6, the starter solenoid. The voltmeter should indicate voltage is present. If voltage is not present, the starter solenoid is defective and must be replaced.

Testing Throttle Advance Cut-Out Switch

◆ See Figure 299

1. Remove the existing wire from the No. 1 switch terminal (in the accompanying illustration).
2. Connect 1 probe lead of an ohmmeter to the terminal.
3. Connect the other test probe lead to a good ground.
4. Depress the switch button and the ohmmeter should indicate continuity. If continuity is not indicated, the switch is defective and must be replaced. Connect the heavy cable at No. 5, the starter motor.

Autolite Starters

REMOVAL & INSTALLATION

◆ See Figures 300, 301 and 302

1. Before beginning any work on the starter motor, disconnect the positive (+) lead from the battery terminal. Remove the hood.

2. Disconnect the red cable at the starter motor terminal.
3. Remove the 2 thru-bolts securing the starter motor. These bolts pass through the starter motor and thread into the drive gear housing. On some engines, particularly smaller ones, a starter motor support bracket is installed under the starter motor securing the starter motor to the engine. If the bracket is installed, remove the 7/16 in. bolts (or nuts) securing the bracket to the powerhead. Remove the starter motor and bracket together.
4. To remove the drive gear assembly, remove the 7/16 in. screw on the base of the drive gear assembly, and lift assembly free of the bracket on the powerhead.

■ If the only motor repair necessary is replacement of the brushes, the drive gear does not have to be removed. All starter motors have thru-bolts securing the upper and lower cap to the field frame assembly. In all cases, both caps have some type of mark or boss. These marks are used to properly align the caps with the field frame assembly.

To Install:
On models with a bracket:
5. Mount the starter motor and bracket onto the engine. Align the top bolt above the carburetor, and then thread it into the block about half-way. Align the other 3 bolts on the starboard side or start the nuts onto the studs, depending on the model engine being serviced. Tighten the 3 on the side evenly and alternately until they are secure.
6. If the starter motor has the mounting flanges permanently attached, then position the motor in place and start the 3 bolts to attach the motor to the engine. Now, tighten the 3 bolts alternately and evenly until all bolts are tight. The bolts MUST be tightened alternately to prevent binding and possibly bending the flanges.
On models without a bracket:
7. On the starboard side of the engine, set the gear shaft into the bronze bushing and start the 7/16 in. bolt through the housing into the engine bracket. DO NOT tighten the bolt at this time.
8. If a bottom bracket is used, the thru-bolts must pass through the bracket and then through the starter motor.

■ Installation of the starter motor to an outboard engine is not the easiest of tasks. It may take a little time and patience to work the motor up into the drive gear bracket until it is seated properly. If difficulty is encountered, cease the work, have a cup of tea or coffee, and then try it again.

9. Use a rubber band around each thru-bolt after the bolt has been installed through the starter to hold them in place as the motor is lifted into place. Once the thru-bolts are started into the drive gear housing, break and remove the rubber bands from the bolts. Tighten the bolts securely.
On all models now:
10. Connect the positive red lead to the starter motor.
11. Connect the electrical lead to the battery. Test the completed work by cranking the engine with the starter motor.

■ DO NOT, under any circumstances, start the engine unless it is mounted in an adequate size test tank or body of water.

Fig. 300 Removing the starter on models with a mounting bracket

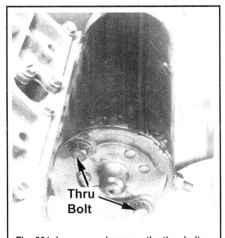
Fig. 301 Loosen and remove the thru-bolts on models without a bracket

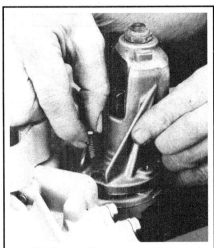
Fig. 302 Remove the drive gear if necessary

※※ **CAUTION**

Water must circulate through the lower unit to the engine any time the engine is run to prevent damage to the water pump in the lower unit. Just 5 seconds without water will damage the water pump.

DISASSEMBLY & ASSEMBLY

◆ **See Figures 303 thru 307**

1. Observe the caps and find the identifying mark or boss on each. If the marks are not visible, make an identifying mark prior to removing the thru-bolts as an essential aid during assembling.
2. Remove the thru-bolts from the starter motor.
3. Use a small hammer and carefully tap the lower cap free of the starter motor.

■ **The brushes are mounted in the end cap.**

4. Pull on the armature shaft from the drive gear end and remove it from the field frame assembly.
5. Remove the positive brush from its holder.

To assemble:

6. Clamp the armature in a vise equipped with soft jaws, with the drive end down.

7. Slide the thrust washers onto the armature shaft.

8. Lower the field assembly down over the armature. The spring action against the brush is built into the retainer; therefore, a separate brush spring is not required.

■ **With one brush attached to the end cap and the other to the frame assembly, positioning the brushes properly is not the easiest task, but it can be done with patience.**

9. Insert the negative brush into the retainer and push it in until the back part of the spring rests on the side of the brush. This force secures the brush in the retainer in a retracted position.

10. Lower the end cap over the frame assembly.

11. Install the positive brush into the retainer and push the brush backward until the spring part of the retainer is on the side of the brush.

12. Work the cap over the end of the armature shaft and the brushes over the commutator. Just before the cap is completely into place against the frame, use a punch and push on the back side of each brush and the brush will snap in to ride against the commutator. The spring will then be behind the brush.

13. Align the end cap notch or mark with the mark on the frame and the upper cap mark with its matching mark. Now, install the thru-bolts through the end cap and hold it all together with a nut on each bolt. The nuts will later be removed when the starter motor is installed onto the powerhead.

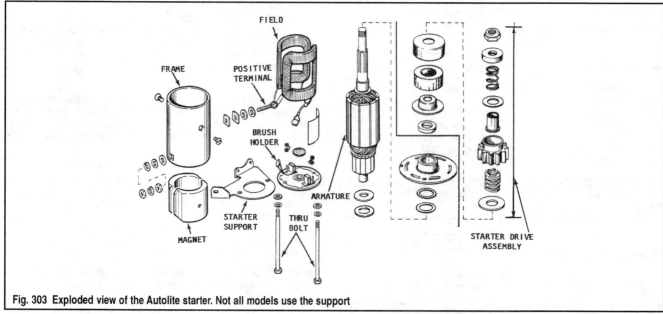

Fig. 303 Exploded view of the Autolite starter. Not all models use the support

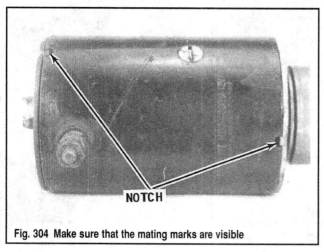

Fig. 304 Make sure that the mating marks are visible

Fig. 305 Tap out the lower cap while holding in the brush

Fig. 306 Pull out the shaft

BRUSH

BRUSH
SPRING

Fig. 307 A good look at the brush and spring

CLEANING & INSPECTION

◆ **See Figure 303**

Clean the field coils, armature, commutator, armature shaft, brush-end plate and drive-end housing with a brush or compressed air. Wash all other parts in solvent and blow them dry with compressed air.

Inspect the insulation and the unsoldered connections of the armature windings for breaks or burns.

Perform electrical tests on any suspected defective part, according to the procedures outlined earlier in this section.

Check the commutator for run-out. Inspect the armature shaft and both bearings for scoring.

Turn the commutator in a lathe if it is out-of-round by more than 0.005 in. (1.27mm).

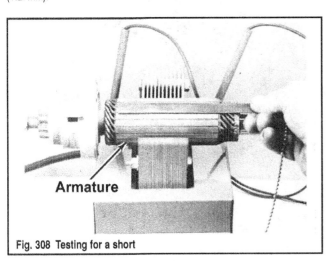

Fig. 308 Testing for a short

Check the springs in the brush holder to be sure none are broken. Check the spring tension and replace if the tension is not 32-40 ounces.

Check the insulated brush holders for shorts to ground. If the brushes are worn down to 1/4 in. (6.35mm) or less, they must be replaced.

Check the field brush connections and lead insulation. A brush kit and a contact kit are available at your local marine dealer, but all other assemblies must be replaced rather than repaired.

The armature, fields, and brush holders, must be checked before assembling the starter motor.

■ **See the Testing section for detailed procedures on testing the starter motor.**

TESTING

Testing Armature for a Short

◆ **See Figure 308**

1. Position the armature on a growler, then hold a hacksaw blade over the armature core.
2. Turn the growler switch to the **ON** position.
3. Slowly rotate the armature. If the hacksaw blade vibrates, the armature or commutator has a short. Clean the grooves between the commutator bars on the armature.
4. Perform the test again. If the hacksaw blade still vibrates during the test; the armature has a short and must be replaced.

Testing for a Ground

◆ **See Figure 309**

1. Obtain a test lamp or continuity meter.
2. Make contact with 1 probe lead on the armature core and the other probe lead on the commutator bar. If the lamp lights or the meter indicates continuity, the armature is grounded and must be replaced.

Checking the Commutator Bar

◆ **See Figure 310**

Check between or check bar-to-bar as shown in the accompanying illustration.

The test light should light, or the meter should indicate continuity. If the commutator fails the test, the armature must be replaced.

Turning the Commutator

◆ **See Figure 311**

1. True the commutator, if necessary, in a lathe. NEVER undercut the mica because the brushes are harder than the insulation.

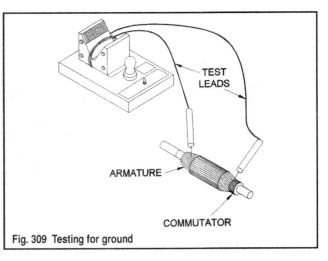

TEST
LEADS

ARMATURE

COMMUTATOR

Fig. 309 Testing for ground

2. Undercut the insulation between the commutator bars 1/32 in. (0.79mm) to the full width of the insulation and flat at the bottom. A triangular groove is not satisfactory.

3. After the undercutting work is completed, clean out the slots carefully to remove dirt and copper dust.

4. Sand the commutator lightly with No. 00 sandpaper to remove any burrs left from the undercutting.

5. Check the armature a second time on the growler for possible short circuits.

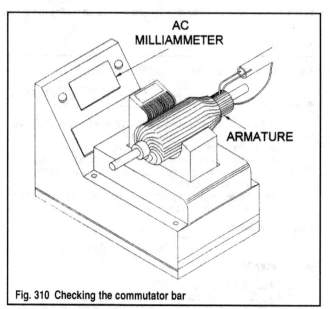

Fig. 310 Checking the commutator bar

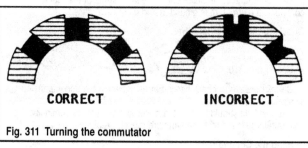

Fig. 311 Turning the commutator

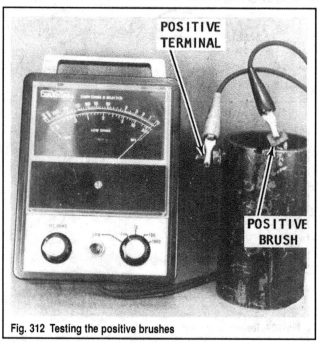

Fig. 312 Testing the positive brushes

Positive Brushes

◆ See Figure 312

The positive brush can always be identified as the brush with the lead connected to the field coil.

1. Obtain an ohmmeter.

2. Connect 1 lead of the meter to the end of the brush and the other lead to the terminal. The ohmmeter must indicate continuity between the brush and the terminal.

3. If the meter indicates any resistance, check the lead to the brush and the lead to the positive terminal solder connection. If the connection cannot be repaired, the brush must be replaced.

Negative Brush

◆ See Figure 313

The negative brush can always be identified because the lead is connected to the starter motor end cap.

1. Obtain an ohmmeter.

2. Make contact with 1 lead on the negative brush and make contact with the other lead on the starter end cap. If the meter does not indicate continuity, the brush or brush retainer is not grounded to the end cap.

Positive Brush Installation

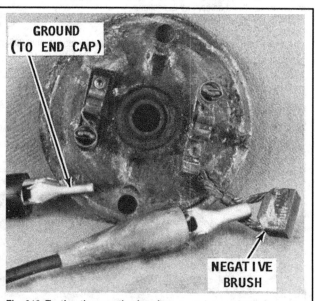

◆ See Figure 314

1. First, cut the old field coil brush free.

2. Next, attach the lead of the new brush to the stiff wire lead on the field coil.

3. Wrap a fine piece of copper wire around the lead and the stiff wire from the coil to hold the brush lead in place while it is soldered. If the wrapped wire becomes soldered also, no problem, leave it in place. If it did not become soldered, pull it free after the soldering is complete.

4. Use rosin flux and solder the leads to the back sides of the wire to prevent any excess solder from rubbing against the armature. Be sure the leads are in the right position to reach the brush holders. Do not overheat the leads, because the solder will run onto the lead and the lead will lose its flexibility.

5. Check to be sure none of the soldered connections are touching the frame. The fields would be grounded if the connections make contact with the frame.

Fig. 313 Testing the negative brushes

Negative Brush Installation

◆ See Figure 315

1. Cut the old brush free from the end cap.
2. Clean the surface thoroughly.
3. Next, solder the lead of the new brush to the end cap. If the retainer is no longer fit for service, the entire end cap must be replaced.

■ **New brushes are not included with the end cap.**

Delco-Remy Starters

REMOVAL & INSTALLATION

■ **Please refer to the Autolite Starter section for all removal and installation procedures.**

DISASSEMBLY & ASSEMBLY

◆ See Figures 304 and 316 thru 322

Two models of Delco-Remy starter motors have been used over the years, a short model and a longer, more powerful model. The short model has 1 positive and 1 negative brush. The longer model has 2 positive and 2

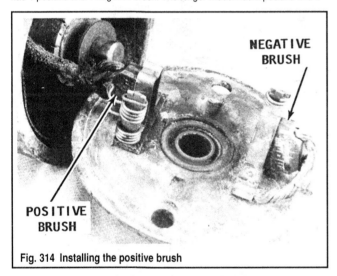

NEGATIVE BRUSH

POSITIVE BRUSH

Fig. 314 Installing the positive brush

BRUSH SPRING

BRUSH LEAD

Fig. 315 Installing the negative brush

negative brushes. The negative brushes can always be identified as the brushes with the lead connected to the frame. The other brush or brushes are connected to the field coil.

1. Observe the caps and find the identifying mark or boss on each. If the marks are not visible, make an identifying mark prior to removing the thru-bolts as an essential aid during assembling.
2. Remove the thru-bolts from the starter motor.
3. Use a small hammer and carefully tap the lower cap free of the starter motor.

■ **The brushes are mounted in the end cap.**

4. Pull on the armature shaft from the drive gear end and remove it from the field frame assembly.
5. Remove the brushes from there holder and then remove the springs.

To assemble:
Negative Brushes and Retainer
The following procedures apply to brushes mounted to the field frame assembly.

6. Remove the old set of ground brushes by cutting off the rivets with a chisel or by drilling them out. Replacement brush holder kits are available at marine outlets. These kits are complete with screws, washers, and nuts for attachment to the frame. Replacement brush springs are also available.

7. The brush spring is removed from the holder by compressing one side of the spring with a small screwdriver until the spring flips out of its seat. After the spring pops out, turn the spring clockwise until it is free. Replacement brush sets are available and usually contain the following parts:
   - One insulated brush, with flexible lead attached.
   - One ground brush holder with brush and lead attached.
   - Necessary attaching screws, washers, and nuts.

Positive Brushes
8. Cut off the old brush lead where it is attached to the field coil.
9. Prepare the ends of the coil for soldering the new brush lead assembly. Clean the ends of the coil by filing or grinding off the old brush lead connection. Remove the varnish only as far back as necessary to enable a good soldered connection to be made.

10. Use rosin flux and solder the leads to the back sides of the coil to prevent any excess solder from rubbing against the armature. Be sure the leads are in the right position to reach the brush holders. Do not overheat the leads, because the solder will run onto the lead and the lead will lose its flexibility.

Assembling the Motor
11. Clamp the drive gear in a vise equipped with soft jaws and with the drive gear down.
12. Insert the brush springs into the brush holders, and then install the brushes in place.
13. Take the time to whittle the end of 2 or 4 match sticks in the shape of a tiny spade. Now push 1 brush outward, and then wedge 1 of the match sticks in between the brush and the lip of the retainer. The match stick will hold the brush in the retracted position during installation of the armature. Retract each of the remaining brushes and hold them with a match stick.

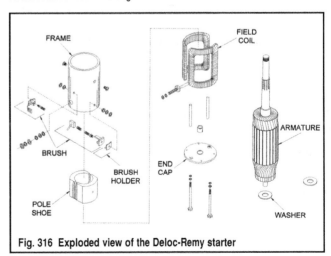

FRAME   **FIELD COIL**   **ARMATURE**   **BRUSH**   **END CAP**   **BRUSH HOLDER**   **POLE SHOE**   **WASHER**

Fig. 316 Exploded view of the Deloc-Remy starter

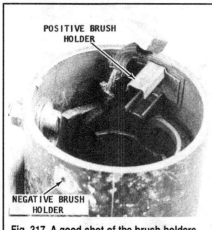

Fig. 317 A good shot of the brush holders

Fig. 318 A new negative brush and holder

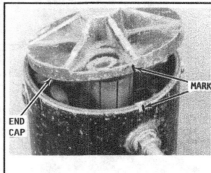

Fig. 319 Installing the negative brush

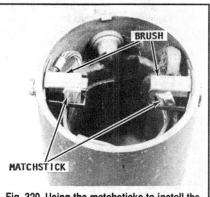

Fig. 320 Using the matchsticks to install the brushes in the retracted position

Fig. 321 Lower the frame into position. . .

Fig. 322 . . .and install end cap

14. After all the brushes have been retracted and held in place with the match sticks, lower the field frame assembly until sticks, and lower the field frame assembly until the brushes make contact with the commutator. Now, remove the match sticks. Align the mark on the upper cap with the matching mark on the field frame.

15. Place the washer onto the commutator shaft, and then place the cap onto the end of the field frame assembly.

16. Align the mark on the lower cap with the mark on the field frame. Install the thru-bolts through the end cap and hold it all together with a nut on the end of each bolt. These nuts will be removed later.

CLEANING & INSPECTION

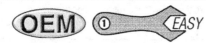

◆ See Figure 316

■ Please refer to the Autolite Starter section for all cleaning and inspection procedures.

TESTING

■ Please refer to the Autolite Starter section for all testing procedures except those detailed below.

Positive Brushes

◆ See Figures 323

The positive brush or brushes can always be identified as the brushes connected to the field coil.
1. Obtain an ohmmeter.
2. Connect 1 lead of the meter to the end of the brush and the other

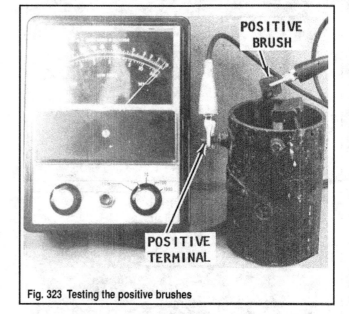

Fig. 323 Testing the positive brushes

lead to the terminal. The ohmmeter must indicate continuity between the brush and the terminal. If the meter indicates any resistance, check the lead to the brush and the lead to the positive terminal solder connection.

3. If the connection cannot be repaired, the brushes MUST be replaced.

4. If the unit being tested has a double set of positive brushes, repeat the test for the other positive brush.

5. Move the test lead from the brush to a good ground on the frame. If continuity is indicated, the field coil is grounded to the case.

Negative Brush

◆ See Figure 324

The negative brushes can always be identified because the lead is connected to the brush retainer as a ground to the frame.

1. Obtain an ohmmeter.
2. Make contact with 1 lead on the negative brush and make contact with the other lead on the starter frame. If the meter does not indicate continuity, the brush or brush retainer is not grounded to the frame.
3. If the unit being tested has a double set of negative and positive brushes, move the test lead from the 1 negative brush to the other negative brush lead and again check for continuity. If the meter does not indicate continuity, the brush or brush retainer is not grounded to the frame.
4. Check to be sure none of the soldered connections are touching the frame. The fields would be grounded if the connections make contact with the frame.

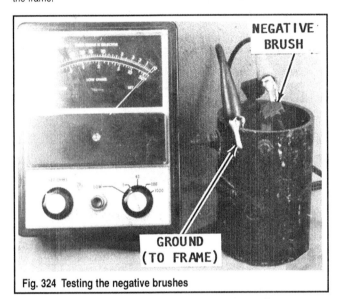

Fig. 324 Testing the negative brushes

Prestolite Starters

Two different methods of installation were used for the Prestolite starter motors used on the engines covered here. If the starter motor is installed on the starboard side of the engine, it is secured to a bracket attached to the powerhead. If the unit is installed on the port side of the engine, a bracket is welded to the starter and the bracket is then secured to the powerhead.

REMOVAL & INSTALLATION

Starboard Installation With Separate Drive Gear

■ Please refer to the Autolite Starter section for all removal and installation procedures.

Port Side Installation Drive Gear On Armature Shaft

◆ See Figure 325

1. Remove the 3 attaching bolts from the powerhead. One is partially hidden just behind the carburetor and the other 2 are visible. Lift the starter motor free of the powerhead.

■ If the only motor repair necessary is replacement of the brushes, the drive gear does not have to be removed. All starter motors have thru-bolts securing the upper and lower cap to the field frame assembly. In all cases both caps have some type of mark or boss. These marks are used to properly align the caps with the field frame assembly.

Fig. 325 Remove the 3 bolts on models with the starter on the port side

To Install:

2. Position the motor in place and start the 3 bolts to attach the motor to the engine. The hole for one bolt is hidden behind the carburetor. Tighten the bolts alternately and evenly until all bolts are tight. The bolts MUST be tightened alternately to prevent binding and possibly bending the flanges.
3. Connect the positive red lead to the starter motor.
4. Connect the electrical lead to the battery.
5. Test the completed work by cranking the engine with the starter motor. DO NOT, under any circumstances, start the engine unless it is mounted in an adequate size test tank or body of water.

✳✳ CAUTION

Water must circulate through the lower unit to the engine any time the engine is run to prevent damage to the water pump in the lower unit. Just 5 seconds without water will damage the water pump.

DISASSEMBLY & ASSEMBLY

◆ See Figures 326 thru 331

1. Look at the caps and find the identifying mark or boss on each. If the marks are not visible, make an identifying mark prior to removing the thru-bolts as an essential aid during assembling.
2. Remove the thru-bolts from the starter motor.
3. Use a small hammer and *carefully* tap the lower cap free of the starter motor.
4. Pull on the armature shaft from the drive gear end and remove it from the field frame assembly.
5. Remove the brushes from their holders and then remove the brush springs.
6. Lift the white plastic retainer free from the frame. Observe the location of the notch on the retainer in relation to the frame. The retainer must be installed in the same position.

To assemble:

7. Slide the plastic terminal and brush lead retainer into the groove in the frame with the small protrusion on one side facing downward. Continue pushing the retainer into the groove until it is fully seated.

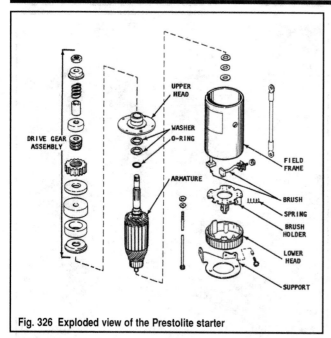

Fig. 326 Exploded view of the Prestolite starter

8. Work the brush retainer down on top of the frame with the positive lead through the cutaway in the retainer plate. Check to be sure the field coil negative brush passes through the cutaway in the plate.

9. Install the spring into the retainer.

10. Push the negative brush into its retainer and then wrap a fine piece of wire around the front side of the brush and the back side of the retainer. Tighten the wire snugly. This wire will hold the brush in the retainer. Repeat the procedure for the positive brush.

11. Check to be sure the plate is secured onto the frame and the cutaway is over the protrusion of the positive plastic terminal.

12. Clamp the armature in a vise equipped with soft jaws with the drive gear facing downward.

13. Install the thrust washers onto the end of the armature shaft.

14. Lower the frame assembly down over the armature until the brushes are over the commutator.

15. After the armature is in place, cut and remove the wire wrapped around the brushes to hold them in place. The brushes should then make firm contact with the commutator.

16. Install the end cap onto the end of the starter motor. Observe the 3 small nipples on the inside of the end cap, these nipples must index with matching dimples in the retaining plate. Align the mark on the side of the end cap with the terminal. Lower the cap onto the frame, and seat it gently. Never tap with a hammer or other tool, because the nipples may not be indexed with the dimples and the tapping may cause damage.

17. Align the end cap notch or mark with the mark on the frame, and the upper cap mark with its mark.

18. Install the thru-bolts.

CLEANING & INSPECTION

◆ See Figure 326

■ Please refer to the Autolite Starter section for all cleaning and inspection procedures.

TESTING

■ Please refer to the Autolite Starter section for all testing procedures except those detailed below.

Positive Brushes

◆ See Figure 332

Notice how the positive brush lead is attached to the terminal on the end of the frame. This is the same terminal to which the heavy battery cable is attached. The terminal may be removed from the frame. Pull the terminal free of the frame.

1. Obtain an ohmmeter.

Fig. 327 Make sure the matchmarks are there...

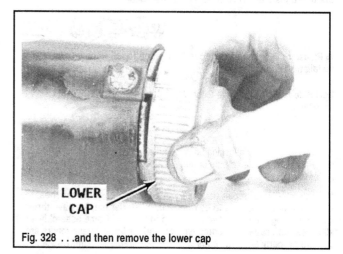

Fig. 328 ...and then remove the lower cap

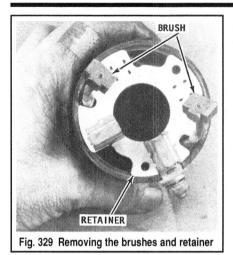

Fig. 329 Removing the brushes and retainer

Fig. 330 Hold the brushes in position while installing the retainer

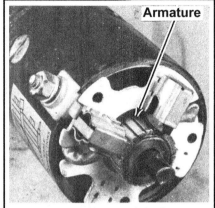

Fig. 331 Once the armature is in place you can cut the wire and remove it

2. Connect 1 test lead of an ohmmeter to the brush and the other test lead to the terminal. Continuity should be indicated on the ohmmeter. If continuity is not indicated, the brush must be replaced. The brush and terminal are sold as an assembly, eliminating the necessity for soldering.

Negative Brushes

◆ See Figure 333

The complete terminology for Prestolite negative brushes is Field Coil - Negative Brush.

    1. Obtain an ohmmeter.

    2. Make contact with 1 test lead to the negative brush and make contact with the other lead to the starter frame. If the meter does not indicate continuity, the field coils are open and must be replaced.

    3. Check to be sure the soldered connections are NOT touching the frame. The fields must not be grounded. If the connections make contact with the frame, the fields would be grounded.

Starter Drive Gear

IDENTIFICATION

Three types of drive gear arrangements are used on Johnson/Evinrude outboard starter motors covered here. One has a spring and spring retainer installed above the drive gear, and then the a castellated or lock nut securing these parts on the drive shaft. This unit is very simple in construction and

therefore, the service procedures; including disassembling and assembling are not difficult or involved. This drive gear is referred to as Type I.

The 2nd type is very similar except for the arrangement of parts on the armature shaft. This second unit is referred to as Type II.

An finally, the 3rd style is called Type III and only used on 1958-68 V4 engines and can be identified by its unique housing over the assembly.

To determine which starter motor drive type is being serviced observe the unit and make a comparison with the exploded illustrations in this section, especially the pinion gear and the screw shaft.

DISASSEMBLY & ASSEMBLY

Type I Drive Gear

◆ See Figures 334 thru 339

    1. Prevent the armature from turning by holding it with the proper size wrench on the hex nut provided for this purpose on the opposite end from the shaft nut. If the hex nut is not provided, hold the drive assembly with a pair of water pump pliers.

    2. Remove the shaft nut, spring retainer, spring, and then the drive assembly. The shaft nut should be replaced and NOT used a second time. Remember, some models will use a castellated nut with a cotter pin, while others will use a more conventional lock nut.

■ **The manufacturer strongly recommends against using any type of self-locking nut on the shaft.**

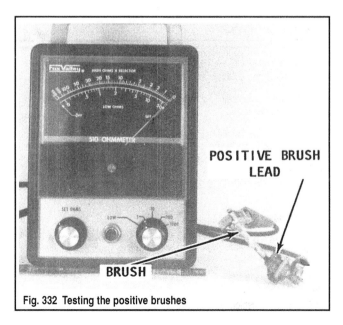

Fig. 332 Testing the positive brushes

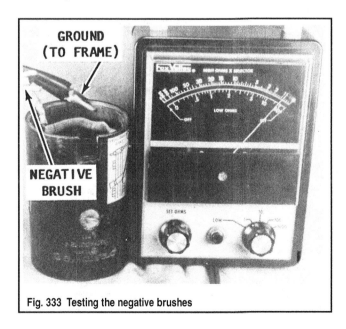

Fig. 333 Testing the negative brushes

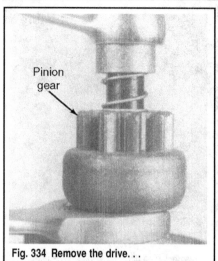

Fig. 334 Remove the drive. . .

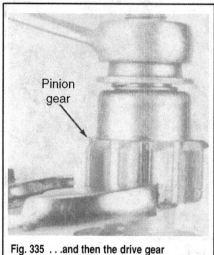

Fig. 335 . . .and then the drive gear

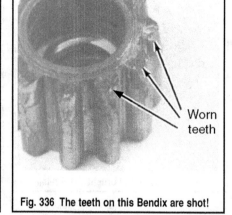

Fig. 336 The teeth on this Bendix are shot!

Fig. 337 A good shot of all the component parts

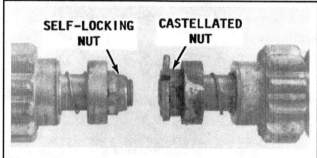

Fig. 338 Some models will use a lock nut, and others will use a castellated nut with a pin

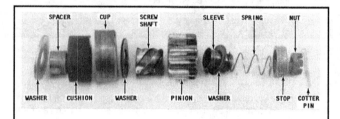

Fig. 339 Exploded view of a typical Type I gear assembly (this one using the nut/cotter pin)

3. The exploded view will be helpful in assembling the starter motor in the proper sequence.

4. Inspect the drive gear teeth for chips, cracks, or a broken tooth. Check the spline inside the drive gear for burrs and to be sure the drive gear moves freely on the armature shaft. Check to be sure the return spring is flexible and has not become distorted. Clean the armature shaft with crocus cloth.

To assemble:

5. Begin by assembling the following parts in the order given - first, slide the drive gear onto the shaft, then the spring, spring retainer, and then a NEW lock nut.

6. Prevent the armature shaft from turning by holding it with the proper size wrench on the hex nut provided for this purpose on the opposite end from the shaft nut. If the armature hex nut is not provided, hold the drive assembly with a pair of water pump pliers.

7. Tighten the shaft nut securely.

8. Now test and/or install the starter motor, as detailed in this section.

Type II Drive Gear

◆ See Figure 340

1. Remove the nut from the end of the armature shaft.
2. Scratch a mark on the top of the screw shaft and one on the top of the pinion as an aid during assembling. These marks will identify the top of both parts.

3. Next, remove the following parts in the sequence given: the pinion stop; anti-drift spring; sleeve; screw shaft cup; screw shaft; pinion; thrust washer; cushion cap; cushion; and the cushion retainer.

■ **The exploded view will be helpful in assembling the starter motor in the proper sequence.**

4. Inspect the drive gear teeth for chips, cracks, or a broken tooth. Check the spline inside the drive rear for burrs and to he sure the drive gear moves freely on the armature shaft. Check to be sure the return spring is flexible and has not become distorted. Inspect the rubber cushion for cracks and for signs of oil on the cushion. Clean the armature shaft with crocus cloth.

To assemble:

5. Begin by assembling the following parts in the sequence given: First, slide the cushion retainer down onto the drive end cap, with the shoulder facing upward.

6. Next, slide the cushion down the armature shaft and seat it over the shoulder of the retainer.

7. Install the cushion cap over the cushion.

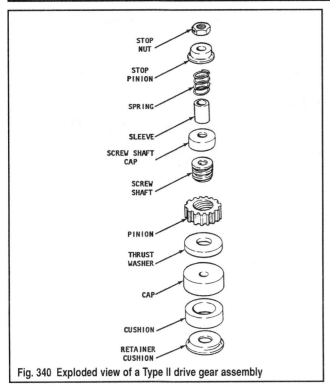

Fig. 340 Exploded view of a Type II drive gear assembly

STOP NUT
STOP PINION
SPRING
SLEEVE
SCREW SHAFT CAP
SCREW SHAFT
PINION
THRUST WASHER
CAP
CUSHION
RETAINER CUSHION

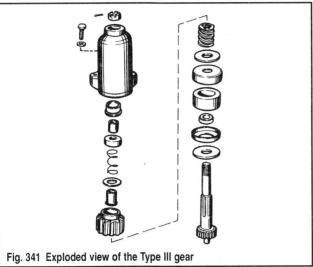

Fig. 341 Exploded view of the Type III gear

8. Slide thrust washer down the armature shaft onto the top of the cap. Rotate the screw shaft clockwise into the pinion, and then slide the pinion and screw shaft down the armature shaft onto the thrust washer. Install the cap over the end of the screw shaft.

9. Slide the following parts onto the armature shaft in the order given: the sleeve; spring; pinion stop washer; and finally thread the nut onto the end of the shaft. Tighten the nut securely.

10. Now test and/or install the starter motor as detailed in this section.

Type III

② ◁MODERATE

◆ **See Figures 341 thru 347**

The starter drive assembly has a nut on top that is easier to loosen before the assembly is removed from the engine. A special tool can be made from an old screwdriver which will greatly assist in this task. After the screwdriver has been modified as shown in the accompanying illustration, it will fit onto the pinion gear.

1. Using the special tool to prevent the pinion gear from rotating, a socket wrench can be used to loosen the nut.

2. After the nut is loose, remove the 7/16 in. screw on the base of the drive gear assembly and lift assembly free of the bracket on the powerhead.

3. With the assembly on the bench, remove and discard the nut from the top of the housing. Pull the shaft and attaching parts free of the housing from the other end.

4. Remove the component parts from the shaft and note their order as an aid to assembling. The top of the housing has a shaft bushing which may remain in the housing as the shaft is removed. If the bushing came out with the shaft, remove it first. Next, slide the pinion stop off the shaft, then the anti-drift spring, a thrust washer, anti-drift spring sleeve, the pinion gear, screw shaft, another thrust washer, upper cushion cup, cushion, spacer, lower cushion cup, and finally another thrust washer free, in that order leaving a bare shaft.

To assemble:

The most important items in the assembly are the shaft and the gear. The gear meshes with the splined gear on the armature shaft. These gears must remain in good condition for the assembly to function properly.

Inspect the teeth of both gears to be sure they are free of any type of damage or wear.

If the shaft is to be replaced, the armature should also be replaced. The armature is an expensive part, but if a new shaft is installed with the old armature, one or the other will most likely wear excessively in a very short time.

Inspect the pinion gear teeth to be sure they are clean and free of any damage. Check to be sure the screw shaft moves freely inside the pinion gear with no sign of binding.

Clean the screw portion of the shaft. Check the rubber cushion for wear, but DO NOT clean it with solvent.

If the bushing in the housing needs to be replaced, it may be pushed or pressed out. If, at all possible, put the new sleeve in a freezer or cold refrigerator overnight to shrink it for easier installation. After the new bushing

Fig. 342 Craft a special holder out of a screwdriver. . .

Pinion Gear

Fig. 343 . . .and use it to stop the gear from rotating

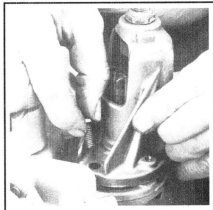

Fig. 344 Remove the bolt at the base of the assembly

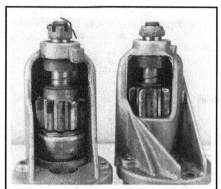

Fig. 345 The gear on the left is used with a magneto ignition system, while the one on the right is used with a battery ignition system

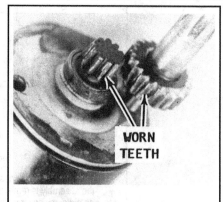

Fig. 346 The teeth on these gear are worn. Worn teeth on either gear will necessitate replacing both

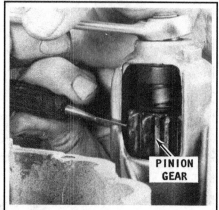

Fig. 347 Using the special screwdriver again

has been installed, insert the shaft through the housing to be sure it turns freely.

5. Apply just a drop or two of oil onto the bushing in the top of the housing. The remainder of the components will be assembled dry - do not use any oil or lubricant on any of the other parts.

6. Assembling the drive gear unit is simply a case of sliding the component parts onto the shaft in the reverse order from which they were removed. First, slide the thrust washer down into place, then the small cushion cup with the lip on the side, then the cushion spacer, the cushion, then the large cushion cup with the flange facing downward, the second thrust washer, the screw shaft, starter drive pinion gear, anti-drift spring

sleeve, another washer, anti-drift spring down over the sleeve, and finally, the pinion stop with the concave side down.

7. Insert the assembled shaft up into the housing and hold it in place by starting a NEW nut onto the shaft. It is easier to tighten the nut after the assembly has been installed using the modified screwdriver employed during loosening of the nut.

8. Install the unit onto the powerhead and secure it in place with the 7/16 in. bolt.

9. Use the modified screwdriver and socket wrench to tighten the nut until it stops at the end of the shaft threads. After the nut is tightened, loosen the 7/16 in. bolt slightly to assist with the starter motor installation and the alignment of the thru-bolts.

ELECTRICAL ACCESSORIES (GAUGES, LIGHTS, HORNS & SWITCHES)

General Information

◆ See Figure 348

Gauges or lights are installed to warn the operator of a condition in the cooling and lubrication systems that may need attention. The fuel gauge gives an indication of the amount of fuel in the tank. If the engine overheats, a warning light will come on or a horn sound advising the operator to shut down the engine and check the cause of the warning before serious damage is done.

In order for gauges to register properly, they must be supplied with a steady voltage. The voltage variations produced by the engine charging system would cause erratic gauge operation, too high when the generator or alternator voltage is high and too low when the generator or alternator is not charging. To remedy this problem, a constant-voltage system is used to reduce the 12-14 volts of the electrical system to an average of 5 volts. This steady 5 volts ensures the gauges will read accurately under varying conditions from the electrical system.

Temperature Gauge and/or Warning Light

TEMPERATURE GUAGE

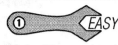

The body of temperature gauges must be grounded and they must be supplied with 12 volts. Many gauges have a terminal on the mounting bracket for attaching a ground wire. A tang from the mounting bracket makes contact with the gauge - always check to be sure the tang does make good contact with the gauge.

Ground the wire to the sending unit and the needle of the gauge should move to the full right position indicating the gauge is in serviceable condition.

WARNING LIGHTS

If a problem arises on a boat equipped with water and temperature lights or warning horn, the first area to check is the light assembly for loose wires or burned-out bulbs. Check the horn in the shift box for loose connections and proper grounding. When the ignition key is turned on, the light assembly is supplied with 12 volts and grounded through the sending unit mounted on the engine. When the sending unit makes contact because the water temperature is too hot, the circuit to ground is completed and the lamp should light or the horn sound.

Check the bulb - turn the ignition switch to the ON position. Disconnect the wire at the engine sending unit, and then ground the wire. The lamp on the dash should light or the horn sound. If it does not light, check for a burned-out bulb or a break in the wiring to the light.

If the horn does not sound, check inside the shift box. Disconnect the horn wires, and then connect a good direct wire from the battery to the horn and another wire from the horn to a good ground. The horn should sound. If

Fig. 348 Gauges should be kept clean and as elry as possible

the horn fails to sound, the horn is defective. If the horn does sound, the wires or connections from the engine to the horn need attention.

THERMOMELT STICKS

◆ See Figure 349

Thermomelt sticks are an easy, inexpensive, and fairly accurate method of determining if the engine is running at approximately the temperature recommended by the manufacturer. Thermomelt sticks are available at your local marine dealer or at an automotive parts house.

To check powerhead operating temperatures proceed as follows:

1. Start the engine with the propeller in the water and run it for about 5 minutes at roughly 3000 rpm.

✳✳ CAUTION

Water must circulate through the lower unit to the engine any time the engine is run to prevent damage to the water pump in the lower unit. Just 5 seconds without water will damage the water pump.

The 140° stick should melt when you touch it to the lower thermostat housing or on the top cylinder. If it does not melt, the thermostat is stuck in the open position and the engine temperature is too low.

Touch the 170° stick to the same spot on the lower thermostat housing or on the top cylinder. The stick should not melt. If it does, the thermostat is stuck in the closed position or the water pump is not operating properly because the engine is running too hot.

Fuel Gauge

◆ See Figure 350

The fuel gauge is intended to indicate the quantity of fuel in the tank. As the experienced boatman has learned, the gauge reading is seldom an accurate report of the fuel available in the tank. The main reason for this false reading is because the boat is rarely on an even keel. A considerable difference in fuel quantity will be indicated by the gauge if the bow or stern is heavy, or if the boat has a list to port or starboard. Therefore, the reading is usually low.

The amount of fuel drawn from the tank is dependent on the location of the fuel pickup tube in the tank. The engine may cutout while cruising because the pickup tube is above the fuel level. Instead of assuming the tank is empty, shift weight in the boat to change the trim and the problem may be solved until you are able to take on more fuel.

FUEL GAUGE HOOK-UP

◆ See Figure 350

The Boating Industry Association recommends the following color coding be used on all fuel gauge installations:

Black - for all grounded current-carrying conductors.

Pink - insulated wire for the fuel gauge sending unit to the gauge.

Red - insulated wire for a connection from the positive side of the battery to any electrical equipment.

1. Connect one end of a pink insulated wire to the terminal on the gauge marked **TANK** and the other end to the terminal on top of the tank unit.

2. Connect one end of a black wire to the terminal on the fuel gauge marked **IGN** and the other end to the ignition switch.

3. Connect one end of a second black wire to the fuel gauge terminal marked **GRD** and the other end to a good ground. It is important for the fuel gauge case to have a good common ground with the tank unit. Aboard an all-metal boat, this ground wire is not necessary. However, if the dashboard is insulated, or made of wood or plastic, a wire MUST be run from the gauge ground terminal to one of the bolts securing the sending unit in the fuel tank, and then from there to the Negative side of the battery.

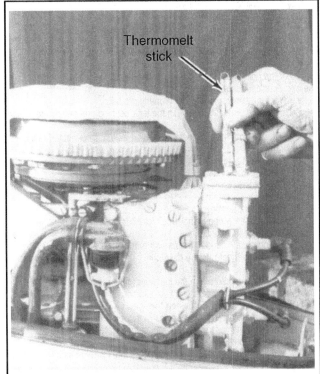

Fig. 349 Thermomelt sticks are a simple, inexpensive means of determining engine running temperature

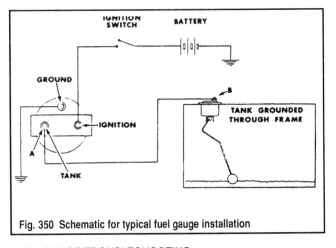

Fig. 350 Schematic for typical fuel gauge installation

FUEL GAUGE TROUBLESHOOTING

◆ See Figure 350

In order for the fuel gauge to operate properly the sending unit and the receiving unit must be of the same type and preferably of the same make.

The following symptoms and possible corrective actions will be helpful in restoring a faulty fuel gauge circuit to proper operation.

1. If you suspect the gauge is not operating properly, the first area to check is all electrical connections from one end to the other. Be sure they are clean and tight.

2. Next, check the common ground wire between the negative side of the battery, the fuel tank, and the gauge on the dash.

3. If all wires and connections in the circuit are in good condition, remove the sending unit from the tank. Run a wire from the gauge mounting flange on the tank to the flange of the sending unit.

4. Now, move the float up-and-down to determine if the receiving unit operates. If the sending unit does not appear to operate, move the float to the midway point of its travel and see if the receiving unit indicates half full.

5. If the pointer does not move from the **EMPTY** position one of four faults could be to blame:
- The dash receiving unit is not properly grounded.
- No voltage at the dash receiving unit.
- Negative meter connections are on a positive grounded system.
- Positive meter connections are on a negative grounded system.

6. If the pointer fails to move from the **FULL** position, the problem could be one of three faults.
- The tank sending unit is not properly grounded.
- Improper connection between the tank sending unit and the receiving unit on the dash.
- The wire from the gauge to the ignition switch is connected at the wrong terminal.

7. If the pointer remains at the 3/4 full mark, it indicates a 6-volt gauge is installed in a 12-volt system.

8. If the pointer remains at about 3/8 full, it indicates a 12-volt gauge is installed in a 6-volt system.

Preliminary Inspection

Inspect all of the wiring in the circuit for possible damage to the insulation or conductor. Carefully check:
- Ground connections at the receiving unit on the dash.
- Harness connector to the dash unit.
- Body harness connector to the chassis harness.
- Ground connection from the fuel tank to the tank floor pan.
- Feed wire connection at the tank sending unit.

Gauge Always Reads Full when the ignition switch is ON:

9. Check the electrical connections at the receiving unit on the dash; the body harness connector to chassis harness connector; and the tank unit connector in the tank.

10. Make a continuity check of the ground wire from the tank to the tank floor pan.

11. Connect a known good tank unit to the tank feed wire and the ground lead. Raise and lower the float and observe the receiving unit on the dash. If the dash unit follows the arm movement, replace the tank sending unit.

Gauge Always Reads Empty when the ignition switch is ON:

12. Disconnect the tank unit feed wire and do not allow the wire terminal to ground.

13. The gauge on the dash should read FULL.

If Gauge Reads Empty:

14. Connect a spare dash unit into the dash unit harness connector and ground the unit.

15. If the spare unit reads FULL, the original unit is shorted and must be replaced.

16. A reading of EMPTY indicates a short in the harness between the tank sending unit and the gauge on the dash.

If Gauge Reads Full:

17. Connect a known good tank sending unit to the tank feed wire and the ground lead.

18. Raise and lower the float while observing the dash gauge.

19. If dash gauge follows movement of the float, replace the tank sending unit.

Gauge Never Indicates Full:

■ **This test requires shop test equipment.**

20. Disconnect the feed wire to the tank unit and connect the wire to a good ground through a variable resistor or through a spare tank unit.

21. Observe the dash gauge reading. The reading should be FULL when resistance is increased to about 90 ohms. This resistance would simulate a full tank.

22. If the check indicates the dash gauge is operating properly, the trouble is either in the tank sending unit rheostat being shorted, or the float is binding. The arm could be bent, or the tank may be deformed. Inspect and correct the problem.

Tachometer

◆ **See Figure 351**

An accurate tachometer can be installed on any engine. Such an instrument provides an indication of engine speed in revolutions per minute (rpm). This is accomplished by measuring the number of electrical pulses per minute generated in the primary circuit of the ignition system. The proper tachometer MUST be installed. Be sure to check with your local marine dealer to ensure the proper unit is being installed. The wrong tachometer will cause serious damage to the ignition system.

The meter readings range from 0 to 6,000 rpm, in increments of 100. Tachometers have solid-state electronic circuits which eliminate the need for relays or batteries and contributes to their accuracy. The electronic parts of the tachometer susceptible to moisture are coated to prolong their life. Take time at the marine dealer to specify the unit being serviced when purchasing a new tachometer to ensure the correct unit is installed.

Horns

◆ **See Figure 352**

The only reason for servicing a horn is because it fails to operate properly or because it is out of tune. In most cases, the problem can be traced to an open circuit in the wiring or to a defective relay.

Cleaning:

Crocus cloth and carbon tetrachloride should be used to clean the contact points. NEVER force the contacts apart or you will bend the contact spring and change the operating tension.

Checking the Relay and Wiring:

Connect a wire from the battery to the horn terminal. If the horn operates, the problem is in the relay or in the horn wiring. If both of these appear satisfactory, the horn is defective and needs to be replaced. Before replacing the horn however, connect a second jumper wire from the horn frame to ground to check the ground connection.

Test the winding for an open circuit, faulty insulation, or poor ground. Check the resistor with an ohmmeter, or test the condenser for capacity, ground, and leakage. Inspect the diaphragm for cracks.

Fig. 351 Never use your helm-installed tach for anything other than an estimate

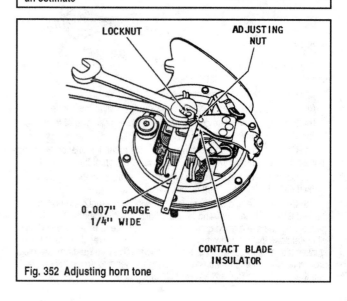

Fig. 352 Adjusting horn tone

Adjusting Horn Tone:

Loosen the lock-nut, and then rotate the adjusting screw until the desired tone is reached. On a dual horn installation, disconnect one horn and adjust each, one at a time. The contact point adjustment is made by inserting a 0.007 in. (1.78mm) feeler gauge blade between the adjusting nut and the contact blade insulator. Take care not to allow the feeler gauge to touch the metallic parts of the contact points because it would short them out. Now, loosen the lock-nut and turn the adjusting nut down until the horn fails to sound. Loosen the adjusting nut slowly until the horn barely sounds. The lock-nut MUST be tightened after each test. When the feeler gauge is withdrawn the horn will operate properly and the current draw will be satisfactory.

Choke Circuit

♦ **See Figure 353**

The choke is activated by a solenoid. This solenoid attracts a plunger to close the choke valves. The solenoid is energized when the ignition key is turned to the **START** position and the choke button is depressed. When using the electric choke, the manual choke MUST be in the Neutral position.

Only the electric choke is used on the engines covered in this service.

On some newer model engines, the choke will be activated if the key is pushed inward while it is being rotated to the **START** position.

This short section provides instructions to test the choke circuit. If the

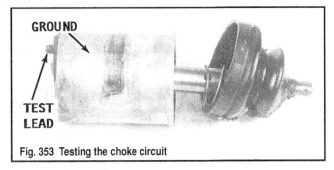

Fig. 353 Testing the choke circuit

system fails the test, the attaching hardware can be removed and the choke assembly replaced.

Choke Circuit Testing

The choke circuit may be quickly tested to determine if it is functioning properly as follows:

1. Obtain an ohmmeter.
2. Connect the black meter lead to an unpainted portion of the engine block for a good ground.
3. Connect the red meter lead to the choke terminal.
4. Test the circuit using the R x 1 scale of the ohmmeter. A satisfactory reading is approximately 3 ohms.
5. After the test is completed, check to be sure the choke plunger is pulled into the choke solenoid.

SPECIFICATIONS

Generator Specifications

Component		Value
Part No.		GJG-4001M
		GJG-4002M
Rotor Direction		Clockwise
Ground Polarity		Negative
Brush Spring Tension		12-24 oz.
Field Coil Draw		
		10.0 volts
		1.7-1.9 amps
Monitoring Draw		
		10.0 volts
		5.0-6.0 amps
Generator		
	Output	15.0 volts
	Max Amps	10.0 amps
	Max RPM	7,000

Condenser Specifications

Engine Size	Model	Condenser Part No.	MFD Capacity
1.5 Hp	1968-70	580321	18-22
2 Hp	1971-72	-	-
3 Hp	1958-68	580321	18-22
4 Hp	1969-72	580321	18-22
5 Hp	1965-68	580321	18-22
5.5 Hp	1958-64	580321	18-22
6 Hp	1965-72	580321	18-22
7.5 Hp	1958	580321	18-22
9.5 Hp	1964-72	580321	18-22
10 Hp	1958-63	580321	18-22
18 Hp	1958-61	580321	18-22
	1962-72	580419	25-29
20 Hp	1966-72	580419	25-29
25 Hp	1969-72	580419	25-29
28 Hp	1962-64	580419	25-29
33 Hp	1965-70	580419	25-29
35 Hp	1958-59	580321	18-22
40 Hp	1960-61	580321	18-22
	1962-72	580419	25-29
50 Hp	1971-72	-	-

Regulator Specifications

Component		Value
Part No.		VRU-6101A
System Voltage		12
Ground Polarity		Negative
Armature Air Gap	Circuit Breaker	0.031-0.034 in.
	Voltage Regulator	0.048-0.052 in.
Current Regulator		0.048-0.052 in.
Current Regulator Setting		9.0-11.0 amps
Shunt Winding	Circuit Breaker	107-121 ohms
	Voltage Regulator	43.7-49.3 ohms
Circuit Breaker	Close	12.6-13.6 volts
	Open, Discharge	3.0-5.0 amps
Regulator Operating Voltage A	50° F	15.2 volts
	80° F	15.0 volts
	110° F	14.8 volts
	140° F	14.6 volts

NOTE: Figures are for a unit in normal operation while charging at half the rated output, or with 1/4 ohm fixed resistor in series with the battery.

A Hot, tolerance +/- 0.4 volts

Starter Motor Specifications

Model	Brush Spring Tension Oz.	Armature End Play In.	Volts	Max Amps	Min RPM	Volts	Max Amps	Min ft. lbs.
MDO-0	42-66	0.005 Min	10	38	10,000	4	170	1.5
MDO-1	42-66	0.010-0.035	10	38	10,000	4	170	1.5
MDW-0	42-66	0.010-0.035	10	26	8,500	4	160	2.1
MDW-1	42-66	0.005 Min	10	26	8,500	4	160	2.1

Pinion position: 1 25/32 +/- 1/16 in. from face of mounting flange to edge of pinion.

MDO-4002M: use test MDO-0, counterclockwise rotation
MDO-4003M: use test MDO-1, clockwise rotation
MDW-4001M: use test MDO-0, clockwise rotation
MDW-4002M: use test MDO-1, counterclockwise rotation

WIRING DIAGRAMS

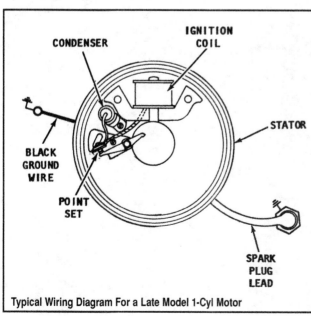

Typical Wiring Diagram For a Late Model 1-Cyl Motor

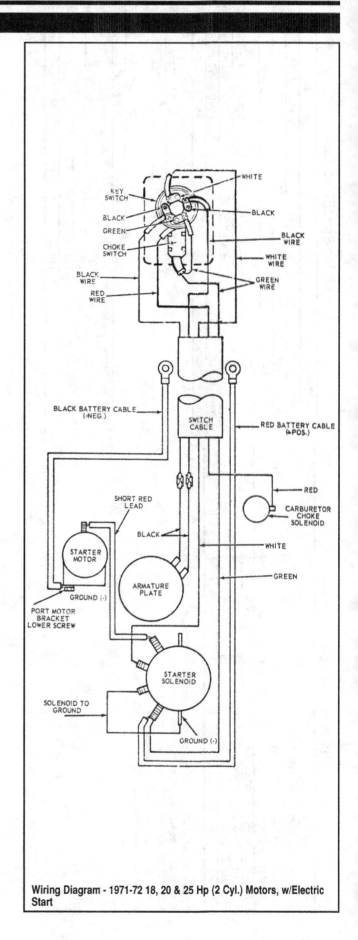

Wiring Diagram - 1971-72 18, 20 & 25 Hp (2 Cyl.) Motors, w/Electric Start

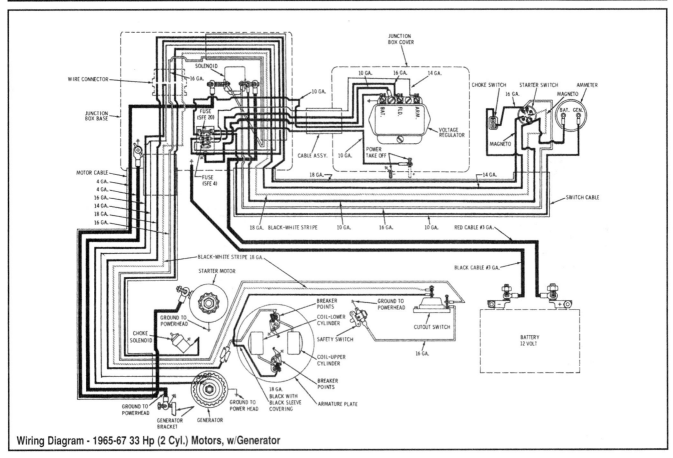

Wiring Diagram - 1965-67 33 Hp (2 Cyl.) Motors, w/Generator

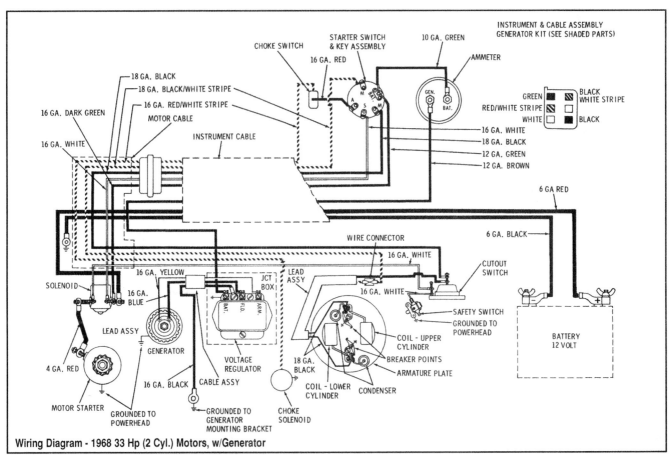

Wiring Diagram - 1968 33 Hp (2 Cyl.) Motors, w/Generator

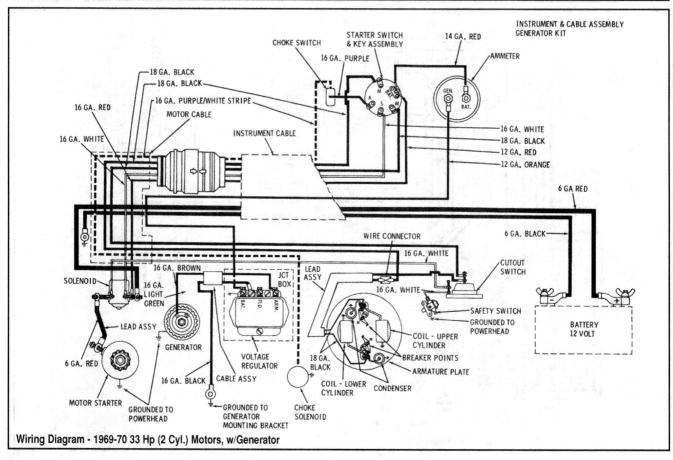

Wiring Diagram - 1969-70 33 Hp (2 Cyl.) Motors, w/Generator

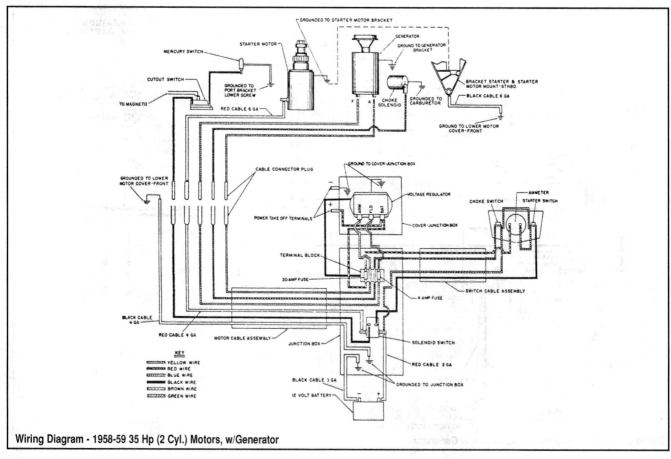

Wiring Diagram - 1958-59 35 Hp (2 Cyl.) Motors, w/Generator

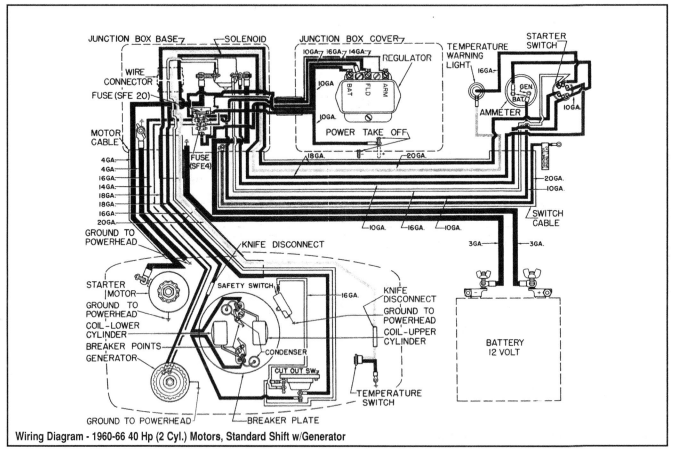

Wiring Diagram - 1960-66 40 Hp (2 Cyl.) Motors, Standard Shift w/Generator

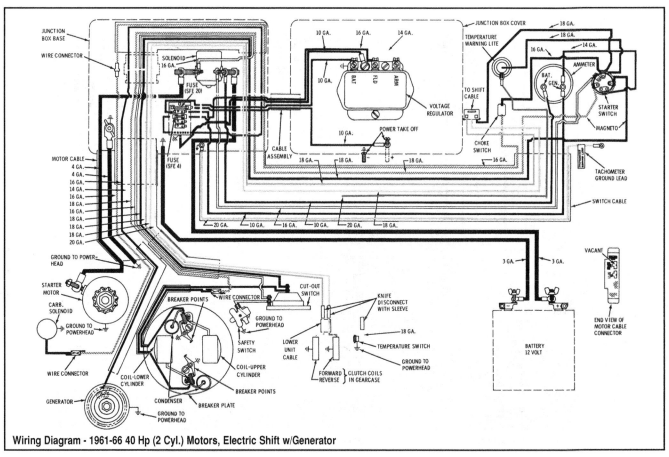

Wiring Diagram - 1961-66 40 Hp (2 Cyl.) Motors, Electric Shift w/Generator

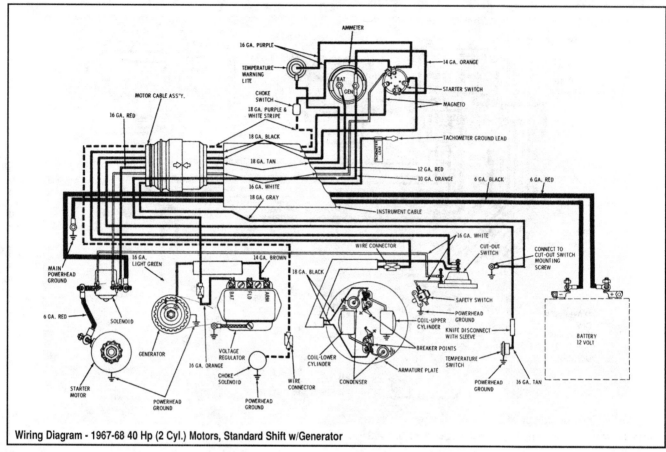

Wiring Diagram - 1967-68 40 Hp (2 Cyl.) Motors, Standard Shift w/Generator

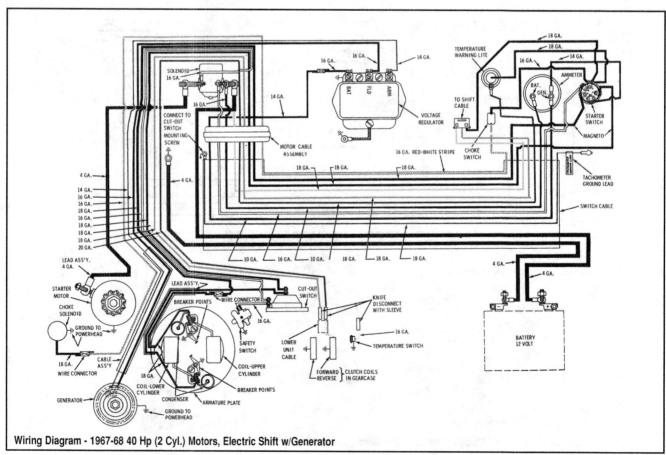

Wiring Diagram - 1967-68 40 Hp (2 Cyl.) Motors, Electric Shift w/Generator

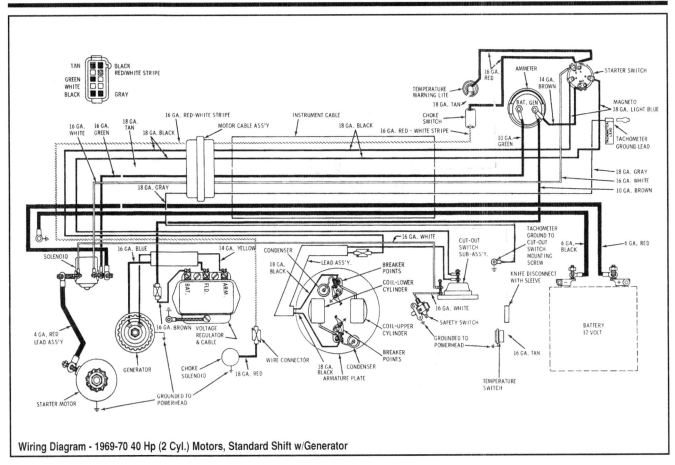

Wiring Diagram - 1969-70 40 Hp (2 Cyl.) Motors, Standard Shift w/Generator

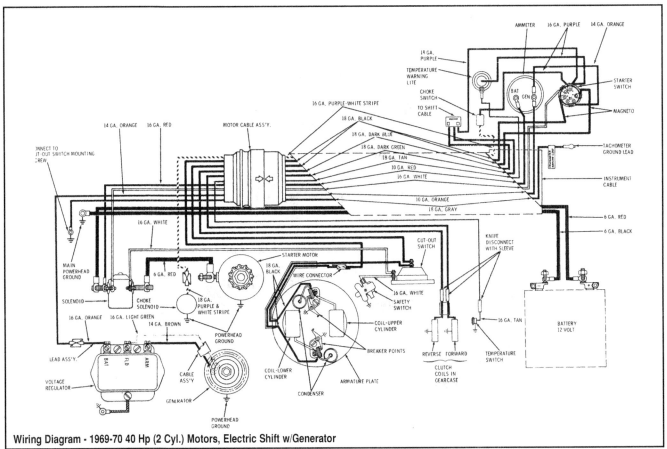

Wiring Diagram - 1969-70 40 Hp (2 Cyl.) Motors, Electric Shift w/Generator

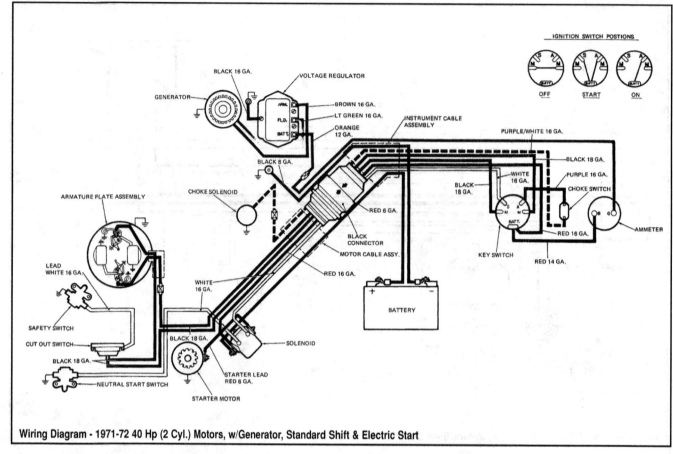

Wiring Diagram - 1971-72 40 Hp (2 Cyl.) Motors, w/Generator, Standard Shift & Electric Start

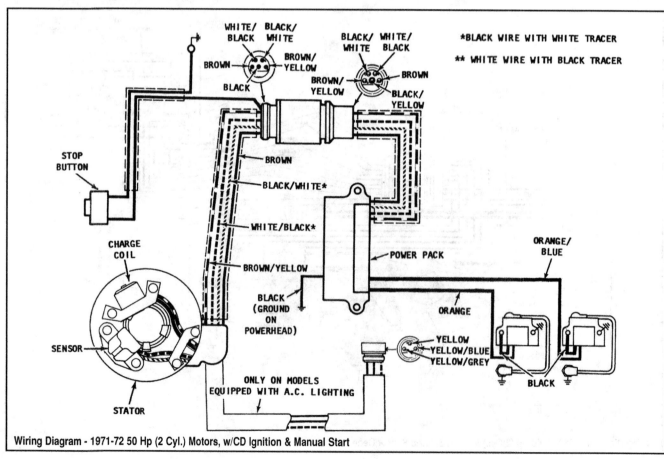

Wiring Diagram - 1971-72 50 Hp (2 Cyl.) Motors, w/CD Ignition & Manual Start

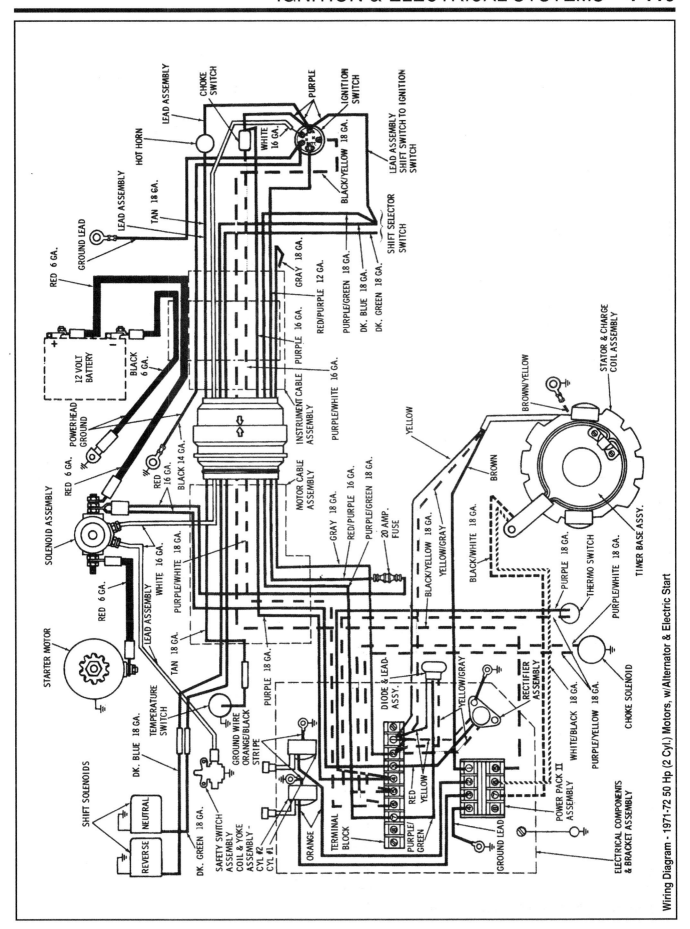

Wiring Diagram - 1971-72 50 Hp (2 Cyl.) Motors, w/Alternator & Electric Start

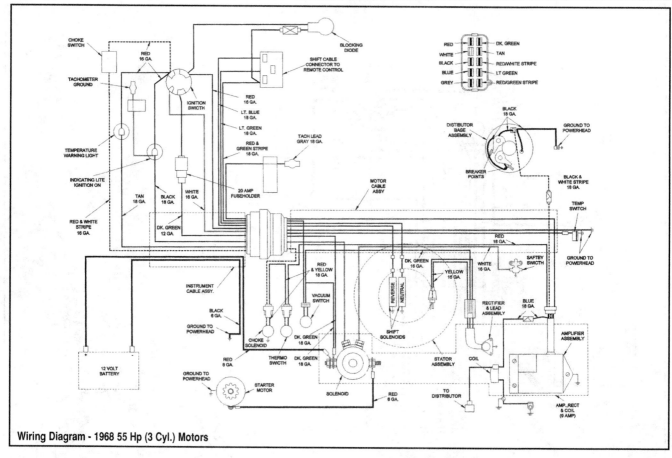

Wiring Diagram - 1968 55 Hp (3 Cyl.) Motors

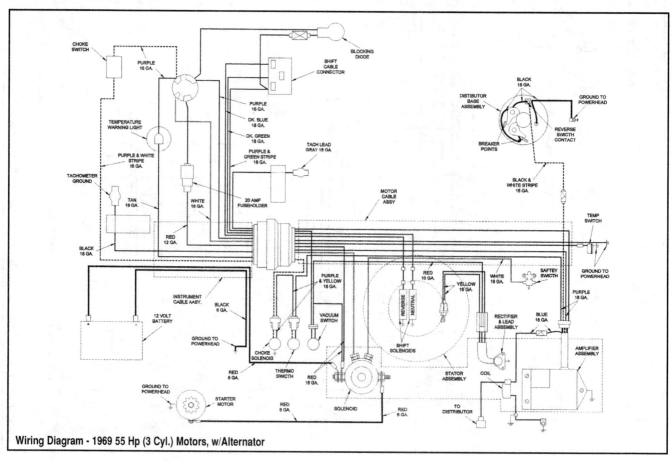

Wiring Diagram - 1969 55 Hp (3 Cyl.) Motors, w/Alternator

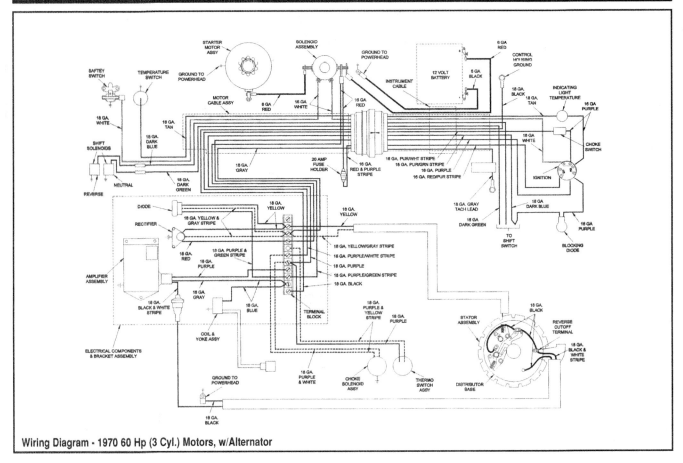

Wiring Diagram - 1970 60 Hp (3 Cyl.) Motors, w/Alternator

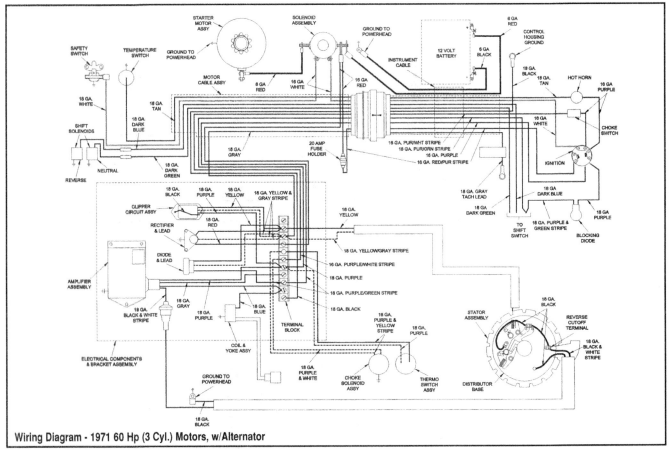

Wiring Diagram - 1971 60 Hp (3 Cyl.) Motors, w/Alternator

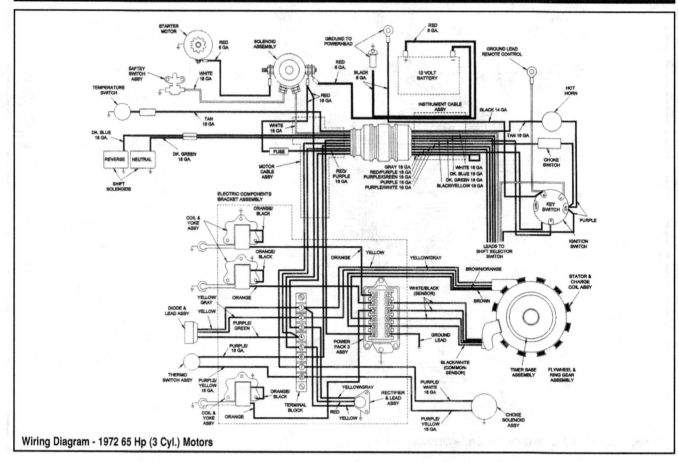

Wiring Diagram - 1972 65 Hp (3 Cyl.) Motors

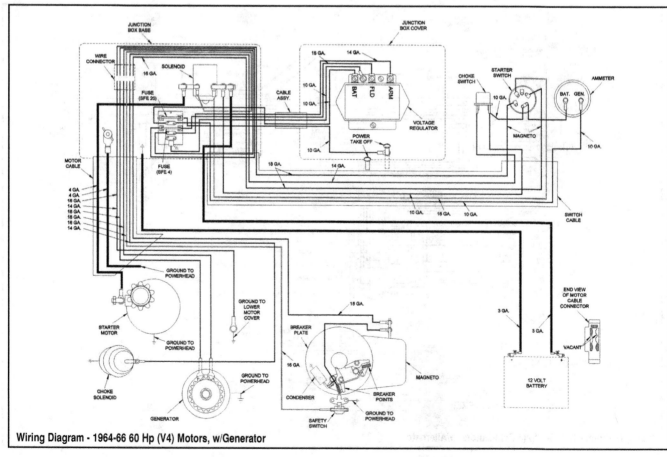

Wiring Diagram - 1964-66 60 Hp (V4) Motors, w/Generator

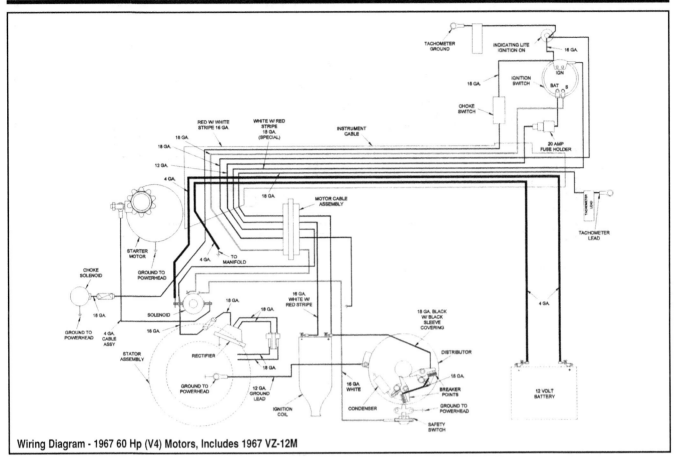

Wiring Diagram - 1967 60 Hp (V4) Motors, Includes 1967 VZ-12M

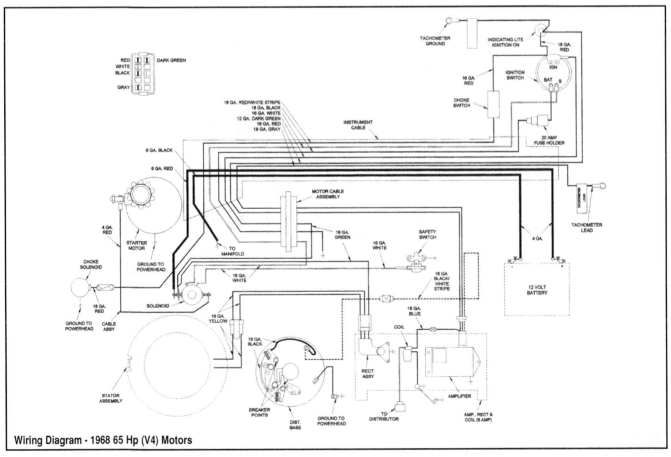

Wiring Diagram - 1968 65 Hp (V4) Motors

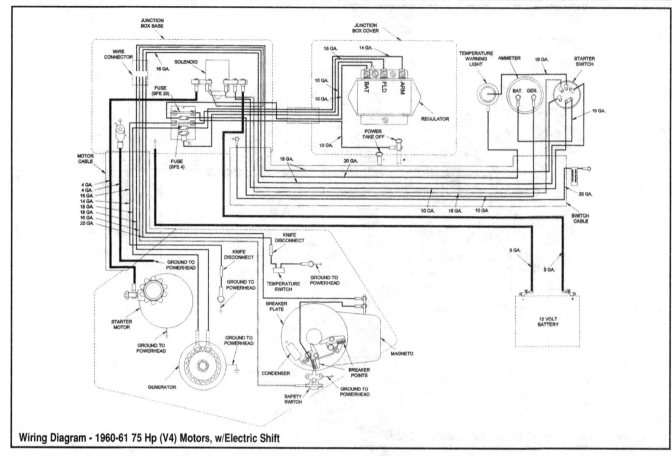

Wiring Diagram - 1960-61 75 Hp (V4) Motors, w/Electric Shift

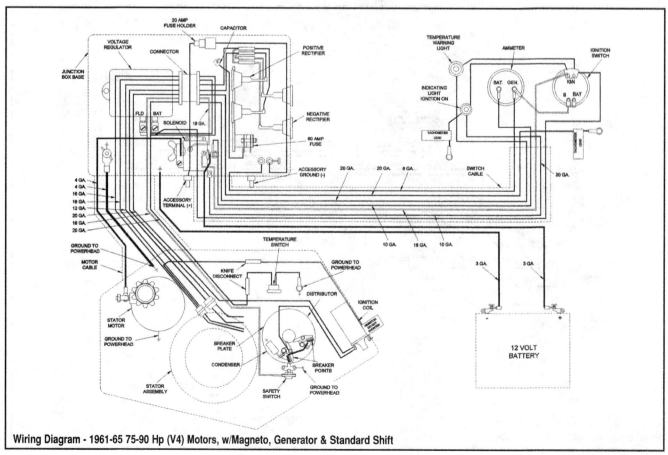

Wiring Diagram - 1961-65 75-90 Hp (V4) Motors, w/Magneto, Generator & Standard Shift

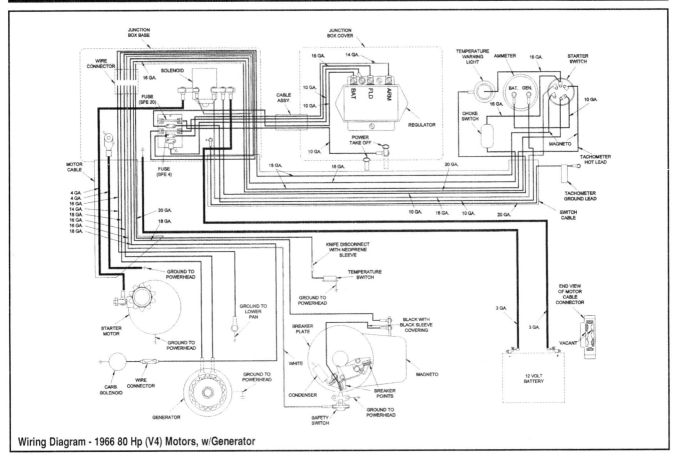

Wiring Diagram - 1966 80 Hp (V4) Motors, w/Generator

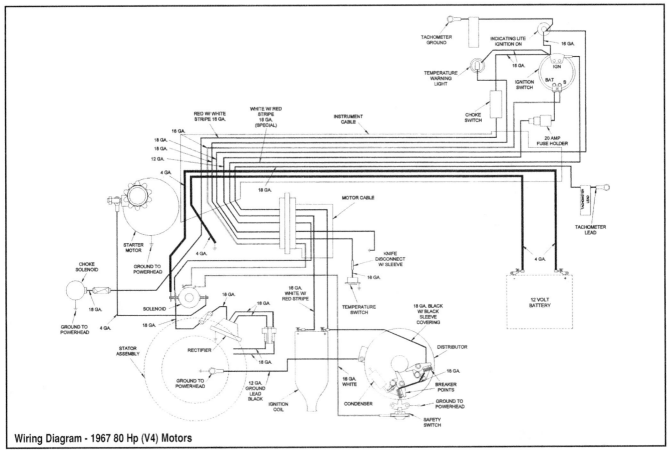

Wiring Diagram - 1967 80 Hp (V4) Motors

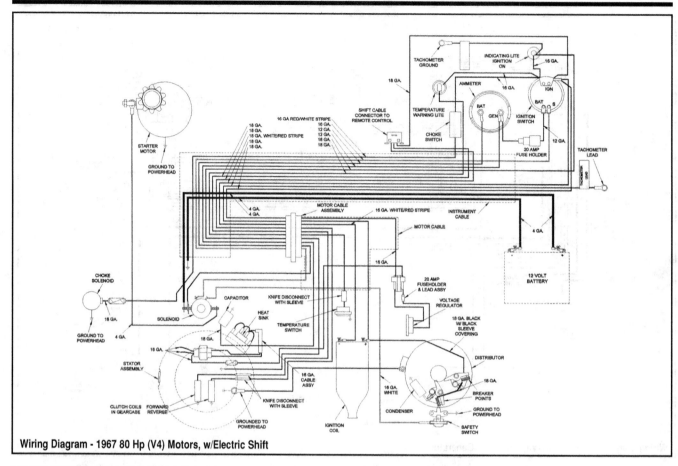

Wiring Diagram - 1967 80 Hp (V4) Motors, w/Electric Shift

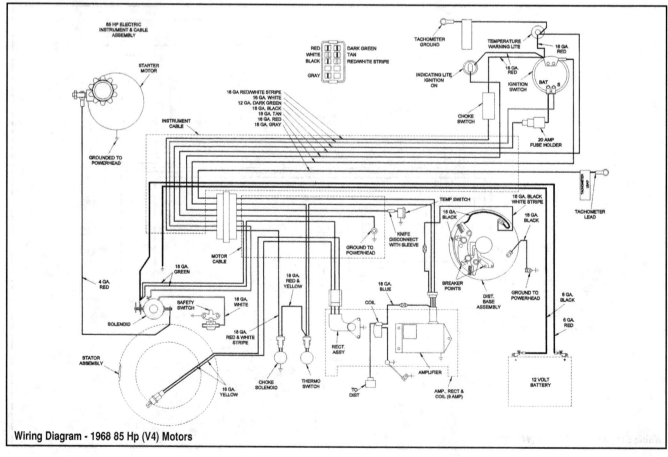

Wiring Diagram - 1968 85 Hp (V4) Motors

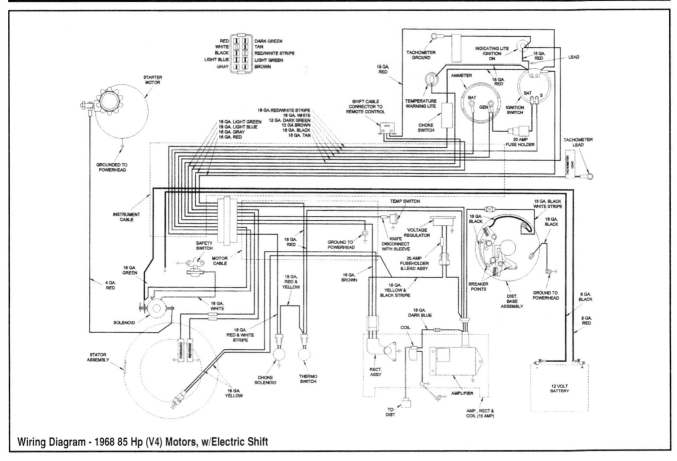

Wiring Diagram - 1968 85 Hp (V4) Motors, w/Electric Shift

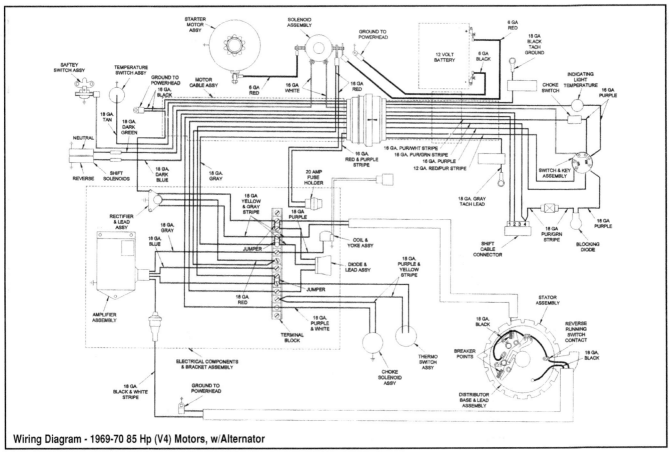

Wiring Diagram - 1969-70 85 Hp (V4) Motors, w/Alternator

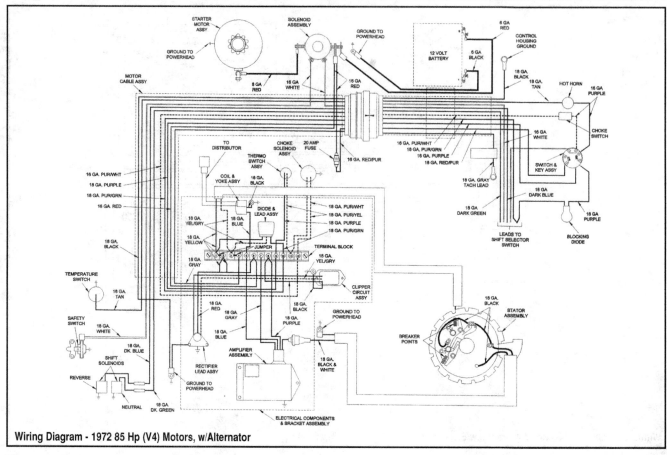

Wiring Diagram - 1972 85 Hp (V4) Motors, w/Alternator

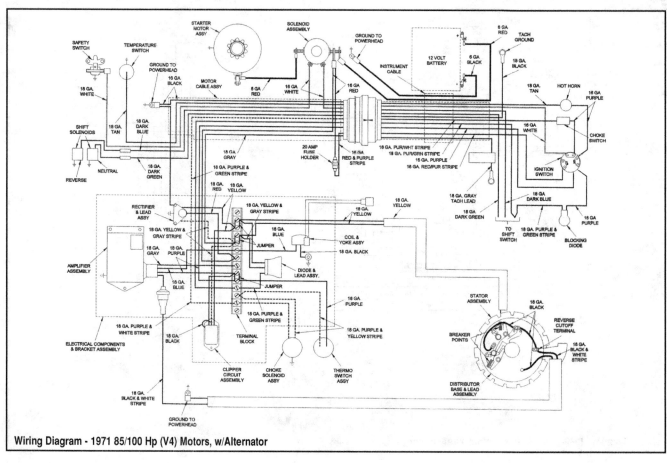

Wiring Diagram - 1971 85/100 Hp (V4) Motors, w/Alternator

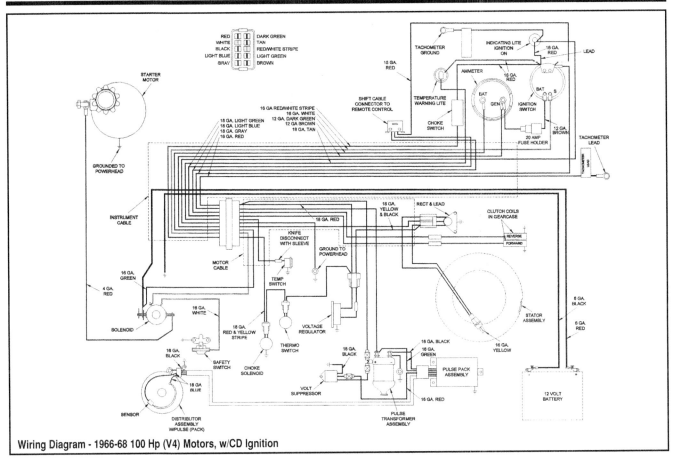

Wiring Diagram - 1966-68 100 Hp (V4) Motors, w/CD Ignition

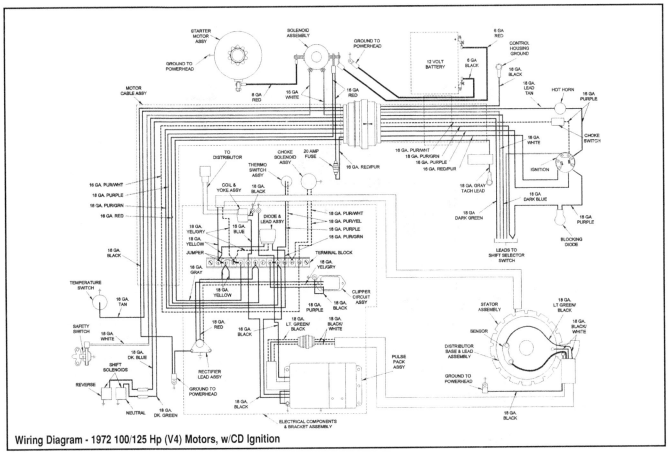

Wiring Diagram - 1972 100/125 Hp (V4) Motors, w/CD Ignition

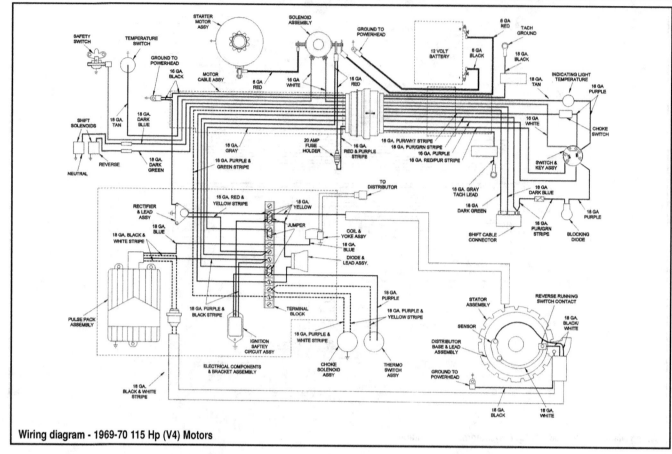

Wiring diagram - 1969-70 115 Hp (V4) Motors

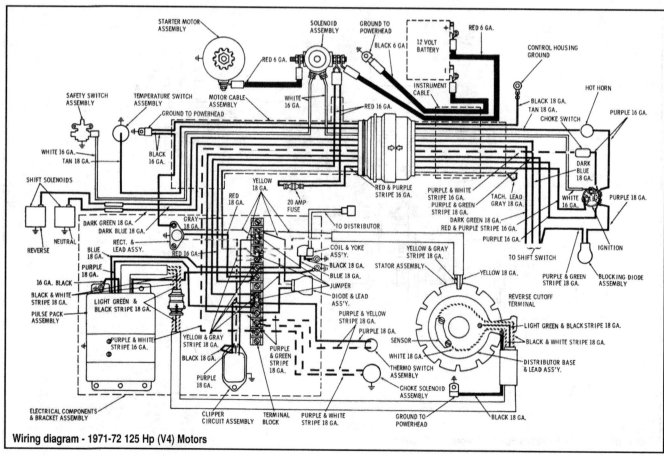

Wiring diagram - 1971-72 125 Hp (V4) Motors

5

COOLING SYSTEM

COOLING SYSTEM

General Information

◆ See Figure 1

All Johnson/Evinrude outboard engines are equipped with a raw water cooling system, meaning that sea, lake or river water is drawn through a water intake in the gearcase lower unit and pumped through the powerhead by a water pump impeller. The exact mounting and location of the pump/impeller varies slightly, but on these years and models, it is mounted to the lower unit along the gearcase-to-intermediate section split line (in a position to be directly driven by the driveshaft).

For many boaters, annual replacement of the water pump impeller is considered cheap insurance for a trouble-free boating season. This is probably a bit too conservative for most people, but after a number of trouble-free seasons, an impeller doesn't owe you anything and you should consider taking the time and a little bit of money necessary to replace it. Remember if an impeller fails, you'll be stranded. Worse, a worn impeller will simply supply less cooling water than required by specification, allowing the powerhead to run hot placing unnecessary stress on components and best or risking overheating the powerhead at worst.

Most of the motors covered here are equipped with a thermostat that restricts the amount of cooling water allowed into the powerhead until the powerhead reaches normal operating temperature. The purpose of a thermostat is to increase engine performance and reduce emissions by making sure the engine warms as quickly as possible to operating temperature and remains there during use under all conditions. Running a motor without a thermostat may prevent it from fully warming, not only increasing emissions and reducing fuel economy, but it will likely lead to carbon fouling, stumbling and poor performance in general. It can even damage the motor, especially if the motor is then run under load (such as full-throttle operation) without allowing it to thoroughly warm. A restricted thermostat can promote engine overheating. The good news is that should you be caught on the water with a restricted thermostat, you should be able to easily remove it and get back to shore, just make sure you replace it before the next outing.

Some of the larger motors may also be equipped with a water-pressure valve (also known as a blow-off valve because it opens in response to high-pressure regardless of engine temperature). The purpose of the pressure valve is to prevent possible damage to the cooling system should pressure rise above a certain point.

The water intake grate and cooling passages throughout the powerhead and gearcase comprise the balance of the cooling system. Both components require the most simple, but most frequent maintenance to ensure proper cooling system operation. The water intake grate should be inspected before and after each outing to make sure it is not clogged or damaged. A damaged grate could allow debris into the motor that could clog passages or damage the water pump impeller (both conditions could lead to overheating the powerhead). Cooling passages have the tendency to become clogged gradually over time by debris and corrosion. The best way to prevent this is to flush the cooling system **after each use** regardless of where you boat (salt or freshwater). But obviously, this form of maintenance is even more important on vessels used in salt, brackish or polluted waters that will promote internal corrosion of the cooling passages.

Description & Operation

◆ See Figures 2 thru 5

The water pump uses an impeller driven by the driveshaft, sealing between an offset housing and lower plate to create a flexing of the impeller blades. The rubber impeller inside the pump maintains an equal volume of water flow at most operating speeds.

At low speeds the pump acts like a full displacement pump with the longer impeller blades following the contour of the pump housing. As pump speed increases, and because of resistance to the flow of water, the impellers bend back away from the pump housing and the pump acts like a centrifugal pump. If the impeller blades are short, they remain in contact throughout the full RPM range, supplying full pressure.

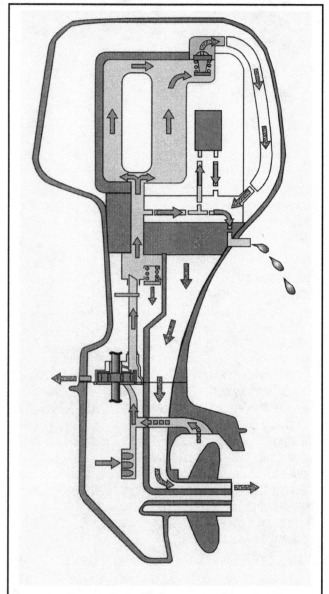

Fig. 1 Cut-away view of a typical outboard cooling system showing water flow

✳✳ WARNING

The outboard should never be run without water, not even for a moment. As the dry impeller tips come in contact with the pump housing or insert, the impeller will be damaged. In most cases, damage will occur to the impeller in seconds.

On many larger powerheads, if the powerhead overheats, a warning circuit is triggered by a temperature switch to signal the operator of an overheat condition. This should happen before major damage can occur. Reasons for overheating can be as simple as a plastic bag over the water inlet, or as serious as a leaking head gasket.

Whenever the powerhead is started and the cooling system begins pumping water through the powerhead, a water indicator stream (or mist on some early models) will appear from a cooling system indicator in the engine cover. The water stream fitting commonly becomes blocked with debris (especially when lazy operators fail to flush the system after each use, yes we said LAZY, does this mean YOU?) and ceases flowing. This leads one to suspect a cooling system malfunction. Clean the opening in the fitting using a stiff piece of wire before testing or inspecting other cooling system components.

Some motors are equipped with a water pressure relief valve that allows additional water flow at higher engine speeds by providing an additional exit passage. Increased pump flow and pressure at higher engine speeds causes the valve to open.

Whenever water is pumped through the powerhead it absorbs and removes excessive heat. This means that anytime a motor begins to overheat, there must be not enough (or no) water flowing to the powerhead (or not enough heat is being exchanged with the water that is flowing). This can happen for various reasons, including a damaged or worn impeller, clogged intake or passages or a stuck closed/restricted thermostat. A sometimes overlooked cause of overheating is the inability of the linings of the cooling passage to conduct heat. Over time, large amounts of corrosion deposits will form, especially on engines that have not received sufficient maintenance. Corrosion deposits can insulate the powerhead passages from the raw water flowing through them.

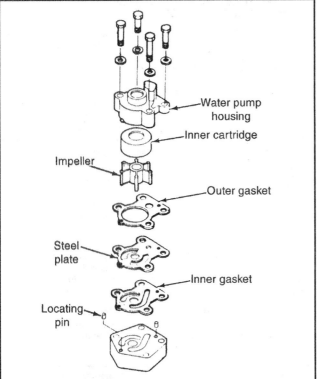

Fig. 2 Exploded view of a typical water pump assembly (note the size/shape of the housing and types of gaskets or seals used will vary by model/gearcase)

Troubleshooting

GENERAL

◆ See Figures 2 thru 5

■ When troubleshooting the cooling system, especially for overheat conditions, the motor should be run in a test tank or on a launched vessel (to simulate normal running conditions.). Running the motor on a flushing device may provide both a higher volume of water than the system would deliver (and often, water that is much colder than water that would normally be drawn into the system).

A water-cooled powerhead has a lot of problems to consider when talking about overheating. The most overlooked tends to be the simplest, clogged cooling passages or water intake grate. Although a visual inspection of the intake grate will go a long way, the cooling passage condition can really only be checked by operating the motor or disassembling it to observe the passages.

Damaged or worn cooling system components tend to cause most other problems. And since there are relatively few components, they are easy to discuss. The most obvious is a thermostat that is damaged or corroded will often cause the motor to run hot or cold (depending on the position in which the thermostat is stuck, closed or open). The water pump impeller is really the heart of the cooling system and it is easy to check, easy that is once it is accessed.

Periodic inspection and replacement of the water pump impeller is a mainstay for many mariners. There are those that wouldn't really consider launching their vessel at the beginning of a season without first replacing (or at least checking) the impeller. If the water pump is removed for inspection, check the impeller and housing for wear, grooves or scoring that might prevent proper sealing. Check for grooves in the driveshaft where the seal rides. Any damage in these areas may cause air or exhaust gases to be drawn into the pump, putting bubbles into the water. In this case, air does not aid in cooling. When inspecting the pump, consider the following:

• Is the pump inlet clear and clean of foreign material or marine growth? Check that the inlet screen is totally open. How about the impeller?

• Try and separate the impeller hub from the rubber. If it shows signs of loosening or cracking away from the hub, replace the impeller.

• Has the impeller taken a set, and are the blade tips worn down or do they look burned? Are the side sealing rings on the impeller worn away? If so, replace the impeller.

Remember that the life of the powerhead depends on this pump, so don't reuse any parts that look damaged. Are any parts of the impeller missing! If so, they must be found. Broken pieces will migrate up the water tube into the water jacket passages and cause a restriction that could block a water passage. It can be expensive or time consuming to locate the broken pieces in the water passages, but they must be found, or major damage could occur.

The best insurance against breaking the impeller is to replace it at the beginning of each boating season (are you sensing a pattern here?), and to NEVER run it out of the water. If installing a metal-bodied pump housing, coat all screws with non-hardening sealing compound to retard galvanic corrosion.

Fig. 3 Most water pumps are mounted to the top of the gearcase lower unit

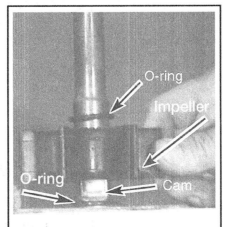

Fig. 4 The water pump impeller is the heart of the cooling system

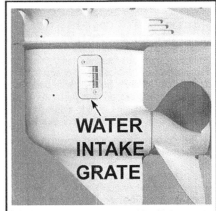

Fig. 5 The water intake grate and cooling passages should be checked and cleaned with each use

The water tube carries the water from the pump to the powerhead. Grommets seal the water tube to the water pump and exhaust housing at each end of the tube, and can deteriorate. Also, the water tube(s) should be checked for holes through the side of the tube, for restrictions, dents, or kinks.

Overheating at high RPM, but not under light load, may indicate a leaking head gasket. If a head gasket it leaking, water can go into the cylinder, or hot exhaust gases may go into the water jacket, creating exhaust bubbles and excessive heat. Remember that aluminum heads have a tendency to warp, and usually need to be surfaced each time they are removed. If necessary, they can be resurfaces by using emery paper and a surface block moving in a figure-eight motion. Also, inspect the cylinders and pistons for damage. Other areas to consider are the exhaust cover gaskets and plate. Look for corrosion pin holes. This is rare, but if the outboard has been operated in salt water over the years, there may just be a problem.

If the outboard is mounted too high on the transom, air may be drawn into the water inlet or sufficient water may not be available at the water inlet. When underway the outboard anti-ventilation plate should be running at or near the bottom of the boat and parallel to the surface of the water. This will allow undisturbed water to come to the lower unit, and the water pick-up should be able to draw sufficient water for proper cooling.

Whenever the outboard has been run in polluted, brackish or saltwater, the cooling system should be flushed. Follow the instructions provided under Flushing the Cooling System in the Maintenance & Tune-Up section for more details. But in most cases, the outboard must be flushed for at least 5 minutes. This will wash the salt from the castings and reduce internal corrosion.

There is no need to run in gear during the flushing operation. After the flushing job is done, rinse the external parts of the outboard off to remove the salt spray. As a matter of fact, many of these motors contain some form of flush fitting, which when used, allows the motor to be properly flushed without running the engine at all.

When service work is done on the water pump or lower unit, all the bolts that attach the lower unit to the exhaust housing, and bolts that hold the water pump housing (unless otherwise specified), should be coated with non-hardening gasket sealing compound to guard against corrosion. If this is not done, the bolts may become seized by galvanic corrosion and may become extremely difficult to remove the next time service work is performed.

Last but not least, check to be sure that the overheat warning system is working properly. By grounding the wire at the sending unit, the horn should sound, and/or a light should illuminate. More details can be found under Electrical Accessories (Gauges, Lights, Horns And Switches) in the Ignition & Electrical System section.

TESTING COOLING SYSTEM EFFICIENCY

◆ See Figure 6

If trouble is suspected, you can check cooling system efficiency by running the motor in a test tank or on a launched boat (while an assistant navigates) and monitoring cylinder head temperatures. There are 2 common methods available to monitor cylinder head temperature, the use of a heat sensitive marker or an electronic pyrometer.

■ **We want to take a moment here to sing the praises of an electric pyrometer. The MiniTemp® in the accompanying illustration is simply too handy to pass up. A laser pyrometer such as this will give you an instant surface temperature reading anywhere you point it. You can use it to check outboard running temperatures, to look for hot spots automotive cooling systems, to check air conditioning output in your home or car, to check fish tank temperatures in the living room, to search for drafts or insulation problems in your house. We could go on, but probably shouldn't need to at this point. Suffice it to say that once you one, you'll wonder why it took you so long to buy it.**

The Stevens Instrument company markets a product known as the Markal Thermomelt Stik®. This is a physical marker that can be purchased to check different heat ranges. The marker is designed to leave a chalky mark behind on a part of the motor that will remain chalky until it is warmed to a specific temperature, at which point the mark will melt appearing liquid and glossy.

When using a Thermomelt Stik or equivalent indicator, markers of 2 different heat specifications are necessary for this test. On all models you will want a 163°F (73°C) or 170°F (77°C) marker to check for overheating and a 125° (52°C) or 140°F (60°C) marker to determine if the motor is failing to reach normal operating temperature.

Alternately, an electronic pyrometer may be used. Many DVOMs are available with thermosensor adapters that can be touched to the cylinder head in order to get a reading. Also, some instrument companies are now producing relatively inexpensive infra-red or laser pyrometers (such as the Raytek® MiniTemp®) of a point-and-shoot design. These units are simply pointed toward the cylinder head while holding down the trigger and the electronic display will give cylinder head temperature. For ease of use and relative accuracy of information, it is hard to beat these infra-red pyrometers. Be sure to follow the tool manufacturer's instructions closely when using any pyrometer to ensure accurate readings.

To test the cooling system efficiency, obtain either a Thermomelt Stik (or equivalent temperature indicating marker) or a pyrometer and proceed as follows:

1. If available, install a shop tachometer to gauge engine speed during the test.
2. Make sure the proper propeller or test wheel is installed on the motor.
3. Place the motor in a test tank or on a launched craft.

■ **In order to ensure proper readings, water temperature must be approximately 60-80°F (18-24°C).**

4. Start and run the engine at 3000 rpm for *at least* 5 minutes.
5. Reduce engine speed to about 900 rpm as proceed as follows depending on the test equipment:
   • If using Thermomelt Stiks, make 2 marks on the top of each cylinder head (on most models, there should be a thermostat pocket on which you should make the marks), one with the low-range marker and one with the high-range marker. Continue to operate the motor at 900 rpm. The low-range mark must turn liquid and glossy or the engine is being overcooled (check the thermostat for a stuck open condition). The high-range mark must remain chalky, or the motor is overheating (check the thermostat for a stuck closed condition and then check the cooling system passages and the water pump impeller).
   • If using a pyrometer, take temperature readings on the top of each cylinder head (on most models, there should be a thermostat pocket on which to take the reading). Temperature readings must be 125-160°F (53-72°C) for all motors otherwise the engine is being over/under cooled. Check the thermostat first for either condition and then suspect the cooling water passages and/or the impeller.

■ **When checking the engine at speed (5000 rpm) expect temperatures to vary slightly from the idle test. Some models will run slightly hotter and some slightly cooler due to the differences in volume of water delivered by the cooling system when compared with engine load.**

6. Increase engine speed to 5000 rpm and continue to watch the markers or the reading on the pyrometer. The engine must not overheat at this speed either or the system components must be examined further. Examine the cooling system further if temperatures rise above 170°F (77°C).

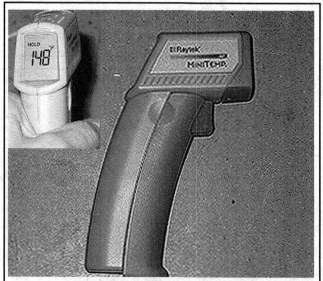

Fig. 6 By far the easiest way to check cylinder head temperature is with a hand-held pyrometer like the MiniTemp® from Raytek® pictured here

Thermostat (& Pressure Relief Valve)

Many (but not all) motors are equipped with a thermostat that restricts the amount of cooling water allowed into the powerhead until the powerhead reaches normal operating temperature. The purpose of the thermostat is to prevent cooling water from reaching the powerhead until the powerhead has warmed to normal operating temperature. In doing this the thermostat will increase engine performance and reduce emissions.

When equipped, the thermostat components are mounted in a cooling passage, under an access cover that is sealed using a gasket or an O-ring. On most models the thermostat is mounted directly into the top of the cylinder head. Although V4 motors are an exception and usually found in a Bakelite housing between the 2 cylinder heads.

Some of the larger models are also equipped with a pressure relieve valve that is usually mounted under the same cover as the thermostat. However, in a few cases they may be mounted separately under a cover secured by 2 or more bolts.

TESTING

◆ See Figure 7

Most motors are equipped with a thermostat that restricts the amount of cooling water allowed into the powerhead until the powerhead reaches normal operating temperature. The purpose of the thermostat is to prevent cooling water from reaching the powerhead until the powerhead has warmed to normal operating temperature. In doing this the thermostat will increase engine performance and reduce emissions.

However, this means that the thermostat is vitally important to proper cooling system operation. A thermostat can fail by seizing in either the open or closed positions, or it can, due to wear or deterioration, open or close at the wrong time. All failures would potentially affect engine operation.

A thermostat that is stuck open or will not fully close, may prevent a powerhead from ever fully warming, this could lead to carbon fouling, stumbling, hesitation and all around poor performance. Although these symptoms could occur at any speed, they are more likely to affect most motors at idle when high water flow through the open orifice will allow for more cooling than the lower production of heat in the powerhead requires.

A thermostat that is stuck closed will usually reveal itself right away as the engine will not only come up to temperature quickly, but the temperature warning circuit should be triggered shortly thereafter. However, thermostat that is stuck partially closed may be harder to notice. Cooling water may reach the powerhead and keep it within normal operating temperature range at various engine rpm, but allow heat to build up at other rpm. Generally speaking, engines suffering from this type of thermostat failure will show symptoms at part or full throttle, but problems can occur at idle as well. Symptoms, besides overheating, may include hesitation, stumbling, increased noise and smoke from the motor and, general, poor performance.

Testing a thermostat is a relatively easy proposition. Simply remove the thermostat from the powerhead and suspend it in a container of water, then heat the water watching for the thermostat element to move (open) and noting at what temperature it accomplishes this. Unfortunately, a few of the thermostats used on Johnson/Evinrude engines are assembled in the thermostat housing on the powerhead. On some of these motors, changes in the thermostat element (or vernatherm) may not be obvious. If you suspect a faulty or inoperable thermostat and cannot seem to verify proper opening/closing temperatures, it may be a good idea (especially since you've already gone through the trouble of removing the thermostat) to simply replace it (it's a relatively low cost part, that performs an important function). Doing so should remove it from suspicion for at least a couple of seasons.

1. Locate and remove the thermostat from the powerhead.

2. Suspend the thermostat and a thermometer in a container of water. For most accurate test results, it is best to hang the thermostat and a thermometer using lengths of string so that they are not touching the bottoms or sides of the container (this ensures that both components remain at the same temperature as the water and not the container).

3. Slowly heat the water while observing the thermostat vernatherm for movement. The moment you observe movement, check the thermometer and note the temperature. If the water begins to boil (reaches about 212°F/100°C at normal atmospheric pressure) and NO movement has occurred, discontinue the test and throw the thermostat away (if you are SURE there was no movement from the vernatherm).

4. Remove the source of heat and allow the water to cool (you can speed this up a little by adding some cool water to the container, but if you're using a glass container, don't add too much or you'll risk breaking the container). Observe the vernatherm again for movement as the water cools. When movement occurs, check the thermometer and record the temperature.

5. In most cases, the thermostat should open by 125°F (53°C) and it MUST be fully open by about 136-144°F (58-62°C).

6. Specifications for closing temperatures are not specifically provided by the manufacturer, but typically a thermostat must close a temperature close to, but below the temperature for the opening specification. A slight modulation (repeated opening and closing) or the thermostat can occur at borderline temperatures to make sure the powerhead remains in the proper operating range.

7. Replace the thermostat if it does not operate as described, or if you are unsure of the test results and would like to eliminate the thermostat as a possible problem.

REMOVAL & INSTALLATION

◆ See Figures 8 and 9

When equipped, the thermostat assembly is mounted under a cover on the powerhead or in the exhaust housing adapter. Most covers are bolted into position and sealed with a gasket (lightly coated with sealant). However, on a few late-model powerheads covers may be threaded into position and are sealed using an O-ring (which is installed dry, though we fail to see how a very light coating of marine grade grease would hurt). In either case, the thermostat itself MAY also contain its own seal or gasket that is mounted

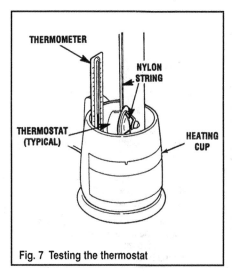

Fig. 7 Testing the thermostat

Fig. 8 Thermostats are usually located under a bolted cover. . .

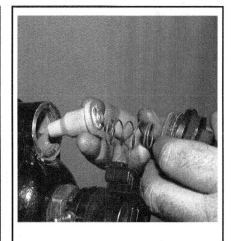

Fig. 9 . . .and are easily removed

between the thermostat lip and cylinder head. The size, shape and location of this cover, including the number of components and seals found underneath may vary greatly by model.

■ **Keep in mind that components may vary in design as well as direction of orientation. For this reason it is important that you take note of the order and orientation of each component as you disassemble them to ensure proper installation and operation.**

1. Disconnect the negative battery cable for safety.
2. Remove the engine top cover for access.
3. Locate the thermostat housing for the motor undergoing service.
4. On some motors, there may be a hose attached to the housing cover. You can disconnect the hose and move it out of the way, or in many cases, simply leave it attached to the cover.

■ **In a few cases, covers that are bolted in position can be installed facing different directions. For these models, matchmark the cover to the mating surface or otherwise make a note of cover orientation to ensure installation facing the proper direction. This is especially important for covers to which hoses attach (that have been removed).**

5. Loosen and remove the thermostat cover as follows:
• On models where the cover is bolted into place, carefully loosen and remove the cover bolts. If necessary, tap around the outside of the cover using a rubber or plastic mallet to help loosen the seal, then remove the cover from the powerhead.
• On models where the cover is threaded into position, there is normally a large flat or hex on the center of the cover. Use a suitably-sized wrench (or a large adjustable or large pair of slip-joint pliers) to loosen the cover and then unthread it from the cylinder head. When the cover has an integral hose nipple, you'll have to grip the flats from the side and not above (meaning a socket usually won't cut it).

■ **Remove the thermostat cover slowly, on most (but not all) models there is a spring located under the cover that may come loose when the cover is removed. Keep track of all components, including the order and the orientation of all components as they are removed.**

6. Check if the seal or gasket was removed with the cover. When a composite gasket and/or sealant was used, make sure all traces of gasket and sealant material are removed from the cover and the powerhead mounting surface.
7. Remove the thermostat and any mounting components. On most motors, that means removing the spring usually mounted above the thermostat. On some motors, that also includes removing the pressure relief valve and spring (mounted under the valve).

■ **If the motor utilizes an assembled thermostat, carefully remove the components, noting the orientation of each, including the cup, vernatherm (thermostat), spring and housing. Lay each component out on the worksurface in order to ensure installation facing the same directions.**

8. If the thermostat itself was sealed to the cylinder head with its own gasket, be sure to remove and discard that gasket or seal as well.
9. Visually inspect the thermostat for obvious damage including corrosion, cracks/breaks or severe discoloration from overheating. Make sure any springs have not lost tension. If necessary, refer to the Testing in this section for details concerning using heat to test thermostat function.

To install:

10. Install each of the thermostat components in the reverse of the removal procedure. Replace any gaskets, seals and/or O-rings. Pay close attention to the direction each component is installed.

■ **On all models that utilize a gasket and a cover that is bolted in position, lightly coat both sides of the new gasket using Johnson/Evinrude Gasket Sealing Compound or equivalent sealant before installation.**

11. Assemble the thermostat and/or pressure relief valve components as noted during removal.
12. Tighten the retaining bolts (or the threaded housing) securely to prevent leaks.
13. If equipped, and removed, reconnect the hose to the cover.
14. Connect the negative battery cable and verify proper cooling system operation.

Water Pump

◆ **See Figures 10, 11 and 12**

On all Johnson/Evinrude motors, the water pump is attached to the gearcase. The pump itself usually consists of a multi-vane composite material impeller attached to a portion of the driveshaft that runs through a pump housing and cover. The housing is located on the top of the gearcase lower unit (along the gearcase-to-midsection or adapter section split line) and because of this, the lower unit must be removed for access. There are multiple differences in water tubes or sleeves/grommets that may be attached to the water pump grommet depending on the year, model and gearcase, however the basic design and water pump removal/installation procedure remains the same for all motors.

■ **We have included this basic section here for quick help on your water pump - the photos and procedures are representative and may not match your pump specifically. For more detailed procedures, photos and diagrams of the water pump on your specific engine, please refer to the Gearcase Overhaul procedures in the Lower Unit section.**

Fig. 10 A wooden fixture is easily fabricated to hold the lower unit during service

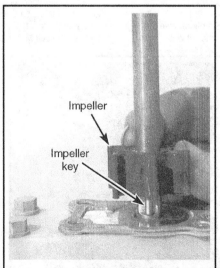

Fig. 11 The impeller (the heart of the water pump) is turned by the driveshaft using an impeller key

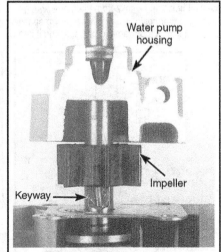

Fig. 12 This cutaway shows the difference in size between the impeller and housing. The impeller blades must bend to one side to fit within the housing

REMOVAL & INSTALLATION

◆ See Figures 10 and 13 thru 21

■ We have included this basic section here for quick help on your water pump - the photos and procedures are representative and may not match your pump specifically. For more detailed procedures, photos and diagrams of the water pump on your specific engine, please refer to the Gearcase Overhaul procedures in the Lower Unit section.

■ Replace the impeller, gaskets and any O-rings/seal whenever the water pump is removed for inspection or service. There is no reason to use questionable parts. Keep in mind that damage to the powerhead caused by an overheating condition (and the subsequent trouble that can occur from becoming stranded on the water) quickly overtakes the expense of a water pump service kit.

For ease of service mount the gearcase in a support such as a homemade cavitation plate holder. To fabricate a cavitation plate holder, cut a groove in a short piece of 2" x 6" or 2" x 8" piece of wood. Cut the groove so it can accommodate the lower unit with the cavitation plate resting on top of the wood. Clamp the wood in a vise to hold the lower unit securely during service.

1. Remove the gearcase lower unit from the intermediate housing. For details, refer to the Lower Unit section.

■ On electric shift models, take time to notice how the shift wires are routed and anchored in position with a clamp on top of the water pump. It is extremely important for the shift wires to be routed and secured in the same position during installation. If you have a digital camera, take a picture.

2. If equipped, remove the O-ring from the top of the driveshaft and/or the water tube or grommet from the top of the pump housing.

3. Remove the bolts, usually 4, securing the water pump to the lower unit housing. On electric shift models, one or more of the bolts will secure a wire clamp, note the position of the clamp and the wire routing before removal to ensure proper positioning during installation.

■ On some models the water pump housing is mounted to a gearcase cover through which the shift rod is inserted and which is bolted to the top of the gearcase using its own fasteners (usually 3, two behind the pump and one in front of the pump). On these models DO NOT remove the gearcase cover bolts unless you are planning on overhauling the gearcase. Removal of the bolts and any repositioning of the cover will allow the shifter to dislodge, requiring gearcase disassembly to reset it.

4. Slide or work the housing up and off the driveshaft.

■ Depending on whether or not the impeller pulls off the driveshaft with the housing, either remove it from the housing using a pair of needle-nose pliers or gently pry it upward from the shaft using a small prybar.

5. If the impeller did not come off with the cover, work the impeller off the driveshaft.

6. Pop the impeller Woodruff key out of the driveshaft keyway.

7. Slide the water pump base plate up and off of the drive shaft. Certain models will utilize a gasket here as well - if yours does, remove this too.

8. Carefully remove any traces of sealant (if present) from the mating surfaces.

9. Thoroughly inspect the impeller, cover, insert (also known as an impeller cup) and impeller plate for signs of damage or wear and replace, as necessary. For more details, please refer to Inspection & Overhaul, in this section. If the insert is damaged or if the driveshaft O-ring (located between the seal and housing) is to be replaced, pull the insert out of the cover with the needle nose pliers.

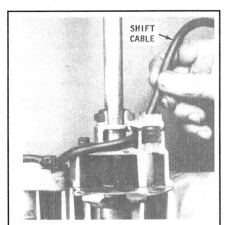

Fig. 13 Always pay attention to shift cable routing on models so equipped (usually electric shift units)

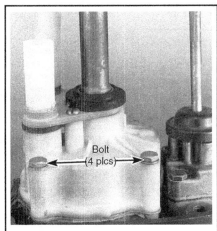

Fig. 14 Find the housing mounting bolts. . .

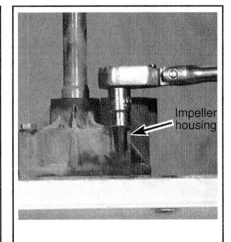

Fig. 15 . . .and then remove them

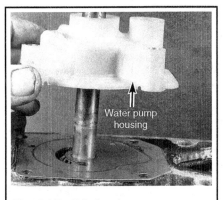

Fig. 16 Lift off the housing. . .

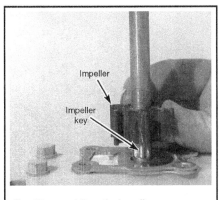

Fig. 17 . . .and then the impeller. . .

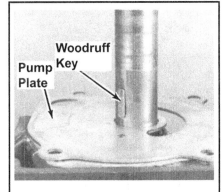

Fig. 18 . . .and key

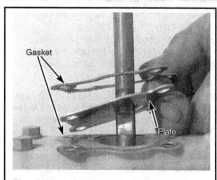

Fig. 19 You might as well remove the gasket (if equipped) and plate while you're at it

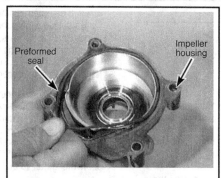

Fig. 20 Turn over the housing and remove all seals and O-rings. . .

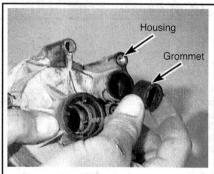

Fig. 21 . . .and then remove any grommets from the other side if equipped

■ If you've taken the trouble to go this far (to service a water pump) it is always a good idea to replace both the impeller and the insert to ensure proper pump and cooling system operation.

10. If equipped, remove the irregularly shaped O-ring from the underside of the water pump housing.

To Install:

◆ See Figures 22 thru 32

11. With the insert removed from the housing, inspect it carefully for any signs of damage, cracks, wear or melting and replace, if necessary.

■ An improved water pump is available as a replacement for many later models. If the old water pump housing is unfit for further service, only the new pump housing can be purchased. It is strongly recommended to replace the water pump with the improved model while the lower unit is disassembled (that is assuming it was not already done considering the age of the unit). The new pump must be assembled before it is installed. Therefore, the following steps outline procedures for both pumps.

Fig. 22 There is a new style pump available for later models, the original is on the left and the new one on the right

12. To pre-assemble and install the improved pump housing, proceed as follows:

a. Remove the water pump parts from the container.

b. Insert the plate into the housing. The tang on the bottom side of the plate MUST index into the short slot in the pump housing.

c. Slide the pump liner into the housing with the 2 small tabs on the bottom side indexed into the 2 cut-outs in the plate.

d. Coat the inside diameter of the liner with light-weight oil.

e. Work the impeller into the housing with all of the blades bent back to the right. In this position, blades will rotate properly when the pump housing is installed. Remember, the pump and the blades will be rotating CLOCKWISE when the housing is turned over and installed in place on the lower unit.

f. Coat the mating surface of the lower unit with a suitable sealer.

g. Slide the water pump base plate down the driveshaft and into place on the lower unit. Insert the Woodruff key into the key slot in the drive shaft.

h. Lay down a very thin bead of a suitable sealer into the irregular shaped groove in the housing.

i. Insert the seal into the groove and then coat the seal using the sealer.

j. Begin to slide the water pump down the driveshaft, and at the same time observe the position of the slot in the impeller. Continue to work the pump down the driveshaft, with the slot in the impeller indexed over the Woodruff key. The pump must be fairly well aligned before the key is covered because the slot in the impeller is not visible as the pump begins to come close to the base plate.

k. Install the short forward bolt through the pump and into the lower unit. DO NOT tighten this bolt at this time. Insert the grommet into the pump housing.

✳✳ CAUTION

On electric shift models, make sure the wiring is routed as noted during removal.

l. Install the grommet retainer and water tube guide onto the pump housing. Install the remaining pump attaching bolts. On electric shift models, install the solenoid cable bracket with the same bolt and in the same position from which it was removed. On some units the solenoid cable fits into a

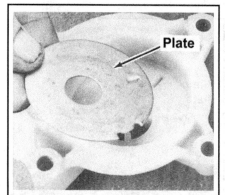

Fig. 23 Insert the plate (if equipped) into the housing

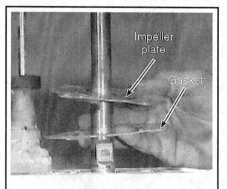

Fig. 24 To prepare for installation, install the impeller plate using a new gasket. . .

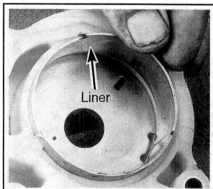

Fig. 25 . . .install a new insert (liner) to the pump housing . . .

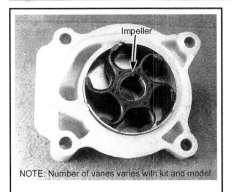

Fig. 26 . . . then install the impeller while rotating counterclockwise (so the blades face clockwise, the normal direction of rotation)

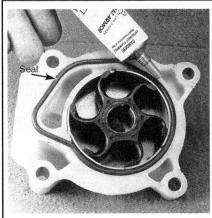

Fig. 27 Apply a thin bead of sealant, then install the pump housing O-ring seal

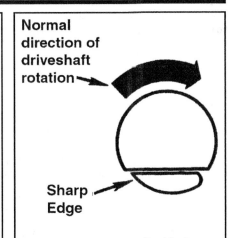

Fig. 28 Driveshaft and impeller drive key installation viewed from **ABOVE** the housing

recess of the water pump and is held in place in that manner. Tighten the bolts alternately and evenly, and at the same time rotate the driveshaft CLOCKWISE. If the driveshaft is not rotated while the attaching bolts are being tightened, it is possible to pinch one of the impeller blades underneath the housing.

m. Slide the large grommet down the driveshaft and seat it over the pump collar. This grommet does not require sealer. Its function is to prevent exhaust gases from entering the water pump.

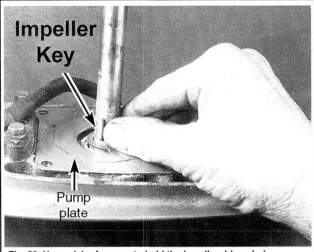

Fig. 29 Use a dab of grease to hold the impeller drive pin in position . . .

13. To install the original equipment housing, proceed as follows:

a. Coat the water pump plate mating surface of the lower unit with a suitable sealer.

b. Slide the water pump plate down the driveshaft and *before* it makes contact with the sealer, check to be sure the bolt holes in the plate will align with the holes in the housing. The plate will only fit one way. If the holes will not align, remove the plate, turn it over and again slide the plate down the driveshaft and into place on the housing. This checking will prevent accidentally getting the sealer on both sides of the plate. If by chance sealer does get on the top surface it MUST be removed before the water pump impeller is installed.

c. Slide the water pump impeller down the driveshaft. Just before the impeller covers the cut-out for the Woodruff key, install the key, and then work the impeller on down, with the slot in the impeller indexed over the Woodruff key. Continue working the impeller down until it is firmly in place on the surface of the pump plate.

d. Check to be sure NEW seals and O-rings have been installed in the water pump. Lubricate the inside surface of the water pump with light-weight oil.

e. Lower the water pump housing down the driveshaft and over the impeller. Always rotate the driveshaft slowly CLOCKWISE as the housing is lowered over the impeller to allow the impeller blades to assume their natural and proper position inside the housing. Continue to rotate the driveshaft and work the water pump housing downward until it is seated on the plate.

✳✳ CAUTION

On electric shift models, make sure the wiring is routed as noted during removal.

f. Coat the threads of the water pump attaching screws with sealer, and then secure the pump in place with the screws. On electric shift models,

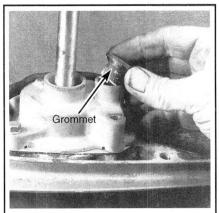

Fig. 30 If not done already, the water tube grommet should be replaced. . .

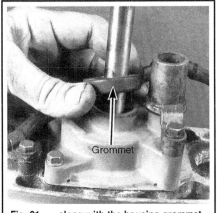

Fig. 31 . . .along with the housing grommet (secure both with adhesive sealant). . .

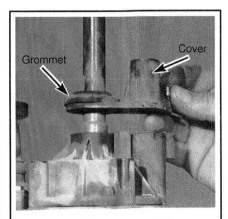

Fig. 32 . . . and the cover must be in position (if equipped)

install the solenoid cable bracket with the same bolt and in the same position from which it was removed. On some units, the solenoid cable fits into a recess of the water pump and is held in place in that manner. Tighten the screws alternately and evenly.

14. Install the gearcase as detailed in the Lower Unit section.

INSPECTION & OVERHAUL

Let's face it, you've gone through the trouble of removing the water pump for a reason. Either, you've already had cooling system problems and you're looking to fix it, or you are looking to perform some preventive maintenance. Although the truth is that you can just remove and inspect the impeller, replacing only the impeller (or even reusing the impeller if it looks to be in good shape) WHY would you? The cost of a water pump rebuild kit is very little when compared with the even the time involved to get this far. If you've misjudged a component, or an O-ring (which by the way, never reseal quite the same way the second time), then you'll be taking this lower unit off again in the very near future to replace these parts. And, at best this will be because the warning system activated or you noticed a weak coolant indicator stream (spray) or, at worst, it will because you're dealing with the results of an overheated powerhead.

In short, if there is one way to protect you and your engine, it is to replace the impeller and insert (if used) along with all O-ring seals, grommets and gaskets, anytime the pump housing is removed. If not, take time to thoroughly clean and inspect the old impeller, housing and related components before assembly and installation. New components should be checked against the old. Seek explanations for differences with your parts supplier (but keep in mind that some rebuild kits may contain upgrades or modifications)

1. Remove and disassemble the water pump assembly as detailed in this section.

2. Carefully remove all traces of gasket or sealant material from components. If some material is stubborn, use a suitable solvent in the next steps to help clean material. Avoid scraping whenever possible, especially on plastic components whose gasket surfaces are easily scored and damaged.

3. Clean all metallic components using a mild solvent (such as Simple Green® cut with water or mineral spirits), then dry using compressed air (or allow them to air dry).

4. Clean plastic components using isopropyl alcohol or Johnson/Evinrude Cleaning Solvent.

■ Skip the next step, we really mean it, don't INSPECT the impeller, REPLACE IT. Ok, we've been there before, if you absolutely don't want to replace the impeller. Let's say it's only been used one season or so and you're here for another reason, but you're just being thorough and checking the pump, then perform the next step.

5. Check the impeller for missing, brittle or burned blades. Inspect the impeller side surfaces and blade tips for cracks, tears, excessive wear or a glazed (or melted) appearance. Replace the impeller if these defects are found. Next, squeeze the vanes toward the hub and release them. The vanes should spring back to the extended position. Replace the impeller if the vanes are set in a curled position and do not spring back when released.

6. The water pump impeller should move smoothly up or downward on the driveshaft. If not, it could become wedged up against the housing or down against the impeller plate, causing undue wear to the top or bottom of the blades and hub. Check the impeller on the driveshaft and, if necessary, clean the driveshaft contact surface (inside the impeller hub) using emery cloth.

7. Inspect the water pump body or the lining inside the water pump body, as applicable, for burned, worn or damaged surfaces. Replace the impeller lining, if equipped, or the water pump body if any defects are noted.

8. Visually check the water pump housing (and insert, if equipped), along with the impeller wear plate for signs of overheating including warpage (especially on the plate) or melted plastic. Some wear is expected on the impeller plate and, if equipped, on the housing insert, but deep grooves (with edges) are signs of excessive wear requiring component replacement.

■ A groove is considered deep or edged if it catches a fingernail.

9. Check the water tube grommets and seals for a burned appearance or for cracked or brittle surfaces. Replace the grommets and seals if any of these defects are noted.

6

POWERHEAD

POWERHEAD

Introduction

■ **Please refer to the Exploded Views section for exploded views of the powerhead and its components.**

You can compare the major components of an outboard with the engine and drivetrain of your car or truck. In doing so, the powerhead is the equivalent of the engine and the gearcase is the equivalent of your drivetrain (the transmission/transaxle). The powerhead is the assembly that produces the power necessary to move the vehicle, while the gearcase is the assembly that transmits that power via gears, shafts and a propeller (instead of tires).

Speaking in this manner, the powerhead is the "engine" or "motor" portion of your outboard. It is an assembly of long-life components that are protected through proper maintenance. Lubrication, the use of high-quality 2-stroke oil and proper fuel/oil ratios are the most important ways to preserve powerhead condition. Similarly, proper tune-ups that help maintain proper air/fuel mixture ratios and prevent pinging, knocking or other potentially damaging operating conditions are the next best way to preserve your motor. But, even given the best of conditions, components in a motor begin wearing the first time the motor is started and will continue to do so over the life of the powerhead.

Eventually, all powerheads will require some repair. The particular broken or worn component, plus the age and overall condition of the motor may help dictate whether a small repair or major overhaul is warranted. The complexity of the job will vary with 2 major factors. As much as you can generalize about mechanical work:

• The age of the motor (the older OR less well maintained the motor is) the more difficult the repair

• The larger and more complex the motor, the more difficult the repair.

Again, these are generalizations and, working carefully, a skilled do-it-yourself boater can disassemble and repair a 2 hp or 125 hp powerhead, as well as a seasoned professional. But both DIYers and professionals must know their limits. These days, many professionals will leave portions of machine work (from cylinder block and piston disassembly, cleaning and inspection to honing and assembly up to a machinist). This is not because they are not capable of the task, but because that's what a machinist does day in and day out. A machinist is naturally going to be more experienced (and efficient) with the procedures.

If a complete powerhead overhaul is necessary on your outboard, we recommend that you find a local machine shop that has both an excellent reputation and that specializes in marine work. This is just as important and handy a resource to the professional as a DIYer. If possible, consult with the machine shop before disassembly to make sure you follow procedures or mark components, as they would desire. Some machine shops would prefer to perform the disassembly themselves. In these cases, you can usually remove the powerhead from the gearcase and deliver the entire unit to the shop for disassembly, inspection, machining and assembly.

If you decide to perform the entire overhaul yourself, proceed slowly, taking care to following instructions closely. Consider using a digital camera (if available) to help document assemblies during the removal and disassembly procedures. This can be especially helpful if the overhaul or rebuild is going to take place over an extended amount of time. If this is your first overhaul, don't even *think* about trying to get it done in one weekend, YOU WON'T. It is better to proceed slowly, asking help when necessary from your trusted parts counterman or a tech with experience on these motors.

Keep in mind that anytime pistons, rings and bearings have been replaced, the powerhead must be broken-in again, as if it were a brand-new motor. Once a major overhaul is completed, refer to the section on Powerhead Break-In for details on how to ensure the rings set properly without damage or scoring to the new cylinder wall or the piston surfaces. Careful break-in or a properly overhauled motor will ensure many years of service for the trusty powerhead.

■ **The information in this section is organized from start to finish in the logical order of a complete teardown and rebuild. Procedures for disassembly are first, followed by various cleaning and inspection procedures and finally, ending with complete installation. Obviously to remove or service just one portion of the powerhead you should only follow the appropriate removal procedures then skip to the relevant cleaning/inspection and installation procedures.**

The carburetion and ignition principles of 2-cycle engine operation MUST be understood in order to perform a proper tune-up on an outboard motor.

Therefore, it would be well worth the time to study the principles of 2-cycle engines as outlined in this section.

Theory Of Operation

◆ **See Figures 1 thru 6**

The 2-cycle engine differs in several ways from a conventional four-cycle (automobile) engine.

1. The method by which the fuel-air mixture is delivered to the combustion chamber.
2. The complete lubrication system.
3. In most cases, the ignition system.
4. The frequency of the power stroke.

These differences will be discussed briefly and compared with four-cycle engine operation.

INTAKE/EXHAUST

◆ **See Figure 1**

2-cycle engines utilize an arrangement of port openings to admit fuel to the combustion chamber and to purge the exhaust gases after burning has been completed. The ports are located in a precise pattern in order for them to be open and closed off at an exact moment by the piston as it moves up and down in the cylinder. The exhaust port is located slightly higher than the fuel intake port. This arrangement opens the exhaust port first, as the piston starts downward, and therefore, the exhaust phase begins a fraction of a second before the intake phase.

Actually, the intake and exhaust ports are spaced so closely together that both open almost simultaneously. For this reason, the pistons of most 2-cycle engines have a deflector-type top. This design of the piston top serves two purposes very effectively.

First, it creates turbulence when the incoming charge of fuel enters the combustion chamber. This turbulence results in more complete burning of the fuel than if the piston top were flat. The second effect of the deflector-type piston crown is to force the exhaust gases from the cylinder more rapidly.

This system of intake and exhaust is in marked contrast to individual valve arrangement employed on 4-cycle engines.

LUBRICATION

◆ **See Figure 3**

A 2-cycle engine is lubricated by mixing oil with the fuel. Therefore, various parts are lubricated as the fuel mixture passes through the crankcase and the cylinder. Four-cycle engines have a crankcase containing oil. This oil is pumped through a circulating system and returned to the crankcase to begin the routing again.

POWER STROKE

The combustion cycle of a 2-cycle engine has 4 distinct phases.
• Intake
• Compression
• Power
• Exhaust

Three phases of the cycle are accomplished with each stroke of the piston, and the fourth phase, the power stroke occurs with each revolution of the crankshaft. Compare this system with a 4-cycle engine. A stroke of the piston is required to accomplish each phase of the cycle and the power stroke occurs on every other revolution of the crankshaft. Stated another way, two revolutions of the 4-cycle engine crankshaft are required to complete 1 full cycle, the 4 phases.

PHYSICAL LAWS

◆ **See Figure 2**

The 2-cycle engine is able to function because of 2 very simple physical laws.

One: Gases will flow from an area of high pressure to an area of lower pressure. A tire blow-out is an example of this principle. The high-pressure

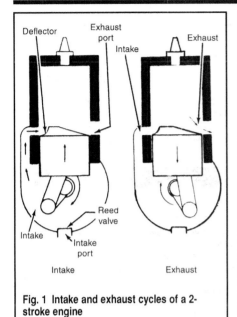

Fig. 1 Intake and exhaust cycles of a 2-stroke engine

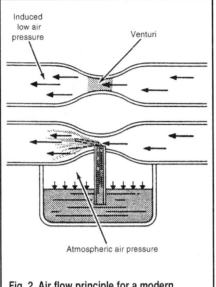

Fig. 2 Air flow principle for a modern carburetor

Fig. 3 Adding oil to the fuel tank

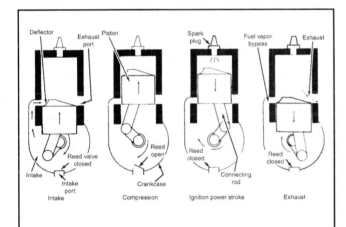

Fig. 4 Complete piston cycle of a 2-stroke engine

air escapes rapidly if the tube is punctured.

Two: If a gas is compressed into a smaller area, the pressure increases, and if a gas expands into a larger area, the pressure is decreased.

If these 2 laws are kept in mind, the operation of the 2-cycle engine will be easier understood.

ACTUAL OPERATION

◆ See Figures 4 thru 6

Beginning with the piston approaching top dead center on the compression stroke: The intake and exhaust ports are closed by the piston; the reed valve is open; the spark plug fires; the compressed fuel-air mixture is ignited; and the power stroke begins. The reed valve was open because as the piston moved upward, the crankcase volume increased, which reduced the crankcase pressure to less than the outside atmosphere.

As the piston moves downward on the power stroke, the combustion chamber is filled with burning gases. As the exhaust port is uncovered, the gases, which are under great pressure, escape rapidly through the exhaust ports. The piston continues its downward movement. Pressure within the

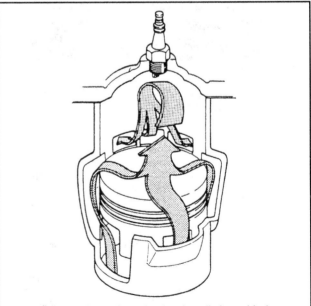

Fig. 5 Fuel flow of the loop charge while the piston is on the down stroke

Fig. 6 Depiction of the exhaust leaving the cylinder and fuel entering through the three ports in the cylinder

crankcase increases, closing the reed valves against their seats. The crankcase then becomes a sealed chamber. The air-fuel mixture is compressed ready for delivery to the combustion chamber. As the piston continues to move downward, the intake port is uncovered. Fresh fuel rushes through the intake port into the combustion chamber striking the top of the piston where it is deflected along the cylinder wall. The reed valve remains closed until the piston moves upward again.

When the piston begins to move upward on the compression stroke, the reed valve opens because the crankcase volume has been increased, reducing crankcase pressure to less than the outside atmosphere. The intake and exhaust ports are closed and the fresh fuel charge is compressed inside the combustion chamber.

Pressure in the crankcase decreases as the piston moves upward and a fresh charge of air flows through the carburetor picking up fuel. As the piston approaches top dead center, the spark plug ignites the air-fuel mixture, the power stroke begins and one complete cycle has been completed.

Cross Fuel Flow Principle

Johnson/Evinrude pistons are a deflector dome type. The design is necessary to deflect the fuel charge up and around the combustion chamber. The fresh fuel mixture enters the combustion chamber through the intake ports and flows across the top of the piston. The piston design contributes to clearing the combustion chamber, because the incoming fuel pushes the burned gases out the exhaust ports.

Loop Scavenging

The 1971-72 50 hp 2 cylinder motor and all 3 cylinder powerheads have what is commonly known as a loop scavenging system. The piston dome is relatively flat on top with just a small amount of crown. Pressurized fuel in the crankcase is forced up through the skirt of the piston and out through irregular shaped openings cut in the skirt. After the fuel is forced out the piston skirt openings it is transferred upward through long deep grooves molded into the cylinder wall. The fuel then enters the combustion portion of the cylinder and is compressed, as the piston moves upward.

This particular powerhead does not have intake cover plates, because the intake passage is molded into the cylinder wall as described in the previous paragraph. Therefore, if these engines are being serviced, disregard the sections covering intake cover plates.

Timing

The exact time of spark plug firing depends on engine speed. At low speed the spark is retarded - fires later than when the piston is at or beyond top dead center. Therefore, the timing is advanced as the throttle is advanced.

At high speed, the spark is advanced - fires earlier than when the piston is at top dead center.

Procedures for making the timing and synchronization adjustment will be found in the Maintenance & Tune-Up section.

POWERHEAD OVERHAUL

Powerhead Rebuilding Preliminary Work - Read This First

◆ See Figure 7

Although really quite simple, before the powerhead can be disassembled:
- the battery must be disconnected;
- fuel lines disconnected;
- the carburetor removed;
- the alternator and starter removed;
- the flywheel and magneto/ignition components all removed.

Take time to identify all hoses and wires to ensure they are connected and routed properly during installation. You may elect want to take a series of digital photographs of the engine with the flywheel removed: one from the top, and a couple from the sides showing the wiring and arrangement of parts.

■ **If in doubt as to how these items are to be removed, refer to the appropriate section.**

After the accessories have been removed, remove the bolts in the front and rear of the powerhead securing the powerhead to the exhaust housing. Lift the powerhead free of the exhaust housing. On 3 cyl/V4 motors, there are 3 bolts on each side of the powerhead, the 2 in the front, and the 2 nuts from the studs at the rear of the powerhead. . .1 & 2 cyl motors are similar.

SUMMARY

More than one phase of the cycle occurs simultaneously during operation of a 2-cycle engine. On the downward stroke, power occurs above the piston while the ports are closed. When the ports open, exhaust begins and intake follows. Below the piston, fresh air-fuel mixture is compressed in the crankcase.

On the upward stroke, exhaust and intake continue as long as the ports are open. Compression begins when the ports are closed and continues until the spark plug ignites the air-fuel mixture. Below the piston, a fresh air-fuel mixture is drawn into the crankcase ready to be compressed during the next cycle.

Rebuilding Tips

TORQUE VALUES

All torque values must be met when they are specified. Many of the outboard castings and other parts are made of aluminum. The torque values are given to prevent stretching the bolts, but more importantly to protect the threads in the aluminum. It is extremely important to tighten the connecting rods to the proper torque value to ensure proper service. The head bolts are probably the next most important torque value.

POWERHEAD COMPONENTS

Service procedures for the carburetors, fuel pumps, starter, and other powerhead components are given in their respective sections.

REED INSTALLATION

All reeds on Johnson/Evinrude engines covered here are installed just behind the carburetor behind the intake manifold.

CLEANLINESS

Make a determined effort to keep parts and the work area as clean as possible. Parts MUST be cleaned and thoroughly inspected before they are assembled, installed, or adjusted. Use proper lubricants, or their equivalent, whenever they are recommended.

Keep rods and rod caps together as a set to ensure they will be installed as a pair and in the proper sequence.

Needle bearings must remain as a complete set. NEVER mix needles from one set with another. If only one needle is damaged, the complete set must be replaced.

If the unit is several years old, or if it has been operated in salt water, or has not had proper maintenance, or shelter, or any number of other factors, then separating the powerhead from the exhaust housing may not be a simple task. An air hammer may be required on the studs to shake the corrosion loose; heat may have to be applied to the casting to expand it slightly; or other devices employed in order to remove the powerhead.

One very serious condition would be the driveshaft "frozen" with the crankshaft. In this case, a circular plug-type hole must be drilled and a torch used to cut the driveshaft. Let's assume the powerhead will come free on the first attempt.

The following procedures in this section pick up the work after these preliminary tasks have been completed.

Cylinder Head

REMOVAL & SERVICE

◆ See Figures 8 thru 17

■ **Please refer to the Exploded Views section for exploded views of the powerhead and its components.**

Fig. 7 Cleaning the pistons while they remain in the powerhead

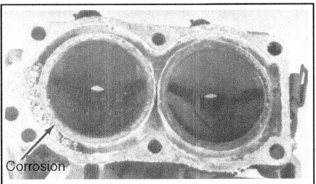

Fig. 8 The water passages on this block are corroded, preventing water circulation

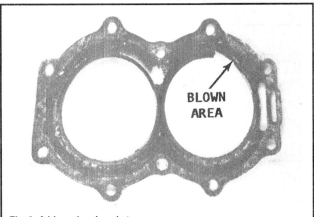

Fig. 9 A blown head gasket

Usually the head is removed and an examination of the cylinders made to determine the extent of overhaul required. However, if the head has not been removed, back out all of the head bolts and lift the head free of the powerhead.

A thermostat is installed in the cylinder head of all 3 cylinder engines and most (but not all) 1 & 2 cyl engines. On V4 engines, the thermostat is usually installed in a Bakelite housing between the 2 heads just below the exhaust housing plate. The thermostat has 3 parts, a top, middle, and bottom. A hose connects the thermostat with the/each head. In addition to the thermostat, the engine may very well have a thermostat bypass valve. These items are easily removed, inspected and cleaned.

Normally, if a thermostat is not functioning properly, it is almost always stuck in the open position. An engine operating at too low a temperature is almost as much a problem as an engine running too hot.

Therefore, during a major overhaul, good shop practice dictates to replace the thermostat and eliminate this area as a possible problem at a later date.

Lay a piece of fine sandpaper or emery paper on a flat surface (such as a piece of glass) with the abrasive side facing up. With the machined face of the head on the sandpaper, move the head in a circular motion to dress the surface. This procedure will also indicate any "high" or "low" spots.

Check the spark plug openings to be sure the threads are not damaged. Most marine dealers can insert a heli-coil into a spark plug opening if the threads have been damaged.

On many engines, a sending unit is installed in the head to warn the operator if the engine begins to run too hot. The light on the dash can be checked by turning the ignition switch to the **ON** position, and then grounding the wire to the sending unit. The light should come on. If it does not, replace the bulb and repeat the test. Many later models changed to a horn in place of the light, but it is unlikely that anything covered here will have this unless it has been retro-fitted.

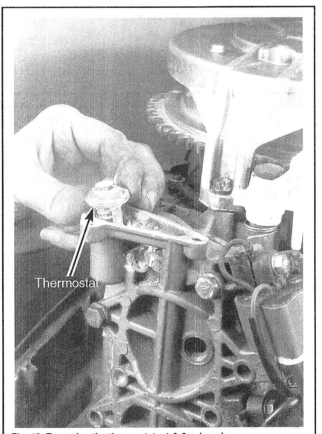

Fig. 10 Removing the thermostat - 1 & 2 cyl engine

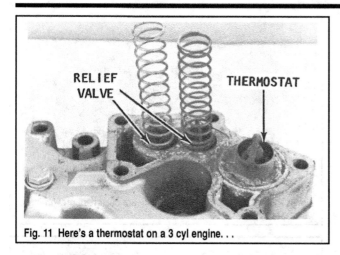

Fig. 11 Here's a thermostat on a 3 cyl engine. . .

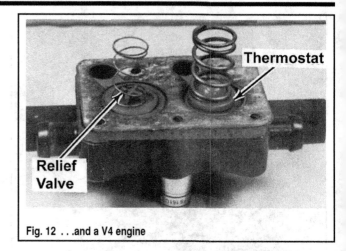

Fig. 12 . . .and a V4 engine

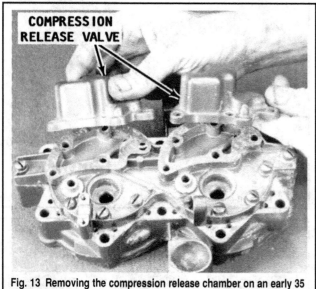

Fig. 13 Removing the compression release chamber on an early 35 hp model

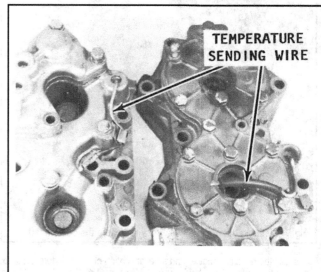

Fig. 14 The water temperature sending unit wires on a 3 cylinder (L) and V4 (R) engine

Fig. 15 Loosen the cylinder head bolts. . .

Fig. 16 . . .and remove the head (2 cyl shown)

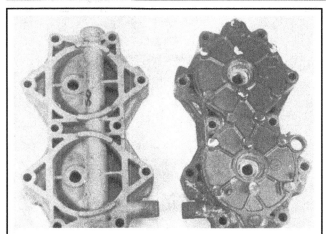

Fig. 17 The head on the left is used on 1958-59 50 hp V4 engines, the one on the right is used on later engines. Note that the one on the right has a built-in water jacket, while the one on the left does not

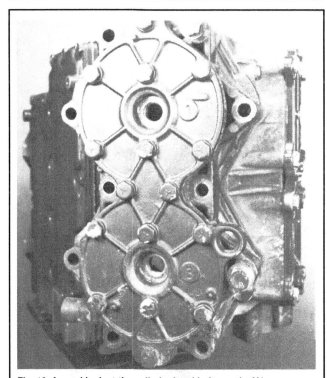

Fig. 18 A good look at the cylinder head bolts on the V4. . .

Fig. 19 . . .and the 3 cylinder

Fig. 20 Tightening the head bolts on the V4 - others similar

To Install:

◆ See Figures 18, 19 and 20

1. Position a NEW head gasket in place on the cylinder block. Usually, the head gaskets for late model powerheads are clearly identified as to which side faces the block and the head. Take an extra minute to be sure the gasket is properly positioned. If the gasket does *not* have any identification, it may be installed with either side facing the block or head. Check to be sure all bolt holes are properly aligned, indicating the gasket is positioned properly end for end. NEVER use automotive type head gasket sealer. The chemicals in the sealer will cause electrolytic action and eat the aluminum faster than you can get to the bank for money to buy a new cylinder block.

2. Carefully place the head in position. Install the head bolts and tighten them finger tight, then just a bit more. Now, tighten the bolts alternately and *evenly* in 3 rounds to the torque value specified in the Torque Specifications chart. On the first round, tighten the bolts to 1/2 the total torque value; on the second round, to 3/4 the total torque value; and to the full torque value on the third and final round.

■ If working on a V4 motor, repeat the last step for the other cylinder bank.

3. Install the assembled powerhead to the exhaust housing and tighten the attaching bolts alternately and evenly until secure.

4. Install all powerhead accessories including the flywheel, carburetor, magneto and ignition components, starter, etc. If any doubts or difficulties are encountered, follow the procedures outlined in the sections covering the particular component.

5. Connect the fuel lines, wiring and, if applicable, battery cables.

6. The complete outboard unit is now ready to be started and "broken in" according to the procedures outlined in the that section.

Reed Valves

DESCRIPTION & OPERATION

◆ **See Figures 21 thru 26**The fuel delivery may be one of several types. All 1, 2 and 3 cylinder engines are equipped with 1, 2 or 3 carburetors, 1 for each cylinder. V4 engines have 2 carburetors, each with double barrels, each barrel serving a single cylinder.

The 3 cylinder engines covered in this manual are equipped with an upper seal and bearing, a lower seal and bearing, and 2 main bearings between. A labyrinth seal is used at the 2 center main bearings and just above the bottom seal, to provide an effective seal between the cylinders.

On the V4 engines covered here, a top and bottom seal is used with sealing rings around the crankshaft in the center between each cylinder, in each bank, to seal for pressure and vacuum. The V4 engines use 6 rings.

Therefore, with a 2 cylinder powerhead, two sets of reeds are installed; one for each cylinder. These reeds may be installed on a reed plate or with a reed box, depending on the model engine. One carburetor provides fuel to both sets of reeds.

The reed arrangement operates in much the same manner as the reed in a saxophone or other wind instrument. At rest, the reed is closed and seals the opening to which it is attached. In the case of an outboard engine, this opening is between the crankcase and the carburetor. The reeds are mounted in the intake manifold, just behind the carburetor

On a 3 cylinder powerhead, 3 sets of reeds are used, 1 for each cylinder. On a V4 powerhead, 4 sets of reeds are installed. These reeds are installed on a reed plate with a reed box. Fuel is delivered to the reeds as described in the paragraph above, Fuel Delivery.

The reed arrangement operates in much the same manner as the reed in a saxophone or other wind instrument. At rest, the reed is closed and seals the opening to which it is attached. In the case of an outboard engine, this opening is between the crankcase and the carburetor. The reeds are mounted in the intake manifold, just behind the carburetor.

ACTUAL OPERATION

The piston creates vacuum and pressure in the crankcase as it moves up and down in the cylinder. As the piston moves upward, a vacuum is created in the crankcase. This vacuum "lifts" the reed off its seat, allowing fuel to pass. On the compression stroke, when the piston moves downward, pressure is created and the reed is forced closed.

REED DESIGNS

A wide range of reed boxes may be found on an outboard unit, due to the varying designs of the engines. All installations employ the same principle and there is no difference in their operation.

BROKEN REED

◆ **See Figure 27**

A broken reed is usually caused by metal fatigue over a long period of time. The failure may also be due to the reed flexing too far because the reed stop has not been adjusted properly or the stop has become distorted. If the reed is broken, the loose piece MUST be located and removed, before the engine is returned to service. The piece of reed may have found its way into the crankcase, into the passage leading to the cylinder, or in the cylinder. If the broken piece cannot be located, the powerhead must be completely disassembled until it is located and removed.

An excellent check for a broken reed on an operating engine is to hold an ordinary business card approximately 2 in. (9.08 cm) in front of the carburetor. Under normal operating conditions, a very small amount of fine mist will be noticeable, but if fuel begins to appear rapidly on the card from the carburetor, one of the reeds is broken and causing the back flow through the carburetor and onto the card.

A broken reed will cause the engine to operate roughly and with a "pop" back through the carburetor.

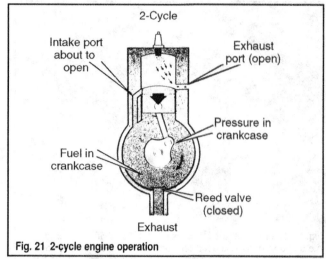

Fig. 21 2-cycle engine operation

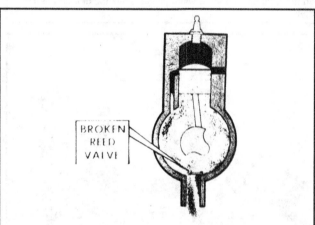

Fig. 22 This cross-section illustrates a broken reed in the crankcase

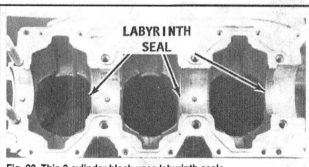

Fig. 23 This 3 cylinder block uses labyrinth seals

Fig. 24 This V4 block uses sealing rings

Fig. 25 A good look at the reed box on 75 hp and larger motors

Fig. 26 Here's one on the early 50 hp V4 engine (1958-59)

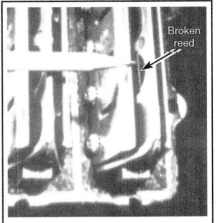

Fig. 27 This cross-section illustrates a broken reed in the crankcase

REED STOPS

◆ See Figure 28

If the reed stops have become distorted, the most effective corrective action is to replace the complete reed box as an assembly.

V-TYPE REED BOXES

◆ See Figure 29

As the name implies, these reed boxes are shaped in a "V" with a set of reeds and stops on both arms of the "V". If a problem develops with this type reed box, it is strongly recommended that you replace the complete assembly - reeds, box, and stops. The assembly may usually be purchased as a complete unit and the cost will usually not exceed the time, effort, and problems encountered in an attempt to replace only one part.

REED-TO-REED BOX CHECK

◆ See Figures 30 and 31

The specified clearance of the reed from the base plate, when the reed is at rest, is 0.010 in. (0.254mm) at the tip of the reed.

An alternate method of checking the reed clearance is to hold the reed up to the sunlight and look through the back side. Some air space should be visible, but not a great amount. If in doubt, check the reed at the tip with a feeler gauge. The maximum clearance should not exceed 0.010 in. (0.254mm). If the clearance is excessive, the reed box must be replaced as a complete assembly.

The reeds must NEVER be turned over in an attempt to correct a problem. Such action would cause the reed to flex in the opposite direction and the reed would break in a very short time.

REED VALVE ADJUSTMENT

◆ See Figure 32

In many instances, the reed is placed on the reed plate in such a manner to cover the openings in the plate. As shown in the accompanying illustration, a small indent is manufactured into the face of the plate. The leaves of the reed should be centered on this indentation for proper operation. If the reed is being replaced, both reeds AND the reed stops should be replaced as a set.

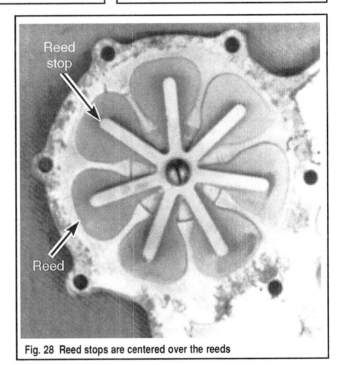

Fig. 28 Reed stops are centered over the reeds

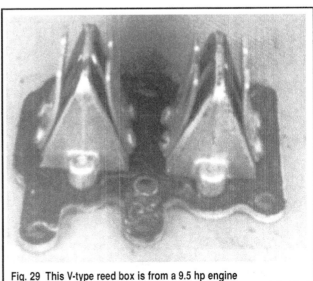

Fig. 29 This V-type reed box is from a 9.5 hp engine

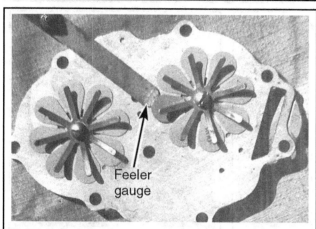

Fig. 30 Measure the clearance between the reed tip and the reed plate here. . .

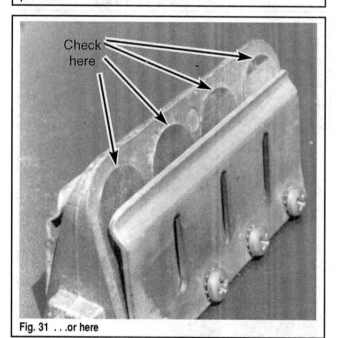

Fig. 31 . . .or here

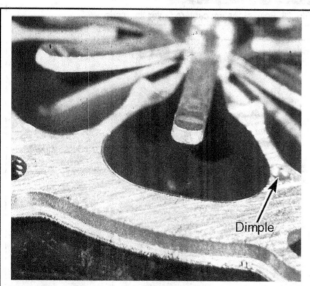

Fig. 32 The reed leaves must straddle the dimple and be centered over the openings

CLEANING & SERVICING

◆ See Figures 33, 34 and 35

■ Please refer to the Exploded Views section for exploded views of the powerhead and its components.

✳✳ WARNING

Always handle the reeds with the utmost care. Rough treatment will result in the reeds becoming distorted and will affect their performance.

1. Wash the reeds in solvent, and blow them dry with compressed air from the BACK SIDE ONLY. DO NOT blow air through the reed from the front side. Such action would cause the reed to open and fly up against the reed stop. Wipe the front of the reed dry with a lint free cloth.

2. Clean the reed box thoroughly by removing any old gasket material.

3. Secure the reed blocks to the reed plate and tighten the screws securely (25-35 ft. lbs. {3-4 Nm} on 3 cyl/V4 models).

4. Check for chipped or broken reeds. Observe that the reeds are not preloaded or standing open. Satisfactory reeds will not adhere to the reed block surface, but still there is not more than 0.010 in. (0.254mm) clearance between the reed and the block surface.

■ DO NOT remove the reeds, unless they are to be replaced. ALWAYS replace reeds in sets. NEVER turn used reeds over to be used a second time.

5. Check the reed location over the reed block, or plate openings, to be sure the reed is centered.

6. The reed assemblies are then ready for installation.

Small Engines

Disassemble the reed block by first removing the screws securing the reed stops and reeds to the reed block, and then lifting the reed stops and reeds from the block.

Clean the gasket surfaces of the reed block or plate. Check the surfaces for deep grooves, cracks, or any distortion that could leakage. Replace the reed block or plate if it is damaged.

After new reeds have been installed, and the reed stop and attaching screws have been tightened to the required torque value, check the new reeds as outlined in the following paragraphs.

Check to be sure the reeds are not preloaded. They should not adhere to the block or plate, and still the clearance between the reed and the block surface, should not be more than 0.010 in. (0.254mm).

■ DO NOT remove the reeds, unless they are to be replaced. ALWAYS replace reeds in sets. NEVER turn used reeds over to be used a second time.

Lay the reeds on a flat surface and measure all the reed stops. If there is a great difference between the stops, the entire reed stop assembly should be replaced. Any attempt to bend and get all the stops equal and level would be almost impossible.

REED BOX INSTALLATION

◆ See Figures 36 thru 41

Install the reed box and intake manifold onto the cylinder block. A gasket is usually installed on both sides of the reed box. The reeds and reed stops face inward toward the cylinder.

On certain later 15-40 hp powerheads, a screw is installed in the center of the reed box into the cylinder block. This center screw is installed first, and then the intake manifold is installed and secured in place.

Fig. 33 The Phillips screws must be removed first before the box can be removed from the plate

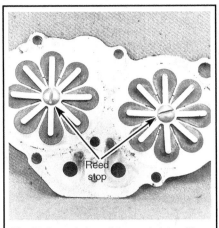

Fig. 34 A good view of the reed plate with the two reeds and stops in place

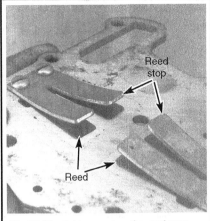

Fig. 35 The flat bars act as the reed stop and are the same width as the reed

Fig. 36 On 2 cylinder engines, install the gasket and reed plate to the powerhead. . .

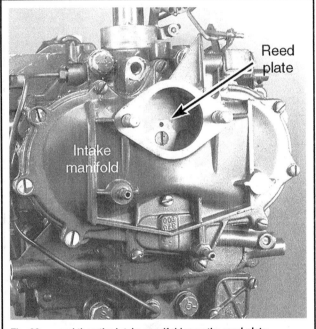

Fig. 38 . . .and then the intake manifold over the reed plate

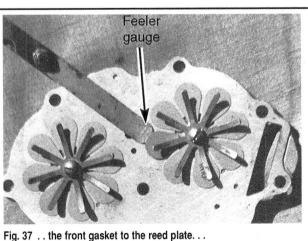

Fig. 37 . . the front gasket to the reed plate. . .

Bypass Cover

REMOVAL & INSTALLATION

② MODERATE

◆ See Figure 42 and 43

■ Please refer to the Exploded Views section for exploded views of the powerhead and its components.

■ On some small horsepower units the powerhead does not contain bypass covers.

The bypass cover actually covers the passageway the fuel travels from the crankcase up the side of the powerhead and into the cylinder. Seldom does a bypass cover cause any problem.

On some models, a fuel pump may be attached to one of the bypass covers.

During a normal overhaul, the bypass covers should be removed cleaned, and new gaskets installed. Identify the covers to ensure installation in the same location from which they are removed.

1. Coat both sides of a NEW gasket with sealer, and then place it in position on the cylinder block.

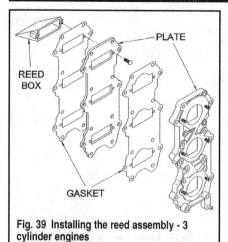

Fig. 39 Installing the reed assembly - 3 cylinder engines

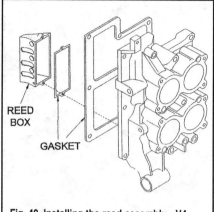

Fig. 40 Installing the reed assembly - V4 engines

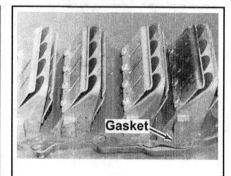

Fig. 41 Position a new gasket on 3 cyl/V4 engines

Fig. 42 Removing the bypass cover - 1 & 2 cylinder engines

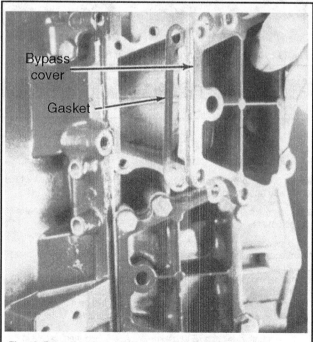

Fig. 43 Removing the bypass cover - 3 cyl & V4 engines

2. Install the bypass covers and in the SAME location from which they were removed. Secure the covers in place with the attaching hardware. If a fuel pump is used, be sure the same bypass cover is installed in the position from which it was removed.

Exhaust Cover

REMOVAL & INSTALLATION

◆ See Figures 44 thru 50

■ Please refer to the Exploded Views section for exploded views of the powerhead and its components.

The exhaust covers are one of the most neglected items on any outboard engine. Seldom are they checked and serviced. Many times an engine may be overhauled and returned to service without the exhaust covers ever having been removed.

The exhaust manifold is located on the port side of 3 cylinder engines. On V4 engines, the exhaust manifold is located at the rear of the powerhead between the 2 cylinder heads.

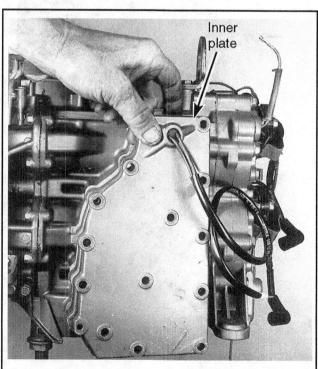

Fig. 44 Removing the inner plate - 40 and 50 hp 2 cyl powerheads

POWERHEAD **6-13**

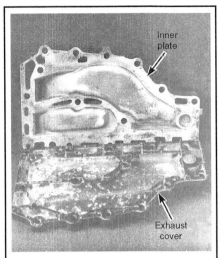
Fig. 45 The inner exhaust plate and cover - 40 and 50 hp 2 cyl powerheads

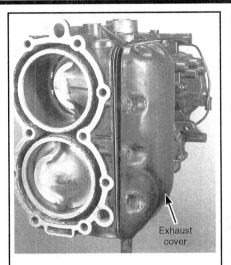
Fig. 46 Removing the exhaust cover on an earlier 2 cyl powerhead

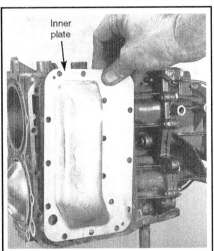
Fig. 47 Removing the exhaust plate on an earlier 2 cyl powerhead

Fig. 48 The exhaust manifold is located on the port side of 3 cylinder engines. . .

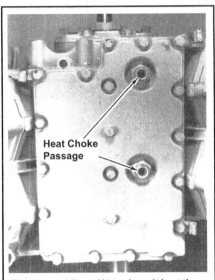
Fig. 48 . . .while on V4 engines, it is at the rear, between the cylinder banks

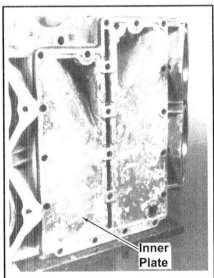
Fig. 50 Many models have an inner plate installed

One reason the exhaust covers are not removed is because the attaching bolts usually become corroded in place. This means they are very difficult to remove, but the work should be done. Heat applied to the bolt head and around the exhaust cover will help in removal. However, some bolts may still be broken. If the bolt is broken it must be drilled out and the hole tapped with new threads.

The exhaust covers are installed over the exhaust ports to allow the exhaust to leave the powerhead and be transferred to the exhaust housing. If the cover was the only item over the exhaust ports, they would become so hot from the exhaust gases they might cause a fire or a person would be severely burned if they came in contact with the cover.

Therefore, an inner plate is installed to help dissipate the exhaust heat. Two gaskets are installed - one on either side of the inner plate. Water is channeled to circulate between the exhaust cover and the inner plate. This circulating water cools the exhaust cover and prevents it from becoming a hazard.

On some early model outboards, the inner plate was constructed of aluminum. Unfortunately, the aluminum would corrode through, especially in a salt water environment, and then water could enter the lower cylinder and cause a powerhead failure. In some cases, the inner plate may even corrode all the way through, allowing water to enter the lower cylinder. To correct this corrosion problem, the inner plate is now made of stainless steel material.

A thorough cleaning of the inner plate behind the exhaust covers should be performed during a major engine overhaul. If the integrity of the exhaust cover assembly is in doubt, replace the complete cover including the inner plate.

On powerheads equipped with the heat/electric choke, a baffle is installed on the inside surface of the inner plate. This baffle is heated from the engine exhaust gases. Air passing through the baffle heats the choke and allows the choke to open as engine temperature rises.

1. Coat both sides of a NEW gasket with sealer, and then place the gasket in position on the exhaust side of the cylinder block. Install the inner plate. Coat both sides of another NEW gasket with sealer, and then install the gasket and exhaust cover.

2. Secure the exhaust cover in place with the attaching hardware.

CLEANING

◆ See Figure 51

Clean any gasket material from the cover and inner plate surfaces. Check to be sure the water passages in the cover end plate are clean to permit adequate passage of cooling water.

Inspect the inlet and outlet holes in the powerhead to be sure they are clean and free of corrosion. The openings in the powerhead may be cleaned with a small size screwdriver.

Clean the area around the exhaust ports and in the webs running up to the exhaust ports. Carbon has a habit of forming in this area.

Fig. 51 The exhaust area of a powerhead open for inspection and cleaning

Top Seal

REMOVAL & INSTALLATION

1 & 2 Cylinder Engines

◆ See Figures 52, 53 and 54

■ Please refer to the Exploded Views section for exploded views of the powerhead and its components.

■ The top seal maintains vacuum and pressure in the crankcase at the top cylinder.

1. This seal can only be removed using one of two methods.

a. The 1st is by using a special puller while the powerhead is still assembled. If the puller is used, thread the end of the puller into the seal. After the puller is secured to the seal, remove the seal from the powerhead by tightening the center screw on the puller. DO NOT attempt to use any other type of tool to remove this seal or the powerhead flanges will be damaged. If the flanges are damaged, the block must be replaced.

b. The 2nd method is to remove the seal during powerhead disassembling. After the crankcase cover has been removed, the seal will be loose and can be easily removed by holding onto the bearing and prying the seal out.

To Install:

2. To install the seal with the powerhead assembled, coat the outside diameter of the seal with Johnson/Evinrude Seal Compound. Use the special tool to tap the seal EVENLY into place around the crankshaft.

3. If the powerhead has been disassembled, a socket or other similar type tool of equal diameter as the seal may be used to tap it into the bearing. If the powerhead being serviced does not have the seal in the bearing, lay the seal in the recess of the block. When the crankcase cover is installed, the seal will be in place.

3 Cylinder & V4 Engines

◆ See Figures 55 and 56

■ Please refer to the Exploded Views section for exploded views of the powerhead and its components.

On early model engines, the top seal maintains vacuum and pressure in the crankcase at the top cylinder. Engineering changes installed extra rings around the crankshaft to seal the crankcase; on these engines the top seal simply prevents oil from escaping around the bearing.

1. On V4 engines, remove the bolts securing the upper cap in place. Carefully work the bearing cap up and free of the block. Drive the seal from the bearing cap using a punch and hammer. Remove the O-rings from the outside edge of the bearing cap.

2. Install a NEW seal into the bearing cap, with the hard side facing UP. Position a NEW O-ring in place around the bearing cap. Some models have 1 O-ring, and others have 2. Do not attempt to install 2 rings if there is only 1 groove. Coat the O-ring/s with oil. Set the assembly aside for installation later.

3. On 3 cylinder engines, in order to replace the top seal, the crankcase cover must be removed, the crankshaft raised, and the top bearing removed, before the seal can be removed. Refer to Crankshaft procedures for more details.

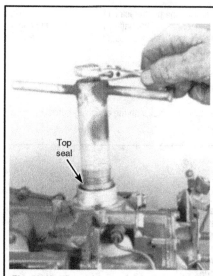

Fig. 52 You'll need a special tool to remove the top seal. . .

Fig. 53 . . .and also to install it

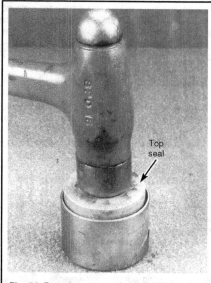

Fig. 54 Pressing a new seal into the bearing

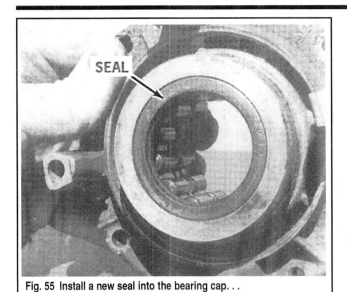

Fig. 55 Install a new seal into the bearing cap. . .

Fig. 56 . . .and then install new O-rings

4. Install a new seal into the bearing cap so the hard side is facing UP and then install a new O-ring(s) into the groove(s) in the cap. Make sure that you coat the O-rings with clean engine oil before installation.

■ Some models may use 2 O-rings, while others use only 1.

Bottom Seal

REMOVAL & INSTALLATION

1 & 2 Cylinder Engines w/Seal Mounted on the Driveshaft

◆ See Figures 54 and 57

■ Please refer to the Exploded Views section for exploded views of the powerhead and its components.

The bottom seal has equal importance as the top seal. This seal is installed to maintain vacuum and pressure in the lower half of the crankcase for the lower cylinder.

The bottom seal will vary, depending on the model engine being serviced. These procedures and accompanying illustrations cover the most common Johnson/Evinrude bottom seal installed.

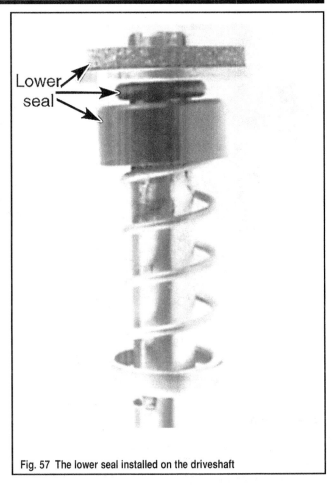

Fig. 57 The lower seal installed on the driveshaft

When the powerhead is removed, observe around the driveshaft at the lower end and the seal will be visible. The seal consists of a gasket, plate, an O-ring, lower seal bearing, spring, washer, and a pin. The pin is installed through the driveshaft and holds the seal upward and in place.

As the powerhead is lowered down over the driveshaft during installation, the seal is held in place and will hold the vacuum and pressure created when the engine is operating.

With the powerhead assembled, it is a simple matter to reach into the exhaust housing and remove the seal and then replace the gasket and O-ring. Check the spring, to be sure it is not distorted, and the washer for damage.

1 & 2 Cylinder Engines w/Seal Mounted on the Crankshaft

◆ See Figures 58, 59 and 60

■ Please refer to the Exploded Views section for exploded views of the powerhead and its components.

The bottom seal has equal importance as the top seal. This seal is installed to maintain vacuum and pressure in the lower half of the crankcase for the lower cylinder.

The bottom seal will vary, depending on the model engine being serviced. These procedures and accompanying illustrations cover the most common Johnson/Evinrude bottom seal installed.

This bottom seal prevents exhaust fumes from entering the crankcase, and holds pressure and vacuum inside. The seal consists of a quadrant ring, O-ring, retainer washer, spring, another washer, and a snapring.

To remove this seal from the lower end of the crankshaft, use a pair of Tru-arc pliers and *carefully* remove the snapring. Be careful not to lose any of the parts due to the spring pressure against the snapring. Notice how the quadrant O-ring fits inside the seal. This ring is also removable. Observe how the seal has a raised edge on one side. This raised edge *must* face upward when the seal is installed.

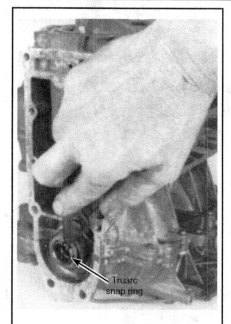

Fig. 58 Remove the snap ring from the crankshaft. . .

Fig. 59 . . .and then remove the O-ring from the quadrant ring

Fig. 60 The new O-ring and quadrant ring are installed onto the crankshaft with the raised edge of the quadrant ring facing upward

1 & 2 Cylinder & All 3 Cylinder Engines w/Seal in a Cap

◆ See Figure 61

 MODERATE

■ **Please refer to the Exploded Views section for exploded views of the powerhead and its components.**

The bottom seal has equal importance as the top seal. This seal is installed to maintain vacuum and pressure in the lower half of the crankcase for the lower cylinder.

The bottom seal will vary, depending on the model engine being serviced. These procedures and accompanying illustrations cover the most common Johnson/Evinrude bottom seal installed.

The seal on these engines is installed in a cap which is bolted to the bottom of the powerhead. The 4 bolts securing the cap are removed before the crankcase cover is removed. The seal can be punched and a new seal installed without difficulty. Remove the O-ring from the cap.

If the powerhead is not to be overhauled, install the seal with the hard side DOWN. Drive the seal and O-ring into place using the correct size socket and hammer.

Align the holes in the cap with the holes in the powerhead. Secure the cap in place with the attaching bolts.

If the powerhead is to be overhauled, refer to the Crankshaft Installation procedures for details on installing the bottom seal and cap assembly.

1958-68 V4 Engines w/Seal Mounted on the Crankshaft

◆ See Figures 62 and 63

 MODERATE

■ **Please refer to the Exploded Views section for exploded views of the powerhead and its components.**

This bottom seal prevents exhaust fumes from entering the crankcase, and holds pressure and vacuum inside. The seal consists of a quadrant ring, O-ring, retainer washer, spring, another washer and a snap ring.

To remove this seal from the lower end of the crankshaft, use a pair of Truarc pliers and carefully remove the snap ring. Take care not to lose any of the parts due to the spring pressure against the snap ring. Notice how the quadrant O-ring fits inside the seal. This ring is also removable. Observe how the quadrant ring has a raised edge on one side - this raised edge MUST face upward when the seal is installed.

Fig. 61 Removing the seal and the O-ring from the cap

Fig. 62 Notice how the quadrant fits inside the seal

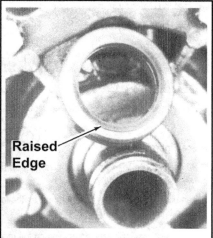

Fig. 63 The raised edge on the quadrant must face upward

Check to be sure the spring has good tension. Check to be sure the washers are not distorted.

The quadrant O-ring should be discarded and a new one installed.

Good shop practice dictates the quadrant O-ring seal be replaced each time the lower seal is serviced.

Check the groove in the lower end of the crankshaft where the Truarc ring fits. If the groove is not clean, the ring will snap out and the lower sealing qualities will be lost. If the groove is badly corroded, the crankshaft must be replaced.

If the powerhead is not to be overhauled, refer to the Crankshaft Installation procedures for details on installing the bearing cap and seal.

1969 & Later V4 Engines w/Seal Installed In Bearing Cap

◆ See Figures 64 and 65

■ **Please refer to the Exploded Views section for exploded views of the powerhead and its components.**

The seal on these engines is installed inside the lower bearing cap. Its function is to prevent exhaust fumes from entering the crankcase, and maintain pressure and vacuum inside the crankcase.

If the seal is to be replaced without overhauling the powerhead, proceed as follows:

1. Remove the outside perimeter bolts on the bearing cap.
2. Remove the 4 bolts closer to the crankshaft.
3. Use a pair of screwdrivers and work the bearing cap free of the lower bearing.
4. Remove the O-rings from the outside edge of the bearing cap.
5. Use a punch and drive the seal out of the bearing cap.

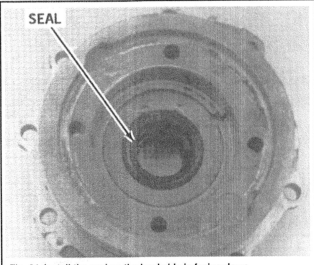

Fig. 64 Install the seal so the hard side is facing down. . .

Fig. 65 . . .and then install new O-rings

6. Install the new seal with the hard side DOWN. Drive the seal into place using the correct size socket and a hammer.

7. Install the outside O-ring/s around the cap. Some models use 1 O-ring and others use 2.

If the powerhead is to be overhauled, set the assembly aside for later installation.

If the powerhead is not to be overhauled, refer to the Crankshaft Installation procedures for details on installing the bearing cap and seal.

INSPECTION

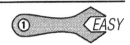

Check to be sure the spring has good tension. Check to be sure the washers are not distorted. The quadrant ring should be discarded and a new one installed.

Good shop practice dictates the quadrant seal be replaced each time the lower seal is serviced.

Check the groove in the lower end of the crankshaft where the truarc ring fits. If the groove is not clean, the ring will snap out and the lower sealing qualities will be lost. If the groove is badly corroded, the crankshaft must be replaced.

Centering Pin

REMOVAL

◆ See Figures 66 and 67

■ **Please refer to the Exploded Views section for exploded views of the powerhead and its components.**

All Johnson/Evinrude outboard engines have at least one, and in most cases two, centering pins installed through the crankcase cover. These pins index into matching holes in the powerhead block when the crankcase cover is installed. These pins center the crankcase cover on the powerhead block.

The centering pins are tapered. The pins must be carefully checked to determine how they are to be removed from the cover. In most cases the pin is removed by using a center punch and tapping the pin towards the carburetor or intake manifold side of the crankcase.

When removing a centering pin, hold the punch securely onto the pin head, and then strike the punch a good hard forceful blow. DO NOT keep beating on the end of the pin, because such action would round the pin head until it would not be possible to drive it out of the cover.

Centering pins are the first item to be installed in the cover when replacing the crankcase cover, thus installation will be covered in that section.

Main Bearing Bolts & Crankcase Side Bolts

REMOVAL & INSTALLATION

◆ See Figures 68 thru 73

■ **Please refer to the Exploded Views section for exploded views of the powerhead and its components.**

The main bearing bolts are installed through the crankcase cover into the powerhead block. Most engines have 2 bolts installed for the top main bearing, 2 for the center main bearing, and 2 for the lower main bearing.

In many cases the upper and lower main bearing bolts are DIFFERENT lengths. Therefore, take time to tag and identify the bolts to ensure they will be installed in the same location from which they were removed.

The crankcase side bolts are installed along the edge of the crankcase cover to secure the cover to the cylinder block. These bolts usually have a 7/16 in. head and all must be removed before the crankcase cover can be removed.

1. Remove the crankcase side bolts.
2. Remove the main bearing bolts. On 1 & 2 cylinder engines, 2 bolts installed in the center are behind the reeds. Normally these 2 are not actually bolts, but Allen head screws. All 6 main bearing bolts must be removed before the crankcase cover can be removed. On 3 cyl and V4 engines, be

sure to remove any bolts or Allen screws from the intake manifold in the reed box areas. Naturally, all main bearing bolts must be removed before the crankcase cover will come free

To Install:

3. Apply a coating of sealer to the threads of the main bearing bolts. Install and tighten the main bearing bolts finger-tight, and then just a bit more.

4. Tighten the main bearing bolts alternately and evenly in 3 rounds to the torque value given in the Torque Specifications chart. Tighten the bolts to 1/2 the total torque value on the 1st round, to 3/4 the total torque value on the 2nd round, and to the full torque value on the third and final round.

■ **As an example: if the total torque value specified is 200 inch lbs., the bolts should be tightened to 100 inch lbs. on the 1st go-around; to 150 inch lbs. on the 2nd round; and to the full 200 inch lbs. on the 3rd round.**

5. Install and tighten the crankcase side bolts to the torque value given in the Torque Specifications chart.

6. On 3 cylinder engines, install the other 2 attaching bolts through the lower bearing cap and tighten them securely.

7. On V4 engines, proceed as follows:

a. Install the attaching bolts to secure the lower and upper bearing caps. Tighten the bolts evenly and securely.

b. Install 2 bolts through the lower bearing cap into the retainer plate. These are special bolts with sealing qualities built in; never use the bolts a second time. Discard the old bolts that were removed and install new bolts.

c. Remove the two 1/4 x 28 guide pins and install the other 2 sealing bolts. Tighten the bolts evenly and alternately.

8. Install the Woodruff key in the crankshaft. Slide the flywheel onto the crankshaft. Rotate the flywheel through several revolutions and check to be sure all moving parts indicate smooth operation without evidence of binding or "rough" spots.

9. Remove the flywheel (and the key to make sure it is not lost before final flywheel installation).

Fig. 66 Cylinder block with the 2 centering pins installed

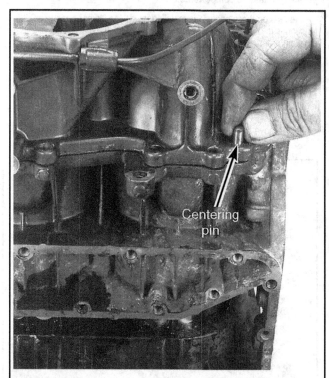

Fig. 67 Removing a centering pin from the block

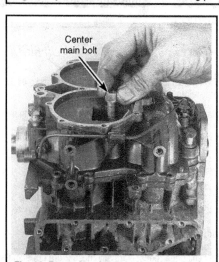

Fig. 68 Removing the main bearing bolt - 2 cyl models

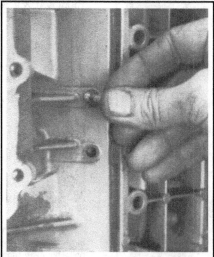

Fig. 69 Remove the crankcase side bolts. . .

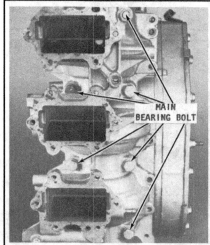

Fig. 70 . . .and then the main bearing bolts on the 3 cylinder. . .

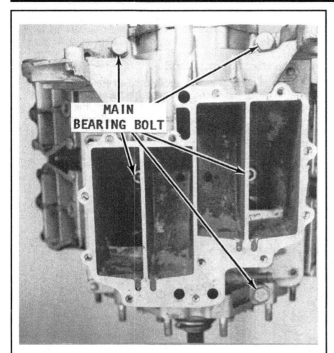

Fig. 71 . . .or V4 engine

Fig. 72 Special bolts are used on the V4

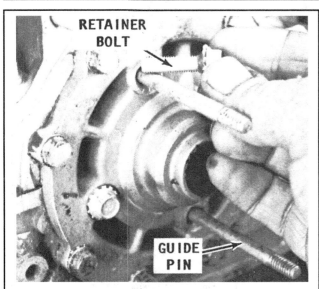

Fig. 73 Make sure that you remove the 2 guide pins (V4)

Crankcase Cover

REMOVAL & INSTALLATION

 **②** **MODERATE**

◆ See Figures 74 thru 84

■ **Please refer to the Exploded Views section for exploded views of the powerhead and its components.**

1. Remove all bolts securing the upper bearing cap and all bolts securing the lower bearing cap in place, if this has not been done.

2. After all side bolts and main bearing bolts have been removed, use a soft-headed mallet and tap on the bottom side of the crankshaft. A soft, hollow sound should be heard indicating the cover has broken loose from the crankcase. If this sound is not heard, check to be sure all the side bolts and main bearing bolts have been removed.

✳✳ CAUTION

NEVER pry between the cover and the crankcase or the cover will surely be distorted. If the cover is distorted, it will fail to make a proper seal when it is installed. Such an installation would damage both crankcase halves and render them unfit for service - an expensive replacement.

3. Once the crankshaft has been tapped, as described, and the proper sound heard, the cover will be jarred loose and may be removed.

4. Clean and thoroughly inspect the cover. Wash the cover with solvent, and then dry it thoroughly. Check the mating surface to the cylinder block for damage that may affect the seal.

5. Inspect the labyrinth seal grooves at the center main bearing area to be sure they are clean and not damaged in any manner.

To Install:

6. First, check to be sure the mating surfaces of the crankcase cover and the cylinder block are clean. Pay particular attention to the labyrinth seal grooves in the center main bearing area. The mating surfaces and the seal grooves MUST be free of any old sealing compound or other foreign material.

Fig. 74 A good shot of the block after the cover has been removed (2 cyl)

7. On 3 cylinder engines, proceed as follows:

a. Check to be sure the main bearings are fully seated into the powerhead. The bearings can be checked by attempting to "rock" the bearing in place - the attempt should fail. The bearing should be locked in place with the pin indexed into the hole in the block.

b. Remove the wooden block from the studs at the upper bearing area. After the wooden block has been removed, do not rotate the crankshaft. Rotating the crankshaft will cause it to "lift" and the bearings indexed over the pins in the block will become misaligned.

8. On V4 engines, proceed as follows:

a. Check to be sure the sealing rings are all down in place and that the ring openings are NOT close to the split between the 2 crankcase halves.

b. Also, double check to be sure the pins in the block are still indexed into the holes in the center main bearing. This can be checked by attempting to rotate the bearing - the attempt should fail.

■ The remainder of the cylinder block installation work should be performed WITHOUT interruption. Do not begin the work if a break in the sequence is expected, coffee, tea, lunch, whatever. The sealer will begin to set almost immediately, therefore, the crankcase cover installation, main bearing bolt installation and tightening, and the side bolt installation and tightening, must move along rapidly.

Fig. 75 Removing the cover on 40 and 50 hp 2 cyl motors

Fig. 76 Cleaning the powerhead surface - 40 and 50 hp 2 cyl motors; notice the seals

Fig. 77 An even better shot of the seal

Fig. 78 Inspect the sealing ring groove

Fig. 79 Installing sealer on the spaghetti seal - 2 cyl engines

Fig. 80 Installing sealer on the block if the seal was not used on 2 cyl engines

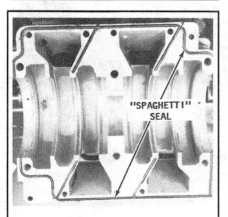

Fig. 81 On 3 cyl/V4 engines, find the spaghetti seal. . .

Fig. 82 . . .and apply sealer

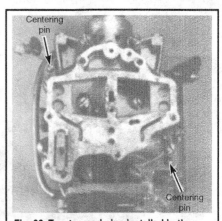

Fig. 83 Two tapered pins installed in the block on 2 cyl engines

Fig. 84 Don't forget to tap in the centering pins (3 cyl shown)

9. Apply just a small amount of suitable sealer into the groove in the cylinder block to hold the seal in place (if used). Install a new spaghetti seal into the groove.

10. After the seal on both sides of the cylinder block has been installed, apply a light coating of sealer to the outside edge of the spaghetti seal. This seal is not used on some models. If the seal is not used double check to be sure the 2 surfaces are clean and smooth. Eject about a 3/8 in. dab of the seal at very close intervals along the surface of the block. Repeat this procedure on the mating surface of the crankcase cover.

11. Next, lower the crankcase cover into place on the cylinder block. Install the 2 guide centering pins through the cover and into the block. The centering pins are tapered, therefore, check the crankcase and notice which side has the large hole and which has the small hole. The pin must be inserted into the large hole first. If the pin is installed into the small hole first, the crankcase cover or the cylinder block will break.

Connecting Rods, Pistons & Rings

REMOVAL & INSTALLATION

◆ See Figures 85 thru 91

■ Please refer to the Exploded Views section for exploded views of the powerhead and its components.

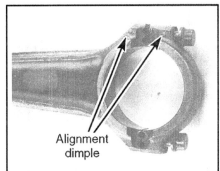
Fig. 85 A good look at the alignment dimples. . .

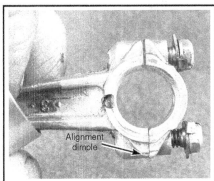
Fig. 86 . . .the alignment marks. . .

Fig. 87 . . .and the and the hill and valley

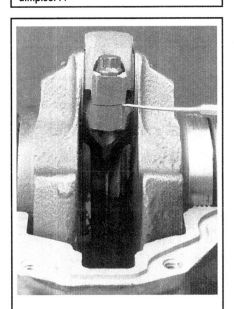
Fig. 88 This punch shows a fracture between the rod and cap

Fig. 89 Removing the bolts from the rod cap. . .

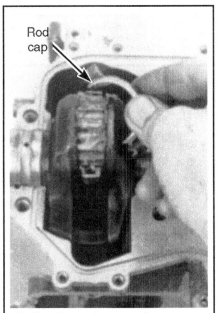
Fig. 90 . . .the cap from the rod. . .

Fig. 91 . . .and finally, the needle bearing and cage

■ The connecting rods and their rod caps are a MATCHED set. They absolutely MUST be identified, kept, and installed as a set. Under no circumstances should the connecting rod and caps be interchanged with another cylinder or another motor. Therefore, take time and care to tag each rod and rod cap; to keep them together as a set while they are on the bench; and to install them into the same cylinder from which they were removed as a set and facing the same direction (meaning you still have to pay attention on single cylinder motors).

The connecting rod and its cap on 15 hp to early 40 hp engines, and all 3 cyl and V4 engines, are manufactured as a set - as a single unit. After the complete rod and cap have been made, 2 holes are drilled through the side of the cap and rod, and the cap is then fractured from the rod. Therefore, the cap must always be installed with its original rod. The cap half of the break can *only* be matched with the other half of the break on the original rod (and only facing the same direction).

The rods and caps on the smaller horsepower engines are made of aluminum with Babbitt inserts. These rods and caps are manufactured as two separate items

Inspect the rod and the rod cap before removing the cap from the crankshaft. Under normal conditions, a line or a dot is visible on the side of the rod and the cap. This identification is an assist to assemble the parts together and in the proper location.

Look into the block and notice how the rods have a "trough". Also notice the hole in the rod in the area where the wrist pin passes through the piston. This hole must always face toward the upper end of the powerhead during installation. A cut-away area at the lower end of the rod serves to "throw" lubrication onto the main bearing. If the hole at the upper end faces correctly,

this cut-away section will also face in the proper direction and the bearing will be adequately lubricated.

1. To remove the rod bolts from the cap, it is recommended to loosen each bolt just a little at a time and alternately. This procedure will prevent one bolt from being completely removed while the other is still tightened to its recommended torque value. Such action may very likely warp the cap.

2. Remove the bolts as described in the previous step, and then *carefully* remove the rod cap.

3. Remove the needle bearings and cages from around the crankshaft. Count the needle bearings and insert them into a separate container - 1 container for each rod on twins, with the container clearly identified to ensure they will be installed with the proper rod at the crankshaft journal from which they were removed.

4. Tap the piston out of the cylinder from the crankshaft side. Immediately attach the proper rod cap to the rod and hold it in place with the rod bolts. The few minutes involved in securing the cap with the rod will ensure the matched cap remains with its mating rod during the cleaning and assembling work.

5. Fill the piston skirt with a rag, towel, shop cloths, or other suitable material. The rag will prevent the rod from coming in contact with the piston skirt while it is lying on the bench. If the rod is allowed to strike the piston skirt, the skirt may become distorted.

6. Identify the rod to ensure it will be installed into the cylinder from which it was removed.

7. Remove and identify the other rod cap, needle bearings and cages, and rod with piston, in the same manner.

To Install:

◆ See Figures 92 thru 107

If new rings are to be installed, each ring from the package MUST be checked in the cylinder. Errors happen. Men and machines can make mistakes. The wrong size ring can be included in a package with the proper part number.

Therefore, the end gap of each ring must be checked, one at a time. If the ring end gap is too LARGE - when the powerhead reaches operating temperature, the ring will fail to bear against the cylinder wall properly - resulting in loss of compression. If the ring end gap is too small - when the powerhead reaches operating temperature, the ring will expand and have excessive friction against the cylinder wall resulting in rapid cylinder wall wear.

8. Turn the ring sideways and lower it a couple of inches into the proper cylinder bore. Now, make the ring "square" in the cylinder bore, using a piston to position the ring properly. Next, use a feeler gauge and measure the distance (the gap), between the ends of the ring. The maximum and minimum allowable ring gap is listed in the Powerhead Specifications chart. On most engines it will be 0.07-0.17 in. (1.8-4.3mm), but check the chart for those few that differ.

9. Turn the piston upside down and slide it in and out of the cylinder. The piston should slide without any evidence of binding.

10. Several different methods are possible to install the piston and rod assembly into the cylinder. The following procedures are outlined for the do-it-yourselfer, working at home without the advantage of special tools.

Fig. 92 Check the ring gap clearance by inserting the ring into the cylinder. . .

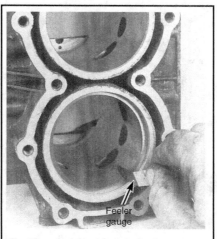
Fig. 93 . . .and then use a feeler gauge to measure it

Fig. 94 Use a hose clamp like this if a ring compressor is not available

11. First, purchase a special hose clamp with a strip of metal inside the clamp. This piece of metal on the inside allows the outside portion of the clamp to slide on the inner strip without causing the ring to rotate.

12. Actually, when this procedure was authored a Mercruiser dealer was the only place such a clamp may be purchased (that might still be true). At the Mercruiser marine dealer, ask for an exhaust bellows hose clamp. The design of this hose clamp prevents the clamp and the piston ring from turning as the clamp is tightened. DO NOT attempt to use an ordinary hose clamp from an automotive parts house because such a clamp will cause the piston ring to rotate as the clamp is tightened. The ring MUST NOT rotate, because the ring ends must remain on either side of the dowel pin in the ring groove.

13. Next, coat the inside surface of the cylinder with a film of light-weight oil. Coat the exterior surface of the piston with the oil.

14. Take just a minute to notice how the piston rings are manufactured. Each end of the ring has a small cutout on the inside circumference. Now, visualize the ring installed in the piston groove. The ring ends must straddle the pin installed in each piston groove. As the ring is tightened around the piston, the ends will begin to come together. When the piston is installed into the cylinder bore, the two ends of the ring will come together and the cutout edge will be up against the pin. For this reason, care must be exercised when installing the rings onto the piston and when the piston is installed into the cylinder.

15. Install only the bottom ring into the bottom piston groove. Do not expand the ring any further than necessary, as a precaution against breaking it.

16. Install the ring into the piston groove with the ends of the ring straddling the pin in the groove. Notice how the ring pins on most motors are staggered from one groove to the next, by 180°. The ring ends *must* straddle the pin to prevent the ring from rotating during engine operation. In a 2-cycle engine, if the ring is permitted to rotate, at one point the opening between the ring ends would align with either the intake or exhaust port in the cylinder, the ring would expand very slightly, catch on the edge of the port, and break.

■ **The following items must be checked at this point in the assembly work:**

• The piston and rod are being installed into the same cylinder from which they were removed.

• The hole in the rod near the wrist pin opening, and at the lower end of the rod, is facing UPWARD.

• The slanted side of the piston (if applicable, some smaller engines and all V4s) is TOWARD the exhaust side of the cylinder.

• The ends of the bottom ring straddle the pin in the piston groove.

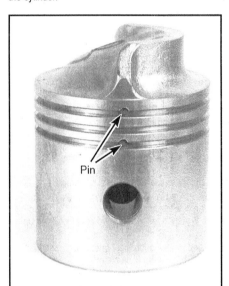

Fig. 95 The end of the piston ring must straddle the groove pins

Fig. 96 Install the clamp over the ring before moving the piston further into the cylinder (2 cylinder shown)

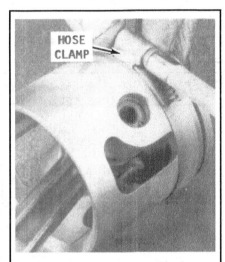

Fig. 97 Install the clamp over the rings. . .

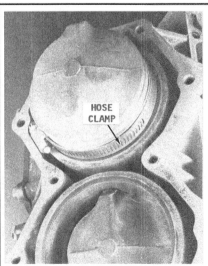

Fig. 98 And then insert the piston until the ring just touches the top of the cylinder (3 cyl and V4 shown)

Fig. 99 You may need a soft mallet to persuade the piston into the cylinder

Fig. 100 Here's another shot of the 'persuasion' on the 40/50 hp 2 cylinder, and most 3 cylinder and V4 motors

Fig. 101 Using a ring installer to expand the ring during installation into the groove

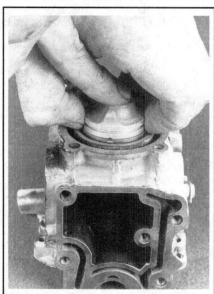

Fig. 102 On small engines, you can actually use your fingers to compress the rings when installing the piston

Fig. 103 The slanted edge of each piston should face toward the exhaust port (V4 engines and if applicable on all others)

Fig. 104 Check the flexibility of the rings through the intake port. . .

• On looper powerheads using the flat dome piston, check to be sure the cut-outs in the piston align with "windows" in the cylinder wall.

17. Install the hose clamp over the piston and bottom ring. Tighten the hose clamp with one hand and at the same time rotate the clamp back and forth slightly with the other hand. This "rocking" motion of the clamp as it is tightened will convince you the ring ends are properly positioned on either side of the pin. Continue to tighten the clamp, and "rocking" the clamp until the clamp is against the piston skirt. At this point, the ring ends will be together and the cutout on each ring end will be against the pin.

18. Carefully insert the rod and the piston skirt down into the cylinder. Push the piston into the cylinder until the bottom ring clamp, just installed, touches the top of the cylinder. Watch to be sure the rod does not hang-up on one of the cylinder ports.

19. Tap the piston with the end of a wooden tool handle, or plastic mallet, until the ring enters the cylinder. Remove the hose clamp.

20. Install the remaining rings in the same manner; one at a time, making sure the ends of each ring straddle the pin in the piston groove.

21. After the last ring has been installed and the clamp removed, tap the piston into the bore until the crown is about even with the cylinder block surface.

22. On twins, install the other piston in exactly the same manner.

Fig. 105 . . .and then through the exhaust port

Fig. 106 The word UP must be at the top of the cylinder on the 40/50 hp 2 cylinder motors (and any others where applicable)

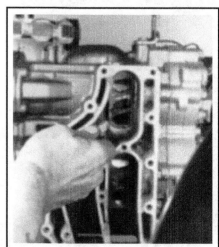

Fig. 107 Checking the ring tension through the exhaust port - 40/50 hp 2 cylinder motors

23. Turn the cylinder block upside down with the top of the block to your LEFT. Remove the bolts and rod caps from each rod. Set each rod cap in a definite position to ensure each will be installed onto the rod from which it was removed.

DISASSEMBLY & ASSEMBLY

 DIFFICULT

◆ **See Figures 108 thru 118**

■ **Please refer to the Exploded Views section for exploded views of the powerhead and its components.**

1. If not already done, remove the piston rings from the piston as detailed previously.

■ **Before separating the piston from the rod, notice the location of the piston in relation to the rod. Observe the hole in the rod trough on one side of the rod near the wrist pin opening. This hole must face toward the top of the engine during installation.**

2. On engines so equipped, note the slanted edge and the sharp edge of the dome-type piston. The slanted edge MUST face toward the exhaust side of the cylinder and the sharp edge toward the intake side during installation.

3. If servicing a 40 or 50 hp 2 cylinder powerhead, or later model pistons; carefully observe the hole in the rod near the wrist pin and the relationship of the irregular cut-outs in the piston skirt. . .only in this position will this relationship exist. The rod and piston MUST be assembled in this manner or the engine will run poorly.

■ **When the rod is installed to the piston, the relationship of the rod can only be one way. The rod holes must face upward and the piston must face as described in the previous paragraph.**

4. Look into the piston skirt. On most model pistons, notice the **L** stamped on the boss through which the wrist pin passes. The letter mark identifies the "loose" side of the piston and indicates the side of the piston from which the wrist pin must be driven out without damaging the piston. Some pistons may have the full word **LOOSE** stamped on the inside of the piston skirt.

■ **If the piston does not have the "L" or the word "LOOSE" stamped, the wrist pin may be driven out in either direction.**

■ **It may be necessary to heat the piston in a container of boiling water in order to press the wrist pin free.**

5. Remove the retaining clips from each end of the wrist pin. Some clips are spring wire type and may be worked free of the piston using a screwdriver or small awl. Other model pistons have a truarc snapring. This type of ring can only be successfully removed using a pair of truarc pliers.

6. Place the piston in an arbor press using the correct size cradle for the piston being serviced, and with the LOOSE side of the piston facing UPWARD.

7. The wrist pin must be driven out from the loose side. This may not seem reasonable, but there is a very simple explanation. By placing the piston in the arbor press cradle with the tight side down, and the arbor ram pushing from the loose side, the piston has good support and will not be distorted. If the piston is placed in the arbor press with the loose side down, the piston would be distorted and unfit for further service.

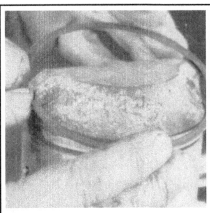

Fig. 108 Remove the piston rings if not done already

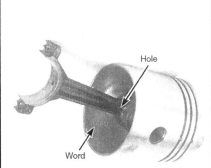

Fig. 109 Find the word LOOSE (or an 'L') on the inside of the skirt and the hole in the rod at the wrist pin end. The pin must be driven from the loose side of the piston out the tight sid

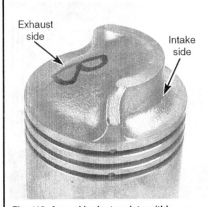

Fig. 110 A good look at a piston (this one slanted)

Fig. 111 Make sure that the word UP is at the top of the cylinder when installing pistons (if applicable)

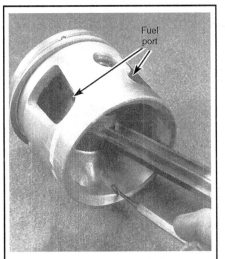

Fig. 112 Note the fuel ports, later model 2 cylinder and all 3 cylinder motors

Fig. 113 A broken rod

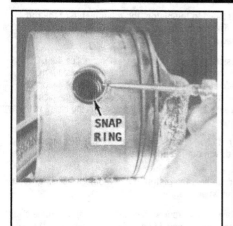

Fig. 114 The wrist pin may use a spring-loaded clip. . .

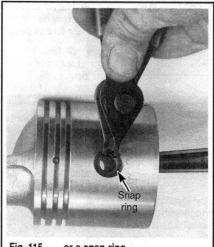

Fig. 115. . . .or a snap ring

Fig. 116 Sometimes it will be necessary to heat the piston

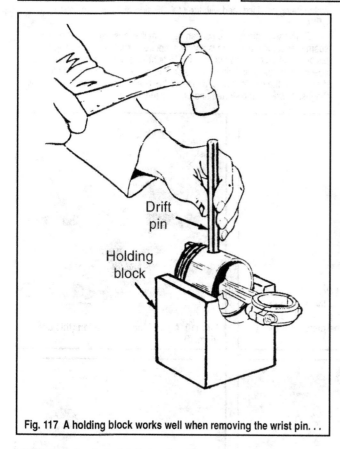

Fig. 117 A holding block works well when removing the wrist pin. . .

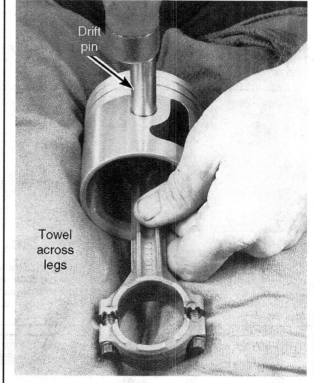

Fig. 118 . . .or you can use a drift pin with the piston between the legs

■ **Many rods have a wrist pin bearing. Some are caged bearings and others are not. Take care not to lose any of the bearings when the wrist pin is driven free of the piston.**

Alternate Removal Method

8. If an arbor press or cradle is not available, proceed as follows:

a. Heat the piston in a container of very hot water for about 10 minutes (heating the piston will cause the metal to expand ever so slightly, but ease the task of driving the pin out); assume a sitting position in a chair, on a box, whatever.

b. Next, lay a couple towels over your legs. Hold your legs tightly together to form a cradle for the piston above your knees.

c. Set the piston between your legs with the LOOSE side of the piston facing upward. Now, drive the wrist pin free using a drift pin with a shoulder. The drift pin will fit into the hole through the wrist pin and the shoulder will ride on the edge of the wrist pin.

d. Use sharp hard blows with a hammer. Your legs will absorb the shock without damaging the piston. If this method is used on a regular basis during the busy season, your legs will develop black and blue areas, but no problem, the marks will disappear in a few days.

To assemble:

◆ **See Figures 119 thru 125**

Two conditions absolutely MUST exist when the piston and rod assembly are installed into the cylinder block.

• Except for looper motors (with domed pistons), the slanted side of the piston must face toward the exhaust side of the cylinder. On loopers, make sure the openings in the pistons are aligned with the cutaways in the cylinder wall.

• The hole in the rod near the wrist pin opening and at the lower end of the rod must face UPWARD.

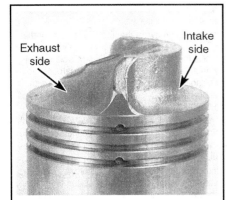

Fig. 119 The slanted side of the piston MUST face the exhaust port and the sharp edge must face the intake port

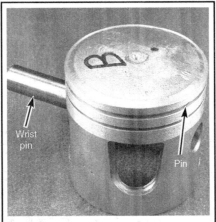

Fig. 120 Ready to install the wrist pin

Fig. 121 Remember, it must go in from the LOOSE side

Fig. 122 Heating the piston in hot water sometimes helps when installing the wrist pin

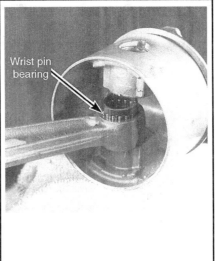

Fig. 123 Final installation

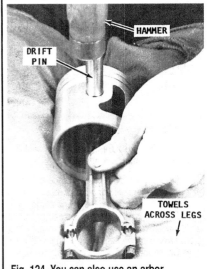

Fig. 124 You can also use an arbor press/holding block if you prefer

Therefore, the rod and piston MUST be assembled correctly in order for the assembly to be properly installed into the cylinder. Soak the piston in a container of very hot water for about 10 minutes. Heating the piston will cause it to expand ever so slightly, but enough to allow the wrist pin to be pressed through without difficulty.

9. Before pressing the wrist pin into place, hold the piston and rod near the cylinder block and check to be sure both will be facing in the right direction when they are installed.

10. Pack the wrist pin needle bearing cage with needle bearing grease, or a good grade of petroleum jelly. Load the bearing cage with needles and insert it into the end of the rod.

11. Slide the rod into the piston boss. On cross-flow motors (non-domed pistons), check a second time to be sure the slanted side of the piston is facing toward the exhaust side of the cylinder and the hole in the rod is facing upward. On looper motors, check a second time to be sure the holes in the piston align with the cut-away in the cylinder wall.

12. If an arbor press is available, proceed as follows:

a. Place the piston and rod in the arbor press with the **LOOSE** or stamped **L** side of the piston facing UPWARD (if applicable, remember not all pistons contain a loose side and marking).

b. Press the wrist pin through the piston and rod.

c. Continue to press the wrist pin through until the groove in the wrist pin for the lock ring is visible on both ends of the pin.

d. Remove the assembly from the arbor press. Install the retaining ring onto each end of the wrist pin. Some models have a wire ring, and others have a Truarc ring. Use a pair of Truarc pliers to install the Truarc ring.

13. If an arbor press is not available, the piston may be assembled to the rod in much the same manner as described for disassembling, proceed as follows:

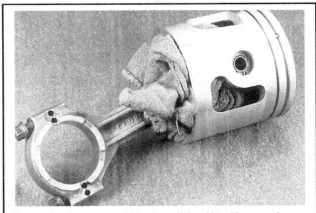

Fig. 125 Its always a good idea to stuff the skirt with a rag when the assembly is out of the cylinder

a. First, soak the piston in a container of very hot water for about 10 minutes.

b. Before pressing the wrist pin into place, hold the piston and rod near the cylinder block and check to be sure both will be facing in the right direction when they are installed.

c. Pack the wrist pin needle bearing cage with needle bearing grease, or a good grade of petroleum jelly. Load the bearing cage with needles and insert it into the end of the rod.

d. Slide the rod into the piston boss and check a second time to be sure the slanted side of the piston is facing toward the exhaust side of the cylinder and the hole in the rod is facing upward.

e. Now, assume a sitting position and lay a couple towels over your lap. Hold your legs tightly together to form a cradle for the piston above your knees. Set the piston between your legs with the LOOSE side of the piston facing upward.

f. Drive the wrist pin through the piston using a drift pin with a shoulder. The drift pin will fit into the hole through the wrist pin and the shoulder will ride on the end of the wrist pin. Use sharp hard blows with a hammer. Your legs will absorb the shock without damaging the piston. If this method is used on a regular basis during the busy season, your legs will develop black and blue areas, but no problem, the marks will disappear in a few days.

g. Continue to drive the wrist pin through the piston until the groove in the wrist pin for the lock-ring is visible at both ends. Install the retaining spring wire or Truarc ring onto each end of the wrist pin.

14. Fill the piston skirt with a rag, towel, shop cloths, or other suitable material. The rag will prevent the rod from coming in contact with the piston skirt while it is lying on the bench. If the rod is allowed to strike the piston skirt, the skirt may become distorted.

15. On multi-cylinder motors, assemble the other pistons, rods, and wrist pins in the same manner. Fill the skirt with rags as protection until the assembly is installed.

16. Piston and rod assembly installation procedures are found under Piston & Rod Assembly Installation.

CLEANING, INSPECTION & SERVICE

Piston & Ring

② MODERATE

◆ See Figures 126 thru 131

■ Please refer to the Exploded Views section for exploded views of the powerhead and its components.

1. Inspect each piston for evidence of scoring, cracks, metal damage, cracked piston pin boss, or worn pin boss. Be especially critical during inspection if the engine has been submerged.

2. Carefully check each wrist pin to be sure it is not the least bit bent. If a wrist pin is bent, the pin and piston *must* be replaced as a set, because the pin will have damaged the boss when it was removed.

3. Check the wrist pin bearings. If the bearing is the pressed in type, use your finger and determine the bearing is in good condition with no indication of binding or "rough" spots. If the wrist pin bearing is the removable type, the needle should be replaced.

4. Grasp each end of the ring with either a ring expander or your thumbnails, open the ring and remove it from the piston. Many times, the ring may be difficult to remove because it is "frozen" in the piston ring groove. In

such a case, use a screwdriver and pry the ring free. The ring may break, but if it is difficult to remove, it must be replaced.

5. Observe the pin in each ring groove of the piston. The ends of the ring *must* straddle this pin. The pin prevents the ring from rotating while the engine is operating. This fact is the direct opposite of a four cycle engine where the ring must rotate. In a 2-cycle engine, if the ring is permitted to rotate, at one point, the opening between the ring ends would align with either the intake or exhaust port in the cylinder. At that time, the ring would expand very slightly, catch on the edge of the port, and break.

Therefore, when checking the condition of the piston, ALWAYS check the pin in each groove to be sure it is tight. If one pin is the least bit loose, the piston MUST be replaced, without question. Never attempt to replace the pin, it is NEVER successful.

6. Check the piston ring grooves for wear, burns, distortion or loose locating pins. During an overhaul, the rings should be replaced to ensure lasting repair and proper engine performance after the work has been completed.

7. Clean the piston dome, ring grooves and the piston skirt. Clean the piston skirt with a crocus cloth.

8. Clean carbon deposits from the top of the piston using a soft wire brush, carbon removal solution, or by sand blasting. If a wire brush is used, take care not to burr or round machined edges.

9. Wear a pair of good gloves for protection against sharp edges, and clean the piston ring grooves using the recessed end of the proper broken ring as a tool. NEVER use a rectangular ring to clean the groove for a tapered ring, or use a tapered ring to clean the groove for a rectangular ring.

✳✳ CAUTION

NEVER use an automotive-type ring groove cleaner to clean piston ring grooves, because this type of tool could loosen the piston ring locating pins. Take care not to burr or round the machined edges.

10. Inspect the piston ring locating pins to be sure they are tight. There is one locating pin in each ring groove. If one locating pin is loose, the piston must be replaced. Never attempt to replace the pin, it is NEVER successful.

Oversize Pistons & Rings

Scored cylinder blocks can be saved for further service by re-boring and installing oversize pistons and piston rings.

Oversize pistons and rings may not available for all engines. At the time of original authoring, the sizes listed in the charts were available. Check with the parts department at your local dealer for the model engine you are servicing, and to be sure the factory has not deleted a size from their stock.

The pistons should always be ordered and received BEFORE the block is re-bored, to ensure a proper piston fit for each cylinder, after the work is accomplished.

Fig. 126 This piston is badly scored and unfit for service

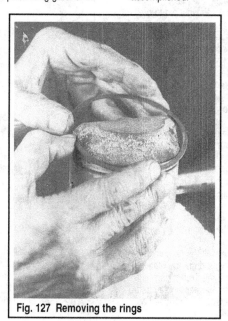

Fig. 127 Removing the rings

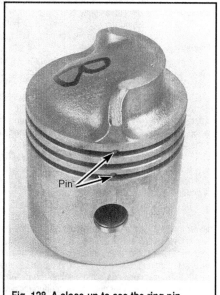

Fig. 128 A close-up to see the ring pin

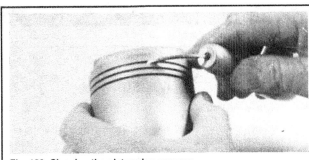

Fig. 129 Cleaning the piston ring grooves

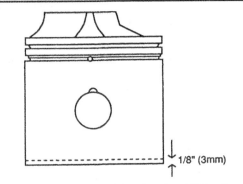

↓ 1/8" (3mm)
↑

Fig. 130 Always measure the piston diameter at 1/8 in. (3mm) above the bottom edge

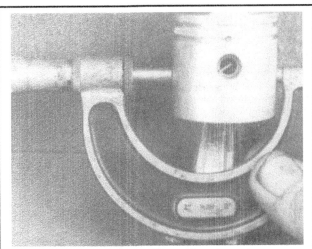

Fig. 131 Check piston skirt roundness with a micrometer

Connecting Rod

◆ See Figures 132 thru 139

■ Please refer to the Exploded Views section for exploded views of the powerhead and its components.

If the rod has needle bearings, the needles should be replaced anytime a major overhaul is performed. It is not necessary to replace the cages, but a complete NEW set of needles should be purchased and installed.

1. Position each connecting rod on a surface plate and check the alignment. If light can be seen under any portion of the machined surfaces, or if the rod has a slight wobble on the plate, or if a 0.002 in. (0.05mm) feeler gauge can be inserted between the machined surface and the surface plate, the rod is bent and unfit for further service.

2. Inspect the connecting rod bearings for rust or signs of bearing failure. NEVER intermix new and used bearings. If even one bearing in a set needs to be replaced, all bearings at that location must be replaced.

3. Inspect the bearing surface of the rod and the rod cap for rust and pitting.

4. Inspect the bearing surface of the rod and the rod cap for water marks, which form when the bearing surfaces are subjected to water contamination, which in turn causes "etching". The "etching" will worsen rapidly.

5. Inspect the bearing surface of the rod and rod cap for signs of spalling (a loss of bearing surface that resembles flaking or chipping). The spalling condition will be most evident on the thrust portion of the connecting rod in line with the I-beam. Bearing surface damage is usually caused by improper lubrication.

6. Check the bearing surface of the rod and rod cap for signs of chatter marks. This condition is identified by a rough bearing surface resembling a tiny washboard. It is caused by a combination of low-speed low-load operation in cold water, and is aggravated by inadequate lubrication and improper fuel. When this occurs, the crankshaft journal is hammered by the connecting rod. As ignition occurs in the cylinder, the piston pushes the connecting rod with tremendous force, and this force is transferred to the connecting rod journal.

Since there is little or no load on the crankshaft, it bounces away from the connecting rod. The crankshaft then remains immobile for a split second, until the piston travel causes the connecting rod to catch up to the waiting crankshaft journal, then hammers it.

In some instances, the connecting rod crank-pin bore becomes highly polished.

While the engine is running, a "whirr" and/or "chirp" sound may be heard when the engine is accelerated rapidly from idle speed to about 1500 rpm, then quickly returned to idle. If chatter marks are discovered, the crankshaft and the connecting rods should be replaced.

7. Inspect the bearing surface of the rod and rod cap for signs of uneven wear and possible overheating. Uneven wear is usually caused by a bent connecting rod. Overheating is identified as a bluish bearing surface color and is caused by inadequate lubrication or operating the engine at excessively high rpm.

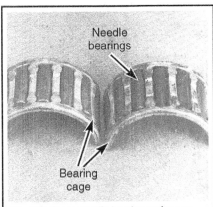

Fig. 132 These needle bearing and cages are unfit for service

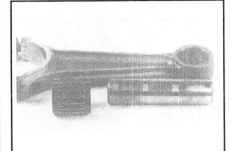

Fig. 133 A good look at the components after removal

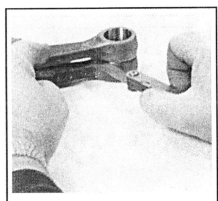

Fig. 134 Testing 2 rods for warpage at the wrist pin end...

8. Inspect the needle bearings. A bluish color indicates the bearing became very hot and the complete set for the rod must be replaced, no question.

9. Service the connecting rod bearing surfaces according to the following procedures and precautions:

a. Align the etched marks on the knob side of the connecting rod with the etched marks on the connecting rod cap.

b. Tighten the connecting rod cap attaching bolts securely.

c. Use only crocus cloth to clean bearing surface at the crankshaft end of the connecting rod. NEVER use any other type of abrasive cloth.

d. Insert the crocus cloth in a slotted 3/8 in. diameter shaft. Chuck the shaft in a drill press and operate the press at high speed and at the same time, keep the connecting rod at a 90° angle to the slotted shaft.

e. Clean the connecting rod only enough to remove marks. DO NOT continue once the marks have disappeared.

f. Clean the piston pin end of the connecting rod using the method described earlier to clean the crankshaft end, but using 320 grit Carborundum cloth instead of crocus cloth.

g. Thoroughly wash the connecting rods to remove abrasive grit. After washing, check the bearing surfaces a second time.

h. If the connecting rod cannot be cleaned properly, it should be replaced.

i. Lubricate the bearing surfaces of the connecting rods with light-weight oil to prevent corrosion.

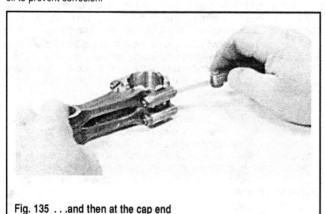

Fig. 135 . . .and then at the cap end

Fig. 136 This crank came from a submerged engine

Scored area

Crankshaft & Bearing

REMOVAL & DISASSEMBLY

1 & 2 Cylinder Motors

◆ See Figures 140 thru 145

■ Please refer to the Exploded Views section for exploded views of the powerhead and its components.

1. Lift the crankshaft assembly from the block. On some models, especially the larger horsepower engines, it may be necessary to use a soft-headed mallet and tap on the bottom side of the crankshaft to jar it loose.

■ As the crankshaft is lifted, work the center main bearing loose. This center bearing is a split bearing held together with a snap wire ring. On some models, the bottom half of the bearing may be stuck in the cylinder block. Therefore, the crankshaft and the center main bearing must be worked free of the block together.

2. If servicing a 15-40 hp powerhead, observe how the center main bearing, and the top and bottom main bearings all have a hole in the outside circumference. Notice the locating pins in the cylinder block. The purpose of this arrangement is to prevent the bearing shell from rotating. During assembly, the holes in the bearings *must* index with the pins in the block. Also notice the grooves in the block on one side of the center main bearing. Observe the grooves in the crankcase cover. This arrangement of grooves forms what is commonly known as a "labyrinth" seal. The grooves fill with oil and/or fuel creating a seal between the cylinders.

3. The 50 hp 2 cylinder powerheads have a pressed-in-place lower roller bearing. A clamp-type puller is required to remove this bearing. The bearing need not be removed for cleaning and inspection. Remove this bearing ONLY if the determination has been made that it is unfit for further service.

4. To install a new bearing, place the bearing onto the shaft and press it into place using an arbor press. If an arbor press is not available a socket large enough to fit over the crankshaft could be used to drive the bearing into place.

■ On the smaller horsepower engines, babbitt bearings are used for the center main with needle bearings installed for the upper and lower main bearings.

3 Cylinder & V4 Motors

◆ See Figures 146 thru 151

■ Please refer to the Exploded Views section for exploded views of the powerhead and its components.

The 3 cylinder engines are equipped with an upper seal and bearing, a lower seal and bearing, and 2 main bearings between. A labyrinth seal is used at the 2 center main bearings and just above the bottom seal, to provide an effective seal between the cylinders.

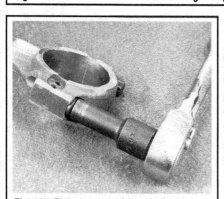

Fig. 137 The cap must always be kept with its matching rod

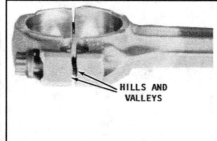

HILLS AND VALLEYS

Fig. 138 Notice the matching hills and valleys

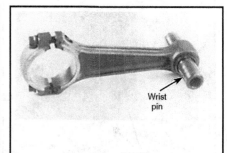

Wrist pin

Fig. 139 Testing the wrist pin before installation

Fig. 140 Here's a crank with the mains bearing ready to be removed

Fig. 141 A good shot of the labyrinth seal

Fig. 142 Removing the crankshaft from the block - 40 and 50 hp motors shown

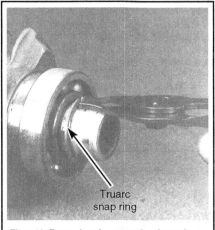
Fig. 143 Removing the snap ring from the lower bearing - 40 and 50 hp motors shown

Fig. 144 Use a puller when removing the lower bearing - 40 and 50 hp motors shown

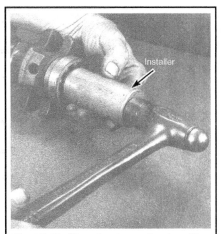
Fig. 145 You'll need a sleeve when installing the bearing

On the V4 engines, a top and bottom seal is used with sealing rings around the crankshaft in the center between each cylinder in each bank to seal for pressure and vacuum.

All models have a "6-ring" or "8-ring" arrangement, which eliminates the necessity of having the labyrinth seals. It is not possible to exchange a 2-ring crankshaft for a 3-ring crankshaft or the other way around. The same type crankshaft must always be installed as a replacement.

✳✳ WARNING

The rings on the V4 engines are brittle and break easily if twisted during removal of the crankshaft. Therefore, exercise care when removing the crankshaft to prevent damaging the rings.

1. Check to be sure the bolts have been removed securing the upper and lower bearing caps.

2. Lift the crankshaft assembly from the block. On some models, it may be necessary to use a soft-headed mallet and tap on the bottom side of the crankshaft to jar it loose. As the crankshaft is lifted, take care to work the center main bearings loose, and the sealing rings, if servicing a V4 engine. The center bearings are a split bearing held together with a snap wire ring.

■ **The bottom half of the bearing may be stuck in the cylinder block. Therefore, the crankshaft and the center main bearing must be worked free of the block together.**

Crankshaft Bearings

On V4 engines, observe how the center main bearing has a hole in the outside circumference.

On 3 cylinder engines, the 2 center main bearings and the top main bearing also have a hole.

On V4 engines, observe how the center main bearing has a hole in the outside circumference. Notice the locating pin in the cylinder block. The

Fig. 146 Lift out the crankshaft

purpose of this arrangement is to prevent the bearing shell from rotating. During assembling, the holes in the bearings must index with the pins in the block. Notice that the hole is not in the center of the bearing. If the bearing is to be removed from the crankshaft, make a suitable mark or identification to ensure the bearing is installed in the same position from which it was removed. Only in this manner can the hole be properly indexed with the pin.

Labyrinth Seal

If servicing a 3 cylinder engine, notice the grooves in the block on one side of the center main bearing. Observe the grooves in the crankcase cover. This arrangement of grooves forms what is commonly known as a labyrinth seal. The grooves fill with oil and/or fuel creating a seal between the cylinders. Also notice that the top bearing and the 2 center main bearings all have a locating pin.

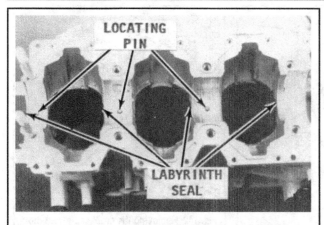

Fig. 147 The bearings are held in place with a pin

Fig. 148 The lower roller bearing is pressed on and held in place with a snap ring

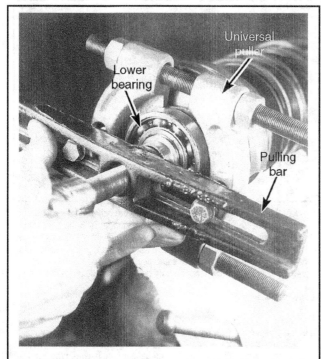

Fig. 149 Once the snap ring is removed, you'll need to press off the bearin

Lower Bearing

The powerheads of engines covered in this manual have a lower roller bearing that is pressed onto the crankshaft. If this bearing requires replacement, it is removed by first removing the snap ring, and then "pulling" the bearing from the crankshaft. A clamp-type puller is required to remove this bearing. The bearing need not be removed for cleaning and inspection. Remove this bearing only if the determination has been made that it is unfit for further service.

ASSEMBLY & INSTALLATION - 1 & 2 CYLINDER MOTORS

◆ See Figures 152 and 153

■ Please refer to the Exploded Views section for exploded views of the powerhead and its components.

1. Insert the proper number of needle bearings into the center main bearing cage.
2. Install the outer sleeve over the bearing cage. Check to be sure the 2 halves of the outer sleeve are matched. Again, these 2 halves are manufactured as a single unit and then broken. Therefore, the hills and valleys of the break absolutely MUST match during installation.
3. Snap the retaining ring into place around the bearing.
4. Slide the upper bearing onto the crankshaft journal at the upper end and the lower bearing onto the lower end.
5. Rotate the installed bearings to be sure there is no evidence of binding or rough spots. The crankshaft is now ready for installation.
6. Installation procedures for the crankshaft and bearing assembly follow:

Fig. 150 Remove the retaining wire. .

Fig. 151 . . .and then the race and bearing

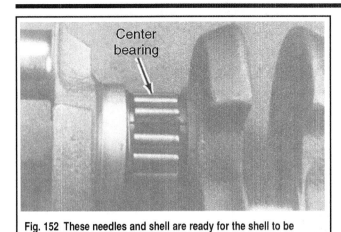

Fig. 152 These needles and shell are ready for the shell to be installed

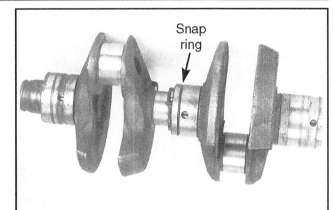

Fig. 153 Make sure the snap ring is installed on the center main bearing

Units With All Needle Main & Rod Bearings

◆ See Figures 154 thru 165

■ Please refer to the Exploded Views section for exploded views of the powerhead and its components.

The following procedures outline steps to install a crankshaft with needle upper, center, and lower main bearings. The upper and lower mains are complete bearings and cannot be disassembled. The center bearing is caged. Installation procedures for small horsepower crankshafts with Babbitt

upper lower and center main bearings and with Babbitt rod bearings are given in the following sections.

1. Observe the pin installed in each main bearing recess. Notice the hole in each main bearing outer shell. During installation, the hole in each bearing shell must index over the pin in the cylinder block.

2. Hold the crankshaft over the cylinder block with the upper end to your left. Now, lower the crankshaft into the block, and at the same time, align the hole in each bearing to enable the pin in the block to index with the hole. Rotate each bearing slightly until all pins are properly indexed with the matching bearing hole. Once all pins are indexed, the crankshaft will be properly seated.

Fig. 154 Note the bearing locating pins in the cylinder block. . .

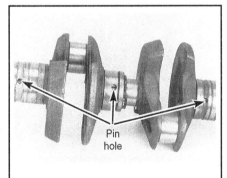

Fig. 155 . . .which must index with holes in the bearings

Fig. 156 Installing the crankshaft - 40 and 50 hp motors shown

Fig. 157 Cage and needle bearings installed into the lower portion of the rod

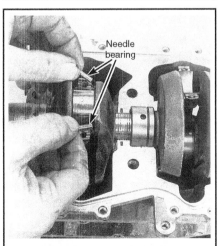

Fig. 158 Installing the needle bearing on each side of the crank

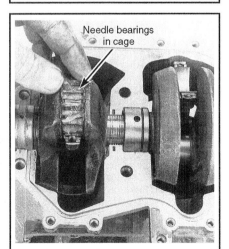

Fig. 159 Lowering the cage and bearings over the top of the crank

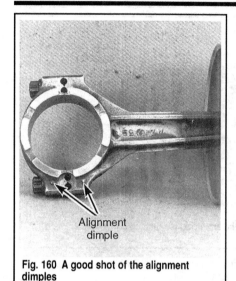

Fig. 160 A good shot of the alignment dimples

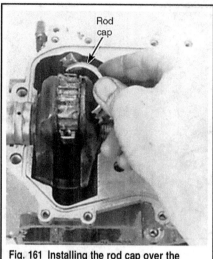

Fig. 161 Installing the rod cap over the bearings and cage

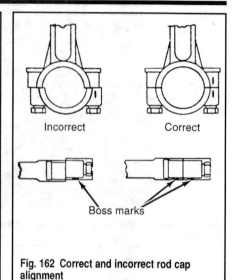

Fig. 162 Correct and incorrect rod cap alignment

Fig. 163 Install the rod cap bolts, and at the same time checking the cap alignment with the rod (using a pick)

Fig. 164 Check the flexibility of the rings through the exhaust port (small hp motor shown)

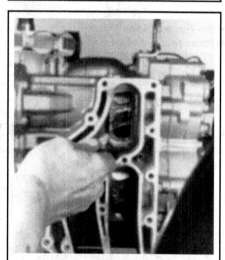

Fig. 165 Check the flexibility of the rings through the exhaust port - 40 and 50 hp motors shown

3. Apply needle bearing grease to each bearing cage. Coat the rod half of the bearing areas with needle bearing grease. Needle bearing grease must be used because other types of grease will not thin out and dissipate. The grease must dissipate to allow the gasoline and oil mixture to enter and lubricate the bearing. If needle bearing grease is not available, use a good grade of petroleum jelly (Vaseline).

4. Insert the proper number of needle bearings into each cage. Set the bearing cage into the bottom half of the rod. With your fingers on each side of the rod, pull up on the rod and bring the rod up to the bottom side of the crankshaft. Put one needle bearing on each side of the crankshaft. Using needle bearing grease load the other cage and install the needle bearings into the cage. Lower the cage onto the crankshaft journal.

5. Install the proper rod cap to the rod with the identifying mark or dimple properly aligned to ensure the cap is being installed in the same position from which it was removed. Tighten the rod bolts finger tight, and then just a bit more.

6. Use a "scratch-all", pick, or similar tool and move it back and forth on the outside surface of the rod and cap. Make the movement across the mating line of the rod and cap. The tool should not catch on the rod or on the cap. The rod cap must seat squarely with the rod. If not, tap the cap until the "scratch-all" will move back and forth on the rod and cap across the mating line without any feeling of catching. Any step on the outside will mean a step on the inside of the rod and cap. Just a whisker of a lip, will cause one of the needle bearings to catch and fail to rotate. The needle will quickly flatten, and the rod will begin to "knock". Needle bearings must rotate or the function of the bearing is lost.

7. Tighten the rod cap bolts alternately and evenly in three rounds to the value given in the Torque Specifications chart. Tighten the bolts to 1/2 the torque value on the first round, to 3/4 the torque value on the second round, and to the full torque value on the third and final round. On each round, check with the pick to be sure the cap remains seated squarely.

8. On twins, install the other rod cap in the same manner.

9. After the rod(s) has (have) been connected to the crankshaft, rotate the crankshaft until the rings on one cylinder are visible through the exhaust port. Use a screwdriver and push on each ring to be sure it has spring tension. It will be necessary to move the piston slightly, because all of the rings will not be visible at one time. If there is no spring tension, the ring was broken during installation - the piston must be removed and a new ring installed. Repeat the tension test at the intake port. On twins, check the other cylinder in the same manner.

Units With Babbit Or Babbit & Needle Bearings

◆ See Figures 166 thru 169

■ Please refer to the Exploded Views section for exploded views of the powerhead and its components.

This section provides detailed instructions to install a small horsepower crankshaft with any of the bearing combinations listed in the heading. Some of the powerheads covered in these paragraphs have rod liners, others do not.

The procedures pick up the work after the pistons have been installed, as described earlier in this section.

■ **Before installing the crankshaft, check to be sure each "throw" is clean and shiny. There should be no evidence of corrosion that might damage the "throw" during engine operation.**

1. Lower the crankshaft into place in the cylinder block with the long threaded shank end at the top of the cylinder block. (It is a known fact, in more than just a few shops around the country, because of haste; the crankshaft installation work has proceeded with the short end at the top.)

■ **The hole in the upper and lower main bearing must index into the pin in the cylinder block.**

2. Some engines may have a lining arrangement as listed in the heading of this section. The lining is made in two parts. Install the liner half into the rod, and then install the bearings as described in the next paragraph. The matching liner is to be installed into the rod cap.

3. Coat the rod half, of the bearing area, with needle bearing grease. Needle bearing grease MUST be used because other types of grease will not thin out and dissipate. The grease must dissipate to allow the gasoline and oil mixture to enter and lubricate the bearing. If needle bearing grease is not available, use a good grade of petroleum jelly (Vaseline).

4. Load the rod half of the rod bearing with needle bearings. Next, bring the rod up to the crankshaft rod journal. Coat the crankshaft journal with needle bearing grease. Place the needle bearings around the crankshaft journal.

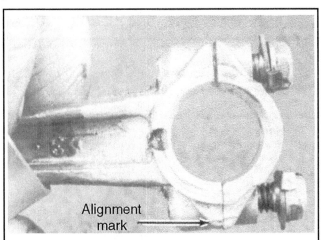

Fig. 166 A good look at the rod and cap alignment marks

5. Position the rod cap, with the liners (if used) over the needle bearings. Install the rod cap bolts and lock washers. Bring the bolts up finger tight and then just a bit more.

6. If the liners are used, the cap and rod automatically align properly. If liners are not used, a dowel pin is installed in the rod cap. This pin will index into a hole in the rod for proper alignment.

7. Tighten the rod cap bolts alternately and evenly in three rounds to the value given in the Torque Specifications chart. Tighten the bolts to 1/2 the torque value on the 1st round, to 3/4 the torque value on the 2nd round, and to the full torque value on the 3rd and final round. On each round, check with the pick to be sure the cap remains seated squarely.

8. After the rod cap bolts have been tightened to the required torque value and the installation appears satisfactory, bend the bolt locking tabs upward to prevent the bolts from loosening.

9. On twins, install the other rod cap in the same manner.

10. After the rod(s) has(have) been connected to the crankshaft, rotate the crankshaft until the rings on one cylinder are visible through the exhaust port. Use a screwdriver and push on each ring to be sure it has spring tension. It will be necessary to move the piston slightly, because all of the rings will not be visible at one time. If there is no spring tension, the ring was broken during installation. The piston must be removed and a new ring installed.

11. Repeat the tension test at the intake port. On twins, check the other cylinder in the same manner.

Units With All Babbit Main & Rod Bearings

◆ **See Figures 166, 170, 171 and 172**

■ **Please refer to the Exploded Views section for exploded views of the powerhead and its components.**

This section provides detailed instructions to install a small horsepower crankshaft with babbitt upper, lower, and center main bearings, and with babbitt rod bearings.

The procedures pickup the work after the pistons have been installed, as described earlier in this section.

1. Lower the crankshaft into place in the cylinder block with the long threaded shank end at the top of the cylinder block. It is a known fact, in more than just a few shops around the country, because of haste; the crankshaft installation work has proceeded with the short end at the top.

2. Pull the rod up to the crankshaft journal. Position the rod cap over the crankshaft journal. Install the rod cap bolts and lock-washers. Bring the bolts up finger tight and then just a bit more.

3. Tighten the rod cap bolts alternately and evenly in 3 rounds to the value given in the Torque Specifications chart. Tighten the bolts to 1/2 the torque value on the 1st round, to 3/4 the torque value on the 2nd round, and

Fig. 167 Needle bearings installed in the rod cap liner and around the crank

Fig. 168 Rod cap with liner ready for installation

Fig. 169 The locking tabs must be bent upward after the rod cap bolts have been tightened

Fig. 170 Tightening the rod cap bolts

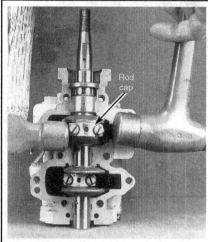

Fig. 171 Use two hammers when fitting the rod cap to the crankshaft

Fig. 172 Install the flywheel temporarily to check the piston and crankshaft movement

to the full torque value on the 3rd and final round. On each round, check with the pick to be sure the cap remains seated squarely.

4. On twins, repeat the procedure for the other rod and cap.

5. After the other rod cap (if applicable) has been installed and the bolts tightened to the proper torque value, hold one hammer on one side of the rod and cap, and at the same time tap the other side of the rod and cap with the other hammer. Tap lightly on the top of the cap. Reverse the hammer positions and tap the opposite sides of the rod and cap. This procedure will "fit" the rod and cap to the crankshaft journal.

6. Repeat the "fitting" procedure for the other rod and cap.

7. Once the installation procedure appears satisfactory and all work has been completed, bend the bolt locking tabs upward to prevent the bolts from loosening.

ASSEMBLY & INSTALLATION - 3 CYLINDER & V4 MOTORS

◆ See Figures 173 thru 190

■ Please refer to the Exploded Views section for exploded views of the powerhead and its components.

■ The V4 engines have a single center main bearing, while the 3 cylinder engines have 2 center main bearings.

1. Insert the proper number of needle bearings into the center main bearing cage.

2. Install the outer sleeve over the bearing cage. Check to be sure the 2 halves of the outer sleeve are matched. Again, these 2 halves are manufactured as a single unit and then broken. Therefore, the hills and valleys of the break absolutely must match during installation. Double check to be sure the marks made during bearing removal match to ensure the hole in the bearing will index with the pin in the block. Remember, the hole in the bearing is not in the center; therefore the bearing can only be installed properly one way.

3. Snap the retaining ring into place around the bearing.

4. To install a new lower bearing, place the bearing onto the shaft and press it into place using an arbor press. If an arbor press is not available a socket large enough to fit over the crankshaft can usually be used to drive the bearing into place.

5. Install the Truarc ring to secure the bearing in place.

■ On V4 engines, the upper bearing is pressed into the upper bearing cap. On the 3 cylinder engines, the upper bearing is simply a "snug" fit onto the crankshaft.

6. Rotate the installed bearings to be sure there is no evidence of binding or rough spots. The crankshaft is now ready for installation.

The following procedures outline steps to install a crankshaft with a single main bearing; the lower bearings pressed onto the crankshaft; and the upper bearing pressed into the upper bearing cap, as for the V4 engine.

The section also contains procedures for installation of the 3 cylinder crankshaft with 2 center main bearings; the lower bearing pressed onto the crankshaft; and the upper bearing simply pushed onto the crankshaft.

The text clearly identifies the procedures for the type of crankshaft being serviced.

7. To install a 3 cylinder crankshaft, start as follows:

a. Observe the pin installed in each main bearing recess. Notice the hole in each main bearing outer shell. During installation, the hole in each bearing shell must index over the pin in the cylinder block.

b. Hold the crankshaft over the cylinder block with the upper end to your left. Now, lower the crankshaft into the block and at the same time, align the hole in each bearing to enable the pin in the block to index with the hole. Rotate each bearing slightly until all pins are properly indexed with the matching bearing hole. Once all pins are indexed, the crankshaft will be properly seated.

c. Slide the lower bearing cap into place and just start the mounting bolts. Do not tighten them at this time. Starting the bolts will hold the crankshaft in place in the block while the work continues.

■ The reason for installing the bearing caps at this time is to prevent the crankshaft from lifting as it is turned during rod bearing and rod cap installation. If servicing a 3 cylinder engine, drill two 7/16 in. holes in a small piece of wood. Now, slide the block of wood down the 2 studs at the upper bearing. Secure the block of wood in place with the nuts. This block of wood will prevent the crankshaft from lifting as it is turned during rod cap installation.

8. To install a V4 crankshaft, start as follows:

a. Lower the crankshaft into place in the block with the center main bearing pin indexed into the hole in the block. At the same time as the crankshaft is being lowered, work the sealing rings around the crankshaft into the recesses in the block. Rotate the rings until the openings are staggered across the crankshaft. Be careful, because the rings are brittle and will break if twisted or distorted.

■ After the crankshaft is in place, it is not possible to rotate the rings. Check to be sure the pins in the block are indexed into the holes in the center main bearings. It should not be possible to rotate the bearings after the crankshaft is in place.

b. Slide the lower and upper bearing caps into place and just start the mounting bolts.

c. On V4 engines, the bearing cap at both ends of the crankshaft can only align with the bolt hole pattern in the powerhead one way. Continue to rotate the caps until all 8 bolt holes are properly aligned. Do not tighten the bolts at this time. Starting the bolts will hold the crankshaft in place in the block while the work continues.

d. The lower bearing retainer plate has 2 ears which index into 2 recesses in the lower bearing cap. Therefore, rotate the retainer until the ears on the retainer plate index with the 2 recesses in the bearing cap, then install two 1/4 x 28 guide pins through the lower bearing cap and into the retainer plate. These guide pins will hold the bearing cap and retainer plate in place while the work continues and until the retaining bolts are installed.

9. Now, for all engines, apply needle bearing grease to each bearing cage. Coat the rod half of the bearing areas with needle bearing grease. Needle bearing grease must be used because other types of grease will not thin out and dissipate. The grease must dissipate to allow the gasoline and oil mixture to enter and lubricate the bearing. If needle bearing grease is not available, use a good grade of petroleum jelly (Vasoline).

10. Insert the proper number of needle bearings into each cage. Set the bearing cage into the bottom half of the rod. With your fingers on each side of the rod, pull up on the rod and bring the rod up to the bottom side of the crankshaft. Place 1 needle bearing on each side of the crankshaft. Using needle bearing grease load the other cage and install the needle bearings into the cage. Lower the cage onto the crankshaft journal.

11. Install the proper rod cap to the rod with the identifying mark or dimple properly aligned to ensure the cap is being installed in the same position from which it was removed. Tighten the rod bolts finger tight, and then just a bit more.

12. Use a "scratch-all", pick, or similar tool and move it back and forth on the outside surface of the rod and cap. Make the movement across the

Fig. 173 Insert the needles into the cage

Fig. 174 Proper bearing installation

Fig. 175 Drive the bearing into position. . .

Fig. 176 . . .and then install the snap ring

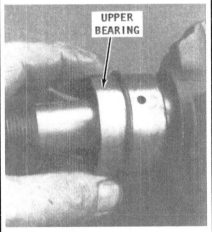

Fig. 177 Press the upper bearing into the cup on the V4 and V6

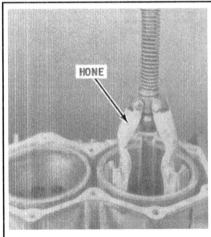

Fig. 178 On the 3-cylinder engine, the upper bearing is a press-fit

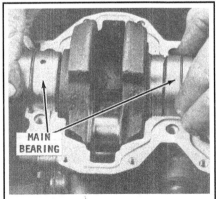

Fig. 179 Lower the crankshaft into the cylinder block so the holes are aligned - 3 cylinder

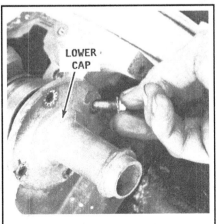

Fig. 180 Thread in the mounting bolts

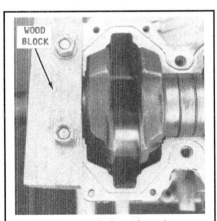

Fig. 181 Use a small piece of wood to secure the crankshaft on the 3 cylinder

mating line of the rod and cap. The tool should not catch on the rod or on the cap. The rod cap must seat squarely with the rod. If not, tap the cap until the "scratch-all" will move back and forth on the rod and cap across the mating line without any feeling of catching. Any step on the outside will mean a step on the inside of the rod and cap. Dust a whisker of a lip will cause one of the needle bearings to catch and fail to rotate. The needle will quickly flatten and the rod will begin to "knock". Needle bearings must rotate or the function of the bearing is lost.

13. Tighten the rod cap bolts alternately and evenly in 3 rounds to the torque value given in the Torque Chart. Tighten the bolts to 1/2 the torque value on the first round, to 3/4 the torque value on the second round, and to the full torque value on the third and final round. On each round, check with the pick to be sure the cap remains seated squarely.

14. Install the other rod caps in the same manner.

15. After the rods have been connected to the crankshaft, rotate the crankshaft until the rings on 1 cylinder are visible through the exhaust port. Use a screwdriver and push on each ring to be sure it has spring tension. It will be necessary to move the piston slightly, because all of the rings will not be visible at one time. If there is no spring tension, the ring was broken during installation. The piston must be removed and a new ring installed. Repeat the tension test at the intake port. Check the other cylinders in the same manner.

Fig. 182 Lower the crankshaft. . .

Fig. 183 . . .and then install the retainer plate on the V4

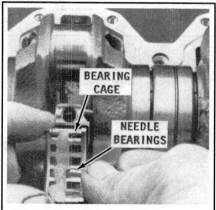
Fig. 184 Install the needle bearings into the cage. . .

Fig. 185 . . .and then position the cage in the bottom half of the connecting rod

Fig. 186 Position a single needle on each side

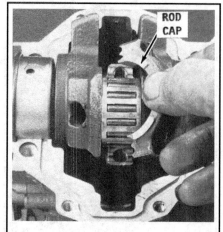
Fig. 187 Installing the rod cap

Fig. 188 Snug up the rod bolts

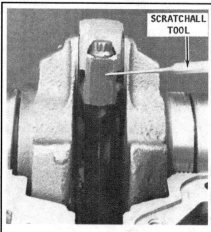
Fig. 189 Mark the rod and cap

Fig. 190 Check the spring tension on each ring

CLEANING & INSPECTION

◆ See Figures 152 and 191 thru 194

■ **Please refer to the Exploded Views section for exploded views of the powerhead and its components.**

Inspect the splines for signs of abnormal wear. Check the crankshaft for straightness. Inspect the crankshaft oil seal surfaces to be sure they are not grooved, pitted or scratched. Replace the crankshaft if it is severely damaged or worn. Check all crankshaft bearing surfaces for rust, water marks, chatter marks, uneven wear or overheating. Clean the crankshaft surfaces with crocus cloth.

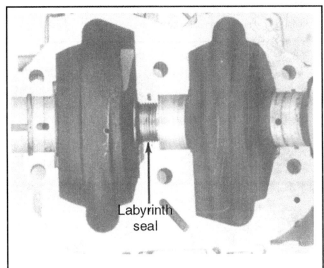

Fig. 191 The labyrinth seal area is clearly visible here

Fig. 192 This crank came from a submerged motor and is no longer fit for service

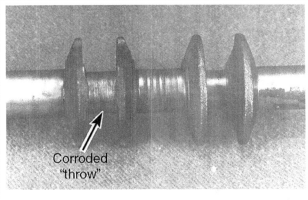

Fig. 193 The 'throw' on this crank is badly corroded and not fit for service

Carefully check the grooves around the crankshaft for the sealing rings. Check the surfaces of block halves to be sure the sealing rings will affect an effective seal.

Clean the crankshaft and crankshaft bearing with solvent. Dry the parts, but NOT the bearing, with compressed air. Check the crankshaft surfaces a second time. Replace the crankshaft if the surfaces cannot be cleaned properly for satisfactory service. If the crankshaft is to be installed for service, lubricate the surfaces with light oil.

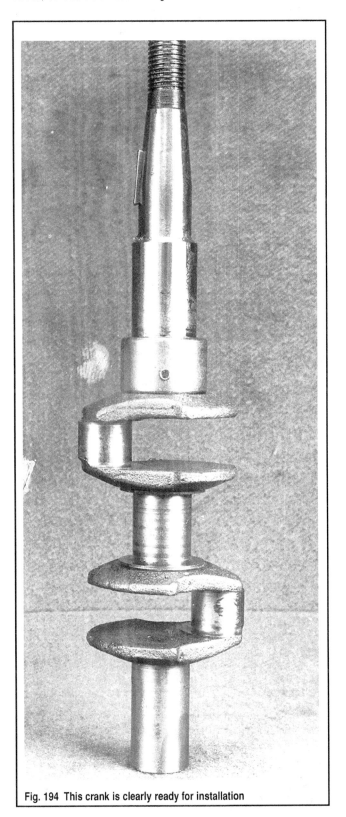

Fig. 194 This crank is clearly ready for installation

On 1 and 2 cylinder motors, the top and lower bearing may be easily removed from the crankshaft. The center main bearing has a spring steel wire securing the two halves together. Remove the wire, and then the outer sleeve, then the needle bearings. Be careful not to lose any of the needles. The outer shell is a fractured break type unit; therefore, the 2 halves of the shell must be kept as a set.

On V4 engines, the center main bearing has a spring steel wire securing the 2 halves together. The 3 cylinder engine has 2 center main bearings. Remove the wire, and then the outer sleeve, then the needle bearings. Take care not to lose any of the needles. The outer shell is a fractured break type unit. Therefore, the 2 halves of the shell must absolutely be kept as a set.

Check the crankshaft bearing surfaces to be sure they are not pitted or show any signs of rust or corrosion. If the bearing surfaces are pitted or rusted, the crankshaft and bearings must be replaced.

During an engine overhaul to this degree, it is a good practice to remove the seal from the top main bearing. If the same type of seal is used in the bottom main bearing, remove that seal also.

Cylinder Block

SERVICE

 DIFFICULT

◆ See Figures 195 and 196

■ Please refer to the Exploded Views section for exploded views of the powerhead and its components.

1. Inspect the cylinder block and cylinder bores for cracks or other damage.
2. Remove carbon with a fine wire brush on a shaft attached to an electric drill or use a carbon remover solution.
3. Use an inside micrometer or telescopic gauge and micrometer to check the cylinders for wear. Check the bore for out of round and/or oversize bore. If the bore is tapered, out of round or worn more than 0.003-0.004 in. (0.076-0.102mm) the cylinders should be re-bored and oversize pistons and rings installed.

■ Oversize piston weight is approximately the same as a standard size piston; therefore, it is NOT usually necessary to re-bore all cylinders in a block just because 1 cylinder requires re-boring. The APBA (American Power Boat Association) accepts and permits the use of 0.015 in. (0.381mm) oversize pistons.

4. Hone the cylinder walls lightly to seat the new piston rings, as outlined in the Honing Procedures in this section. If the cylinders have been scored, but are not out of round or the sleeve is rough, clean the surface of the cylinder with a cylinder hone as described in Honing Procedures.

■ If only one cylinder is damaged, a cylinder sleeve may be installed on some models, but the cost is very high. Installation of the sleeve will bring the powerhead back to standard and permit re-boring of that cylinder at a later date.

HONING PROCEDURES

 DIFFICULT

◆ See Figures 195 and 196

To ensure satisfactory engine performance and long life following the overhaul work, the honing work should be performed with patience, skill, and in the following sequence:
1. Follow the hone manufacturer's recommendations for use of the hone and for cleaning and lubricating during the honing operation.
2. Pump a continuous flow of honing oil into the work area. If pumping is not practical, use an oil can. Apply the oil generously and frequently on both the stones and work surface.

■ If honing a block from a looper motor, use a long-size hone as a precaution against the hone becoming stuck in one of the oblong openings intake ports, in the cylinder wall and breaking. If the special long hone is not available, the block should be taken to a machine shop.

3. Begin the stroking at the smallest diameter (usually the top of the cylinder). Maintain a firm stone pressure against the cylinder wall to assure fast stock removal and accurate results.
4. Expand the stones as necessary to compensate for stock removal and stone wear. The best cross-hatch pattern is obtained using a stroke rate of 30 complete cycles per minute. Again, use the honing oil generously.
5. Hone the cylinder walls ONLY enough to de-glaze the walls.
6. After the honing operation has been completed, clean the cylinder bores with hot water and detergent. Scrub the walls with a. stiff bristle brush and rinse thoroughly with hot water. The cylinders must be cleaned well as prevention against any abrasive material remaining in the cylinder bore. Such material will cause rapid wear of new piston rings, the cylinder bore, and the bearings.
7. After cleaning, swab the bore(s) several times with engine oil and a clean cloth, and then wipe them dry with a clean cloth. NEVER use kerosene or gasoline to clean the cylinder(s).
8. Clean the remainder of the cylinder block to remove any excess material spread during the honing operation.

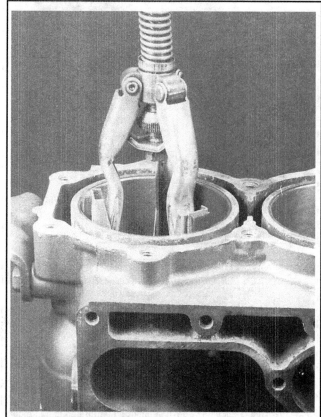

Fig. 195 Honing the cylinder

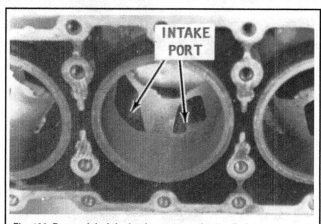

Fig. 196 Be careful of the intake ports on the 3-cylinder

POWERHEAD BREAK-IN

General Information

Anytime a new or rebuilt powerhead is installed (this includes a powerhead whose wear components such as pistons and rings and bearings have been replaced), the motor must undergo proper break-in.

By following break-in procedures largely consisting of specific engine operating limitations during the first 10 hours of operation, you will help you will help ensure a long and trouble-free life. Failure to follow these recommendations may allow components to seat improperly, causing accelerated wear and premature powerhead failure.

On all motors, special attention is required to the fuel/oil mixture requirements.

During break-in, pay close attention to all pre- and post- operation checks. This goes double when checking for fuel, oil or water leaks. At each start-up and frequently during operation, check for presence of the cooling indicator stream.

At the completion of break-in, double-check the tightness of all exposed engine fasteners.

Although some engines require slightly different steps for powerhead break-in, one procedure is common. During the entire first 10 hours of engine operation, **vary** the engine speed and perform a complete service at the end of the first 20 hours of operation. Varying engine speed allows parts to wear in under conditions throughout the power band, not just at idle or mid-throttle. Conducting the 20 hour service makes sure that you find and correct any problems or change any settings that may have occurred/changed during break-in (or replaced parts expected to wear during break-in).

■ During break-in, check your hourmeter or a watch frequently and be sure to change the engine speed at least every 15 minutes (that means between every 2-3 tenths on the hourmeter).

Be sure to **always** allow the engine to reach operating temperature before setting the throttle anywhere above idle. This means you should always start and run the motor for at least 5 minutes before advancing the throttle, but be especially aware of engine warm-up time during break-in.

✳✳ WARNING

NEVER run the engine when it's out of the water, unless a flush fitting is used to provide a source of cooling. Remember that the water pump can be destroyed in less than a minute just from a lack of water. The powerhead will suffer damage in very little time as well, but even if it is not overheated out of water, reduced cooling from a damaged water pump impeller could destroy it later. Don't risk it. And keep in mind, since you must run the engine at varying speeds and loads during break-in, it cannot be done on a flush fitting.

Breaking In A Powerhead

For the first 10 hours of engine operation, you **must** use a 25:1 pre-mix fuel/oil mixture in the primary fuel tank. For more details, please refer to the information for Engine Oil in the Maintenance & Tune-Up section.
During break-in, observe the following time-table and limitations on engine operation.

■ Check the cooling indicator repeatedly to ensure proper engine cooling. Prior to 1977, a water mist should discharge from the exhaust relief holes at rear of driveshaft housing.

• During the first 10 minutes, operate the engine in gear at **only** fast idle.
• During the next 50 minutes, operate the engine in gear **below** 3500 rpm. If the boat planes easily, use **full** throttle to quickly bring the boat on plane, then immediately reduce throttle to 1/2 or less, but making sure the boat remains on plane. Vary the engine speed at least every 15 minutes.
• During the second hour of break-in, use full throttle to quickly plane the boat, then immediately reduce throttle to 3/4 or less, but make sure the boat remains on plane. Continue to vary the engine speed at least every 15 minutes. At various intervals, operate the engine at **full throttle** for 1-2 minutes, then reduce the throttle to 3/4 for an additional minute or two in order to allow the pistons to cool off slowly. Don't just drop from wide open throttle (WOT) to idle.
• For the next 8 hours, continue to vary the engine speed. Be sure to avoid continuous full-throttle operation for long periods.
• After the first 20 hours, the powerhead should be fully broken-in, perform a complete service. Be sure properly re-torque the cylinder head bolts (this should be done after the engine is run, but only after the cylinder head has cooled to the touch).

EXPLODED VIEWS

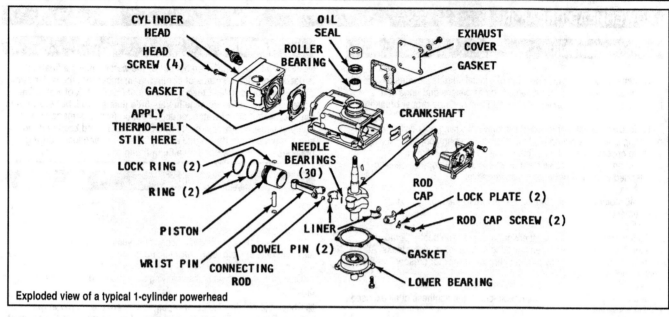

Exploded view of a typical 1-cylinder powerhead

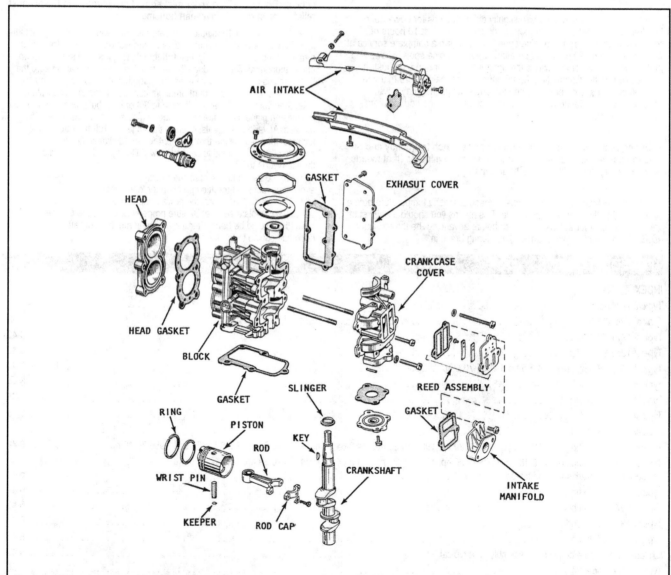

Exploded view of a typical 3 hp and 1969-70 4 hp powerhead. The 4 hp uses a bearing and seal on the crankshaft instead of the oil slinger

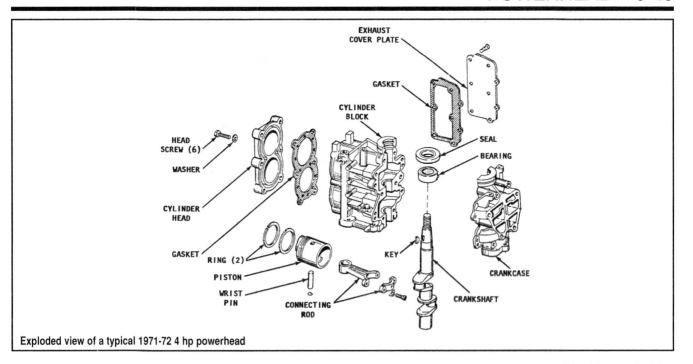

Exploded view of a typical 1971-72 4 hp powerhead

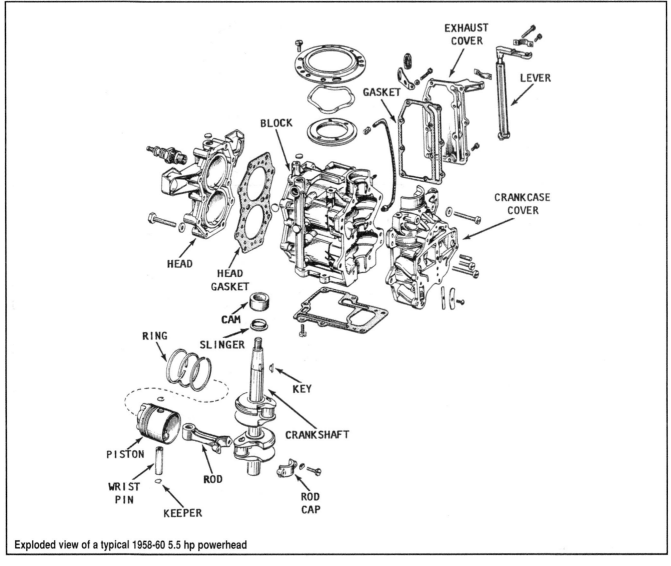

Exploded view of a typical 1958-60 5.5 hp powerhead

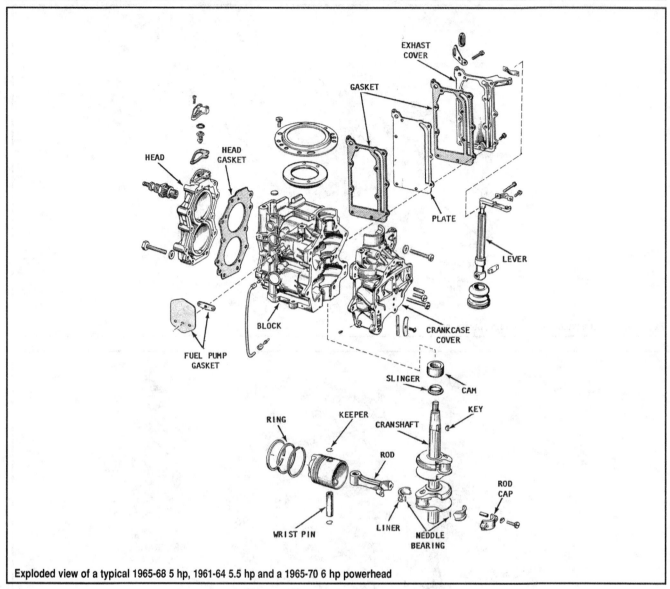

Exploded view of a typical 1965-68 5 hp, 1961-64 5.5 hp and a 1965-70 6 hp powerhead

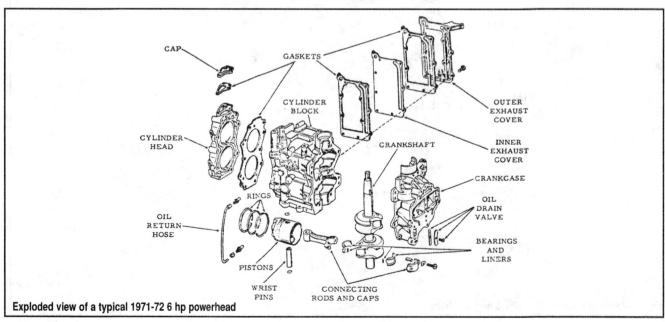

Exploded view of a typical 1971-72 6 hp powerhead

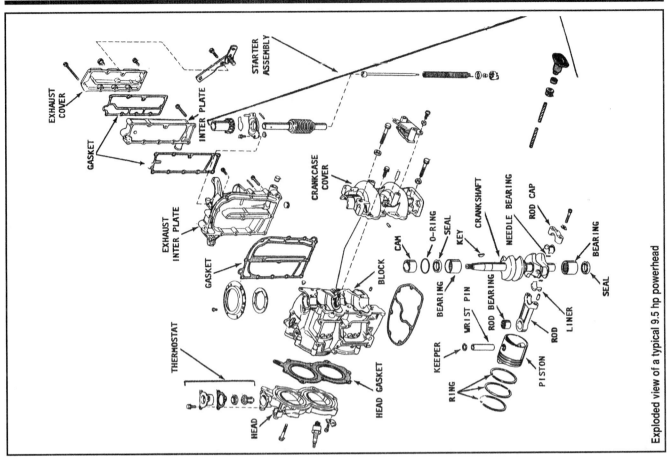

Exploded view of a typical 9.5 hp powerhead

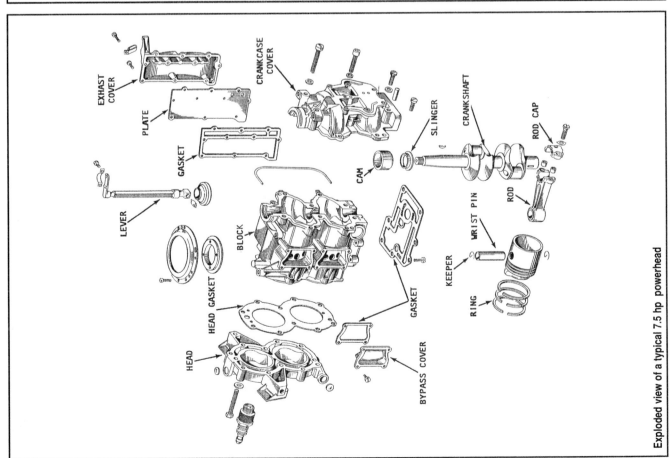

Exploded view of a typical 7.5 hp powerhead

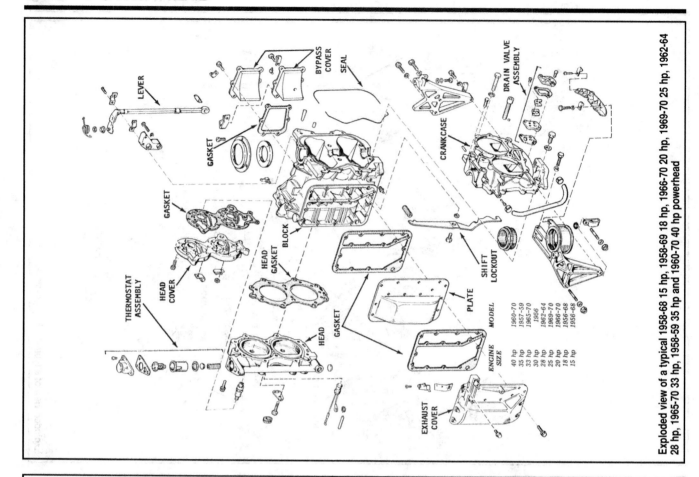

Exploded view of a typical 1958-68 15 hp, 1958-69 18 hp, 1966-70 20 hp, 1969-70 25 hp, 1962-64 28 hp, 1965-70 33 hp, 1958-59 35 hp and 1960-70 40 hp powerhead

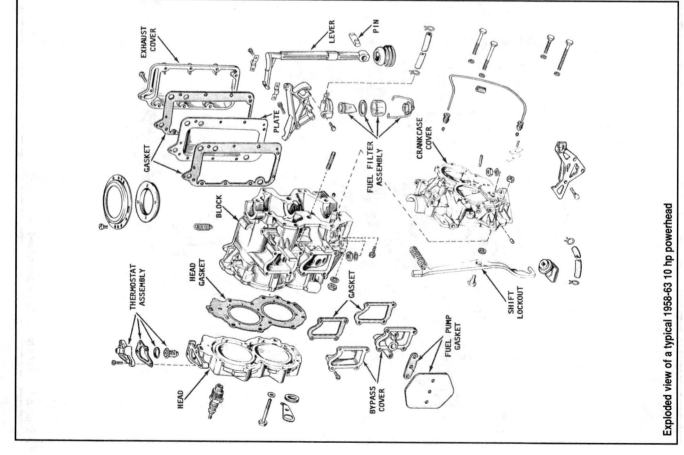

Exploded view of a typical 1958-63 10 hp powerhead

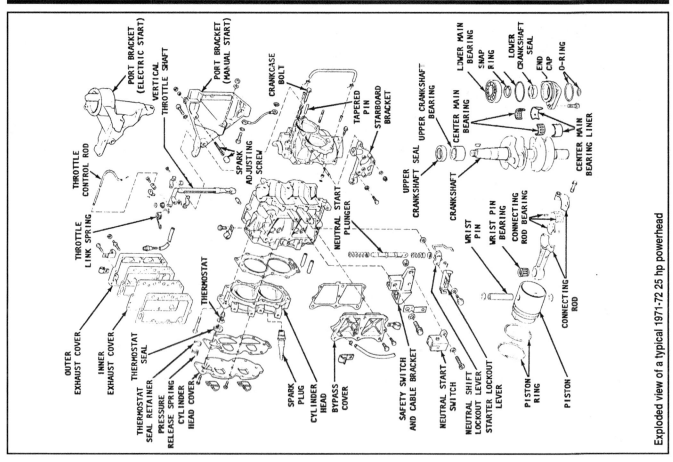

Exploded view of a typical 1971-72 25 hp powerhead

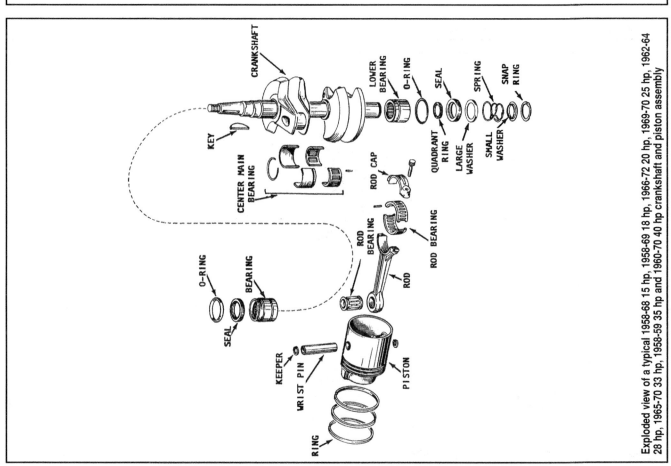

Exploded view of a typical 1958-68 15 hp, 1958-69 18 hp, 1966-72 20 hp, 1969-70 25 hp, 1962-64 28 hp, 1965-70 33 hp, 1958-59 35 hp and 1960-70 40 hp crankshaft and piston assembly

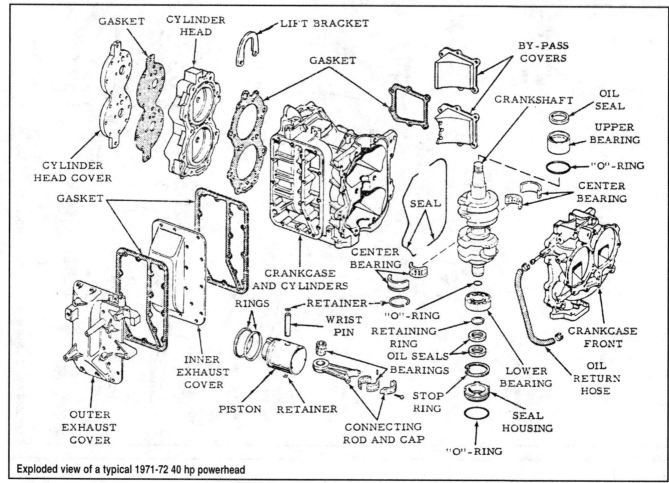

Exploded view of a typical 1971-72 40 hp powerhead

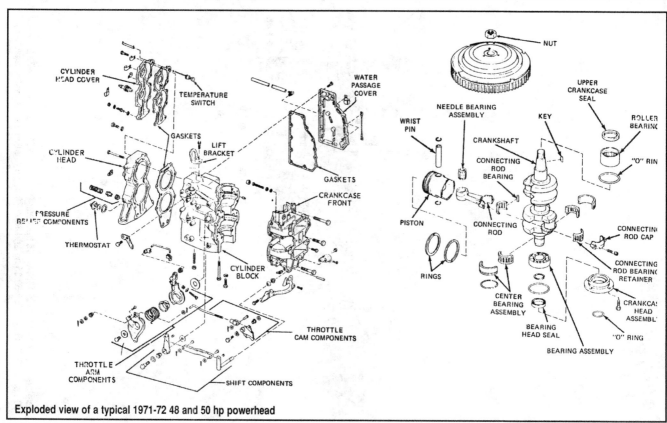

Exploded view of a typical 1971-72 48 and 50 hp powerhead

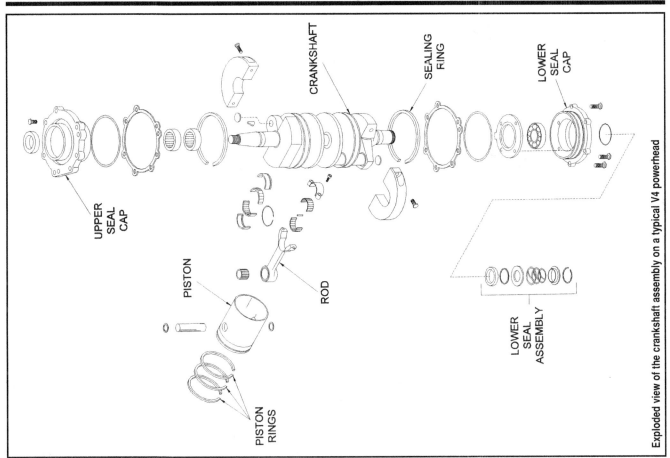

Exploded view of the crankshaft assembly on a typical V4 powerhead

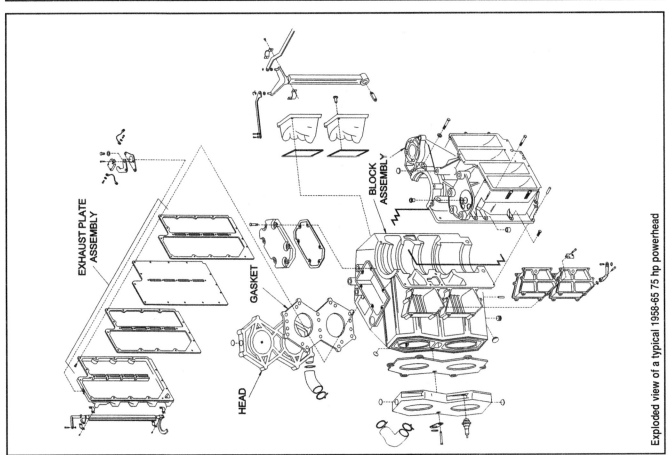

Exploded view of a typical 1958-65 75 hp powerhead

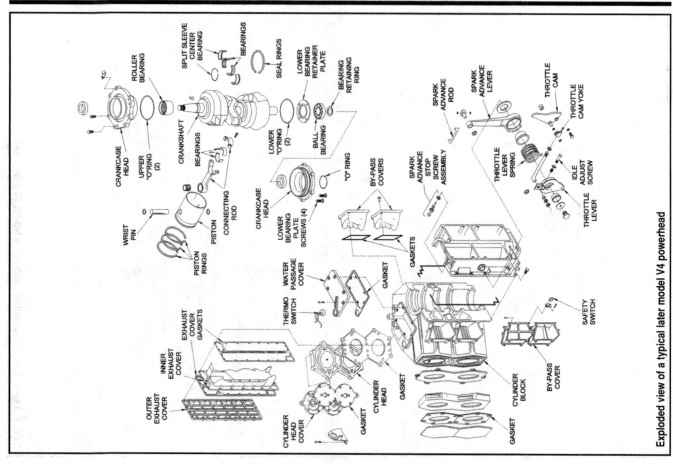

Exploded view of a typical later model V4 powerhead

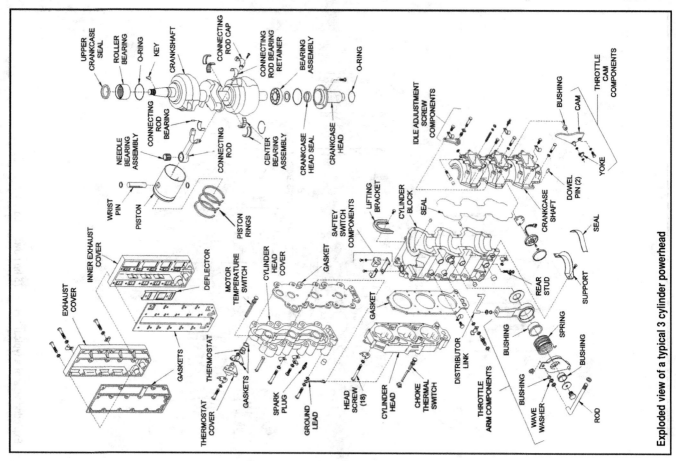

Exploded view of a typical 3 cylinder powerhead

Powerhead Specifications

Year	Model (Hp)	No. of Cyl	Displace cu. in. (cc)	Bore & Stroke in.	Bore 1st O/S in.	Bore 2nd O/S in.	Piston to Cylinder Clearance Max in.	Min in.	Piston Ring Groove Max in.	Min in.	Piston Ring Width Max in.	Min in.	Piston Ring End Gap Max in.	Min in.
1958	3	2	5.28 (85.5)	1.56 x 1.37	0.020	-	0.0020	0.0013	0.0035	0.0010	-	-	0.015	0.005
	5.5	2	8.84 (145)	1.94 x 1.50	0.020	-	0.0025	0.0013	0.0035	0.0010	0.0935	0.0925	0.015	0.005
	7.5	2	12.4 (203)	2.125 x 1.75	-	-	0.0015	0.0030	0.0035	0.0010	-	-	0.015	0.005
	10	2	16.6 (281)	2.37 x 1.875	-	-	0.0035	0.0030	0.0030	0.0015	-	-	0.017	0.007
	18	2	22 (361)	2.50 x 2.25	0.025	0.040	0.0035	0.0040	0.0030	0.0015	0.0935	0.0925	0.017	0.007
	35	2	40.5 (664)	3.06 x 2.75	0.020	0.040	0.0040	0.0035	0.0065	0.0050	-	-	0.017	0.007
	50	V4	70.7 (1159)	3.00 x 2.50	-	-	0.0040	0.0035	0.0065	0.0050	0.0935	0.0925	0.017	0.007
1959	3	2	5.28 (85.5)	1.56 x 1.37	0.020	-	0.0020	0.0013	0.0035	0.0010	-	-	0.015	0.005
	5.5	2	8.84 (145)	1.94 x 1.50	0.020	-	0.0025	0.0013	0.0035	0.0010	0.0935	0.0925	0.015	0.005
	10	2	16.6 (281)	2.37 x 1.875	-	-	0.0035	0.0030	0.0030	0.0015	-	-	0.017	0.007
	18	2	22 (361)	2.50 x 2.25	0.025	0.040	0.0035	0.0040	0.0030	0.0015	0.0935	0.0925	0.017	0.007
	35	2	40.5 (664)	3.06 x 2.75	0.020	0.040	0.0040	0.0035	0.0065	0.0050	-	-	0.017	0.007
	50	V4	70.7 (1159)	3.00 x 2.50	-	-	0.0040	0.0035	0.0065	0.0050	0.0935	0.0925	0.017	0.007
1960	3	2	5.28 (85.5)	1.56 x 1.37	0.020	-	0.0020	0.0013	0.0035	0.0010	-	-	0.015	0.005
	5.5	2	8.84 (145)	1.94 x 1.50	0.020	-	0.0025	0.0013	0.0035	0.0010	0.0935	0.0925	0.015	0.005
	10	2	16.6 (281)	2.37 x 1.875	-	-	0.0035	0.0030	0.0030	0.0015	-	-	0.017	0.007
	18	2	22 (361)	2.50 x 2.25	0.025	0.040	0.0035	0.0040	0.0030	0.0015	0.0935	0.0925	0.017	0.007
	40	2	43.9 (719)	3.19 x 2.75	-	0.025	0.0040	0.0035	0.0065	0.0050	0.0935	0.0925	0.017	0.007
	75	V4	89.5 (1467)	3.375 x 2.50	-	0.040	0.0060	0.0045	0.0070	0.0045	0.0935	0.0925	0.017	0.007
1961	3	2	5.28 (85.5)	1.56 x 1.37	0.020	-	0.0020	0.0013	0.0035	0.0010	-	-	0.015	0.005
	5.5	2	8.84 (145)	1.94 x 1.50	0.020	-	0.0025	0.0013	0.0035	0.0010	0.0935	0.0925	0.015	0.005
	10	2	16.6 (281)	2.37 x 1.875	-	-	0.0035	0.0030	0.0030	0.0015	-	-	0.017	0.007
	18	2	22 (361)	2.50 x 2.25	0.025	0.040	0.0035	0.0040	0.0030	0.0015	0.0935	0.0925	0.017	0.007
	40	2	43.9 (719)	3.19 x 2.75	-	0.025	0.0040	0.0035	0.0065	0.0050	0.0935	0.0925	0.017	0.007
	75	V4	89.5 (1467)	3.375 x 2.50	-	0.040	0.0060	0.0045	0.0070	0.0045	0.0935	0.0925	0.017	0.007
1962	3	2	5.28 (85.5)	1.56 x 1.37	0.020	-	0.0020	0.0013	0.0035	0.0010	-	-	0.015	0.005
	5.5	2	8.84 (145)	1.94 x 1.50	0.020	-	0.0025	0.0013	0.0035	0.0010	0.0935	0.0925	0.015	0.005
	10	2	16.6 (281)	2.37 x 1.875	-	-	0.0035	0.0030	0.0030	0.0015	-	-	0.017	0.007
	18	2	22 (361)	2.50 x 2.25	0.025	0.040	0.0035	0.0040	0.0030	0.0015	0.0935	0.0925	0.017	0.007
	28	2	35.7 (585)	2.875 x 2.75	0.020	0.040	0.025	0.0040	0.0070	0.0045	0.0935	0.0925	0.017	0.007
	40	2	43.9 (719)	3.19 x 2.75	-	0.025	0.0045	0.0030	0.0070	0.0045	0.0935	0.0925	0.017	0.007
	75	V4	89.5 (1467)	3.375 x 2.50	-	0.040	0.0060	0.0045	0.0070	0.0045	0.0935	0.0925	0.017	0.007
1963	3	2	5.28 (85.5)	1.56 x 1.37	0.020	-	0.0020	0.0013	0.0035	0.0010	-	-	0.015	0.005
	5.5	2	8.84 (145)	1.94 x 1.50	0.020	-	0.0025	0.0013	0.0035	0.0010	0.0935	0.0925	0.015	0.005
	10	2	16.6 (281)	2.37 x 1.875	-	-	0.0035	0.0030	0.0030	0.0015	-	-	0.017	0.007
	18	2	22 (361)	2.50 x 2.25	0.025	0.040	0.0035	0.0040	0.0030	0.0015	0.0935	0.0925	0.017	0.007
	28	2	35.7 (585)	2.875 x 2.75	0.020	0.040	0.025	0.0040	0.0070	0.0045	0.0935	0.0925	0.017	0.007
	40	2	43.9 (719)	3.19 x 2.75	-	0.025	0.0045	0.0030	0.0070	0.0045	0.0935	0.0925	0.017	0.007

Powerhead Specifications

Year	Model (Hp)	No. of Cyl	Displace cu. in. (cc)	Bore & Stroke in.	Bore 1st O/S in.	Bore 2nd O/S in.	Piston to Cylinder Clearance Max in.	Piston to Cylinder Clearance Min in.	Piston Ring Groove Max in.	Piston Ring Groove Min in.	Piston Ring Width Max in.	Piston Ring Width Min in.	Piston Ring End Gap Max in.	Piston Ring End Gap Min in.
1963	75	V4	89.5 (1467)	3.375 x 2.50	-	.040	0.0060	0.0045	0.0070	0.0045	0.0935	0.0925	0.017	0.007
1964	3	2	5.28 (85.5)	1.56 x 1.37	.020	-	0.0020	0.0013	0.0035	0.0010	-	-	0.015	0.005
	5.5	2	8.84 (145)	1.94 x 1.50	.020	-	0.0025	0.0013	0.0035	0.0010	0.0935	0.0925	0.015	0.005
	9.5	2	15.2 (249)	2.31 x 1.81	.020	-	0.0045	0.0030	0.0035	0.0010	0.0935	0.0925	0.017	0.007
	18	2	22 (361)	2.50 x 2.25	.025	.040	0.0035	0.0040	0.0030	0.0015	0.0935	0.0925	0.017	0.007
	28	2	35.7 (585)	2.875 x 2.75	.020	.040	0.025	0.0040	0.0070	0.0045	-	-	0.017	0.007
	40	2	43.9 (719)	3.19 x 2.75	-	.025	0.0045	0.0030	0.0070	0.0045	0.0935	0.0925	0.017	0.007
	60	V4	70.7 (1159)	3.00 x 2.50	.020	.040	0.0050	0.0035	0.0070	0.0045	0.0935	0.0925	0.017	0.007
	75	V4	89.5 (1467)	3.375 x 2.50	-	.040	0.0060	0.0045	0.0070	0.0045	0.0935	0.0925	0.017	0.007
	90	V4	89.5 (1467)	3.375 x 2.50	.020	.040	0.0060	0.0045	0.0070	0.0045	0.0935	0.0925	0.017	0.007
1965	3	2	5.28 (85.5)	1.56 x 1.37	.020	-	0.0020	0.0013	0.0035	0.0010	-	-	0.015	0.005
	5	2	8.84 (145)	1.94 x 1.50	.020	-	0.0030	0.0018	0.0035	0.0010	0.0935	0.0925	0.015	0.005
	6	2	8.84 (145)	1.94 x 1.50	.020	-	0.0030	0.0018	0.0035	0.0010	0.0935	0.0925	0.015	0.005
	9.5	2	15.2 (249)	2.31 x 1.81	.020	-	0.0045	0.0030	0.0035	0.0010	0.0935	0.0925	0.017	0.007
	18	2	22 (361)	2.50 x 2.25	.020	.040	0.0045	0.0030	0.0035	0.0045	0.0935	0.0925	0.017	0.007
	33	2	40.5 (664)	3.06 x 2.75	.020	.040	0.0045	0.0030	0.0070	0.0045	0.0935	0.0925	0.017	0.007
	40	2	43.9 (719)	3.19 x 2.75	-	.025	0.0045	0.0030	0.0070	0.0045	0.0935	0.0925	0.017	0.007
	60	V4	70.7 (1159)	3.00 x 2.50	.020	.040	0.0050	0.0035	0.0070	0.0045	0.0935	0.0925	0.017	0.007
	75	V4	89.5 (1467)	3.375 x 2.50	-	.040	0.0060	0.0045	0.0070	0.0045	0.0935	0.0925	0.017	0.007
	90	V4	89.5 (1467)	3.375 x 2.50	.020	.040	0.0060	0.0045	0.0070	0.0045	0.0935	0.0925	0.017	0.007
1966	3	2	5.28 (85.5)	1.56 x 1.37	.020	-	0.0020	0.0013	0.0035	0.0010	-	-	0.015	0.005
	5	2	8.84 (145)	1.94 x 1.50	.020	-	0.0030	0.0018	0.0035	0.0010	0.0935	0.0925	0.015	0.005
	6	2	8.84 (145)	1.94 x 1.50	.020	-	0.0030	0.0018	0.0035	0.0010	0.0935	0.0925	0.015	0.005
	9.5	2	15.2 (249)	2.31 x 1.81	.020	-	0.0045	0.0030	0.0035	0.0010	0.0935	0.0925	0.017	0.007
	18	2	22 (361)	2.50 x 2.25	.020	.040	0.0045	0.0030	0.0035	0.0010	0.0935	0.0925	0.017	0.007
	20	2	22 (361)	2.50 x 2.25	.020	.040	0.0045	0.0030	0.0035	0.0010	0.0625	0.0615	0.017	0.007
	33	2	40.5 (664)	3.06 x 2.75	.020	.040	0.0045	0.0030	0.0070	0.0045	0.0935	0.0925	0.017	0.007
	40	2	43.9 (719)	3.19 x 2.75	-	.025	0.0045	0.0030	0.0070	0.0045	0.0935	0.0925	0.017	0.007
	60	V4	70.7 (1159)	3.00 x 2.50	.020	.040	0.0050	0.0035	0.0070	0.0045	0.0935	0.0925	0.017	0.007
	80	V4	89.5 (1467)	3.375 x 2.50	-	.040	0.0060	0.0045	0.0070	0.0045	0.0935	0.0925	0.017	0.007
	100	V4	89.5 (1467)	3.375 x 2.50	.020	.040	0.0060	0.0045	0.0070	0.0045	0.0935	0.0925	0.017	0.007
1967	3	2	5.28 (85.5)	1.56 x 1.37	.020	-	0.0020	0.0013	0.0035	0.0010	-	-	0.015	0.005
	5	2	8.84 (145)	1.94 x 1.50	.020	-	0.0030	0.0018	0.0035	0.0010	0.0935	0.0925	0.015	0.005
	6	2	8.84 (145)	1.94 x 1.50	.020	-	0.0030	0.0018	0.0035	0.0010	0.0935	0.0925	0.015	0.005
	9.5	2	15.2 (249)	2.31 x 1.81	.020	-	0.0045	0.0030	0.0035	0.0010	0.0935	0.0925	0.017	0.007
	18	2	22 (361)	2.50 x 2.25	.020	.040	0.0045	0.0030	0.0035	0.0010	0.0935	0.0925	0.017	0.007
	20	2	22 (361)	2.50 x 2.25	.020	.040	0.0045	0.0030	0.0035	0.0010	0.0625	0.0615	0.017	0.007
	33	2	40.5 (664)	3.06 x 2.75	.020	.040	0.0045	0.0030	0.0070	0.0045	0.0935	0.0925	0.017	0.007

Powerhead Specifications

Year	Model (Hp)	No. of Cyl	Displace cu. in. (cc)	Bore & Stroke in.	Bore 1st O/S in.	Bore 2nd O/S in.	Piston to Cylinder Clearance Max in.	Min in.	Piston Ring Groove Max in.	Min in.	Piston Ring Width Max in.	Min in.	Piston Ring End Gap Max in.	Min in.
1967 (cont'd)	40	2	43.9 (719)	3.19 x 2.75	-	0.025	0.0045	0.0030	0.0070	0.0045	0.0935	0.0925	0.017	0.007
	60	V4	70.7 (1159)	3.00 x 2.50	0.020	0.040	0.0050	0.0035	0.0070	0.0045	0.0935	0.0925	0.017	0.007
	80	V4	89.5 (1467)	3.375 x 2.50	0.020	0.040	0.0040	0.0025	0.0070	0.0045	0.0935	0.0925	0.017	0.007
	100	V4	89.5 (1467)	3.375 x 2.50	0.020	0.040	0.0060	0.0045	0.0070	0.0045	0.0935	0.0925	0.017	0.007
1968	1.5	1	2.64 (43.3)	1.56 x 1.37	0.020	-	0.0025	0.0013	0.0035	0.0010	0.0935	0.0925	0.015	0.005
	3	2	5.28 (85.5)	1.56 x 1.37	0.020	-	0.0025	0.0013	0.0035	0.0010	0.0935	0.0925	0.015	0.005
	5	2	8.84 (145)	1.94 x 1.50	0.020	-	0.0030	0.0018	0.0035	0.0010	0.0935	0.0925	0.015	0.005
	6	2	8.84 (145)	1.94 x 1.50	0.020	-	0.0030	0.0018	0.0035	0.0010	0.0935	0.0925	0.015	0.005
	9.5	2	15.2 (249)	2.31 x 1.81	0.020	-	0.0045	0.0030	0.0035	0.0010	0.0935	0.0925	0.017	0.007
	18	2	22 (361)	2.50 x 2.25	0.020	0.040	0.0045	0.0030	0.0035	0.0010	0.0935	0.0925	0.017	0.007
	20	2	22 (361)	2.50 x 2.25	0.020	0.040	0.0045	0.0030	0.0035	0.0010	0.0625	0.0615	0.017	0.007
	33	2	40.5 (664)	3.06 x 2.75	0.020	0.040	0.0045	0.0030	0.0070	0.0045	0.0935	0.0925	0.017	0.007
	40	2	43.9 (719)	3.19 x 2.75	-	0.025	0.0045	0.0030	0.0070	0.0045	0.0935	0.0925	0.017	0.007
	55	3	49.7 (814)	3.00 x 2.34	0.020	-	0.0055	0.0040	0.0070	0.0045	0.0935	0.0925	0.017	0.007
	65	V4	70.7 (1159)	3.00 x 2.50	0.020	0.040	0.0050	0.0035	0.0070	0.0045	0.0935	0.0925	0.017	0.007
	85	V4	89.5 (1467)	3.375 x 2.50	0.020	0.040	0.0040	0.0025	0.0070	0.0045	0.0935	0.0925	0.017	0.007
	100	V4	89.5 (1467)	3.375 x 2.50	0.020	0.040	0.0040	0.0025	0.0070	0.0045	0.0935	0.0925	0.017	0.007
1969	1.5	1	2.64 (43.3)	1.56 x 1.37	0.020	-	0.0055	0.0043	0.0035	0.0010	0.0625	0.0615	0.015	0.005
	4	2	5.28 (85.5)	1.56 x 1.37	0.020	-	0.0049	0.0014	0.0035	0.0010	0.0625	0.0615	0.015	0.005
	6	2	8.84 (145)	1.94 x 1.50	0.020	-	0.0030	0.0018	0.0035	0.0010	0.0935	0.0925	0.015	0.005
	9.5	2	15.2 (249)	2.31 x 1.81	0.020	-	0.0050	0.0035	0.0035	0.0010	0.0935	0.0925	0.017	0.007
	18	2	22 (361)	2.50 x 2.25	0.020	0.040	0.0047	0.0032	0.0040	0.0020	0.0625	0.0615	0.017	0.007
	20	2	22 (361)	2.50 x 2.25	0.020	0.040	0.0047	0.0032	0.0040	0.0020	0.0625	0.0615	0.017	0.007
	25	2	22 (361)	2.50 x 2.25	0.020	0.040	0.0045	0.0033	0.0040	0.0020	0.0625	0.0615	0.017	0.007
	33	2	40.5 (664)	3.06 x 2.75	0.020	0.040	0.0045	0.0030	0.0070	0.0045	0.0935	0.0925	0.017	0.007
	40	2	43.9 (719)	3.19 x 2.75	-	0.025	0.0045	0.0030	0.0045	0.0020	0.0935	0.0925	0.017	0.007
	55	3	49.7 (814)	3.00 x 2.34	0.020	-	0.0055	0.0040	0.0045	0.0010	0.0625	0.0615	0.017	0.007
	85	V4	92.6 (1517)	3.375 x 2.59	0.020	0.040	0.0055	0.0040	0.0070	0.0045	0.0935	0.0925	0.017	0.007
	115	V4	96.1 (1575)	3.44 x 2.59	0.020	-	0.0055	0.0040	0.0070	0.0045	0.0620	0.0620	0.017	0.007
1970	1.5	1	2.64 (43.3)	1.56 x 1.37	0.020	-	0.0055	0.0043	0.0040	0.0020	0.0625	0.0615	0.015	0.005
	4	2	5.28 (85.5)	1.56 x 1.37	0.020	-	0.0080	0.0020	0.0040	0.0020	0.0625	0.0615	0.015	0.005
	6	2	8.84 (145)	1.94 x 1.50	0.020	-	0.0030	0.0018	0.0035	0.0010	0.0935	0.0925	0.015	0.005
	9.5	2	15.2 (249)	2.31 x 1.81	0.020	-	0.0050	0.0035	0.0035	0.0010	0.0935	0.0925	0.017	0.007
	18	2	22 (361)	2.50 x 2.25	0.020	0.040	0.0047	0.0032	0.0040	0.0020	0.0625	0.0615	0.017	0.007
	20	2	22 (361)	2.50 x 2.25	0.020	0.040	0.0047	0.0032	0.0040	0.0020	0.0625	0.0615	0.017	0.007
	25	2	22 (361)	2.50 x 2.25	0.020	0.040	0.0045	0.0033	0.0040	0.0020	0.0625	0.0615	0.017	0.007
	33	2	40.5 (664)	3.06 x 2.75	0.020	0.040	0.0045	0.0030	0.0070	0.0045	0.0935	0.0925	0.017	0.007
	40	2	43.9 (719)	3.19 x 2.75	-	0.025	0.0045	0.0030	0.0045	0.0020	0.0935	0.0925	0.017	0.007

Powerhead Specifications

Year	Model (Hp)	No. of Cyl	Displace cu. in. (cc)	Bore & Stroke in.	Bore 1st O/S in.	Bore 2nd O/S in.	Piston to Cylinder Clearance Max in.	Min in.	Piston Ring Groove Max in.	Min in.	Piston Ring Width Max in.	Min in.	Piston Ring End Gap Max in.	Min in.
1970 (cont'd)	60	3	49.7 (814)	3.00 x 2.34	0.020	-	0.0050	0.0035	0.0040	0.0015	0.0900 ①	0.0895 ①	0.017	0.007
	85	V4	92.6 (1517)	3.375 x 2.59	0.020	-	0.0040	0.0025	0.0070	0.0045	0.0930	0.0930	0.017	0.007
	115	V4	96.1 (1575)	3.44 x 2.59	0.020	-	0.0050	0.0035	0.0040	0.0015	0.0620	0.0620	0.017	0.007
1971	2	1	2.64 (43.3)	1.56 x 1.37	0.030	-	0.0055	0.0043	0.0040	0.0020	0.0625	0.0615	0.025	0.015
	4	2	5.28 (85.5)	1.56 x 1.37	0.030	-	0.0030	0.0018	0.0040	0.0020	0.0625	0.0615	0.015	0.005
	6	2	8.84 (145)	1.94 x 1.50	0.030	-	0.0030	0.0018	0.0035	0.0010	0.0935	0.0925	0.015	0.005
	9.5	2	15.2 (249)	2.31 x 1.81	0.030	-	0.0050	0.0035	0.0035	0.0010	0.0935	0.0925	0.017	0.007
	18	2	22 (361)	2.50 x 2.25	0.030	-	0.0048	0.0033	0.0040	0.0020	0.0900 ①	0.0895 ①	0.017	0.007
	20	2	22 (361)	2.50 x 2.25	0.030	-	0.0048	0.0033	0.0040	0.0020	0.0900 ①	0.0895 ①	0.017	0.007
	25	2	22 (361)	2.50 x 2.25	0.030	-	0.0048	0.0033	0.0040	0.0015	0.0900 ①	0.0895 ①	0.017	0.007
	40	2	43.9 (719)	3.19 x 2.75	0.030	-	0.0050	0.0030	0.0040	0.0015	0.0900 ①	0.0895 ①	0.017	0.007
	50	2	41.5 (680)	3.19 x 2.82	0.030	-	0.0065	0.0045	0.0040	0.0015	0.0900 ①	0.0895 ①	0.017	0.007
	60	3	49.7 (814)	3.00 x 2.34	0.020	-	0.0050	0.0035	0.0045	0.0015	0.0625	0.0615	0.017	0.007
	85	V4	92.6 (1517)	3.375 x 2.59	0.020	-	0.0040	0.0025	0.0070	0.0045	0.0935	0.0925	0.017	0.007
	100	V4	92.6 (1517)	3.375 x 2.59	0.020	0.040	0.0040	0.0025	0.0070	0.0045	0.0935	0.0925	0.017	0.007
	125	V4	99.6 (1632)	3.50 x 2.59	0.020	-	0.0045	0.0030	0.0040	0.0020	0.0900 ①	0.0895 ①	0.017	0.007
1972	2	1	2.64 (43.3)	1.56 x 1.37	0.030	-	0.0055	0.0043	0.0040	0.0020	0.0625	0.0615	0.025	0.015
	4	2	5.28 (85.5)	1.56 x 1.37	0.030	-	0.0030	0.0018	0.0040	0.0020	0.0625	0.0615	0.015	0.005
	6	2	8.84 (145)	1.94 x 1.50	0.030	-	0.0030	0.0018	0.0035	0.0010	0.0935	0.0925	0.015	0.005
	9.5	2	15.2 (249)	2.31 x 1.81	0.030	-	0.0050	0.0035	0.0035	0.0010	0.0935	0.0925	0.017	0.007
	18	2	22 (361)	2.50 x 2.25	0.030	-	0.0048	0.0033	0.0040	0.0020	0.0900 ①	0.0895 ①	0.017	0.007
	20	2	22 (361)	2.50 x 2.25	0.030	-	0.0048	0.0033	0.0040	0.0020	0.0900 ①	0.0895 ①	0.017	0.007
	25	2	22 (361)	2.50 x 2.25	0.030	-	0.0048	0.0033	0.0040	0.0015	0.0900 ①	0.0895 ①	0.017	0.007
	40	2	43.9 (719)	3.19 x 2.75	0.030	-	0.0050	0.0030	0.0040	0.0015	0.0900 ①	0.0895 ①	0.017	0.007
	50	2	41.5 (680)	3.19 x 2.82	0.030	-	0.0065	0.0045	0.0040	0.0015	0.0900 ①	0.0895 ①	0.017	0.007
	65	3	49.7 (814)	3.00 x 2.34	0.020	0.030	0.0055	0.0035	0.0040	0.0015	0.0900 ①	0.0895 ①	0.017	0.007
	85	V4	92.6 (1517)	3.375 x 2.59	0.020	-	0.0040	0.0025	0.0070	0.0045	0.0935	0.0925	0.017	0.007
	100	V4	92.6 (1517)	3.375 x 2.59	0.020	-	0.0040	0.0025	0.0070	0.0045	0.0935	0.0925	0.017	0.007
	125	V4	99.6 (1632)	3.50 x 2.59	0.020	-	0.0045	0.0025	0.0040	0.0020	0.0900 ①	0.0895 ①	0.017	0.007

① Figure for upper & center ring, lower ring is 0.0615-0.0625 in.

Torque Specifications

Year	Model (Hp)	No. of Cyl	Displace cu. in. (cc)	Flywheel Nut ft. lbs.	Flywheel Nut inch lbs.	Connecting Rod Bolt ft. lbs	Connecting Rod Bolt inch lbs.	Cylinder Head Bolts ft. lbs.	Cylinder Head Bolts inch lbs.	Crankcase to Cylinder Bolts Upper & Lower ft. lbs.	Crankcase to Cylinder Bolts Upper & Lower inch lbs.	Crankcase to Cylinder Bolts Center ft. lbs.	Crankcase to Cylinder Bolts Center inch lbs.
1958	3	2	5.28 (85.5)	40-45	-	-	60-96	-	60-84	-	60-84	-	60-84
	5.5	2	8.84 (145)	40-45	-	-	60-66	-	60-84	-	60-84	-	60-84
	7.5	2	12.4 (203)	40-45	-	-	60-66	-	60-84	-	60-84	-	60-84
	10	2	16.6 (281)	40-45	-	15-15.5	180-186	8-10	96-120	10-12	120-144	10-12	120-144
	18	2	22 (361)	40-45	-	15-15.5	180-186	8-10	96-120	10-12	120-144	10-12	120-144
	35	2	40.5 (664)	60-65	-	18-18.5	216-22	18-20	216-240	13.5-14	162-168	13.5-14	162-168
	50	V4	70.7 (1159)	80	-	22	264	15	180	13	156	14	168
1959	3	2	5.28 (85.5)	40-45	-	-	60-96	-	60-84	-	60-84	-	60-84
	5.5	2	8.84 (145)	40-45	-	-	60-66	-	60-84	-	60-84	-	60-84
	10	2	16.6 (281)	40-45	-	15-15.5	180-186	8-10	96-120	10-12	120-144	10-12	120-144
	18	2	22 (361)	40-45	-	15-15.5	180-186	8-10	96-120	10-12	120-144	10-12	120-144
	35	2	40.5 (664)	60-65	-	18-18.5	216-22	18-20	216-240	13.5-14	162-168	13.5-14	162-168
	50	V4	70.7 (1159)	80	-	22	264	15	180	13	156	14	168
1960	3	2	5.28 (85.5)	40-45	-	-	60-96	-	60-84	-	60-84	-	60-84
	5.5	2	8.84 (145)	40-45	-	-	60-66	-	60-84	-	60-84	-	60-84
	10	2	16.6 (281)	40-45	-	15-15.5	180-186	8-10	96-120	10-12	120-144	10-12	120-144
	18	2	22 (361)	40-45	-	15-15.5	180-186	8-10	96-120	10-12	120-144	10-12	120-144
	40	2	43.9 (719)	100-105	-	29-31	-	14-16	168-192	12.5-14	150-170	13.5-14	162-168
	75	V4	89.5 (1467)	80	-	22	264	15	180	13	156	14	168
1961	3	2	5.28 (85.5)	40-45	-	-	60-96	-	60-84	-	60-84	-	60-84
	5.5	2	8.84 (145)	40-45	-	-	60-66	-	60-84	-	60-84	-	60-84
	10	2	16.6 (281)	40-45	-	15-15.5	180-186	8-10	96-120	10-12	120-144	10-12	120-144
	18	2	22 (361)	40-45	-	15-15.5	180-186	8-10	96-120	10-12	120-144	10-12	120-144
	40	2	43.9 (719)	100-105	-	29-31	-	14-16	168-192	12.5-14	150-170	13.5-14	162-168
	75	V4	89.5 (1467)	80	-	22	264	15	180	13	156	14	168
1962	3	2	5.28 (85.5)	40-45	-	-	60-96	-	60-84	-	60-84	-	60-84
	5.5	2	8.84 (145)	40-45	-	-	60-66	-	60-84	-	60-84	-	60-84
	10	2	16.6 (281)	40-45	-	15-15.5	180-186	8-10	96-120	10-12	120-144	10-12	120-144
	18	2	22 (361)	40-45	-	15-15.5	180-186	8-10	96-120	10-12	120-144	10-12	120-144
	28	2	35.7 (585)	100-105	-	29-31	-	14-16	168-192	13.5-14	162-168	13.5-14	162-168
	40	2	43.9 (719)	100-105	-	29-31	-	14-16	168-192	12.5-14	150-170	13.5-14	162-168
	75	V4	89.5 (1467)	80	-	30	360	15	180	13	156	14	168
1963	3	2	5.28 (85.5)	40-45	-	-	60-96	-	60-84	-	60-84	-	60-84
	5.5	2	8.84 (145)	40-45	-	-	60-66	-	60-84	-	60-84	-	60-84
	10	2	16.6 (281)	40-45	-	15-15.5	180-186	8-10	96-120	10-12	120-144	10-12	120-144
	18	2	22 (361)	40-45	-	15-15.5	180-186	8-10	96-120	10-12	120-144	10-12	120-144
	28	2	35.7 (585)	100-105	-	29-31	-	14-16	168-192	13.5-14	162-168	13.5-14	162-168
	40	2	43.9 (719)	100-105	-	29-31	-	14-16	168-192	12.5-14	150-170	13.5-14	162-168
	75	V4	89.5 (1467)	80	-	30	360	15	180	13	156	14	168

Torque Specifications

Year	Model (Hp)	No. of Cyl	Displace cu. in. (cc)	Flywheel Nut ft. lbs.	Flywheel Nut inch lbs.	Connecting Rod Bolt ft. lbs	Connecting Rod Bolt inch lbs.	Cylinder Head Bolts ft. lbs.	Cylinder Head Bolts inch lbs.	Crankcase Upper & Lower ft. lbs.	Crankcase Upper & Lower inch lbs.	Crankcase Center ft. lbs.	Crankcase Center inch lbs.
1964	3	2	5.28 (85.5)	40-45	-	-	60-96	-	60-84	-	60-84	-	60-84
	5.5	2	8.84 (145)	40-45	-	-	60-66	-	60-84	-	60-84	-	60-84
	9.5	2	15.2 (249)	40-45	-	-	90-100	8-10	96-120	10-12	120-144	10-12	120-144
	18	2	22 (361)	40-45	15-15.5	180-186	-	8-10	96-120	10-12	120-144	10-12	120-144
	28	2	35.7 (585)	100-105	29-31	-	14-16	168-192	13.5-14	162-168	13.5-14	162-168
	40	2	43.9 (719)	100-105	29-31	-	14-16	168-192	12.5-14	150-170	13.5-14	162-168
	60	V4	70.7 (1159)	80	-	30	360	15	180	13	156	14	168
	75	V4	89.5 (1467)	80	-	30	360	15	180	13	156	14	168
	90	V4	89.5 (1467)	80	-	30	360	15	180	13	156	14	168
1965	3	2	5.28 (85.5)	40-45	-	-	60-96	-	60-84	-	60-84	-	60-84
	5	2	8.84 (145)	40-45	-	-	60-66	-	60-80	-	60-80	-	60-80
	6	2	8.84 (145)	40-45	-	-	60-66	-	60-80	-	60-80	-	60-80
	9.5	2	15.2 (249)	40-45	-	-	90-100	8-10	96-120	10-12	120-144	10-12	120-144
	18	2	22 (361)	40-45	15-15.5	180-186	-	8-10	96-120	10-12	120-144	10-12	120-144
	33	2	40.5 (664)	100-105	29-31	-	14-16	168-192	12.5-14	150-170	13.5-14	162-168
	40	2	43.9 (719)	100-105	29-31	-	14-16	168-192	12.5-14	150-170	13.5-14	162-168
	60	V4	70.7 (1159)	80	-	30	360	15	180	13	156	14	168
	75	V4	89.5 (1467)	80	-	30	360	15	180	13	156	14	168
	90	V4	89.5 (1467)	80	-	30	360	15	180	13	156	14	168
1966	3	2	5.28 (85.5)	40-45	-	-	60-96	-	60-84	-	60-84	-	60-84
	5	2	8.84 (145)	40-45	-	-	60-66	-	60-80	-	60-80	-	60-80
	6	2	8.84 (145)	40-45	-	-	60-66	-	60-80	-	60-80	-	60-80
	9.5	2	15.2 (249)	40-45	-	-	90-100	8-10	96-120	10-12	120-144	10-12	120-144
	18	2	22 (361)	40-45	15-15.5	180-186	-	8-10	96-120	10-12	120-144	10-12	120-144
	20	2	22 (361)	40-45	15-15.5	180-186	-	8-10	96-120	10-12	120-144	9-11	110-130
	33	2	40.5 (664)	100-105	29-31	-	14-16	168-192	12.5-14	150-170	13.5-14	162-168
	40	2	43.9 (719)	100-105	29-31	-	14-16	168-192	12.5-14	150-170	13.5-14	162-168
	60	V4	70.7 (1159)	80	-	30	360	15	180	13	156	14	168
	80	V4	89.5 (1467)	80	-	30	360	15	180	13	156	14	168
	100	V4	89.5 (1467)	80	-	30	360	15	180	13	156	14	168
1967	3	2	5.28 (85.5)	40-45	-	-	60-96	-	60-84	-	60-84	-	60-84
	5	2	8.84 (145)	40-45	-	-	60-66	-	60-80	-	60-80	-	60-80
	6	2	8.84 (145)	40-45	-	-	60-66	-	60-80	-	60-80	-	60-80
	9.5	2	15.2 (249)	40-45	-	-	90-100	8-10	96-120	10-12	120-144	10-12	120-144
	18	2	22 (361)	40-45	15-15.5	180-186	-	8-10	96-120	10-12	120-144	10-12	120-144
	20	2	22 (361)	40-45	15-15.5	180-186	-	8-10	96-120	9-11	110-130	9-11	110-130
	33	2	40.5 (664)	100-105	29-31	-	14-16	168-192	12.5-14	150-170	13.5-14	162-168
	40	2	43.9 (719)	100-105	29-31	-	14-16	168-192	12.5-14	150-170	13.5-14	162-168
	60	V4	70.7 (1159)	80	-	30	360	15	180	13	156	14	168
	80	V4	89.5 (1467)	80	-	30	360	15	180	13	156	14	168
	100	V4	89.5 (1467)	80	-	30	360	15	180	13	156	14	168

Torque Specifications

Year	Model (Hp)	No. of Cyl	Displace cu. in. (cc)	Flywheel Nut ft. lbs	Flywheel Nut inch lbs.	Connecting Rod Bolt ft. lbs	Connecting Rod Bolt inch lbs	Cylinder Head Bolts ft. lbs.	Cylinder Head Bolts inch lbs.	Crankcase to Cylinder Bolts — Upper & Lower ft. lbs.	Crankcase to Cylinder Bolts — Upper & Lower inch lbs.	Crankcase to Cylinder Bolts — Center ft. lbs.	Crankcase to Cylinder Bolts — Center inch lbs.
1968	1.5	1	2.64 (43.3)	22-25	-	-	60-66	-	60-80	-	60-80	-	60-80
	3	2	5.28 (85.5)	40-45	-	-	60-96	-	60-84	-	60-84	-	60-84
	5	2	8.84 (145)	40-45	-	-	60-66	-	60-80	-	60-80	-	60-80
	6	2	8.84 (145)	40-45	-	-	60-66	-	60-80	-	60-80	-	60-80
	9.5	2	15.2 (249)	40-45	-	90-100	-	8-10	96-120	10-12	120-144	10-12	120-144
	18	2	22 (361)	40-45	-	15-15.5	180-186	8-10	96-120	10-12	120-144	10-12	120-144
	20	2	22 (361)	40-45	-	15-15.5	180-186	8-10	96-120	9-11	110-130	9-11	110-130
	33	2	40.5 (664)	100-105	-	29-31	-	14-16	168-192	12.5-14	150-170	13.5-14	162-168
	40	2	43.9 (719)	100-105	-	29-31	-	14-16	168-192	12.5-14	150-170	13.5-14	162-168
	55	3	49.7 (814)	100	-	30	360	15	180	13	156	13	156
	65	V4	70.7 (1159)	100	-	30	360	15	180	13	156	14	168
	85	V4	89.5 (1467)	80	-	30	360	15	180	13	156	14	168
	100	V4	89.5 (1467)	100	-	30	360	15	180	13	156	14	168
1969	1.5	1	2.64 (43.3)	22-25	-	-	60-66	-	60-80	-	60-80	-	60-80
	4	2	5.28 (85.5)	30-40	-	-	60-66	-	60-80	-	60-80	-	60-80
	6	2	8.84 (145)	40-45	-	-	60-66	-	60-80	-	60-80	-	60-80
	9.5	2	15.2 (249)	40-45	-	90-100	-	8-10	96-120	10-12	120-144	10-12	120-144
	18	2	22 (361)	40-45	-	15-15.5	180-186	8-10	96-120	10-12	120-144	10-12	120-144
	20	2	22 (361)	40-45	-	15-15.5	180-186	8-10	96-120	9-11	110-130	9-11	110-130
	25	2	22 (361)	40-45	-	15-15.5	180-186	8-10	96-120	10-12	120-144	10-12	120-144
	33	2	40.5 (664)	100-105	-	29-31	-	14-16	168-192	12.5-14	150-170	13.5-14	162-168
	40	2	43.9 (719)	100-105	-	29-31	-	14-16	168-192	12.5-14	150-170	13.5-14	162-168
	55	3	49.7 (814)	100	-	30	360	15	180	13	156	13	156
	85	V4	92.6 (1517)	100	-	30	360	15	180	13	156	14	168
	115	V4	96.1 (1575)	100	-	30	360	15	180	13	156	14	168
1970	1.5	1	2.64 (43.3)	22-25	-	-	60-66	-	60-80	-	60-80	-	60-80
	4	2	5.28 (85.5)	30-40	-	-	60-66	-	60-80	-	60-80	-	60-80
	6	2	8.84 (145)	40-45	-	-	60-66	-	60-80	-	60-80	-	60-80
	9.5	2	15.2 (249)	40-45	-	90-100	-	8-10	96-120	10-12	120-144	10-12	120-144
	18	2	22 (361)	40-45	-	15-15.5	180-186	8-10	96-120	10-12	120-144	10-12	120-144
	20	2	22 (361)	40-45	-	15-15.5	180-186	8-10	96-120	9-11	110-130	9-11	110-130
	25	2	22 (361)	40-45	-	15-15.5	180-186	8-10	96-120	10-12	120-144	10-12	120-144
	33	2	40.5 (664)	100-105	-	29-31	-	14-16	168-192	12.5-14	150-170	13.5-14	162-168
	40	2	43.9 (719)	100-105	-	29-31	-	14-16	168-192	12.5-14	150-170	13.5-14	162-168
	60	3	49.7 (814)	100	-	30	360	15	180	13	156	13	156
	85	V4	92.6 (1517)	100	-	30	360	15	180	13	156	14	168
	115	V4	96.1 (1575)	100	-	30	360	15	180	13	156	14	168

Torque Specifications

Year	Model (Hp)	No. of Cyl	Displace cu. in. (cc)	Flywheel Nut ft. lbs	Flywheel Nut inch lbs.	Connecting Rod Bolt ft. lbs	Connecting Rod Bolt inch lbs.	Cylinder Head Bolts ft. lbs.	Cylinder Head Bolts inch lbs.	Crankcase to Cylinder Bolts Upper & Lower ft. lbs.	Crankcase to Cylinder Bolts Upper & Lower inch lbs.	Crankcase to Cylinder Bolts Center ft. lbs.	Crankcase to Cylinder Bolts Center inch lbs.
1971	2	1	2.64 (43.3)	20-25	-	-	60-66	-	60-80	-	60-80	-	60-80
	4	2	5.28 (85.5)	30-40	-	-	60-66	-	60-80	-	60-80	-	60-80
	6	2	8.84 (145)	40-45	-	-	60-66	-	60-80	-	60-80	-	60-80
	9.5	2	15.2 (249)	40-45	-	-	90-100	8-10	96-120	10-11	120-132	10-12	120-144
	18	2	22 (361)	40-45	-	15-15.5	180-186	8-10	96-120	9-11	110-130	10-11	120-130
	20	2	22 (361)	40-45	-	15-15.5	180-186	8-10	96-120	9-11	110-130	10-11	120-130
	25	2	22 (361)	40-45	-	15-15.5	180-186	8-10	96-120	10-12	120-144	10-12	120-144
	40	2	43.9 (719)	100-105	-	29-31	-	14-16	168-192	12.5-14	150-170	13.5-14	162-168
	50	2	41.5 (680)	100-105	-	29-31	-	14-16	168-192	18-20	216-240	18-20	216-240
	60	3	49.7 (814)	100	-	30	360	15	180	13	156	13	156
	85	V4	92.6 (1517)	100	-	30	360	15	180	13	156	14	168
	100	V4	92.6 (1517)	100	-	30	360	15	180	13	156	14	168
	125	V4	99.6 (1632)	100	-	30	360	15	180	13	156	14	168
1972	2	1	2.64 (43.3)	20-25	-	-	60-66	-	60-80	-	60-80	-	60-80
	4	2	5.28 (85.5)	30-40	-	-	60-66	-	60-80	-	60-80	-	60-80
	6	2	8.84 (145)	40-45	-	-	60-66	-	60-80	-	60-80	-	60-80
	9.5	2	15.2 (249)	40-45	-	-	90-100	8-10	96-120	10-11	120-132	10-12	120-144
	18	2	22 (361)	40-45	-	15-15.5	180-186	8-10	96-120	9-11	110-130	11-Oct	120-130
	20	2	22 (361)	40-45	-	15-15.5	180-186	8-10	96-120	9-11	110-130	10-11	120-130
	25	2	22 (361)	40-45	-	15-15.5	180-186	8-10	96-120	10-12	120-144	10-12	120-144
	40	2	43.9 (719)	100-105	-	29-31	-	14-16	168-192	12.5-14	150-170	13.5-14	162-168
	50	2	41.5 (680)	100-105	-	29-31	-	14-16	168-192	18-20	216-240	18-20	216-240
	65	3	49.7 (814)	100	-	30	360	19	228	-	90	-	90
	85	V4	92.6 (1517)	100	-	30	360	15	180	13	156	14	168
	100	V4	92.6 (1517)	100	-	30	360	15	180	13	156	14	168
	125	V4	99.6 (1632)	100	-	30	360	15	180	13	156	14	168

7

LOWER UNIT

LOWER UNIT - GEARCASE BASICS

General Information

◆ See Figures 1 thru 4

■ **Model identification should be included with each section, and also in the Gearcase Applications chary at the end of this section. You may still have a different gearcase than suggested so please view and read the different description if necessary.**

The lower unit is considered as that part of the outboard below the exhaust housing. The unit contains the propeller shaft, the driven and pinion gears, the driveshaft from the powerhead and the water pump. On models equipped with shifting capabilities, the forward and reverse gears, together with the clutch, shift assembly, and related linkage, are all housed within the lower unit.

The lower unit is removed by 1 of nine methods depending on the model year and the engine horsepower.

• The lower unit does not have shifting capabilities; therefore, removal of the lower unit is not an involved procedure.

• The lower unit has shifting capabilities. The upper end of the shift rod indexes into the shift handle gear and the lower end of the rod indexes into the gear in the lower unit.

• The lower unit is lowered a couple inches and the shift connector removed.

• A window in the exhaust housing is opened and the shift connector disconnected.

• A window in the lower unit is opened to disconnect the shift rod.

• Green and blue shift wires are disconnected on the port side of the engine (electric shift models).

• The shift rod is disconnected at the linkage under and to the rear of the bottom carburetor.

WATER PUMP

◆ See Figure 5

Water pump service work is by far the most common reason for removal of the lower unit. Each lower unit service section contains complete detailed procedures to rebuild the water pump. The instructions given to prepare for the water pump work must be performed as listed. However, once the pump is ready for installation, if no other work is to be performed on the lower unit, the reader may jump to the pump assembling procedures and proceed with installation of the water pump.

Each section is presented with complete detailed instructions for removal, disassembly, cleaning and inspecting, assembling, adjusting, and installation of only one type unit.

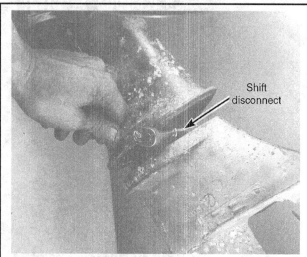

Fig. 1 Disconnecting the shift connector after the lower unit has been separated slightly from the housing

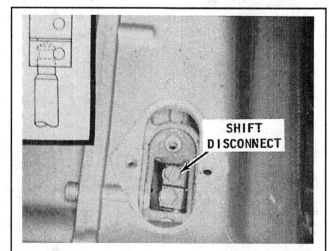

Fig. 2 The window on this unit has been removed to gain access to the shift connector. The drawing in the upper left corner illustrates the relationship of the bolt to the rod

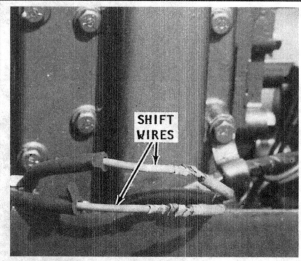

Fig. 3 These wires must be disconnected on many units

Fig. 4 This shift rod is disconnected under the carburetor

Fig. 5 Typical water pump - 4.5 and 7.5 hp shown

Fig. 6 A neglected lower unit cannot be expected to perform to maximum efficiency, compared with a unit receiving TLC (tender loving care)

Maintenance & Service

◆ See Figure 6

The single most important task for proper gearcase maintenance is inspecting it for signs of leakage after each use and properly maintaining the oil level/condition. For starters, remember that if oil can get out, then water can get in. And, water, mixing with or replacing the oil in the gearcase will wreak havoc with the bearings, shafts and gears contained within the housing. Besides a visual inspection after each use, periodically check and maintain the oil inside the case. Not only is it important to make sure the oil is at the proper level (not above or below), but it is important to check the oil for signs of contamination from moisture. Water entering the gearcase will usually cause the oil to turn a slightly milky-white color. Also, significant amounts of water mixed with the oil will give the appearance of an overfilled condition.

If you suspect water in the gearcase, start by draining and closely inspecting the fluid. Then, refill the unit with fresh oil and test the outboard (by using it!). Watch the fluid level closely after the test, and for the first few outings. If any oil leaks out or water enters, either the propeller shaft seal must be replaced or the gearcase must be disassembled, inspected and completely overhauled. To be honest, a complete overhaul is recommended, because corrosion and damage may have occurred if moisture was in the gearcase long enough. But, in some cases, if the leak was caught in time, and there is no significant wear, damage or corrosion in the gearcase, the propeller shaft seal can usually be replaced with the gearcase still installed to the outboard.

The next most important task you can perform to help keep your gearcase in top shape, is to flush the inside and outside of the gearcase after each use. Rinse the outside of the unit with a hose to remove any sea life, salt, chemicals or other corrosion inducing substances that you may have picked up in the water. Cleaning the gearcase will also help you spot potential trouble, such as gearcase oil leaks, cracks or damage that may have occurred during use. Remove any sand, silt or dirt that could potentially damage seals or clog passages. Once you've rinsed the outside, hook up a flushing device and do the same for the inside. Details are found in the Maintenance & Tune-Up or Cooling System sections, look under Flushing the Cooling System.

Important Gearcase Service Tips:

• All threaded parts on these units are right-handed unless otherwise indicated in the text.

• If the presence of any water or metal particles is discovered in the gear lubricant, the lower unit should be completely disassembled, cleaned, and inspected. All defective and/or excessively worn components must be replaced to restore the unit to maximum performance.

• Use "soft jaws" in a vise to prevent damage to expensive parts. Usually a couple pieces of scrap wood will suffice.

• Take time to obtain a suitable size mandrel, which will contact only the bearing race or seal casing, when it is necessary to press or drive bearings into place.

• Keep a record of all shim material removed, as an aid during installation.

Problems in the lower unit can be classified into 3 broad areas:

1. Lack of proper lubrication in the lower unit. Most often such a condition is caused by failure of the operator to check the gear oil level frequently and to add lubricant when required.

2. Faulty seals, allowing water to enter the lower unit. Water allowed to remain in the lower unit over a period of non-use time will corrode the finish on bearings, gears and bushings causing premature failure.

3. Excessive clutch dog and clutch ear wear on the forward and reverse gears. This condition is caused by excessive wear in the bellcrank under the powerhead. A worn bellcrank will result in "sloppy" shifting of the lower unit and cause the clutch components to wear and develop shifting problems.

■ **Excessive high idle speeds will cause the clutch dog and clutch ears on the forward and reverse gears to wear extremely fast. Continued service over a long period of time will cause parts to wear and require replacement.**

Propeller

■ **Please refer to the Maintenance & Tune-Up section for detailed information on Description & Operation, Inspection and Removal & Installation of your propeller.**

Lower Unit (Gearcase) Oil

■ **For detailed information on Oil Recommendations, Checking Level & Condition and Draining & Filling, please refer to the Maintenance & Tune-Up section.**

TYPE I - NON-SHIFTING LOWER UNIT

Description & Operation

A non-shifting gearcase is used on all 1.5-4 hp models.

This is a very simple direct drive unit without any shift capabilities. Reverse is obtained by rotating the engine 180° and holding that position while the boat is moved stern ward. Therefore, no shift rod disconnects are necessary.

Troubleshooting

The first item to check whenever loss of boat movement is encountered is the shear pin. The next area to check is the rubber hub in the propeller, if one is installed. A worn hub will give an indication the unit is not in gear.

The splines in the crankshaft or on the driveshaft may be damaged or worn and thus prevent rotation from the crankshaft to reach the propeller shaft. If the splines in the crankshaft are destroyed, the crankshaft will have to be replaced. See Powerhead. If the splines on the driveshaft have been destroyed, the driveshaft must be replaced.

Frozen Powerhead

This condition is suggested when the operator unsuccessfully attempts to crank the engine with a hand starter. The fly wheel will not rotate. Do not assume the engine is "frozen" until the lower unit has been removed and thoroughly checked. If the lower unit is "locked" (the driveshaft or propeller shaft will not rotate), the powerhead will have the indication of being "frozen" (failure to rotate the flywheel).

The first step to perform under these conditions is to "pull" the lower unit, and then again attempt to crank the engine. If the attempt is successful with the lower unit disconnected, the problem is in the lower unit. If the attempt to crank the engine is still unsuccessful, the problem is in the powerhead.

Lower Unit

REMOVAL & INSTALLATION

◆ See Figures 7 thru 10

■ If the only work to be performed is service of the water pump, be extremely careful to prevent the driveshaft from being pulled up and free of the pinion gear in the lower unit. NEVER carry the lower unit by the driveshaft. If the shaft should be released from the pinion, the lower unit must be disassembled to align the pinion gear and driveshaft, then the driveshaft installed.

1. Disconnect the spark plug wire from the plug.
2. Remove the retaining bolts securing the lower unit to the exhaust housing. Carefully pull directly downward, to prevent damage to the water tube, and remove the lower unit.

To Install:

3. Clean and shine the water pump tube with lightweight sandpaper, and then coat it with oil as an aid to installation.
4. Apply oil to the grommet in the water pump housing, also as an aid to installation. This tube is very small in size and will bend easily during installation if it has even a little difficulty passing through the rubber grommet in the water pump housing.
5. Bring the lower unit together to mate with the exhaust housing. Guide the water tube into the water pump housing grommet, and at the same time rotate the propeller shaft *clockwise*. Rotating the propeller shaft will also

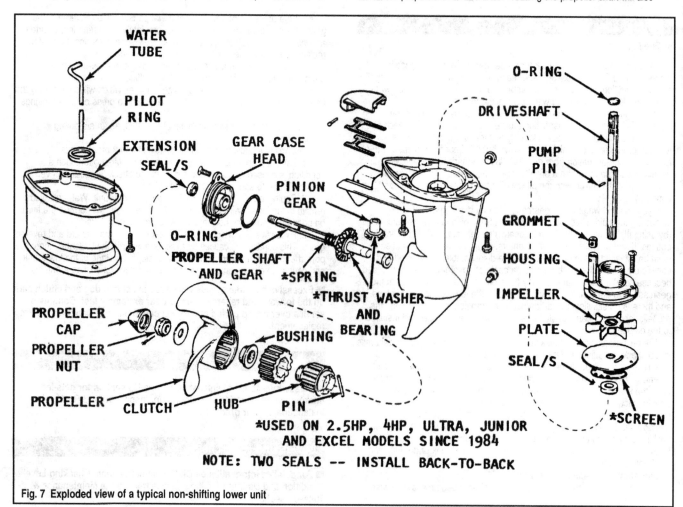

Fig. 7 Exploded view of a typical non-shifting lower unit

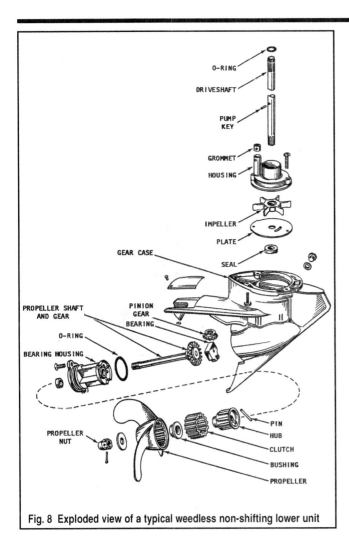

Fig. 8 Exploded view of a typical weedless non-shifting lower unit

O-RING
DRIVESHAFT
PUMP KEY
GROMMET
HOUSING
IMPELLER
PLATE
GEAR CASE
SEAL
PINION GEAR
BEARING
PROPELLER SHAFT AND GEAR
O-RING
BEARING HOUSING
PROPELLER NUT
PIN
HUB
CLUTCH
BUSHING
PROPELLER

rotate the driveshaft and allow the splines on the driveshaft to index with the splines of the engine crankshaft. Continue to work the lower unit closer to the exhaust housing until the mating surfaces make contact.

6. Coat the retaining screws with sealer or a thread locking compound to prevent corrosion, and then start them in place. Tighten the retaining screws EVENLY and ALTERNATELY until secure.

DISASSEMBLY & ASSEMBLY

◆ See Figures 7, 8 and 11 thru 15

1. Remove the gearcase as detailed earlier in this section.
2. Remove the gearcase head and the 2 screws.
3. Pull on the propeller shaft or tap on the gearcase head to separate the gearcase head from the lower unit housing.

■ The driven gear is pressed onto the propeller shaft. Therefore, the propeller shaft and gear are considered as a complete assembly. If either is damaged and requires replacement, the two are purchased as an assembly.

4. Pull upward on the driveshaft and at the same time, reach inside the lower unit and remove the pinion gear.
5. On the weedless-type lower unit, a thrust bearing is installed under the pinion gear. This thrust bearing can only be removed by tapping it out - turn the lower unit so the propeller shaft opening is facing downward. Now, gently rap the unit on a work bench or block of wood. The thrust bearing and pinion gear will be dislodged and fall free.
6. If the seal at the top of the lower unit housing under the water pump is to be replaced, remove the seal using any type seal remover. To remove the seal in the gearcase head, work the seal free by using a punch and mallet from the back side. Remove the O-ring.

To assemble:

7. Perform the procedures detailed under Cleaning & Inspection.
8. Tap a NEW seal into place on top of the lower unit housing.
9. Tap a NEW seal into place in the gearcase head. Install a NEW O-ring into the groove in the gearcase.
10. Install the pinion gear into the recess in the lower unit housing.
11. If the unit being serviced is the weedless type gearcase, install the

Fig. 9 Removing the propeller. . .

Fig. 10 . . .separating the case

Fig. 11 Remove the propeller shaft and gear assembly

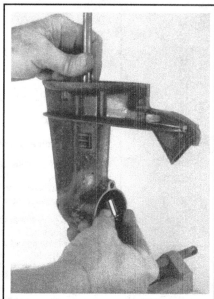

Fig. 12 Pull up on the driveshaft while lifting out the pinion gear

Fig. 13 Removing the upper seal

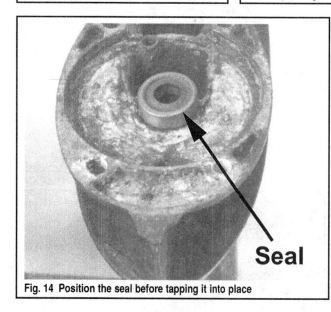

Fig. 14 Position the seal before tapping it into place

Fig. 15 Tap a new seal into place here as well

thrust bearing with the bosses on the bearing indexed between the 2 bosses in the gearcase.

12. Hold the pinion gear in place with 1 hand and with the other hand install the driveshaft down into the lower unit. Continue to hold the pinion gear, and at the same time, rotate the driveshaft slightly after it makes contact with the pinion gear to allow the splines on the shaft to index with the splines in the gear.

■ **After the driveshaft is installed, care must be exercised not to allow the driveshaft to slip out of position in the pinion gear. This is especially important during water pump installation work. If the driveshaft should come free, the lower unit must be disassembled in order to install the driveshaft back into the pinion gear.**

13. Coat the propeller shaft and the gearcase O-ring with oil as an aid to installation.

14. Install the gearcase head over the propeller shaft. Slide the propeller shaft through the lower unit with the driven gear teeth indexed with the teeth of the pinion gear. It may be necessary to rotate the propeller shaft slightly in order to index the driven and pinion gear teeth. The teeth must engage fully and properly or the gear-case head will be damaged when the attaching screws are installed.

15. Coat the screws securing the head to the lower unit with sealer or a thread locking compound, and then install the screws.

■ **If the unit being serviced is the weedless-type, 2 sets of matching marks on the gearcase head and the lower unit must be aligned when the head is installed.**

■ **If the unit being serviced is the highthrust-type, the gearcase has a hole which must face upward when the head is installed.**

16. CAREFULLY tap on the gearcase head with a soft mallet and tighten the screws EVENLY and ALTERNATELY.

✳✳ WARNING

If the screws are not tightened evenly, or the driven gear and pinion gear teeth are not fully and properly engaged, the gearcase head will be thrown out of line just a whisker and the ears through which the bolts pass may snap off. Not good. . .a new gearcase head would have to be purchased.

17. Install the gearcase as detailed earlier in this section.

CLEANING & INSPECTION

◆ **See Figures 7 and 8**

1. Clean all water pump parts with solvent, and then dry them with compressed air.

2. Inspect the water pump cover and base for cracks and distortion, possibly caused from overheating.

3. Inspect the face plate and water pump insert for grooves and/or rough surfaces.

■ **If possible, ALWAYS install a complete new water pump while the lower unit is disassembled. A new impeller will ensure extended satisfactory service and give peace of mind to the owner.**

✳✳ CAUTION

If the old impeller must be returned to service, NEVER install it in reverse to the original direction of rotation. Installation in reverse will cause premature impeller failure.

4. Inspect the impeller side seal surfaces and the ends of the impeller blades for cracks, tears, and wear. Check for a glazed or melted appearance, caused from operating without sufficient water. If any question exists, and as previously stated, install a new impeller if at all possible.

5. Clean all parts with solvent and dry them with compressed air. Discard all O-rings and gaskets.

6. Inspect and replace the driveshaft if the splines are worn.

7. Inspect the gearcase and exhaust housing for damage to the machined surfaces. Remove any nicks and refurbish the surfaces on a surface plate. Start with a No. 120 Emery paper and finish with No. 180.

8. Check the water intake screen and passages.

9. Inspect the drive gear, pinion gear, and thrust washers. Replace these items if they appear worn.

Water Pump

REMOVAL & INSTALLATION

◆ **See Figures 7, 8 and 16 thru 20**

1. Remove the gearcase for access to the water pump, as detailed earlier in this section.

2. Remove the screws securing the water pump to the lower unit housing.

■ **It is very possible corrosion will cause the screw heads to break-off when an attempt to remove them is made. If this should happen, use a chisel and break away the water pump housing from the lower unit. EXERCISE CARE not to damage the lower unit housing.**

3. After the screws have been removed, slide the water pump and impeller upward and free of the driveshaft.

4. Remove the Woodruff key, and then the lower water pump plate,

Fig. 16 Lift off the water pump after loosening the bolts

Fig. 17 Lift off the impeller. . .

Fig. 18 . . .remove the woodruff key. . .

Fig. 19 . . .and lift off the plate

Fig. 20 Always rotate the driveshaft clockwise while tightening the water pump bolts

To Install:

5. Lay down a bead of sealer No. 1000 or equivalent onto the lower unit surface.

6. Slide the water pump plate down the driveshaft and onto the lower unit surface, or screen if used.

7. Insert the Woodruff key into the driveshaft groove.

8. Slide the water pump impeller down the driveshaft and into place on top of the water pump base plate with the pump pin indexed in the impeller. Lubricate the inside surface of the water pump with light weight oil.

9. Lower the water pump housing down the driveshaft and over the impeller. Rotate the driveshaft CLOCKWISE as the water pump housing is lowered to allow the impeller blades to assume their natural and proper position inside the housing. Continue to rotate the driveshaft and work the water pump housing downward until it is seated on the lower unit upper housing surface.

10. Rotate the driveshaft CLOCKWISE while the screws are tightened to prevent damaging the impeller vanes. If the impeller is not rotated, the housing could damage or cut the end of the vanes as the screws are brought up tight. The rotation allows them to spring back into a natural position.

11. Place a *new* grommet into the water pump housing for the water pickup tube. If a new water pump was installed, this seal will already be in place. Install a *new* O-ring on the top of the driveshaft.

12. Install the Gearcase as detailed earlier in this section.

TYPE II - MECHANICAL SHIFT LOWER UNIT, SPLIT GEARCASE (5-25 HP)

◆ See Figures 21, 22 and 23

A mechanical shift, split gearcase is used on most 5-25 hp motors.

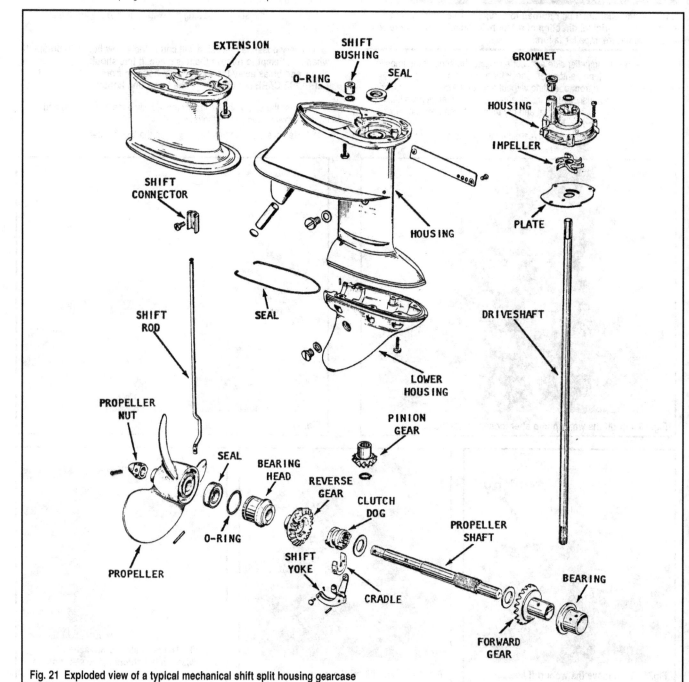

Fig. 21 Exploded view of a typical mechanical shift split housing gearcase

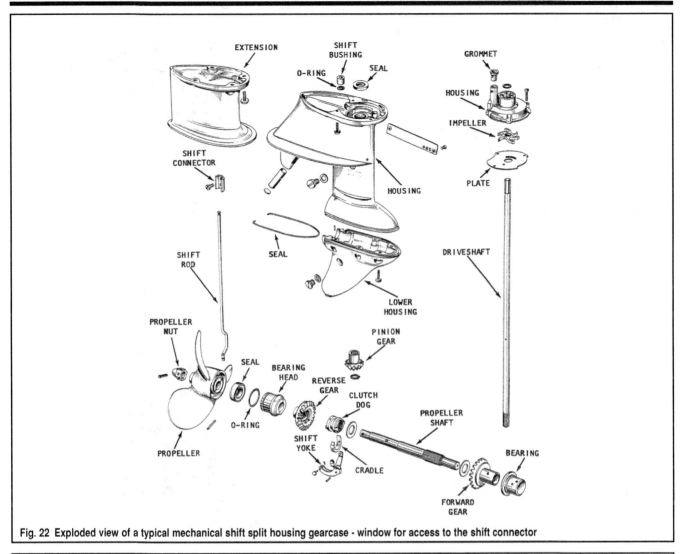

Fig. 22 Exploded view of a typical mechanical shift split housing gearcase - window for access to the shift connector

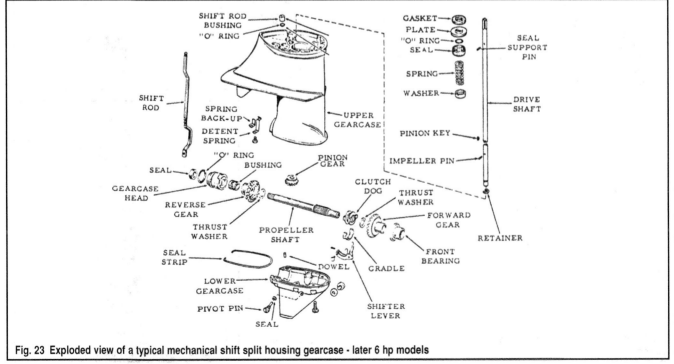

Fig. 23 Exploded view of a typical mechanical shift split housing gearcase - later 6 hp models

Shift Rod Disconnect

◆ See Figures 24 thru 28

Any one of 5 different shift rod connection arrangements may be encountered on the engines covered here. The following paragraphs describe the connections and how they are to be handled for removal of the lower unit. The horsepower and model years are also given for each.

No Shift Unit

This type is used on the direct drive engines, without a reverse gear, The engine is rotated 180° with the steering lever to move the boat sternward. Because a shift rod is not used, there is, obviously, nothing to disconnect.

Pin in Upper Driveshaft

■ A few of the other types may also use this in conjunction with their main disconnect.

"Pin in the upper driveshaft," means that the pin holds and pushes the seal and spring assembly against the powerhead and thus provides a bottom seal for the powerhead.

After the lower unit attaching bolts have been removed, the flywheel must be rotated (to rotate the driveshaft) until the pin is aligned with 2 slots in the upper portion of the exhaust housing. The lower housing can then be separated from the exhaust housing.

If an attempt is made to force the lower unit from the exhaust housing without aligning the driveshaft pin, as just described, the pin may be broken and other items damaged.

Loosen the attaching screws securing the lower unit to the exhaust housing. Allow the lower unit to drop approximately 1 inch, and then remove the bottom bolt. The lower unit may then be completely separated from the exhaust housing.

Shift Disconnect Under Powerhead

Remove the powerhead - the shift rod is attached to the shift lever underneath the powerhead. Disconnect the rod from the shift lever, remove the mounting bolts and then lower the lower unit.

Shift Disconnect Connector

These units do not have the pin in the driveshaft.

Loosen the attaching screws securing the lower unit to the exhaust housing. Allow the lower unit to drop approximately 1 in., and then remove the bottom bolt in the shift connector. The lower unit may then be completely separated from the exhaust.

Window Removal To Gain Access

Remove the metal plate from the port side of the engine. Access to the shift connector is gained through the opening.

Disconnect the shift rod from the exhaust housing by removing the bottom bolt from the shift connector.

■ In most cases, if any unit being serviced has the 6-inch extension, it is NOT necessary to remove the extension in order to "drop" the lower unit. However, as in most things in life, there are rare exceptions and here is one. If the lower unit is separated from the extension and the driveshaft connection is not accessible, then the extension will have to be removed to gain access to the coupler.

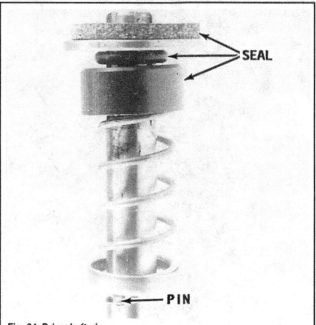

Fig. 24 Driveshaft pin

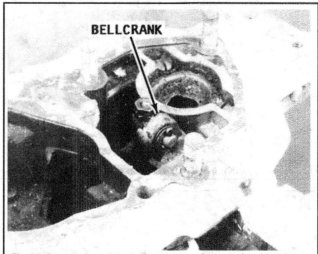

Fig. 25 A good shot of the bellcrank once the powerhead has been removed

Fig. 26 Window removal to access the disconnect

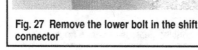

Fig. 27 Remove the lower bolt in the shift connector

Fig. 28 A good shot of the shift disconnect

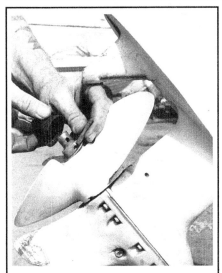

Fig. 29 Checking for a broken shear pin

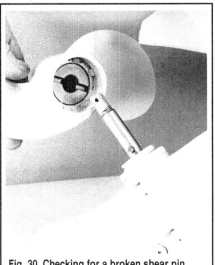

Fig. 30 Checking for a broken shear pin (behind prop)

Fig. 31 Checking for a worn bellcrank

Troubleshooting

Troubleshooting must be done *before* the unit is removed from the powerhead to permit isolating the problem to one area. Always attempt to proceed with troubleshooting in an orderly manner. The shotgun approach will only result in wasted time, incorrect diagnosis, replacement of unnecessary parts, and frustration.

The following procedures are presented in a logical sequence with the most prevalent, easiest, and less costly items to be checked listed first.

UNABLE TO SHIFT INTO FORWARD OR REVERSE

◆ See Figures 29, 30 and 31

Remove the propeller and check to determine if the shear pin has been broken. If the unit being serviced has the shear pin at the rear of the propeller, the propeller should be removed and the shear pin checked at the rear of the propeller shaft.

Access the shift linkage. If the unit has a window in the exhaust housing, then the window must be removed. If procedures indicate the powerhead must be removed in order to check the shift mechanism, then the powerhead must be removed. Hold the shift rod with a pair of pliers and at the same time attempt to move the shift lever on the starboard side of the engine. If it is possible to move the shift lever, the bell crank is worn.

If the engine is the type requiring the lower unit to be lowered slightly to gain access to the shift rod, proceed as follows: Lower the lower unit slightly, and then hold the shift rod with a pair of pliers and attempt to move the shift lever on the starboard side of the engine. If the lever can move, the bell crank is worn and must be repaired.

WATER IN THE LOWER UNIT

◆ See Figure 32

Water in the lower unit is usually caused by fish line becoming entangled around the propeller shaft behind the propeller and damaging the propeller seal. If the line is not removed, it will cut the propeller shaft seal and allow water to enter the lower unit.

Fish line has also been known to cut a groove in the propeller shaft. The propeller should be removed each time the boat is hauled from the water at the end of an outing and any material entangled behind the propeller removed before it can cause expensive damage. The small amount of time and effort involved in pulling the propeller is repaid many times by reduced maintenance and service work, including the replacement of expensive parts.

SLIPPAGE IN THE LOWER UNIT

If the shift seems to be slipping as the boat moves through the water: Check the propeller and the rubber hub.

If the propeller has been subjected to many strikes against underwater objects, it could slip on its hub.

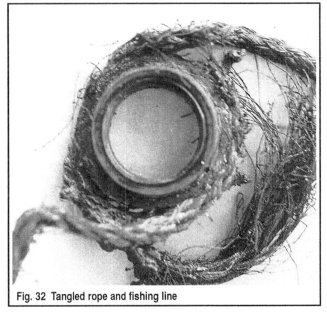

Fig. 32 Tangled rope and fishing line

If the hub is damaged or excessively worn on the small propellers, it is not economical to have the hub or propeller rebuilt.

A new propeller may be purchased for considerably less than meeting the expense of rebuilding an old worn propeller.

DIFFICULT SHIFTING

◆ See Figures 33 and 34

1. Verify that the ignition switch is **OFF**, or better still, disconnect the spark plug wires from the plugs, to prevent possible personal injury, should the engine start.

2. Shift the unit into Reverse at the shift control box, and at the same time have an assistant turn the propeller shaft to ensure the clutch is fully engaged. If the shift handle is hard to move, the trouble may be in the lower unit, with the shift cable, or in the shift box, if used.

3. Isolate the Problem:

   a. Disconnect the shift cable, if used, at the engine.

   b. Operate the shift lever.

   c. If shifting is still hard, the problem is in the shift cable or control box.

   d. If the shifting feels normal with the shift cable disconnected, the problem must be in the lower unit. To verify the problem is in the lower unit, have an assistant turn the propeller and at the same time move the shift cable back and forth. Determine if the clutch engages properly.

JUMPING OUT OF GEAR

◆ See Figures 31, 33 and 35

If a loud thumping sound is heard at the transom while the boat is underway, the unit is jumping out of gear, the propeller does not have a load, therefore the rushing water under the hull forces the lower unit in a backward direction.

The unit jumps back into gear; the propeller catches hold; the lower unit is forced forward again, and the result is the thumping sound as the action is repeated. Normally this type of action occurs perhaps once a day, and then more frequently each time the clutch is operated, until finally the unit will not stay in gear for even a short time.

The following areas must be checked to locate the cause:

1. Access the shift linkage. If the unit has a window in the exhaust housing, then the window must be removed. If procedures indicate the power-head must be removed in order to check the shift mechanism, then the powerhead must be removed. Hold the shift rod with a pair of pliers and at the same time attempt to move the shift lever on the starboard side of the engine. If it is possible to move the shift lever, the bell crank is worn.

If the engine is the type requiring the lower unit to be lowered slightly to gain access to the shift rod, lower the lower unit slightly, and then hold the shift rod with a pair of pliers and attempt to move the shift lever on the starboard side of the engine. If the lever can move, the bell crank is damaged and must be repaired.

2. Disconnect the shift cable at the engine. Attempt to shift the unit into forward gear with the shift lever on the starboard side of the engine and at the same time rotate the propeller in an effort to shift into gear. Shift the control lever at the control box into forward gear. Move the shift cable at the engine up to the shift handle and determine if the cable is properly aligned. The control lever may have jumped a tooth on the slider or on the shift lever arc. If a tooth has been jumped, the cable would lose its adjustment and the unit would fail to shift properly. If the inner cable should slip on the end cable guide, the adjustment would be lost.

3. Move the shift lever at the engine into the neutral position and the shift lever at the control box to the neutral position. Now, move the shift cable up to the shift lever and see if it is aligned. Shift the unit into reverse at the engine and shift the control lever at the control box into reverse. Move the cable up and see if it is aligned. If the cable is properly aligned, but the unit still jumps out of gear when the cable is connected, one of three conditions may exist.

• The bell crank is worn excessively or damaged.
• The shift rod connector is misaligned. This connector is used to link the upper shift rod with the lower rod. If the connector has not been installed properly, any shifting will be difficult.
• Parts in the lower unit are worn from extended use.

FROZEN POWERHEAD

◆ See Figure 36

This condition is suggested when the operator unsuccessfully attempts to crank the engine, either with a hand starter or with a starter motor. The flywheel will not rotate.

Do not assume the engine is "frozen" until the lower unit has been removed and thoroughly checked. If the lower unit is "locked" (the driveshaft or propeller shaft will not rotate), the powerhead will have the indication of being "frozen" (failure to rotate the flywheel).

The first step to perform under these conditions is to pull the lower unit, and then again attempt to crank the engine. If the attempt is successful with the lower unit disconnected, the problem is in the lower unit. If the attempt to crank the engine is still unsuccessful, the problem is in the powerhead.

Lower Unit

REMOVAL & INSTALLATION

 MODERATE

◆ See Figures 21, 22, 23 and 37 thru 43

1. After the shift rod has been disconnected, as described earlier under Shift Rod Disconnect, remove the bolts securing the lower unit to the housing. Some units may have an additional bolt on each side and 1 at the rear of the engine.

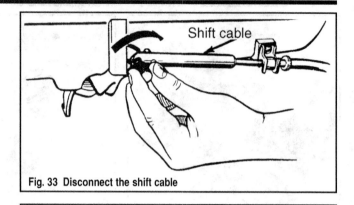

Fig. 33 Disconnect the shift cable

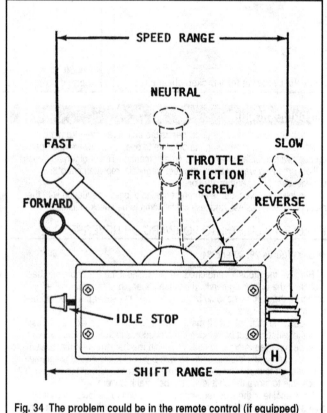

Fig. 34 The problem could be in the remote control (if equipped)

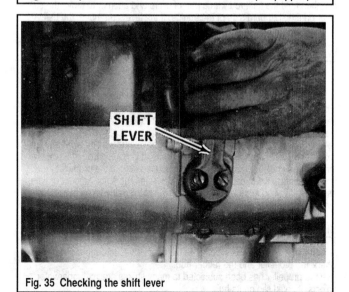

Fig. 35 Checking the shift lever

Fig. 36 A good look at a frozen piston

2. Work the lower unit loose from the exhaust housing. It is not uncommon for the water tube to be stuck in the water pump making separation of the lower unit from the exhaust housing difficult. However, with patience and persistence, the tube will come free of the pump and the lower unit separated from the exhaust housing.

■ Position the lower unit in a vertical position on the edge of the work bench resting on the cavitation plate. Secure the lower unit in this position with a C-clamp. The lower unit will then be held firmly in a favorable position for further service work. An alternate method is to cut a groove in a short piece of 2" x 6" wood to accommodate the lower unit with the cavitation plate resting on top of the wood. Clamp the wood in a vise and service work may then be performed with the lower unit erect (in its normal position), or inverted (upside down). In both positions, the cavitation plate is the supporting surface.

3. Remove the O-ring from the top of the driveshaft. Some units may have a pin installed in this location, instead of an O-ring. In this case, remove the pin from the driveshaft. The washer springs, and other parts will have remained in the exhaust housing.

To Install:

If the unit being serviced uses the shift rod connector arrangement either through the window or before the lower unit and exhaust housings are fully mated, these words are extremely critical:

Connecting the shift rod with the connector is not an easy task but can be accomplished as follows. First, notice the cut-out area on the end of the shift rod. This area permits the bolt to pass through the connector past the shift rod, and into the other side of the connector. It is this bolt that holds the shift rod in the connector.

Now, in order for the bolt to be properly installed, the cut-out area on the shift rod must be aligned in such a manner to allow the bolt to be properly installed. Therefore, as the lower unit is mated with the exhaust housing, exercise patience as the two units come together, to enable the bolt to be

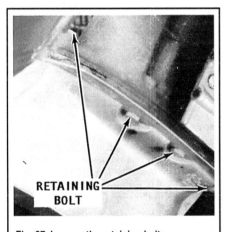
Fig. 37 Loosen the retaining bolts

Fig. 38 Some units may use a pin

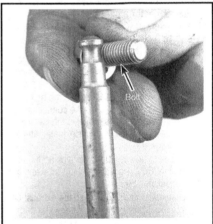
Fig. 39 Shift rod and bolt

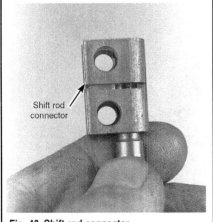
Fig. 40 Shift rod connector

Fig. 41 Connector and bolt installed on shift rod end

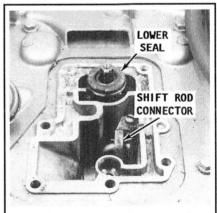
Fig. 42 On some motors, the connector is under the powerhead. . .

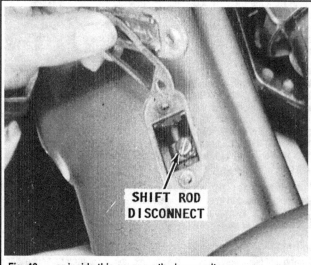

Fig. 43 . . .or inside this cover on the lower unit

installed at the proper time. If the rod is allowed to move too far into the connector before the bolt is installed, it may be possible to force the bolt into place, past the shift rod. The threads on the bolt will be stripped, and the shift rod will eventually come out of the connector.

4. Install the connector onto the lower unit shift rod, with the NO THREAD section facing towards the window. With the connector in this position, the bolt may be inserted through the connector and "catch" the threads on the far side. Install the connector bolt in the manner described in the previous paragraphs.

■ If the unit being serviced uses the shift rod connector arrangement, then the connector must be connected to the shift rod *before* the lower unit is fully mated with the exhaust housing, as described in the previous step and paragraphs.

5. If the unit being serviced uses the driveshaft with the pin, extreme care must be exercised as the shaft is guided into the exhaust housing to allow the pin to index with the groove in the housing.

6. If the unit being serviced uses the bolt through the window arrangement, insert the bolt into the connector. Take time to read and understand the paragraphs earlier in this section before making this connection. After the bolt is in place, install and secure the window with the attaching hardware.

7. Check to be sure the water pick-up tube is clean, smooth, and free of any corrosion. Coat the water pick-up tube and grommet with lubricant as an aid to installation.

8. Guide the lower unit up into the exhaust housing with the water tube sliding into the rubber grommet of the water pump. If the shift rod must be connected before the lower unit makes contact with the exhaust housing, make the connection at this time.

9. Continue to work the lower unit towards the exhaust housing, and at the same time rotate the propeller shaft as an aid to indexing the driveshaft splines with the crankshaft.

10. Start the bolts securing the lower unit to the exhaust housings together. Tighten the bolts ALTERNATELY and EVENLY until secured.

11. If drained, refill the lower unit with lubricant.

12. Final adjustment for remote control units:

a. Shift the lower unit into Neutral.

b. At the shift box, move the shift lever to the Neutral position. If the pin on the end of the shift cable does not align with the shift handle, move the adjusting knob until the pin aligns and will move into the shift handle.

c. With the shift cable removed, move the lower unit into Forward gear and at the same time rotate the propeller CLOCKWISE to ensure the gears are fully indexed.

d. At the control box, move the shift lever into the Forward position. Again check to be sure the pin on the end of the shift cable aligns with the hole in the shift lever.

e. Adjust the knob on the shift cable until the pin does align with the hole in the shift lever.

FUNCTIONAL CHECK

Perform a functional check of the completed work by mounting the engine in a test tank, in a body of water, or with a flush attachment connected to the lower unit. If the flush attachment is used, NEVER operate the engine above an idle speed, because the no-load condition on the propeller would allow the engine to RUNAWAY resulting in serious damage or destruction of the engine.

✳✳ CAUTION

Water must circulate through the lower unit to the engine any time the engine is run to prevent damage to the water pump in the lower unit. Just 5 seconds without water will damage the water pump.

Start the engine and observe the tattle-tale flow of water from idle relief in the exhaust housing. The water pump installation work is verified. If a "Flushette" is connected to the lower unit, very little water will be visible from the idle relief port. Shift the engine into the 3 gears and check for smoothness of operation and satisfactory performance.

DISASSEMBLY & ASSEMBLY

DIFFICULT

◆ See Figures 21, 22, 23 and 44 thru 64

■ One of two type of driveshafts may be installed in the lower units covered in this section. One has a spline on the lower end of the shaft to index with the splines in the pinion gear. The other type driveshaft has a key and keyway in the lower end of the driveshaft. The key indexes with a matching key way in the pinion gear.

1. Remove the gearcase as detailed earlier in this section.

2. Carefully pull upward on the driveshaft. If the driveshaft comes free easily, the unit is the type with the splines on the end of the shaft. If the shaft

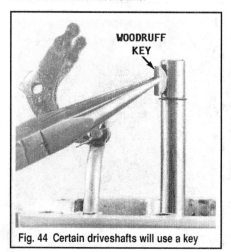

Fig. 44 Certain driveshafts will use a key

Fig. 45 Remove the Phillips screw in the cap. . .

Fig. 46 . . .and then remove the cap retaining screws

will not come free, it is the type with the key and keyway. Therefore, the driveshaft will be removed later when the lower unit is disassembled. If the driveshaft comes free, remove it at this time.

3. Turn the lower unit upside down and again clamp it in the vise or slide it into the wooden block, if one is used. Carefully examine the lower portion of the unit. The cap is considered that part below the split with the skeg attached. The cap has a Phillips screw installed. Remove the screw.

4. Remove the attaching screws around the cap. These screws may be slotted type or Phillips screws. Carefully tap the cap to jar it loose, and then separate it from the lower unit housing. If the cap did not have a Phillips screw on the outside, observe the 2 slots inside the cap.

■ Before proceeding with the disassembly work, take time to study the arrangement of parts in the lower unit. You may wish to take a couple of digital pictures of the unit as an aid during the assembly work. Several engineering and production changes were made to the lower unit over the years. Therefore, the positioning of the gears, shims, bearings, and other parts may vary slightly from one unit to the next.

To show each and every arrangement with a picture in this manual would not be practical. Even if it were done, the ability to associate the unit being serviced with the illustration would be almost impossible. Therefore, take time to make notes, scribble out a sketch, or take a couple photographs.

5. Lift the shift lever out of the cradle, and then remove the cradle from the shift dog.

6. Raise the propeller shaft and at the same time, tap with a soft-headed mallet on the bottom side to jar it loose; and then remove the shaft assembly from the lower unit. The forward and reverse gear including the bearings will all come out with the propeller shaft. The forward gear is the gear at the opposite end of the shaft from the propeller.

■ Notice the bearing split on the back side of the forward gear. Also observe the pin in the housing and a matching slot in the bearing. The pin must index in the slot during installation. The reverse gear has a tab protruding from the bearing head. This tab indexes in a slot in the housing during installation. By taking note at this time of the particular type of installation for the unit being serviced, the task of installation will progress more smoothly. If the installation work is not performed properly, the lower unit housing will quickly be damaged requiring the purchase of a new unit.

7. If the unit being serviced is the type with a keyway in the driveshaft, remove the pinion gear from the shaft, then the key, and snapring, before attempting to remove the driveshaft.

8. Slide the forward gear, Babbitt bearing, and washer, free of the propeller shaft. Remove the clutch dog. Remove the reverse gearcase head, reverse gear, and washer, from the shaft.

9. Turn the lower unit housing right side up and again clamp it in the vise. Remove the bearing carrier and bearing assembly - use a bearing carrier puller, as shown. An alternate method is to use 2 screwdrivers to

Fig. 47 Always take a good look at the assembled unit before removing components

Fig. 48 Lift the shift lever out of the cradle

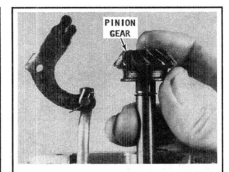

Fig. 49 Remove the pinion gear on units with a keyed driveshaft

Fig. 50 Pull the components off of the propeller shaft

Fig. 51 Use these 2 screws for alignment during installation

Fig. 52 Use a puller to remove the carrier

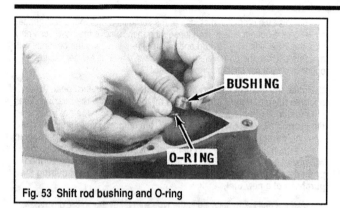

Fig. 53 Shift rod bushing and O-ring

Fig. 54 Use a slide hammer to pull out the bearing carrier

remove the carrier from the lower unit. Sometimes the bearing carrier is difficult to remove. One effective method to release a stubborn bearing carrier is to heat the lower unit housing while attempting to remove the carrier. If this method is employed, be careful not to overheat the lower unit. Excessive heat may damage internal parts.

10. Remove the gasket from underneath the bearing carrier housing.

11. Clean the upper part of the shift rod as an aid to pulling it through the bushing and O-ring. Pull the shift rod from the lower unit housing. The shift rod passes through an O-ring and bushing in the lower unit housing. These two items prevent water from entering the lower unit.

12. A special tapered punch is required to remove the bushing from the lower unit housing. Obtain the special punch, and then remove the bushing, and the O-ring.

To assemble:

◆ **See Figures 55 thru 66**

13. Place the lower unit on the workbench with the water pump recess facing upward.

14. Install a NEW O-ring into the shift cavity. Work the bushing into place on top of the O-ring with a punch and mallet. Inject just a couple drops of oil into the bushing and O-ring as an assist during installation of the shift rod.

15. Turn the exhaust housing upside down.

16. If the unit being serviced uses a Woodruff key, to secure the pinion gear to the driveshaft, install the driveshaft through the housing.

17. Install the snapring into the groove near the end of the driveshaft, and then the Woodruff key, and finally the pinion gear onto the driveshaft.

18. Lower the assembled driveshaft into place in the lower unit housing. If the driveshaft is the splined type, lower the pinion gear into place at this time. The driveshaft will be installed later.

Assembling the Propeller Shaft

19. Slide the clutch dog onto the propeller shaft splines. Apply a light coating of lubricant to the washer and then insert it into the center of the forward gear. Slide the forward gear onto the end of the propeller shaft. Slide the forward gear bearing onto the shaft and into the forward gear.

Fig. 55 Slide on the clutch dog, and then the forward gear

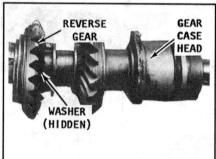

Fig. 56 A good look at the assembled shaft

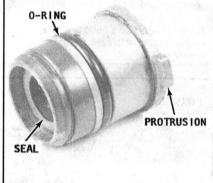

Fig. 57 There is a protrusion on the head that will need to index along with the locating pin

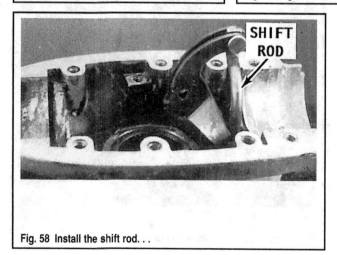

Fig. 58 Install the shift rod. . .

Fig. 59 . . .and swing up the lever

20. Apply a light coating of lubricant to the washer, and then insert it into the center of the reverse gear. Slide the reverse gear onto the propeller shaft from the propeller end. Check to be sure a new O-ring and bearing seal has been installed into the gearcase head, and then install the gearcase head assembly onto the propeller shaft.

✳✳ CAUTION

Look into the front part of the lower unit housing. Notice the pin protruding up from the housing. Now, observe the slot in the forward gear bearing. When the propeller shaft assembly is installed into the lower unit, this pin MUST index into the hole in the forward gear bearing. Also notice the protrusion on the end of the gearcase head. This protrusion *must* index with the slot in the housing when the propeller shaft assembly is installed.

21. Check to be sure the pinion gear is properly located. Check to be sure the shift rod is clean and smooth (free of any burrs or corrosion).

22. Coat the shift rod and the O-ring with oil as an aid to installation.

23. Slide the shift rod down through the O-ring and bushing into the gearcase.

24. Slide the propeller shaft assembly in to the lower unit housing. Check to be sure the slot in the forward gear bearing indexes with the pin in the lower unit, and the protrusion on the end of the gearcase head indexes in the slot in the housing. On some models, a pin in the lower unit housing must index with a hole in the gearcase head.

25. Lubricate the cradle, and then slip it into the clutch dog groove.

26. Bring the shift lever down over the cradle and snap the fingers of the lever into the cradle. Check to be sure the clutch dog is in the Neutral position. Push or pull on the shift rod to move it up or down until the clutch dog is in the center between the forward and reverse gears.

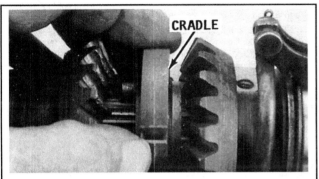

Fig. 60 Position the cradle into the clutch dog groove. . .

Fig. 61 . . .and then swivel over the shift lever

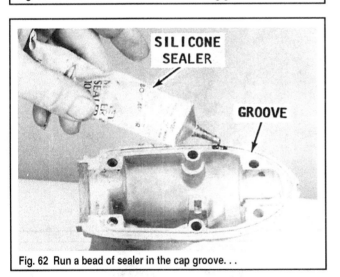

Fig. 62 Run a bead of sealer in the cap groove. . .

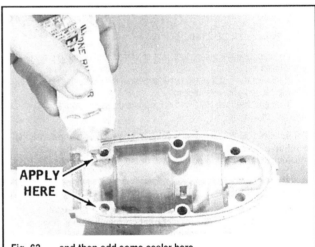

Fig. 63 . . .and then add some sealer here

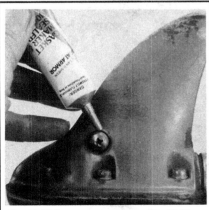

Fig. 64 Make sure use sealer on the Phillips screw

Fig. 65 Install a new gasket. . .

Fig. 66 . . .and then position the bearing housing base plate

27. Lay down a bead of No. 1000 Sealer into the groove of the cap in preparation to installing the seal.

28. Place a NEW seal in the lower cap and hold the seal in the groove with sealer. Apply a small amount of silicon sealer on each side of the bearing gearcase head. This sealer will form a complete seal when the lower unit cap is installed.

■ **If time is taken to grind the end of the screw to a short point, it will make the task of installation much easier. If the cap and shift lever are not aligned exactly, the screw will "seek" and make the alignment as it passes through. However, do not make a long point or the screw will not have enough support and would bend during operation of the shift lever.**

29. Position the lower unit cap over the gear assembly onto the lower unit housing. If the unit being serviced uses a shift lever pin, work the cap until the pin indexes into the recess of the cap.

30. Apply a drop of sealer into the opening for each cap retaining screw to ensure a complete seal between the cap and the lower unit housing. Install the screws securing the cap to the lower unit housing. Tighten the screws ALTERNATELY and EVENLY.

31. A Phillips screw is used in the side of the cap. Use a flashlight and align the hole in the cap with the hole in the shift lever. Install the tapered Phillips screw into the housing and through the lever.

32. Apply a drop of good grade sealer to the threads, and then tighten the screw securely.

33. Turn the lower unit right side up. Install a NEW gasket onto the upper surface of the lower unit.

34. Install the bearing housing and bearing assembly by sliding a couple of bolts through the housing to align the base gasket.

35. Install the Gearcase, as detailed earlier in this section.

CLEANING & INSPECTION

◆ **See Figures 21, 22, 23 and 67 thru 75**

1. Clean all water pump parts with solvent, and then dry them with compressed air.

2. Inspect the water pump cover and base for cracks and distortion, possibly caused from overheating.

3. Inspect the face plate and water pump insert for grooves and/or rough surfaces.

■ **If possible, ALWAYS install a complete new water pump while the lower unit is disassembled. A new impeller will ensure extended satisfactory service and give "peace of mind" to the owner.**

✳✳ CAUTION

If the old impeller must be returned to service, NEVER install it in reverse to the original direction of rotation. Installation in reverse will cause premature impeller failure.

4. Inspect the impeller side seal surfaces and the ends of the impeller blades for cracks, tears, and wear. Check for a glazed or melted appearance, caused from operating without sufficient water. If any question exists, and as previously stated, install a new impeller if at all possible.

5. Discard all O-rings and gaskets.

6. Inspect and replace the driveshaft if the splines are worn.

7. Inspect the gearcase and exhaust housing for damage to the machined surfaces. Remove any nicks and refurbish the surfaces on a surface plate. Start with a No. 120 Emery paper and finish with No. 180.

8. Check the water intake screen and passages by removing the bypass cover, if one is used. Inspect the clutch dog, drive gears, pinion gear, and thrust washers. Replace these items if they appear worn. If the clutch dog and drive gear arrangement surfaces are nicked, chipped, or the edges rounded, the operator may be performing the shift operation improperly or the controls may not be adjusted correctly. These items *must* be replaced if they are damaged.

9. Inspect the dog ears on the inside of the forward and reverse gears. The gears must be replaced if they are damaged.

10. Check the cradle that rides on the inside diameter of the clutch dog. The sides of the cradle must be in good condition, free of any damage or signs of wear. If damage or wear has occurred, the cradle must be replaced.

11. Check the shift lever and the 2 prongs that fit inside the cradle. Check to be sure the prongs are not worn or rounded. Damage or wear to the prongs indicates the lever must be replaced.

Incorrect Assembly

The 3 accompanying illustrations clearly show a lower unit that has been assembled INCORRECTLY.

The 1st illustration: Inspect very closely, the bushing bearing on the forward gear for any kind of indication the pin was missed when the housing was installed during the last repair work. Check the reverse gearcase head and if there is any indication of a pin mark, then check the housing for evidence the pin has been driven into the housing. Some bearing carriers have a lip that indexes with a slot in the lower unit housing. Check for evidence the lip did not index properly.

The 2nd illustration: If the pin has been driven down into the lower unit housing, it must be drilled out as described in the next paragraph. The accompanying illustration compares a proper and an improper installation.

The 3rd illustration: If the pin must be drilled, be careful not too drill too deeply. If a hole is drilled deeper than necessary, then insert a couple drops of melted solder into the hole, and set the new pin in place with the large portion of the pin flush with the housing. If the pin is not flush, remove it, drop more soldered into the hole and make another test. Continue to drop

Fig. 67 Compare these two cradles - the new one is on the right

Fig. 68 And yet another bearing that is shot!

Fig. 69 The cage on this ball bearing has been destroyed

Fig. 70 This pinion gear is badly worn. . .

Fig. 71 . . .while this one suffers from a bad case of corrosion

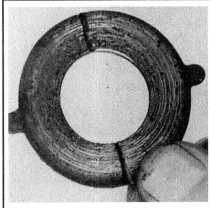

Fig. 72 This Babbitt bearing is unfit for service

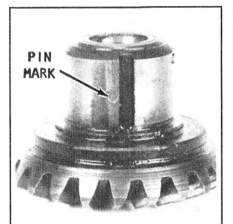

Fig. 73 Check the forward gear *closely* for evidence that the pin was missed when the housing was last opened (1)

Fig. 74 If the pin was misaligned with the bearing slot, then it would be driven down deep into the housing (2). . .

Fig. 75 . . .and it must be carefully drilled out so it can be replaced (3)

solder into the hole and test until the pin is flush with the housing when it is installed. If more solder is inserted into the hole than necessary, the pin may be tapped into the solder while it is still warm and the pin made flush with the housing.

Water Pump

REMOVAL & INSTALLATION

② ◁MODERATE

◆ See Figures 21, 22, 23 and 76 thru 79

1. Remove the gearcase for access to the water pump, as detailed earlier in this section.

2. Remove the screws securing the water pump to the lower unit housing. It is very possible corrosion will cause the screw heads to break off when an attempt to remove them is made. If this should happen, use a chisel and breakaway the water pump housing from the lower unit. Be careful not to damage the lower unit housing.

3. After the screws have been removed, slide the water pump, impeller, the impeller key, and the lower water pump plate, upward and free of the driveshaft.

■ If the only work to be performed is service of the water pump, be extremely careful to prevent the driveshaft from being pulled up and free of the pinion gear in the lower unit. NEVER carry the lower unit by the driveshaft. If the shaft should be released from the pinion, the lower unit MUST be disassembled to align the pinion gear and the driveshaft, then the driveshaft installed.

Fig. 76 Remove the housing. . .

To Install:

4. Apply a coating of sealer to the upper surface of the lower unit. Install the water pump base plate.

5. If the unit being serviced uses the splined type driveshaft, slide the driveshaft into the lower unit, and then rotate the shaft very slowly. When the splines of the driveshaft index with the pinion gear, the shaft will drop slightly. Install the water pump pin or key.

6. Slide the water pump impeller down the driveshaft and into place on top of the water pump base plate with the pump pin or key indexed in the impeller. Lubricate the inside surface of the water pump with light weight oil.

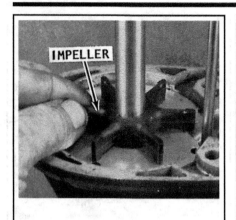

Fig. 77 . . .and lift off the impeller

Fig. 78 Always use a new grommet. . .

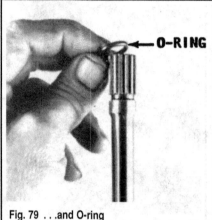

Fig. 79 . . .and O-ring

7. Lower the water pump housing down the driveshaft and over the impeller. Rotate the driveshaft clockwise as the water pump housing is lowered to allow the impeller blades to assume their natural and proper position inside the housing. Continue to rotate the driveshaft and work the water pump housing downward until it is seated on the lower unit upper housing.

8. Rotate the driveshaft clockwise while the screws are tightened to prevent damaging the impeller vanes. If the impeller is not rotated, the housing could damage or cut the end of the vanes as the screws are brought up tight. The rotation allows them to spring back in a natural position. Place a NEW grommet into the water pump housing for the water pickup. If a new water pump was installed, this seal will already be in place.

9. Install a NEW O-ring on the top of the driveshaft.

10. If the lower unit being serviced uses a pin on the top of the driveshaft, install the pin at this time. Shift the lower unit into Forward gear and at the same time rotate the propeller shaft clockwise. The lower unit assembling is now complete and ready to mate with the exhaust housing.

11. Install the gearcase.

TYPE III - MECHANICAL SHIFT LOWER UNIT, SPLIT GEARCASE (28-40 HP)

Description & Operation

◆ See Figures 80 and 81

A unique mechanical shift, split gearcase is used on many 28-40 hp motors.

Only one type of shift mechanism and removal procedures are used on the engines covered in this section. Access to the shift disconnect is through a window on the side of the engine.

Troubleshooting

■ For all troubleshooting procedures, please refer to Troubleshooting in the Type II - Mechanical Shift Lower Unit W/Split Gearcase (5-25 Hp) section.

Lower Unit

REMOVAL & INSTALLATION

② MODERATE

◆ See Figures 26, 37, 39, 40, 41, and 80 thru 83

1. Remove the metal plate from the port side of the engine. Access to the shift connector is gained through the opening.

2. Disconnect the shift rod from the exhaust housing by removing the bottom bolt from the shift connector.

■ In most cases, if any unit being serviced has the 6-inch extension, it is not necessary to remove the extension in order to "drop" the lower unit.

3. Remove the attaching hardware securing the rear exhaust housing cover. This cover must be removed to gain access to one of the bolts securing the lower unit to the exhaust housing.

4. Remove the bolts securing the lower unit to the housing. Some units may have an additional bolt on each side and 1 at the rear of the engine.

5. Work the lower unit loose from the exhaust housing. It is not uncommon for the water tube's to be stuck in the water pump making separation of the lower unit from the exhaust housing difficult. However, with patience and persistence, the tubes will come free of the pump and the lower unit separated from the exhaust housing.

6. Position the lower unit in a vertical position on the edge of the work bench resting on the cavitation plate. Secure the lower unit in this position with a C-clamp. The lower unit will then be held firmly in a favorable position for further service work. An alternate method is to cut a groove in a short piece of 2" x 6" wood to accommodate the lower unit with the cavitation plate resting on top of the wood. Clamp the wood in a vise and service work may then be performed with the lower unit erect (in its normal position), or inverted (upside down). In both positions, the cavitation plate is the supporting surface.

7. Remove and discard the O-ring from the top of the driveshaft.

To Install:

■ Connecting the shift rod with the coupler is not an easy task but can be accomplished as follows:

a. First, notice the cutout area on the end of the shift rod. This area permits the bolt to pass through the connector, past the shift rod, and into the other side of the connector. It is this bolt that holds the shift rod in the connector.

b. Now, in order for the bolt to be properly installed, the cut-out area on the shift rod MUST be aligned in such a manner to allow the bolt to be properly installed. Therefore, as the lower unit is mated with the exhaust housing, exercise patience as the two units come together, to enable the bolt to be installed at the proper time. If the rod is allowed to move too far into the connector before the bolt is installed, it may be possible to force the bolt into place, past the shift rod. The threads on the bolt will be stripped, and the shift rod will eventually come out of the connector.

8. Install the connector, onto the lower unit shift rod, with the NO THREAD section facing towards the window. With the connector in this position, the bolt may be inserted through the connector and "catch" the threads on the far side. Install the connector bolt in the manner described in the previous paragraph.

9. Check to be sure the water pick-up tubes are clean, smooth, and free of any corrosion. Coat the water pick-up tubes and grommets with lubricant as an aid to installation.

10. Guide the lower unit up into the exhaust housing with the water tube sliding into the rubber grommet of the water pump. Continue to work the lower unit towards the exhaust housing, and at the same time rotate the propeller shaft as an aid to indexing the driveshaft splines with the crankshaft.

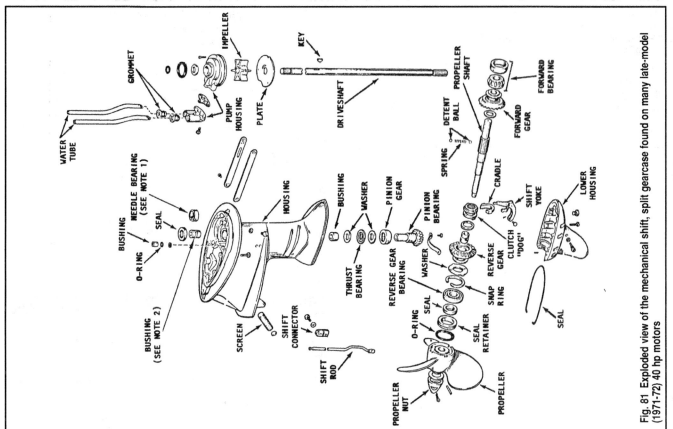

Fig. 81 Exploded view of the mechanical shift, split gearcase found on many late-model motors (1971-72) 40 hp motors

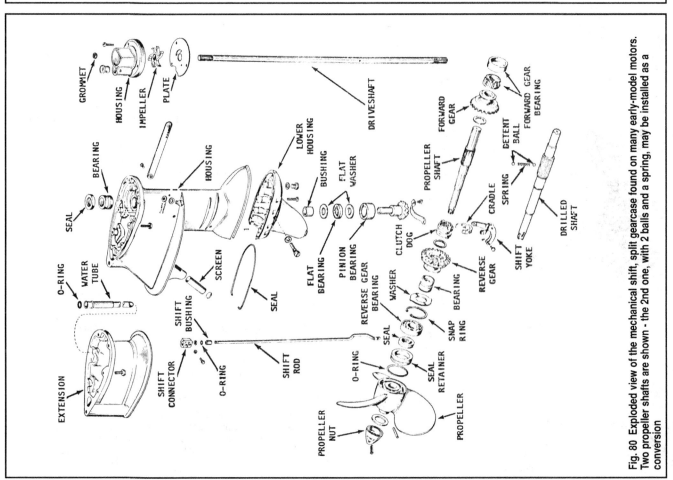

Fig. 80 Exploded view of the mechanical shift, split gearcase found on many early-model motors. Two propeller shafts are shown - the 2nd one, with 2 balls and a spring, may be installed as a conversion

11. Insert the bolt into the water connector. Take time to reread and understand the paragraphs earlier in this section, before making this connection. After the bolt is in place, install and secure the window with the attaching hardware.

12. Start the bolts securing the lower unit to the exhaust housing. Tighten the bolts EVENLY and ALTERNATELY until secure.

13. Install the rear cover over the exhaust housing. When the cover is installed, check to be sure the idle relief rubber tube on the upper side underneath the powerhead fits into the recess of the cover. Secure the cover in place with the attaching hardware.

Fig. 82 You can fabricate your own stand

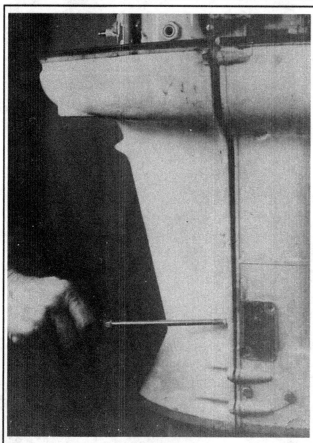

Fig. 83 Performing the final adjustment

14. If drained, properly refill the lower unit with lubricant.

15. If removed, install the propeller.

16. Final adjustment for remote control units:

a. Shift the lower unit into Neutral gear.

b. At the shift box, move the shift lever to the Neutral position. If the pin on the end of the shift cable does not align with the shift handle, move the adjusting knob until the pin aligns and will move into the shift handle.

c. With the shift cable removed, move the lower unit into Forward gear and at the same time rotate the propeller CLOCKWISE to ensure the gears are fully indexed.

d. At the control box, move the shift lever into the Forward position. Again check to be sure the pin on the end of the shift cable aligns with the hole in the shift lever. Adjust the knob on the shift cable until the pin does align with the hole in the shift lever.

FUNCTIONAL CHECK

Perform a functional check of the completed work by mounting the engine in a test tank, in a body of water, or with a flush attachment connected to the lower unit. If the flush attachment is used, NEVER operate the engine above an idle speed, because the no-load condition on the propeller would allow the engine to runaway resulting in serious damage or destruction of the engine.

✳✳ CAUTION

Water must circulate through the lower unit to the engine any time the engine is run to prevent damage to the water pump in the lower unit. Just 5 seconds without water will damage the water pump.

Start the engine and observe the tattle-tale flow of water from idle relief in the exhaust housing. The water pump installation work is verified. If a "Flushette" is connected to the lower unit, very little water will be visible from the idle relief port. Shift the engine into the 3 gears and check for smoothness of operation and satisfactory performance.

DISASSEMBLY & ASSEMBLY

 DIFFICULT

◆ See Figures 80, 81 and 84 thru 98

1. Remove the gearcase as detailed earlier in this section.

2. Carefully pull upward on the driveshaft and remove it from the lower unit.

3. Turn the lower unit upside down and again clamp it in the vise or slide it into the wooden block, if one is used. Carefully examine the lower portion of the unit. The cap is considered that part below the split with the skeg attached. Remove the Phillips screw from the starboard side of the lower housing. This screw passes through the shift yoke and threads into the other side of the housing.

4. Remove the attaching screws around the cap. These screws may be slotted type or Phillips screws. Carefully tap the cap to jar it loose, and then separate it from the lower unit housing.

■ Before proceeding with the disassembly work, take time to study the arrangement of parts in the lower unit. You may elect to take a couple of digital pictures of the unit as an aid during the assembly work. Several engineering and production changes have been made to the lower unit over the years. Therefore, the positioning of the gears, shims, bearings, and other parts may vary slightly from one unit to the next.

To show each and every arrangement with a picture in this manual would not be practical. Even if it were done, the ability to associate the unit being serviced with the illustration would be almost impossible. Therefore, take time to make notes, scribble out a sketch, or take a couple photographs.

5. Lift the shift lever out of the cradle, and then remove the cradle from the shift dog. Raise the propeller shaft and at the same time tap with a soft-headed mallet on the bottom side to jar it loose, and then remove the shaft assembly from the lower unit. The forward and reverse gear including the bearings will all come out with the propeller shaft. The forward gear is the gear at the opposite end of the shaft from the propeller.

■ Notice that the forward gear bearing is a tapered bearing with a race and that the taper faces outward, AWAY from the gear. Also observe the seal retainer on the propeller end of the propeller shaft. Now, notice the matching pin in the lower unit housing. During the installation work, the retainer must be installed with the pin indexed in the hole. Take note of the snapring installed between the thrust washer and the reverse gear bearing. One more item of particular interest. Notice the two sides of the thrust washer. One side is as a normal washer, but the other side is a babbitt. The babbitt side MUST face toward the reverse gear during installation. The washer also has 2 dog ears, one facing upward and the other downward.

By taking note at this time of these items and exactly how they are installed, the task of assembling and installation will progress more smoothly.

6. Remove the attaching screws, and then the U-shaped bracket from the top of the pinion gear

7. Reach into the lower housing and remove the pinion gear.

8. Slide the tapered bearing, forward gear, washer, and clutch dog, off the propeller shaft.

9. Remove the seal retainer, reverse bearing, snapring, washer, reverse gear and bearing, and the washer, from the propeller end of the shaft.

10. Turn the lower unit housing right side up and again clamp it in the vise. Remove the upper seal using a seal puller. An alternate method, if the puller is not available, is to use 2 screwdrivers and work the seal out of the housing. Be careful not to damage the seal recess as the seal is being removed. A Babbitt bearing is installed under the seal. Late model units may have caged needle bearings installed. Normally, it is not necessary to remove this bearing. However, check the bearing surface with a finger and if any roughness is felt, the bearing must be replaced.

11. Clean the upper portion of the shift rod as an aid to pulling it through the bushing and O-ring. Pull the shift rod from the lower unit housing.

12. The shift rod passes through an O-ring and bushing in the lower unit housing. These 2 items prevent water from entering the lower unit. A special tapered punch is required to remove the bushing from the lower unit housing. Obtain the special punch, and then remove the bushing, and the O-ring.

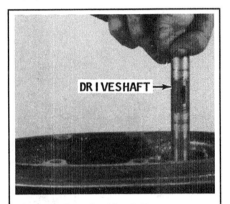
Fig. 84 Pull out the driveshaft

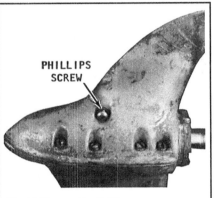
Fig. 85 Remove the Phillips screw. . .

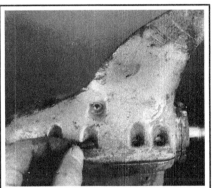
Fig. 86 . . .and then remove the cap retaining screws

Fig. 87 Take the time to check out component orientation before pulling everything apart

Fig. 88 This pin will be important during assembly

Fig. 89 A good look at the propeller shaft assembly

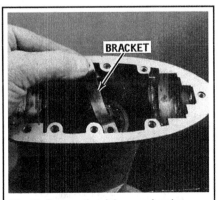
Fig. 90 Remove the pinion gear bracket. . .

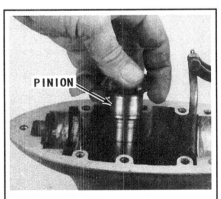
Fig. 91 . . .and then lift out the pinion

Fig. 92 Slide off the bearing, forward gear and clutch dog. . .

Fig. 93 . . .and the remove the reverse gear

Fig. 94 Removing the upper seal

Fig. 95 Use a slide hammer when removing the seals/bearing from under the water pump plate

Fig. 96 Newer models may use a needle bearing

Fig. 97 Pull out the shift rod. . .

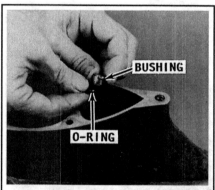

Fig. 98 . . .without losing the bushing and O-ring

To assemble:

◆ See Figures 99 and 100

13. Place the lower unit on the work bench with the water pump recess facing upward. Install a NEW O-ring into the shift cavity. Work the bushing into place on top of the O-ring with a punch and mallet. Inject just a couple drops of oil into the bushing and O-ring as an assist during installation of the shift rod.

14. Lower the pinion gear into the housing. Check to be sure it seats properly.

15. Lower the U-shaped pinion gear retaining bracket into position and secure it in place with the attaching screws.

16. Check to be sure the pinion gear is properly located. Check to be sure the shift rod is clean and smooth (free of any burrs or corrosion). Coat the shift rod and the O-ring with oil as an aid to installation.

17. Slide the shift rod down through the O-ring and bushing into the gearcase.

Assembling the Propeller Shaft

18. Apply a light coating of lubricant to the washer, and then insert it into the center of the reverse gear.

19. Slide the reverse gear onto the propeller shaft from the propeller end.

20. Install the thrust washer with the babbitt side *toward* the reverse gear.

21. Install the snapring, the bearing.

22. Check to be sure a NEW seal and O-ring has been installed into the seal retainer, and then install the retainer.

23. Slide the clutch dog onto the propeller shaft splines.

24. Apply a light coating of lubricant to the washer and then insert it into the center of the forward gear.

25. Slide the forward gear onto the end of the propeller shaft. Slide the forward gear bearing onto the shaft with the large end of the taper *toward* the forward gear. Move the bearing into place on the forward gear.

26. Check to be sure a new O-ring and bearing seal has been installed into the gearcase head, and then install the gearcase head assembly onto the propeller shaft.

✳✳ CAUTION

The seal retainer has a hole and the lower housing of the lower unit has a pin. This pin must index into the hole in the retainer when the propeller shaft is installed. If the pin is not seated properly in the hole, the seal retainer will work part way out of the housing and the lubricant in the lower unit will be lost.

27. Slide the propeller shaft assembly into the lower unit housing. Check to be sure the forward and reverse gear index with the pinion gear and the hole in the seal retainer indexes with the pin in the lower unit housing. Lubricate the cradle, and then slip it into the clutch dog groove.

28. Bring the shift lever down over the cradle and snap the fingers of the lever into the cradle. Check to be sure the clutch dog is in the Neutral position. Push or pull on the shift rod to move it up or down until the clutch dog is in the center between the forward and reverse gears.

29. Lay down a bead of No. 1000 Sealer into the groove of the cap in preparation to installing the seal.

30. Place a NEW seal in the lower cap and hold the seal in the groove with sealer. Apply a small amount of Silicon sealer on each side of the bearing gearcase head. This sealer will form a complete seal when the lower unit cap is installed. Position the lower unit cap over the gear assembly onto the lower unit housing.

31. Apply a drop of sealer into the opening for each cap retaining screw to ensure a complete seal between the cap and the lower unit housing. Install the screws securing the cap to the lower unit housing. Tighten the screws ALTERNATELY and EVENLY.

■ If time is taken to grind the end of the screw to a *short* point, it will make the task of installation much easier. If the cap and shift lever are not aligned exactly, the screw will "seek" and make the alignment as it passes through. However, do not make a long point or the screw will not have enough support and would bend during operation of the shift lever.

32. Use a flashlight and align the hole in the cap with the hole in the shift lever. Install the tapered Phillips screw into the housing and through the lever. Tighten the screw securely.

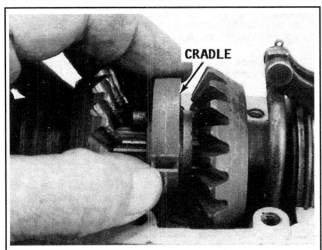

Fig. 99 Make sure that you lubricate the cradle before positioning it over the dog

Fig. 100 Move the shift lever down and into position

33. Install the babbitt or needle bearing, if it was removed. The babbitt bearing may be installed using the proper size socket and hammer. If the caged needle bearing is installed tap on the numbered side of the bearing.

34. Coat the outside edge of a NEW seal with No. 1000 sealer, and then tap the seal into place in the top of the upper lower unit housing.

35. Install the gearcase.

CLEANING & INSPECTION

◆ See Figures 68 thru 72, 80, 81, 101 and 102

1. Clean all water pump parts with solvent, and then dry them with compressed air.

2. Inspect the water pump cover and base for cracks and distortion, possibly caused from overheating.

3. Inspect the face plate and water pump insert for grooves and/or rough surfaces.

■ If possible, ALWAYS install a complete new water pump while the lower unit is disassembled. A new impeller will ensure extended satisfactory service and give "peace of mind" to the owner.

✳✳ WARNING

If the old impeller must be returned to service, never install it in reverse to the original direction of rotation. Installation in reverse will cause premature impeller failure.

Fig. 101 This is what happens when water gets into the unit and stays there

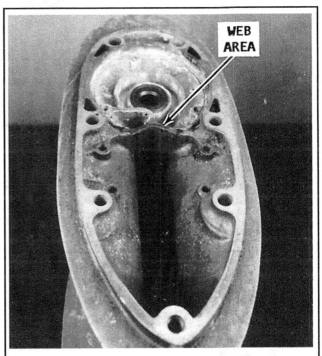

Fig. 102 The web area should be inspected carefully for even the slightest sign of a crack

4. Inspect the impeller side seal surfaces and the ends of the impeller blades for cracks, tears, and wear. Check for a glazed or melted appearance, caused from operating without sufficient water. If any question exists, and as previously stated, install a new impeller if at all possible.

5. Discard all O-rings and gaskets. Inspect and replace the driveshaft if the splines are worn.

6. Inspect the gearcase and exhaust housing for damage to the machined surfaces. Remove any nicks and refurbish the surfaces on a surface plate. Start with a No. 120 Emery paper and finish with No. 180.

7. Check the water intake screen and passages by removing the bypass cover, if one is used. Inspect the clutch dog, drive gears, pinion gear, and thrust washers. Replace these items if they appear worn. If the clutch dog and drive gear arrangement surfaces are nicked, chipped, or the edges rounded, the operator may be performing the shift operation improperly or the controls may not be adjusted correctly. These items must be replaced if they are damaged.

8. Inspect the dog ears on the inside of the forward and reverse gears. The gears must be replaced if they are damaged.

9. Check the cradle that rides on the inside diameter of the clutch dog. The sides of the cradle must be in good condition, free of any damage or signs of wear. If damage or wear has occurred, the cradle must be replaced.

10. Check the shift lever and the 2 prongs that fit inside the cradle. Check to be sure the prongs are not worn or rounded. Damage or wear to the prongs indicates the lever must be replaced.

Water Pump

REMOVAL & INSTALLATION

◆ See Figures 80, 81 and 103 thru 109

1. Remove the gearcase for access to the water pump, as detailed earlier in this section.

2. Remove the O-ring from the top of the water pump.

3. Remove the screws securing the water pump to the lower unit housing. It is very possible corrosion will cause the screw heads to break-off when an attempt to remove them is made. If this should happen, use a chisel and breakaway the water pump housing from the lower unit. Be careful not to damage the lower unit housing.

4. After the screws have been removed, slide the water pump, impeller, the impeller key, and the lower water pump plate upward and free of the driveshaft.

To Install:

5. Apply a coating of sealer to the upper surface of the lower unit.

6. Install the water pump base plate. Slide the driveshaft into the lower unit, and then rotate the shaft very slowly. When the splines of the driveshaft index with the pinion gear, the shaft will drop slightly. Install the water pump pin or key.

7. Slide the water pump impeller down the driveshaft and into place on top of the water pump base plate with the pump pin or key indexed in the impeller. Lubricate the inside surface of the water pump with light weight oil.

8. Lower the water pump housing down the driveshaft and over the impeller. Rotate the driveshaft CLOCKWISE as the water pump housing is lowered to allow the impeller blades to assume their natural and proper position inside the housing. Continue to rotate the driveshaft and work the water pump housing downward until it is seated on the water pump plate.

9. ALWAYS rotate the driveshaft clockwise while the screws are tightened to prevent damaging the impeller vanes. If the impeller is not rotated, the housing could damage or cut the end of the vanes as the screws are brought up tight. The rotation allows them to spring back in a natural position.

10. Place NEW grommets into the water pump housing for the water pickup. If a new water pump was installed, this seal will already be in place.

11. Install a NEW O-ring on the top of the driveshaft.

12. Install the gearcase as detailed earlier in this section.

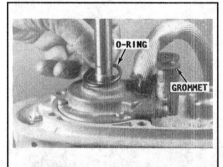

Fig. 103 Remove the O-ring and grommet. . .

Fig. 104 . . .and then remove the pump housing

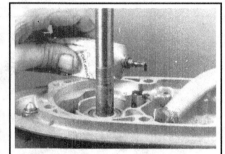

Fig. 105 Coat the upper surface of the unit with sealant

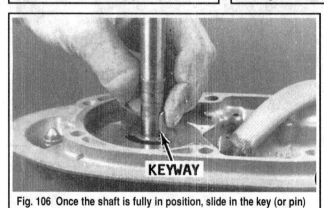

Fig. 106 Once the shaft is fully in position, slide in the key (or pin)

Fig. 107 An old style pump on the left and a newer one on the right

Fig. 108 Position the impeller and slide on the pump

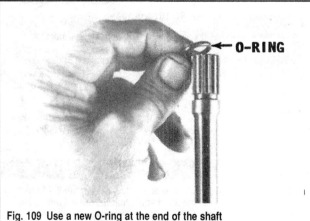

Fig. 109 Use a new O-ring at the end of the shaft

TYPE IV - MANUAL SHIFT LOWER UNIT, SPLIT LOWER UNIT

◆ **See Figure 110**

The Type IV manual shift lower unit with a split lower unit can be found on:
- 1958-59 50 hp (V4)
- 1964-67 60 hp
- 1968 65 hp
- 1964-65 75 hp

The manufacturer did not make things easy to service the 1956 75 hp engine lower unit - because the unit may have one of three different type lower units. Therefore, it is extremely important to determine the type unit for the engine being serviced.

One has a split lower unit, one is a single unit type, and the third is an electric shift and each is detailed in its appropriate section.

Troubleshooting

Troubleshooting must be done before the unit is removed from the powerhead to permit isolating the problem to one area. Always attempt to proceed with troubleshooting in an orderly manner. The shotgun approach will only result in wasted time, incorrect diagnosis, replacement of unnecessary parts and frustration.

The following procedures are presented in a logical sequence with the most prevalent, easiest, and less costly items to be checked listed first.

■ **One contributing factor to lower unit problems can be blamed on the helmsman's operation. If the operator attempts to "ease" the unit into gear, he is causing problems instead of preventing them. Any time the unit is shifted into, or out of gear, it must always be done with a definite and deliberate action.**

SLIPPAGE IN THE LOWER UNIT

◆ **See Figure 111**

If the shift seems to be slipping as the boat moves through the water, check the propeller and the rubber hub.

If the propeller has been subjected to many strikes against underwater objects, it could slip on its hub. If the hub is damaged or excessively worn, it is not economical to have the hub or propeller rebuilt.

A new propeller may be purchased for considerably less than meeting the expense of rebuilding an old worn propeller.

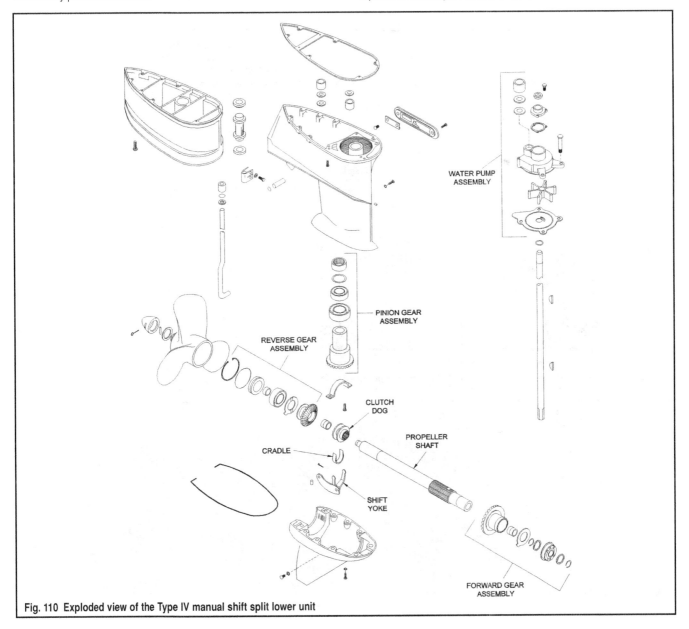

WATER PUMP
ASSEMBLY

PINION GEAR
ASSEMBLY

REVERSE GEAR
ASSEMBLY

CLUTCH
DOG

PROPELLER
SHAFT

CRADLE

SHIFT
YOKE

FORWARD GEAR
ASSEMBLY

Fig. 110 Exploded view of the Type IV manual shift split lower unit

SHIFT LINKAGE

◆ **See Figure 112**

A definite problem area, especially on a V4 unit, is in the linkage from the shift handle to the lower unit. The connection at the end of the shift handle consists of a bellcrank.

Over a long period of time and operation, the bellcrank and the rod fittings wear, developing slack in the linkage to the lower unit. Without tight fittings, free of slack, the lower unit cannot be shifted fully into gear as the design engineers intended. Therefore, this area should be checked early in the troubleshooting work, as follows.

UNABLE TO SHIFT INTO FORWARD OR REVERSE

◆ **See Figure 113**

Remove the propeller and check to determine if the shear pin has been broken.

Check the bellcrank under the powerhead. This is accomplished by removing the outer and inner windows in the exhaust housing.

Hold the shift rod with a pair of needle-nose pliers and at the same time attempt to move the shift lever on the starboard side of the engine. If it is possible to move the shift lever, the bellcrank is worn.

SHIFT ADJUSTMENT

◆ **See Figures 114 and 115**

1. Remove the hood.
2. Remove the rear cowling around the back of the engine.
3. Observe the 2 boss marks on the cam below the head, on the starboard side of the engine. Notice the boss or an arrow mark on the block.
4. Shift the unit into Forward and at the same time rotate the propeller until it stops, indicating the gear is against the clutch dog.

5. Ease the handle back out of Forward gear, and then rotate the propeller a very small amount. Shift back into Forward. The clutch dog and gear should be on top of each other. The clutch dog should be closest to going into gear without actually going into gear. The boss on the cam should now be aligned with the mark on the block.
6. If it is not, loosen the nut below the indent and make the adjustment. Repeat the procedure for Reverse gear. The other boss mark on the cam should align with the mark on the block. If it is not possible to adjust both Forward and Reverse exactly, it is best to have Forward gear nearer perfect than Reverse.
7. Install the exhaust cover and secure it in place with the attaching screws.

CRANKSHAFT & DRIVESHAFT SPLINES

The splines in the crankshaft or on the driveshaft may be damaged or worn and thus prevent rotation from the crankshaft to reach the propeller shaft.

If the splines in the crankshaft are destroyed, the crankshaft will have to be replaced as detailed in the Powerhead section.

If the splines on the driveshaft have been destroyed, the driveshaft must be replaced. Procedure to replace the driveshaft is included in each of the individual procedures of this chapter.

DIFFICULT SHIFTING

Verify that the ignition switch is in the **OFF** position, or better still, disconnect the spark plug wires from the plugs to prevent possible personal injury should the engine start.

Shift the unit into Reverse gear at the shift control box, and at the same time have an assistant turn the propeller shaft to ensure the clutch is fully engaged.

If the shift handle is hard to move, the trouble may be in the lower unit, with the shift cable, or in the shift box, if used.

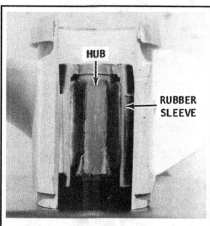

Fig. 111 Always check the rubber hub

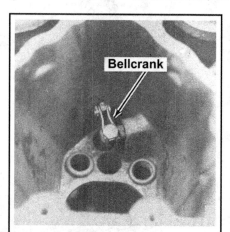

Fig. 112 The bellcrank is always a weak link

Fig. 113 Remove the window to access the shift rod

Fig. 114 A good look at the marks on the cam and block in Forward. . .

Fig. 115 . . .and in Reverse

WATER IN THE LOWER UNIT

Water in the lower unit is usually caused by fish line becoming entangled around the propeller shaft behind the propeller and damaging the propeller seal. If the line is not removed, it will cut the propeller shaft seal and allow water to enter the lower unit. Fish line has also been known to cut a groove in the propeller shaft.

The propeller should be removed each time the boat is hauled from the water at the end of an outing and any material entangled behind the propeller removed before it can cause expensive damage. The small amount of time and effort involved in pulling the propeller is repaid many times by reduced maintenance and service work, including the replacement of expensive parts.

ISOLATE THE PROBLEM

1. Disconnect the shift cable at the engine.
2. Operate the shift lever.
3. If shifting is still hard, the problem is in the shift cable or control box.
4. If the shifting feels normal with the shift cable disconnected, the problem must be in the lower unit, or in the shift lever rod passing through the exhaust housing.
5. To verify the problem is in the lower unit, have an assistant turn the propeller and at the same time move the shift cable back-and-forth. Determine if the clutch engages properly.

JUMPING OUT OF GEAR

If a loud thumping sound is heard at the transom, while the boat is underway, the unit is jumping out of gear. The propeller does not have a load; therefore the rushing water under the hull forces the lower unit in a backward direction. The unit jumps back into gear, the propeller catches hold, the lower unit is forced forward again, and the result is the thumping sound as the action is repeated. Normally this type of action occurs perhaps once a day, and then more frequently each time the clutch is operated, until finally the unit will not stay in gear for even a short time.

The following areas must be checked to locate the cause:
1. Check the bellcrank under the powerhead. This is accomplished by outer and inner windows in the exhaust housing. Hold the shift rod with a pair of needle-nose pliers and at the same time attempt to move the shift lever on the starboard side of the engine. If it is possible to move the shift lever, the bellcrank is damaged.
2. Disconnect the shift cable at the engine. Attempt to shift the unit into forward gear with the shift lever on the starboard side of the engine and at the same time rotate the propeller in an effort to shift into gear. Shift the control lever at the control box into forward gear. Move the shift cable at the engine up to the shift handle and determine if the cable is properly aligned. The control lever may have jumped a tooth on the slider or on the shift lever arc. If a tooth has been jumped, the cable would lose its adjustment and the unit would fail to shift properly. If the inner cable should slip on the end cable guide, the adjustment would be lost.
3. Move the shift lever at the engine into the neutral position and the shift lever at the control box to the Neutral position.
4. Now, move the shift cable up to the shift lever and see if it is aligned.

5. Shift the unit into reverse at the engine and shift the control lever at the control box into reverse. Move the cable up and see if it is aligned. If the cable is properly aligned, but the unit still jumps out of gear when the cable is connected, one of three conditions may exist:
   a. The bellcrank is worn excessively or damaged.
   b. The coupler at the connector at the shift rod is misaligned. This coupler is used to connect the upper shift rod with the lower rod. If the coupler has not been installed properly, any shifting will be difficult.
   c. Parts in the lower unit are worn from extended use.

Lower Unit

REMOVAL & INSTALLATION

◆ **See Figures 116 thru 121**

 ② 〈MODERATE

1. Remove the outer metal plate from the starboard side of the engine housing. Remove the inner plate from the exhaust housing. Access to the shift coupler is gained through this opening - disconnect the shift rod from the exhaust housing by removing the bottom bolt from the shift coupler.

■ **In most cases, if any unit being serviced has the 6 in. extension, it is not necessary to remove the extension in order to drop the lower unit.**

2. Remove the 4 bolts from each side of the lower unit. Work the lower unit loose from the exhaust housing. It is not uncommon for the water tube/s to be stuck in the water pump making separation of the lower unit from the exhaust housing difficult. However, with patience and persistence, the tube/s will come free of the pump and the lower unit separated from the exhaust housing.

3. Position the lower unit in a vertical position on the edge of the work bench resting on the cavitation plate. Secure the lower unit in this position with a C-clamp. The lower unit will then be held firmly in a favorable position for further service work. An alternate method is to cut a groove in a short piece of 2 x 6" wood to accommodate the lower unit with the cavitation plate resting on top of the wood. Clamp the wood in a vise and service work may then be performed with the lower unit erect (in its normal position), or inverted (upside down). In both positions, the cavitation plate is the supporting surface.

4. Remove and discard the O-ring from the top of the driveshaft.

To Install:

5. Connecting the shift rod with the coupler is not an easy task but can be accomplished as follows. Notice the cut-out area on the end of the shift rod, this area permits the bolt to pass through the coupler, past the shift rod, and into the other side of the coupler. It is this bolt that holds the shift rod in the coupler. Now, in order for the bolt to be properly installed, the cut-out area on the shift rod must be aligned in such a manner as to allow the bolt to be properly installed. As the lower unit is mated with the exhaust housing, exercise patience as the 2 units come together in order to enable the bolt to be installed at the proper time. If the rod is allowed to move too far into the coupler before the bolt is installed, it may be possible to force the bolt into place, past the shift rod. The threads on the bolt will be stripped, and the shift rod will eventually come out of the coupler.

Fig. 116 Disconnect the shift coupler. . .

Fig. 117 . . .and then remove the lower unit bolts

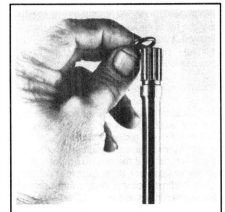

Fig. 118 Always get rid of the O-ring at the top of the driveshaft

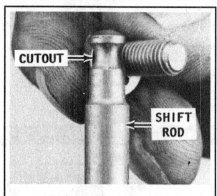

Fig. 119 There is a cut-out on the shaft that rides on the bolt. . .

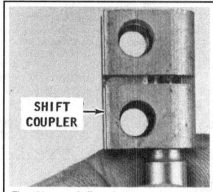

Fig. 120 . . .and allows it to pass through the coupler

Fig. 121 Make sure that the shift coupler window is closed and secured

■ Normally the coupler is not removed. In most cases it stays with the upper shift rod. If, however, the coupler was removed, for any number of reasons, perform the procedures following.

6. Install the coupler onto the upper unit shift rod, with the NO THREAD section towards the window. With the coupler in this position, the bolt may be inserted through the coupler and will catch the threads on the far side. Install the coupler bolt in the manner described in the previous step.

7. Check to be sure the water pick-up tubes are clean, smooth and free of any corrosion. Coat the water pick-up tubes and grommets with lubricant as an aid to installation.

8. Shift the lower unit into Forward or Reverse gear as an aid to rotating the shaft and the lower unit housing is worked towards the exhaust housing. Guide the lower unit up into the exhaust housing with the water tube sliding into the rubber grommet of the water pump. Continue to work the lower unit towards the exhaust housing, and at the same time rotate the propeller shaft as an aid to indexing the driveshaft splines with the crankshaft.

Start the bolts securing the lower unit to the exhaust housing. Tighten the bolts evenly and alternately.

9. Insert the bolt into the connector as detailed previously. After the bolt is in place, install and secure the gasket and inner window with the attaching hardware.

10. Install the outer window and gasket.

DISASSEMBLY & ASSEMBLY

◆ See Figures 110 and 122 thru 126

1. Turn the lower unit upside down and again clamp it in the vise or slide it into the wooden block, if one is used.

2. Carefully examine the lower portion of the unit. The cap is considered that part below the split with the skeg attached. Remove the Phillips screw from the starboard side of the lower housing. This screw passes through the shift yoke and threads into the side of the housing. Remove the attaching screws around the cap. These screws may be slotted-type or Phillips screws. Carefully tap the cap to jar it loose, and then separate it from the lower unit housing.

■ Take some time to study the arrangement of parts in the lower unit. You may elect to follow the practice of many professional mechanics and take a digital picture of the unit as an aid during the assembly work. Several engineering and production changes have been made to the lower unit over the years. Therefore, the positioning of the gears, shims, bearings and other parts may vary slightly from one unit to the next. To show each and every arrangement with a picture in this manual would not be practical. Even if it were done, the ability to associate the unit being serviced with the illustration would be almost impossible. Therefore, take time to make notes, scribble out a sketch, or take a couple photographs.

■ Take a look at the seal retainer on the propeller end of the propeller shaft. Now, notice the matching pin in the lower unit housing. During the installation work, the retainer must be installed with the pin indexed in the hole. Look at the snap ring installed in front of the seal retainer. One more item of particular interest; notice the 2 sides of the thrust washers, one side is as a normal washer, but the other side is a babbitt. The babbitt side must face toward the reverse gear during installation. The washer also has a dog ear which must face upward when it is installed. By taking note, at this time, of these items and exactly how they are installed, the task of assembling and installation will progress more smoothly.

3. Lift the shift lever out of the cradle and then remove the cradle from the shift dog.

4. Raise the propeller shaft while at the same time tapping with a soft-headed mallet on the bottom side to jar it loose; then remove the shaft assembly from the lower unit. The forward and reverse gear including the bearings will all come out with the propeller shaft. The forward gear is the gear at the opposite end of the shaft from the propeller.

5. Remove the attaching screws and then the U-shaped bracket from the top of the pinion gear. Reach into the lower housing and remove the pinion gear.

6. The pinion gear bearing, thrust bearing and driveshaft bearing do not have to be removed unless they are unfit for further service. Feel carefully with your fingers to determine their condition. If they do have to be removed, use a slide hammer with fingers. An alternate method is to drive them out with a drift punch from above.

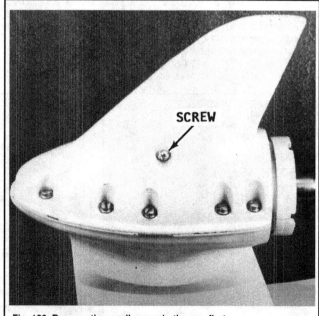

Fig. 122 Remove the small screw in the cap first

Fig. 123 A good look at the thrust washers

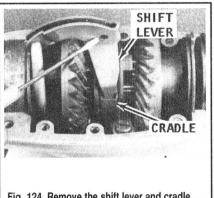

Fig. 124 Remove the shift lever and cradle. . .

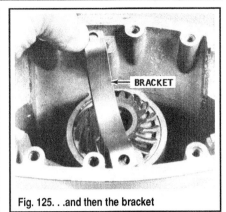

Fig. 125. . .and then the bracket

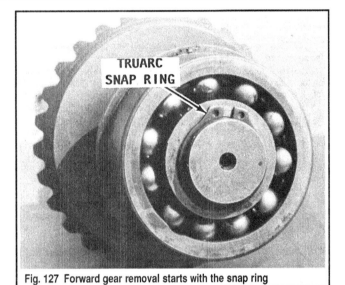

Fig. 126 Removing the pinion, thrust and driveshaft bearing is not necessary unless they are damaged

Fig. 127 Forward gear removal starts with the snap ring

Forward Gear Removal

◆ See Figures 127 and 128

7. Remove the Truarc snap ring from the propeller shaft.

8. Remove the small thrust washer, roller bearing, another small thrust washer and then the large thrust washer.

9. Remove the second Truarc snap ring, the forward gear bushing and finally the forward gear.

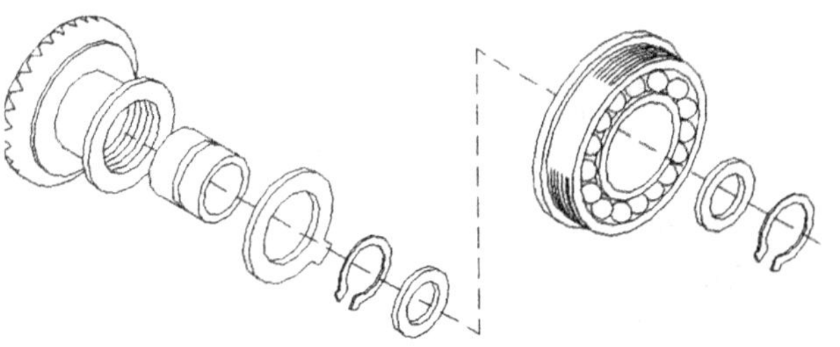

Reverse Gear Removal

◆ See Figures 129 thru 132

10. Remove the snap ring, the oil retainer, the roller bearing, thrust washer, reverse gear and bushing, and finally the clutch dog.

11. Turn the lower unit housing right side up and again clamp it in the vise. Remove the upper seal using a seal puller. An alternate method, if the puller is not available, is to use 2 small prybars and work the seal out of the housing. Be very careful not to damage the seal recess as the seal is being removed.

12. A babbitt bearing is installed under the seal. Normally, it is not necessary to remove this bushing. However, check the bushing surface with a finger and if any roughness is felt, the bearing must be replaced.

13. Clean the upper portion of the shift rod as an aid to pulling it through the bushing and O-ring. Pull the shift rod from the bottom portion of the lower unit housing. The shift rod passes through an O-ring and bushing in the lower unit housing; these 2 items prevent water from entering the lower unit. A special tapered punch is required to remove the bushing from the lower unit housing. Obtain the special punch, and then remove the bushing, and the O-ring.

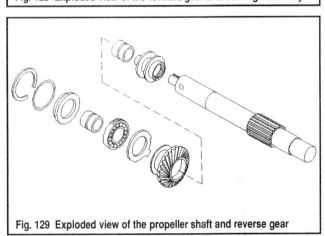

Fig. 128 Exploded view of the forward gear and bearing assembly

Fig. 129 Exploded view of the propeller shaft and reverse gear

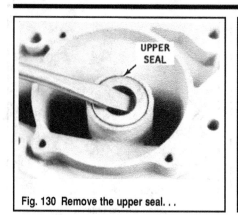

Fig. 130 Remove the upper seal. . .

Fig. 131 . . .and then the babbitt bearing

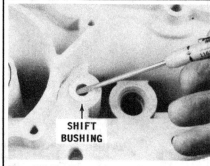

Fig. 132 You'll need a special punch to remove the shift bushing

To assemble:

◆ **See Figures 133 thru 137**

14. Place the lower unit on the work bench with the water pump recess facing upward.

15. Install a new O-ring into the shift cavity. Work the bushing into place on top of the O-ring with a punch and mallet. Squirt just a couple drops of oil into the bushing and O-ring as an assist during installation of the shift rod.

16. Install the babbitt bearing if it was removed. The babbitt bearing may be installed using the proper size socket and hammer.

17. Coat the outside edge of a new seal with No.1000 sealer and then tap the seal into place in the top of the upper lower unit housing.

18. Turn the unit over and then install the driveshaft bearing. Press against the lettered side of the bearing from the bottom side of the lower unit housing. Use the proper size socket and be careful not to distort the bearing.

19. Lower in the same number of shims that were removed during disassembly. If the shims are distorted or damaged, use a micrometer to measure the thickness and then purchase new ones of exactly the same size. If the shims were accidentally lost, a new bearing must be purchased and new shims will be included in the package.

20. Install the first bearing by pressing on the lettered side. Install the pinion gear bearing.

21. Lower the pinion gear into the housing. Check to be sure it seats properly. Lower the U-shaped pinion gear retaining bracket into position and secure it in place with the attaching screws. Check to be sure the pinion gear is properly located.

22. Check to be sure the shift rod is clean and smooth (free of any burrs or corrosion). Coat the shift rod and the O-ring with oil and then slide the shift rod through the O-ring and bushing into the gear case; from the bottom side of the lower unit housing.

23. Install a new seal into the seal retainer. Install a new O-ring around the outside perimeter of the retainer.

Forward Gear Installation

24. Slide the forward gear and forward gear bushing onto the end of the propeller shaft. Secure the bushing in place by installing the Truarc snap ring. Install the thrust washer with the babbitt side facing towards the forward gear.

25. Install the large thrust washer, roller bearing, small thrust washer and, finally, the Truarc snap ring.

Reverse Gear Installation

26. Slide the clutch dog onto the propeller shaft so the splines index with the splines on the shaft.

27. Install the reverse gear and reverse gear bushing.

Fig. 133 Pressing in the driveshaft bearing. . .

Fig. 134 . . .and then position the shims

Fig. 135 Press on the lettered side of the 1st bearing

Fig. 136 Coat the shift rod with oil

Fig. 137 Install a new seal and O-ring

28. Slide the thrust washer onto the shaft with the Babbitt side facing towards the reverse gear.

29. Slide the roller bearing onto the shaft, and then the oil retainer seal. Check to be sure a new seal and O-ring has been installed into the seal retainer, and then install the retainer.

Propeller Shaft Installation

◆ See Figure 138

■ **The seal retainer has a hole and the lower housing of the lower unit has a pin. This pin must index into the hole in the retainer when the propeller shaft is installed. If the pin is not seated properly in the hole, the seal retainer will work part way out of the housing and the lubricant in the lower unit will be lost.**

30. Lower the assembled propeller shaft into the lower unit housing, and, at the same time, check to be sure the oil retainer hole indexes with the pin in the housing.

31. Install the snap ring in front of the seal retainer with the open end towards the top. Check to be sure the tabs on the thrust washers are facing upward.

32. Place the cradle into place in the clutch dog recess. Bring the shifting forks to the cradle, check to be sure the forks seat properly in the cradle.

33. Work the shift rod up and down to be sure the unit is in Neutral.

Assembling Lower Gear Case

◆ See Figures 139 thru 141

34. Lay down a bead of No.1000 Sealer into the groove of the cap in preparation to installing the seal. Apply a small amount of silicone sealer on each side of the bearing gear case head. This sealer will form a complete seal when the lower unit cap is installed.

35. Position the lower unit cap over the gear assembly onto the lower unit housing. Apply a drop of sealer into the opening for each cap retaining screw to ensure a complete seal between the cap and the lower unit housing. Install the screws securing the cap to the lower unit housing. Tighten the screws alternately and evenly.

Fig. 138 Make sure the retainer hole indexes with the pin

■ **If time is taken to grind the end of the screw to a short point, it will make the task of installation much easier. If the cap and shift lever are not aligned exactly, the screw will seek and make the alignment as it passes through. However, do not make a long point or the screw will not have enough support and would bend during operation of the shift lever.**

36. Use a flashlight and align the hole in the cap with the hole in the shift lever. Install the tapered Phillips screw into the housing and through the lever. Always use a new washer with this screw and 1000 sealer on the screw to ensure a leak proof seal. The lubricant in the lower unit must not escape and water must not enter. Tighten the screw securely.

CLEANING & INSPECTION

◆ See Figures 110 and 142, 143 and 144

Clean all water pump parts with solvent, and then dry them with compressed air. Inspect the water pump cover and base for cracks and distortion, possibly caused from overheating. Inspect the face plate and water pump insert for grooves and/or rough surfaces. If possible, always install a complete new water pump while the lower unit is disassembled. A new impeller will ensure extended satisfactory service and give peace of mind to the owner. If the old impeller must be returned to service, never install it in reverse to the original direction of rotation. Installation in reverse will cause premature impeller failure.

Inspect the impeller side seal surfaces and the ends of the impeller blades for cracks, tears, and wear. Check for a glazed or melted appearance, caused from operating without sufficient water. If any question exists, and as previously stated, install a new impeller if at all possible.

Clean all parts with solvent and dry them with compressed air. Discard all O-rings and gaskets.

Inspect and replace the driveshaft if the splines are worn.

Inspect the gearcase and exhaust housing for damage to the machined surfaces. Remove any nicks and refurbish the surfaces on a surface plate. Start with a No.120 Emery paper and finish with No.180.

Check the water intake screen and passages by removing the bypass cover. Inspect the clutch dog, drive gears, pinion gear and thrust washers. Replace these items if they appear worn. If the clutch dog and drive gear arrangement surfaces are nicked, chipped or the edges rounded, the operator may be performing the shift operation improperly or the controls may not be adjusted correctly. These items MUST be replaced if they are damaged.

Inspect the dog ears on the inside of the forward and reverse gears. The gears must be replaced if they are damaged.

Check the cradle that rides on the inside diameter of the clutch dog. The sides of the cradle must be in good condition, free of any damage or signs of wear. If damage or wear has occurred, the cradle must be replaced.

Check the shift lever and the two prongs that fit inside the cradle. Check to be sure the prongs are not worn or rounded. Damage or wear to the prongs indicates the lever must be replaced.

Fig. 139 Lay a bead of sealer around the cap

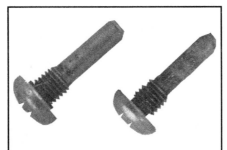

Fig. 140 Grinding down the taper on the cap screw will make things easier. . .

Phillips Screw

Fig. 141 . . .before installing it

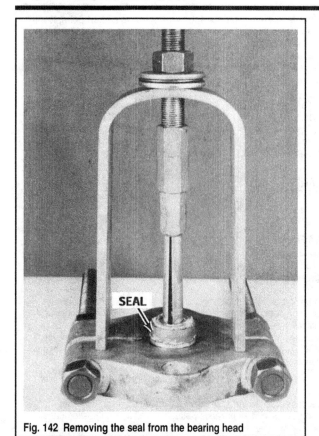

Fig. 142 Removing the seal from the bearing head

Water Pump

REMOVAL & INSTALLATION

◆ **See Figures 145 thru 150**

1. Remove the lower unit.

2. Remove the bolts (should be 4) securing the water pump to the lower unit housing. It is very possible corrosion will cause the bolt heads to break-off when an attempt to remove them is made. If this should happen, use a chisel and break away the water pump housing from the lower unit. Be very careful not to damage the lower unit housing.

3. After the screws have been removed, slide the water pump, impeller, the impeller key and the lower water pump plate upward and free of the driveshaft. Carefully pull upward on the driveshaft and remove it from the lower unit (if necessary).

To Install:

4. Apply a coating of sealer to the upper surface of the lower unit and install the water pump base plate.

5. Slide the driveshaft into the lower unit and then rotate the shaft very slowly. When the splines of the driveshaft index with the pinion gear, the shaft will drop slightly. Install the water pump key (a little Vaseline may help).

6. Slide the water pump impeller down over the driveshaft and into position on top of the water pump base plate with the pump key indexed in the impeller. Lubricate the inside surface of the water pump with lightweight oil.

7. Lower the water pump housing down over the driveshaft and over the impeller. Rotate the driveshaft CLOCKWISE as the water pump housing is lowered over the impeller to allow the blades to assume their natural and proper position inside the housing. Continue to rotate the driveshaft and work the water pump housing downward until it is seated on the lower unit upper housing.

Always rotate the driveshaft clockwise while the screws are tightened to prevent damaging the impeller vanes. If the impeller is not rotated, the

Fig. 143 The ears on this clutch dog are worn beyond repair

Fig. 144 Water in the lower unit has badly damaged this forward gear assembly

Fig. 145 Remove the water pump cover. . .

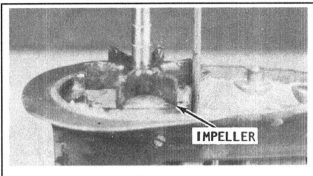

Fig. 146 .. and then the impeller

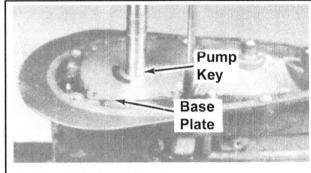

Fig. 147 Install the base plate and pump key. . .

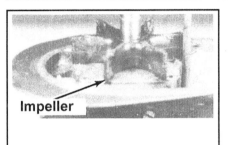

Fig. 148 . . .and then slide on the impeller

Fig. 149 Lower the pump housing down over the impeller. . .

Fig. 150 . . .and then replace the seal and grommet after tightening the bolts

housing could damage or cut the end of the vanes as the screws are brought up tight. The rotation allows them to spring back in a natural position.

8. Place new grommets into the water pump housing for the water tubes. install a new seal ring on top of the water pump.

9. Install a new O-ring on the top of the driveshaft.

Functional Check

Perform a functional check of the completed work by mounting the engine in a test tank, in a body of water, or with a flush attachment connected to the lower unit. If a flush attachment is used, never operate the engine above an idle speed, because the no-load condition on the propeller would allow the engine to runaway resulting in serious damage or destruction of the engine.

✳✳ CAUTION

Water must circulate through the lower unit to the engine any time the engine is run to prevent damage to the water pump in the lower unit. Just 5 seconds without water will damage the water pump.

Start the engine and observe the tattletale flow of water from the idle relief in the exhaust housing. The water pump installation work is verified.

If a flush attachment is connected to the lower unit, very little water will be visible from the idle relief port. Shift the engine into the 3 gears and check for smoothness of operation and satisfactory performance.

TYPE V - MANUAL SHIFT LOWER UNIT, SINGLE ENCLOSED HOUSING

◆ See Figure 151

- 1964-67 60 hp
- 1968 65 hp
- 1960-65 75 hp
- 1966-67 80 hp
- 1968 85 hp

On the 1960 75 hp engine, the pinion gear/bearing assembly and the driveshaft bearing assembly were different from the other lower units covered in this section. Where these differences occur, they will be noted in the procedures and supported with accompanying illustrations.

Clutch Dog Modification

DIFFICULT

◆ See Figures 152 thru 155

A standard unit from the manufacturer included a spline for the clutch dog to operate into forward and reverse gears. Design of the clutch dog caused it to chatter, or the unit to slide slightly out of gear during normal operation. This action caused the ears of the clutch dog to wear very rapidly.

As a result of this condition, a service bulletin, fixture, new clutch dog and associated parts, were dispatched from the factory, Illustration. Service centers were instructed to modify the lower unit any time it was disassembled. A hole was drilled through the propeller shaft where a spring

and 2 balls, provided in the package, were installed on the propeller shaft. One on either side.

A groove had been cut into the inside diameter of the new clutch dog. When the new clutch dog is placed in the Neutral position, the groove is over the balls. The spring exerts a force on the balls to hold the clutch dog in position.

Two ramps are manufactured on the inside surface of the clutch dog, one towards the forward gear and the other towards the reverse gear. As the clutch dog is moved toward either forward or reverse gear, the spring and balls exert a small force on the clutch dog to prevent the chattering.

If the unit being serviced has not had this clutch dog modification incorporated, the authors strongly recommend the propeller shaft be taken to an authorized repair facility and the work performed. DO NOT attempt to drill the hole without the special fixture.

The following procedures and supporting illustrations cover service of the original factory delivered lower unit without the spring and balls. However, during the assembling sequence, extra procedures and supporting illustrations are included as an aid to installation.

Troubleshooting

■ Please refer to the Type IV section for all troubleshooting procedures on this lower unit.

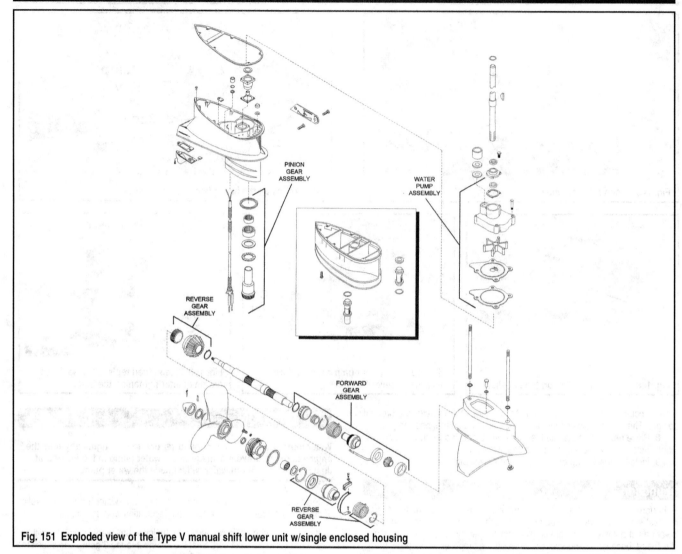

Fig. 151 Exploded view of the Type V manual shift lower unit w/single enclosed housing

PINION GEAR ASSEMBLY

WATER PUMP ASSEMBLY

REVERSE GEAR ASSEMBLY

FORWARD GEAR ASSEMBLY

REVERSE GEAR ASSEMBLY

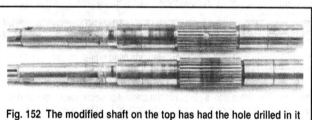

Fig. 152 The modified shaft on the top has had the hole drilled in it

FIXTURE

Fig. 153 Here's the new kit

Lower Unit

REMOVAL & INSTALLATION

 MODERATE

◆ See Figures 116 thru 121

1. Remove the propeller and drain the lower unit.
2. Disconnect the cables from the battery.
3. Remove the hood.
4. Disconnect the shift cable from the shift handle.
5. On the starboard side of the engine, remove the egg shaped outer window from the exhaust housing about halfway down. After the outer is removed, remove the inner elongated window.
6. Observe the shift rod connection through the windows. If the engine does not have the long extension, the connector will be about 1 1/4 (3.18 cm) long. Remove the lower screw in the connector. After the screw is loose, it may be necessary to grasp it with a pair of needle-nose pliers in order to withdraw it from the connector.
7. If the unit being serviced has the long shaft extension, remove the 8 bolts securing the lower unit to the extension.
8. Work the lower unit loose and separate it from the exhaust housing. If difficulty is encountered separating the lower unit from the exhaust housing, the cause is usually the water pump tubes being stuck in the water pump, and/or the shift rod being tight in the connector, or the driveshaft frozen in the powerhead crankshaft. Use patience, a rubber mallet, more patience, and perhaps a break for coffee or tea, until the lower unit is free.

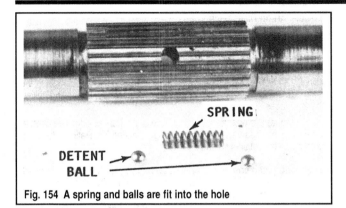

Fig. 154 A spring and balls are fit into the hole

Fig. 155 The new dog has 2 ramps

To Install:

■ **Connecting the shift rod with the coupler is not an easy task but can be accomplished as follows:**

9. Notice the cut-out area on the end of the shift rod. This area permits the bolt to pass through the coupler, past the shift rod, and into the other side of the coupler. It is this bolt that holds the shift rod in the coupler. Now, in order for the bolt to be properly installed, the cut-out area on the shift rod must be aligned in such a manner to allow the bolt to be properly installed. Therefore, as the lower unit is mated with the exhaust housing, exercise patience as the 2 units come together, to enable the bolt to be installed at the proper time. If the rod is allowed to move too far into the coupler before the bolt is installed, it may be possible to force the bolt into place, past the shift rod. The threads on the bolt will be stripped, and the shift rod will eventually come out of the coupler.

■ **Normally the coupler is not removed. In most cases it stays with the upper shift rod. If, however, the coupler was removed, for any number of reasons, perform the following.**

10. Install the coupler onto the upper unit shift rod, with the NO THREAD section facing towards the window. With the coupler in this position, the bolt may be inserted through the coupler and catch the threads on the far side. Install the coupler bolt in the manner described previously.

11. Check to be sure the water pick up tubes are clean, smooth and free of any corrosion. Coat the water pick-up tubes and grommets with lubricant as an aid to installation.

12. Shift the lower unit into Forward or Reverse gear as an aid to rotating the shaft and the lower unit housing is worked towards the exhaust housing. Guide the lower unit up into the exhaust housing with the water tube sliding into the rubber grommet of the water pump. Continue to work the lower unit towards the exhaust housing and at the same time rotate the propeller shaft as an aid to indexing the driveshaft splines with the crankshaft.

13. Start the bolts securing the lower unit to the exhaust housing. Tighten the bolts evenly and alternately.

14. Insert the bolt into the connector. Take the time to read and understand the earlier Note before making this connection.

15. After the bolt is in place, install and secure the gasket and inner window with the attaching hardware. Install the outer window and gasket.

16. Install the propeller and fill the unit with oil.

DISASSEMBLY

◆ **See Figures 151 and 156 thru 162**  ③ *DIFFICULT*

1. Remove the lower unit and set it in a fixture, or clamp the anti-cavitation plate to the edge of a work bench with the lower unit in its normal upright position.

2. Remove the O-ring from the groove at the top of the driveshaft.

3. Remove the water pump.

4. Now clamp the driveshaft in a vise so as not to damage the splines. Place a wooden block on the upper surface of the housing and use a hammer to drive the lower unit from the driveshaft. This action will pull the seal free of the cap and the driveshaft free of the lower unit housing. Check the bearing surface on the driveshaft. If the surface is rough, corroded, or damaged in any way, the cap must be removed and the bearing replaced.

5. Lay the lower unit on the bench, on its side. Remove the two 9/16 in. nuts on the front and rear section of the upper gear housing.

6. Tap on the torpedo end of the lower unit with a rubber mallet. Continue to tap with the mallet until the lower section separates from the upper section. Coat the shift rod with lightweight oil as an aid for it to slip through the housing. As the 2 parts of the lower unit are separated, the shift rod will then come through the upper housing. As the halves are separated, the needle bearings will usually fall free of the bearing carrier so be prepared to keep track of them as they come out.

7. To remove the driveshaft bearing installed just below the seal that was removed with the driveshaft, remove the four 7/16 in. bolts and then the bearing assembly.

8. On 1960 75 hp units, a roller bearing is installed at this location. This bearing is pressed onto the driveshaft and secured with a washer and snap ring. The bearing need not be removed unless it is unfit for further service; water has entered the lower unit. The driveshaft on these units is in 2 sections, one above the water pump and the other below the pump.

Fig. 156 Hold the driveshaft and tap off the lower unit

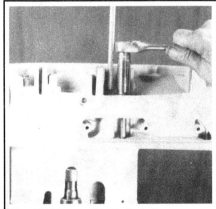

Fig. 157 Remove the 2 bolts in the upper section. . .

Fig. 158 . . .and then tap the units free of themselves

■ If the 2 halves of the lower unit are frozen and refuse to separate, the cause is usually corrosion around the 2 long bolts passing through the halves; therefore, it may be necessary to cut the bolts. If one of the two halves is to be damaged in the separation process, let it be the less expensive upper half. After the bolts have been cut, again tap with the mallet and the 2 halves should separate. Drive the 2 studs out of the upper part. Have the cut marks welded, sand the surface smooth, apply a new coat of paint, and the section can be restored to service.

Shift Rod Removal
◆ See Figures 163 and 164

9. Pull upward on the shift rod and rotate the propeller shaft. Remove the 7/16 in. bolt through the shift lever and shift rod bolt, and then the shift rod can then be removed from the 2 levers.

10. Work the 2 levers out of the lower unit housing.

Fig. 159 Remove the 4 bolts and lift out the driveshaft bearing

Fig. 160 Removing the snap ring on the 1960 75 hp engine. . .

Lower Portion Disassembly
◆ See Figures 165 thru 175

11. Remove the 4 screws from the propeller shaft bearing cap. Use a hammer and chisel and remove the bearing carrier cap from the lower unit housing - DO NOT use the chisel in the groove between the housing and the bearing carrier cap as such action would probably destroy the sealing ability of the 2 surfaces.

12. Note that this washer has a plain side and a babbitt side. The babbitt side must be installed towards the forward gear during installation. The 2 ears fit inside the retainer washer.

■ The bearing cap contains a seal and a bearing for the propeller shaft. The seal prevents lubricant in the lower unit from escaping and prevents water from entering. Special tools are required to remove the bearing and seals from the cap; therefore, if the bearing or seal is unfit for further service, the purchase of a new bearing cap may be more practical than attempting to remove and replace the individual parts. The new cap will have the bearing, seal, and O-ring properly installed and ready for service.

Fig. 161 . . .before removing the driveshaft

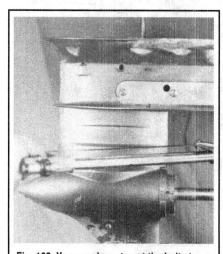

Fig. 162 You may have to cut the bolts to get the units separated

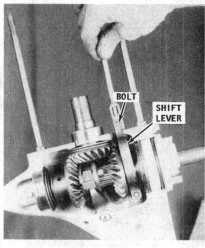

Fig. 163 Remove the connecting bolt. . .

Fig. 164 . . .and then pull out the shift rod and lever

✳✳ WARNING

The next step involves a dangerous procedure and should be executed with care while wearing safety glasses. The retaining ring is under tremendous tension in the groove. While it is being removed, it could slip off the Truarc pliers and will travel with incredible speed causing personal injury if it should strike a person. Therefore, continue to hold the ring and pliers firmly after the ring is out of the groove and clear of the lower unit. Place the ring on the floor and hold it securely with one foot before releasing the grip on the pliers. An alternate method is to hold the ring inside a trash barrel, or other suitable container, before releasing the pliers.

13. Obtain a pair of Truarc pliers. Insert the tips of the pliers into the holes of the retaining ring inside the housing. Now, carefully remove the retaining ring from the groove and gearcase without allowing the pliers to slip. Release the grip on the pliers in the manner described in the previous Warning.

14. Remove the retainer washer with the 4 holes and the thrust washer. Notice the 2 cut-outs in the washer; during installation, the ears of the babbitt washer index into these cut-outs.

15. The reverse gear will come with the shaft. Slide the reverse gear free of the propeller shaft. Notice that the reverse gear has a washer on the inside and a removable bushing. The bushing and washer must be purchased as separate items, they are not included with the purchase of a new reverse gear.

Fig. 165 Remove the bearing cap. . .

Fig. 166 . . .the thrust washer. . .

Fig. 167 . . .and then the snap ring

Fig. 168 Pull out the retainer plate. . .

Fig. 169 . . .and then the reverse gear

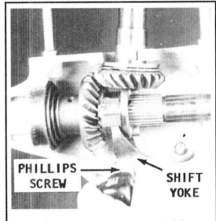

Fig. 170 Remove the small screw holding the shift yoke in position. . .

Fig. 171 . . .and then pull out the propeller shaft assembly

Fig. 172 Remove the pinion gear, washer and thrust bearing. . .

Fig. 173 . . .before pulling out the forward gear

16. On the starboard side of the lower unit, remove the Phillips screw. This screw holds the shift yoke in place.

17. Grasp the propeller shaft firmly and pull it free of the lower unit. Reach inside the lower unit and remove the clutch dog, clutch cradle and yoke.

18. Remove the pinion gear, the pinion flat washer and the pinion thrust bearing.

■ **On the 1960 75 hp lower unit, the pinion gear does not use a flat washer or a thrust bearing.**

19. Reach in and withdraw the forward gear. Notice that the forward gear also has a flat washer and sometimes this washer fails to come with the gear and remains on the shaft. In that case, the washer is removed with the propeller shaft. Also notice that the bushing is quite different from the bushing in the reverse gear. This bushing is pressed in with a flange on the back side and cannot be replaced. A new forward gear will have the bushing pressed in place and ready for service, so if the bearing is worn and unfit for further service, a new forward gear must be purchased.

20. Reach in and withdraw the forward gear bearing.

21. The bearing race does not have to be removed unless the bearing is being replaced. A special tool is available to remove the race; however, the torpedo end of the lower unit may be heated with a torch and then tapped with a soft-head mallet to help the race come free. There are no shims installed behind the race.

Upper Portion Disassembly

◆ See Figure 176

■ **The upper pinion gear bearing shell usually does not require replacement. Do not remove this item unless the bearing is unfit for further service. The bearing and the shell must be replaced as a set.**

22. Use a long punch inserted through the top of the upper lower unit section and drive the pinion gear bearing shell free of the housing. To remove the upper driveshaft bearing, drive it free from the top side. Remove this bearing ONLY if it is unfit for further service.

ASSEMBLY

Forward Gear & Pinion Gear Installation

◆ See Figures 177 thru 180

1. Drive the forward gear bearing race into place using a bearing installer, until it bottoms in the housing.

2. Install the forward bearing into the race.

3. Install the forward gear. Check to be sure it is the gear with the babbitt boss on the back side. Also, make sure that the washer is in place inside the gear. This washer may be held in place within the gear, with a small amount of grease.

4. Install the flat bearing and flat washer on top of the pinion gear. If the unit being serviced is a 1960 75 hp engine, the flat bearing and flat washer is not used. Install the pinion gear assembly into the lower unit housing. The back side of the pinion gear will lay on the forward gear.

■ **The parts installed thus far, except the forward gear bearing race, are all just laying in place without being secured at this time.**

5. Lay the shifting yoke into position in the bottom of the lower unit.

■ **If the propeller has had the modification described in the Clutch Dog Modification procedures at the beginning of this section, perform the next step. Otherwise, skip it.**

6. Insert the spring into the hole in the propeller shaft. Set a ball in place on the spring on each side of the shaft. Slide the clutch dog down the propeller shaft until the balls index into the groove in the inside surface of the clutch dog. Move the clutch dog until it snaps into the Neutral position.

7. Coat the groove in the outside surface of the clutch dog with OMC Grease, or equivalent. Install the cradle to the clutch dog and slide the clutch dog onto the propeller shaft with the splines of the clutch dog indexed with the splines on the shaft.

8. Slide the propeller shaft into the lower unit with the forward end of the shaft entering the forward bearing. Push the shaft forward as far as possible to seat it properly in the forward bearing.

■ **The following step is tricky, requiring patience and possibly a break for coffee or tea; but it can be done.**

9. Reach into the opening of the lower unit with a screwdriver and work the fingers of the yoke until they index into the cradle around the clutch dog. After the yoke fingers are in position, move the yoke very slightly until the hole on the starboard side of the housing is aligned with the hole in the yoke. Once the holes are aligned, thread the Phillips screw, with a new O-ring coated with 1000 sealer, through the housing and into the yoke. Tighten the screw securely. A difficult task has been completed and if you did it on the first try, you're good!.

10. Check to be sure the bushing and thrust washer have been installed into the reverse gear. Slide the assembled reverse gear onto the propeller shaft until the gear teeth index with the teeth of the pinion gear.

11. Install the large retainer washer with the 4 holes. Slide the washer up against the reverse gear.

✳✳ WARNING

This next step can be dangerous. The snap ring is placed under tremendous tension with the Truarc pliers while it is being placed into the groove. Wear safety glasses and exercise care to prevent the snap ring from slipping out of the pliers. If the snap ring should slip out, it would travel with incredible speed and cause personal injury if it struck a person.

12. Notice how one edge of the snap ring is square and the other edge is rounded. When the snap ring is installed, the square edge must face towards

Fig. 174 Pull out the forward gear bearing

Fig. 175 If you have to remove the forward gear bearing race, heating the end of the unit may help

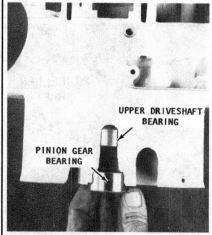

Fig. 176 Removing the pinion gear and upper driveshaft bearings

Fig. 177 Drive the forward gear bearing race in until it seats

Fig. 178 Positioning the shift yoke

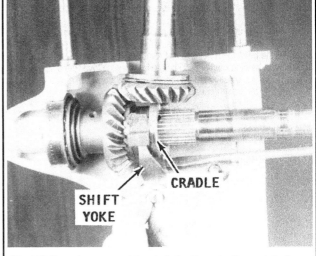

Fig. 179 A look at the modified shaft

you. Carefully install the snap ring into the groove in the lower unit following the precautions given in the previous Warning.

13. Install 2 guide pins into the flat washer with the 4 holes; these guide pins will assist to install the bearing cap. Hold the guide pins to prevent the washer from sliding out of the retainer. Slide the thrust washer onto the propeller shaft with the Babbitt side facing toward the gear and the 2 ears indexed into the cut-outs of the large retainer washer.

14. Install a new O-ring, coated with 1000 sealer, onto the propeller shaft bearing cap. Slide the cap onto the propeller shaft and at the same time hold the guide pins. Work the bearing cap over the guide pins. Hold the guide pins with needle nose pliers and tap the bearing cap into place with a soft-headed mallet. Start 2 attaching screws, with new O-rings coated with 1000 sealer, through the 2 holes in the cap without guide pins. After the 2 screws have been well started, remove the 2 guide pins and install the other 2 screws with new O-rings coated with 1000 sealer. Tighten all 4 screws evenly and alternately.

■ If you think some of the other assembling procedures were tricky on this unit, stand by! This next one will put you to the test.

Two Lever Installation

15. Lower the 2 shift levers into the lower unit and then index the arms to the shift yoke. Insert the shift rod with the bend towards the driveshaft.

16. Install the bolt through the arms and the shift rod. Rotate the propeller shaft and at the same time pull upward on the shift rod. This action will shift the unit into Forward gear and the bolt will be high enough to permit tightening it.

17. Move the shift rod downward to the Neutral position and at the same time rotate the propeller shaft. Watch the pinion gear as the propeller shaft is rotated - it should not move.

18. Continue to push the shift rod down as far as possible and at the same time turn the propeller shaft. As the propeller shaft is rotated clockwise, the pinion gear should turn counterclockwise. Pull upward on the shift rod and engage the clutch dog into the forward gear. Turn the propeller shaft clockwise and the pinion gear should turn clockwise.

Driveshaft Bearing Installation

◆ See Figure 181

19. Turn the upper half of the lower unit upside down and press the driveshaft caged needle bearing into place.

Fig. 180 Use a long screwdriver to index the yoke fingers into the cradle around the dog

Lower Pinion Gear Bearing Installation

◆ See Figures 182 and 183

If this bearing and shell was removed, perform the next step.

20. Push the lower pinion gear bearing shell into the housing recess with the lettered side facing outward. Coat the inside of the shell with needle bearing grease and insert the individual needle bearings into the shell. The count must be the same as noted during disassembly. The proper number allows the needles to rotate and perform their bearing function properly.

21. Apply just a small amount of OMC 1000 sealer around the shell and in the recess. Work a new O-ring into the recess.

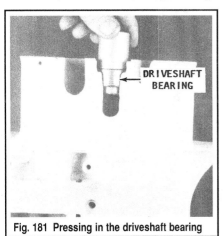

Fig. 181 Pressing in the driveshaft bearing

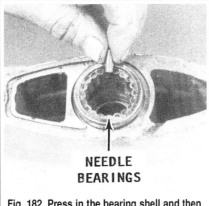

Fig. 182 Press in the bearing shell and then insert the needles

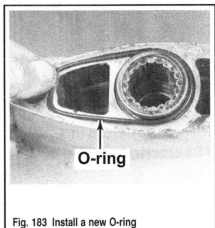

Fig. 183 Install a new O-ring

Lower Unit Mating

◆ See Figures 184 thru 187

22. Clamp the lower portion of the lower unit to the work bench with a C-clamp on the anti-cavitation plate, or clamp the unit in a vise.

23. Lower the upper portion of the lower unit down over the shift rod, but do not make contact with the surface of the lower portion. With a little distance between the 2 parts, clamp a pair of vise grip pliers onto 1 of the studs to hold the units apart.

24. Coat the lower portion mating surface with OMC 1000 sealer. Remove the vise grip pliers and lower the upper portion until the surfaces of the 2 lower unit sections make contact. Thread a new nut onto each stud; the nuts are self-locking and should never be used a second time because after the first use, they lose their locking ability. Tighten the nuts evenly and alternately.

■ Some bearing carriers use a gasket and an O-ring, while others just use the O-ring. Check the horsepower and model year of the unit being serviced, and then determine the type installation from the parts book.

25. If the upper bearing carrier was removed, install the driveshaft down into the housing.

26. Slide the upper bearing carrier down the driveshaft and into the housing. Install the 4 retaining bolts.

27. If the bearing carrier was not removed, install the driveshaft down through the bearing carrier and into the lower unit. Cover the splines of the driveshaft with protective tape. Work a new seal down the driveshaft and tap the seal into the bearing head.

CLEANING & INSPECTION

◆ See Figures 151 and 188 thru 195

Clean all water pump parts with solvent, and then dry them with compressed air. Inspect the water pump cover and base for cracks and distortion, possibly caused from overheating. Inspect the face plate and water pump insert for grooves and/or rough surfaces. If possible, install a complete new water pump while the lower unit is disassembled. A new impeller will ensure extended satisfactory service and give peace of mind to the owner. If the old impeller must be returned to service, never install it in reverse to the original direction of rotation. Installation in reverse will cause premature impeller failure.

Inspect the impeller side seal surfaces and the ends of the impeller blades for cracks, tears, and wear. Check for a glazed or melted appearance, caused from operating without sufficient water. If any question exists, as previously stated, install a new impeller if at all possible.

Clean all parts with solvent and dry them with compressed air. Discard all O-rings and gaskets.

Inspect and replace the driveshaft if the splines are worn.

Inspect the gearcase and exhaust housing for damage to the machined surfaces. Remove any nicks and refurbish the surfaces on a surface plate. Start with a No.120 Emery paper and finish with No.180.

Check the water intake screen and passages by removing the bypass cover. Inspect the clutch dog, drive gears, pinion gear and thrust washers.

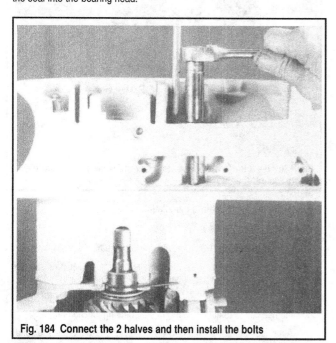

Fig. 184 Connect the 2 halves and then install the bolts

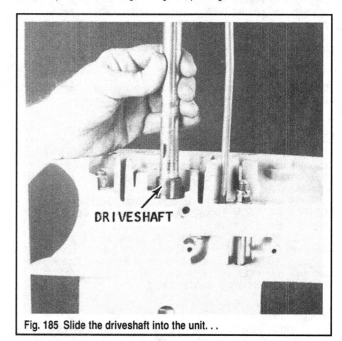

Fig. 185 Slide the driveshaft into the unit. . .

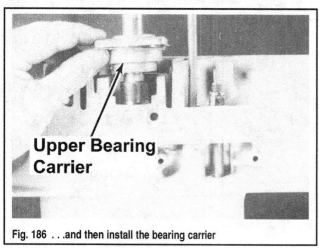

Fig. 186 . . .and then install the bearing carrier

Fig. 187 Pressing on a new seal

Replace these items if they appear worn. If the clutch dog and drive gear arrangement surfaces are nicked, chipped or the edges rounded, the operator may be performing the shift operation improperly or the controls may not be adjusted correctly. These items must be replaced if they are damaged.

Inspect the dog ears on the inside of the forward and reverse gears. The gears must be replaced if they are damaged.

Check the cradle that rides on the inside diameter of the clutch dog. The sides of the cradle must be in good condition, free of any damage or signs of wear. If damage or wear has occurred, the cradle must be replaced.

Check the shift lever and the 2 prongs that fit inside the cradle. Check to be sure the prongs are not worn or rounded. Damage or wear to the prongs indicates the lever must be replaced.

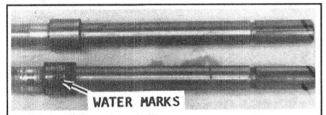

Fig. 188 Look at the water damage on the lower shaft. A new one is on the top

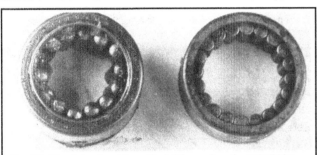

Fig. 189 The pinion bearing on the left is used on the 1960 75 hp engine, while the one on the right is used on all others

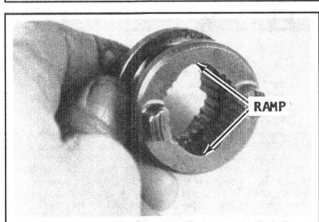

Fig. 190 The newer modified clutch dog has ramps for the balls installed in the shaft

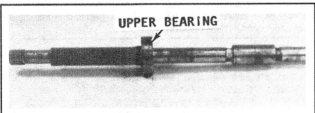

Fig. 191 The upper bearing is pressed into place on this driveshaft from a 1960 75 hp engine

Fig. 192 There are some needle bearings missing from this thrust washer and it will require replacement

Fig. 193 This snap ring is what holds the driveshaft in position

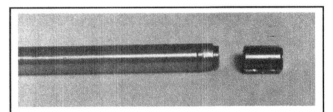

Fig. 194 Shifting with the engine at a high rpm can break the driveshaft

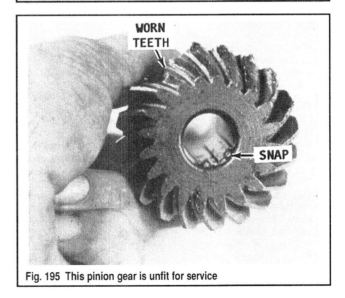

Fig. 195 This pinion gear is unfit for service

Water Pump

REMOVAL & INSTALLATION

 MODERATE

◆ See Figures 145 thru 150 and 196

1. Remove the lower unit.
2. Remove the four 7/16 in. bolts from the water pump housing. Lift the water pump housing upward off the pump impeller and then free of the driveshaft.
3. Slide the water pump impeller up and free of the driveshaft.
4. Remove the impeller Woodruff key from the driveshaft. Slide the water pump base plate up and free of the driveshaft.

■ Notice how the shift rod extends out of the housing. Also notice how the shift rod has a bend near the top towards the driveshaft. This is the normal configuration for the shift rod. Do not consider this bend as damage and attempt to straighten it. When the shift rod is installed during assembling of the lower unit, make certain the bend is towards the driveshaft.

■ After the water pump base plate has been removed you'll be able to see the upper bearing cap in the lower unit housing. Do not remove the cap or bearing unless there is no doubt the bearing is unfit for further service. The screws are very difficult to remove and many times the housing will be damaged in the attempt to remove the bearing.

To Install:

5. Coat the upper housing surface with OMC 1000 Sealer, or equivalent.
6. Slide a new water pump gasket down onto the housing surface. Coat the top surface of the gasket with sealer.
7. Slide the water pump plate down into position on the gasket. Install a new Woodruff key into the driveshaft.
8. Slide the water pump impeller down the driveshaft and into place on top of the water pump base plate with the pump key indexed in the impeller. Lubricate the inside surface of the water pump with lightweight oil.

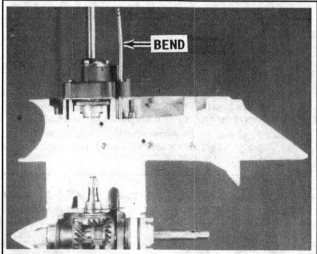

Fig. 196 The shift rod has a slight bend and you should not mistake this for a problem

9. Lower the water pump housing down the driveshaft and over the impeller. Rotate the driveshaft clockwise as the water pump housing is lowered to allow the impeller blades to assume their natural and proper position inside the housing. Continue to rotate the driveshaft and work the water pump housing downward until it is seated on the lower unit upper housing.
10. Always rotate the driveshaft clockwise while the screws are being tightened to prevent damaging the impeller vanes. If the impeller is not rotated, the housing could damage or cut the end of the vanes as the screws are brought up tight. The rotation allows them to spring back in a natural position.
11. Place new grommets into the water pump housing for the water tubes. Install a new seal ring on top of the water pump.
12. Install a new O-ring on the top of the driveshaft.

TYPE VI - ELECTROMATIC LOWER UNIT (ALL EXC. 100 HP)

◆ See Figures 197 and 198

- 1962-70 40 hp
- 1962-65 75 hp
- 1966-67 80 hp
- 1968 85 hp
- 1964-65 90 hp

Description & Operation

◆ See Figure 199

When the unit is shifted to the Forward position, an electric switch in the shift box closes the circuit to the forward electromagnetic coil in the gearcase. After the coil is energized, magnetism attracts and anchors the free end of the clutch spring to the flange of the clutch hub. The revolving gear causes the spring to wrap around the hub, creating a direct coupling with the propeller shaft.

Power is transmitted through the pinion gear, forward gear and propeller shaft to the propeller.

When the lower unit is shifted to the Reverse position, the reverse coil is energized, and the same sequence of events takes place. The reverse gear assembly is ALWAYS the one nearest the propeller.

The boat battery provides 12-volt power for operation. Therefore, all engines covered here are equipped with an alternator to maintain battery amperage and voltage for efficient operation of the shift mechanism.

When the key is in the **ON** position, power moves through the ignition switch to the switch in the shift box, and on to the lower unit.

The necessary wiring is routed from the dash to the shift box, then to the rear of the engine to a knife-disconnect fitting, and then down to the lower unit. The forward Shift wire is green and the reverse wire is blue. An easy way to remember the color code is green for go; forward that is.

Troubleshooting

  MODERATE

In order to prevent unnecessary service work, specific troubleshooting should be performed. The following steps present a logical sequence of tests and checks to pinpoint problems in an Electromatic lower unit.

1. Check the quantity of lubricant in the lower unit and top it off if necessary. The unit will not operate properly if a lubricant other than OMC Type C or Premium Lube is used. If any doubt exists as to the type of lubricant in the lower unit, drain the unit, refill with Type C (now known as Premium Lube material) and then check operation of the shift mechanism. At the same time the quantity of lubricant is being checked, observe the material carefully for any sign of water. Position a suitable container under the lower unit, and then remove the FILL plug and the VENT plug.

✳✳ CAUTION

Do not remove the plugs if the engine has been operated recently or if the unit has been sitting exposed to the hot sun. If one of the plugs should be removed when the lubricant is hot, the material will squirt out under considerable pressure.

2. Allow the lubricant to drain into the container. As the lubricant drains, catch some with your fingers, from time-to-time, and rub it between your thumb and finger to determine if any metal particles are present. If metal is detected in the lubricant, the unit must be completely disassembled, inspected, the cause of the problem determined, and then corrected.

3. Check for a broken shear pin by removing the propeller. First pull the cotter key, and then the propeller nut, drive pin and washer. Because the drive pin is not a tight fit, the propeller is able to move on the pin and

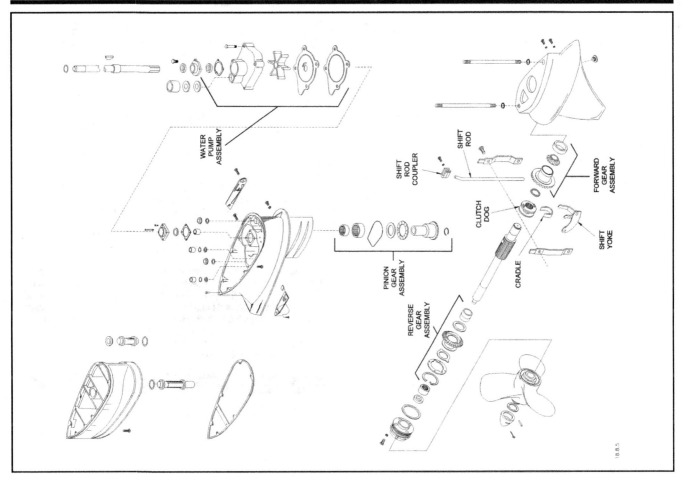

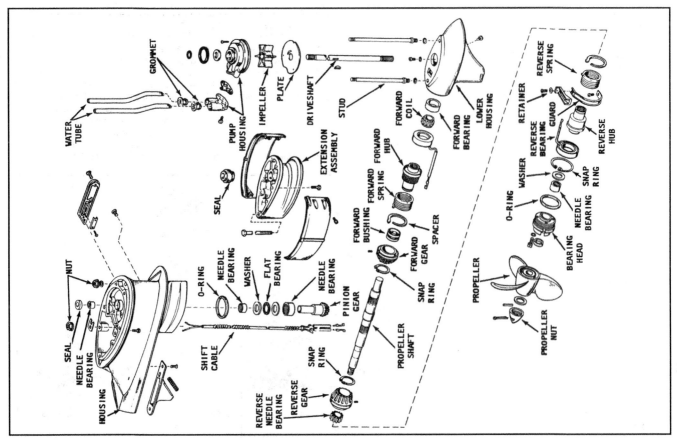

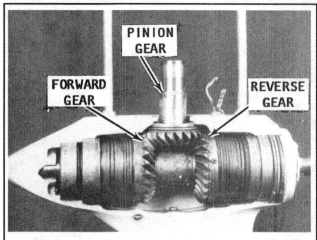

Fig. 199 The reverse gear is the one closest to the propeller

cause burrs on the hole. Propeller removal may be difficult because of these burrs. To overcome this problem, the propeller hub has 2 grooves running the full length of the hub. Hold the shaft from turning and then rotate the propeller 1/4 turn to position the grooves over the drive pin holes. The propeller can then be pulled straight off the shaft. After the propeller has been removed, file the drive pin holes on both sides of the shaft to remove the burrs.

4. Check the propeller and the rubber hub. See if the hub is shredded. If the propeller has been subjected to many strikes against underwater objects it could slip on its hub. For this size engine, in most cases, it is less expensive to purchase a new propeller instead of having it rebuilt.

5. Begin with a thorough check of the battery. Measure the gravity of the electrolyte in each cell by withdrawing only enough to lift the float. Take the reading at eye level. A fully charged battery cell should read 1.280; at half-charge, 1.210; and a dead battery will read only 1.150. If the electrolyte level is low, bring it up to full level with clean clear water. NEVER ADD ACID to a battery cell. If water is added, it is not possible to take an accurate reading until the battery has been charged for a few hours.

6. Check the total battery voltage for a full 12 volts. Clean any corrosion from, on, or, around the cables and terminals. Remove the cables; clean the posts until bright metal is visible. Scrape out the inside of the battery terminals, then connect and tighten them securely.

7. Turn the ignition switch to the **ON** position and observe the amperage reading on the dash ammeter. If an ammeter is not installed on the dash, one must be temporarily connected to the system for this test, by first removing the wire marked **BAT** from the key switch, and then connecting the ammeter in series with this wire and the key switch terminal marked **BAT**.

Check the current draw, if the draw exceeds 2.5 amps in either gear, disconnect the shift wires at the shift wire disconnects on the port side of the

engine. Again, check the current draw. A higher reading than 2.5 amps indicates a short in the wiring, in the shift switch, or in the shift box. If the readings are within acceptable limits, reconnect the shift wires at the engine, shift the unit into forward and then reverse gear. Check the current draw in each gear. A high-amp draw indicates a shorted wire to the lower unit, or a short in one of the coils.

8. A broken driveshaft from the power head to the lower unit indicates both forward and reverse gears were energized at the same time. Check the shift box, shift switch and the wiring to the lower unit.

9. Leave the wire marked **BAT** disconnected from the key switch for this test. Check the wiring leading to the ignition switch, and from the switch for 12 volts. If the reading is less than 12 volts, the key switch is defective and should be replaced.

10. Shift Box and Coil Tests. Disconnect the blue and green shift wires at the rear of the engine. Connect 1 lead of a voltmeter to the green wire of the shift box, and the other lead to a good ground. Turn the ignition switch to the **ON** position and move the shift lever into forward gear. The voltmeter must indicate 12 volts. Next, connect the voltmeter to the blue wire, shift into reverse gear, and the voltmeter should indicate 12 volts. If the voltmeter fails to indicate 12 volts during either one of these tests, the shift box requires service.

11. Leave the shift wires disconnected. Turn the ignition switch to the **OFF** position and move the shift lever to the Neutral position. Connect one lead of an ohmmeter to the green (forward) wire leading from the rear of the engine to the lower unit, and the other lead to a good ground. The ohmmeter should indicate from 4.5-6.5 ohms. Make the same test for reverse gear, the blue wire leading from the rear of the engine to the lower unit, and check for the same reading. If the ohmmeter fails to indicate the required resistance, a wire is broken, or the coil in the lower unit is shorted.

Lower Unit

REMOVAL & INSTALLATION

◆ See Figures 117, 118 and 200 thru 202

1. Drain the lower unit and remove the propeller.
2. Remove the lower unit.
3. Disconnect the cables from the battery. Remove the hood.
4. At the rear of the engine, slide the insulating sleeve back on the shift cable wires and disconnect the shift and engine shift terminals.
5. Remove the attaching hardware from the exhaust plate on the starboard side of the engine. Remove the 2 screws from the inner plate and clamp on the shift wire. Pull the shift wire down through the exhaust cover. Apply oil or soap onto the cable and remove the inner plate from the cable.
6. If the lower unit being serviced has the 6-inch extension above the lower unit, remove the screws from the bottom side of the extension. It is not necessary to remove the extension in order to remove the lower unit.
7. Remove the screws securing the lower unit to the exhaust housing or to the extension. Work the lower unit away from the exhaust housing or

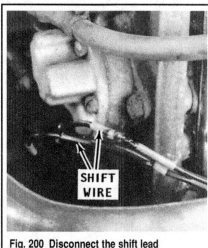

Fig. 200 Disconnect the shift lead connectors on the 40 hp models. . .

Fig. 201 . . .and all others

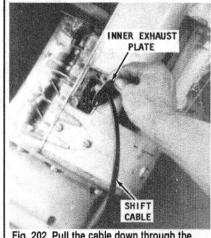

Fig. 202 Pull the cable down through the inner plate

extension. The water tubes may be stuck in the water pump. . .if difficulty is encountered in freeing the lower unit, force a wide blade chisel, stiff scraper, or other suitable tool, between the 2 surfaces and work each side of the lower unit away from the exhaust housing.

To Install:

8. Position the assembled lower unit under the exhaust housing.

9. Work the electric shift cable up through the exhaust housing and out the hole on the starboard side.

10. Slowly lift the lower unit into place with the driveshaft indexing with the crankshaft splines and the water tubes entering the exhaust housing grommets. It may be necessary to have an assistant rotate the flywheel ever so slowly clockwise and to pull the electric shift cable through while the lower unit is being mated with the exhaust housing. Rotating the flywheel will permit the driveshaft to index with the splines of the crankshaft.

11. Coat the threads of the attaching screws with sealer. After the mating surfaces of the exhaust housing and the lower unit have made contact, start the screws securing the 2 units together, tightening them alternately and evenly.

12. Place a new gasket in position on the exhaust housing.

■ **The exhaust cover has a small short section of pipe on one side. This pipe must face inward and downward and the shift cable passed through.**

13. Feed the electric shift cable through the inner exhaust cover until the cover is in place on the surface of the exhaust housing.

14. Secure the cover in place with the 2 attaching screws. Don't forget - the upper screw also holds the clamp used to secure the shift cable in place.

15. Position the outer exhaust cover and gasket in place on the starboard side and secure the cover with the attaching hardware.

16. Connect the shift wires to the harness at the back of the engine, Blue-to-Blue and Green-to-Green. Slide the rubber protective sleeves in place over the connectors.

17. Install the propeller and fill the unit with oil.

DISASSEMBLY & ASSEMBLY

◆ **See Figures 117, 118 and 203 thru 229**

1. Remove the lower unit.

2. Set the cavitation plate on the edge of the work bench or other suitable surface, and secure it firmly with a C-clamp. An alternate method is to cut a deep "V" in a 2 x 6 piece of wood and then slide the lower unit into the "V", resting it on the cavitation plate.

3. Remove the O-ring from the top of the driveshaft.

4. On 40 hp models, remove the water pump as detailed later in this section.

   a. Remove the pin, or key, from the driveshaft and then lift off the pump base plate.

   b. Lift the driveshaft up and out of the lower unit.

5. On all other models:

■ **The upper bearing and seal assembly is a very tight fit into the upper gear housing. Usually one or more of the mounting bolts will break during removal. After all the bolts have been removed, the bearing carrier will still be difficult to remove.**

   a. Clamp the driveshaft in a vise about midway. Tap on the housing with a softheaded mallet and pull the shaft from the housing. As the shaft is

Fig. 203 On the 40 hp, lift off the water pump base plate. . .

Fig. 204 . . .and then slide out the driveshaft

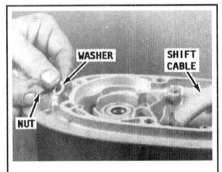

Fig. 205 Remove the stud nuts and then the cable retainer

Fig. 206 Remove the bearing cover if need be

Fig. 207 Loosen the attaching bolts. . .

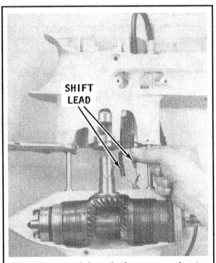

Fig. 208 . . .and drop the lower case about 3 in.

removed, a doughnut on the end of the shaft will pull the seal with it. After the seal and shaft have been removed, a check of the bearing can be made to determine if it must be replaced.

b. If the bearing is defective and unfit for further service, remove the 4 screws, and then work 2 small prybars under the bearing head and remove the bearing. The O-ring around the outside edge of the bearing must be replaced with the bearing.

6. Working on all models now, lay the lower unit flat on the bench because when the gear case nuts are removed, the lower section could drop to the floor and be severely damaged. Use a 9/16 in. deep well socket and remove the lower unit gearcase stud nuts, then the washers. Discard the nuts because they are the self-locking type and must not be used a second time.

7. Tap the front cone with a softheaded mallet to separate the lower housing from the upper housing a distance of about 3 in. After the upper and lower housings have been separated slightly, slide the terminal sleeves on the forward and reverse wires back and disconnect the quick-disconnects of the shift cable to the coil.

■ The upper housing has a bearing to accommodate the pinion gear. This bearing is comprised of 20 individual needles. Be careful not to lose any of the needles to the bearing set, as the needles are removed, or they fall out.

■ If the lower housing cannot be dislodged from the upper housing because the long bolts extending through the upper housing into the lower housing are badly corroded, a decision must be made. Something will be destroyed in order to proceed with the work. In almost all cases the sacrificed piece is the less expensive upper housing. Therefore, cut through the upper housing on both sides, as shown. The lower housing can then be separated from the upper housing. The studs can then be pressed out of the upper housing and the hacksaw cut welded shut and the housing returned to service. If the studs cannot be pressed free, the upper housing must be replaced.

8. Clamp the lower unit by the skeg in a vise equipped with soft jaws.

9. Remove the retainer screw and washer, then lift the nylon coil lead retainer from the lower housing and at the same time work the forward and reverse wires free of the retainer.

10. Remove the 4 screws in the bearing head. This is the cap the propeller shaft passes through. Notice how each screw has an O-ring behind the head? These O-rings must be in place during installation to maintain a water-tight unit.

11. With a small chisel and mallet, work the cap free of the propeller shaft. The chisel is to be worked on the cap, NOT in the groove between the cap and the lower housing. If the cap is damaged it may be replaced without great expense, but damage to the lower unit is bad news.

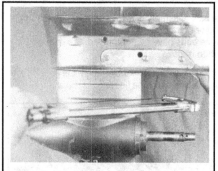

Fig. 209 You may have to cut through the upper case and bolts. . .

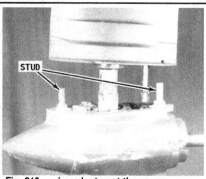

Fig. 210 . . .in order to get the cases separated

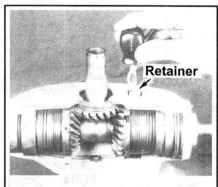

Fig. 211 Remove the small retainer screw

Fig. 212 Remove the bearing head bolts. . .

Fig. 213 . . .and then carefully pry the cap off

Fig. 214 Remove the snap ring. . .

Fig. 215 . . .and then the reverse coil

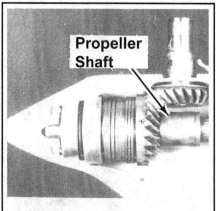

Fig. 216 Pull the propeller shaft out of the housing. . .

Fig. 217 . . .and then reach in and remove the pinion gear assembly

☀☀ WARNING

The next step involves a dangerous procedure and should be executed with care while wearing safety glasses. The retaining ring is under tremendous tension in the groove. If it should slip off the Truarc pliers, it will travel with incredible speed causing personal injury if it should strike a person. Therefore, continue to hold the ring and pliers firm after the ring is out of the groove and clear of the lower unit. Place the ring on the floor and hold it securely with one foot before releasing the grip on the pliers. An alternate method is to hold the ring inside a trash barrel, or other suitable container, before releasing the pliers.

12. Obtain a pair of Truarc pliers. Insert the tips of the pliers into the holes of the retaining ring. Now, carefully remove the retaining ring from the groove and gear case without allowing the pliers to slip. Release the grip on the pliers in the manner described in the Warning just before this step.

13. Install the 2 screws into the coil. These are the 2 screws that were removed from the gear case earlier. Rock the coil out and at the same time feed the reverse coil blue wire down into the recess in the lower unit.

Continue working the coil out and down the propeller shaft until it is clear. The thrust washer will come off with the coil.

14. Hold the propeller shaft firmly and pull it free from the lower unit housing. While the shaft is being pulled, feed the blue wire down into the recess of the lower unit. The reverse gear, spring, hub and babbitt thrust washer will come out with the propeller shaft.

15. Reach into the lower housing and remove the pinion gear.

16. Remove the forward gear, spring and hub assembly.

■ Rarely does the bearing in back of the coil require replacement. If the lower unit has contained water over a period of time, or sustained other damage, the bearing may be defective. The next 2 steps need only be performed IF the coil and the bearing are unfit for further service.

17. Notice the metal guard extending down the inside of the lower unit. This guard protects the forward coil wire. Identify the Phillips screw in the bottom of the lower housing and remove the screw and metal tab in the bottom of the lower housing.

18. Using a special coil remover (#79784 and Kit #384414), remove the forward coil. If these tools are not available, you may heat the outside edge

Fig. 218 Remove the forward gear assembly

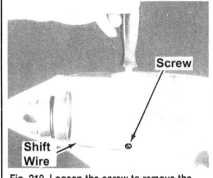

Fig. 219 Loosen the screw to remove the shift wire guard

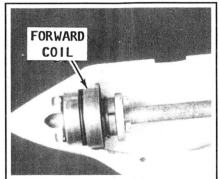

Fig. 220 You'll need a special tool to remove the forward coil. . .

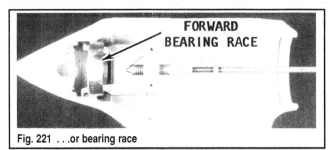

Fig. 221 . . .or bearing race

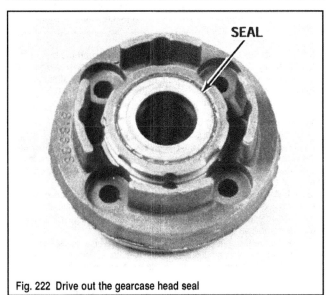

Fig. 222 Drive out the gearcase head seal

Fig. 223 Removing the upper pinion bearing. . .

Fig. 224 Remove the upper pinion bearing and components on the 40 hp. . .

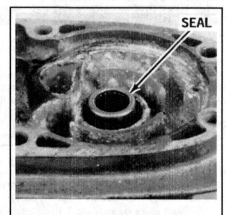

Fig. 225 . . .the needle bearing. . .

Fig. 226 . . .and then the seal

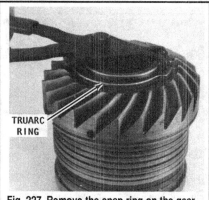

Fig. 227 Remove the snap ring on the gear assemblies. . .

Fig. 228 . . .and then separate from the hub

Fig. 229 Removing the spring assembly

of the lower housing with a torch. A considerable amount of heat will be required. After the housing is heated, grasp the studs with a gloved hand and at the same time tap the housing on a board. The forward coil and bearing race will be released from the housing. If the tools are available, the coil and race can be removed with no sweat.

19. Remove the seal from the gearcase head by driving it out from the backside with a punch and mallet. This method is very effective and adjacent parts are not harmed. Remove the O-ring from the bearing case head.

■ Professional mechanics have discovered that when the lower unit is being rebuilt, a new gearcase head should be purchased and installed. The new head will have a new bearing, O-ring and seal installed as an assembly.

20. On the upper housing of the lower unit, use a slide hammer with the adaptor fingers fitting behind the pinion gear bearing, and remove the bearing from the housing.

21. On 40 hp models, reach in and remove the washers and thrust bearing from the upper bearing recess. Pull out the upper driveshaft seal using #377565, or a standard seal removal tool and then remove the needle bearing with the special removal tool (#309916)

■ In most cases, it is not necessary to remove the lower driveshaft bearing. A couple of quick checks can be made to determine the condition of the bearing. One is to feel the driveshaft in the area where it passes through the bearing. If the shaft is smooth with no indication of roughness, the bearing is usually considered fit for further service. Another method is to insert a finger into the center of the bearing and determine if the needles roll freely and smoothly. If there is no evidence of binding or roughness, the bearing does not have to be removed and replaced. In other words, let a sleeping dog lie. If the bearing is unfit for further service, use a punch and hammer, and drive it free. Further damage to the bearing is of no consequence, because a new one is to be installed.

■ The forward and reverse gear assemblies look almost identical. However, there is a difference. The forward gear uses a babbitt bearing and the hub is knurled to provide more positive engagement. The reverse gear assembly uses needle bearings and the hub is smooth. The greatest percentage of motor operation is in forward gear with the reverse gear turning in the opposite direction. The needle bearings are used in the reverse gear hub for more satisfactory operation. Proper gear installation is extremely critical for satisfactory performance. The hubs are different and are easily identified. After the gears are assembled it is very easy in a moment of haste, to pick a gear assembly from the bench and install it in the wrong location. Therefore, after the service work on the gears has been completed and they are ready for installation, identify one with a felt pen marking or with a tag to ensure proper installation.

■ If the clutch has been slipping, replace the spring and the hub. It is very difficult to accurately determine what is considered *excessive* wear on these two items. Therefore, if the clutch has been slipping, the modest cost of the spring and hub is justified in eliminating this area as a possible source of shifting problems.

Forward & Reverse Gear Disassembly

22. Please put on a pair of safety glasses while working with the Truarc pliers in this step. Practice the same safety precautions given previously and in the Warning. Disassemble the forward and reverse gear assembly using a pair of Truarc pliers and carefully remove the snap ring from the hub on the front of the gear.

23. Lift the gear and spring assembly from the hub. As mentioned earlier, the forward gear has a babbitt bearing and the reverse gear a needle bearing arrangement. Exercise care when lifting the reverse gear from the hub so as not to lose any of the needles.

24. To remove the springs from the forward and reverse gears, first remove the Allen screws around the outside diameter of each gear. Next, pull the spring from the gear. Notice the nylon tapered washer installed under the spring; this washer must be installed properly to permit the spring to seat level in the gear. Repeat the last couple steps for the other gear.

To assemble:

◆ See Figures 230 thru 243

Forward Gear

25. Insert the spacer, with its key, into the slot in the cupped end of the Forward gear in such a way that it encircles the gear in the opposite direction to the normal winding of the spring coils.

26. Place the spring in the gear with the spring key beside the spacer key. Now, shift both keys to the side of the slot against which they will pull.

27. Coat the new setscrews with Loctite TL-242, and install them in the forward gear. Tighten the setscrews in rotation, to 30-35 inch lbs., beginning with the one nearest the spring. Bake the assembly in a 300° oven for 1/2 hour. If an oven is not available, apply Locquic Primer "T" to the screws before tightening them, and then allow them to cure for 4 hours.

28. The tolerance between the clutch hub and the bushing is very close. Carefully slide the gear and spring assembly onto the hub.

This next step can be dangerous. The snap ring is placed under tremendous tension with the Truarc pliers while it is being placed into the hub groove. Wear safety glasses and exercise care to prevent the snap ring from slipping out of the pliers. If the snap ring should slip out it would travel with incredible speed and cause personal injury if it struck a person.

29. Install the Truarc snap ring into the groove of the forward gear hub.
Reverse Gear

30. Assemble the reverse gear by first installing the nylon spacer into the cupped end of the reverse bevel gear, with its key in the slot of the bevel gear. The spacer must be positioned to encircle the cupped area in the opposite direction to the normal winding of the spring coil.

31. Install the spring with the key indexed in the slot beside the spacer key. Slide both keys against the side of the slot they will pull against when the operator selects reverse gear.

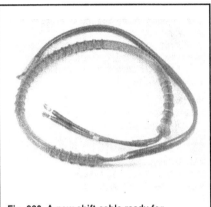

Fig. 230 A new shift cable ready for installation

Fig. 231 Position the spacer before installing the spring

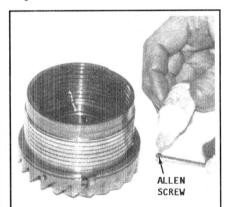

Fig. 232 Make sure that you use Loctite on the Allen screws

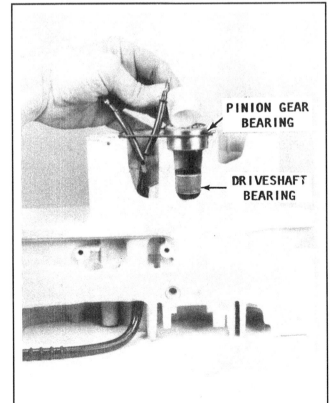

Fig. 233 You'll need special tools to install the lower driveshaft bearing

Fig. 234 Press the forward bearing into the race until it is fully seated

Fig. 235 Rest the pinion gear against the forward gear

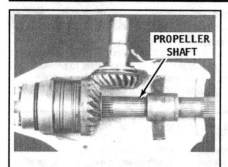

Fig. 236 Pull the pinion gear up while inserting the propeller shaft

Fig. 237 Always use a new O-ring on the bearing head screw and make sure that it seats itself in the recess of the screw head

Fig. 238 Make sure that the coils have not been grounded to the housing

32. Coat the Allen head cup-point setscrews with Loctite, and then install them to secure the spring to the bevel gear. Tighten the setscrews in rotation, to 30-35 inch lbs., beginning with the one nearest the spring. Bake the assembly in a 300° oven for 1/2 hour. If an oven is not available, apply Locquic Primer "T" to the screws before tightening them, and then allow them to cure for 4 hours.

33. Look at the ring at the top of the reverse hub used to retain the needle bearings. On some lower unit models, the needle bearings are held in a cage. On other units the needle bearings merely fit around the hub. The correct number of needles will fill the cage. This ring is the only visible difference between the forward and reverse hubs. Coat the needle bearings with grease or Vaseline to hold them in place. Always count and take care to be sure the total number of needle bearings are replaced during installation. Never use a grease to hold the needles in place which will not dissolve quickly, or the parts will be ruined due to lack of initial lubrication. After all needle bearings are all in place, carefully slide the gear and spring assembly down over the hub.

Fig. 239 There are 20 needle bearings in the outer race

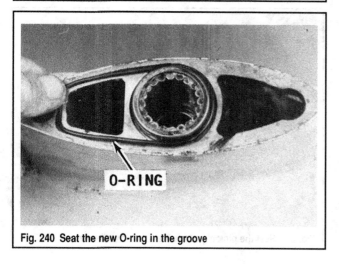

Fig. 240 Seat the new O-ring in the groove

34. A Truarc snap ring secures the gear to the hub; use a pair of Truarc snap ring pliers to install this snap ring with the chamfered edge against the bevel gear.

■ The next step involves assembling the bearing head, because either the seal or bearing, or both, were unfit for further service. In this one case, it may prove more profitable and efficient to purchase and install a new bearing head. The new unit will have the bearing and seal installed, ready for service. If the decision is made to rebuild the head, proceed with the next step, otherwise skip it.

Assembling the Bearing Head

35. If the bearing head seal was removed, press a new seal into the head. The seal can be installed using a block of wood and a mallet. If the bearing was removed, install a new bearing from the back side of the head. A special bearing installer tool (#308104 for the 40 hp, #308119 on all others) is required to install the bearing. Press against the lettered side of the bearing until the bearing is flush with the head surface. If the special bearing tool is not available, a new head must be purchased with the bearing installed. Install a new O-ring around the head seal.

Installing the Upper Bearing

36. On 40 hp models, press the upper bearing (if removed) into the housing using #378737 and #308099, making certain that the tools presses

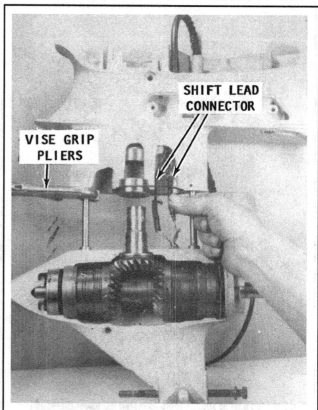

Fig. 241 Connect the shift wires before fully mating the two halves

against the lettered side of the bearing. Now, with the unit right side up, position the seal around the bearing and lightly (and carefully!) tap it into position.

37. On other models, if the lower driveshaft bearing was removed, now is the time to install the new bearing. Obtain the special tool(s) (#308102 and #378737 from Kit #378446) and install the bearing. Install the bearing race by tapping it on the lettered side until it is fully seated. The needle bearings will not be installed at this time.

Propeller Shaft Front Bearing Race

38. On all models now, install the front propeller shaft bearing race using a driveshaft guide plate and bearing race installer (#379248 from kit #384415, exc. on the 40 hp where you'll need #309033, #309932 and #379247). Place the tapered front portion of the lower unit on a block of wood and drive the race into the housing. If the special tools are not available, the outside diameter of the lower unit may be heated and the race tapped in with a wooden block and mallet.

39. Install the cone-shaped roller bearing into the bearing race; naturally, the tapered end of the bearing enters the race first.

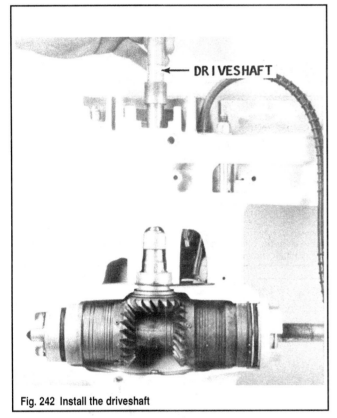

Fig. 242 Install the driveshaft

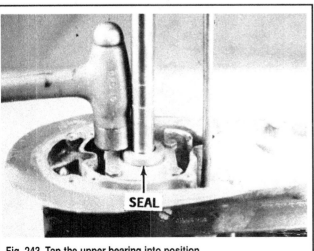

Fig. 243 Tap the upper bearing into position

40. Install the forward coil in the housing using a special tool (#379245 from kit #384415, or on the 40 hp - #379320). Feed the coil wire into the groove in the bottom of the housing.

41. Install the small washer, and screw securing the wire, in the recess in the bottom of the housing. The wire has a plastic covering; this covering must be located behind the metal guard. The guard clamp must fit alongside the housing. Install the Phillips screw in the bottom of the housing with part of the metal tab protruding out of the square hole on top of the lower unit and also with the wire coming out the hole.

42. Lower the forward gear, hub and spring assembly into the lower unit and over the forward coil.

43. Install the thrust bearing over the shank of the pinion gear, then the thrust washer. Position the pinion gear into its recess, resting against the forward gear.

44. Hold the pinion gear up, and at the same time, install the propeller shaft. As the propeller shaft is moved into the lower unit, turn the shaft slowly clockwise to allow the splines on the propeller shaft to engage in the forward gear.

45. Install the reverse gear with the splines of the reverse gear hub engaged with the splines of the propeller shaft.

46. Install the reverse coil and then feed the blue lead through the opening in the lower housing. The lead on the back side of the coil must be on top.

✳✳ WARNING

This next step can be dangerous. The snap ring is placed under tremendous tension with the Truarc pliers while it is being placed into the groove. Therefore, Wear safety glasses and exercise care to prevent the snap ring from slipping out of the pliers. If the snap ring should slip out, it would travel with incredible speed and cause personal injury if it struck a person.

47. Use a pair of Truarc pliers and install the Truarc snap ring into the groove just forward of the coil. Check to be sure the coil leads are correctly positioned and will not be damaged by any moving part in the lower unit. Double check to be sure the green lead is well protected by the metal guard.

48. Slide the thrust washer onto the propeller shaft and into the recess of the coil with the babbitt side of the washer facing the coil.

■ Alignment of the gearcase head holes with the holes in the reverse coil is very difficult because once the head is in place, the O-ring prevents the head from turning. Therefore, before installing the gearcase head, insert a guide pin into opposite corner holes in the reverse coil. Check to be sure the large O-ring is properly positioned in the groove in the gearcase head.

49. Install the gearcase head with the holes in the head indexing over the pins protruding from the reverse coil holes as described in the previous Note. Slide an O-ring onto each screw, and then dip the gearcase head screws in Perfect Seal No.4, or equivalent. Now, start the screws through the holes that do not contain pins. Do *not* tighten the screws at this time. Remove the 2 pins and start the other 2 screws. Tighten the 4 screws alternately and evenly. Rotate the propeller shaft and check to be sure it turns without excessive drag.

50. Set the wires into the retainer and secure them in place with the screw and washer. The washer will hold the screw.

51. Check to be sure the coils have not grounded to the lower unit. This can be accomplished by using an ohmmeter to check the forward and reverse coil for resistance. The meter should indicate 4.5-6.5 ohms resistance.

52. Turn the upper section of the lower unit upside down. Apply a heavy coating of OMC needle bearing lubricant into the pinion gear race. Install the 20 needle bearings in the outer race.

53. Coat the O-ring groove around the pinion gear bearing with sealer and install the O-ring into the groove.

54. Clamp the skeg of the lower section in a vise. Attach a pair of vise grip pliers about 3 in. (7.62 cm) up on one of the lower section studs. Lower the upper section down over the studs until the section rests on the vise grip pliers. Connect the wire cable, green-to-green and blue-to-blue. Pull the sleeves down over the connectors. Apply sealer around the surface of the lower section O-ring. Remove the vise grip pliers from the stud and lowly lower the upper housing down onto the lower section.

55. Check to be sure the shift cable retaining fork (actually a washer-type item) is over the hole as the stud passes through the retainer washer. Continue to lower the upper housing and at the same time work the shift

wires up into the cavity of the upper housing to prevent the wires from being pinched when the upper housing makes contact with the lower section. Place a washer over the other stud. Thread new nuts onto the studs.

■ **These nuts are the self-locking type and should not be used a second time. Once they have been tightened, the locking ability is lost and they should be replaced with new ones.**

56. Tighten the nuts alternately and evenly.

57. With an ohmmeter, check the forward and reverse wires for 4.5-6.5 ohms resistance. If the ohmmeter does not indicate the proper resistance on each lead, the lower unit must be separated again and the wiring checked for a short or broken wire.

58. Lower the driveshaft down into the lower unit. As the driveshaft is lowered, rotate the shaft slowly to permit the splines on the shaft to index with the pinion gear.

■ **If the upper bearing was not removed, proceed as follows:**

59. Cover the splines of the driveshaft with electrical tape, pulled as tight as possible to allow the driveshaft to pass through the bearing and to allow the seal to pass over the tape.

60. Slide a new seal down the driveshaft and into place in the bearing housing. Tap the bearing into place.

61. If the upper driveshaft bearing was removed, cover the splines on the driveshaft with electrical tape, pulled tight. Install the new seal into the bearing housing using a new gasket and after applying sealer under the gasket. Secure the housing with the 4 screws.

CLEANING & INSPECTION

◆ **See Figures 197, 198 and 244 thru 246**

Clean the parts with solvent and blow them dry with compressed air. Remove all seal and gasket material from mating surfaces. Blow all water and oil passages, and screw holes clean with air.

After the parts are clean and dry, apply a coating of lightweight engine oil to the bearings and bright mating surfaces of the shafts and gears as prevention against corrosion.

Inspect the shaft bearing surfaces, splines and keyways for wear and burrs. Check for evidence of an inner bearing race turning on the shaft. Check for damaged threads.

Measure the run-out on all shafts to reveal any bent condition. If necessary, turn the shaft in a lathe as a check for out-of-round.

Carefully check the inside and outside surfaces of the gearcase, housing and covers for cracks. Pay special attention to the areas around screw and shaft holes. Verify that all traces of old gasket material have been removed from mating surfaces.

Check O-ring grooves for sharp edges which could cut a new seal.

Inspect gear teeth and shaft holes for wear and burrs.

Hold the center race of each bearing and turn the outer race to be sure it turns freely without any evidence of rough spots or binding. Inspect the rollers and balls for any sign of pits or flat spots.

Inspect the outside diameter of the outer races and the inside diameter of the inner races for evidence of turning in the housing or on the shaft. Any sign of discoloration or scores is evidence of overheating.

Check the thrust washers for wear and distortion. If they do not have uniform thickness and lay flat, they must be replaced.

Inspect all springs for tension, distortion, corrosion or discoloration.

Inspect the shift cables for broken leads or damaged insulation. Use an ohmmeter and test for continuity. Use the ohmmeter to test the coil resistance which should indicate 4.5-6.5 ohms. Check the coil leads for breaks and damaged insulation.

Check the water pick-up screen on the upper housing. Blow air through the screen to dislodge any debris. Clean the area behind the screen.

Inspect the propeller for cracks, gouged, bent or broken blades.

Replace all bent, worn, corroded or damaged parts. Burrs can be removed with a file.

Always install new O-rings, gaskets, and seals during assembling and installation to prevent leaks.

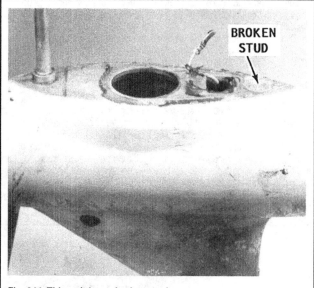

Fig. 244 This unit has a broken stud

Fig. 245 The thrust bearing on the left is new and the one on the right has a broken cage

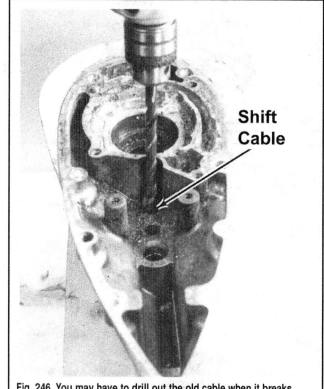

Fig. 246 You may have to drill out the old cable when it breaks

Functional Check

Perform a functional check of the completed work by mounting the engine in a test tank, in a body of water, or with a flush attachment connected to the lower unit. If the flush attachment is used, never operate the engine above an idle speed, because the no-load condition on the propeller would allow the engine to runaway resulting in serious damage or destruction of the engine.

✳✳ CAUTION

Water must circulate through the lower unit to the engine any time the engine is run to prevent damage to the water pump in the lower unit. Just 5 seconds without water will damage the water pump.

Start the engine and observe the telltale flow of water from idle relief in the exhaust housing. The water pump installation work is verified.

TYPE VII - ELECTROMATIC LOWER UNIT (100 HP)

◆ See Figures 247 and 248

• 1966-68 100 hp

The Type VII lower unit is almost identical to the Type VI unit except for much heavier construction of the housing and component parts. Although the service procedures do differ from the Type VI, the description and all troubleshooting steps are the same.

If a flush attachment is connected to the lower unit, very little water will be visible from the idle relief port.

Shift the engine into the three gears and check for smoothness of operation and satisfactory performance.

Water Pump

REMOVAL & INSTALLATION

■ Please refer to the Type V Water Pump section for all water pump procedures. You will note that in the pictures shown in that section, there is a shift rod that is not utilized on your lower unit - disregard this rod as all other images and procedures are identical between the two units.

Troubleshooting

■ Please refer to the Type VI section for all Troubleshooting procedures on this unit.

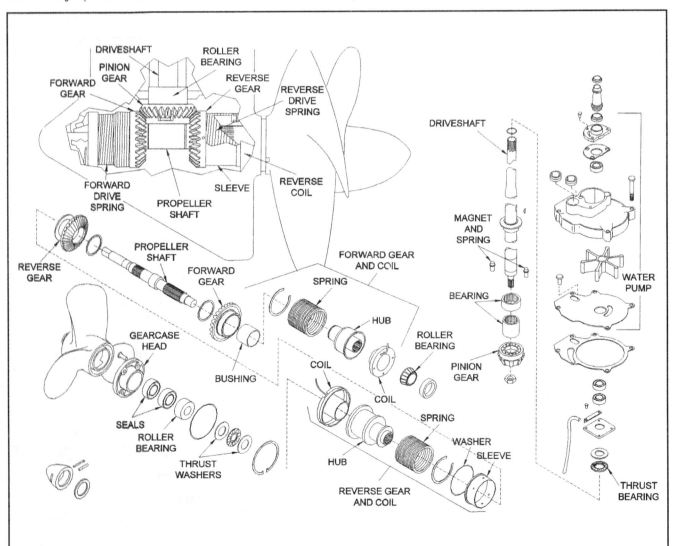

Fig. 247 Exploded view of the Type IV Electromatic lower unit (inside)

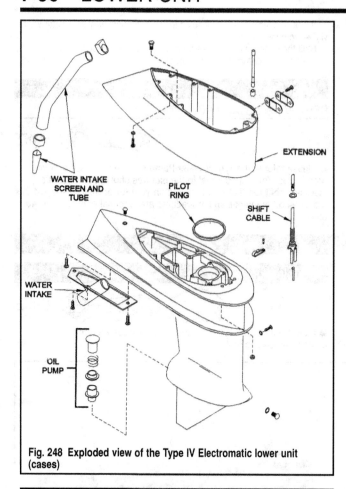

Fig. 248 Exploded view of the Type IV Electromatic lower unit (cases)

Lower Unit

REMOVAL & INSTALLATION

MODERATE

◆ See Figures 249, 250 and 251

1. Drain the lower unit and remove the propeller.
2. At the rear of the engine, disconnect the blue and green wires leading to the lower unit.
3. Free the shift cable by first bending back the clip located under the thermostat housing at the rear of the engine.
4. Attach about a 5 in. (152 cm) piece of wire to the green and blue wires leading to the lower unit. Tape the connections to allow the wire to feed

through the exhaust housing as the lower unit is separated from the exhaust housing. These wires will prove very useful during installation of the lower unit to the exhaust housing.

■ **The extension will separate with the lower unit when it is separated from the exhaust housing.**

5. Remove the 6 nuts, 3 on each side of the anti-cavitation plate. Remove the 2 bolts extending up through the lower unit into the exhaust housing. Work to separate the lower unit from the exhaust housing. If the engine is still attached to the boat, it may be necessary to lower the bow in order to obtain enough clearance for the lower unit. It may also be necessary to wedge a tool between the surfaces of the lower unit and the exhaust housing in order to jar the 2 housings apart.

To Install:

6. Position the assembled lower unit under the exhaust housing.
7. Connect the jumper wire from the exhaust housing, left when the lower unit was removed, to the blue and green wires from the lower unit. Tape the connections smoothly to allow the wires to pass through the exhaust housing.
8. As the lower unit is moved toward the exhaust housing, pull on the jumper wires to pull the lower unit wires through. Slowly lift the lower unit into place with the driveshaft indexing with the crankshaft splines and the water tubes entering the exhaust housing grommets. It may be necessary to have an assistant rotate the flywheel ever so slowly clockwise and to pull the electric shift cable through while the lower unit is being mated with the exhaust housing. Rotating the flywheel will permit the driveshaft to index with the splines of the crankshaft.
9. Coat the threads of the attaching screws with sealer. After the mating surfaces of the exhaust housing and the lower unit have made contact, start the screws securing the 2 units together. Tighten the screws alternately and evenly.
10. Remove the jumper wire that was attached to shift lead in order to feed the lead through the exhaust housing. Connect the shift wires to the harness at the back of the engine, blue-to-blue and green-to-green. Slide the rubber protective sleeves in place over the connectors.
11. Install the propeller and fill the unit with oil.

DISASSEMBLY & ASSEMBLY

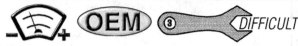
OEM ③ DIFFICULT

◆ See 227 thru 229, 247, 248 and 252 thru 270

1. Remove the lower unit and the water pump.
2. Turn the lower unit upside down and remove the screws and the water pick-up.
3. On units with the long shaft extensions, there is 1 more bolt inside where the water pick-up was removed. Remove this bolt on these units.
4. Raise the 6 in. extension about 4 in. (10.2 cm). Clamp a pair of vise grip pliers onto the driveshaft to hold the extension above the lower unit housing. Pull back the sleeves on the green and blue wires and disconnect the wiring leading into the lower unit. Notice the ring on the extension, used to properly locate the extension during installation.

Fig. 249 Disconnect the blue and green wires. . .

Fig. 250 . . .and then splice in an extension to each

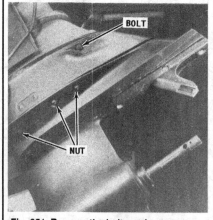

Fig. 251 Remove the bolts and nuts

Thrust Plate Removal

5. Remove the 4 screws and then the thrust plate, bracket and oil line extending into the lower unit. Pay particular attention to how the oil line sets just in front of the opening to the upper thrust washer.

6. Remove the 4 screws securing the gearcase head to the lower unit and then slide the gearcase head off the propeller shaft. Observe the reverse coil wire at the top of the coil - do not remove the coil at this time.

7. Remove the 2 thrust washers and the thrust bearing from the shaft. Bear in mind the order of the washers, and the bearing, as an aid to installation. Notice the difference in the washer thickness. The thinner washer must be installed first against the reverse coil.

✳✳ WARNING

The next step involves a dangerous procedure and should be executed with care while wearing safety glasses. The retaining ring is under tremendous tension in the groove and while it is being removed. If it should slip off the Truarc pliers, it will travel with incredible speed causing personal injury if it should strike a person. Therefore, continue to hold the ring and pliers firm after the ring is out of the groove and clear of the lower unit. Place the ring on the floor and hold it securely with one foot before releasing the grip on the pliers. An alternate method is to hold the ring inside a trash barrel, or other suitable container, before releasing the pliers.

8. Remove the Truarc ring securing the reverse coil in place.

9. Carefully pull the reverse coil lead down and out the hole prior to removing the reverse coil in the next step.

10. Obtain the special tool (#380658) and thread the tool into the reverse coil; and then pull the coil with the tool. If the special tool is not available, use a long bolt with a nut in the manner of a slide hammer.

11. The propeller shaft has a cam. In order to remove the shaft, the high side of the cam must be downward to prevent damaging the oil pump in the bottom of the lower unit when the shaft is removed. Alright, to position the cam downward, hold the propeller shaft with feeling (as if you were turning the knob on a safe) and rotate the shaft, ever so slowly, until you "feel" the cam depressing the pump. Once in this position, grasp the propeller shaft firmly, and then jerk hard to pull the shaft free of the lower unit. The propeller shaft, reverse gear, spring and hub assembly will come out with the shaft.

12. Reach in and lift the oil pump free of the lower unit.

13. A pinion gear and locknut hold the driveshaft in the gearcase. Hold the nut with a 1-1/16 in. wrench and at the same time, use a torque bar and special tool (#312752) to turn the driveshaft. Discard the locknut after it is free because its locking ability is ruined once the nut has been removed.

14. Carefully lift the driveshaft out of the lower unit with one hand and catch the pinion gear with the other hand. Remove the pinion gear. Remove the thrust washer and the thrust bearing from the driveshaft.

15. Reach inside the housing and remove the pinion roller bearings. If the lower unit has been operated, when the level of the gear oil is low, the upper bearing on the driveshaft will be destroyed.

16. Reach inside the housing, insert 2 fingers into the center of the forward gear assembly, and then withdraw the assembly.

■ **The coil and the rear bearing do not have to be removed unless they are unfit for further service.**

17. Use a slide hammer (#380658) to remove the forward coil and wire lead. Take care to pull the coil out evenly to prevent it from binding in the housing.

18. Reach in and remove the forward bearing installed behind the coil.

19. Use a special tool (#380657) and a slide hammer to remove the forward bearing cup from the lower gearcase.

20. If the upper bearing is unfit for further service, you'll need to obtain a special tool (#380657). Insert the special tool, "puller", into the bearing and pull the bearing free of the lower unit.

21. If the lower bearing sleeve is unfit for further service, it must be removed using special tool #380659. A special handle (#311885), must be

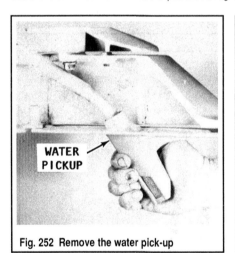

Fig. 252 Remove the water pick-up

Fig. 253 Long shaft extension units have 1 additional bolt

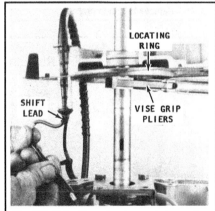

Fig. 254 Raise the extension and clamp vise grips on the driveshaft

Fig. 255 Pay close attention to how the oil tube is situated

Fig. 256 Remove the gearcase head...

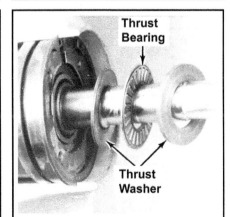

Fig. 257 ...and then the thrust washers and bearing

used with the #380659 tool. Insert the tool down through the top of the housing and drive the sleeve free of the lower unit.

■ The forward and reverse gear assemblies look almost identical. However, there is a difference. The forward gear uses a babbitt bearing and the hub is knurled to provide more positive engagement. The reverse gear assembly uses needle bearings and the hub is smooth. The greatest percentage of motor operation is in forward gear with the reverse gear turning in the opposite direction. The needle bearings are used in the reverse gear hub for more satisfactory operation. Proper gear installation is extremely critical for satisfactory performance. The hubs are different and are easily identified. After the gears are assembled it is very easy, in a moment of haste, to pick a gear assembly from the bench and install it in the wrong location. Therefore, after the service work on the gears has been completed and they are ready for installation, identify one with a felt pen marking or with a tag to ensure proper installation.

■ If the clutch has been slipping, replace the spring and the hub. It is very difficult to accurately determine what is considered "excessive" wear on these 2 items. Therefore, if the clutch has been slipping, the modest cost of the spring and hub is justified in eliminating this area as a possible source of shifting problems.

Forward and Reverse Gear Disassembly

22. Wear a pair of safety glasses while working with the Truarc pliers in this step. Practice the same precautions given previously. Disassemble the forward and reverse gear assembly using a pair of Truarc pliers and carefully remove the snap ring from the hub on the front of the gear.

23. Lift the gear and spring assembly from the hub. As mentioned earlier in the Note, the forward gear has a babbitt bearing and the reverse gear a needle bearing arrangement. Exercise care when lifting the reverse gear from the hub, not to lose any of the needles.

24. To remove the springs from the forward and reverse gears, first remove the Allen screws around the outside diameter of each gear. Next, pull

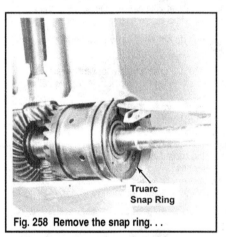

Fig. 258 Remove the snap ring. . .

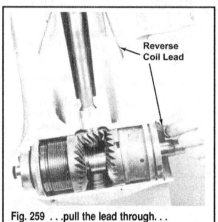

Fig. 259 . . .pull the lead through. . .

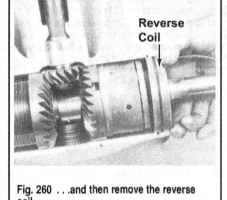

Fig. 260 . . .and then remove the reverse coil

Fig. 261 Now you can pull out the reverse gear assembly

Fig. 262 Remove the oil pump

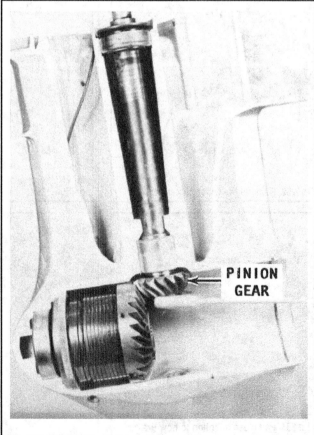

Fig. 263 Remove the pinion nut. . .

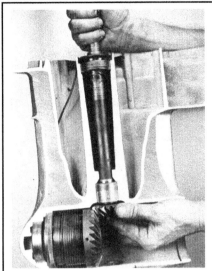

Fig. 264 . . .and then lift up on the driveshaft to remove the pinion gear

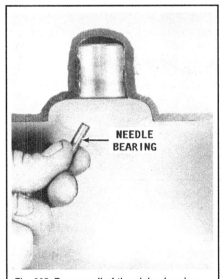

Fig. 265 Remove all of the pinion bearing needles

Fig. 266 Pull out the forward gear. . .

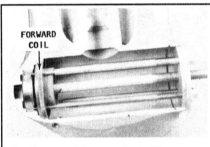

Fig. 267 . . .and then use the tool to remove the forward coil

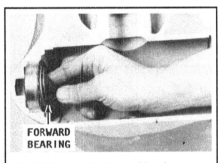

Fig. 268 Remove the forward bearing. . .

Fig. 269 . . .and then pull out the race

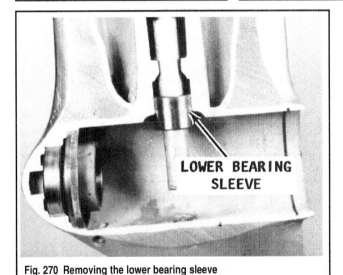

Fig. 270 Removing the lower bearing sleeve

the spring from the gear. Notice the nylon tapered washer installed under the spring. This washer MUST be installed properly to permit the spring to seat level in the gear. Forty (40) needle bearings are used on the reverse clutch hub. The forward hub has a bronze bearing - NEVER interchange these bearings.

To assemble:

◆ See Figures 230, 231, 232 and 271 thru 289

25. Insert the spacer, with its key, into the slot in the cupped end of the forward gear in such a way that it encircles the gear in the opposite direction

to the normal winding of the spring coils. The forward gear of this lower unit does not use a sleeve. Place the spring in the gear with the spring key beside the spacer key. Now, shift both keys to the side of the slot against which they will pull.

26. Coat the new setscrews with Loctite TL-24-2 and install them in the forward gear. Tighten the set screws, in rotation, to 30-35 inch lbs.; beginning with the one nearest the spring. Bake the assembly in a 300 deg. oven for 1/2 hour. If an oven is not available, apply Locquic Primer "T" to the screws before tightening them, and then allow them to cure for 4 hours.

27. The tolerance between the clutch hub and the bushing is very close. Carefully slide the gear and spring assembly onto the hub. Install the Truarc snap ring with the chamfered (beveled) side of the retaining ring facing toward the gear. This arrangement places the square side of the Truarc ring in the groove of the hub.

Reverse Gear

28. Assemble the reverse gear by first installing the spacer into the cupped end of the reverse bevel gear, with its key in the slot of the bevel gear. The spacer must be positioned to encircle the cupped area in the opposite direction to the normal winding of the spring coil.

29. Install the spring with the key indexed in the slot beside the spacer key. Slide both keys against the side of the slot they will pull against when reverse gear is selected.

30. Coat the Allen head cup-point set screws with Loctite and then install them to secure the spring to the bevel gear. Tighten the set screws, in rotation, to 30-35 inch lbs., beginning with the one nearest the spring. Bake the assembly in a 300° oven for 1/2 hour. If an oven is not available, apply Locquic Primer "T" to the screws before tightening them, and then allow them to cure for 4 hours.

31. Slide the sleeve over the spring with the flanged end pointing away from the bevel gear. Take note of the ring at the top of the reverse hub used to retain the 40 needle bearings - this ring is the only visible difference between the forward and reverse hubs.

32. Coat the needle bearings with grease or Vaseline to hold them in place. Always count and take care to be sure the total number of needle

bearings are replaced during installation. Never use a grease to hold the needles in place which will not dissolve quickly, or the parts will be ruined due to lack of initial lubrication. After the forty needle bearings are all in place, carefully slide the gear-and-spring assembly down over the hub.

✳✳ WARNING

The next step involves a dangerous procedure and should be executed with care while wearing safety glasses. The retaining ring is under tremendous tension while in the grip of the Truarc pliers and after it is in place in the groove. If the retaining ring should slip off the Truarc pliers, it will travel with incredible speed causing personal injury if it should strike a person. Therefore, continue to hold the ring and pliers firmly after the ring is in the groove, and then release the pliers carefully.

33. A Truarc snap ring secures the bevel gear to the hub. Use a pair of Truarc No.6 snap ring pliers to install this snap ring with the chamfered edge against the bevel gear.

Forward Bearing

34. Tilt the gearcase slightly, and then attach a Drive Handle (#311880) to a Bearing Cup Installer (#311872), and then start to seat the cup at the front of the gearcase. Use extra care not to move the bearing cup.

35. After the cup is properly seated, install the forward bearing into the bearing cup.

36. Before coil installation, mark the front of the coil opposite the wire WITH A LEAD PENCIL (nothing else) to enable you to see the mark as the coil is installed into the rear of the housing.

37. Mark a line to indicate the hole location in the gear housing. When the coil is properly installed, these marks must be aligned.

38. The following special tools must be obtained to properly install the coil: OMC #380691, #311759 and #311880. Now, install the forward coil into the housing and at the same time feed the shift wire through the hole in the rear of the housing. As you push the coil inward, pull on the wire from inside the top of the lower unit.

39. Next, align the mark you made on the coil with the mark on the housing. Insert the coil installer tool into the lower housing and into the coil. Press the forward coil inward and at the same time pull on the shift wire until the coil is fully seated in the housing. Remove the installer tool.

40. After the coil is installed, use an ohmmeter to check that the wire behind the coil is not shorted. The coil resistance should be 4.5-6.5 ohms. An infinite reading indicates an open circuit, such as a broken wire. A reading of less than 4.5 ohms indicates a short circuit, such as a bare wire making contact with the case. Replace the coil if the resistance is too high because the clutch will slip.

Lower Pinion Gear Bearing & Upper Driveshaft Bearing Installation

41. Obtain the special OMC tools #311876, #380758 and #312019. Insert the lower bearing sleeve in the housing and then install the adaptor (#380758) inside the race. Screw the long rod bolt through the top into the

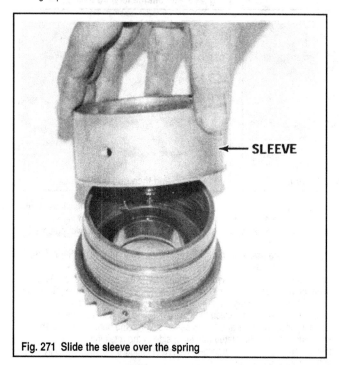

Fig. 271 Slide the sleeve over the spring

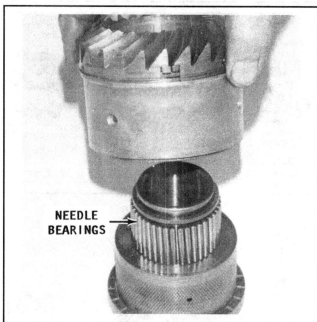

Fig. 272 Always make sure that you have the correct number of needle bearings

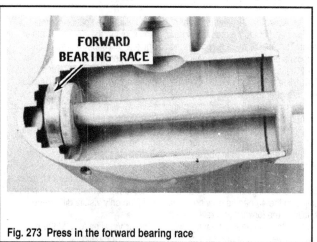

Fig. 273 Press in the forward bearing race

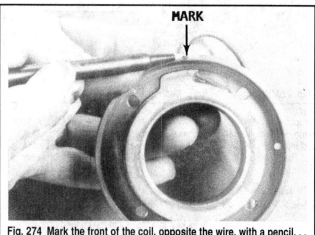

Fig. 274 Mark the front of the coil, opposite the wire, with a pencil. . .

Fig. 275 . . .and then mark inside the housing. . .

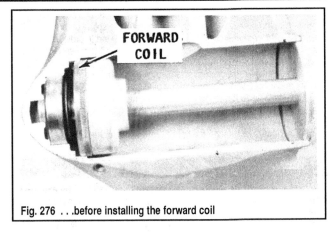

Fig. 276 . . .before installing the forward coil

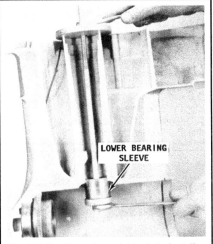

Fig. 277 You'll need special tools to install the lower bearing sleeve. . .

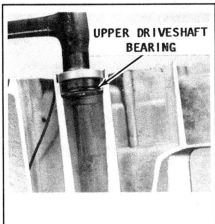

Fig. 278 . . .but you can just tap in the upper shaft bearing

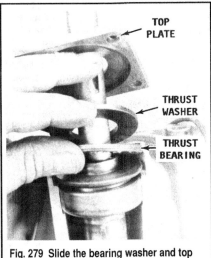

Fig. 279 Slide the bearing washer and top plate over the driveshaft

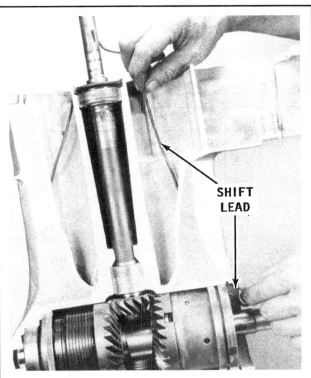

Fig. 280 Feed the lead through the hole

special tool (#312019). Insert the upper driveshaft bearing into place, and then use the special tool (#311876) with a nut on the tool stud. Tighten the nut, and at the same time hold the lower installer with a wrench to prevent it from turning. Continue to tighten the nut until the lower bearing race and the upper bearing are in place in the recesses. Remove the special tools.

42. Install the upper driveshaft bearing into the lower unit housing. The proper size socket may be used. Press against the lettered side of the bearing.

43. Place the forward gear assembly in the gearcase, as far forward as possible and with the gear towards the propeller.

44. Coat the race at the bottom of the pinion driveshaft bore with needle bearing grease No. 378642. Install the 18 needle bearings. The grease will hold them in place.

45. Hold the pinion gear in place inside the lower unit with one hand and with the other, install the driveshaft. Rotate the driveshaft as it moved into place to allow the splines on the shaft to index with the splines in the pinion gear. Take care not to dislodge the needles in the lower pinion gear bearing as the drive shaft is lowered and engaged with the pinion gear.

46. Install a new locknut onto the end of the driveshaft finger-tight. Never use an old locknut because its locking ability is lost once it is removed. Tighten the locknut to 70-80 ft. lbs. using a Driveshaft Holding Socket Tool (#312752) and a 1-1/16 in. open-end wrench.

47. Slide the thrust bearing onto the driveshaft, then the thrust washer. Slide the top plate down the driveshaft.

48. Slide the oil line through the retainer and then down through the opening just behind the driveshaft and down into the hole in the lower unit housing. Set the bracket on top of the top plate and secure the bracket and plate to the lower unit housing with the attaching screws.

49. Install the oil pump into the lower unit housing. After the pump is in place, work it up and down to be sure it functions properly, without evidence of binding.

50. Insert the propeller shaft through the forward gear and coil into the bearing, with the propeller shaft spline engaged in the spline of the forward clutch hub, AND with the cam on the shaft facing upward to provide clearance for the shaft to pass over the oil pump. Check to be sure the cam is positioned over the oil pump.

51. Place the reverse gear assembly over the propeller shaft with the gear teeth engaged with the teeth of the pinion gear.

52. Carefully slide the reverse coil over the propeller shaft and into the gearcase past the ring groove with the lead at the top. This lead must not be twisted or kinked as it leaves the coil.

53. Feed the lead through the hole, then reach into the cavity at the top rear of the gearcase, and pull the lead through.

✳✳ WARNING

The next step involves a dangerous procedure and should be executed with care while wearing safety glasses. The retaining ring is under tremendous tension while in the grip of the Truarc pliers and after it is in place in the groove. If the retaining ring should slip off the Truarc pliers, it will travel with incredible speed causing personal injury if it should strike a person. Therefore, continue to hold the ring and pliers firmly after the ring is in the groove, and then release the pliers carefully.

54. Using a pair of Truarc pliers, replace the snap ring into the recess and against the reverse coil. Take care not to damage the reverse coil wire. The snap ring opening must be at the top of the housing to allow clearance for the shift wire to route into the hole.

55. Check to be sure the coil and leads were not damaged during installation by making a resistance test. Set the ohmmeter selector switch to the LO OHMS position, and then zero the meter. Make contact with the black

meter lead to a clean metal surface of the gear case and with the red ohmmeter lead to the connector of the green forward coil lead. The meter reading must indicate 4.5-6.5 ohms.

56. Conduct the same test on the reverse coil. If either coil is damaged, the unit must be repaired or replaced.

57. Install the snap ring with the chamfered side facing in, and the open part of the ring to the top, to allow the reverse coil wire room to pass into the upper hole.

58. Install the thinner thrust washer over the propeller shaft, and then the thrust bearing, and the remaining thrust washer. This arrangement is necessary to position the reverse gear properly and to control propeller shaft end play.

Bearing Head Seal Installation

■ Several special tools are required to install the seals and bearing in the bearing head. The bearing must be removed before the seals can be installed. If a new bearing head is purchased, the seals, bearing and O-ring will be included and installed. This may be a cheaper route than buying the special tools and attempting to install these items.

The seals are installed back-to-back with special grease between the 2 surfaces. One seal prevents lubricant in the lower unit from escaping, and the other seal prevents water from entering.

59. Special tools (#311877 and #311880) are required. If the special tools are not available or purchase cannot be made, a proper size socket can be used. Coat the outside surface of both seals with OMC Seal Compound, or equivalent. Drive the first seal into place using the special tools. Coat the surface of the seal with Triple Guard Grease. Install the second seal back-to-back against the first seal, using the special tools.

60. Obtain special tools #311869 and #311880 and drive the bearing into the bearing head from the lettered side of the bearing.

Fig. 281 Make sure the snap ring opening is at the top

Fig. 282 Drive the seals into place. . .

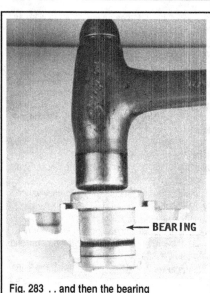

Fig. 283 . . and then the bearing

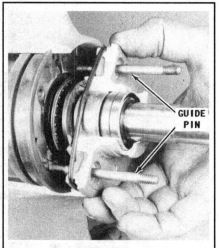

Fig. 284 It's tough aligning the holes in the gear head with those in the coil, so use the pins

Fig. 285 Install the screws after removing the pins

Fig. 286 Connecting the reverse coil lead

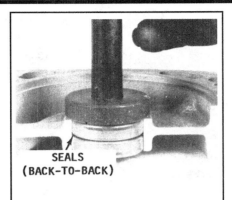

Fig. 287 Installing the lower seals in the extension

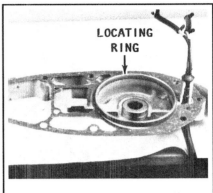

Fig. 288 Seat the locating ring in the extension

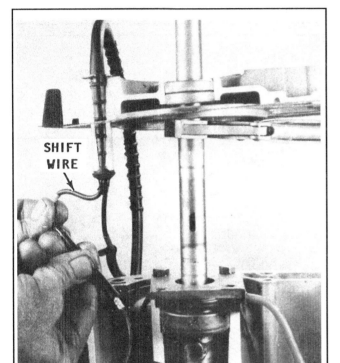

Fig. 289 Make sure you reconnect the leads before mating the two case halves

61. Alignment of the holes in the gear head with the holes in the coil is very difficult because once the head is in place, the O-ring prevents the head from turning. Therefore, before installing the gear case head, insert a guide pin into opposite corner holes of the reverse coil. Coat the O-ring with light-weight oil or gasket seal compound, as an aid to installation. Now, install the gearcase head with the holes in the head sliding over the pins in the reverse coil holes. Dip the retaining screws in sealing compound, and then install the two retaining screws in the holes without the pins.

62. Remove the pins, install the other 2 screws, and then tighten the 4 retaining screws to 5-7 ft. lbs. Rotate the propeller shaft and check to be sure there is no evidence of drag.

63. Restrain the reverse coil wire with the clamp and screw at the top of the gearcase. Install and push the 2 spring-and-magnet assemblies into the cavities at the top of the lower gear case until they are firmly seated.

Lower Seal Installation - Extension

64. Coat the outside surface of the seals with OMC Seal Compound. The seals are installed back-to-back with Triple Guard Grease between the 2 surfaces. The inner seal prevents lubricant in the lower unit from escaping and the outer seal prevents water from entering. Install the seals back-to-back with Triple Guard Grease between the 2 surfaces.

65. Slide the locating ring down the driveshaft and seat it into place in the extension. Place a new gasket in position on the extension.

66. Coat the mating surfaces of the lower unit and the extension with OMC 1000 Sealer, or equivalent. Tape the splines of the upper driveshaft. Lower the extension down over the driveshaft, but stop about 5 in. (12.7 cm) from the lower unit surface. Connect the shift wires, green-to-green and blue-to-blue. Slide the protective sleeves over the connections. Lower the extension until the mating surfaces make contact.

67. Install the 2 Phillips screws alongside the shift wires and the 2 bolts on the opposite side extending alongside the water tube. Install the 1 bolt extending up from the lower unit into the extension. This bolt is not used with the 2 in. (5.1 cm) extension.

68. From the bottom side of the lower unit, install the bolt into the extension. This bolt is only used with the 5 in. (12.7 cm) extension.

69. Place a new grommet into the water pick-up tube opening. Insert the water pick-up tube through the grommet and then work the grommet into the groove. Install the water pick-up plate containing a grommet in place. A bit of oil on the grommet will ease installation. Secure the pick-up plate with the attaching screws.

CLEANING & INSPECTION

◆ See Figures 237 and 248

■ Please refer to the Type VI section for all cleaning and inspection procedures.

Water Pump

REMOVAL & INSTALLATION

◆ See Figures 290, 291 and 292

1. Remove the lower unit.
2. Set the cavitation plate on the edge of the work bench or other suitable surface, and secure it firmly with a C-clamp. An alternate method is to cut a deep "V" in a 2 x 6 in. piece of wood, and then slide the lower unit into the "V" resting it on the cavitation plate.
3. Remove the O-ring from the top of the driveshaft.
4. Remove the extension tube.
5. Remove the 4 screws through the water pump housing, and then pull the water pump housing, impeller and Woodruff key from the driveshaft.
6. Remove the short screw securing the water pump base plate, and then remove the base plate.
7. Remove the 2 Phillips head screws from just behind the water pump. Remove the 2 bolts from alongside the water tube. Pull the water tube free of the lower unit.

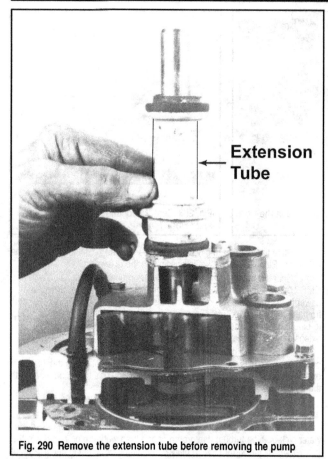

Fig. 290 Remove the extension tube before removing the pump

To Install:

8. Coat the mating surface of the lower unit with OMC 1000 sealer, or equivalent. Place a new water pump gasket in position. Apply more sealer on the water pump gasket.

9. Lower the water pump plate down the driveshaft. Secure the plate in place with the 1 short screw on the port side of the plate. Do not tighten this screw at this time.

10. Install the Woodruff key in the driveshaft and then slide the water pump impeller down the driveshaft with the slot in the impeller indexed over the Woodruff key. Set the impeller onto the water pump plate.

11. Coat the inside surface of the water pump housing with light-weight oil. Slide the water pump housing down over the driveshaft and impeller. At the same time rotate the driveshaft clockwise to allow the impeller fins to assume their normal position.

12. Continue working the housing downward until it is seated on top of the plate. Install and tighten the 4 screws securing the water pump housing to the lower unit housing.

13. Install the seal ring on top of the water pump. The ends of the extension tube are of different sizes. The word **UP** is stamped on the tube to

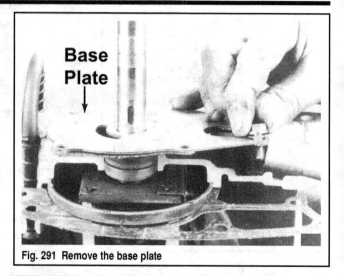

Fig. 291 Remove the base plate

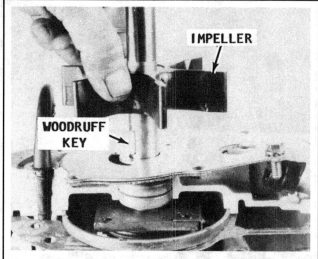

Fig. 292 Slide on the impeller

indicate which end of the tube is the upper end. Install the extension tube down over the driveshaft with the proper end of the tube at the top. Check to be sure the extension tube has a new seal installed.

14. Install a new O-ring onto the end of the driveshaft.

■ Clean both lower unit water tubes with sandpaper. These tubes should be clean and shiny as an aid to mating the lower unit to the exhaust housing. If these tubes are not thoroughly clean, great difficulty may be encountered in mating the lower unit with the exhaust housing. After the water tubes have been cleaned, apply a light coating of oil or lubricant to the outside surface of the tubes. Apply a light coating of oil to the electric shift cable.

TYPE VIII - ELECTRIC SHIFT LOWER UNITS, 2 SOLENOIDS

Description & Operation

◆ See Figures 293 and 294

The dual-solenoid electric shift gearcase was normally found on:
• 1971-72 50 hp models (2 cyl)
• 1968-69 55 hp
• 1970-71 60 hp
• 1972 65 hp
• 1969-72 85 hp
• 1971-72 100 hp
• 1969-70 115 hp
• 1971-72 125 hp

The lower unit covered in this section is a 3 shift position, hydraulic activated, solenoid controlled, propeller exhaust unit. A hydraulic pump

mounted in the forward portion of the lower unit provides the force required to shift the unit. Two solenoids installed in the lower unit, above the pump, control and operate the pump valve. The pump valve directs the hydraulic force to place the clutch dog in the desired position for neutral, forward, or reverse gear position. One solenoid controls the valve for the neutral position. Both solenoids control the valve for the reverse position. When neither solenoid is activated, the unit is at rest in the forward gear position.

In simple terms, something must be done (a solenoid activated and hydraulic pressure applied) to move the unit into neutral or reverse gear position. If no action is taken (shift mechanism at rest) the unit is in the forward gear position.

A full 12-volt is required to activate the solenoids. This means shifting is not possible if the battery should become low for any number of reasons. A potentially dangerous condition could exist because the unit could not be

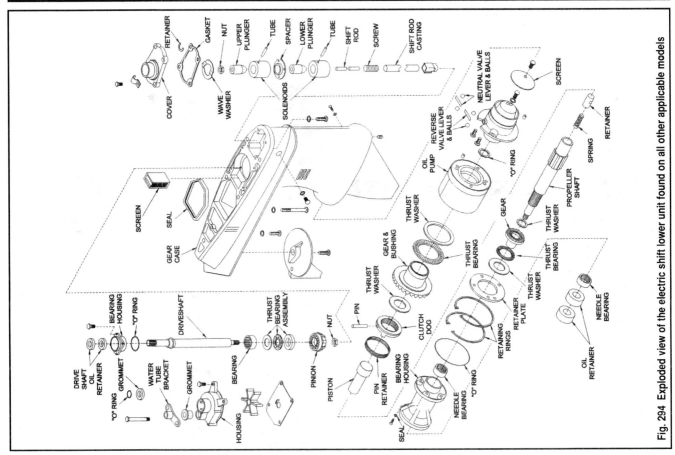

Fig. 294 Exploded view of the electric shift lower unit found on all other applicable models

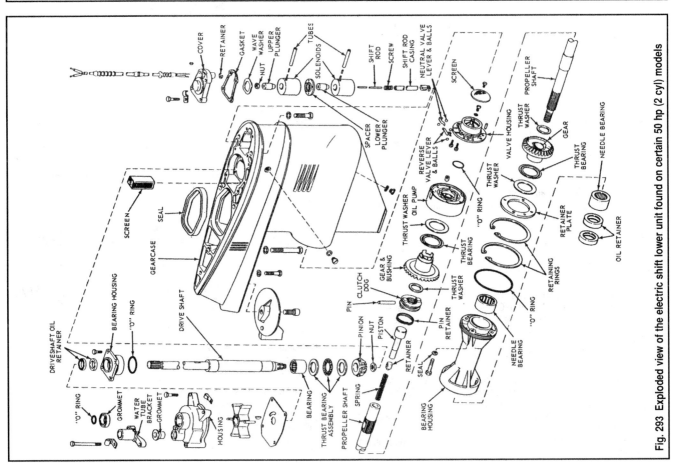

Fig. 293 Exploded view of the electric shift lower unit found on certain 50 hp (2 cyl) models

taken out of forward gear. Therefore, the only way to stop forward boat movement would be to shut the engine down. A low battery would also mean the electric starter motor would fail to crank the engine properly for engine start. Such a condition would require hand starting in an emergency, if a second battery were not available.

The lower unit houses the driveshaft and pinion gear, the forward and reverse driven gears, the propeller shaft, clutch dog, hydraulic pump, 2 solenoids, and the necessary shims, bearings, and associated parts to make it all work properly.

■ **A useful piece of information to remember is that the green wire carries current for the neutral position; and the green and blue wires carry current for the reverse gear operation.**

Troubleshooting

PRELIMINARY CHECKS

Whenever the lower unit fails to shift properly the first place to check is the condition of the battery. Determine if the battery contains a full charge. Check the condition of the battery terminals, the battery leads to the engine, and the electrical connections.

The second area to check is the quantity and quality of the lubricant in the lower unit. If the lubricant level is low, contaminated with water, or is broken down because of overuse, the shift mechanism may be affected. Water in the lower unit is very bad news for a number of reasons, particularly when the lower unit contains electrical or hydraulic components. Electrical parts short out and hydraulic units will not function with water in the system.

Before making any tests, remove the propeller. Check the propeller carefully to determine if the hub has been slipping and giving a false indication the unit is not in gear. If there is any doubt, the propeller should be taken to a shop properly equipped for testing, before the time and expense of disassembling the lower unit is undertaken. The expense of the propeller testing and possible rebuild is justified.

BY SYMPTOM

The following troubleshooting procedures are presented on the assumption the battery, including its connections, the lower unit lubricant, and the propeller have all been checked and found to be satisfactory.

Lower Unit Locked

◆ See Figure 36

Determine if the problem is in the powerhead or in the lower unit. Attempt to rotate the flywheel. If the flywheel can be moved even slightly in either direction, the problem is most likely in the lower unit. If it is not possible to rotate the flywheel, the problem is a "frozen" powerhead. To absolutely verify the powerhead is "frozen", separate the lower unit from the exhaust housing and then again attempt to rotate the flywheel. If the attempt is successful, the problem is definitely in the lower unit. If the attempt to rotate the flywheel, with the lower unit removed, still fails, a "frozen" powerhead is verified.

Unit Fails to Shift (Neutral, Forward, or Reverse)

Disconnect the green and blue electrical wires from the lower unit at the engine in order to test the shift circuitry.

Voltmeter Tests

◆ See Figure 295

Separate the green and blue wires at the engine by first sliding the sleeve back, and then making the disconnect. Connect 1 lead of the voltmeter to the green wire to the dash, and the other lead to a good ground on the engine. Turn the ignition key to the **ON** position. With the shift box handle in the forward position, the voltmeter should indicate NO voltage. Move the test lead from the green wire to the blue wire to the dash. With the shift lever still in the forward position, the voltmeter should indicate NO voltage.

Move the shift lever to the Neutral position. With the voltmeter still connected to the blue wire, NO voltage should be indicated. Move the test lead to the green wire. Voltage should be indicated.

Move the shift lever to the Reverse position. Voltage should be indicated on the green wire AND on the blue wire.

If the desired results are not obtained on any of these tests, the problem is in the shift box switch or the wiring under the dashboard.

Fig. 295 Performing the voltmeter tests

Ohmmeter Test

◆ See Figure 296

Set the ohmmeter to the low scale. Connect 1 lead to the green wire to the lower unit, and the other lead to a good ground. The meter should indicate 5-7 ohms. Connect the meter to the blue wire to the lower unit and ground. The meter should again indicate from 5-7 ohms.

■ **If the unit fails the voltmeter and ohmmeter tests just outlined, the only course of action is to disassemble the lower unit to determine and correct the problem.**

Lower Unit

REMOVAL & INSTALLATION

◆ See Figures 293, 294 and 296 thru 301

■ **If water is discovered in the lower unit and the propeller shaft seal is damaged and requires replacement, the lower unit does not have to be removed in order to accomplish the work.**

The bearing carrier can be removed and the seal replaced without disassembling the lower unit. However, such a procedure is not considered good shop practice, but merely a quick-fix. If water has entered the lower unit, the unit really should be disassembled and a detailed check made to determine if any other seals, bearings, bearing races, O-rings or other parts have been rendered unfit for further service by the water.

Fig. 296 Performing the ohmmeter tests

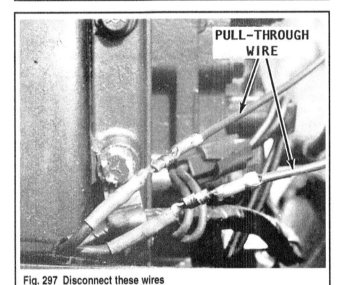

Fig. 297 Disconnect these wires

1. If necessary, remove the propeller and/or drain the lower unit. Neither are necessary for removal of the gearcase assembly, however both are recommended the first for safety and the second to inspect the gearcase oil and help determine potential condition.

2. Disconnect and ground the spark plug wires. Slide back the insulators on the shift wires at the engine. Disconnect the blue wire from the blue and the green wire from the green, at the engine.

■ **Obtain a piece of electrical wire about 5 ft. long. Connect 1 end to the green wire and the other end to the blue wire. Tape the connections. Now, when the lower unit is separated from the exhaust housing the ends of the wire will feed down through the exhaust housing. When the lower unit is free, disconnect the wire ends from the blue and green wires and leave the wire loop in the exhaust housing. When it is time to bring the lower unit together with the exhaust housing, the wire ends will be connected again and the blue and green wires easily pulled back up through the exhaust housing. No sweat! Alright, on with the work.**

3. Scribe a mark on the trim tab and a matching mark on the lower unit to ensure the trim tab will installed in the same position from which it is removed. Remove the attaching hardware, and then remove the trim tab.

4. Use a 1/2 in. socket with a short extension and remove the bolt from inside the trim tab cavity. Remove the 5/8 in. countersunk bolt located just ahead of the trim tab position.

5. Remove the four 9/16 in. bolts, 2 on each side, securing the lower unit to the exhaust housing.

6. Work the lower unit free of the exhaust housing. If the unit is still mounted on a boat, tilt the engine forward to gain clearance between the lower unit and the deck (floor, ground, whatever). Be careful to withdraw the lower unit straight away from the exhaust housing to prevent bending the driveshaft. Once the lower unit is free of the exhaust housing, stop and disconnect the wires installed as described in the "Helpful Word" earlier. Leave the wire loop in the exhaust housing as an aid during installation.

To Install:

7. Check to be sure the water tubes are clean, smooth, and free of any corrosion. Coat the water pick-up tubes and grommets with lubricant as an aid to installation.

8. Check to be sure the spark plug wires are disconnected from the spark plugs (for safety).

9. Bring the lower unit housing together with the exhaust housing, and at the same time, guide the water tube into the rubber grommet of the water pump.

10. Connect the ends of the wire left in the exhaust housing during removal to the blue and green wires from the lower unit. Tape the connections, and then oil the shift cable as an aid to slipping the cable through the exhaust housing.

11. Continue to bring the two units together and at the same time: pull the wires through the exhaust housing; guide the water pick-up tubes into the rubber grommet of the water pump; and, rotate the flywheel slowly to permit

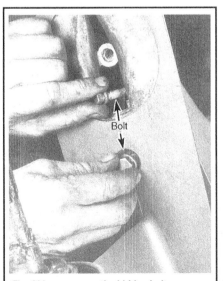

Fig. 298 Matchmark the trim tab to the case before removing it. . .

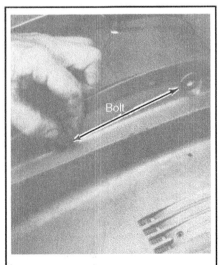

Fig. 300 . . .and then remove the remaining bolts

Fig. 299 . . .remove the hidden bolt. . .

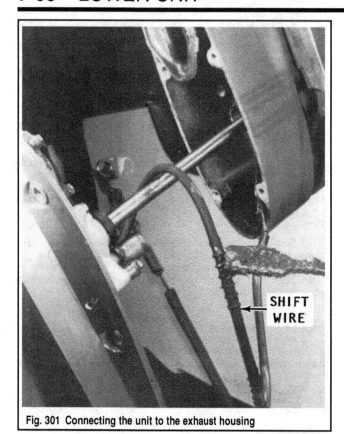

Fig. 301 Connecting the unit to the exhaust housing

the splines of the drives haft to index with the splines of the crankshaft. This may sound like it is necessary to do 4 things at the same time, and so it is. Therefore, make an earnest attempt to secure the services of an assistant for this task.

12. After the surfaces of the lower unit and exhaust housing are close, dip the attaching bolts in Johnson/Evinrude Sealer and then start them in place. Two bolts are used on each side of the two housings.

13. Install the retaining bolt in the recess of the trim tab and another bolt in the Cavitation plate. Tighten the bolts ALTERNATELY and EVENLY.

14. Install the trim tab with the mark made on the tab during disassembling aligned with the mark made on the lower unit housing.

15. Disconnect and remove the "pull" wire used to feed the electrical wires UP through the exhaust housing.

16. Obtain an ohmmeter. Ground the meter and again check the green and blue wires for continuity. The meter should indicate 5-7 ohms. Connect the green wire to the green wire and the blue wire to the blue wire. After the connections have been made, slide the sleeve down over the connections. A bit of oil on the wires will allow the sleeves to slide more easily.

17. If drained, refill the lower unit with oil.

18. Install the propeller.

DISASSEMBLY

◆ See Figures 293, 294 and 302 thru 321

1. Remove the gearcase as detailed earlier in this section.

2. If necessary for access, remove the water pump as detailed earlier in this section. Besides, isn't it time for impeller replacement as long as you're here?

3. Remove the shift solenoid cover located just aft of the water pump position. Do not to lose the wavy washer installed under the cover.

4. Grasp the upper (green) shift solenoid and withdraw the solenoids and shift rod from the lower unit cavity as an assembly.

■ It is not recommended to attempt service of this assembly. If troubleshooting has been performed and the determination made that the unit or any part is faulty, the ONLY satisfactory solution is to purchase and install a new assembly.

Bearing Carrier, Bearings & Seals

It appears as though this procedure can be done with the gearcase assembly still installed IF your only intention is to service the seals in the carrier. However, if driveshaft and propshaft removal is going to be necessary, the gearcase must be removed. Further, on these models the forward gear is held in place by the pinion and is not pressed onto the propeller shaft, so the propshaft itself can be serviced without removing the gearcase, though installation can be tricky with the gearcase and shift assembly still installed.

5. Remove the four 5/16 in. bolts from inside the bearing carrier. Notice how each bolt has an O-ring seal. These O-rings should be replaced each time the bolts are removed. Also observe the word **UP** embossed into the metal rim of some bearing carriers. This word must face UP in relation to the lower unit during installation. Clean the surface and if the word **UP** does not show, the position of the carrier during installation is not important.

6. Remove the bearing carrier using one of the methods described in the following paragraphs.

Several models of bearing carriers are used on the lower units covered in this section. The bearing carriers are a very tight fit into the lower unit opening. Therefore, it is not uncommon to apply heat to the outside surface of the lower unit with a torch at the same time the puller is being worked to remove the carrier. DO NOT overheat the lower unit.

One model carrier has 2 threaded holes on the end of the carrier. These threads permit the installation of 2 long bolts. These bolts will then allow the use of a flywheel puller to remove the bearing carrier.

Another model does not have the threaded screw holes. To remove this type bearing carrier, a special puller with arms must be used. The arms are hooked onto the carrier web area, and then the carrier removed.

✳✳ WARNING

The next step involves a dangerous procedure and should be executed with care while wearing safety glasses. The retaining rings are under tremendous tension in the groove and while they are being removed. If a ring should slip off the Truarc pliers, it will travel with incredible speed causing personal injury if it should strike a person. Therefore, continue to hold the ring and pliers firm after the ring is out of the groove and clear of the lower unit. Place the ring on the floor and hold it securely with 1 foot before releasing the grip on the pliers. An alternate method is to hold the ring inside a trash barrel, or other suitable container, before releasing the pliers.

7. Obtain a pair of Truarc pliers and insert the tips of the pliers into the holes of the first retaining ring. Now, *carefully* remove the retaining ring from the groove and gearcase without allowing the pliers to slip. Release the grip on the pliers in the manner described in the above Warning. Remove the second retaining ring in the same manner. The rings are identical and either one may be installed first.

8. Remove the retainer plate. As the plate is removed, notice which surface is facing into the housing, as an aid during installation.

9. The bearings in the carrier need not be removed unless they are unfit for further service. Insert a finger end rotate the bearing. Check for rough spots or binding. Inspect the bearing for signs of corrosion or other types of damage. If the bearings must be replaced, proceed with the next step.

10. Use a seal remover to remove the 2 back-to-back seals or clamp the carrier in a. vise and use a pry bar to pop each seal out.

11. Use a drift punch to drive the bearings free of the carrier. The bearings are being removed because they are unfit for service; therefore, additional damage is of no consequence.

Propeller Shaft

12. Grasp the propeller shaft firmly and withdraw it from the lower unit. The retainer plate, thrust washer, thrust bearing, reverse gear, reverse gear small thrust washer, and the clutch dog, will come out with the shaft.

On rare occasions, especially if water has been allowed to enter the lower unit, it may not be possible to withdraw the propeller shaft as described in the previous step. The shaft may be "frozen" in the hydraulic pump due to corrosion.

If efforts to remove the propeller shaft after the bearing carrier has been removed are unsuccessful, all is not lost:

13. Thread an adaptor to the end of the propeller shaft. The adaptor has internal threads at both ends.

14. Attach a slide hammer to the adaptor; operate the hammer; and remove the propeller shaft.

■ Bad news! Using the slide hammer, under these conditions, to remove the propeller shaft, will usually result in some internal part being damaged as the shaft is withdrawn. However, the cost of replacing the damaged part is reasonable considering the seriousness of a "frozen" shaft and getting it out.

15. Notice the spring retainer on the outside surface of the clutch dog. Use a small screwdriver and work one end of the spring up onto the shoulder of the clutch dog.

16. Continue working the spring out of the groove until it is free. Be careful not to distort the spring.

17. Place one end of the propeller shaft on the bench and push the pin free of the clutch dog.

18. Raise the propeller end of the shaft upward and the piston, retainer, and spring, will come free of the shaft.

19. Notice how the small end of the retainer fits into the spring. Also notice the hole in the retainer. Turing installation, this hole must align with the hole in the propeller shaft and clutch dog to allow the pin to pass through. Slide the clutch dog free of the propeller shaft.

Pinion Gear Removal

■ A special tool (#316612) is required to turn the driveshaft in order to remove the pinion gear nut.

20. Obtain the special tool and slip it over the end of the driveshaft with the splines of the tool indexed with the splines on the driveshaft.

21. Hold the pinion gear nuts with the proper size wrench and at the same time rotate the driveshaft, with the special tool and wrench COUNTERCLOCKWISE until the nut is free.

22. If the special tool is not available, clamp the driveshaft in a vise equipped with soft jaws, in an area below the splines but not in the water pump impeller area.

23. Now, with the proper size wrench on the pinion gear nut, rotate the complete lower unit COUNTERCLOCKWISE until the nut is free. This procedure will probably require the driveshaft to be loosened in the vise several times and re-clamped in order to affect rotation of the lower unit and wrench.

Driveshaft Removal

24. Remove the 4 bolts from the top of the lower unit securing the bearing housing.

25. Carefully pry the bearing housing upward away from the lower unit, then slide it free of the driveshaft.

26. An alternate method is to again clamp the driveshaft in a vise equipped with soft jaws. Use a soft headed mallet and tap on the top side of the bearing housing. This action will jar the housing loose from the lower unit. Continue tapping with the mallet and the bearing housing, O-rings, shims, thrust washer, thrust bearing, and the driveshaft will all breakaway from the lower unit and may be removed as an assembly.

Fig. 302 Remove the solenoid cover

Fig. 303 Notice that the bolts utilize an O-ring

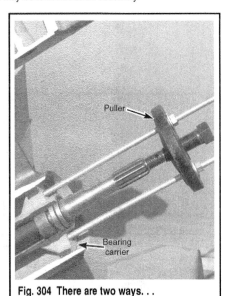

Fig. 304 There are two ways. . .

Fig. 305 . . .to remove the bearing carrier

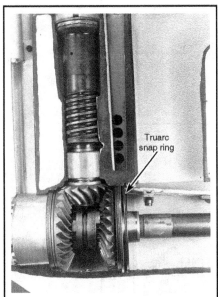

Fig. 306 Removing the retaining ring

Fig. 307 Removing the retaining plate

27. Remove the pinion gear from the lower unit cavity.

28. Remove the forward gear from the hydraulic pump.

Hydraulic Pump

29. Obtain 2 long rods with 1/4 in. x 20 threads on both ends. Thread the rods into the hydraulic pump housing.

30. Attach a slide hammer to the rods and secure it with a nut on the end of each rod. Check to be sure the slide hammer is installed onto the rods evenly to allow an even pull on the pump. If the slide hammer is not installed to the rods properly, the pump may become tightly wedged in the lower unit. Operate the slide hammer and pull the hydraulic pump free. If the pump should happen to become lodged in the lower unit, stop operating the slide hammer *immediately*. Tap the hydraulic pump back into place in the lower unit and start the removal procedure over.

31. Remove the screw from the center of the screen on the back side of the pump. Remove the screen.

32. Remove the screws securing the valve housing to the pump, and then lift the housing free of the pump.

33. Lift the 2 gears out off the pump housing and hold them just as they were removed. Check the face of each gear for an indent mark (a dot, dimple, or similar identification). The identification mark will indicate how the gear must face in the housing. Make a note of how the mark faces, outward or inward, to ensure the gears will be installed properly in the same position from, which they were removed.

Lower Driveshaft Bearing

■ This bearing cannot be removed/installed without the aid of a special tool. Therefore, DO NOT attempt to remove this bearing unless it is unfit for further service. To check the bearing, first use a flashlight and inspect it for corrosion or other damage. Insert a finger into the bearing, and then check for "rough" spots or binding while rotating it.

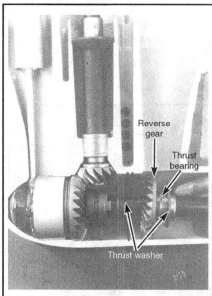

Fig. 308 Reach in and remove the reverse gear

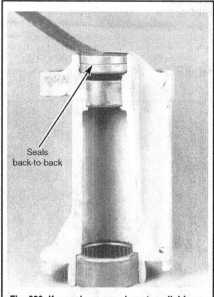

Fig. 309 If a seal remover is not available, use a pry bar

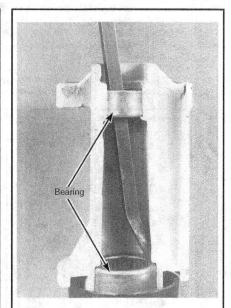

Fig. 310 It will also work for the bearings

Fig. 311 Removing the clutch dog spring

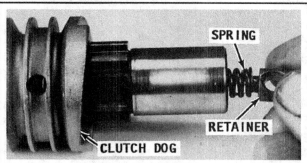

Fig. 312 Removing the retainer and spring

Fig. 313 Removing the pinion gear and nut

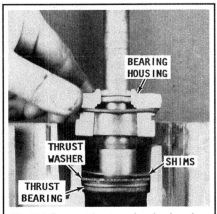

Fig. 314 Remove the upper bearing housing before pulling out the driveshaft

Fig. 315 Pull out the forward gear/bearing

Fig. 316 Pull out the hydraulic pump

34. On 1971 50 hp models ONLY, use a punch or similar tool and drive the lower driveshaft bearing out of position and into the lower unit cavity.

✳✳ CAUTION

On all other models, the bearing must actually be pulled upward to come free. NEVER make an attempt to "drive" it down and out or the lip in the lower unit holding the bearing will be broken off. The lower unit housing would have to be replaced.

35. On all models but the 1971 50 hp, remove the Allen screw from the water pickup slots in the starboard side of the lower unit housing, this screw secures the bearing in place and MUST be removed before an attempt is made to remove the bearing. Obtain the special tool (#385546). Use the special tool and "pull" the bearing from the lower unit.

Fig. 317 Remove the screen screw from the back of the pump

Fig. 318 Separate the pump and housing

CLEANING & INSPECTION

◆ **See Figures 293 and 294**

1. Wash all, except electrical, parts in solvent and dry them with compressed air.

2. Discard all O-rings and seals that have been removed. A new seal kit for this lower unit is normally available from the local dealer. The kit should contain the necessary seals and O-rings to restore the lower unit to service.

3. Inspect all splines on shafts and in gears for wear, rounded edges, corrosion, and damage.

4. Carefully check the driveshaft and the propeller shaft to verify they are straight and true without any sign of damage. A complete check must be performed by turning the shaft in a lathe. This is only necessary if there is evidence to suspect the shaft is not true.

5. Check the water pump housing for corrosion on the inside and verify the impeller and base plate are in good condition. Actually, good shop practice dictates to rebuild or replace the water pump each time the lower unit is disassembled. The small cost is rewarded with "peace of mind" and satisfactory service.

6. Inspect the lower unit housing for nicks, dents, corrosion, or other signs of damage. Nicks may be removed with No. 120 and No. 180 emery cloth. Make a special effort to ensure all old gasket material has been removed and mating surfaces are clean and smooth.

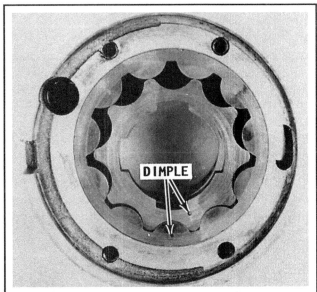

Fig. 319 Make sure you take note of the dimples before removing the gears

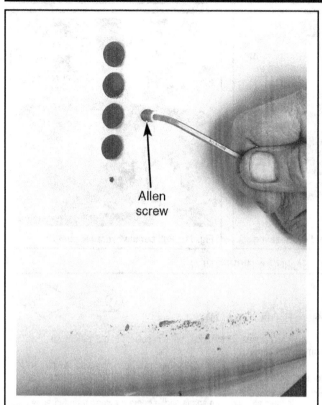

Fig. 320 Remove the Allen screw (if equipped) that holds the bearing in place. . .

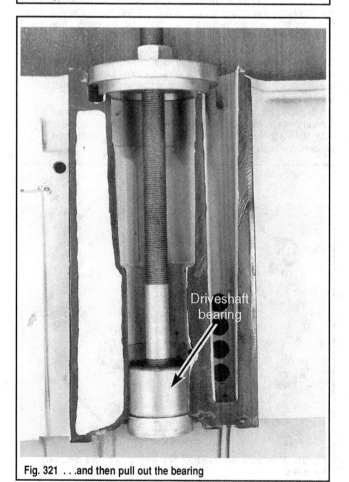

Fig. 321 . . .and then pull out the bearing

7. Inspect the water passages in the lower unit to be sure they are clean. The screen may be removed and cleaned.

8. Check the gears and clutch dog to be sure the ears are not rounded. If doubt exists as to the part performing satisfactorily, it should be replaced.

9. Inspect the bearings for "rough" spots, binding, and signs of corrosion or damage.

10. Test the neutral and reverse solenoids with an ohmmeter. A reading of 5-7 ohms is normal and indicates the solenoid is in satisfactory condition.

ASSEMBLY

◆ See Figures 293, 294 and 322 thru 347

■ The lower unit should never be assembled in a dry condition. Always coat all internal components with OMC Hi-Vis lube oil as they are assembled. All seals should be coated with OMC Gasket Seal compound. Whenever 2 seals are to be installed back-to-back, use Triple Guard Grease between them.

■ Many of the accompanying illustrations may show a hydraulic pump with a snap-ring and rubber seal installed on the front of the pump. These were not used in most 1971-72 models, so please disregard them if you are working on one of these models. Everything else is the same.

■ As many components installation illustrations are identical between disassembly and assembly procedures, make sure you refer back to the Disassembly section for additional images.

Propeller Shaft - Assembly

1. Slide the clutch dog onto the propeller shaft with the face of the dog marked **PROP END** facing toward the propeller end of the shaft. Before the

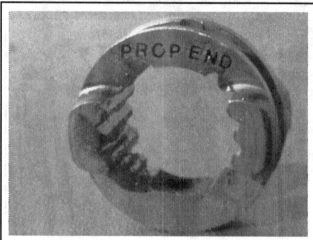

Fig. 322 This end of the clutch dog should face the propeller

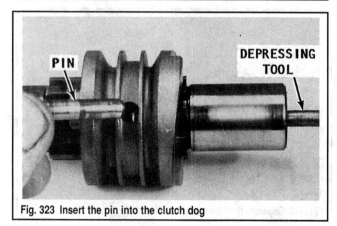

Fig. 323 Insert the pin into the clutch dog

splines of the clutch dog engage the splines of the propeller shaft, rotate the dog until the hole for the pin appears to align with the hole through the propeller shaft. Slide the clutch dog onto the splines until the hole in the dog aligns with the hole in the shaft. If the hole is off just a bit, slide the clutch dog back off the splines, rotate it one spline in the required direction, and then slide it into place.

2. Insert the spring into the end of the propeller shaft. Secure the spring in place with the spring retainer. Install the retainer with the small end going into the propeller shaft *first*.

3. Depress the spring retainer and insert the pin through the clutch dog, the shaft, spring retainer, and out the other side of the shaft and clutch dog. Center the pin through the clutch dog.

4. Install the spring-type pin retainer around the clutch dap to secure the pin in place. Be careful not to distort the pin retainer during the installation process.

Bearing Carrier Bearings & Seals Installation

5. Install the reverse gear bearing into the bearing carrier by pressing against the *lettered* side of the bearing with the proper size socket.

6. Press the forward gear bearing into the bearing carrier in the same manner. - press against the *lettered* side of the bearing.

7. Coat the outside surfaces of the seals with HI-VIS oil. Install the first seal with the flat side facing out.

8. Coat the flat surface of both seals with Triple Guard Grease, and then install the second seal with the flat side going in *first*. The seals are then back-to-back with the grease between the 2 surfaces. The outside seal prevents water from entering the lower unit and the inside seal prevents the lubricant in the lower unit from escaping.

Hydraulic Pump Assembly

9. Check out the note you made during disassembly to determine how the identifying marks (dots, dimples, whatever) on the gears must face - inward or outward. The gears MUST be installed in the same position from which they were removed.

10. Install the rear valve housing with the tang on the outside edge of the housing indexed with the small slot in the pump housing. Secure the valve housing in place with the attaching screws tightened securely.

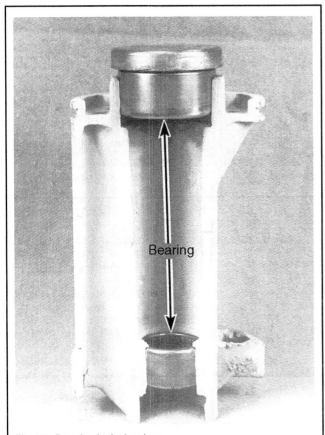

Fig. 324 Pressing in the bearings. . .

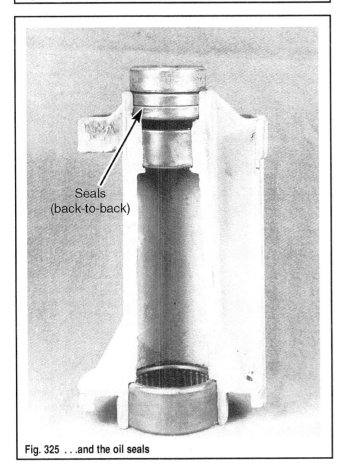

Fig. 325 . . .and the oil seals

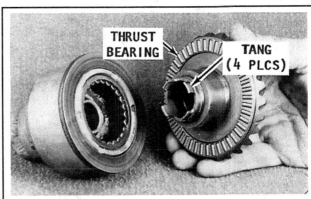

Fig. 326 Press the forward gear assembly into the hydraulic pump

Fig. 327 There are 2 types of pumps. The one on the left has the shift rod passing through a hole in 2 levers on top of the pump, and is the more common. The one on the right has the rod passing through a hole in the pump top

Fig. 328 It will take patience to get the pump indexed into the housing properly

Fig. 329 Make sure that the plunger is seated fully

Fig. 330 Install the forward gear assembly into the pump

11. Place the screen in position on the back side of the valve housing, and then secure it in place with the screw.

12. Install the forward gear into the pump housing. It may be necessary to work the gears around in the pump to permit the tangs on the forward gear shank to index in the slots in the housing.

Lower Driveshaft Bearing Installation

13. Obtain the special tool (#385546) and assemble with the washer, guide sleeve, and remover portion of the tool, in the order given. The shoulder of the tool must face down.

14. Place the bearing onto the end of the tool, with the lettered side of the bearing facing the tool. Drive the bearing down until the large washer on the tool makes contact with the surface of the lower unit. The bearing is then seated to the proper depth.

15. If an Allen screw is used to secure the bearing in place, apply Loctite to the threads, and then install the screw through the lower unit.

Hydraulic Pump Installation

16. Observe the tang on the backside of the pump. This tang must face directly up in relation to the lower unit housing to permit installation of the shift rod into the pump. Also notice the pin on the backside of the pump. This pin must index into a matching hole in the housing to restrain the pump from rotating.

17. Secure the lower unit housing in the horizontal position with the bearing carrier opening facing up.

18. Remove the forward gear from the pump.

19. Obtain 2 long 1/4 x 20 rods with threads on both ends. Thread the rods into the pump and then lower the pump into the lower unit housing. To index the pin on the back of the pump housing into the hole in the lower unit is not an easy task. However, exercise patience and rotate the pump ever so slowly. A helpful hint at this point: As the pump is being lowered into the cavity, align the opening and tang on top of the pump in the approximate position your eye indicates the shift rod may be installed. When the pin indexes, it will not be possible to rotate the pump. The pump must be properly seated to permit installation of the shift rod and the pinion gear. After the pump is in place, remove the 2 rods used during installation.

20. Coat the plunger with oil, and then lower the large end of the plunger into the hydraulic pump. Check to be sure it seats all the way into place.

21. Install the thrust washer and thrust bearing into the pump with the flat side of the bearing facing outward. Lower the forward gear into the pump. Work the gear slowly until the teeth index with the teeth of the pump gear. This should not be too difficult because the forward gear was installed once, and then removed in the previous step. However, it is entirely possible the gears moved when the pump was installed, so use a flashlight and check the position of the gears. If necessary, use a long shank screwdriver and rotate the gears until they are close to center, then install the forward gear.

Driveshaft & Pinion Gear Installation

■ The driveshaft and pinion gear must be assembled prior to installation, and then checked with a special shimming gauge. This shimming must be accomplished properly, the unit disassembled, and then installed into the lower unit. Use of the shimming gauge is the ONLY way to determine the proper amount of shimming required at the upper end of the driveshaft. The following detailed step outlines the procedure.

22. Clamp the driveshaft in a vise equipped with soft jaws and in such a manner that the splines, water pump area, or other critical portions of the shaft cannot be damaged.

23. Slide the pinion gear onto the driveshaft with the bevel of the gear teeth facing toward the lower end of the shaft.

24. Install the pinion gear nut and tighten it to 40-45 ft. lbs. No parts should be installed on the upper end of the driveshaft at this point. Slide the same amount of shim material removed during disassembling, onto the driveshaft and seat it against the driveshaft shoulder.

25. Obtain a special shimming tool (#315767). Slip the special tool down over the driveshaft and onto the upper surface of the top shim. Measure the distance between the top of the pinion gear and the bottom of the tool. The tool should just barely make contact with the pinion gear surface for zero clearance.

26. Add or remove shims from the upper end of the driveshaft to obtain the required zero clearance.

27. Remove the tool and set the shims aside for installation later.

28. Back off the pinion gear nut and remove the pinion gear.

29. Remove the driveshaft from the vise.

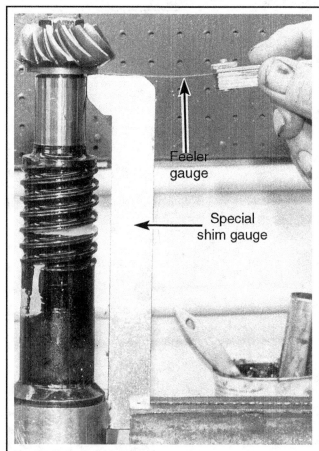

Fig. 331 Using the shim gauge

Fig. 332 Slide the shims and thrust washer over the driveshaft

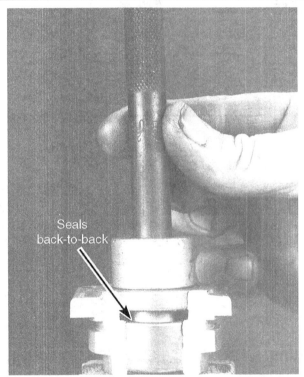

Fig. 333 Press the 2 seals into the top of the bearing housing

30. Insert the pinion gear into the cavity in the lower unit with the flat side of the bearing facing upward. Fold the pinion gear in place and at the same time lower the driveshaft down into the lower unit. As the driveshaft begins to make contact with the pinion gear, rotate the shaft slightly to permit the splines on the shaft to index with the splines of the pinion gear. After the shaft has indexed with the pinion gear, thread the pinion gear nut onto the end of the shaft.

31. Obtain the special tool (#316612) and slide the tool over the upper end of the driveshaft with the splines of the tool indexed with the splines on the shaft. Attach a torque wrench to the special tool. Now, hold the pinion gear nut with the proper size wrench and rotate the driveshaft CLOCKWISE with the special tool until the pinion gear nut is tightened to 40-45 ft. lbs. Remove the special tool.

32. Slide the thrust bearing, thrust washer, and the shims, set aside after the shim gauge procedure, onto the driveshaft.

33. Install the 2 seals back-to-back into the opening on top of the bearing housing. Coat the outside surface of the NEW seals with Johnson/Evinrude Lubricant. Press the first seal into the housing with the flat side of the seal facing outward. After the seal is in place, apply a coating of Triple Guard Grease to the flat side of the installed seal and the flat side of the second seal. Press the second seal into the bearing housing with the flat side of the seal facing inward. Insert a NEW O-ring into the bottom opening of the bearing housing.

34. Wrap friction tape around the splines of the driveshaft to protect the seals in the bearing housing as the housing is installed. Now, slide the assembled bearing housing down the driveshaft and seat it in the lower unit housing. Secure the housing in place with the 4 bolts. Tighten the bolts ALTERNATELY and EVENLY.

Propeller Shaft Installation

35. Secure the lower unit in the horizontal position with the bearing carrier opening facing upward. Check the inside diameter of the forward gear to be sure the small thrust washer in the gear is still in place. Lower the propeller shaft assembly down into the lower unit with the inside diameter of the shaft indexing over the shaft of the piston.

36. Apply a light coating of grease to the inside surface of the reverse gear to hold the thrust washer in place. Insert the thrust washer into the reverse gear.

37. Lower the thrust bearing and thrust washer down onto the shank of the reverse gear. Slide the reverse gear down the propeller shaft with the splines of the reverse gear indexing with the splines of the shaft.

38. Insert the retainer plate into the lower unit against the reverse gear.

✳✳ WARNING

This next step can be dangerous. Each snapring is placed under tremendous tension with the Truarc pliers while it is being placed into the groove. Therefore, wear safety glasses and exercise care to prevent the snapring from slipping out of the pliers. If the snapring should slip out, it would travel with incredible speed and cause personal injury if it struck a person.

39. Install the Truarc snap rings one at a time following the precautions given in the Warning and the advice given in the following paragraph.

Fig. 334 Slide on the bearing retainer and tighten the bolts

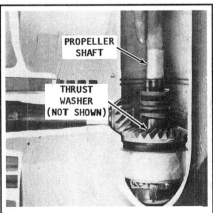

Fig. 335 Position the shaft into the forward gear. . .

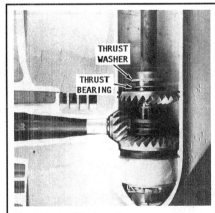

Fig. 336 . . .slide down the reverse gear, thrust bearing and washers. . .

■ The 2 snap rings index into separate grooves in the lower unit housing. As the first ring is being installed, depth perception may play a trick on your eyes. It may appear that the first ring is properly indexed all the way around in the proper groove, when in reality; a portion may be in one groove and the remainder in the other groove. Should this happen, and the Truarc pliers be released from the ring, it is extremely difficult to get the pliers back into the ring to correct the condition. If necessary use a flashlight and carefully check to be sure the first ring is properly seated all the way around *before* releasing the grip on the pliers. Installation of the second ring is not so difficult because the one groove is filled with the first ring.

40. Obtain 2 long 1/4 in. rods with threads on one end. Thread the rods into the retainer plate opposite each other to act as guides for the bearing carrier.

41. Check the bearing carrier to be sure a NEW O-ring has been installed. Position the carrier over the guide pins with the embossed word **UP** on the rim of the carrier facing up in relation to the lower unit housing. Now, lower the bearing carrier down over the guide pins and into place in the lower unit housing.

42. Slide NEW little O-rings onto each bolt, and apply some Johnson/Evinrude Sealer onto the threads. Install the bolts through the carrier and into the retaining plate. After a couple bolts are in place, remove the guide pins and install the remaining bolts. Tighten the bolts EVENLY and ALTERNATELY to the correct torque value.

Shift Cable, Solenoids and Rod

■ The following procedures must be performed exactly as given and in the order presented. Do not attempt any shortcuts or anticipate what will be done next. All parts must be installed and adjusted to the letter for the unit to function properly.

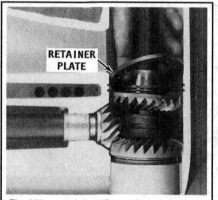
Fig. 337 . . .and then the retainer

Fig. 338 Install the snap-rings

Fig. 339 Thread in the guide rods. . .

Fig. 340 . . .and install the bearing carrier

Fig. 341 Installing the carrier bolts

Fig. 342 Installing the solenoid. . .

Fig. 343 . . .and the plunger

Fig. 344 The lip on the spacer must be facing up. . .

Fig. 345 . . .before installing the green solenoid

43. Separate the upper (green) solenoid from the lower (blue) solenoid, by pulling them apart. An inner shift rod will be released from the shift rod casing. Check to be sure the cap on the bottom of the shift rod casing is in place.

44. Lower the blue solenoid down into the lower unit housing. Continue lowering the solenoid until the cap on the end of the shift rod casing seats on top of the valve lever and check ball assembly. Check to be sure the solenoid is fully seated in the housing. The lower plunger should be flush with the top of the solenoid.

■ If the plunger is not flush with the solenoid, remove the solenoid and screw the lower plunger up or down on the shift rod casing, then install the solenoid again and check the plunger.

45. Install the spacer with the lip on the spacer facing upward.

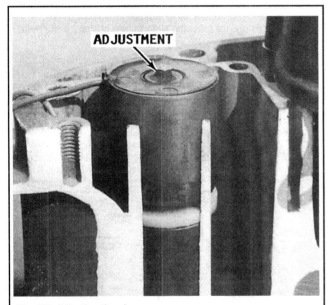
Fig. 346 Adjusting the plunger

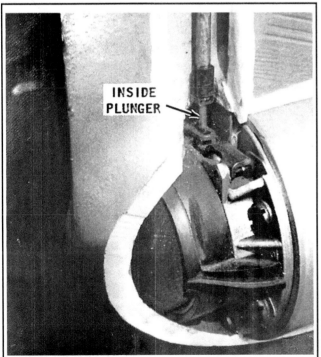
Fig. 347 Make sure the inner plunger indexes to the hole in the pump

46. Lower the green solenoid into the lower unit housing. Insert the shift rod into the shift rod casing.

47. The upper plunger must be flush with the top surface of the solenoid. If it is not flush with the solenoid, remove the solenoid, loosen the nut, and make an adjustment. Install the solenoid again and check the upper plunger.

48. Check to be sure the inside plunger indexes into the hole in the hydraulic pump. Install the wavy washer and a NEW gasket into the lower unit housing. Work the shift wires down into the lower unit cavity. Lower the cover into place in the lower unit housing. The wavy washer will give you a false pressure against the cap. Therefore, it takes a bit of patience to be sure the wavy washer indexes into the recess of the cover.

49. Start the bolts securing the cover. Tighten the bolts ALTERNATELY and EVENLY. As the bolts are tightened, make continuous checks to be sure the wavy washer and the green solenoid fit up into the cover as the bolts are tightened. This may not be accomplished on the first attempt, but keeping cool and working slowly will be rewarded with success.

50. Connect an ohmmeter, ground 1 lead of the meter, and then check the blue and green wires for continuity. The ohmmeter should indicate 5-7 ohms.

51. If removed, install the water pump assembly.

52. Install the gearcase.

Water Pump

REMOVAL & INSTALLATION

◆ See Figures 293, 294 and 348 thru 361

1. Remove the gearcase for access to the water pump as detailed earlier in this section.

2. Position the lower unit in the vertical position on the edge of the work bench resting on the cavitation plate. Secure the lower unit in this position with a C clamp. The lower unit will then be held firmly in a favorable position during the service work. An alternate method is to cut a groove in a short piece of 2 x 6" wood to accommodate the lower unit with the cavitation plate resting on top of the wood. Clamp the wood in a vise and service work may then be performed with the lower unit erect (in its normal position), or inverted (upside down). In both positions, the cavitation plate is the supporting surface.

■ Take the time to notice how the shift wires are routed and anchored in position with a clamp on top of the water pump. It is extremely important for the shift wires to be routed and secured in the same position during installation.

3. Remove the O-ring from the top of the driveshaft.

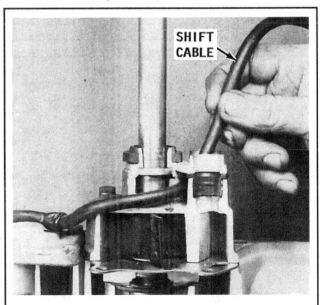
Fig. 348 Make sure that you note the shift cable positioning

Fig. 349 Remove the pump housing bolts

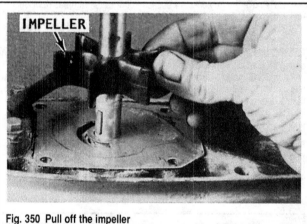

Fig. 350 Pull off the impeller

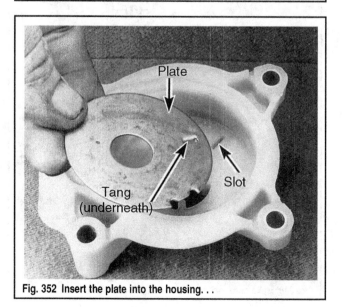

Fig. 352 Insert the plate into the housing. . .

4. Remove the bolts securing the water pump to the lower unit housing and the clamp securing the shift wires in place. Leave the clamp on the shift cable as an aid during installation. Pull the water pump housing up and free of the driveshaft.

5. Slide the water pump impeller up and free of the driveshaft.

6. Pop the impeller Woodruff key out of the driveshaft keyway.

7. Slide the water pump base plate up and off of the driveshaft.

To Install:

8. With the insert removed from the housing, inspect it carefully for any signs of damage, cracks, wear or melting and replace, if necessary.

■ An improved water pump is available as a replacement. If the old water pump housing is unfit for further service, only the new pump housing can be purchased. It is strongly recommended to replace the water pump with the improved model while the lower unit is disassembled (that is assuming it was not already done considering the age of the unit). The new pump must be assembled before it is installed. Therefore, the following steps outline procedures for both pumps.

9. **To pre-assemble and install the improved pump housing, proceed as follows:**

a. Remove the water pump parts from the container.

b. Insert the plate into the housing. The tang on the bottom side of the plate MUST index into the short slot in the pump housing.

c. Slide the pump liner into the housing with the 2 small tabs on the bottom side indexed into the 2 cut-outs in the plate.

d. Coat the inside diameter of the liner with light-weight oil.

e. Work the impeller into the housing with all of the blades bent back to the right. In this position, blades will rotate properly when the pump housing

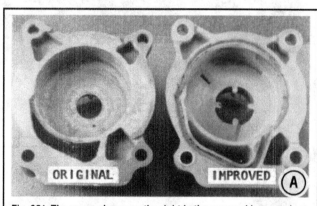

Fig. 351 The pump shown on the right is the new and improved version

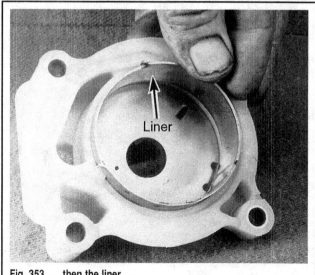

Fig. 353 . . .then the liner. . .

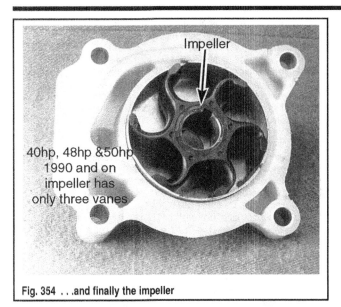

Fig. 354 . . .and finally the impeller

is installed. Remember, the pump and the blades will be rotating CLOCKWISE when the housing is turned over and installed in place on the lower unit.

f. Coat the mating surface of the lower unit with a suitable sealer.

g. Slide the water pump base plate down the driveshaft and into place on the lower unit. Insert the Woodruff key into the key slot in the driveshaft.

h. Lay down a very thin bead of a suitable sealer into the irregular shaped groove in the housing.

i. Insert the seal into the groove and then coat the seal using he sealer.

j. Begin to slide the water pump down the driveshaft, and at the same time observe the position of the slot in the impeller. Continue to work the pump down the driveshaft, with the slot in the impeller indexed over the Woodruff key. The pump must be fairly well aligned before the key is covered because the slot in the impeller is not visible as the pump begins to come close to the base plate.

k. Install the short forward bolt through the pump and into the lower unit. DO NOT tighten this bolt at this time. Insert the grommet into the pump housing.

l. Make sure the electric shift wiring is routed as noted during removal.

m. Install the grommet retainer and water tube guide onto the pump housing. Install the remaining pump attaching bolts. On electric shift models, install the solenoid cable bracket with the same bolt and in the same position from which it was removed. On some units the solenoid cable fits into a

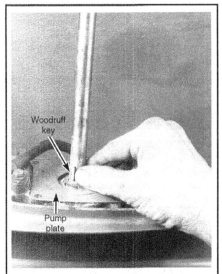

Fig. 355 Slide on the pump plate and install the key

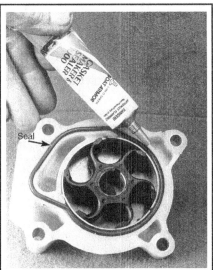

Fig. 356 Use sealer before installing the new seal

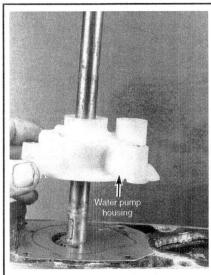

Fig. 357 Slide on the pump housing assembly. . .

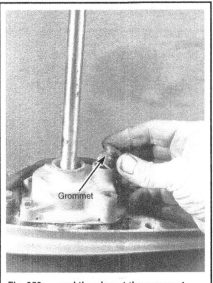

Fig. 358 . . .and then insert the grommet

Fig. 359 Tighten the mounting bolts. . .

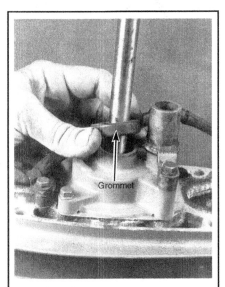

Fig. 360 . . .and then insert the grommet

PUMP HOUSING

Fig. 361 Installing the pump housing on original equipment units

recess of the water pump and is held in place in that manner. Tighten the bolts ALTERNATELY and EVENLY, and at the same time rotate the driveshaft clockwise. If the driveshaft is not rotated while the attaching bolts are being tightened, it is possible to pinch 1 of the impeller blades underneath the housing.

n. Slide the large grommet down the driveshaft and seat it over the pump collar. This grommet does not require sealer. Its function is to prevent exhaust gases from entering the water pump.

TYPE IX - MECHANICAL SHIFT LOWER UNIT, SLIDING CLUTCH DOG

Description & Operation

◆ See Figures 362 and 363

The mechanical shift and sliding clutch gearcase (without hydraulic assist) is normally found on the following models:
- 1972 50 hp (2 Cyl)

The lower unit covered in this section is complete mechanical shift unit.

A shift cable connects the shift box to the shift linkage at the engine.

A shift rod extends from the engine down through the exhaust housing to the lower unit.

Shifting into forward, neutral, and reverse, is accomplished directly through mechanical means from the shift control handle through the cable and linkage to the clutch dog in the lower unit.

The lower unit houses the driveshaft and pinion gear forward and reverses driven gears, the propeller shaft, shift lever, cradle, shift shaft, clutch dog, shift rod and the necessary shims, bearings, and associated parts to make it all work properly. A detent ball and spring" is installed on some models.

The water pump is considered a part of the lower unit.

Two different lower units are covered in this section. One unit is used with the electric start model and the other with manual start. The model years actually overlap. The upper driveshaft bearing and the shift mechanism differ between the two units. These differences are clearly indicated in the procedural steps and illustrations.

Troubleshooting

PRELIMINARY CHECKS

◆ See Figure 364

1. At rest, and without the engine running, the lower unit is in forward gear, Mount the engine in a test tank, in a body of water, or with a flush attachment connected to the lower unit. If the flush attachment is used, NEVER operate the engine above an idle speed, because the no-load condition on the propeller would allow the engine to run-away, resulting in serious damage or destruction of the engine.

10. **To install the original equipment housing, proceed as follows:**

a. Coat the water pump plate mating surface of the lower unit with a suitable sealer.

b. Slide the water pump plate down the driveshaft and *before* it makes contact with the sealer, check to be sure the bolt holes in the plate will align with the holes in the housing. The plate will only fit one way. If the holes will not align, remove the plate, turn it over and again slide the plate down the driveshaft and into place on the housing. This checking will prevent accidentally getting the sealer on both sides of the plate. If by chance sealer does get on the top surface it must be removed before the water pump impeller is installed.

c. Slide the water pump impeller down the driveshaft. Just before the impeller covers the cutout for the Woodruff key, install the key, and then work the impeller on down, with the slot in the impeller indexed over the Woodruff key. Continue working the impeller down until it is firmly in place on the surface of the pump plate.

d. Check to be sure NEW seals and O-rings have been installed in the water pump. Lubricate the inside surface of the water pump with light-weight oil.

e. Lower the water pump housing down the driveshaft and over the impeller. Always rotate the driveshaft slowly CLOCKWISE as the housing is lowered over the impeller to allow the impeller blades to assume their natural and proper position inside the housing. Continue to rotate the driveshaft and work the water pump housing downward until it is seated on the plate.

f. On electric shift models, make sure the wiring is routed as noted during removal.

g. Coat the threads of the water pump attaching screws with sealer, and then secure the pump in place with the screws.

h. On electric shift models, install the solenoid cable bracket with the same bolt and in the same position from which it was removed. On some units the solenoid cable fits into a recess of the water pump and is held in place in that manner. Tighten the screws ALTERNATELY and EVENLY.

11. Install the gearcase.

✳✳ CAUTION

Water must circulate through the lower unit to the engine any time the engine is run to prevent damage to the water pump in the lower unit. Just 5 seconds without water will damage the water pump.

2. Attempt to shift the unit into Neutral and Reverse. It is possible the propeller may turn very slowly while the unit is in neutral, due to "drag" through the various gears and bearings. If difficult shifting is encountered, the problem is in the shift linkage or in the lower unit.

3. The second area to check is the quantity and quality of the lubricant in the lower unit. If the lubricant level is low, contaminated with water, or is broken down because of overuse, the shift mechanism may be affected. Water in the lower unit is very bad news for a number of reasons, particularly when the lower unit contains hydraulic components. Hydraulic units will not function with water in the system.

4. Before making any tests, remove the propeller. Check the propeller carefully to determine if the hub has been slipping and giving a false indication the unit is not in gear. If there is any doubt, the propeller should be taken to a shop properly equipped for testing, before the time and expense of disassembling the lower unit is undertaken. The expense of the propeller testing and possible rebuild is justified.

BY SYMPTOM

The following troubleshooting procedures are presented on the assumption the lower unit lubricant and the propeller have been checked and found to be satisfactory.

Lower Unit Locked

◆ See Figure 36

Determine if the problem is in the power-head or in the lower unit. Attempt to rotate the flywheel. If the flywheel can be moved even slightly in either direction, the problem is most likely in the lower unit. If it is not possible to rotate the flywheel, the problem is a "frozen" powerhead. To absolutely verify the overhead is "frozen", separate the lower unit from the exhaust housing and then again attempt to rotate the flywheel. If the attempt is successful, the problem is definitely in the lower unit. If the attempt to rotate the flywheel, with the lower unit removed, still fails, a "frozen" overhead is verified.

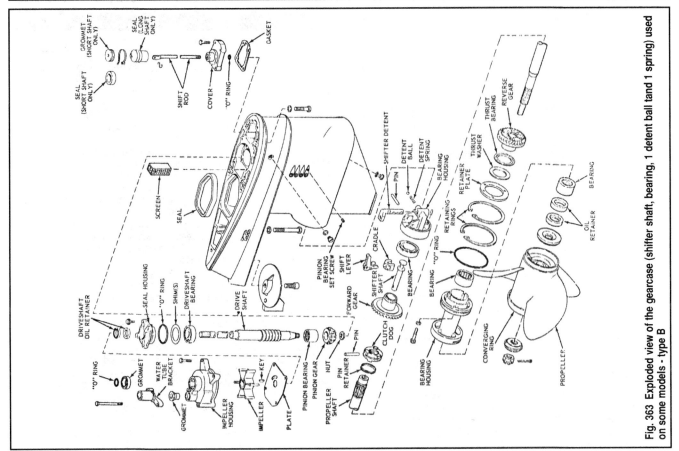

Fig. 363 Exploded view of the gearcase (shifter shaft, bearing, 1 detent ball tand 1 spring) used on some models - type B

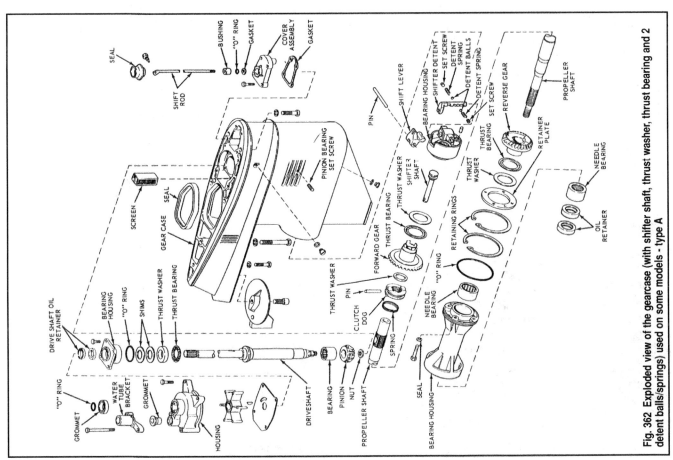

Fig. 362 Exploded view of the gearcase (with shifter shaft, thrust washer, thrust bearing and 2 detent balls/springs) used on some models - type A

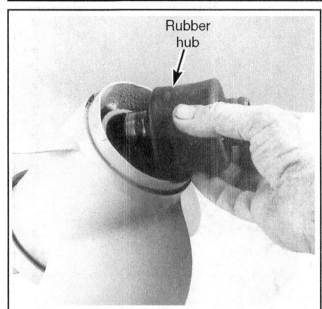

Fig. 364 The rubber hub on this unit was found to be slipping and a new one is being installed

Unit Fails to Shift, Neutral, Forward, or Reverse

◆ See Figure 365

With the outboard mounted on a boat, in a test tank, or with a flush attachment connected, disconnect the shift lever at the engine. Attempt to manually shift the unit into Neutral, Reverse, Forward. At the same time move the shift handle at the shift box and determine that the linkage and shift lever are properly aligned for the shift positions.

If the alignment is not correct, adjust the shift cable at the trunnion. It is also possible the inner wire may have slipped in the connector at the shift box. This condition would result in a lack of inner cable to make a complete "throw" on the shift handle. Check to be sure the shift handle is moved to the full shift position and the linkage to the lower unit is moved to the full shift position.

If an adjustment is required at the shift box, see the section on the Remote Controls.

If it is not possible to shift the unit into gear by manually operating the shift rod while the engine is running, the lower unit requires service as described in this section.

Fig. 365 A good look at the shift and throttle linkage

Lower Unit

REMOVAL & INSTALLATION

 **MODERATE**

◆ See Figures 298 thru 300, 362, 363, and 366

If water is discovered in the lower unit and the propeller shaft seal is damaged and requires replacement, the lower unit does not have to be removed in order to accomplish the work.

The bearing carrier can be removed and the seal replaced without disassembling the lower unit. However, such a procedure is not considered good shop practice, but merely a quick-fix. If water has entered the lower unit, the unit should be disassembled and a detailed check made to determine if any other seals, bearings, bearing races, O-rings or other parts have been rendered unfit for further service by the water.

1. If necessary, remove the propeller and/or drain the lower unit. Neither are necessary for removal of the gearcase assembly, however both are recommended the first for safety and the second to inspect the gearcase oil and help determine potential condition.

2. Disconnect and ground the spark plug wires.

3. On electric start motors, remove the starter motor for access.

4. Remove the bolt or bolts from the shift disconnect coupler under the lower carburetor.

5. Scribe a mark on the trim tab and a matching mark on the lower unit to ensure the trim tab will be installed in the same position from which it is removed. Use an Allen wrench and remove the trim tab.

6. On some units, an attaching bolt is installed inside the trim tab cavity. Use a 1/2 in. socket with a short extension and remove this bolt. Failure to remove this bolt from inside the trim tab cavity may result in an expensive part being broken in an attempt to separate the lower unit from the exhaust housing. Remove the 5/8 in. counter sunk bolt located just ahead of the trim tab position.

7. Remove the four 9/16 in. bolts, 2 on each side, securing the lower unit to the exhaust housing. Work the lower unit free of the exhaust housing. If the unit is still mounted on a boat, tilt the engine forward to gain clearance between the lower unit and the deck (floor, ground, whatever). Be careful to withdraw the lower unit straight away from the exhaust housing to prevent bending the driveshaft.

To Install:

8. Check to be sure the water tubes are clean, smooth, and free of any corrosion. Coat the water pick-up tubes and grommets with lubricant as an aid to installation.

9. Check to be sure the spark plug wires are disconnected from the spark plugs.

Fig. 366 Remove the shift disconnect

10. Bring the lower unit housing together with the exhaust housing, and at the same time, guide the water tube into the rubber grommet of the water pump.

11. Continue to bring the two units together, at the same time guiding the water pickup tubes Into the rubber grommet of the water pump, and, simultaneously rotating the flywheel slowly to permit the splines of the driveshaft to index with the splines of the crankshaft. This may sound like it is necessary to do 4 things at the same time, and so it is. Therefore, make an earnest attempt to secure the services of an assistant for this task.

12. After the surfaces of the lower unit and exhaust housing are close, dip the attaching bolts in Johnson/Evinrude Sealer and then start them in place. Install the 2 bolts on each side of the two housings. Install the bolt in the recess of the trim tab. Install the bolt just forward of the trim tab that extends up through the Cavitation plate. Tighten the bolts ALTERNATELY and EVENLY.

13. Install the trim tab with the mark made on the tab during disassembling aligned with the mark made on the lower unit housing.

14. The shift rod is visible beneath and to the rear of the lower carburetor. Install the shift connector with a bolt or pin, depending on the model being serviced.

15. If drained, refill the lower unit with oil.

16. Install the propeller.

DISASSEMBLY

◆ See Figures 303 thru 310, 320, 321, 362, 363 and 367 thru 379

Bearing Carrier - Removal & Disassembly

It appears as though this procedure can be done with the gearcase assembly still installed if your only intention is to service the seals in the carrier. However, if driveshaft and propshaft removal is going to be necessary, the gearcase must be removed.

1. If not done already, remove the propeller.

2. If desired for ease of service, or to service other components, remove the gearcase assembly, as detailed in this section.

3. Remove the four 5/16 in. bolts from inside the bearing carrier. Notice how each bolt has an O-ring seal. These O-rings should be replaced each time the bolts are removed. In most cases, the word **UP** is embossed into the metal of the bearing carrier rim. This word must face up in relation to the lower unit during installation.

4. Remove the bearing carrier using one of the methods described in the following paragraphs.

Several models of bearing carriers are used on the lower units covered in this section. The bearing carriers fit very tightly into the lower unit opening. Therefore, it is not uncommon to apply heat to the outside surface of the lower unit with a torch; at the same time the puller is being worked to remove the carrier. Be careful not to overheat the lower unit.

One model carrier has 2 threaded holes on the end of the carrier. These threads permit the installation of 2 long bolts. These bolts will then allow the use of a flywheel puller to remove the bearing carrier.

Another model does not have the threaded screw holes. To remove this type bearing carrier, a special puller with arms must be used. The arms are hooked onto the carrier web area, and then the carrier removed.

✳✳ WARNING

The next step involves a dangerous procedure and should be executed with care while wearing safety glasses. The retaining rings are under tremendous tension in the groove and while they are being removed. If a ring should slip off the Truarc pliers, it will travel with incredible speed causing personal injury if it should strike a person. Therefore, continue to hold the ring and pliers firm after the ring is out of the groove and clear of the lower unit. Place the ring on the floor and hold it securely with 1 foot before releasing the grip on the pliers. An alternate method is to hold the ring inside a trash barrel, or other suitable container, before releasing the pliers.

5. Obtain a pair of Truarc pliers. Insert the tips of the pliers into the holes of the first retaining ring. Now, carefully remove the retaining ring from the groove and gearcase without allowing the pliers to slip. Release the grip on the pliers in the manner described in the above Warning. Remove the second retaining ring in the same manner. The rings are identical and either one may be installed first.

6. Remove the retainer plate, thrust washer, thrust bearing, and reverse gear
from the propeller shaft.

7. Remove the 4 bolts from the driveshaft bearing housing. Pull up on the shift rod. This action will place shift dog into the forward gear position. Rotate the propeller shaft a bit as a check to be sure the unit is in forward gear.

8. The bearings in the carrier need not be removed unless they are unfit for further service. Insert a finger and rotate the bearing. Check for "rough" spots or binding. Inspect the bearing for signs of corrosion or other types of damage. If the bearings must be replaced, proceed with the next step.

9. Use a seal remover to remove the 2 back-to-back seals or clamp the carrier in a vise and use a pry bar to pop each seal out.

10. Use a drift punch to drive the bearings free of the carrier. The bearings are being removed because they are unfit for service; therefore, additional damage is of no consequence.

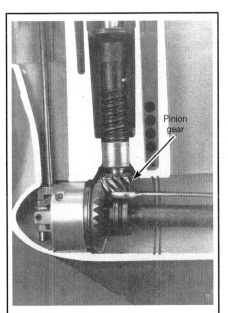

Fig. 367 Removing the pinion gear

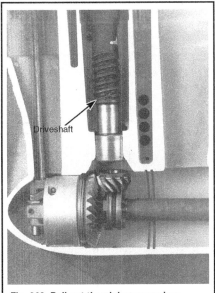

Fig. 368 Pull out the pinion gear when removing the driveshaft

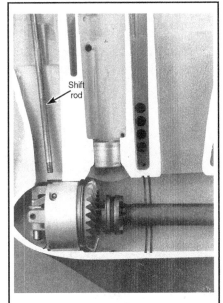

Fig. 369 Removing the shift rod. . .

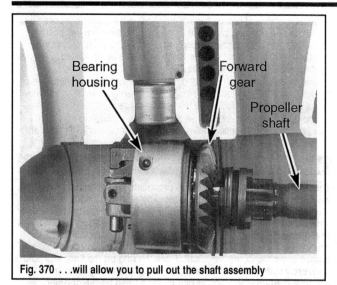

Fig. 370 . . .will allow you to pull out the shaft assembly

Pinion Gear Removal

A special tool (#316612) is required to turn the driveshaft in order to remove the pinion gear nut.

11. Obtain the special tool and slip it over the end of the driveshaft with the splines of the tool indexed with the splines on the driveshaft.

12. Hold the pinion gear nut with the proper size wrench and at the same time rotate the driveshaft, with the special tool and wrench COUNTERCLOCKWISE until the nut is free.

13. If the special tool is not available, clamp the driveshaft in a vise equipped with soft jaws, in an area below the splines but not in the water pump impeller area.

14. Now, with the proper size wrench on the pinion gear nut, rotate the complete lower unit COUNTERCLOCKWISE until the nut is free. This procedure will require the lower unit to be rotated, then the wrench released, the lower unit turned back, the wrench again attached to the nut, and the unit rotated again. Continue with this little maneuver until the nut is free.

15. After the nut is free, proceed with the next step. The driveshaft will be withdrawn from the pinion gear.

Driveshaft Removal & Disassembly

■ **The driveshaft bearing housing bolts were removed at the end of the bearing carrier removal procedure (right before disassembly).**

16. Carefully pry the bearing housing upward away from the lower unit, then slide it free of the driveshaft.

17. An alternate method is to again clamp the driveshaft in a vise equipped with soft jaws. Use a soft headed mallet and tap on the top side of the bearing housing. This action will jar the housing loose from the lower unit. Continue tapping with the mallet and the bearing housing, O-rings, shims, thrust washer, thrust bearing, and the driveshaft will all breakaway from the lower unit and may be removed as an assembly. As the driveshaft is removed, reach into the lower unit, catch, and remove the pinion gear.

18. Push on the shift rod to move the unit into the reverse gear position.

19. Remove the 4 bolts securing the shift rod cover.

20. Rotate the shift rod COUNTERCLOCKWISE until it is free of the shifter detent in the lower unit. Remove the shift rod and cover as an assembly. As the plate is removed, notice which surface is facing into the housing, as an aid during installation.

21. The driveshaft on some models may be disassembled. To do so, proceed as follows:

a. The driveshaft upper bearing and cone are replaced as an assembly. If replacement is required, obtain the special tool (#387131). Clamp the tool below the bearing, and then place the unit in an arbor press equipped with a deep throat pedestal. Press the shaft from the bearing.

b. An alternate method is to clamp the special tool (#387206) to the driveshaft just above the shoulder.

c. Now, invert the driveshaft and place it in an arbor press with the special tool (#387206) seated on the press.

d. Place a piece of pipe over the driveshaft, and then press against the tool (#387131) to remove the bearing.

Propeller Shaft Removal & Disassembly

22. Grasp the propeller shaft firmly with 1 hand and remove the propeller shaft from the lower unit. The forward gear and bearing housing will come out with the shaft as an assembly.

On rare occasions, especially if water has been allowed to enter the lower unit, it may not be possible to withdraw the propeller shaft as described. The shaft may be "frozen" in the hydraulic pump due to corrosion.

If efforts to remove the propeller shaft after the bearing carrier has been removed fail, all is not lost, here are two possible methods of getting it out:

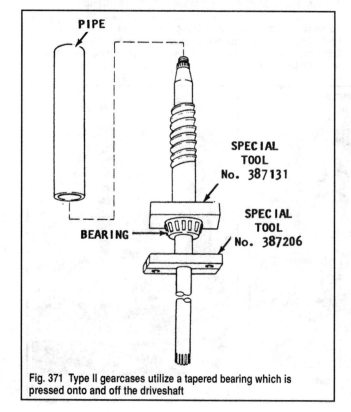

Fig. 371 Type II gearcases utilize a tapered bearing which is pressed onto and off the driveshaft

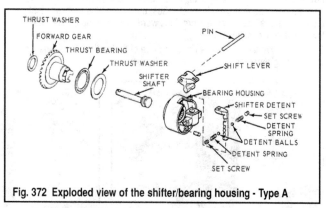

Fig. 372 Exploded view of the shifter/bearing housing - Type A

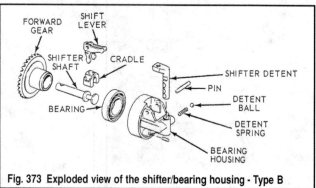

Fig. 373 Exploded view of the shifter/bearing housing - Type B

If efforts to remove the propeller shaft after the bearing carrier has been removed fail, all is not lost:

23. Obtain a block of wood 2 x 4" approx. 1 ft. in length. Drill a hole in the center of the flat side, large enough for the propeller shaft to pass through.

24. Place the block over the shaft. Slide some thick large washers over the shaft and thread the propeller nut onto the shaft.

25. With the skeg clamped securely in a vise equipped with soft jaws, attempt to pull the shaft free. If necessary hammer on the wood, rotating the block at intervals to prevent wedging the shaft in any one direction.

The second and less desirable method is as follows:

26. Clamp the propeller shaft horizontally in a vise equipped with soft jaws. Two blocks of wood may be substituted for the soft jaws, but NEVER clamp the shaft in the vise without protection, to prevent damage to the threads or splines.

27. Use a soft head mallet and strike the lower unit with quick sharp blows midway between the anti-cavitation plate and the propeller shaft. This action will drive the lower unit from the propeller shaft. Make sure you hold the lower unit to keep it in line with the propeller shaft and prevent wedging the shaft in any one direction. Holding the lower unit will also prevent the housing from falling to the floor when the housing comes free of the shaft.

■ **Bad news.** Using either method to remove a frozen propeller shaft will often result in some internal part being damaged as the shaft is withdrawn. Usually, the parts damaged will be the shift cylinder, piston, and/or the push rod. However, the cost of replacing the damaged parts is reasonable considering the seriousness of a "frozen" shaft and the problem of getting it out.

28. Remove the coil spring from the outside groove of the clutch dog. Do not not to distort the spring as it is removed.

29. Remove the clutch dog pin by pushing it through the clutch dog and propeller shaft. This pin is not a tight fit; therefore, it is not difficult to remove.

30. Grasp the propeller shaft and pull it free of the bearing housing.

Two different shifter arrangements on the propeller shaft are used on the units covered in this section. Exploded views are included of both types. Type A has a shifter shaft, thrust washer, thrust bearing and 2 detent balls and springs. Type B has a shifter shaft, bearing, 1 detent ball, and 1 spring. Compare the propeller shaft from the unit being serviced with the two drawings.

■ **The photos are of a Type A shifter.**

31. On Type A shifters, proceed as follows:

a. Remove the 2 Allen screws from the back side of the bearing housing and at the same time be prepared to catch the 2 detent balls and springs.

b. Remove the shift lever pin. Remove the shift lever and shift detent, from the bearing housing. Remove the shifter shaft.

32. On Type B shifters, proceed as follows:

a. Remove the shift lever pin and disengage the shift lever from the cradle in the shift shaft.

b. Remove the shifter shaft and cradle. Remove the shift lever.

✳✳ WARNING

The next step is dangerous. The detent balls are under tension from the springs. The balls are released with pressure. Therefore, safety glasses should be worn as personal protection for the eyes.

---

c. Observe the short arm at the upper end of the shifter detent. Now, rotate the shifter detent until this arm is 180° from its original position (facing toward the opposite direction). Rotating the shifter detent 180° will depress the detent ball and spring. From this position, work the shifter detent up out of the housing.

Please refer to the Propeller Shaft & Driveshaft Assembly/Installation procedure for details on installing the propeller shaft.

Fig. 374 Remove the coil spring and then the clutch dog pin - Type A

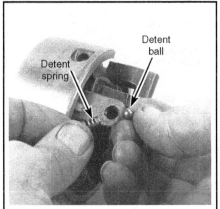

Fig. 375 Remove the detent ball and spring after removing the Allen screws - Type A

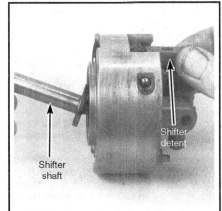

Fig. 376 Remove the shifter shaft and detent - Type A

Fig. 377 Remove the shift lever pin and then disengage the lever - Type B

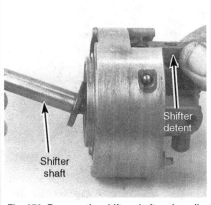

Fig. 378 Remove the shifter shaft and cradle and then the lever - Type B

Fig. 379 Removing the shifter detent

Lower Driveshaft Bearing Removal

Two different type bearings are used on the lower unit models covered in this section. Exploded drawings of both type bearings are provided. To determine which type bearing is used on the unit being serviced, check the shifter arrangement on the propeller shaft and compare it with the two illustrations.

■ **A special tool must be used to remove either type bearing. Therefore, DO NOT attempt to remove this bearing unless it is unfit for further service.**

33. To check the bearing, first use a flashlight and inspect it for corrosion or other damage. Insert a finger into the bearing, and then check for "rough" spots or binding while rotating it.

34. Remove the Allen screw from the water pick-up slots in the starboard side of the lower unit housing. This screw secures the bearing in place and MUST be removed before an attempt is made to remove the bearing.

✳✳ CAUTION

The bearing must actually be pulled upward to come free. NEVER make an attempt to drive it down and out.

35. If servicing a unit shown with a Type A shifter, obtain the special tool (#385546). If servicing a Type B unit, obtain another special tool (#391257). Use the special tool and "pull" the bearing from the lower unit.

CLEANING & INSPECTION

◆ **See Figuresâ 362 and 363**

1. Wash all parts in solvent and dry them with compressed air. Discard all O-ring's and seals that have been removed. A new seal kit for this lower unit is available from the local dealer. The kit will contain the necessary seals and O-rings to restore the lower unit to service.

2. Inspect all splines on shafts and in gears for wear, rounded edges, corrosion, and damage.

3. Carefully check the driveshaft and the propeller shaft to verify they are straight and true without any sign of damage. A complete check must be performed by turning the shaft in a lathe. This is only necessary if there is evidence to suspect the shaft is not true.

4. Check the water pump housing for corrosion on the inside and verify the impeller and base plate are in good condition. Actually, good shop practice dictates to rebuild or replace the water pump each time the lower unit is disassembled. The small cost is rewarded with "peace of mind" and satisfactory service.

5. Inspect the lower unit housing for nicks, dents, corrosion, or other signs of damage. Nicks may be removed with No. 120 and No. 180 emery cloth. Make a special effort to ensure all old gasket material has been removed and mating surfaces are clean and smooth.

6. Inspect the water passages in the lower unit to be sure they are clean. The screen may be removed and cleaned.

7. Check the gears and clutch dog to be sure the ears are not rounded. If doubt exists as to the part performing satisfactorily, it should be replaced.

8. Inspect the bearings for "rough" spots, binding, and signs of corrosion or damage.

ASSEMBLY

◆ **See Figures 324, 325, 331, 362, 363 and 380 thru 394**

■ **The lower unit should never be assembled in a dry condition. Always coat all internal components with OMC Hi-Vis lube oil as they are assembled. All seals should be coated with OMC Gasket Seal compound. Whenever 2 seals are to be installed back-to-back, use Triple Guard Grease between them.**

■ **As many component's installation illustrations are identical between the disassembly and assembly procedures, make sure you refer back to the Disassembly section for additional images if you do not see one here.**

Propeller Shaft & Driveshaft - Assembly & Installation

Two different shifter arrangements on the propeller shaft are used on the units covered in this section. A different type of lower driveshaft bearing is also used. Exploded views of both types are provided. However, it is also possible to tell them apart by examining the shifter components.

Type A has a shifter shaft, thrust washer, thrust bearing, and 2 detent balls and springs. Type B has a shifter shaft, bearing, 1 detent ball, and 1 spring. Compare the propeller shaft from the unit being serviced with the two drawings.

1. On Type A shifters (which utilize a shifter shaft, thrust washer, thrust bearing, and 2 detent balls and springs), proceed as follows:

   a. Install the shift lever and detent.
   b. Install the shifter shaft and then the pin.
   c. Insert the 2 detent balls into the shifter detent, then the springs.
   d. Coat the threads of the set screws with Loctite, and then secure the springs and detent balls in place with the screws. The screws should be tightened until each head is flush with the housing.
   e. Install the thrust bearing and thrust washer onto the shank of the forward gear.
   f. Now, slide the assembled forward gear into the bearing housing.
   g. Slide the small thrust washer onto the propeller shaft.
   h. Insert the propeller shaft into the forward gear with the slot in the shaft aligned with the hole in the clutch dog shaft.
   i. Insert the pin, and then install the coil spring around the outside groove of the clutch dog. Check to be sure 1 coil of the spring does not overlap another coil.

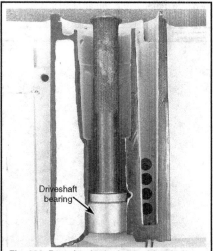

Fig. 380 Pressing in the lower driveshaft bearing

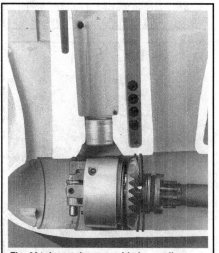

Fig. 381 Insert the assembled propeller shaft into the housing. . .

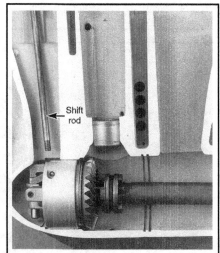

Fig. 382 . . .and then thread in the shift rod

j. Press the shifter down to move the clutch dog into the reverse gear position. Set the propeller shaft and shifter assembly.

k. If removed, install the lower driveshaft bearing. Obtain a bearing installer (#385546). Assemble the washer, plate, guide sleeve, and installer portion of the tool to the screw in the order given, and with the shoulder of the tool facing down.

l. Place the bearing onto the tool with the lettered side of the bearing TOWARD the shoulder. Place the lower driveshaft bearing and tool in position in the lower unit.

Now, drive the bearing into the lower unit until the plate makes contact with the lower unit surface. Coat the threads of the set screw with Loctite and then secure the bearing in place with the sets crew.

2. On Type B shifters (which utilize a shifter shaft, bearing, 1 detent ball, and 1 spring), proceed as follows:

a. Apply a thin coating of Needle Bearing Grease onto the detent ball and spring.

b. Insert the detent ball and spring into the bearing housing.

c. Observe the short arm on the shifter detent. Position the detent with this short arm facing forward. Now, use a punch, or similar tool, and depress the hall and spring while at the same time pressing the shift detent down into the bearing housing. Rotate the shifter detent 180°.

d. Place the cradle onto the shifter shaft and in position in the bearing housing. Install shifter lever with the cradle engaged with the shifter detent. Work the pin through the shifter detent and cradle.

e. Install the clutch dog onto the propeller shaft with the **PROP END** identification on the dog facing the propeller end of the shaft.

f. Align the holes in the dog with the slot in the shaft.

g. Slide the forward gear bearing onto the forward gear shoulder.

h. Install the thrust washer into the bearing housing.

i. Align the hole in the shifter shaft with the hole in the clutch dog. Now, slide the propeller shaft into the assembled forward gear and bearing housing.

j. Insert the clutch dog retaining pin.

k. Install a NEW clutch dog retaining spring. Check to be sure that 1 coil of the spring does not overlap another coil.

l. Move the shifter detent down into the reverse gear position. Set the assembly aside for later installation.

m. If removed, install the lower driveshaft bearing. Obtain a bearing installer (#391257). Assemble the parts as shown in the accompanying illustration. Place the lower driveshaft bearing on the tool with the lettered side of the bearing facing UP towards the driving face of the tool.

n. Use needle bearing grease to hold the bearing on the tool. Now, drive the bearing into place until the washer makes contact with the spacer. Coat the threads of the set screw with Loctite. Secure the bearing in place with the screw.

o. Obtain the special tools (#387131 and #387206), and a short piece of pipe to fit over the driveshaft. Use the 2 special tools and the piece of pipe

in an arbor press to install the upper driveshaft bearing. Press with the tool against the inner race until the bearing is seated on the driveshaft shoulder.

3. Clamp the driveshaft in a vise equipped with soft jaws and in such a manner that the splines, water pump area, or other critical portions of the shaft cannot be damaged.

4. Slide the pinion gear onto the driveshaft with the bevel of the gear teeth facing toward the lower end of the shaft.

5. Install the pinion gear nut and tighten it to 40-45 ft. lbs. No parts should be installed on the upper end of the driveshaft at this point.

6. Slide the shims removed during disassembling onto the driveshaft and seat them against the driveshaft shoulder.

7. If servicing a Type A shifter, obtain a special shimming tool (#315767). If servicing a unit with the Type B shifter, obtain another special shimming tool (#320739).

8. Slip the special tool down over the driveshaft and onto the upper surface of the top shim. Measure the distance between the top of the pinion gear and the bottom of the tool. The tool should just barely make contact with the pinion gear surface for zero clearance.

9. Add or remove shims from the upper end of the driveshaft to obtain the required zero clearance.

10. If servicing the model with a Type B shifter, remove 0.007 in. of shim material AFTER the zero clearance has been obtained.

11. Remove the tool and set the shims aside for installation later.

12. Back off the pinion gear nut and remove the pinion gear.

13. Remove the driveshaft from the vise.

14. Insert the complete assembled propeller shaft into the lower unit housing with pin on the back side of the forward gear housing indexed into hole in the lower unit.

15. Thread the shift rod into the shift detent assembly. Coat both sides of the shift rod gasket with sealer. Place the gasket in position, and then install the shift rod cover onto the housing. The shift rod is adjusted in the next procedure.

16. Slide the shift rod boot down over the shift rod cover.

17. To adjust the shift rod, measure the distance from the top of the lower unit housing to the center of the hole in the shift rod. This distance should be 16-5/32 in. (41.0cm) for a standard long shaft unit. Add exactly 5 in. (12.7cm) for a unit with an extra long shaft.

18. Rotate the shift rod clockwise or counterclockwise until the required dimension is obtained.

19. Insert the pinion gear into the cavity in the lower unit with the flat side of the bearing facing upward.

20. Hold the pinion gear in place and at the same time lower the driveshaft down into the lower unit.

21. As the driveshaft begins to make contact with the pinion gear, rotate the shaft slightly to permit the splines on the shaft to index with the splines of the pinion gear.

22. After the shaft has indexed with the pinion gear, thread the pinion gear nut onto the end of the shaft.

Fig. 383 Tighten the pinion gear nut. . .

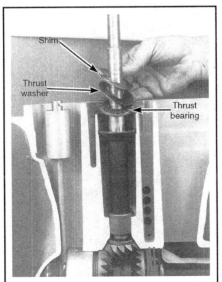

Fig. 384 . . .and then slide on the thrust washer bearing and shim material

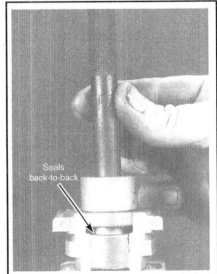

Fig. 385 Drive the 2 seals into the upper bearing housing. . .

23. Obtain the special tool (#316612) and slide the special tool over the upper end of the driveshaft with the splines of the tool indexed with the splines on the upper end of the driveshaft.

24. Attach a torque wrench to the special tool. Now, hold the pinion gear nut with the proper size wrench and rotate the driveshaft CLOCKWISE with the special tool until the pinion gear nut is tightened to 40-45 ft. lbs. (54-61 Nm). Remove the special tool.

25. If servicing a unit with a Type A shifter, slide the thrust bearing, thrust washer, and the shims, set aside after the shim procedure, onto the driveshaft. If servicing a unit with the Type B shifter, slide the shims only onto the driveshaft. (This unit does not have the other items.)

26. Install the 2 seals back-to-back into the opening on top of the bearing housing. Coat the outside surface of the NEW seals with Johnson/Evinrude Lubricant. Press the first seal into the housing with the flat side of the seal facing outward. After the seal is in place, apply a coating of Triple Guard Grease to the flat side of the installed seal and the flat side of the second seal. Press the second seal into the bearing housing with the flat side of the seal facing inward. Insert a NEW O-ring into the bottom opening of the bearing housing.

27. Wrap friction tape around the splines of the driveshaft to protect the seals in the bearing housing as the housing is installed. Now, slide the

assembled bearing housing down the driveshaft and seat it in the lower unit housing. Secure the housing in place with the 4 bolts. Tighten the bolts ALTERNATELY and EVENLY.

Reverse Gear Installation

28. Apply a light coating of grease to the inside surface of the reverse gear to hold the thrust washer in place. Insert the thrust washer into the reverse gear.

29. Install the thrust bearing and thrust washer onto the shank on the backside of the reverse gear.

30. Slide the reverse gear down the propeller shaft and index the teeth of the reverse gear with the teeth of the pinion gear.

31. Insert the retainer plate into the lower unit against the reverse gear.

✳✳ WARNING

This next step can be dangerous. Each snap-ring is placed under tremendous tension with the Truarc pliers while it is being placed into the groove. Therefore, wear safety glasses and exercise care to prevent the snapring from slipping out of the pliers. If the snap-ring should slip out, it would travel with incredible speed and cause personal injury if it struck a person.

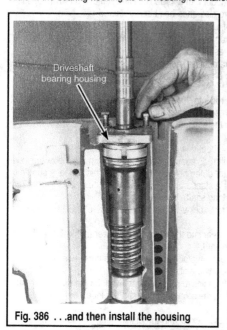

Fig. 386 ...and then install the housing

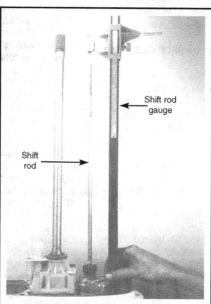

Fig. 387 Setting up the shift rod

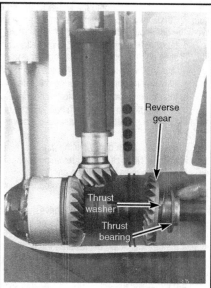

Fig. 388 Installing the reverse gear and associated components

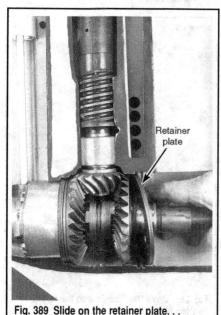

Fig. 389 Slide on the retainer plate...

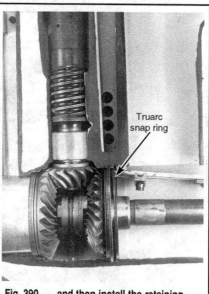

Fig. 390 ...and then install the retaining ring

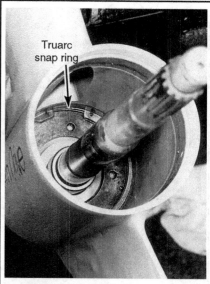

Fig. 391 Make sure the snap ring locks into the grooves properly

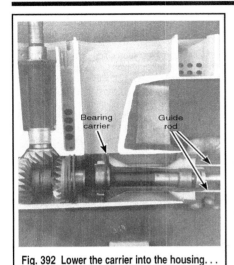

Fig. 392 Lower the carrier into the housing. . .

Fig. 393 . . .making sure that the word UP is at 12 o'clock when installed

Fig. 394 Always use new O-rings on the bolts

32. Install the Truarc snap rings one at a time following the precautions rings in the Warning and the advice given in the following paragraph.

The 2 snap rings index into separate grooves in the lower unit housing. As the first ring is being installed, depth perception may play a trick on your eyes. It may appear that the first ring is properly indexed all the way around in the proper groove, when in reality; a portion may be in one groove and the remainder in the other groove. Should this happen, and the Truarc pliers be released from the ring, it is extremely difficult to get the pliers back into the ring to correct the condition. If necessary use a flashlight and carefully check to be sure the first ring is properly seated all the way around BEFORE releasing the grin on the pliers. Installation of the second ring is not so difficult because the one groove is filled with the first ring.

Bearing Carrier Assembly & Installation

33. Install the reverse gear bearing into the bearing carrier by pressing against the lettered side of the bearing with the proper size socket.

34. Press the forward gear bearing into the bearing carrier in the same manner. Press against the lettered side of the bearing.

35. Coat the outside surfaces of the seals with gearcase oil.

36. Install the first seal with the flat side facing out.

37. Coat the flat surface of both seals with Triple Guard Grease, and then install the second seal with the flat side going in first. The seals are then back-to-back. The outside seal prevents water from entering the lower unit and the inside seal prevents the lubricant in the lower unit from escaping.

38. Obtain 2 long 1/4 in. rods with threads on one end. Thread the rods into the retainer plate opposite each other to act as guides for the bearing carrier.

39. Check the bearing carrier to be sure a NEW O-ring has been installed.

40. Position the carrier over the guide pins with the embossed word **UP** on the rim of the carrier facing up in relation to the lower unit housing.

41. Now, lower the bearing carrier down over the guide pins and into place in the lower unit housing.

42. Slide NEW little O-rings onto each bolt, and apply some Johnson/Evinrude Sealer onto the threads. Install the bolts through the carrier and into the retaining plate. After a couple bolts are in place, remove the guide pins and install the remaining bolts. Tighten the bolts EVENLY and ALTERNATELY.

43. If removed, install the gearcase.

44. Install the propeller assembly.

Water Pump

REMOVAL & INSTALLATION

◆ See Figures 362, 363, 350 thru 360, 395 and 396

1. Remove the gearcase for access to the water pump, as detailed earlier in this section.

2. Position the lower unit in the vertical position on the edge of the work bench resting on the cavitation plate. Secure the lower unit in this position with a C-clamp. The lower unit will then be held firmly in a favorable position during the service work. An alternate method is to cut a groove in a short piece of 2 x 6" wood to accommodate the lower unit with the cavitation plate resting on top of the wood. Clamp the wood in a vise and service work may

Fig. 395 Remove the pump housing bolts

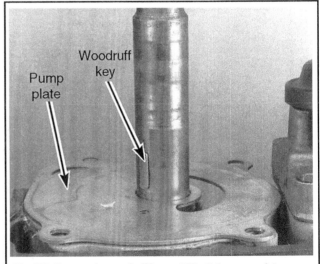

Fig. 396 Remove the woodruff key and lift off the pump plate

then be performed with the lower unit erect (in its normal position), or inverted (upside down). In both positions, the cavitation plate is the supporting surface.

3. Remove the O-ring from the top of the driveshaft.

4. Remove the bolts securing the water pump to the lower unit housing. Pull the water pump housing up and free of the driveshaft.

5. Slide the water pump impeller up and free of the driveshaft.

6. Pop the impeller Woodruff key out of the driveshaft keyway.

7. Slide the water pump base plate up and off of the driveshaft.

To Install:

8. With the insert removed from the housing, inspect it carefully for any signs of damage, cracks, wear or melting and replace, if necessary.

■ **An improved water pump is available as a replacement. If the old water pump housing is unfit for further service, only the new pump housing can be purchased. It is strongly recommended to replace the water pump with the improved model while the lower unit is disassembled (that is assuming it was not already done considering the age of the unit). The new pump must be assembled before it is installed. Therefore, the following steps outline procedures for both pumps.**

9. **To pre-assemble and install the improved pump housing, proceed as follows:**

a. Remove the water pump parts from the container.

b. Insert the plate into the housing. The tang on the bottom side of the plate MUST index into the short slot in the pump housing.

c. Slide the pump liner into the housing with the 2 small tabs on the bottom side indexed into the 2 cut-outs in the plate.

d. Coat the inside diameter of the liner with light-weight oil.

e. Work the impeller into the housing with all of the blades bent back to the right. In this position, blades will rotate properly when the pump housing is installed. Remember, the pump and the blades will be rotating CLOCKWISE when the housing is turned over and installed in place on the lower unit.

f. Coat the mating surface of the lower unit with a suitable sealer.

g. Slide the water pump base plate down the driveshaft and into place on the lower unit. Insert the Woodruff key into the key slot in the driveshaft.

h. Lay down a very thin bead of a suitable sealer into the irregular shaped groove in the housing.

i. Insert the seal into the groove and then coat the seal using he sealer.

j. Begin to slide the water pump down the driveshaft, and at the same time observe the position of the slot in the impeller. Continue to work the pump down the driveshaft, with the slot in the impeller indexed over the Woodruff key. The pump must be fairly well aligned before the key is covered because the slot in the impeller is not visible as the pump begins to come close to the base plate.

k. Install the short forward bolt through the pump and into the lower unit. DO NOT tighten this bolt at this time. Insert the grommet into the pump housing.

l. Install the grommet retainer and water tube guide onto the pump housing. Install the remaining pump attaching bolts. On electric shift models, install the solenoid cable bracket with the same bolt and in the same position from which it was removed. On some units the solenoid cable fits into a recess of the water pump and is held in place in that manner. Tighten the bolts ALTERNATELY and EVENLY, and at the same time rotate the driveshaft CLOCKWISE. If the driveshaft is not rotated while the attaching bolts are being tightened, it is possible to pinch 1 of the impeller blades underneath the housing.

m. Slide the large grommet down the driveshaft and seat it over the pump collar. This grommet does not require sealer. Its function is to prevent exhaust gases from entering the water pump.

10. **To install the original equipment housing, proceed as follows:**

a. Coat the water pump plate mating surface of the lower unit with a suitable sealer.

b. Slide the water pump plate down the driveshaft; and before it makes contact with the sealer, check to be sure the bolt holes in the plate will align with the holes in the housing. The plate will only fit one way. If the holes will not align, remove the plate, turn it over and again slide the plate down the driveshaft and into place on the housing. This checking will prevent accidentally getting the sealer on both sides of the plate. If by chance sealer does get on the top surface it must be removed before the water pump impeller is installed.

c. Slide the water pump impeller down the driveshaft. Just before the impeller covers the cutout for the Woodruff key, install the key, and then work the impeller on down, with the slot in the impeller indexed over the Woodruff key. Continue working the impeller down until it is firmly in place on the surface of the pump plate.

d. Check to be sure NEW seals and O-rings have been installed in the water pump. Lubricate the inside surface of the water pump with light-weight oil.

e. Lower the water pump housing down the driveshaft and over the impeller. Always rotate the driveshaft slowly CLOCKWISE as the housing is lowered over the impeller to allow the impeller blades to assume their natural and proper position inside the housing. Continue to rotate the driveshaft and work the water pump housing downward until it is seated on the plate.

f. Coat the threads of the water pump attaching screws with sealer, and then secure the pump in place with the screws. On electric shift models, install the solenoid cable bracket with the same bolt and in the same position from which it was removed. On some units the solenoid cable fits into a recess of the water pump and is held in place in that manner. Tighten the screws ALTERNATELY and EVENLY.

11. Install the gearcase.

Gearcase Applications

Year	Model (Hp)	No. of Cyl	Displace cu. in. (cc)	Gearcase Type ①
1958	3	2	5.28 (85.5)	I
	5.5	2	8.84 (145)	II
	7.5	2	12.4 (203)	II
	10	2	16.6 (281)	II
	18	2	22 (361)	II
	35	2	40.5 (664)	III
	50	V4	70.7 (1159)	IV
1959	3	2	5.28 (85.5)	I
	5.5	2	8.84 (145)	II
	10	2	16.6 (281)	II
	18	2	22 (361)	II
	35	2	40.5 (664)	III
	50	V4	70.7 (1159)	IV
1960	3	2	5.28 (85.5)	I
	5.5	2	8.84 (145)	II
	10	2	16.6 (281)	II
	18	2	22 (361)	II
	40	2	43.9 (719)	III
	75	V4	89.5 (1467)	V
1961	3	2	5.28 (85.5)	I
	5.5	2	8.84 (145)	II
	10	2	16.6 (281)	II
	18	2	22 (361)	II
	40	2	43.9 (719)	III
	75	V4	89.5 (1467)	V
1962	3	2	5.28 (85.5)	I
	5.5	2	8.84 (145)	II
	10	2	16.6 (281)	II
	18	2	22 (361)	II
	28	2	35.7 (585)	III
	40	2	43.9 (719)	III, VI
	75	V4	89.5 (1467)	V, VI
1963	3	2	5.28 (85.5)	I
	5.5	2	8.84 (145)	II
	10	2	16.6 (281)	II
	18	2	22 (361)	II
	28	2	35.7 (585)	
	40	2	43.9 (719)	VI
	75	V4	89.5 (1467)	VI
1964	3	2	5.28 (85.5)	I
	5.5	2	8.84 (145)	II
	9.5	2	15.2 (249)	II
	18	2	22 (361)	II
	28	2	35.7 (585)	III
	40	2	43.9 (719)	III, VI
	60	V4	70.7 (1159)	IV, V
	75	V4	89.5 (1467)	IV, V, VI
	90	V4	89.5 (1467)	VI
1965	3	2	5.28 (85.5)	I
	5	2	8.84 (145)	II
	6	2	8.84 (145)	II
	9.5	2	15.2 (249)	II
	18	2	22 (361)	II
1965 (cont'd)	33	2	40.5 (664)	III
	40	2	43.9 (719)	III, VI
	60	V4	70.7 (1159)	IV, V
	75	V4	89.5 (1467)	IV, V, VI
	90	V4	89.5 (1467)	VI
1966	3	2	5.28 (85.5)	I
	5	2	8.84 (145)	II
	6	2	8.84 (145)	II
	9.5	2	15.2 (249)	II
	18	2	22 (361)	II
	20	2	22 (361)	II
	33	2	40.5 (664)	III
	40	2	43.9 (719)	III, VI
	60	V4	70.7 (1159)	IV
	80	V4	89.5 (1467)	V, VI
	100	V4	89.5 (1467)	VII
1967	3	2	5.28 (85.5)	I
	5	2	8.84 (145)	II
	6	2	8.84 (145)	II
	9.5	2	15.2 (249)	II
	18	2	22 (361)	II
	20	2	22 (361)	II
	33	2	40.5 (664)	III
	40	2	43.9 (719)	III, VI
	60	V4	70.7 (1159)	IV
	80	V4	89.5 (1467)	V, VI
	100	V4	89.5 (1467)	VII
1968	1.5	1	2.64 (43.3)	I
	3	2	5.28 (85.5)	I
	5	2	8.84 (145)	II
	6	2	8.84 (145)	II
	9.5	2	15.2 (249)	II
	18	2	22 (361)	II
	20	2	22 (361)	II
	33	2	40.5 (664)	III
	40	2	43.9 (719)	III, VI
	55	3	49.7 (814)	VIII
	65	V4	70.7 (1159)	IV, V
	85	V4	89.5 (1467)	V, VI
	100	V4	89.5 (1467)	VII
1969	1.5	1	2.64 (43.3)	I
	4	2	5.28 (85.5)	I
	6	2	8.84 (145)	II
	9.5	2	15.2 (249)	II
	18	2	22 (361)	II
	20	2	22 (361)	II
	25	2	22 (361)	II
	33	2	40.5 (664)	III
	40	2	43.9 (719)	III, VI
	55	3	49.7 (814)	VIII
	85	V4	92.6 (1517)	VIII
	115	V4	96.1 (1575)	VIII

Gearcase Applications

Year	Model (Hp)	No. of Cyl	Displace cu. in. (cc)	Gearcase Type ①		Year	Model (Hp)	No. of Cyl	Displace cu. in. (cc)	Gearcase Type ①
1970	1.5	1	2.64 (43.3)	I		1971 (cont'd)	40	2	43.9 (719)	III
	4	2	5.28 (85.5)	I			50	2	41.5 (680)	VIII
	6	2	8.84 (145)	II			60	3	49.7 (814)	VIII
	9.5	2	15.2 (249)	II			85	V4	92.6 (1517)	VIII
	18	2	22 (361)	II			100	V4	92.6 (1517)	VIII
	20	2	22 (361)	II			125	V4	99.6 (1632)	VIII
	25	2	22 (361)	II		1972	2	1	2.64 (43.3)	I
	33	2	40.5 (664)	III			4	2	5.28 (85.5)	I
	40	2	43.9 (719)	III, VI			6	2	8.84 (145)	II
	60	3	49.7 (814)	VIII			9.5	2	15.2 (249)	II
	85	V4	92.6 (1517)	VIII			18	2	22 (361)	II
	115	V4	96.1 (1575)	VIII			20	2	22 (361)	II
1971	2	1	2.64 (43.3)	I			25	2	22 (361)	II
	4	2	5.28 (85.5)	I			40	2	43.9 (719)	III
	6	2	8.84 (145)	II			50	2	41.5 (680)	VIII, IX
	9.5	2	15.2 (249)	II			65	3	49.7 (814)	VIII
	18	2	22 (361)	II			85	V4	92.6 (1517)	VIII
	20	2	22 (361)	II			100	V4	92.6 (1517)	VIII
	25	2	22 (361)	II			125	V4	99.6 (1632)	VIII

① Your gearcase may differ, please review other gearcase types if you feel yours is different than listed.

Gearcase Capacities

Year	Model (Hp)	No. of Cyl	Displace cu. in. (cc)	Gear Oil Oz. (ml)
1958	3	2	5.28 (85.5)	2.9 (68)
	5.5	2	8.84 (145)	8.5 (200)
	7.5	2	12.4 (203)	8.5 (200)
	10	2	16.6 (281)	10 (236)
	18	2	22 (361)	8.3 (196)
	35	2	40.5 (664)	13.9 (328)
	50	V4	70.7 (1159)	34.8 (820)
1959	3	2	5.28 (85.5)	2.9 (68)
	5.5	2	8.84 (145)	8.5 (200)
	10	2	16.6 (281)	10 (236)
	18	2	22 (361)	8.3 (196)
	35	2	40.5 (664)	13.9 (328)
	50	V4	70.7 (1159)	34.8 (820)
1960	3	2	5.28 (85.5)	2.9 (68)
	5.5	2	8.84 (145)	8.5 (200)
	10	2	16.6 (281)	10 (236)
	18	2	22 (361)	8.3 (196)
	40	2	43.9 (719)	15 (354)
	75	V4	89.5 (1467)	19.5 (460)
1961	3	2	5.28 (85.5)	2.9 (68)
	5.5	2	8.84 (145)	8.5 (200)
	10	2	16.6 (281)	10 (236)
	18	2	22 (361)	8.3 (196)
	40	2	43.9 (719)	15 (354)
	75	V4	89.5 (1467)	19.5 (460)
1962	3	2	5.28 (85.5)	2.9 (68)
	5.5	2	8.84 (145)	8.5 (200)
	10	2	16.6 (281)	10 (236)
	18	2	22 (361)	8.3 (196)
	28	2	35.7 (585)	13.9 (328)
	40	2	43.9 (719)	15 (354)
	75	V4	89.5 (1467)	19.5 (460) ③
1963	3	2	5.28 (85.5)	2.9 (68)
	5.5	2	8.84 (145)	8.5 (200)
	10	2	16.6 (281)	10 (236)
	18	2	22 (361)	8.3 (196)
	28	2	35.7 (585)	13.9 (328)
	40	2	43.9 (719)	15 (354)
	75	V4	89.5 (1467)	19.5 (460) ③
1964	3	2	5.28 (85.5)	2.9 (68)
	5.5	2	8.84 (145)	8.5 (200)
	9.5	2	15.2 (249)	9.7 (229)

Gearcase Capacities

Year	Model (Hp)	No. of Cyl	Displace cu. in. (cc)	Gear Oil Oz. (ml)
1964 (cont'd)	18	2	22 (361)	8.3 (196)
	28	2	35.7 (585)	13.9 (328)
	40	2	43.9 (719)	15 (354)
	60	V4	70.7 (1159)	19.5 (460) ④
	75	V4	89.5 (1467)	19.5 (460) ③④
	90	V4	89.5 (1467)	17.4 (410)
1965	3	2	5.28 (85.5)	2.9 (68)
	5	2	8.84 (145)	2.9 (68)
	6	2	8.84 (145)	8.5 (200)
	9.5	2	15.2 (249)	9.7 (229)
	18	2	22 (361)	8.3 (196)
	33	2	40.5 (664)	13.9 (328)
	40	2	43.9 (719)	15 (354)
	60	V4	70.7 (1159)	19.5 (460) ④
	75	V4	89.5 (1467)	19.5 (460) ③④
	90	V4	89.5 (1467)	17.4 (410)
1966	3	2	5.28 (85.5)	2.9 (68)
	5	2	8.84 (145)	2.9 (68)
	6	2	8.84 (145)	8.5 (200)
	9.5	2	15.2 (249)	9.7 (229)
	18	2	22 (361)	8.3 (196)
	20	2	22 (361)	8.3 (196)
	33	2	40.5 (664)	13.9 (328)
	40	2	43.9 (719)	15 (354)
	60	V4	70.7 (1159)	19.5 (460) ④
	80	V4	89.5 (1467)	19.5 (460) ③
	100	V4	89.5 (1467)	32.2 (759)
1967	3	2	5.28 (85.5)	2.9 (68)
	5	2	8.84 (145)	2.9 (68)
	6	2	8.84 (145)	8.5 (200)
	9.5	2	15.2 (249)	9.7 (229)
	18	2	22 (361)	8.3 (196)
	20	2	22 (361)	8.3 (196)
	33	2	40.5 (664)	13.9 (328)
	40	2	43.9 (719)	15 (354)
	60	V4	70.7 (1159)	19.5 (460) ④
	80	V4	89.5 (1467)	19.5 (460) ③
	100	V4	89.5 (1467)	37.2 (877)
1968	1.5	1	2.64 (43.3)	0.75 (18)
	3	2	5.28 (85.5)	2.9 (68)

Gearcase Capacities

Year	Model (Hp)	No. of Cyl	Displace cu. in. (cc)	Gear Oil Oz. (ml)
1968 (cont'd)	5	2	8.84 (145)	2.9 (68)
	6	2	8.84 (145)	8.5 (200)
	9.5	2	15.2 (249)	9.7 (229)
	18	2	22 (361)	8.3 (196)
	20	2	22 (361)	8.3 (196)
	33	2	40.5 (664)	13.9 (328)
	40	2	43.9 (719)	15 (354)
	55	3	49.7 (814)	25.3 (596)
	65	V4	70.7 (1159)	19.5 (460) ④
	85	V4	89.5 (1467)	19.5 (460) ③
	100	V4	89.5 (1467)	37.2 (877)
1969	1.5	1	2.64 (43.3)	0.75 (18)
	4	2	5.28 (85.5)	1.3 (31) ①
	6	2	8.84 (145)	8.5 (200)
	9.5	2	15.2 (249)	9.7 (229)
	18	2	22 (361)	8.3 (196)
	20	2	22 (361)	8.3 (196)
	25	2	22 (361)	8.3 (196)
	33	2	40.5 (664)	13.9 (328)
	40	2	43.9 (719)	13.9 (328) ②
	55	3	49.7 (814)	25.3 (596)
	85	V4	92.6 (1517)	27.9 (658)
	115	V4	96.1 (1575)	26.9 (634)
1970	1.5	1	2.64 (43.3)	0.75 (18)
	4	2	5.28 (85.5)	1.3 (31) ①
	6	2	8.84 (145)	8.5 (200)
	9.5	2	15.2 (249)	9.7 (229)
	18	2	22 (361)	8.3 (196)
	20	2	22 (361)	8.3 (196)
	25	2	22 (361)	8.3 (196)
	33	2	40.5 (664)	13.9 (328)
	40	2	43.9 (719)	13.9 (328) ②
	60	3	49.7 (814)	25.3 (596)
	85	V4	92.6 (1517)	27.9 (658)
	115	V4	96.1 (1575)	26.9 (634)
1971	2	1	2.64 (43.3)	1.3 (31)
	4	2	5.28 (85.5)	1.3 (31) ①
	6	2	8.84 (145)	8.5 (200)
	9.5	2	15.2 (249)	9.7 (229)
	18	2	22 (361)	8.3 (196)
	20	2	22 (361)	8.3 (196)

Gearcase Capacities

Year	Model (Hp)	No. of Cyl	Displace cu. in. (cc)	Gear Oil Oz. (ml)
1971 (cont'd)	25	2	22 (361)	8.3 (196)
	40	2	43.9 (719)	13.9 (328)
	50	2	41.5 (680)	25.3 (596)
	60	3	49.7 (814)	25.3 (596)
	85	V4	92.6 (1517)	27.9 (658)
	100	V4	92.6 (1517)	26.9 (634)
	125	V4	99.6 (1632)	26.9 (634)
1972	2	1	2.64 (43.3)	1.3 (31)
	4	2	5.28 (85.5)	1.3 (31) ①
	6	2	8.84 (145)	8.5 (200)
	9.5	2	15.2 (249)	9.7 (229)
	18	2	22 (361)	8.3 (196)
	20	2	22 (361)	8.3 (196)
	25	2	22 (361)	8.3 (196)
	40	2	43.9 (719)	13.9 (328)
	50	2	41.5 (680)	25.3 (596)
	65	3	49.7 (814)	19.5 (460)
	85	V4	92.6 (1517)	27.9 (658)
	100	V4	92.6 (1517)	26.9 (634)
	125	V4	99.6 (1632)	26.9 (634)

NOTE: Fuel/Oil ratio based on normal operating conditions, some severe or high performance applications may need higher ratio, refer to Fuel Recommendations in the Maintenance section

① 4.0W: 3.4 (80)
② 40E: 15 (354)
③ Electric: 17.4 (410)
④ HD: 34.8 (820)

8

TRIM & TILT

TRIM/TILT SYSTEMS

Description & Operation

◆ See Figures 1 and 2

■ Although it is unlikely that your engine will have been equipped with a factory trim/tilt system, there were definitely some units that came with them so we have included this information if you are one of the lucky few.

The power trim and tilt unit is a hydraulic/mechanical unit mounted port and starboard of the stern brackets. The unit consists of a single trim cylinder (port) and a single tilt cylinder (starboard), a combination pump, fluid reservoir and bi-directional electric motor in a single unit is installed on the starboard side of the transom brackets, and the necessary tubing, valves, check valves and relief valves, to make it all function properly.

Two versions of this trim/tilt system were used. One manufactured by Prestolite and the other by Calco, as indicated in the accompanying illustrations. The exploded drawing of the trim cylinder and the exploded drawing of the tilt cylinder are valid for both manufacturers.

The upper ends of both the trim and tilt cylinders are attached to a removable bracket mounted to the motor swivel bracket. When the cylinders are extended, they push up on this bracket. When the cylinders retract, they pull down on this bracket.

When the boat operator activates an electric switch on the control panel to the UP position, power is supplied to the electric motor which drives the hydraulic pump. The pump forces hydraulic fluid into the trim cylinder and into the tilt cylinder below the pistons. The trim cylinder moves the outboard motor the first 15 of movement, considered the "tilt range". At this position, the tilt cylinder takes over and moves the outboard through the final 50° of motion, the "tilt range".

The outboard may be operated up to approximately 1500 rpm while it is elevated past the 15 position for boat movement in very shallow water. However, if powerhead speed is increased above 1500 rpm, a relief valve will automatically open causing the outboard to lower to the fully trimmed out position. The outboard cannot be tilted while operating above 1500 rpm.

When the boat operator activates an electric switch on the control panel to the DOWN position, power is supplied to the electric motor which drives the hydraulic pump in the opposite direction. The pump forces hydraulic fluid into the trim cylinder and into the tilt cylinder above the pistons. The outboard is moved in the down direction, movements of the cylinders are all in the opposite direction to the up movement. The tilt cylinder moves first until the swivel bracket is resting on the trim rods. At that point the trim cylinder continues movement in the down direction.

A manual control valve/screw *should* be located at the lower end of the pump. This screw may be rotated COUNTERCLOCKWISE so that you can raise or lower the outboard unit by hand in the event a malfunction in the system prevents movement using the trim/tilt system.

A trim gauge, mounted on the control panel, registers the position of the outboard whenever the key is in the ON position. A variable resistance sending unit, located inside the trim cylinder, is connected to the trim gauge.

The electric motor has a built-in thermal overload protection device, which shuts off the power to the motor in the event of an overload in the circuit. This device cools in approximately a minute to close the switch and allow current to flow to the electric motor.

Power Trim/Tilt Reservoir

RECOMMENDED LUBRICANT

✳✳ WARNING

When equipped with power trim/tilt, proper fluid level is necessary for the built-in impact protection system. Incorrect fluid level could lead to significant lower unit damage in the event of an impact.

The power trim/tilt reservoir must be kept full with Johnson/Evinrude Power Trim/Tilt & Power Steering Fluid.

CHECKING FLUID LEVEL/CONDITION

◆ See Figures 1 and 2

✳✳ WARNING

When equipped with power trim/tilt, proper fluid level is necessary for the built-in impact protection system. Incorrect fluid level could lead to significant lower unit damage in the event of an impact.

The fluid in the power trim/tilt reservoir should be checked periodically to ensure it is full and is not contaminated.

To check the fluid, tilt the motor upward to the full tilt position and manually engage the tilt support, for safety and to prevent damage.

Remove the filler cap and make a visual inspection of the fluid. The cap is usually threaded in position on the outboard side of the reservoir and equipped with a flat to accept a bladed screwdriver. The fluid should seem clear and not milky.

The level is correct if, with the motor at full tilt, the level is even with the bottom of the filler cap hole (but only when the motor is tilted upward, as the oil level will rise above this point when the motor is tilted downward).

The total system capacity is 25 fl. oz. (740mL).

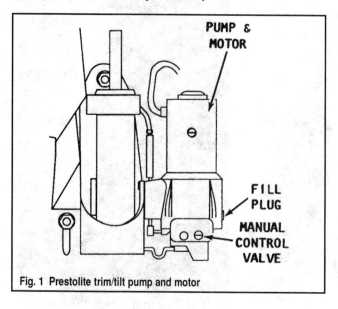

Fig. 1 Prestolite trim/tilt pump and motor

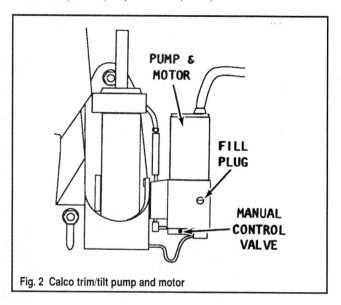

Fig. 2 Calco trim/tilt pump and motor

Troubleshooting Mechanical Components

Troubleshooting must be done *before* the system is opened in order to isolate the problem to one area. Always attempt to proceed with troubleshooting in a definite, orderly, and directed manner. The "shot gun" approach will only result in wasted time, incorrect diagnosis, replacement of unnecessary parts, and frustration.

The following procedures are presented in a logical sequence to check the mechanical components, with the most prevalent, easiest, and less costly items to be checked listed first. All may be checked and corrected without the use of special tools or equipment, with the exception of a torque wrench.

1. Check the battery for a full charge condition and to be sure all connections are clean and secure. Make a visual inspection of all exposed wiring for an open circuit or other damage that might cause a problem.

2. Check the hydraulic fluid level at the reservoir fill plug after all air has been removed from the unit and when the tilt and trim cylinders are fully extended. The fluid level should be even with the bottom of the fill hole with the motor full tilted up. Top off the reservoir to the plug level. Operate the motor and then again check the oil level. Attempt to cycle the unit several times and again check the level when the cylinders are fully extended. The system should be cycled through at least five complete movements to ensure all air has been purged from the system.

3. Add ONLY Johnson/Evinrude Power Trim/Tilt Fluid or GM Dexron Automatic Transmission Fluid, as required. Total capacity of the system is 25 oz. (740 mL).

4. Make an external (outside) inspection of the system for damage or signs of a fluid leak.

5. Seat the manual release valve by tightening the screw to 45-55 inch lbs. (5.1-6.2 Nm).

6. Inspect the stern brackets for signs of binding with the swivel brackets in the thrust rod area. Check the tilt tube nut for a torque value of 24-26 ft. lbs. (32.5-35.2 Nm), then back off (loosen) the nut 1/8 - 1/4 turn. Inspect the trim and tilt cylinders for bent rods.

7. Trailering the boat with the motor in the full tilt position and unsupported can cause a hydraulic "lock-up". To relieve such a "lock-up", loosen the trim cylinder end cap 1/4 turn with a socket type spanner wrench. Operate the unit down then up slightly. Tighten the end cap.

Trim/Tilt Assembly

REMOVAL & INSTALLATION

◆ See Figures 3 and 4

 **DIFFICULT**

1. Raise the outboard unit to the full UP position. If the trim/tilt system is not operative, rotate the manual release valve/screw COUNTERCLOCKWISE, and then lift the unit manually.

2. Secure the outboard in a safe manner using a restraint or support (such as the trailering bracket, if so equipped).

3. Obtain a suitable container to receive the hydraulic fluid from the reservoir.

✳✳ CAUTION

Always use flare wrenches when disconnecting or connecting hydraulic lines at the fittings to prevent "rounding" the corners, which is likely if a standard wrench is used.

4. Begin draining the system by disconnecting the two hydraulic lines at the trim/tilt housing base. Remove the fill plug at the reservoir and allow the hydraulic fluid to drain into the container.

5. After the fluid has drained, temporarily install the fill plug to prevent contaminates from entering the system.

6. Disconnect the trim motor electrical harness at the connector plug.

7. Remove the hardware securing the trim and tilt shaft ends to the outboard. Remove the bolts securing the system to the transom, and then carefully remove the complete unit.

8. Installation is essentially the reverse of the removal procedure. Keeping in mind the following points:

• The oil pump normally cannot be serviced. If defective, it must be replaced. The brushes in the electric motor should be replaced if they are worn to 1/4 in. (6mm) or less.

• Follow the procedures in the following section to service the electric motor.

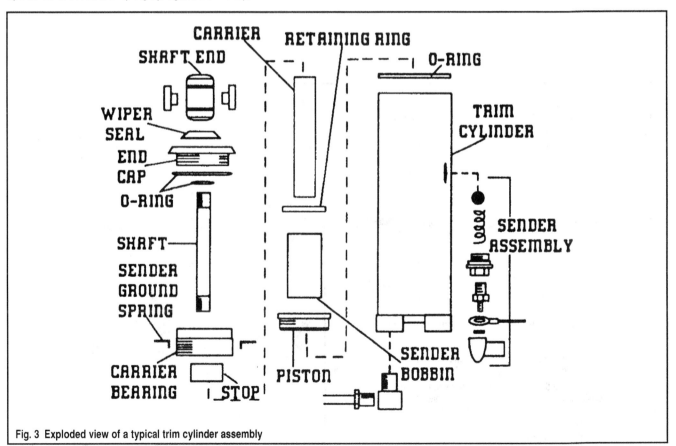

Fig. 3 Exploded view of a typical trim cylinder assembly

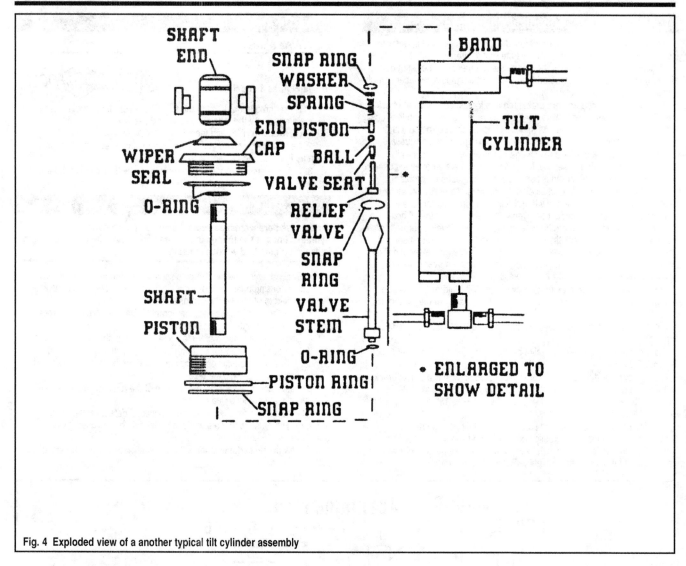

Fig. 4 Exploded view of a another typical tilt cylinder assembly

• Upon installation, use care when tightening the hydraulic lines - they must be snug enough to prevent leaks, but do NOT strip or round them.

In the majority of cases, service of the hydraulic items removed thus far will solve any rare problems encountered with the trim/tilt system.

The reservoir can be removed through the attaching hardware and cleaned if the system is considered to be contaminated with foreign material. The trim/tilt components comprise what is considered a "closed" system. The only route for entry of foreign material is through the fill opening.

Trim/Tilt Motor

TESTING

■ **The condition of the motor can be tested with a current draw test on a no load test.**

On a no load test, the motor should have a maximum current draw of 18 amps (at 12-volts) with a minimum of 8500 rpm at 12-volts.

To test the motor, connect the black wire to negative and the green/white wire (DOWN) to positive. The motor shaft from the drive end should turn in a COUNTERCLOCKWISE direction. Repeat the test with the blue/white (UP) wire to positive and the motor shaft should turn in a CLOCKWISE direction.

If the motor fails either of the tests it must be serviced or replaced.

DISASSEMBLY

◆ See Figure 5

The following procedures pick up the work after the motor has been removed from the trim/tilt assembly.
1. Remove and discard the O-ring.
2. Remove the thru bolts and discard the seals on the bolts.
3. Remove the drive end cap from the motor. Discard the gasket. When removing the seal, exercise care not to scratch the casting surfaces to ensure the new seal will seat properly. Discard the old seal.

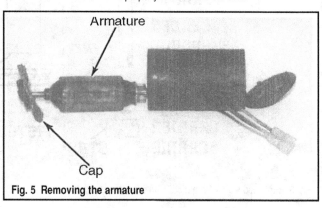

Fig. 5 Removing the armature

4. Remove the armature from the motor housing. Be careful not to lose the fiber washer on each end of the shaft. Tip the end cap free of the motor housing. Discard the springs and the end cap gasket.

■ **The end can is serviced as a complete assembly.**

CLEANING & INSPECTION

Clean all parts with a dry cloth.

✳✳ CAUTION

DO NOT clean either head in solvent, because the solvent will remove the lubricating oils in the armature shaft bushings. DO NOT clean the armature in solvent, because the solvent will leave traces of oil residue on the commutator segments. Oil will cause arcing between the commutator and the brushes.

Brush Replacement
 MODERATE

A new brush head may be purchased from the local Johnson/Evinrude dealer as a complete assembly with new brushes installed.

Checking For a Shorted Armature
 MODERATE

◆ See Figure 6

The armature cannot be checked in the usual manner on a growler, because the internal connections and the low resistance of the windings. If a growler is used, all the coils will check out shorted.

However, the armature can be tested using an AC milli-ammeter, five milli-amperes with 100 scale divisions, and making tests between the commutator segments.

Move from one segment to the next and watch closely for changes in the meter readings. The segments should all check out with almost the same reading. If a test between 2 segments indicates a significant lower reading, the winding is shorted.

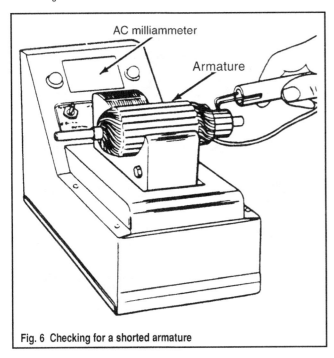
Fig. 6 Checking for a shorted armature

Checking For a Grounded Armature

◆ See Figure 7  MODERATE

1. Connect one lead of a continuity tester to a good ground, and then move the other lead around the entire surface of the commutator. Any indication of continuity means the armature is grounded and must be replaced.
2. If the commutator segments are dirty or show signs of wear (roughness), clean between the bars, and then true it in a lathe. NEVER undercut the mica because the brushes are harder than the insulation.
3. After turning the armature, the insulation between the segments must be undercut to a depth of 1/32 in. (0.79 mm). The undercut must be flat at the bottom and should extend the full width of each insulated groove and beyond the brush contact in both directions. This will prevent the segment insulation from being smeared over the commutator as the segments wear.
4. After undercutting, the commutator should be sanded to remove the ridges left during the undercutting. Now, clean the commutator thoroughly to remove any metal chips or sanding grit.
5. Again perform the shorting and grounding tests on the armature.

Checking End Head Bushings
 MODERATE

Side play of each end head on the armature should be carefully checked. Any side play indicates bearing wear and the end head must be replaced, because the bearings are not serviced separately. If the heads with worn bearings are returned to service, the armature will rub against the pole shoes, or the armature shaft may actually bind.

To replace the commutator end head, first cut the lead connecting it to the field coil as close to the end head as possible. Next, solder the new end lead to the brush holder.

Checking Field Coils
 MODERATE

◆ See Figure 8

The field coils are series-wound and are NOT grounded to the frame.
1. To test the field coils for a short, make contact with one probe of a test light to a good ground on the frame. Make contact with the other probe to the blue/white or green/white tilt motor lead.
2. If the test light comes on, the field coil is grounded and must be replaced.

The field coils are normally only available as a complete field coils and frame assembly.

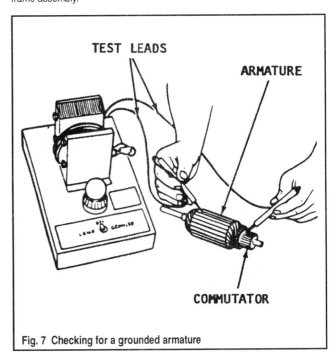
Fig. 7 Checking for a grounded armature

ASSEMBLY

② MODERATE

1. Press a NEW seal into the end cap with the lip and seal spring facing UP, toward the tool. Use a flat ended bar. Continue to press the seal into place until the seal is below the chamber in the end cap.

2. Install a new gasket onto the end cap.

3. Split the end of NEW brush lead, *carefully*, to 1/8 in. (3.18mm) to fit over the piece of brush lead left on the motor head. Fit the end of new split brush leads over the old brush lead ends on the motor head, and then twist them slightly.

4. Hold the new leads with a pair of pliers and solder them using rosin core solder. The pliers will act as a heat barrier, so the solder will not spread and the rest of the brush leads will stay flexible.

5. Install a NEW end cap gasket. Slide the new gasket over the wires and the end head into position.

6. Use two paper clips bent to make brush and spring holders, as shown in the accompanying illustration. Install the new brush springs and insert the brushes into position using the paper clip as a tool to hold the brushes retracted in place.

7. Install the armature into the motor housing and into the motor head, with a fiber washer on the shaft at each end of the armature. Remove the paper clips to permit the brushes to ride on the armature.

8. Install new washer seals on each thru bolt. Tighten the thru-bolts to 20 inch lbs. (2.5 Nm). Apply Johnson/Evinrude Black Neoprene Dip, or equivalent, over the bolt heads and over the two end cap gaskets to prevent any leakage.

9. Run the motor for a few seconds in both directions to seat the brushes.

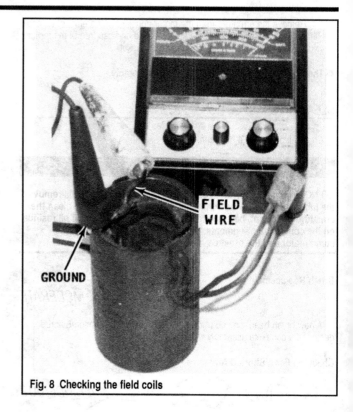

Fig. 8 Checking the field coils

9

REMOTE CONTROLS

REMOTE CONTROLS

General Information

◆ See Figures 1 thru 5

Remote controls are seldom obtained from the original equipment manufacturer, except in the case of the electric shift unit. The electric shift box is considered a part of the new engine. Therefore, unless an owner made a change, the electric shift unit with the engine is probably original engine manufacturer equipment. Mechanical shift units are sold and installed separately.

Shift boxes, steering, and other similar equipment may be added after the boat leaves the plant. Because of the wide assortment, styles, and price ranges of such accessories, the distributor, dealer, or customer has a wide selection from which to draw, when outfitting the boat.

Therefore, the procedures and suggestions in this section are general in nature in order to cover as many units as possible, but still specific and in enough detail to allow troubleshooting, repair, and adjustment for each of these accessories. Proper operation will do much for maximum comfort, performance, and enjoyment.

Complete procedures for removal, installation, and adjustment of shift arrangements are covered in this chapter: manual shift; electric shift; push-button shift mechanism; and the single-lever remote control shift box. These shift boxes are all considered original Johnson/Evinrude equipment.

Description & Operation

◆ See Figures 1 thru 5

Undoubtedly, the most used accessory on any boat is the shift control box. This unit is a remote-control device for shifting the outboard and at the same time controlling the throttle.

Engines equipped with the manual mechanical shift are the only engines sold that do not have a shift box included as part of the complete package. If this engine is to be converted to a shift box operation, a shift box kit must be purchased as an accessory.

Because the cable length requirements cannot be known for each installation, the shift and throttle cables must be purchased separately. When the cables are purchased, the cable ends will have the end fittings attached and ready for installation at the shift box and the engine. A kit is also available enabling the owner of old-style cables to adapt to the new shift boxes.

Johnson/Evinrude equipped boats may be equipped with one of a number of different early-model shift boxes:
- Single lever manual shift
- Double lever manual shift
- Single lever electric shift
- Pushbutton electric shift
- The new improved electric shift box incorporating all of the electrical harness, key switch, and the "hot horn"

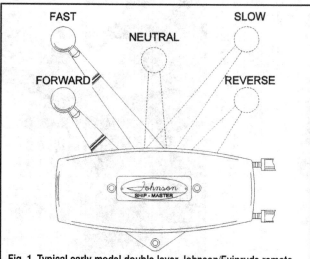

Fig. 1 Typical early model double lever Johnson/Evinrude remote control unit for models with a mechanical shift lower unit

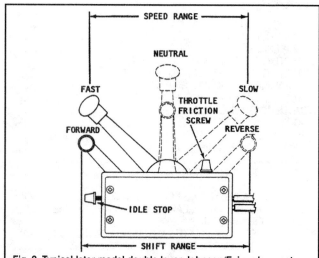

Fig. 2 Typical later model double lever Johnson/Evinrude remote control unit for models with a mechanical shift lower unit

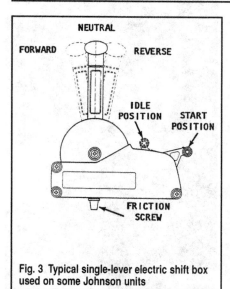

Fig. 3 Typical single-lever electric shift box used on some Johnson units

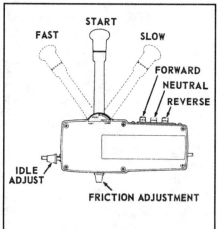

Fig. 4 Typical single-lever electric pushbutton shift box used on some Evinrude units

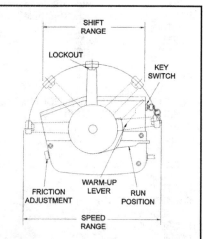

Fig. 5 Typical single-lever Johnson/Evinrude remote control unit with integral key, choke and warning horn

However, any number of late-model shift boxes may have been retrofitted to a number of the mechanical shifting motors.

The mechanical shift box units have 2 levers, a long lever handle and a short lever handle. The long handle controls the throttle and the short one the shift mechanism.

The electric shift units, including the pushbutton models, have only one lever handle for control of the shift and throttle. The shift box installed with Johnson engines has one handle for shifting and throttles control. Another lever, considered a warm-up lever, is installed at the rear, or at the side of the box. This warm-up lever may be adjusted for low and fast idle speeds. The pushbutton type locks out the shift lever to prevent shifting if the throttle is advanced too far while the engine is in neutral.

The single lever remote control box has one lever for both shift and throttle control, plus the warm-up lever on the side of the box, and a safety button on the handle to prevent movement into gear unless the button is depressed.

As the name implies, the 2-lever manual shift box uses the 2 lever principle for the throttle and shift. A friction feature on the throttle mechanism permits the operator to release his grip on the lever handle without the throttle changing position. An idle stop is also built into the shift box. This feature prevents the throttle from being retracted past normal idle to the point where the engine would shut down.

Outboard models are equipped with a cut-out switch in the cranking system to open the circuit to the starter solenoid. This arrangement prevents the cranking system from operating unless the throttle is in the proper idle range. Stating it another way, the throttle must be in the idle position or the starter system will not operate. The position of the shift lever does not affect the starting motor circuit.

All shift box models have a means of advancing the throttle without moving the shift lever into gear. This device is commonly known as the warm-up lever and may be adjusted for low and fast idle speeds.

Double Lever Shift Box

TROUBLESHOOTING

◆ See Figures 1, 2 and 6 thru 10

The following paragraphs provide a logical sequence of tests, checks, and adjustments, designed to isolate and correct a problem in the shift box operation.

The double-lever shift boxes are fairly simple in construction and operation. Seldom do they fail creating problems requiring service in addition to normal lubrication.

Hard Shifting or Difficult Throttle Advance
Checking Throttle Side

Remove the throttle and shift control at the engine. Now, at the shift box, attempt to move the throttle or shift lever.

If the lever moves smoothly, without difficulty, the problem is immediately isolated to the engine. The problem may be in the tower shaft between the connector of the throttle and the armature plate. The armature plate may be "frozen", unable to move properly.

If the problem with shifting is at the engine, the first place to check is the area where the shift lever extends through the exhaust housing. The bushing may be worn, or corroded. If the bushing requires replacement, the engine power head must be removed. Another cause of hard shifting is water entering the lower unit. In this case the lower unit must be disassembled.

If hard shifting is still encountered at the shift box when the controls are disconnected from the engine, the cables may be corroded, and require replacement, or lack of lubrication in the shift box has resulted in excessive wear, or corrosion.

Unable to Obtain Full Shift Movement or Full Throttle

Normally, this type of problem is the result of improper shift box installation. This area includes connection of the shift and throttle cables in the shift box.

If the stainless steel inner wire was not heated and the clamp did not hold the inner cable (wire), the wire could slip inside the sleeve and the cable would be shortened.

Therefore, if it is not possible to obtain full shift or full throttle, the shift box must be removed, opened, and checked for proper installation work. The inner wire could also slip at the engine end of the control, but problems at that end are very rare. Usually if improper installation work has been done at the engine end, the ability to shift at all is lost, or the throttle cannot be actuated.

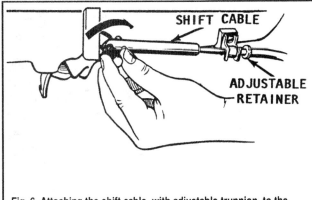
Fig. 6 Attaching the shift cable, with adjustable trunnion, to the shift arm

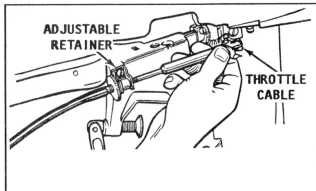
Fig. 7 Connecting the throttle cable with adjustable trunnion

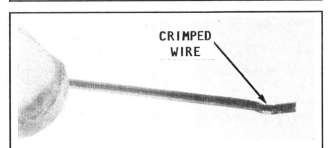
Fig. 8 Here's an inner shift wire after removal. Always make sure it is crimped properly

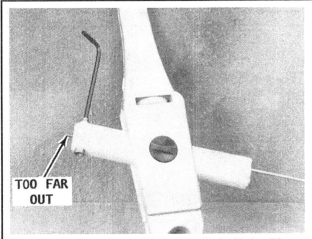
Fig. 9 The wire is too far out and should be flush with the slider surface

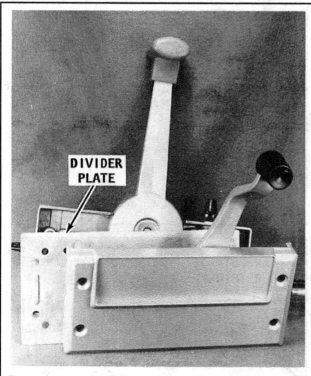

Fig. 10 A good look at the divider plate

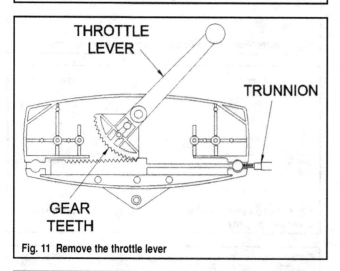

Fig. 11 Remove the throttle lever

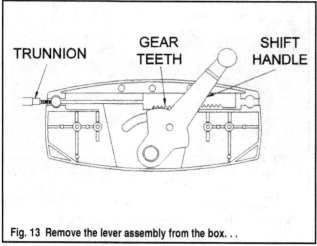

Fig. 13 Remove the lever assembly from the box. . .

DISASSEMBLY & ASSEMBLY

Early Models

◆ **See Figures 1 and 11 thru 15**

　1. Removing the shift box:

　a. Remove the attaching hardware securing the shift box to the side of the boat.

　b. Once the shift box is free, the service work may be performed in the boat. The cables may remain as routed.

　c. Remove the 2 screws, on the side at the rear of the shift box, holding the 2 halves together. Separate the halves.

■ **Observe how one side accommodates the throttle and the other side the shift mechanism. Notice the plastic plate between the 2 halves. This plate prevents any contact between the shift parts and those for the throttle.**

　2. Throttle half disassembly:

　a. Remove the screw attaching the throttle handle to the shift box.

　b. Lift the throttle handle and rocker free of the shift box. If the handle is to be replaced, the shift ball on the end of the handle must be removed and saved because a new ball is not included with a new handle.

　c. Remove the 2 Allen screws securing the ratchet to the end of the shift cable and then pull the ratchet free of the cable end. Take care not to lose the small brass sleeve.

■ **As the ratchet is removed from the cable end, take notice of exactly where the Allen screws made contact with the cable as an aid during installation.**

　d. Remove the small sleeve from the ratchet.

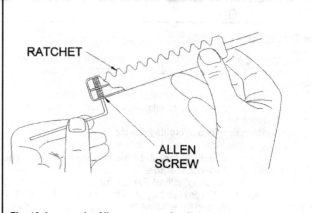

Fig. 12 Loosen the Allen screws and pull the ratchet from the shift cable

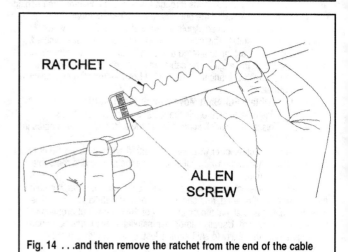

Fig. 14 . . .and then remove the ratchet from the end of the cable

3. Shift half disassembly:

a. Lift the lever handle assembly free of the shift box.

b. Remove the ratchet from the end of the shift cable. This is accomplished by loosening the Allen screws in the sleeve and pulling the ratchet free of the cable. Take care not to lose the small brass sleeve.

Cleaning & Inspection

Check to be sure the teeth on the rocker of the shift and throttle handle are not worn or damaged.

Inspect the ratchets removed from the end of the cables. The teeth should not be damaged or worn excessively.

Wash the outside and inside of the shift box halves with solvent and dry them thoroughly with a cloth or compressed air.

If the throttle or shift cables are not to be replaced, now is an excellent time to lubricate the inner wire. To lubricate the inner wire, remove the casing guide from the cable at both ends. Attach an electric drill to one end of the wire. Momentarily turn the drill on and off to rotate the wire and at the same time allow lubricant to flow into the cable, as shown.

To assemble:

■ **Location of the cable end is of the utmost importance. One Allen screw must be tightened hard, until there is a definite crimp in the wire. If the Allen screw is not tightened enough, the cable will slip in the sleeve and the adjustment will be lost.**

4. Check the end of the cable to determine if the temper has been removed. If the end has a bluish appearance, it has been heated at an earlier date and the temper removed. The temper must be removed to permit the holding screw to make a crimp in the wire to hold an adjustment. If the wire has not been tempered, heat the end, but not enough to melt the wire. Be careful not to overheat the stainless steel wire with a torch because it has a very low melting point.

5. Assembling - wire cable to ratchet on the throttle & shift cables:

a. Working with either cable, insert the small sleeve onto the end of the ratchet with the 2 Allen screws started in the sleeve. Notice how the sleeve has a hole completely through it? The wire must pass through the hole and protrude out the end of the ratchet.

b. Work the ratchet onto the end of the cable. Continue working the ratchet onto the cable until the inner cable end is completely through the ratchet, but just flush with end surface of the ratchet. Tighten 1 Allen screw until a definite crimp is made in the cable, and then tighten the other Allen screw securely.

c. Repeat for the other cable.

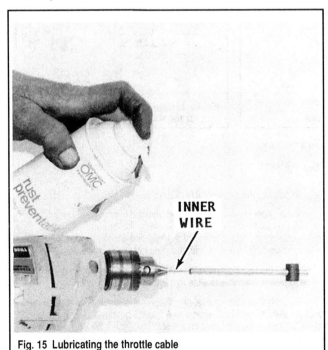

Fig. 15 Lubricating the throttle cable

INNER WIRE

d. Apply a coating of lubricant around the pivot point of the throttle lever bushing. Set the lever in position inside the box half.

e. Install the washer with the concave side of the washer on the same side as the screw. Install the screw and then move the throttle lever back-and-forth to check for freedom of movement.

f. Slip the ratchet and trunnion into the shift box starting the last tooth on the throttle rocker engaged with the last tooth of the slider. Again, move the throttle lever back-and-forth to check for freedom of movement. Check to be sure the same number of teeth are engaging on the rocker as on the slider. The teeth indexing is extremely important and the key to a successful installation. The ratchet and rocker both have the same number of teeth. All teeth must be used on the ratchet and rocker when the lever handle is moved to maximum forward or aft.

6. Shift half assembly:

a. Lay the shift ratchet in position inside the shift half of the box. Place the shift lever down over the top of the ratchet with the last tooth on the lever rocker engaged with the last tooth of the slider.

b. Apply a coating of light lubricant onto the surface of the gear teeth, the bottom and side walls of the ratchet and onto the full length of the gear rack. Do not use grease that will harden because the lubricating qualities will be lost leaving the assembly dry.

c. Install the divider separating the throttle and shift halves of the box. Bring the 2 box halves together, and then install the retaining screws. Double check operation of both levers for smoothness and no evidence of binding.

d. Install the shift box onto the side of the boat and secure it in place with the attaching hardware.

Later Models

MODERATE

◆ See Figures 2 and 16 thru 20

1. Remove the attaching hardware securing the shift box to the side of the boat.

■ **Once the shift box is free, the service work may be performed in the boat. The cables may remain as routed.**

2. Remove the 2 screws, at the rear side of the shift box, holding the 2 halves together. Separate the halves.

3. Observe how one side accommodates the throttle and the other side the shift mechanism. Notice the metal plate between the 2 halves. This plate prevents any contact between the shift parts and those for the throttle. Notice the friction screw and throttle stop on the throttle side of the box. The shift side of the box does not have any adjustments, except for the low idle stop. Observe how the shift lever pivots at the bottom and the throttle lever pivots at the top.

4. To disassemble the throttle half:

a. Remove the screw from the center of the throttle lever. Notice how the washer has a concave side to allow the screw to fit flush with the washer.

b. Loosen the screw on the top side of the shift box to relieve pressure on the anti-friction knob.

c. Lift the lever and throttle cable free of the shift box. Notice how the cable on the trunnion has 2 small caps - one on the underside and the other on top.

d. Loosen the 2 screws securing the gear rack to the end of the throttle cable and remove the end of the cable from the throttle lever. Take care not to lose the small sleeve from the end of the rack to which the screws were attached.

e. Push out the center square button, and then remove the rack from the shift lever.

5. To disassemble the shift half:

a. Remove the screw and washer from the bottom of the shift lever. Notice how this washer also has a concave side to accommodate the screw.

b. Lift the lever and shift cable free of the shift box. Notice how the cable on the trunnion has 2 small caps - one on the underside and the other on top.

c. Loosen the two screws securing the gear rack to the end of the shift cable and remove the end of the cable from the throttle lever. Take care not to lose the small sleeve from the end of the rack to which the screws were attached.

d. Push out the center square button, and then remove the rack from the shift lever.

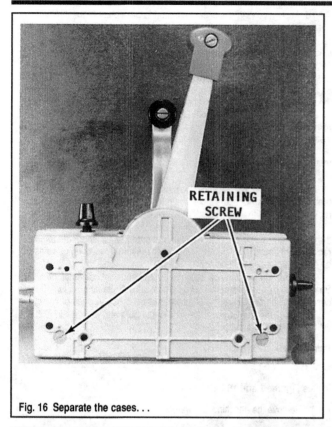

Fig. 16 Separate the cases. . .

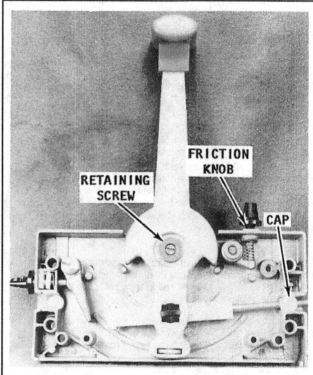

Fig. 17 . . .and then remove the handle bolt and loosen the friction knob

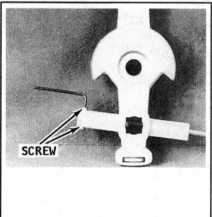

Fig. 18 Loosen the 2 screws on the gear rack for the throttle cable

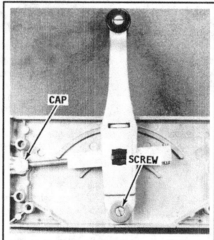

Fig. 19 Removing the shift lever

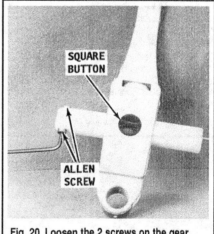

Fig. 20 Loosen the 2 screws on the gear rack for the shift cable

Cleaning & Inspection

◆ See Figure 15

6. Check the nylon wear block on the end of the anti-friction cap. The cap has teeth which index into the inside diameter of the throttle lever. If the teeth are damaged a new block may be purchased and slipped into place.

7. Clean the box halves thoroughly inside and out with solvent, and then dry them with compressed air.

8. Inspect the spring on the anti-friction lever to be sure it is not distorted.

9. Check the screw on the throttle idle stop to ensure it moves in and out freely without any sign of binding.

■ If the throttle or shift cables are not to be replaced, now is an excellent time to lubricate the inner wire.

10. To lubricate the inner wire, remove the casing guide from the cable at both ends. Attach an electric drill to one end of the wire. Momentarily turn the drill on and off to rotate the wire and at the same time allow lubricant to flow into the cable.

To assemble:

◆ See Figure 21

11. To connect the throttle cable to the shift box, proceed as follows:

a. If the slider sleeve was removed from the throttle lever, install the slide rack into the throttle lever with the hole on the end for securing the throttle cable on the opposite end of the hole that accommodates the cable. Position the center of the slide with the center of the throttle lever.

b. Install the square nylon plug with the holes in the plug in a vertical position to permit the cable to slide through.

c. Two different size screws, or possibly Allen screws, are used on each end of the sleeve. Install the short screw into the bottom of the sleeve to prevent the sleeve from rubbing on the shift box. Install the longer screw on the top part of the sleeve. Install the sleeve with the hole in the sleeve aligned with the hole for the cable.

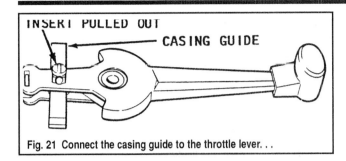

Fig. 21 Connect the casing guide to the throttle lever. . .

✳✳ CAUTION

Check the end of the cable to determine if the temper has been removed. If the end has a bluish appearance, it has been heated at an earlier date and the temper removed. The temper must be removed to permit the holding screw to make a crimp in the wire to hold an adjustment. If the wire has not been tempered, heat the end, but not enough to melt the wire.

d. Slide the cable into the rack. Work the inner wire into the sleeve and out the end of the rack.

e. Push the wire back until the end is flush with the rack surface. Tighten the top holding screw enough to make a definite crimp in the wire. If this screw is not tightened to make the crimp, the wire will slip during operation and the adjustment will be lost.

f. After the top screw has been fully tightened, bring the other screw up tight against the wire. It is not necessary for this second screw to make a crimp in the wire.

g. Work the throttle levers handle down over the friction nylon block and at the same time feed the throttle cable into place in the box half. Check to be sure one of the small caps is on the bottom side of the trunnion.

h. Install the washer and the screw with the concave side of the washer on the same side as the screw. Tighten the screw securely. Install the other trunnion cap on top of the trunnion.

i. Check the throttle lever for ease of movement with no sign of binding.

12. To connect the shift cable to the shift box, proceed as follows:

a. If the slider sleeve was removed from the shift lever, install the slide rack into the shift lever with the hole on the end for securing the shift cable on the opposite end of the hole that will accommodate the cable. Position the center of the slide with the center of the shift lever.

b. Install the square nylon plug with the holes in the plug in a vertical position to permit the cable to slide through.

c. Two different size screws are used on each end of the sleeve. Install the short screw into the bottom of the sleeve to prevent the sleeve from rubbing on the shift box. Install the longer screw on the top part of the sleeve. Install the sleeve with the hole in the sleeve aligned with the hole for the cable.

✳✳ CAUTION

Check the end of the cable to determine if the temper has been removed. If the end has a bluish appearance, it has been heated at an earlier date and the temper removed. The temper must be removed to permit the holding screw to make a crimp in the wire to hold an adjustment. If the wire has not been tempered, heat the end, but not enough to melt the wire.

d. Slide the cable into the rack. Work the inner wire into the sleeve and out the end of the rack.

e. Push the wire back until the end is flush with the rack surface. Tighten the top holding screw enough to make a definite crimp in the wire. If this screw is not tightened to make the crimp, the wire will slip during operation and the adjustment will be lost.

f. After the top screw has been fully tightened, bring the other screw up tight against the wire. It is not necessary for this second screw to make a crimp in the wire.

g. Place the wavy washer and regular washer into the shift box, and then work the shift lever handle down into the shift box with one of the small caps under the shift cable trunnion.

h. Install the bushing into the bottom of the shift handle.

i. Install the washer and the screw with the concave side of the washer on the same side as the screw. Tighten the screw securely.

j. Install the other trunnion cap on top of the trunnion.

13. Check the shift lever for ease of movement with no sign of binding.

14. Place the divider plate between the two halves. Bring the two halves together and secure them with the two screws from the back side.

15. Install the box in the boat and secure it in place with the attaching hardware. Again check the levers for ease of movement and no sign of binding.

Electric Gear Boxes & Single Lever Controls (Johnson)

TROUBLESHOOTING

◆ **See Figure 3**

The accompanying sections provide a logical sequence of tests, checks and adjustments designed to isolate and correct a problem in the Johnson single lever shift box with the warm-up lever to the rear and the Evinrude single lever push button unit.

Difficult Shift Operation

◆ **See Figure 22**

Many times this type of problem is the result of incorrect cable installation - the cable is not the proper length or there are too many bends or kinks in the routing. Such an installation will cause the inner cable to travel much further than necessary and therefore, wear on the outer cable. Over a period of time, inner cable wear will result in difficult shifting or throttle operation.

Be sure to cycle the shift lever to the full position in both directions, when making any test on the shift box. The shift switch may have a dead spot and will not indicate the switch is defective unless the shift lever is fully cycled for each test.

Amp Draw Test

◆ **See Figure 23**

1. Turn the ignition switch to the **ON** position and note the ammeter reading.

2. Now, operate the shift control lever to the Forward, Neutral, and then to the Reverse position. Note how much the ammeter reading increased each time the shift lever was moved.

3. If the reading was more than 2.5 amperes (1.5-2.0 amps for 1971-72 2 cyl models) for any of the 3 shift positions, continue with the following checks.

4. If the boat is not equipped with an ampere gauge, then temporarily disconnect the Green and Brown (or Red) wires from the back side of the

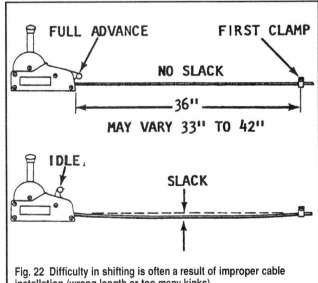

Fig. 22 Difficulty in shifting is often a result of improper cable installation (wrong length or too many kinks)

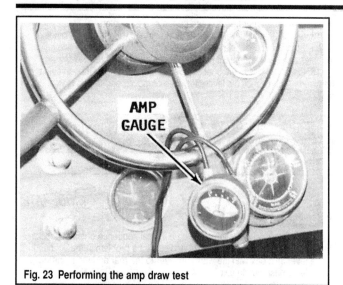

Fig. 23 Performing the amp draw test

key switch and temporarily install an amp gauge for the test. Replace the wires after the test is completed.

5. Disconnect the shift leads at the rear of the engine. Temporarily lay a piece of cloth or other insulating material under the wires to prevent them from shorting out during the following tests.

Standard Prop Models:

a. Again operate the shift lever and note the current loss. If the current draw is still more than 2.5 amperes (1.5-2.0 on 1971-72 2 cyl), then check for a short in the control box switch or wiring.

b. If the current draw is normal with the leads disconnected from the engine, then check for a short in the gear case coil(s) or wiring.

c. If the coil leads are shorted to each other, both shift coils would be energized, stalling the engine or causing serious damage to the driveshaft.

Propeller Exhaust Models:

a. Engines equipped with the prop exhaust system have 2 shift solenoids installed in the lower unit. A check of these solenoids is presented in the next test.

Shift Coil Test

◆ See Figure 24

1. Testing the shifting coils is accomplished by first disconnecting the wires at the rear of the engine.

2. Next, connect an ohmmeter first to one shift coil lead and ground, and then to the other in the same manner. A reading of more than 4.5-6.5 ohms indicates a short in the coil or lead.

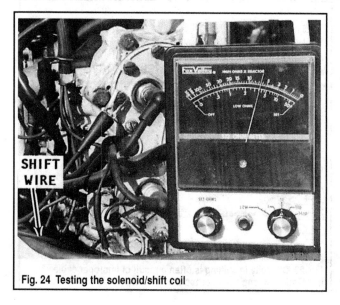

Fig. 24 Testing the solenoid/shift coil

3. No reading at all indicates an open circuit.

4. If the results of this test indicate a short in the circuit, the lower unit must be disassembled, inspected and serviced, to correct the problem.

Shift Solenoid Test

◆ See Figure 24

1. Obtain an ohmmeter. At the rear of the engine, disconnect the Green and Blue wires at the knife disconnect.

2. Set the ohmmeter to the low scale.

3. Connect 1 lead to the Green wire leading to the lower unit, and the other lead to a good ground. The meter should indicate 5-7 ohms.

4. Connect the meter to the Blue wire leading to the lower unit and ground. The meter should again indicate from 5-7 ohms.

■ **If the unit fails the ohmmeter tests just outlined, the only course of action is to disassemble the lower unit to determine and correct the problem**

Testing The Shift Switch - Forward

◆ See Figure 25

■ **Please note that on 1971-72 2 cylinder models, the testing order is reversed. The Forward shift switch is tested after testing of the Neutral and Reverse switches has been completed. Please skip this procedure and move to the Neutral one.**

1. On standard exhaust models:

a. To test the switch for the Forward position, make contact with 1 probe of a continuity meter (or a test light) to the terminal (Purple or Red lead) and to the forward (Green lead) terminal with the other probe.

b. Now, move the shift lever to the Forward position. The meter should indicate continuity (or the test light comes on) when the shift lever is in the Forward position.

c. Move to the Reverse switch test.

2. On propeller exhaust models:

a. To test the switch for the Forward position, make contact with 1 probe of a continuity meter, or test light, to the Purple or Red lead terminal. Make contact with the other meter probe to the Blue or Green wire terminal.

b. Move the shift lever into the Forward position. The test light should not come on, or the meter should indicate no continuity. If the light comes on, the switch has a short.

c. Move to the Reverse switch test.

3. 1971-72 2 cyl. models:

a. After the Neutral and Reverse tests have been completed, check for continuity with the shift lever in the Forward position.

b. Leave the Red lead connected and check the Blue lead (Reverse) and the Green lead (Neutral).

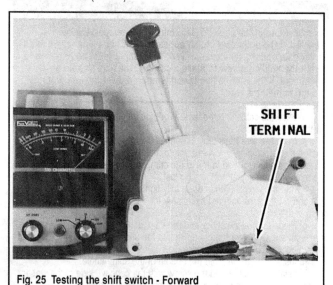

Fig. 25 Testing the shift switch - Forward

Continuity should not be indicated when the shift handle is in Forward. If the switch is defective and requires replacement.

Testing The Shift Switch - Reverse

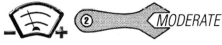

◆ See Figure 26

1. Standard exhaust models:

a. To test the shift switch for Reverse, make contact with 1 probe of a continuity meter (or test light), to the terminal (Purple or Red lead) and to the Reverse terminal (Blue lead) with the other probe.

b. Move the shift lever to the Reverse position. The meter should indicate continuity (or the test light comes on), when the lever is in the Reverse position.

c. Move to the Neutral switch test.

2. Propeller exhaust models:

a. To test the shift switch for Reverse, make contact with 1 probe of a continuity meter (or test light), to the Purple or Red lead terminal and with the other probe to the Blue lead Reverse terminal.

b. Move the shift lever to the Reverse position. The meter should indicate continuity or the test light should come on.

c. Move the 1 probe from the Blue lead terminal to the Green lead terminal. The meter should again indicate continuity or the test light should come on.

d. Move to the Neutral switch test.

3. 1971-72 2 cyl. models:

■ **Please note that on 1971-72 2 cylinder models, the Neutral shift switch is tested first. Please skip back to the procedures for the Reverse terminal after this.**

a. Move the shift lever to the Reverse position.

b. To test the shift switch for reverse, make contact with 1 probe of a continuity meter (or test light), to the Purple or Red lead from the plug.

c. Now, with the other probe, make contact alternately to the Blue and Green wires in the plug.

d. The meter should indicate continuity (or the test light comes on), during both of these tests when the lever is in the Reverse position.

e. Move to the Forward switch test.

Testing The Shift Switch - Neutral

◆ See Figure 27

1. Standard exhaust models:

a. After the Forward and Reverse tests have been completed, check for continuity with the shift lever in the Neutral position.

b. Leave the Red lead connected and check the Blue lead (Reverse) and the green lead (Forward). Continuity should not be indicated when the shift handle is in Neutral.

c. If the switch is defective and requires replacement, procedures are presented in this section under Disassembly.

2. Propeller exhaust models:

a. After the Forward and Reverse tests have been completed, check for continuity with the shift lever in the Neutral position.

Leave 1 probe of the meter or test light connected to the Red lead terminal. Make contact with the other probe to the Green lead coming from the shift box. The meter should indicate continuity or the light should come on.

b. Test the Blue wire in a similar manner. Continuity should NOT be indicated, or the light should not come on.

3. 1971-72 2 cyl. models:

■ **Please note that on 1971-72 2 cylinder models, the Neutral shift switch is tested first. Please skip back to the procedures for the Reverse terminal after this.**

a. To test the switch for the Neutral position, make contact with 1 probe of a continuity meter (or a test light) to the terminal (Purple or Red lead) and to the (Green lead) terminal with the other probe.

b. Now, move the shift lever to the Neutral position. The meter should indicate continuity (or the test light comes on), when the shift lever is in the Neutral position.

c. Move to the Reverse procedure.

Cranking System Inoperative

◆ See Figure 28

1. If the starter fails to crank the engine, check to be sure the throttle lever is in the idle position.

2. If the throttle is advanced more than 1/4 forward, the cut-out switch attached to the armature plate will open the circuit to the starter solenoid. Some later models *may* have the cut-out switch installed in the shift box (but we doubt it). If the cranking system fails to operate the starter properly when the throttle lever is in the idle position, check the 20 amp fuse between the ignition switch **BAT** terminal and the ammeter **GEN** terminal.

3. If the starter operates in full throttle (which it should not do), check for a short between the 2 white leads in the shift box wiring.

4. Further problems in the cranking system may indicate more serious problems with the system itself.

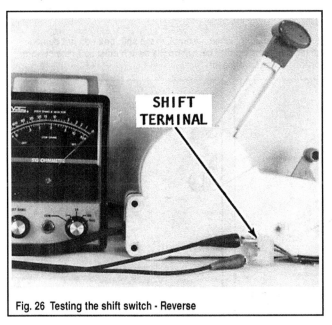

Fig. 26 Testing the shift switch - Reverse

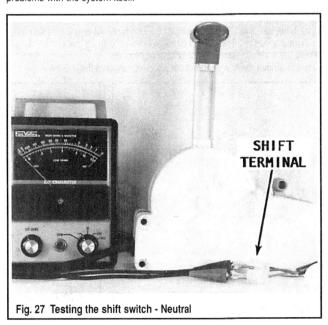

Fig. 27 Testing the shift switch - Neutral

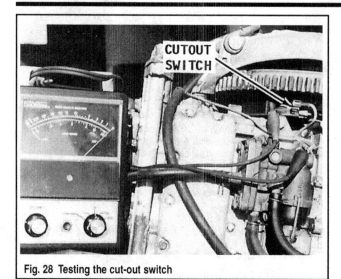

Fig. 28 Testing the cut-out switch

DISASSEMBLY & ASSEMBLY

◆ See Figures 3 15 and 29 thru 34

■ Before starting work on the shift system, disconnect both battery cables at the battery terminals.

■ On many Johnson units, a friction screw is installed in the bottom side of the shift box. This screw allows friction adjustment of the throttle handle. This arrangement prevents the throttle handle from "creeping" after the operator releases his grip on the handle. The box has a maximum advance screw. The adjustment is made through movement of a screw in the warm-up lever. If the engine shuts down when the throttle lever is moved back, then an adjustment must be made at the engine. This is accomplished through an adjustment knob at the engine. Movement of this knob will actually lengthen or shorten the cable slightly for proper operation.

1. Remove the attaching screws; pull the shift box clear; and then disconnect the shift wire under the dash.

2. Remove the screws from the back side of the box.

3. Carefully separate the two halves. Check the shift box for salt water corrosion, worn bushings, and general condition.

4. To disconnect the throttle cable, proceed as follows:

a. Notice how the throttle cable enters the throttle half of the shift box through the idle link. On the top side of the shift box, observe the screw and the concave washer. The washer must be installed with the concave side toward the screw to allow the screw to seat properly.

b. Remove the bushing from the shift rod. Observe the 2 slots in the

bushing and how the flat area without a hole faces toward you. The bushing must be installed in this same position.

c. Remove the screw and washer from the top of the shift box half. This is the screw and washer described in the previous paragraph. Lift out the cam lever and the idle link as an assembly.

d. Remove the screws from the end of the sleeve on the end of the cable, and then pull the throttle cable free of the link and sleeve.

e. If the switch fails to check out, as described in the previous tests, the switch and cable assembly must be replaced. The switch is easily removed by simply removing the attaching screws and lifting the switch free of the shift box.

To assemble:

5. Clean the box halves thoroughly inside and out with solvent and blow them dry with compressed air.

6. Apply a thin coat of engine oil on all metal parts.

7. The 3-position switch installed in the gear box cannot be repaired. Therefore, if a problem is isolated to the switch, it must be replaced.

8. If the throttle cable is not to be replaced, now is an excellent time to lubricate the inner wire.

9. To lubricate the inner wire, remove the casing guide from the cable at both ends.

10. Attach an electric drill to one end of the wire. Momentarily turn the drill on and off to rotate the wire and at the same time allow lubricant to flow into the cable.

✳✳ CAUTION

Check the end of the cable to determine if the temper has been removed. If the end has a bluish appearance, it has been heated at an earlier date and the temper removed. The temper must be removed to permit the holding screw to make a crimp in the wire to hold an adjustment. If the wire has not been tempered, heat the end, but not enough to melt the wire.

It is very easy to shear the wire by applying excessive force when tightening the screw to make the crimp. Therefore, play it cool. Tighten the screw; make one more complete turn to make the crimp; call it good; and then bring the other screw up just tight against the wire.

11. Start the 2 cable retaining screws into the sleeve. These screws are different sizes. On some models, Allen screws are used. Install the short screw on the bottom to prevent the sleeve from rubbing on the shift box. Slide the sleeve onto the cable with the hole aligned with the hole in the plastic sleeve. Feed the wire on through until the end of the wire is flush with the end of the white plastic sleeve.

12. Lower the shift link and throttle link into the box half and secure the throttle link with the screw and washer. Check to be sure the concave side of the washer is facing toward the screw side to permit the screw to seat properly.

13. Slide the throttle cable through the idle link, and then slide the bushing down over the cable.

■ Always be careful when assembling the shift box - do not damage the remote control unit. The arm on the switch must lie in the cut-out portion of the throttle cam.

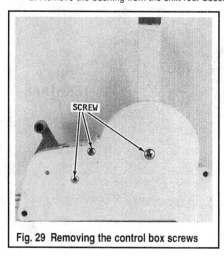

Fig. 29 Removing the control box screws

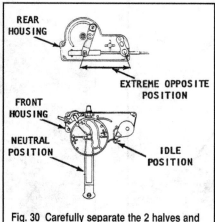

Fig. 30 Carefully separate the 2 halves and inspect the internal components

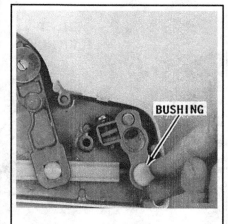

Fig. 31 Remove the bushings. . .

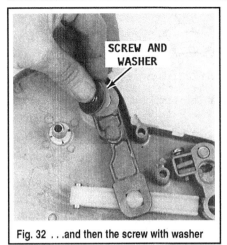

Fig. 32 . . .and then the screw with washer

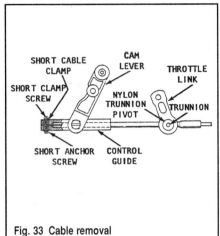

Fig. 33 Cable removal

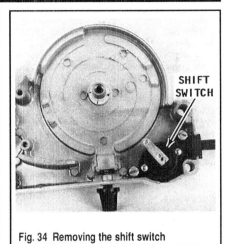

Fig. 34 Removing the shift switch

14. Carefully work the 2 halves of the box together with the cam lever fitting into the recess of the throttle handle and the throttle link fitting into the warm-up lever.

15. Secure the halves together with the screws into the side of the box. Secure the shift box to the side of the boat with the attaching hardware. Bolts with self-locking nuts should always be used because a loose shift box during high speed operation could be extremely dangerous. Connect the shift wire under the dash.

16. The tension of the throttle lever is adjusted by the friction knob under the shift box. Turn the knob clockwise to increase friction and counterclockwise to decrease friction.

REMOTE CONTROL CABLE - INSTALLATION IN THE BOAT

◆ See Figure 22

The remote control cable must be installed properly for satisfactory operation. The clamp nearest the shift box must be positioned correctly as follows:

1. First, move the warm-up lever on the shift box to full advance. Now, measure 36 in. (actually this measurement could range from 33-42 in.) on the cable from the shift box. BE SURE there is no slack in the cable, and then secure the clamp to the boat at the measured position.

2. Next, place the warm-up lever in the slow position and observe the amount of slack in the cable between the shift box and the first clamp. The slack should not be more than 1/2 in. AVOID SHARP TURNS in the cable. The radius of any bend must not be less than 0.5 in.

■ ALWAYS use the correct length of cable when replacing the assembly.

STARTER LOCK-OUT SWITCH ADJUSTMENT

◆ See Figure 35

1. Move the shift lever to the Neutral position and the warm-up lever to the **START** position.

2. Now, turn the ignition switch to the **START** position in an attempt to crank the engine.

3. If the starter fails to crank the engine, move the warm-up lever to the **IDLE** position, and then rotate the set screw counterclockwise 1/2 turn.

4. Move the auxiliary throttle to the **START** position, and then turn the ignition switch to the **START** position again. If the starter still fails to crank the engine with the warm-up lever in the full ADVANCE position, back off the warm-up lever to determine the point at which the starter ceases to operate.

5. Make the adjustment of the set screw clockwise a half turn at a time until the starter operates only with the warm-up lever in the full **START** position.

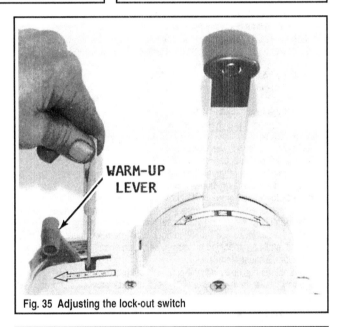

Fig. 35 Adjusting the lock-out switch

Electric Pushbutton Gear Boxes (Evinrude)

DESCRIPTION & OPERATION

◆ See Figure 4

A friction screw is installed in the bottom side of the shift box. This screw allows friction adjustment of the throttle handle. This arrangement prevents the throttle handle from "creeping" after the operator releases his grip on the handle.

A thumbscrew on the front of the box permits adjustment of the throttle handle to prevent movement past a satisfactory idle position and subsequent shutdown of the engine.

If adequate adjustment cannot be made at the shift box, and the engine continues to shut down when the throttle lever is moved back, then an adjustment must be made at the engine. This is accomplished through an adjustment knob at the engine. Movement of the engine knob will actually lengthen or shorten the cable slightly for proper operation.

TROUBLESHOOTING

The accompanying sections provide a logical sequence of tests, checks and adjustments designed to isolate and correct a problem in the Johnson single lever shift box with the warm-up lever to the rear and the Evinrude single lever push button unit.

Difficult Shift Operation

Many times this type of problem is the result of incorrect cable installation - the cable is not the proper length or there are too many bends or kinks in the routing. Such an installation will cause the inner cable to travel much further than necessary and therefore, wear on the outer cable. Over a period of time, inner cable wear will result in difficult shifting or throttle operation.

■ **Be sure to cycle all 3 shift pushbuttons to the full shift position in both directions, when making any test on the shift box. The shift switch may have a dead spot and will not indicate the switch is defective unless the 3 buttons are fully cycled for each test.**

Amp Draw Test

◆ **See Figure 23**

1. Turn the ignition switch to the **ON** position and note the ammeter reading.
2. Now, operate the push button for the Forward, Neutral, and then the Reverse position. Note how much the ammeter reading increased each time the shift button was depressed.
3. If the reading was more than 2.5 amperes (1.5-2.0 amps for 1971-72 2 cyl models) for any of the 3 shift positions, continue with the following checks.
4. If the boat is not equipped with an ampere gauge, then temporarily disconnect the Green and Brown (or Red) wires from the back side of the key switch and temporarily install an amp gauge for the test. Replace the wires after the test is completed.
5. Disconnect the shift leads at the rear of the engine. Temporarily lay a piece of cloth or other insulating material under the wires to prevent them from shorting out during the following tests.

Standard Prop Models:
a. Again operate/depress the shift buttons and note the current loss. If the current draw is still more than 2.5 amperes (1.5-2.0 on 1971-72 2 cyl), then check for a short in the control box switch or wiring.
b. If the current draw is normal with the leads disconnected from the engine, then check for a short in the gear case coil(s) or wiring.
c. If the coil leads are shorted to each other, both shift coils would be energized, stalling the engine or causing serious damage to the driveshaft.

Propeller Exhaust Models:
a. Engines equipped with the prop exhaust system have 2 shift solenoids installed in the lower unit. A check of these solenoids is presented in the next test.

Shift Coil Test

◆ **See Figure 24**

1. Testing the shifting coils is accomplished by first disconnecting the wires at the rear of the engine.
2. Next, connect an ohmmeter first to one shift coil lead and ground, and then to the other in the same manner. A reading of more than 4.5-6.5 ohms indicates a short in the coil or lead.
3. No reading at all indicates an open circuit.
4. If the results of this test indicate a short in the circuit, the lower unit must be disassembled, inspected and serviced, to correct the problem.

Shift Solenoid Test

◆ **See Figure 24**

1. Obtain an ohmmeter. At the rear of the engine, disconnect the Green and Blue wires at the knife disconnect.
2. Set the ohmmeter to the low scale.
3. Connect 1 lead to the Green wire leading to the lower unit, and the other lead to a good ground. The meter should indicate 5-7 ohms.
4. Connect the meter to the Blue wire leading to the lower unit and ground. The meter should again indicate from 5-7 ohms.

■ **If the unit fails the ohmmeter tests just outlined, the only course of action is to disassemble the lower unit to determine and correct the problem**

Testing The Shift Switch - Forward

◆ **See Figure 36**

■ **Please note that on 1971-72 2 cylinder models, the testing order is reversed. The Forward shift switch is tested after testing of the Neutral and Reverse switches has been completed. Please skip this procedure and move to the Neutral one.**

■ **Before the following shift tests are performed, the shift wires under the dash must be disconnected at the quick-disconnect fitting.**

1. On standard exhaust models:
a. To test the switch for the Forward position, make contact with one probe of a continuity meter (or a test light) to the terminal (purple or red lead) and to the forward (green lead) terminal with the other probe.
b. Now, depress each pushbutton for the Forward, Neutral and Reverse positions. The meter should indicate continuity (or the test light comes on), when the Forward button is depressed, and indicates an open circuit, for the other 2 shift positions.
c. Move to the Reverse switch test.
2. On propeller exhaust models:
a. To test the switch for the Forward position, make contact with 1 probe of a continuity meter, or test light, to the Purple or Red lead terminal. Make contact with the other meter probe to the Blue or Green wire terminal.
b. Depress the Forward shift button. The test light should not come on, or the meter should indicate no continuity. If the light comes on, the switch has a short.
c. Move to the Reverse switch test.
3. 1971-72 2 cyl. models:
a. After the Neutral and Reverse tests have been completed, check for continuity by depressing the pushbutton for the Forward position.
b. Leave the red lead connected and check the blue lead (Reverse) and the green lead (Neutral).
Continuity should NOT be indicated when the Forward pushbutton is depressed. If so, the switch is defective and requires replacement.

Testing The Shift Switch - Reverse

◆ **See Figure 37**

1. Standard exhaust models:
a. To test the shift switch for Reverse, make contact with 1 probe of a continuity meter (or test light), to the purple or red lead terminal and to the Reverse blue lead terminal with the other probe.
b. Again, depress the Forward, Neutral and Reverse pushbuttons. The meter should indicate continuity (or, the test light comes on), for the Reverse position and indicate an open circuit, in the other two.
c. Move to the Neutral switch test.
2. Propeller exhaust models:
a. To test the shift switch for Reverse, make contact with one probe of a continuity meter (or test light), to purple or red lead terminal and with the other probe to the blue lead reverse terminal.
b. Depress the Reverse pushbutton. The meter should indicate continuity or the test light should come on.
c. Move the one probe from the blue lead terminal to the green lead terminal. The meter should again indicate continuity or the test light should come on.
d. Move to the Neutral switch test.
3. 1971-72 2 cyl. models:

■ **Please note that on 1971-72 2 cylinder models, the Neutral shift switch is tested first. Please skip back to the procedures for the Reverse terminal after this.**

a. Depress the pushbutton for the Reverse position.
b. To test the shift switch for Reverse, make contact with 1 probe of a continuity meter (or test light), to the purple or red lead from the plug.
c. Now, with the other probe make contact *alternately* to the blue and green wires in the plug.
d. The meter should indicate continuity (or, the test light comes on), during both of these tests when the pushbutton is depressed for the Reverse position.
e. Move to the Forward switch test.

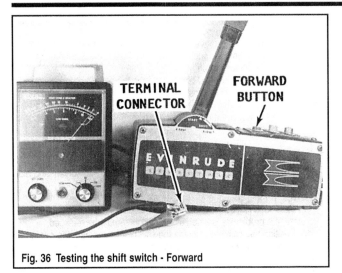

Fig. 36 Testing the shift switch - Forward

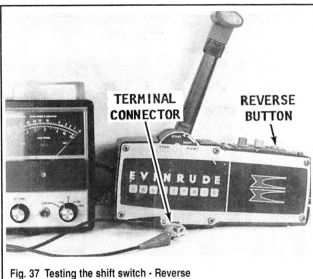

Fig. 37 Testing the shift switch - Reverse

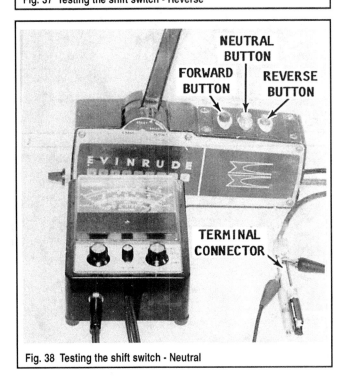

Fig. 38 Testing the shift switch - Neutral

Testing The Shift Switch - Neutral

◆ See Figure 38

1. Standard exhaust models:

a. After the forward and reverse tests have been completed, check for continuity with the Neutral pushbutton.

b. Leave the red lead connected and check the blue lead (Reverse) and the green lead (Forward). Continuity should not be indicated when the Neutral pushbutton is depressed.

c. If the switch is defective and requires replacement, procedures are presented in this section under Disassembly.

2. Propeller exhaust models:

a. After the forward and reverse tests have been completed, check for continuity with the pushbutton depressed for the Neutral position. Leave one probe of the meter or test light connected to the red lead terminal.

b. Make contact with the other probe to the green lead coming from the shift box. The meter should indicate continuity or the test light should come on.

c. Continuity should not be indicated when the probe is making contact with the blue wire.

3. 1971-72 2 cyl. models:

■ Please note that on 1971-72 2 cylinder models, the Neutral shift switch is tested first. Please skip back to the procedures for the Reverse terminal after this.

a. To test the switch for the Neutral position, make contact with 1 probe of a continuity meter (or a test light) to the terminal (purple or red lead) and to the (green lead) terminal with the other probe. Now, depress the Neutral pushbutton.

b. The meter should indicate continuity (or, the test light comes on). When the Forward button is depressed, the meter should indicate NO continuity or the test light should NOT come on.

c. Move to the Reverse procedure.

Cranking System Inoperative

◆ See Figure 39

1. If the starter fails to crank the engine, check to be sure the throttle lever is in the idle position.

2. If the throttle is advanced more than 1/4 forward, the cut-out switch attached to the armature plate will open the circuit to the starter solenoid. Some later models *may* have the cut-out switch installed in the shift box (but we doubt it). If the cranking system fails to operate the starter properly when the throttle lever is in the idle position, check the 20 amp fuse between the ignition switch **BAT** terminal and the ammeter **GEN** terminal.

3. If the starter operates in full throttle (which it should not do), check for a short between the 2 white leads in the shift box wiring.

4. Further problems in the cranking system may indicate more serious problems with the system itself.

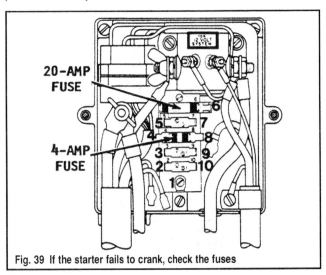

Fig. 39 If the starter fails to crank, check the fuses

DISASSEMBLY & ASSEMBLY

MODERATE

◆ See Figures 4, 15 and 40 thru 47

■ Before starting work on the shift system, disconnect both battery cables at the battery terminals.

The throttle cable and switch box may be replaced without removing the shift box from the boat. If the only service to be performed is replacement of the cable, leave the shift box in place.

1. Remove the 6 Phillips screws on the side plate of the shift box. Remove the front side cover.

2. Notice the screw and retainer at the forward end of the casing guide and just below the throttle lever. Remove the screw and retainer.

3. Pull the throttle cable and casting guide free of the shift box. Take care not loose the trunnion caps, one on the top and another on the bottom.

4. Remove the screws from the end of the casing guide and then pull the throttle cable free. Be careful not to lose the screws and sleeve from the end of the guide.

5. Remove the 4 Phillips screws from the top of the shift box, and then lift off the shift box cover around the push buttons.

6. Pull upward and remove the 3 push buttons. New buttons are not supplied with replacement switches. Therefore, SAVE the 3 buttons for installation with the new switch.

7. Pull the red, green, and blue wires from the bottom of the switch box.

8. Notice the 2 small Phillips screws on top of the switch box holding the switch to the retainer. Remove these 2 screws.

9. Work the switch out of the switch box.

To assemble:

◆ See Figures 48 and 49

10. Clean the box halves thoroughly inside and out with solvent and blow them dry with compressed air.

11. Apply a thin coat of engine oil on all metal parts.

12. The 3-position switch installed in the gear box cannot be repaired. Therefore, if a problem is isolated to the switch, it must be replaced.

13. Position the switch box inside the shift box underneath the retainer and slider and secure it in place with the 2 Phillips screws. Check to be sure the 2 terminals on the bottom side of the switch are towards you. This will place the forward button closest to the throttle handle.

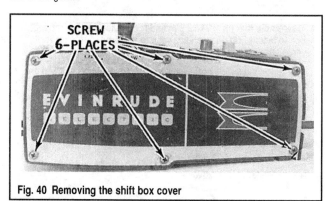

Fig. 40 Removing the shift box cover

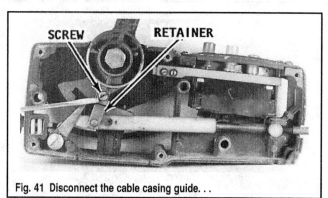

Fig. 41 Disconnect the cable casing guide. . .

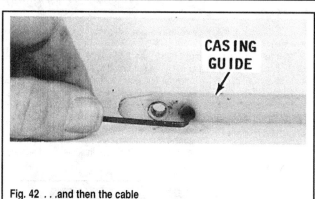

Fig. 42 . . .and then the cable

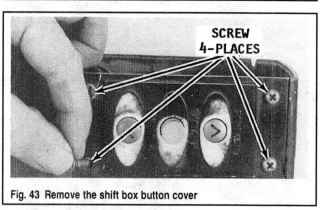

Fig. 43 Remove the shift box button cover

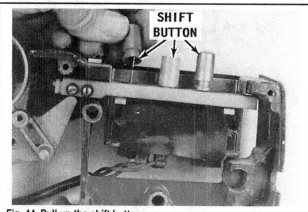

Fig. 44 Pull up the shift buttons. . .

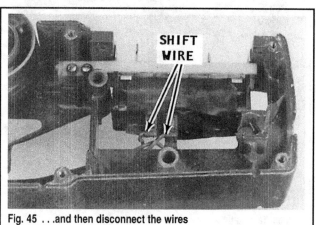

Fig. 45 . . .and then disconnect the wires

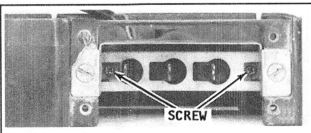

Fig. 46 Remove the screws. . .

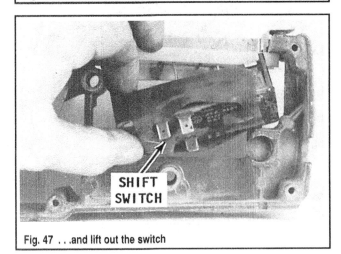

Fig. 47 . . .and lift out the switch

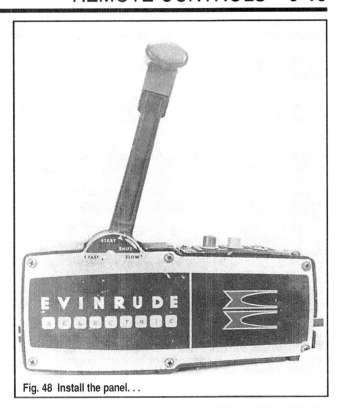

Fig. 48 Install the panel. . .

14. Install the 2 small Phillips screws into the top of the shift box. These 2 screws secure the switch box to the retainer.

15. Connect the wires to the bottom of the switch. Connect the red wire to the Positive terminal; the green to the Neutral terminal; and the blue wire to the Reverse terminal.

16. Slide the buttons down over the protrusions of the switch and seat them in place.

17. Temporarily install the side plate, and then move the throttle hand forward until the boss mark on the bottom of the throttle hand aligns with the mark on the side of the shift box panel.

18. Remove the panel and depress the three buttons one at a time. If it is not possible to depress the buttons, loosen the 2 screws on the selector bracket. Move the bracket forward or aft until the buttons can be depressed. Tighten the screws to secure the bracket in the proper position.

19. Install the shift box cover and secure it in place with the 4 screws.

■ **If the throttle adjustment is not properly performed, the circuit to the starter solenoid will be opened preventing the starter motor from cranking the engine. Adjustment is made by moving the throttle cable adjustment knob in the trunnion on the side of the engine.**

20. Check the end of the cable to determine if the temper has been removed. If the end has a bluish appearance, it has been heated at an earlier date and the temper removed. The temper must be removed to permit the holding screw to make a crimp in the wire to hold an adjustment. If the wire has not been tempered, heat the end, but not enough to melt the wire.

21. Feed the inner cable into the casing guide and align it with the hole in the sleeve. Tighten the 2 Allen screws in the sleeve until one screw makes a crimp in the wire. The screw must be tightened to this degree to prevent the wire from slipping during operation. Bring the other Allen screw up tight against the wire.

22. Install the cable and cable end into the shift box with the trunnion cap on the bottom side.

23. Lower the cable trunnion retainer into the recess and at the same time install the end of the shift cable sleeve over the end of the protrusion.

24. Install the other trunnion cap over the top of the cable retainer.

25. Slide the retaining clip over the end of the guide. The guide slips over a pin and the retainer has a hole in the end. The retainer fits over the pin and holds the end of the throttle cable onto the pin. Install the side plate with the attaching Phillips screws.

Fig. 49 . . .and use the adjusting screws

✳✳ **CAUTION**

Water must circulate through the lower unit to the engine any time the engine is run to prevent damage to the water pump in the lower unit. Just five seconds without water will damage the water pump.

26. Start the engine and run it at 700 rpm.

27. Now, adjust the slide yoke to allow the pushbuttons to be depressed at 700 rpm, but not at 750 rpm. If it is not possible to depress the buttons at 700 rpm, remove the side panel and loosen the 2 screws on the selector bracket. Move the bracket forward or aft until the buttons can be depressed.

28. To adjust the friction knob under the shift box, turn the knob clockwise to increase friction and counterclockwise to decrease friction.

29. Refer to Lower Unit section for details on how to properly adjust the shift cable and to adjust the throttle cable.

Single Lever Remote Control Shift Box

DESCRIPTION & OPERATION

◆ See Figure 5

This unit is a single throttle and gear shift lever shift box with a warm-up lever on the side of the box. The unit has the ignition key switch, and chokes built-in. Some models may have additional built-in features such as a motor overheat horn, a start-in neutral only switch, and an ignition ON light.

The control cables connected between the motor and the remote control lever at the shift box open the throttle after the desired gear is engaged. A throttle friction adjustment is provided to permit the operator to release his grip on the control lever without a change in engine speed.

The warm-up lever mounted on the side of the shift box opens the throttle enough to start the engine and to control the fast idle speed for warm-up after the engine has started.

On models equipped with the start-in-neutral only switch, the starting circuit is completed only when the control lever is in the Neutral position. The switch opens the circuit when the control lever is in either Forward or Reverse positions, making the ignition key switch inactive.

To start the engine, the control lever should be moved to the Neutral position and the warm-up lever to the **START** position. After the engine has started and been allowed to warm to normal operating temperature, the warm-up lever should be moved to the **RUN** position.

A lock-out knob is installed under the control lever handle. This knob must be depressed to permit the control lever to move to the Forward, or to the Reverse, position. The control lever handle must be moved approximately 45° of its total travel for complete shift movement in the lower unit. If the control handle is moved past the 45° point, the throttle is advanced and engine speed increases.

A throttle friction adjustment knob installed on the front of the control box can be adjusted to permit the operator to release his grasp on the handle without the throttle "creeping" and thus changing engine speed. The friction knob should be adjusted only to the point to prevent the throttle from "creeping".

TROUBLESHOOTING

② ◁MODERATE

◆ See Figures 5, 50 and 51

The following paragraphs provide a logical sequence of tests, checks, and adjustments, designed to isolate and correct a problem in the shift box operation.

The single lever remote control shift boxes are fairly simple in construction and operation. Seldom do they fail creating problems requiring service in addition to normal lubrication.

Hard Shifting or Difficult Throttle Advance (Checking Throttle Side)
  1. Remove the throttle and shift control at the engine.
  2. Now, at the shift box, attempt to move the throttle or shift lever. If the lever moves smoothly, without difficulty, the problem is immediately isolated to the engine. The problem may be in the tower shaft between the connector of the throttle and the armature plate. The armature plate may be "frozen", unable to move properly. On certain late model units, the "lever advance arm" located on the starboard side of the engine may be "frozen" and require disassembly and lubrication.
  3. If the problem with shifting is at the engine, the first place to check is the area where the shift lever extends through the exhaust housing. The bushing may be worn or corroded. If the bushing requires replacement, the engine powerhead must be removed.
  4. Another cause of hard shifting is water entering the lower unit. In this case the lower unit must be disassembled.
  5. If hard shifting is still encountered at the shift box when the controls are disconnected from the engine, one of two areas may be causing the problem: the cables may be corroded and require replacement; or, the teeth on the plastic shift lever assembly in the shift box may be worn or broken. This is a common area for problems.

Unable to Obtain Full Shift Movement or Full Throttle "Long Life" Cables Only
Normally, this type of problem is the result of improper shift box installation. This area includes connection of the shift and throttle cables in the shift box. If the stainless steel inner wire was not heated and the clamp did not hold the inner cable (wire), the wire could slip inside the sleeve and

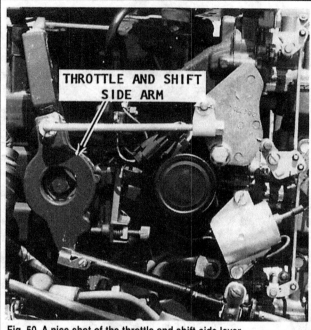
Fig. 50 A nice shot of the throttle and shift side lever

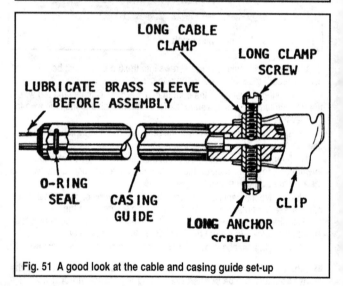
Fig. 51 A good look at the cable and casing guide set-up

the cable would be shortened. Therefore, if it is not possible to obtain full shift or full throttle, the shift box must be removed, opened, and checked for proper installation work. The inner wire could also slip at the engine end of the control, but problems at that end are very rare.

Usually if improper installation work has been done at the engine end, the ability to shift at all is lost, or the throttle cannot be actuated. If the lower unit has previously been removed, the shift rod may not have been adjusted properly.

SHIFT/THROTTLE CABLE

② ◁MODERATE

Removal and Installation

◆ See Figures 5, 15 and 52 thru 62

  1. Disconnect the leads at the battery terminals and disconnect the spark plug wires at the plugs, as a safety precaution to prevent possible personal injury during the work. Some mechanics can service the cables, including replacement, without removing the box from the side of the boat. However, the job is made much easier if the box is removed and laid on its side on the boat seat.

2. Use the proper size Allen wrench and loosen the screw by backing it out about 3 complete turns. Set a small center punch in the center of the screw, and then strike the punch with a quick hard blow. The shock will break the handle loose from the splined shaft. Remove the Allen screw, and then the handle.

3. It is not necessary to remove the shift handle, however the work will progress easier if the handle is removed and out of the way.

4. Remove the 3 Phillips screws on the bottom, the side, or both ends of the panel on the lower section of the shift box. Look inside the box and notice the attaching screws securing the box to the side of the boat. Remove the screws and lay the shift box on the beat seat.

5. Lift the electrical cable and grommet up out of the slot at the rear of the shift box. Loosen the anchor screws on the end of the casing guides.

6. On the back side of the shift box, 2 holes are provided to permit inserting an Allen wrench to hold the underneath screw while the upper

outside Allen screws are removed. After the screws have been loosened, pull the shift cable out of the cable casing.

7. To remove the cables at the engine end:
   a. Remove the self-locking nuts securing the cables to the engine.
   b. Remove the clip on the trunnion.
   c. Slip the end of the cable out of the engine retainer, and then remove the cable from the boat.
   d. To remove the guide casings at the engine end, loosen the Allen screws on both sides and pull the casing free.

To Install:

If the throttle cable is NOT to be replaced, now is an excellent time to lubricate the inner wire.

To lubricate the inner wire, remove the casing guide from the cable at both ends. Attach an electric drill to one end of the wire. Momentarily turn the

Fig. 52 Remove the handle screw. . .

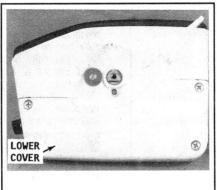

Fig. 53 . . .and the cover screws

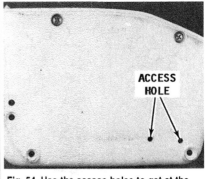

Fig. 54 Use the access holes to get at the inner cable retaining screws

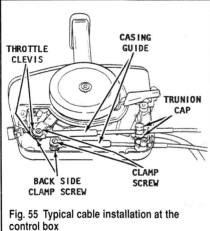

Fig. 55 Typical cable installation at the control box

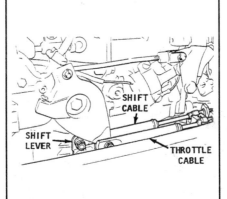

Fig. 56 Typical cable installation at the powerhead

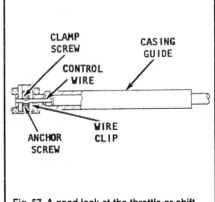

Fig. 57 A good look at the throttle or shift cable connections at the shift box

Fig. 58 Install a trunnion cap and then position the trunnion in the control box

Fig. 59 Don't forget the grommet

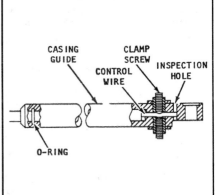

Fig. 60 A good look at the shift cable connections at the shift control lever

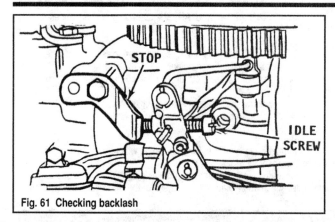

Fig. 61 Checking backlash

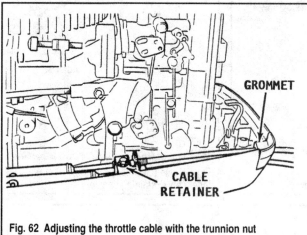

Fig. 62 Adjusting the throttle cable with the trunnion nut

drill on and off to rotate the wire and at the same time allow lubricant to flow into the cable.

■ **The following procedures are to be followed to install either the shift cable or the throttle cable.**

✱✱ CAUTION

Check the end of the cable to determine if the temper has been removed. If the end has a bluish appearance, it has been heated at an earlier date and the temper removed. The temper must be removed to permit the holding screw to make a crimp in the wire to hold an adjustment. If the wire has not been tempered, heat the end, but not enough to melt the wire.

■ **It is very easy to shear the wire by applying too much force when tightening the screw to make the crimp. Therefore, play it cool. Tighten the screw, make one more complete turn to make the crimp; and then bring the other screw up just tight against the wire.**

8. Place the shift control handle in the Neutral position and the warm-up lever in the **START** position.

9. Coat the cable sleeve with anti-corrosive lubricant, and then slide the casing guide over the cable end.

10. Thread one screw into the control cable clamp. Insert the casing guide into the shift control clevis and align the hole in the casing guide with the hole in the clevis.

11. Lubricate and insert the control wire clamp in the clevis hole with the screw toward the back of the control lever.

12. Align the wire holes in the clamp with the wire holes in the casing guide.

13. Feed the control wire through the clamp until the wire is flush to within 1/32 in. (0.8mm) recessed with the end of the casing.

14. Secure the cable wire in the casing guide. The control wire is held in place with 2 screws - the clamp screw already in place and the anchor screw. The wire must be crimped as described earlier to hold the adjustment.

This is accomplished by reaching through the access hole in the rear of the control box with a 3/32 in. Allen wrench and tightening the screw until it is up just snug against the wire.

15. Now, tighten the screw ONLY one complete turn more to make the crimp. Install the 2nd screw, from the front and bring it up just tight against the wire. The wire will now be held securely in the casing guide.

■ **The previous procedure is complete to install either the throttle or the shift cable. If both cables are to be replaced, the complete procedure must be followed again to install the second cable properly.**

16. Snap the nylon trunnion caps onto the cable trunnion, and then position the trunnion in the remote control. If the cable has a spherical trunnion, use the anchor blocks included with the new cable instead of the nylon trunnion caps furnished with the remote control. The nylon caps may be discarded.

17. Insert the electric cable grommet into the remote control.

18. Mount the shift control box in the boat and secure it in place with the attaching screws. Install the shift control handle. Install the access cover with the throttle cable positioned in the machined recess in the cover and the nylon trunnion caps in place.

19. Clamp the control cables to the boat along the run to the engine.

20. Check to be sure the shift control lever is in the Neutral position and the warm-up lever is in the **RUN** position.

21. Insert the clamp in the casing guide. Start the Allen screws into the clamp. Work the casing guide down over the cable (shift or throttle cable) until the wire protrudes out into the inspection hole of the guide.

✱✱ CAUTION

The flat side of the casing guide must face toward the engine. This flat side is necessary to allow the guide to move as the throttle or shift lever is operated.

22. Tighten one of the Allen screws until it makes contact with the wire, and then give it only one more complete turn to make the crimp in the wire.

23. Tighten the other Allen screw just snug against the wire.

■ **The previous steps were complete to install either the throttle or the shift cable at the engine. If both cables are to be replaced, the complete procedure must be followed again to install the second cable properly.**

24. Place the shift or throttle cable onto the shift or throttle lever studs. Secure the cables in place with the washers and lock-nuts.

25. Move the throttle lever on the engine until the idle stop screw makes contact with the stop. Pull firmly on the throttle casing guide and trunnion nut to remove any backlash in the cable run and at the remote control at the shift box. If this backlash is not removed, the engine may not return to a consistent idle speed.

26. Adjust the trunnion adjustment nut on the throttle cable until the cable will slip into the trunnion. Install the throttle or shift cable into the trunnion and install the retaining cover over the top of the trunnion and tighten the screw in the center of the trunnion.

SHIFT BOX

Removal and Installation

◆ See Figures 5, 52, 53 and 63 thru 70

✱✱ WARNING

Always disconnect the electrical leads at the battery terminals to prevent possible personal injury during the work.

1. Remove the throttle and shift handle by first removing the Allen screw in the center of the handle.

2. Remove the 3 Phillips screws on the lower side of the shift box, and then remove the cover.

3. Remove the screws from the top of the shift box, and then the arm rest. Remove the two screws from the back side of the shift box securing the upper panel to the outside of the shift box, and then lift off the panel.

4. Remove the screws from the inside of the shift box securing the shift box to the boat. Use an Allen wrench and working from the back side and

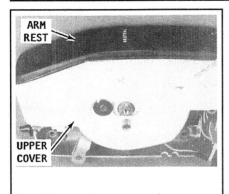

Fig. 63 Remove the armrest and upper cover

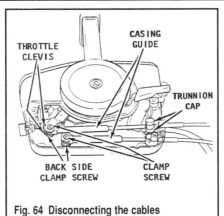

Fig. 64 Disconnecting the cables

Fig. 65 Remove the cam assembly. . .

another Allen wrench from the front side, remove the Allen screws in the control wire on the end of the casing guide. Remove these screws from the shift and the throttle wires.

■ As the panel is lifted, notice how the arm and the mechanism for the throttle are mounted on one side of the panel. Notice the key switch; overheat horn, cam, and start-in-neutral switch installed on the other side of the panel. Also observe the spring and ball bearing under the plate installed under the shift cam.

5. Lift the cam assembly from the panel. Do not lose the spring and ball bearing installed under the plate.

6. Remove the countersunk screw, flat washer, shift lever, and bushing from the housing.

7. Remove the screw and cover plate containing the spring and ball bearing on the right side of the shift box.

To assemble:

8. Disassemble and clean mechanical parts in solvent, and then blow them dry with compressed air.

✳✳ CAUTION

NEVER dip electrical parts in solvent.

9. Check wiring and electrical parts for continuity with a test light or an ohmmeter. Faulty electrical parts must be replaced.

10. Inspect mechanical parts for wear, cracks, or ether damage. Guest parts should be replaced to ensure satisfactory service.

11. Pay special attention to the shift lever teeth. The teeth are made of a hard plastic material. Worn teeth will result in hard shifting.

■ During the assembly work, take time to coat the friction areas of mechanical moving parts with Johnson/Evinrude Multi-purpose grease.

12. Slide the bushing onto the shift lever post in the housing, and then install the lever with the countersunk side of the washer facing up. Tighten the screw securely.

13. Before installing the shift lever, the detent spring and ball must be removed from the retainer.

14. Lower the shift lever down over the shift cam with the center tooth of the shift lever indexes with the center tooth on the cam.

15. Install the ball and detent spring into the recess, and then install the cover over the spring.

16. Place the upper housing onto the control box with the lever cam follower seats in shift lever cam channel. Tighten the screws on the back side of the shift box.

17. Install the arm rest to the top of the shift box and secure it in place with the retaining screws.

18. Slide the throttle and shift cables into the shift box and attach the casing guides in the shift and throttle levers. Tighten the Allen screws from the rear and side of the box. If difficulty is encountered during installation of the cables, see the more detailed instructions under Shift Cable Installation earlier in this section.

19. Install the shift box to the side of the boat. Check to be sure the cables are in their trunnions and the wiring harness is in the recess. Install the lower cover, and then install the shift handle and secure it with the Allen screw.

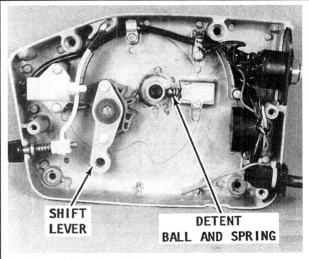

Fig. 66 . . .and then the shift lever

Fig. 67 Notice the difference between the old (left) and new (right) shift levers

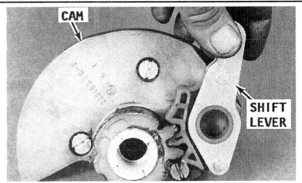

Fig. 68 A good shot of how the shift cam and lever teeth mesh with each other

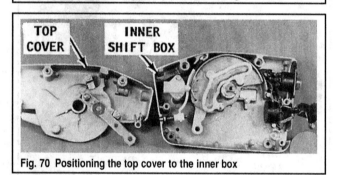

Fig. 69 Install the ball and spring before installing the cover

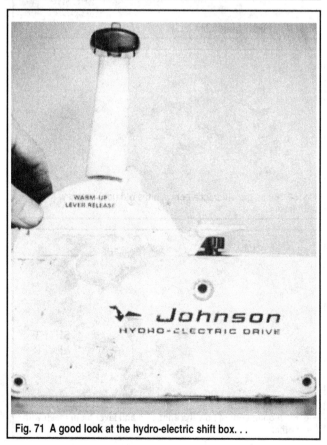

Fig. 70 Positioning the top cover to the inner box

Fig. 71 A good look at the hydro-electric shift box...

Single Lever Hydro-Electric Remote Control Shift Box

DESCRIPTION & OPERATION

◆ See Figures 71 and 72

This shift box is frequently used on the following:
- 1970-71 60 hp
- 1972 65 hp
- 1970-72 85 hp
- 1971-72 100 hp
- 1970 115 hp
- 1971-72 125 hp

The hydro-electric shift box is mounted on the starboard side of the boat. The unit contains a shift select lever, an ignition and choke switch, temperature warning horn, and a warm-up speed control lever.

An electric switch activates the Neutral, Forward and Reverse solenoids for shift movement. With no current to the switch, the lower unit will automatically move into the Forward position. Therefore, current must be present and the unit shifted into Neutral and Reverse. Both the neutral and reverse solenoids are activated to shift into Reverse.

The box is equipped with a friction adjustment to hold the shift lever and throttle in desired engine speed position. A warm-up lever is installed to advance the throttle without the need to move the shift lever. This lever is provided with an adjustment to obtain maximum efficiency during engine startup. The lever has a cutout feature to prevent current from passing to the starter motor circuit once engine speed has reached a predetermined rpm.

The front of the shift box contains a blocking diode. This diode is used to block current to the lower unit when the key switch is in the **OFF** position. Detailed testing procedures of the diode are presented in Lower Unit section.

TROUBLESHOOTING

◆ See Figures 23 and 73 thru 78

The following paragraphs provide a logical sequence of tests, checks and adjustments, designed to isolate and correct a problem in the Johnson single

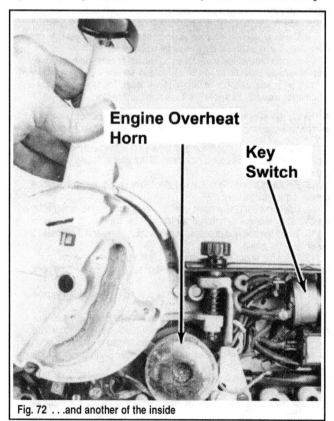

Fig. 72 ...and another of the inside

lever shift box with the warm-up lever to the rear and the Evinrude single lever pushbutton shift box operation.

The procedures and suggestions are keyed by number to matching numbered illustrations as an aid in performing the work.

1. **Difficult Shift Operation**. If difficult shifting is experienced, the problem is usually in the shift box. The friction knob may be adjusted too tightly, the pad that works the friction knob could be excessively worn, or the advance arm on the starboard side of the engine could be inoperative.

2. **Amp Draw Test**. Turn the ignition switch to the **ON** position and note the ammeter reading. There will be no amp draw when the shift lever is in the Forward position. Now, move the shift control lever to the Neutral and then to the Reverse position. Note how much the ammeter reading increased each time the shift lever was moved. If the reading is more than 2.5 amperes for either shift positions, continue with the following checks. If the boat is not equipped with an ampere gauge, then temporarily disconnect the green and brown (or red) wires from the back side of the key switch and temporarily install an amp gauge for the test. Replace the wires after the test is completed.

Disconnect the shift leads at the rear of the engine. Temporarily lay a piece of cloth or other insulating material under the wires to prevent them from shorting out during the following tests.

Again operate the shift lever and note the current loss. If the current draw is still more than 2.5 amperes, then check for a short in the control box switch or wiring. If the current draw is normal with the leads disconnected from the engine, then check for a short in the gear case coil(s) or wiring. If the coil leads are shorted to each other, both shift coils would be energized, stalling the engine or causing serious damage to the driveshaft.

3. **Shift Solenoid Test**. Testing the shift solenoids is accomplished by first disconnecting the blue and green wires to the lower unit at the rear of the engine. Next, connect an ohmmeter first to 1 solenoid lead and ground, and then to the other in the same manner. A reading of more than 5.0-6.0 ohms indicates a short in the solenoid or lead. No reading at all indicates an open circuit. If the results of this test indicate a short in the circuit, the lower unit must be disassembled and inspected.

4. **Testing Shift Switch - Forward**. The shift switch is tested by making meter connections on the blue and green wires running to the key switch.

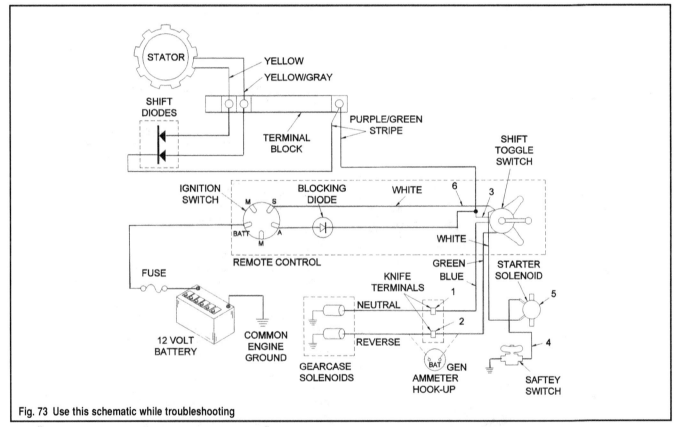

Fig. 73 Use this schematic while troubleshooting

Fig. 74 Checking the shift solenoid. . .

Fig. 75 Checking the shift switch - Forward. . .

Fig. 76 . . .and Reverse. . .

Fig. 77 . . .and finally Neutral

Fig. 78 Checking the starting system

These are the wires that were disconnected in Test No. 3 at the engine terminal. (To test at the shift box would involve cutting wires in order to make the meter connections.)

To test the switch for the Forward position, make contact with 1 probe of a voltmeter to a good ground on the engine. Move the shift lever to the Forward position and the key switch to the **ON** position. Make contact with the other meter probe to first the green and then to the blue wires. The voltmeter should indicate zero volts. If voltage is indicated, the shift switch is defective and must be replaced.

5. **Testing Shift Switch - Reverse.** To test the shift switch for Reverse, make contact with 1 probe of a voltmeter to a good ground on the engine. Move the shift lever to the Reverse position and the key switch to the **ON** position. Make contact with the other meter probe to the green and blue wires. The meter should indicate 12-volts.

If voltage is not present , the shift switch is defective and must be replaced.

6. **Testing Shift Switch - Neutral.** After the Forward and Reverse tests have been completed, check for continuity with the shift lever in the Neutral position. With 1 voltmeter probe still connected to a good ground on the engine and the key switch still at the **ON** position, make contact with the other meter probe to the green and blue wires. When the meter probe makes contact with the green wire, the meter should indicate 12 volts. When the probe makes contact with the blue wire, the meter should indicate zero volts. If these meter readings are not satisfactory, the shift switch is defective and must be replaced.

7. **Cranking System Inoperative.** If the starter fails to crank the engine, check to be sure the throttle lever is in the idle position. If the throttle is advanced more than 1/4 of the way forward, the cut-out switch, attached to the armature plate, will open the circuit to the starter solenoid. If the cranking system fails to operate the starter properly when the throttle lever is in the IDLE position, check the 20 amp fuse between the ignition switch **BAT** terminal and the ammeter **GEN** terminal.

CONTROL UNIT

Disassembly

◆ See Figures 79 thru 86

■ Before starting work on the shift system, disconnect both battery cables at the battery terminals.

1. Remove the 3 attaching bolts and then move the box away from the side of the boat. Remove the 3 screws from the back side of the box. The throttle cable can be removed and serviced without removing it from the boat.

2. Lift off the front cover, but keep the warm-up lever with the back cover. Observe the warm-up lever and the throttle cable in the front cover. Remove the screw from the top of the box securing the arm to the warm-up lever.

3. Lift the arm and the throttle cable from the shift box housing. Take care not to lose the 2 caps from the back side of the cable. One cap is located underneath the cable and the other on top of the cable.

4. Remove the Allen screws from the slider on the cable end. Pull the cable free of the slider.

5. Remove the 2 warm-up lever retaining clips.

6. Depress the release for the warm-up lever and rotate the release up about halfway. Work the warm-up lever upward and at the same time, be careful not to lose the spring and detent located in the warm-up lever. Once the warm-up lever begins to move upward, place a towel or cloth over the lever to prevent the spring and detent from becoming lost. The warm-up lever, detent, and spring are all sold as separate items.

The shift box housing contains the shift lever, shift switch, warning horn, diode, choke, key switch and the wiring harness. To service any of these items, simply remove the attaching hardware and cut or disconnect the attaching wires as shown in the illustration.

The friction knob on top of the shift box may be removed by pulling it free. After the knob is free, the "L" shaped bracket may be removed if the wiring needs service.

Any time the wires from the key switch are removed, the new connection MUST be coated with a Neoprene sealer as a protection against moisture and corrosion.

Cleaning & Inspection

◆ See Figures 15, 79 and 87

1. Clean the box halves thoroughly inside and out with solvent and blow them dry with compressed air. Apply a thin coat of engine oil on all metal parts. If the throttle cable is not to be replaced, now is an excellent time to lubricate the inner wire.

■ This procedure is meant for Long Life cables only, and is not applicable For Snap-in Type cables.

2. To lubricate the inner wire, remove the casing guide from the cable at both ends. Attach an electric drill to one end of the wire. Momentarily turn the drill on and off to rotate the wire and at the same time allow lubricant to flow into the cable, as shown.

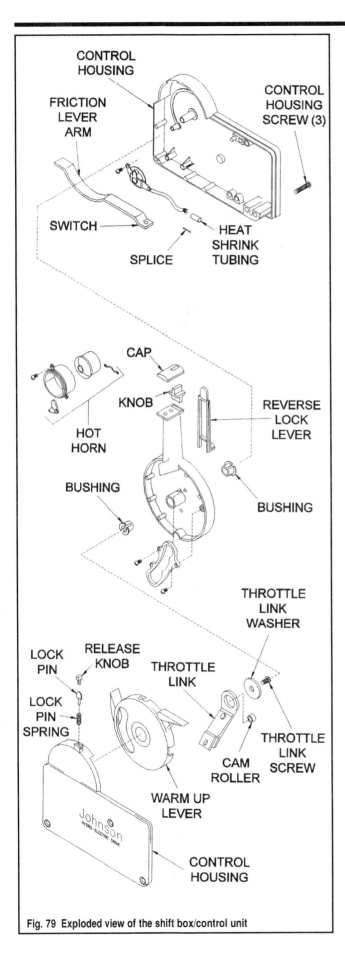

Fig. 79 Exploded view of the shift box/control unit

3. To checking the shift diode:

a. Disconnect both wires from the shift diode. These wires have a rubber seal where they are routed underneath the horn.

b. Peel back the seal and attach 1 lead of an ohmmeter to 1 of the wires, and the other meter lead to the other wire. Observe the ohmmeter for a reading.

c. Reverse the meter leads to the wires. Again observe the meter reading. The meter should indicate continuity when the leads are connected one way to the wires, and no continuity when the leads are connected the other way.

d. If there was a meter reading when the leads were connected both ways, the diode is defective. If the meter did not indicate continuity when the

Fig. 80 Remove the mounting bolts and then the cover screws. . .

Fig. 81 . . .to expose the inside of the unit

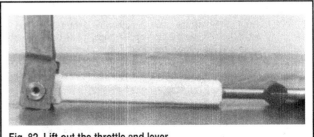

Fig. 82 Lift out the throttle and lever

Fig. 83 Use an Allen wrench to remove the slider

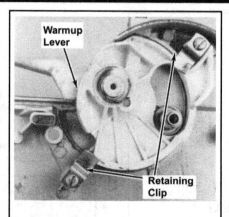

Fig. 84 Remove the warm-up lever retaining clips. . .

Fig. 85 . . .and lift out the lever

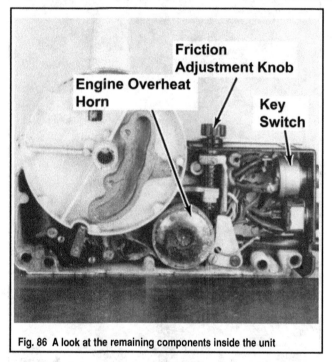

Fig. 86 A look at the remaining components inside the unit

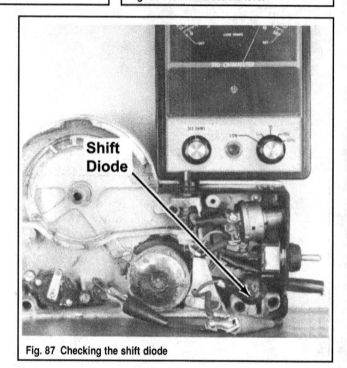

Fig. 87 Checking the shift diode

meter leads were connected either way, the diode is defective. Stating it another way, the meter should indicate continuity when the meter leads are connected to the wires only one way.

Assembly

MODERATE

◆ **See Figures 58 and 88 thru 91**

1. Lubricate the warm-up lever with light-weight oil and then just start it into place in the housing. Install the spring and detent, and then push the warm-up lever fully into place in the housing. Secure the warm-up lever in place with the 2 retaining clips.

■ **Check the end of the throttle cable to determine if the temper has been removed. If the end has a bluish appearance, it has been heated at an earlier date and the temper removed. The temper must be removed to permit the holding screw to make a crimp in the wire to hold an adjustment. If the wire has not been tempered, heat the end, but not enough to melt the wire.**

2. Work the cable end into the slider and through the anchor hole. Adjust the cable end to be flush with the surface of the slider. Tighten the top holding screw enough to make a definite crimp in the wire, as shown. If this screw is not tightened to make the crimp, the wire will slip during operation

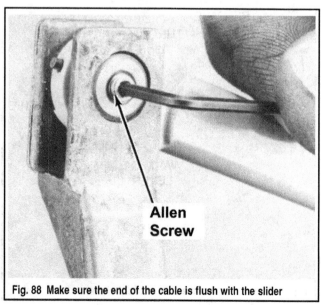

Fig. 88 Make sure the end of the cable is flush with the slider

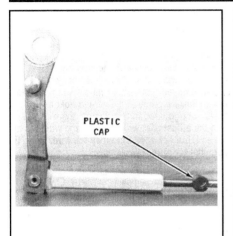

Fig. 89 Installing the slider and cable

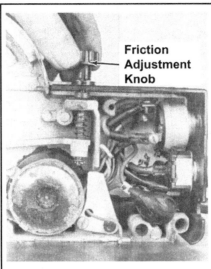

Fig. 90 Releasing the friction adjustment knob

Fig. 91 Always use Neoprene dip on all connections

and the adjustment will be lost. After the top screw has been fully tightened, bring the other screw up tight against the wire. It is not necessary for this second screw to make a crimp in the wire.

3. Place 1 of the plastic caps in place in the shift box housing. Lower the slider and cable into position in the housing. Place the other plastic cap on top of the cable.

4. Secure the arm to the warm-up lever with the washer (having a concave surface on the inside diameter) and the screw.

5. Release the friction adjustment all the way. Set the shift handle over the protrusion in the inner box.

6. Coat all wiring connections and screws with Neoprene Dip as a protection against corrosion and a possible short.

7. Bring the 2 halves of the shift box together. Notice the cut-a-way on the shift handle; it must index with a matching protrusion on the warm-up lever. Work the 2 halves completely together and secure them with the 3 screws in the back. Install the shift box into place in the boat.

REMOTE CONTROL CABLES

◆ **See Figures 92 and 93**

In the early days, the throttle and shift cables were installed using non-adjustable trunnions. The trunnions were installed on the ends of the cables and formed the connection for the cables to the engine. The inner cable (wire) moved in both directions inside the outer cable and actuated the mechanism at the engine.

The anchor on the engine, to which the trunnion is attached, has a **P** and an **S** stamped on the inside diameter or inside edge of the trunnion retainers. These letters identified port and starboard.

As improvements and refinements were incorporated over the years, new cables and trunnions became adjustable through the trunnion.

The non-adjustable unit is totally obsolete and no longer available. Therefore, if the old-style cable with the non-adjustable trunnion requires replacement, the new adjustable type will be installed.

When the new cable and trunnion are to be installed, a new trunnion retainer must be purchased and installed port and starboard on the engine. The new cable and trunnion cannot be connected to the old-style retainer.

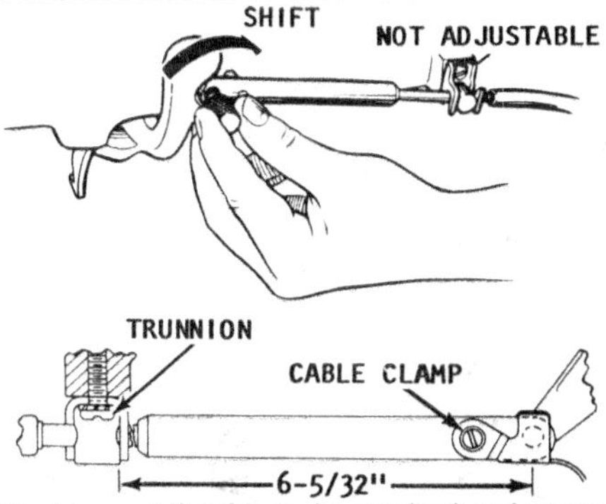

Throttle Cable

INSTALLATION AT THE REMOTE

All cable installation procedures for the remote control are detailed in the Shift Box procedures for the individual remote unit covered elsewhere in this section.

INSTALLATION AT THE ENGINE

① EASY

◆ **See Figures 94 thru 97**

1. Install the throttle lock pin spring over the casing guide.
2. Start the screws into the small cylinder and then slide the cylinder down through the pin spring and into the casing guide. Notice how the cylinder has a hole? This hole should be positioned vertically with the casing to align with the hole in the guide.

Fig. 92 Attaching a shift cable to the arm (top) and connecting the throttle cable to the non-adjustable trunnion (bottom)

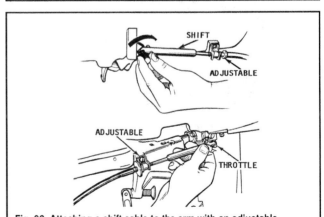

Fig. 93 Attaching a shift cable to the arm with an adjustable trunnion (top) and connecting the throttle cable with an adjustable trunnion (bottom)

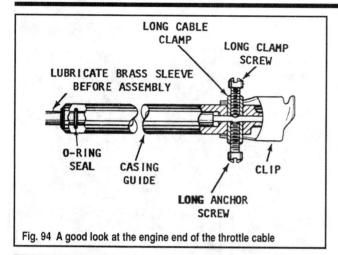

Fig. 94 A good look at the engine end of the throttle cable

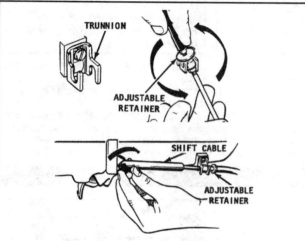

Fig. 95 Attach the shift cable end to the shift lever on the engine by inserting the fitting into the shift control lever, and then pushing inward, and at the same time rotating the fitting 1/2 turn. This action will lock the fitting in the shift lever

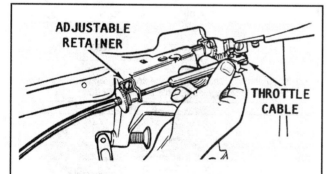

Fig. 96 Slide the guide over the pin onto the engine, and then snap the retainer clip over the end of the guide to lock it in place

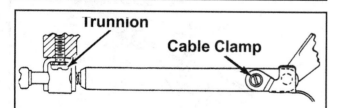

Fig. 97 Connecting the throttle cable with the non-adjustable trunnion

3. Slide the casing guide down over the throttle cable and insert the end of the wire through the sleeve. Tighten the top screw until a definite crimp is made in the wire.

■ **Check the end of the cable to determine if the temper has been removed. If the end has bluish appearance, it has been heated at an earlier date and the temper removed. The temper must be removed to permit the holding screw to make a crimp in the wire to hold an adjustment. If the screw is not tightened to this degree, the wire will slip during operation and the adjustment will be lost. If the wire has not been tempered, heat the end, but not enough to melt the wire. Bring the bottom screw up tight against the wire.**

4. Install the trunnion retainers to the engine if necessary. Check to be sure the retainer with **P** stamped on the inside is installed on the port side of the engine and the retainer with the **S** installed on the starboard side. Connect the trunnion cap to the trunnion retainer. This is accomplished by holding the trunnion in a vertical position, inserting it into the retainer and then turning it to the horizontal position.

5. Slide the guide over the pin onto the engine and then snap the retainer clip over the end of the guide to lock it in place.

Shift Cable

INSTALLATION AT THE REMOTE

All cable installation procedures for the remote control are detailed in the Shift Box procedures for the individual remote unit.

INSTALLATION AT THE ENGINE

◆ **See Figures 98 thru 103**

1. Move the control lever at the shift control box to the Neutral position.

2. Slide the gear shift fitting onto the control wire. Check to be sure the inner wire passes completely through the small holes in the cable clamp. Clamp the anchor screws to prevent twisting the cable. The clamp and the anchor screws must be parallel to the trunnion on the gear shift cable.

3. Notice the flat and rounded areas of the casting guide. The flat edge must face towards the engine; in this position, there is a flat area for the lever to ride during the shifting action.

4. After the cable is in place in the casting guide, tighten the top screw until a definite crimp is made in the cable. If the screw is not tightened enough, the inner wire will slip during operation and the adjustment will be lost.

■ **Check the end of the cable to determine if the temper has been removed. If the end has a bluish appearance, it has been heated at an earlier date and the temper removed. The temper must be removed to permit the holding screw to make a crimp in the wire to hold an adjustment. If the wire has not been tempered, heat the end, but not enough to melt the wire.**

5. Bring the second screw up tight against the wire.

6. Insert the shift cable control vertically into the trunnion bracket and turn the cable to a horizontal position, as indicated by the arrows in the accompanying illustration.

7. Attach the shift cable end to the shift lever on the engine by inserting the fitting into the shift control lever, and then pushing inward, and at the same time rotating the fitting 1/2 turn. This action will lock the fitting in the shift lever.

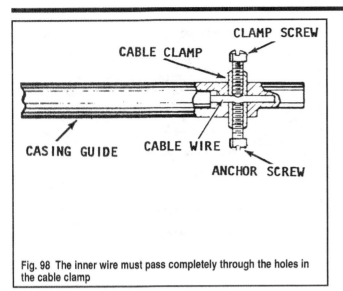

Fig. 98 The inner wire must pass completely through the holes in the cable clamp

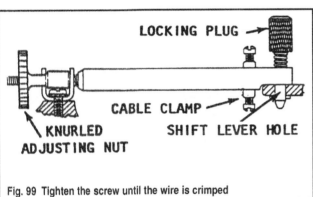

Fig. 99 Tighten the screw until the wire is crimped

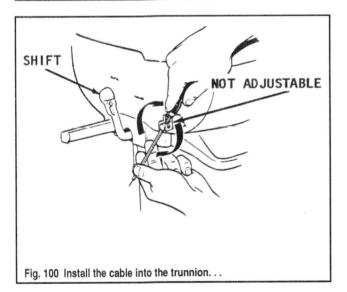

Fig. 100 Install the cable into the trunnion. . .

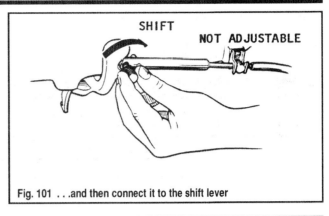

Fig. 101 . . .and then connect it to the shift lever

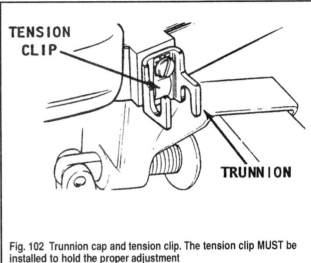

Fig. 102 Trunnion cap and tension clip. The tension clip MUST be installed to hold the proper adjustment

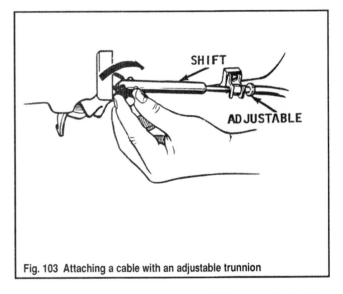

Fig. 103 Attaching a cable with an adjustable trunnion

Rigging

◆ See Figure 104

The control cables should be replaced if inspection reveals any signs of damage, wear or even fraying at the exposed ends. Remember that loss of 1 or more cables while underway could cause loss of control at worst or, at best, strand the boat. Check cable operation frequently and inspect the cables at each service. Replace any that are hard to move or if excessive play is noted. Never replace just 1 cable (unless a freak accident caused

damage to the cable), if 1 cable is worn or has failed, assume the other is in like condition and will soon follow.

Before removal, mark the cable mounting points on the remote control using a permanent marker. This will help ensure easy and proper installation of the replacement. Unless there is a reason to doubt the competence of the person who originally rigged the craft, match the cable lengths as closely as possible. Otherwise, re-measure and determine proper cable lengths as if this was a new rigging.

When rigging an engine to a new boat, determine cable length by measuring from the center point of the motor along the intended cable route to either the side-mount or center-console remote location (refer to the accompanying illustration). Add 3 ft. 4 in. (1.02 M) to the measurement and purchase a cable of the same length (or one that is **slightly** longer than that measurement. Johnson/Evinrude replacement cables are normally available in 1 ft. increments from 5-20 ft. and in 2 ft. increments to 50 ft.

✳✳ CAUTION

To prevent the danger of cable binding or other conditions that could cause a loss of steering control when underway, take care to route all cables with the fewest number and most gentle bends possible. A bend should NEVER have a radius of less than 6 in. (15cm).

Follow the procedures for remote service in this section and the Timing & Synchronization adjustments in the Maintenance & Tune-Up section whenever removing/installing or replacing the throttle and shift cables.

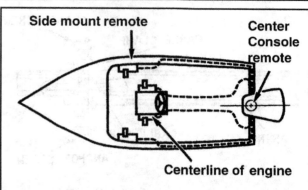

Fig. 104 For cable replacement or rigging, measure the distance from the centerline of the motor to the remote (and add 3 ft. 4 in./1.02 meters to the measurement)

10

HAND REWIND STARTER

HAND REWIND STARTER

General Information

◆ **See Figures 1 thru 7**

The hand rewind starters installed on the Johnson/Evinrude outboards covered here may be one of 3 basic designs:

• Vertical spool type mounted on the side of the powerhead with a drive gear engaging the teeth of the flywheel.

• Disc type which engages the teeth of the flywheel and may be mounted either horizontally or vertically.

• Flat disc type mounted atop the flywheel. This design starter engages a ratchet plate on the flywheel.

Unfortunately, engineering changes have resulted in several models being used for each design. The only logical and practical method of designating different procedures for each model, for each design, was to assign a 'type' number for each different starter. Therefore, the hand rewind starters in this section are designated Type I through Type VII with identification of the outboard models using each type. If in doubt, please refer to the Starter Applications chart at the rear of this section for our best suggestions as to which starter is *supposed* to be on your engine.

DESIGN 1

The first design is a cylinder with a pinion gear arrangement similar to an automotive starter motor. The unit is mounted vertically on the side of the powerhead. When the starter rope is pulled, a nylon drive gear slides upward and engages the flywheel ring gear. After the powerhead starts, the drive gear automatically disengages and retracts to the "rest" position.

Fig. 1 A good look at the Type I starter

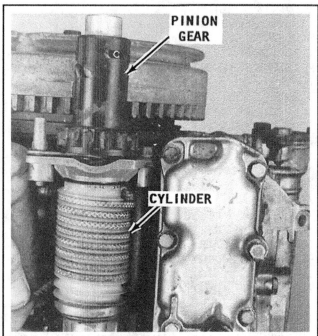

Fig. 2 A good look at the Type II starter

Fig. 3 A good look at the Type III starter

Fig. 4 A good look at the Type IV starter

Fig. 5 A good look at the Type V starter

Fig. 6 A good look at the Type VI starter

Fig. 7 A good look at the Type VII starter

Two models of this hand starter are installed on the outboards covered here and identified as Type I and II.

DESIGN 2

This second design is is mounted on the port side of the powerhead and the drive gear works on an axis. As the rope is pulled, a swing arm moves the drive gear upward to engage with the teeth of the flywheel ring gear. A coil spring winds and tightens as the rope unwinds. The spring then coils the rope around a pulley as the rope handle is returned to the control panel. This model is identified as Type III.

DESIGN 3

This design starter is usually mounted atop the flywheel with 3 mounting legs attached to the powerhead. Four types of this design hand starter are installed on various Johnson/Evinrude powerheads and they will be identified as Type IV thru VII.

Emergency Starting

The most larger powerheads are normally only equipped with an electric starter motor. However, cut-outs are manufactured into the flywheel to permit the use of a rope for emergency starting.

✳✳ WARNING

If an emergency rope is used on ANY powerhead, NEVER wrap the rope end around your hand for a better grip. If the powerhead should happen to backfire, the sudden jerk on the rope would severely injure your hand.

Operation

◆ See Figure 8

Normally, very few problems are encountered with the hand starter. It is strictly a mechanical device to crank the powerhead for starting. The spring will last an incredibly long time, if used properly. The greatest enemy of the spring is the operator.

Three causes contribute to starter failure; two may be prevented, the third cannot.

The most common problem is the result of the operator pulling the starter rope too far outward. If the operator places one hand on the powerhead and pulls the rope with the other hand, it is physically impossible, in this position, to pull the rope too far. Problems develop when the operator uses both hands to pull on the rope, with no control on how far the rope can be extended. The rope may be broken or the knot released from the starter disc. In either case, the spring rewinds with tremendous speed and in almost all cases travels past its normal rewind position bending the end of the spring in reverse. Therefore, more maintenance work is involved than merely replacing the rope.

Another bad habit while using the hand starter is to release the grip on the rope when it is in the extended position, allowing the rope to freely

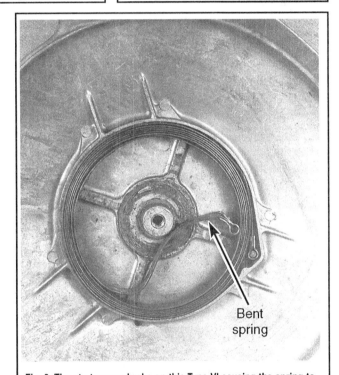

Fig. 8 The starter rope broke on this Type VI causing the spring to rewind with such speed that it actually doubled back in a reverse direction

rewind. The operator should NEVER release his grip, but hold onto the rope, and thus control the rewind. The owner should always be alert to any wear on the rope and replace it long before the possibility of breaking might occur. If the rope should break, the spring would rewind with incredible speed, the same as if the rope were released, causing damage to the spring and other starter parts.

The 3rd cause of spring failure cannot be prevented - age. As the outboard continues to perform year after year, the age of the spring steel will finally take its toll.

The rewind spring is made of spring steel. Depending on the model and the powerhead, from 6-12 feet of spring length is wound into about a 4 inch diameter. This places the spring under unbelievable tension, making it a highly dangerous force. Therefore, any time the hand starter is serviced, especially during work on the spring, safety glasses should be worn and the work performed with the utmost care.

Any time the rope is broken, the starter spring will rewind with incredible speed. Such action will cause the spring to rewind past its normal travel and the end of the spring will be bent back out of shape. Therefore, if the rope has been broken, the starter should be completely disassembled and the spring repaired or replaced.

TYPE I STARTER - CYLINDER W/PINION GEAR (5–6 HP)

◆ **See Figure 1**

The Type I hand rewind starter is used on all 5 and 6 hp models.

This gear-drive starter is a new design employing the principle of an automotive type starter motor. When the starter rope is pulled, the starter rotates, and a nylon pinion gear slides upward and engages the flywheel ring gear. The gear automatically disengages when the powerhead starts.

The ratio between the pinion gear and the ring on the flywheel has been selected to provide maximum cranking speed with minimum pulling effort to ensure fast, easy powerhead start.

Starter Rope

REPLACEMENT

◆ **See Figures 9 and 10**

1. Disconnect the high-tension leads from the spark plugs.

2. Pull the starter rope out slowly until it is fully extended. Now, allow the rope to retract just a little, until the knot end on the spool is facing the port side of the powerhead.

3. Lift the pinion gear to engage the flywheel ring gear. Hold the pinion gear engaged with the ring gear, and at the same time, slide the handles of a pair of pliers under the pinion gear to lock the pinion gear with the ring gear.

4. Remove the handle from the end of the starter rope. Take note of how the rope is wound onto the spool and how the rope is secured by a loop formed in a slot in the spool.

5. Remove the rope from the starter spool.

To Install:

◆ **See Figures 11 thru 14**

■ The length and diameter of the starter rope required will vary depending on the horsepower size of the model being serviced. Therefore, check the Hand Starter Rope Specifications chart, and then purchase a quality nylon piece of the proper length and diameter size. Only with the proper rope will you be assured of efficient operation following installation.

Fig. 9 Lift the pinion gear and lock it in place

Fig. 10 Remove the rope

Fig. 11 Feed the end of the new rope through the spool anchor and make a loop. . .

Fig. 12 . . .and then bring the short end of the rope through the loop

Fig. 13 Now work the short end back through the anchor. . .

Fig. 14 . . .and pull both ends tight

Each end of the nylon rope should be "fused" by burning them slightly with a very small flame (a match flame will do) to melt the fibers together. After the end fibers have been "fused" and while they are still hot, use a piece of cloth as protection and pull the end out flat to prevent a "glob" from forming.

6. Feed one end of the new rope through the spool anchor, make a loop, and then thread the end back through the hole in the anchor. Do not pull it tight at this time - leave a loop.

7. Bring the short end of the rope through the loop just formed.

8. Work the short end back through the anchor.

9. Now, pull both ends of the rope tight. Feed the rope through the front engine cowling, and then install the starter rope handle.

10. Pull and hold tension on the rope and at the same time remove the pliers from under the pinion gear. Allow the starter rope to rewind in a normal manner.

11. After the rope is fully wound onto the starter spool, the rope handle should be up tight against the cowling. If the handle is not up tight, the rope was installed to long or the starter spring is weak and should be replaced.

Starter

REMOVAL & DISASSEMBLY

 MODERATE

◆ See Figures 15 thru 19

■ For photographic clarity, the accompanying pictures were taken while servicing a starter with the powerhead removed from the exhaust housing. The hood need only be removed to work on the starter under normal conditions.

1. Pull approximately 3/4 of the starter rope out, and then form a knot in the rope to prevent it from recoiling. Allow the rope to recoil until the knot is tight against the cowling. Remove the rope handle.

2. The rope may be removed now, or later. To remove the rope now, first pull the rope all the way out. Slide the handles of a pair of pliers under the pinion gear to hold the gear engaged with the flywheel. Remove the rope from the starter spool.

3. Grasp the spool firmly. Remove the pliers and allow the spool to slip a little at a time until the spring is completely unwound.

4. Remove the 2 retaining bolts on top of the starter.

5. Loosen, but do not remove the bolts on the bottom and on each side of the starter spool. When the bottom 2 bolts are loosened, the retainer will separate from the lower cap.

6. Lift the starter from the powerhead. If the spring is still clipped into the lower retainer, release the spring by disengaging the spring tang from the retainer.

To disassemble:
◆ See Figures 20 thru 22

7. Remove the pin in the pinion gear, and then remove the pinion gear from the collar. Slip the bearing head off the spool. Remove the retainer, installed under the pinion gear, from the spool.

8. Remove the spring retainer (the long tube) from the center of the spool.

9. Pull the starter spring from the spool.

CLEANING & INSPECTION

 EASY

1. Wash all parts in solvent, and then dry them with compressed air.

2. Inspect and replace the main spring if it is damaged or worn.

3. Check the bottom end of the spring very carefully to be sure the two tangs (one on the inner and the other on the outer spring) are in good condition with no sign of distortion.

4. Inspect the bushing in the bottom collar of the starter housing. This bushing was not removed in the disassembling procedures. Check for any roughness, burrs, or other evidence of excessive wear or damage. If the bushing is in good condition, it need not be removed.

5. Check the rope condition. If the rope is frayed or shows any sign of weakness, it should be replaced. There will never be an easier time to replace the rope than while the starter is disassembled.

Fig. 15 Remove the rope handle. . .

Fig. 16 . . .and then remove the rope

6. Inspect the teeth of the pinion gear. The teeth will show some signs of normal wear. A broken tooth or excessive wear on one side of the teeth is justification for replacement.

7. Inspect the groove through the pinion gear. This is the groove to accommodate the roll pin. Check to be sure the upper part of the pinion gear is not cracked or distorted.

✳✳ CAUTION

Do not lubricate the pinion gear. Oil applied to the pinion gear will attract dirt causing the gear to bind on the spool.

Fig. 17 Grab the spool and allow it to slip a bit

Fig. 18 Remove the thru-bolts. . .

Fig. 19 . . .and then loosen the bolts at the bottom of the spool

Fig. 20 Removing the pinion gear

Fig. 21 Remove the spring retainer. . .

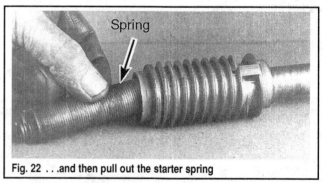

Fig. 22 . . .and then pull out the starter spring

8. Lubricate the upper and lower spool bearing surfaces with just a drop of outboard lubricant.

9. Apply outboard oil to the spring on the pinion gear. Do not oil the pinion gear bearing or the surfaces of the spring.

ASSEMBLY & INSTALLATION

 *MODERATE*

■ Two methods of assembling and installing this starter are presented. The first is the factory suggested procedure. However, an alternate method is also outlined which many professional mechanics feel is much simpler, easier and quicker.

Factory Recommended Method:
◆ See Figures 23 thru 28

1. Install the spring retainer from the bottom side. Slide the spring onto the spring retainer. Work the spring upward until the inner spring tang engages the slot on the bottom of the retainer.

2. Align the hole in the retainer with the hole in the spool sleeve.

3. Slip the bearing head and pinion gear down over the spool shaft. Install the roll pin through the pinion gear and sleeve.

4. If the lower bushing was removed, install a NEW bushing into the bottom collar on the powerhead.

■ As an assist to installation, first soak the bushing in hot water for about 10 minutes, and then lubricate it with just a drop of outboard oil.

5. The tang on the outer spring must hook into the slot of the lower spring retainer plate. Pull on the outer spring to elongate the spring, and at the same time, lower the spring into the spring retainer plate and hook the tang into the slot in the plate.

6. Rotate the spring CLOCKWISE to lock the tang in the plate. Hold upward and turn the spring CLOCKWISE, and at the same time tighten the 2 screws in the lower spring retainer plate.

Fig. 23 Install the spring retainer from the bottom

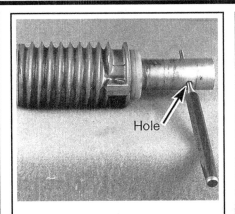

Fig. 24 Make sure that the two holes are aligned

Fig. 25 Slip the bearing head over the gear

Fig. 26 Make sure you use a new bushing if the old one was removed

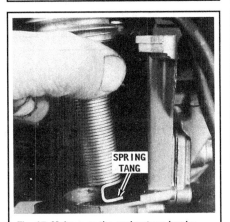

Fig. 27 Make sure the spring tang hooks into the slot

Fig. 28 Start the 2 upper bolts

7. Place the starter assembly in position on the powerhead and start the 2 upper screws through the upper bearing support. Check to be sure the guide is in place in the bottom and top retainers. Tighten the 2 bottom retainer screws.

8. Move to the Final For Both Methods procedure.

Alternate Method:
◆ See Figures 18, 25 and 29 thru 32

a. If the lower bushing was removed, install a NEW bushing into the bottom collar on the powerhead.

■ **As an assist to installation, first soak the bushing in hot water for about ten minutes, and then lubricate it with just a drop of outboard oil.**

b. Install the spring retainer from the bottom side of the spool. Slip the bearing head and pinion gear down the spool shaft.

c. Align the holes and install the roll pin through the pinion gear and spool.

d. Take the spring and lower it into the bottom retainer. Hook the outer spring into the retainer.

e. Lower the spool assembly down over the spring and engage the spring retainer into the tang of the inner spring.

f. Start the two upper bolts through the upper bearing support. Check to be sure the guide is in place in the bottom and top retainers. Tighten the 2 bottom retainer bolts.

g. Move to the Final For Both Methods procedures.

Final For Both Methods:
◆ See Figures 11, 12, 13, 33, 34 and 35

9. Insert a large size screwdriver into the top of the spool, and then rotate the spool, by count, exactly 16 1/2 complete turns.

10. Lift the pinion gear to engage the flywheel ring gear.

11. Hold the pinion gear engaged with the ring gear, and at the same time, slide the handles of a pair of pliers under the pinion gear to lock the pinion gear with the ring gear.

Fig. 29 Make sure you use a new bushing if the old one was removed

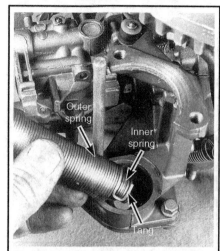

Fig. 30 Lower the spring and hook the outer spring into the retainer

Fig. 31 Lower the spool assembly over the spring. . .

Fig. 32 . . .and install the thru-bolts

Fig. 33 Rotate the spool exactly 16 1/2 turns

Fig. 34 . . .and pull both ends tight

Fig. 35 Install the handle

12. Feed one end of the new rope through the spool anchor, make a loop, and then thread the end back through the hole in the anchor, but DO NOT pull it tight at this time, leave a loop.

13. Bring the short end of the rope through the loop just formed.

14. Work the short end back through the anchor.

15. Now, pull both ends of the rope tight.

16. Feed the rope through the front engine cowling, and then install the starter rope handle. Hold tension on the rope with the handle and at the same time; remove the pliers from underneath the pinion gear. Allow the starter rope to wind onto the spool. After the rope has been wound onto the spool, the starter handle should be up tight against the cowling. If the handle is not up tight against the cowling the rope was installed too long and needs to be shortened.

TYPE II STARTER - CYLINDER W/PINION GEAR (9.5 HP)

The Type II hand rewind starter is used on all 9.5 hp models.

This gear-drive starter is a new design employing the principle of an automotive type starter motor. When the starter rope is pulled, the starter rotates, and a nylon pinion gear slides upward and engages the flywheel ring gear.

The gear automatically disengages when the powerhead starts. The ratio between the pinion gear and the ring on the flywheel has been selected to provide maximum cranking speed with minimum pulling effort to ensure fast, easy start.

Starter Rope

REPLACEMENT

MODERATE

◆ See Figures 36 thru 39

1. Disconnect the high-tension leads from the spark plugs.

2. Pull the starter rope slowly out until it is fully extended. Now, allow the rope to retract just a little until the knot end on the spool is facing the port side of the powerhead.

3. Lift the pinion gear to engage the flywheel ring gear. Hold the pinion gear engaged with the ring gear, and at the same time, slide the handles of a pair of pliers under the pinion gear to lock the pinion gear with the ring gear.

4. Remove the handle from the end of the starter rope. Observe and remember how the rope is wound onto the spool and how the rope is secured by a knot. Pull the knot and the rope from the spool.

To Install:

■ The length and diameter of the starter rope required will vary depending on the horsepower size of the model being serviced. Therefore, check the Hand Starter Rope Specifications chart, and then purchase a quality nylon piece of the proper length and diameter size. Only with the proper rope will you be assured of efficient operation following installation.

5. Each end of the nylon rope should be "fused" by burning them slightly with a very small flame (a match flame will do) to melt the fibers together.

Fig. 36 Lock the pinion gear with a pair of pliers. . .

Fig. 37 . . .remove the handle and pull the rope from the spool

Fig. 38 Feed the knotted rope through the spool and install the handle. . .

Fig. 39 . . .before allowing the spool to take up the rope

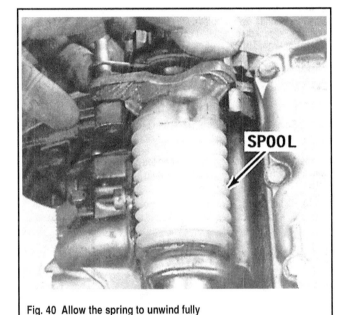

Fig. 40 Allow the spring to unwind fully

After the end fibers have been "fused" and while they are still hot, use a piece of cloth as protection and pull the end out flat to prevent a "glob" from forming.

6. Tie a figure 8 knot in one end of the new rope. Feed the rope through the spool anchor around the back of the spool and out the hole in the cowling.

7. Install the handle on the end of the rope.

8. Pull and hold tension on the rope, and at the same time remove the pliers from under the pinion gear. Allow the starter rope to rewind in a normal manner. After the rope is fully wound onto the starter spool, the rope handle should be up tight against the cowling. If the handle is not up tight, the rope was installed too long or the starter spring is weak and should be replaced.

Starter

REMOVAL & DISASSEMBLY

 MODERATE

◆ See Figures 36, 37, 40 and 41

■ The exhaust shroud has been removed only for photographic clarity in accompanying illustrations, it is not necessary in normal operations.

Fig. 41 Remove the 2 retaining bolts. . .

1. Pull the starter rope out until it is fully extended. Now, allow the rope to retract just a little, until the knot end on the spool is facing the port side of the powerhead.

2. Lift the pinion gear to engage the flywheel ring gear. Hold the pinion gear engaged with the ring gear, and at the same time, slide the handles of a pair of pliers under the pinion gear to lock the pinion gear with the ring gear.

3. Remove the handle from the end of the starter rope. Take note of how the rope is wound onto the spool and how the rope is secured by a knot. Pull the knot and the rope from the spool.

4. Grasp the spool firmly and remove the pliers from under the pinion gear. Now, allow the spool to slip a little at a time until the spring is completely unwound.

5. Remove the 2 retaining bolts on top of the starter. Lift the starter assembly from the powerhead.

To disassemble:

◆ See Figures 42 thru 46

■ As the starter is removed, take special note of how the upper bearing retainer extends over a hole in the powerhead. This hole is a water passage. A gasket is installed under the retainer to form a seal for the water passage. This gasket may remain on the block or come with the starter retainer as the starter is removed. To ensure a good seal, the gasket should be discarded and replaced with a new one at time of installation.

6. Remove the roll pin extending through the pinion gear.

7. Remove the pinion gear, spring and bearing head from the spool.

8. Pull the main spring and upper spring retainer from the spool.

9. Notice the lower spring retainer on the bottom of the spool secured with a set screw. Remove the set screw from the retainer and then pull the retainer and bushing from the spring.

10. Remove the upper spring retainer and outer bearing from the spring.

Fig. 42 . . .and then remove the roll pin after taking out the starter

Fig. 43 Remove the pinion gear and bearing head. .

CLEANING & INSPECTION

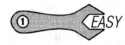

◆ See Figure 47

1. Wash all parts in solvent, and then blow them dry with compressed air.

2. Inspect the main spring. Check the tab on the bottom end of the inner spring to be sure it is not bent or cracked.

3. Inspect the teeth of the pinion gear. The teeth will show some signs of normal wear. A broken tooth or excessive wear on one side of the teeth is justification for replacement.

4. Inspect the groove through the pinion gear. This is the groove to accommodate the roll pin. Check to be sure the upper part of the pinion gear is not cracked or distorted.

✳✳ CAUTION

Do not lubricate the pinion gear, spring, or spool. Oil applied to these parts will attract dirt causing the gear to bind on the spool.

5. Lubricate the upper and lower spool bearing surfaces with just a drop of outboard lubricant.

6. Apply outboard oil to the spring on the pinion gear. Do not oil the pinion gear bearing or the surfaces of the spring.

7. Check the rope condition. If the rope is frayed or shows any sign of weakness, it should be replaced. There will never be an easier time to replace the rope than while the starter is disassembled.

ASSEMBLY & INSTALLATION

◆ See Figures 42, 43, 44, 46, 48 and 49

1. Place the outer bearing and the upper spring retainer onto the main spring.

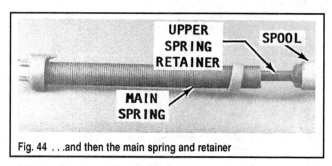

Fig. 44 . . .and then the main spring and retainer

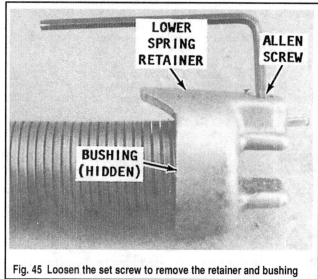

Fig. 45 Loosen the set screw to remove the retainer and bushing

2. Slide the bushing over the main spring. Guide the tang on the main spring through the hole in the main spring retainer.

3. Install the set screw securing the inner spring to the lower retainer. Set the completed spring assembly aside.

4. Slide the outer bearing down over the outer spring.

5. Insert the spring retainer through the center of the inner spring from the top.

6. Rotate the inner spring until the hook on the end of the spring seats in the groove of the spring retainer.

7. Insert the assembled springs and retainer from the previous steps into the bottom of the spool.

8. Install the bearing head, spring, and pinion gear down over the spool shaft.

9. Align the hole in the upper spring retainer with the hole in the starter spool and with the slot in the pinion gear.

To Install:

◆ See Figures 50 thru 53

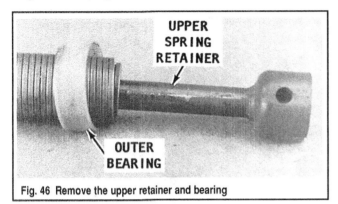

Fig. 46 Remove the upper retainer and bearing

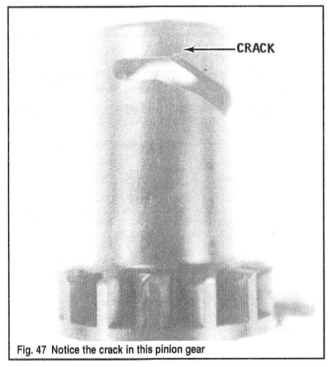

Fig. 47 Notice the crack in this pinion gear

Fig. 50 Use Perfect Seal on the new gasket

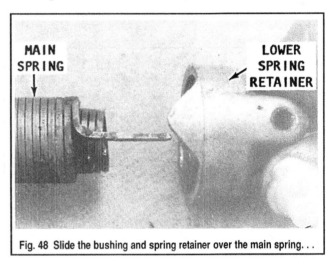

Fig. 48 Slide the bushing and spring retainer over the main spring. . .

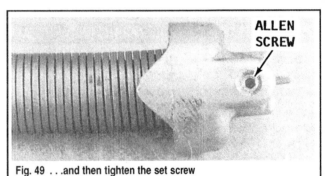

Fig. 49 . . .and then tighten the set screw

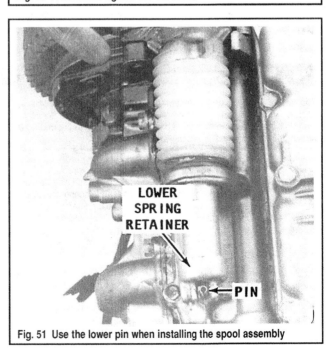

Fig. 51 Use the lower pin when installing the spool assembly

10. Cover both sides of a NEW gasket with Perfect Seal No. 4, and then place the gasket in position on the powerhead over the water passage.

11. Lower the spool assembly down into place on the side of the powerhead with the lower spring retainer indexed over the pin in the bottom portion of the housing.

12. Install the 2 bolts securing the bearing head to the powerhead.

13. Insert a large screwdriver blade into the top of the spool, and then rotate the spool COUNTERCLOCKWISE, by count, 20 1/2 complete turns. After the required numbers of turns have been made, hold pressure on the screwdriver, and at the same time lift the pinion gear to engage with the flywheel ring gear, and insert the handles of a pair of pliers under the pinion gear to hold the gear engaged.

Fig. 52 Install the mounting bolts

14. Tie a figure 8 knot in one end of the new rope. Feed the rope through the spool anchor around the back of the spool and out the hole in the cowling. Install the handle on the end of the rope.

15. Pull and hold tension on the rope, and at the same time remove the pliers from under the pinion gear. Allow the starter rope to rewind in a normal manner. After the rope is fully wound onto the starter spool, the rope handle should be up tight against the cowling. If the handle is not up tight, the rope was installed too long or the starter spring is weak and should be replaced.

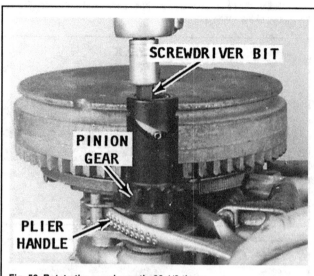

Fig. 53 Rotate the spool exactly 20 1/2 times

TYPE III STARTER - COIL SPRING W/SWING ARM VERTICAL MOUNT

The Type III hand rewind starter is used on 1968 3 hp and 1969-72 4 hp models.

This type hand starter is a flat type mounted on the port side of the powerhead. As the rope is pulled, a swing arm moves the drive gear upward to engage with the teeth of the flywheel ring gear. A coil spring winds and tightens as the rope unwinds. The spring then coils the rope around a pulley as the rope handle is returned to the front of the cowling.

The coil spring consists of a lengthy piece of spring steel (approximately 12 ft. in length), tightly wound inside a housing (the cup and stop assembly).

Movement of the drive gear to the retracted position is accomplished through a second spring.

The starter must be disassembled to replace the rope.

✳✳ WARNING

The rewind spring is under tremendous tension and is a potential hazard. Therefore, safety glasses should be worn and extreme care exercised to follow the procedures carefully during disassembling and assembling work with the starter.

Work on the starter can be very dangerous. Because approximately 12 ft. of spring steel is tightly wound into about 4 in. housing, the spring is placed under tremendous tension - a real tiger in a cage. If the spring should accidentally be released, severe personal injury could result from being struck by the spring with force. Therefore, the service instructions must, and we say again, MUST, be followed closely to prevent release of the spring at the wrong time. Such action would be a bad scene, because serious personal injury could result.

✳✳ CAUTION

The starter rope should NEVER be released from the extended position. Such action would allow the spring to wind with incredible speed resulting in serious damage to the starter mechanism.

Starter

REMOVAL & DISASSEMBLY

◆ See Figures 54 thru 59

1. Remove the spark plug leads.

2. Pull the starter rope slowly out, and then tie a knot in the rope behind the handle. Allow the rope to rewind until the knot is against the cowling. Untie the knot in the end of the rope, and then remove the handle and the rubber bumper.

3. Remove the knot tied in the rope in the previous step. Allow the rope to *slowly* wind into the starter. Before the rope end passes the cowling, firmly grasp the starter pulley, and then allow the starter to unwind.

4. Observe the back side of the starter. Notice the hook of the starter spring protruding out of a hole in the starter. Grasp the spring hook with a pair of needle-nose pliers, and then pull the spring out as far as possible, to relieve tension on the spring.

5. Remove the 3/8 in. bolt from the bracket between the starter and the exhaust housing.

6. Hold the starter together with one hand, and at the same time loosen the large bolt from the center of the starter. DO NOT remove this bolt at this time. Remove the starter from the powerhead.

7. If the starter is only removed in order to accomplish other work, install a 3/8 in. x 16 nut onto the far side of the thru-bolt to hold the starter together and prevent the spring from escaping.

To disassemble:
◆ See Figures 60 thru 63

8. Remove the center bolt, idler gear arm, and the idler gear arm spring.

✳✳ WARNING

The next step could be dangerous. Removing the pulley from the cup must be done with care to prevent personal injury.

9. Lift the pulley *slightly* and then use a screwdriver and work the spring free of the pulley. DO NOT allow the spring to be released from the cup.

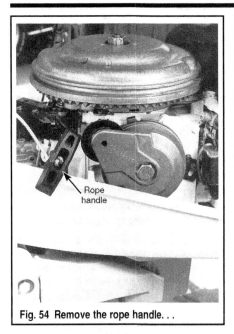

Fig. 54 Remove the rope handle. . .

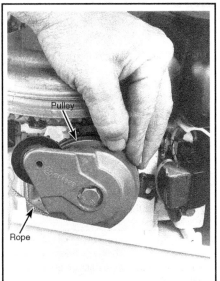

Fig. 55 . . .and then grab the pulley while the starter unwinds

Fig. 56 Grab the spring hook and pull the spring out as far as it will go

Fig. 57 Remove this bolt. . .

Fig. 58 . . .and then remove the starter

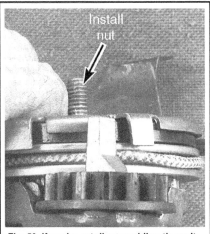

Fig. 59 If you're not disassembling the unit, make sure to run a nut onto the bolt to hold everything together

After the pulley has been removed, notice the position of the spring loop. Remove the rope from the pulley. Notice how the rope unwinds from the pulley COUNTERCLOCKWISE.

10. Remove the bushings from the idler gear arm and the bushings installed one on each side of the pulley.

11. Two different methods are suggested to remove the spring from the starter cup. One method involves pulling continuously on the end of the spring that contains the loop. The second method is to simply toss the cup a safe distance onto carpeting or a lawn, allowing the spring to be released instantly from the cup. If this second method is used, be sure the spring will not cause a threat to any individual in the area when it is released.

CLEANING & INSPECTION

 EASY

1. Wash all parts except the rope in solvent and then blow them dry with compressed air.

2. Remove any trace of corrosion and wipe all metal parts with an oil-dampened cloth.

3. Inspect the starter spring end loops. Replace the spring if it is weak, corroded or cracked.

4. Inspect the rope. Replace the rope if it appears to be weak or frayed. If the rope is frayed, check the hole through which the rope passes for rough

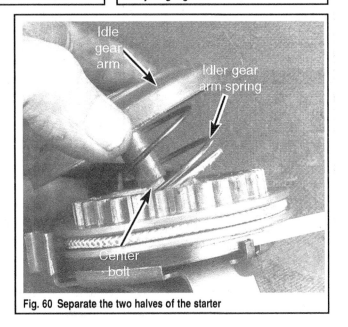

Fig. 60 Separate the two halves of the starter

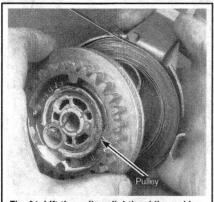

Fig. 61 Lift the pulley slightly while working the spring free

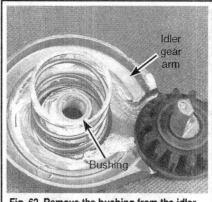

Fig. 62 Remove the bushing from the idler gear arm

Fig. 63 Removing the spring can be an experience!

edges or burrs. Remove the rough edges or burrs with a file, and polish the surface until it is smooth.

5. Inspect the dog ears of the pulley gears to be sure they are not worn and are free of burrs. Check the idler gear for cracks and missing teeth.

ASSEMBLY & INSTALLATION

◆ **See Figures 64 thru 75**

■ **The length and diameter of the starter rope required will vary depending on the horsepower size of the model being serviced. Therefore, check the Hand Starter Rope Specifications chart, and then purchase a quality nylon piece of the proper length and diameter size. Only with the proper rope, will you be assured of efficient operation following installation.**

1. Each end of the nylon rope should be "fused" by burning them slightly with a very small flame (a match flame will do) to melt the fibers together. After the end fibers have been "fused" and while they are still hot, use a piece of cloth as protection and pull the end out fiat to prevent a "glob" from forming.

2. Tie a figure 8 knot in one end of the rope.

3. Feed the other end of the rope through the pulley hole, and then pull the rope tight until the knot is seated in the pulley.

4. Wind the rope CLOCKWISE around the pulley. Use a piece of masking tape or a rubber band to hold the rope in place in the pulley.

5. Coat the bushings with a light film of Johnson/Evinrude Type A lubricant. Insert one bushing into the pulley, another bushing into the idler gear arm, and two more bushings, one on each side of the pulley.

6. If the spring was NOT removed from the cup, lower the pulley down over the spring and insert the end of the spring into the spring anchor post of the pulley. If the spring WAS removed from the cup, hook the end of the

Fig. 64 Tie a knot in one end of the rope. . .

Fig. 65 . . .feed the other end through the hole in the pulley. . .

Fig. 66 . . .and then wind it around the pulley clockwise

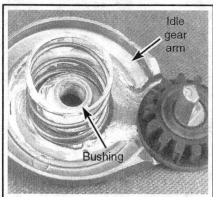

Fig. 67 Lightly lubricate the bushing before installing them

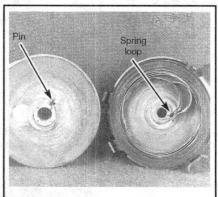

Fig. 68 A good look at everything when properly installed

Fig. 69 Lower the pulley down over the spring

spring into the pulley and then allow the spring to come out the slot in the cup.

7. Turn the pulley and wind the spring into the cup until about 1/2 of the spring length has been wound. Hold the gear and allow it to back off slowly. DO NOT allow the spring to rewind quickly. The remainder of the spring will be installed later.

8. Assemble the idler gear with the shoulder against the idler gear arm. Install the idler gear arm and spring to the pulley and cup with the stop on the underside of the idler gear shaft located between the upper stop and the lower stop on the cup and stop assembly.

9. Hold the assembly together and install the assembly onto the powerhead. Thread the shoulder bolt into place first. This bolt will hold the starter assembly together.

10. Install the bolt through the idler arm and into the exhaust manifold; but do not tighten this bolt at this time. Apply a light coating of Johnson/Evinrude Type A lubricant to the portion of the spring extending out of the starter.

■ Special tools are available to lock-in the starter to the flywheel. However, the tools are usually not available; they are expensive; and professional mechanics have developed an alternate method. The procedure will take time and patience, but it is the only way without the special tools. To work without the special tools proceed as follows:

11. Remove the rubber band or the masking tape from the coiled rope.

12. Working from the rear of the powerhead, pull on the rope and the idler gear will engage with the flywheel ring gear. Continue to pull the rope, and at the same time, work the spring down into the cup. If the rope becomes fully extended, before the spring is installed into the cup, allow the rope to rewind onto the pulley as far as possible and then wind the rope around the pulley again.

13. Now, pull on the rope again from the back side of the rear of the powerhead, and continue to work the spring into the cup until it is completely installed.

14. Ease back on the rope until there is no spring tension on the starter. Thread the rope into the pulley CLOCKWISE around the starter. Two, or possibly more, loops may be required to accomplish the task. Use all of the rope in the pulley with the starter in the relaxed position.

15. After all of the rope has been fed into the pulley, grab the end of the rope in front of the starter and pull it out, and then feed it through the cowling at the front of the powerhead. Continue to pull the rope until about 2 feet is extending out through the cowling. Tie a slip knot in the rope.

16. Install the rubber bumper and handle onto the rope. Tie a figure 8 knot in the end of the rope, and then pull the knot into the handle.

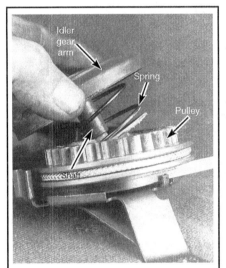

Fig. 70 Positioning the idler gear arm and spring into the pulley

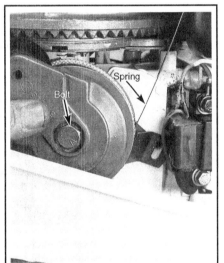

Fig. 71 Position the starter for installation. . .

Fig. 72 . . .and then tighten the bolt

Fig. 73 Work the spring into the cup. . .

Fig. 74 . . .and then wind on the rope

Fig. 75 Slip on the handle and then tie another knot in the free end of the rope

17. Untie the slip knot and ease the rope back into the starter. The starter handle must be up tight against the cowling when the rope is completely rewound on the starter pulley. If the rope is not tight against the cowling, remove the knot and handle from the rope, and then wind the rope around the pulley one complete turn. Tie another knot in the end of the rope as described earlier in this step and then check to be sure the handle is tight against the cowling when the rope is wound onto the pulley.

ADJUSTMENT

◆ **See Figure 76**

1. Hold the idler gear arm stop against the cup stop.
2. Fully engage the idler gear teeth with the teeth in the flywheel ring gear.
3. Tighten the cup and stop assembly screw.
4. Tighten the shoulder screw securely.

BOLT

Fig. 76 Tighten the shoulder bolt

TYPE IV STARTER - MOUNTED ATOP FLYWHEEL W/NO RETURN SPRINGS

The Type IV hand rewind starter is used on:
- 1964 28 hp
- 1965-70 33 hp
- 1964-72 40 hp models

This type starter is installed as original equipment by the manufacturer. However, if the starter pulley was damaged sometime in the past, and the hub replaced, the replacement kit would contain parts modifying the unit. The most noticeable change is the absence of the pawl return springs.

However, it is possible, and has happened quite often that the previous owner may have replaced the starter unit with a unit having the pawl return springs. Possibly this was the only unit available from the local dealer.

✳✳ WARNING

As with other types of hand starters, the rewind spring is a potential hazard. The spring is under tremendous tension when it is wound - a real tiger in a cage. If the spring should accidentally be released, severe personal injury could result from being struck by the spring with force. Therefore, the service instructions must, and we say again MUST, be followed closely to prevent release of the spring at the wrong time. Such action would be a bad scene, because serious personal injury could result.

✳✳ CAUTION

The starter rope should NEVER be released from the extended position. Such action would allow the spring to wind with incredible speed, resulting in serious damage to the starter mechanism.

Any time the rope is broken, the starter spring will rewind with incredible speed. Such action will cause the spring to rewind past its normal travel and the end of the spring will be bent back out of shape. Therefore, if the rope has been broken, the starter must be completely disassembled and the spring repaired or replaced.

Starter

REMOVAL & DISASSEMBLY

◆ **See Figures 77 thru 85**

1. Disconnect any linkage between the starter and the carburetor. Move the linkage out of the way.
2. Remove the starter leg retaining bolts and then lift the complete starter from the powerhead.
3. Pull the rope out far enough, and then tie a knot in the rope. Allow the rope to rewind to the knot.
4. Work the rope anchor out of the rubber-covered handle, and then remove the rope from the anchor. Remove the handle from the rope.

5. Untie the knot in the rope, and then hold the disc pulley, but permit it to turn and thus allow the rope to wind back on to the pulley SLOWLY. Continue to allow the spring in the pulley to unwind *slowly* until all tension has been released.

6. Remove the center nut from the top side of the starter. Some models do not have a center nut. On other models, the nut may have vibrated loose, but if the center bolt has threads showing, a nut must be installed during assembling.

7. Lay the starter on its back on a work surface. Remove the 3 screws securing the pawl retainers to the pulley.

8. Remove the 3 pawls from their retainers.

9. Remove the center bolt from the hub.

10. Hold the pulley, and at the same time, remove the spindle, the wavy washer, friction ring, and the nylon bushing from the center of the pulley.

✳✳ WARNING

The rewind spring is a potential hazard. The spring is under tremendous tension when it is wound - a real tiger in a cage. If the spring should accidentally be released, severe personal injury could result from being struck by the spring with force. Therefore, the following step must be performed with care to prevent personal injury to self and others in the area. If the spring should be accidentally released at the wrong time, such action would be a bad scene, because serious personal injury could result.

11. Lift the pulley straight up and at the same time work the spring free of the pulley. The spring has a small loop hooked into the pulley.

12. An alternate and safe method is to hold the pulley and the housing together tightly and turn the complete assembly so the legs are facing downward. Now, lower the complete assembly to the floor. When the legs make contact with the floor, release the grip on the pulley. The pulley will fall and the spring will be released from the housing, but the three legs will contain the spring and prevent it from traveling across the room. If the spring was not released from the housing, the only safe method is to jar the three legs on the floor to release the spring. Unwind the rope out of the pulley groove, and then pull it free.

✳✳ WARNING

If the spring was not released from the housing, the only safe method is to jar the three legs on the floor a second or third time to release the spring. Check to be sure ALL of the spring has been released. If some of the spring is hung up in the housing, tap on the top of the housing with a mallet while the legs are still on the floor.

CLEANING & INSPECTION

◆ **See Figures 86 and 87**

■ **If the rope was broken and the spring is bent backward, it is a simple matter to bend the spring end back to its normal position.**

1. Wash all parts except the rope in solvent and then blow them dry with compressed air.
2. Remove any trace of corrosion and wipe all metal parts with an oil-dampened cloth.

3. Inspect the rope. Replace the rope if it appears to be weak or frayed. If the rope is frayed, check the hole through which the rope passes for rough edges or burrs. Remove the rough edges or burrs with a file, and polish the surface until it is smooth.

4. Inspect the starter spring end loops. Replace the spring if it is weak, corroded or cracked. Check the spring pin located at the back side of the pulley to be sure it is straight and solid.

5. Check the inside surface of the housing and remove any burrs.

6. Inspect the hub center locating pin to be sure it is straight and tight.

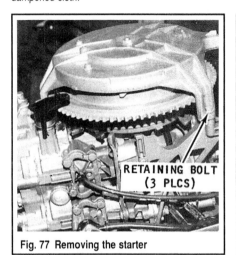

Fig. 77 Removing the starter

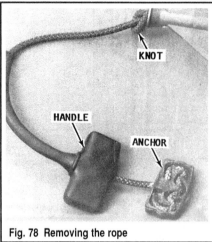

Fig. 78 Removing the rope

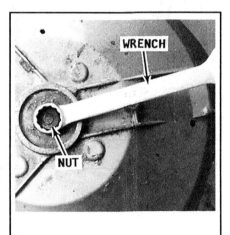

Fig. 79 Remove the center nut. . .

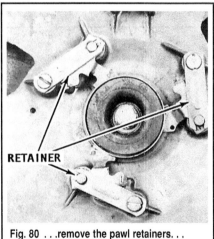
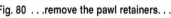

Fig. 80 . . .remove the pawl retainers. . .

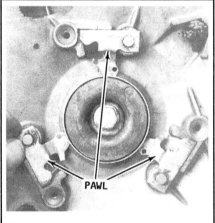

Fig. 81 . . .and then the pawls

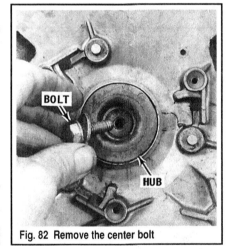

Fig. 82 Remove the center bolt

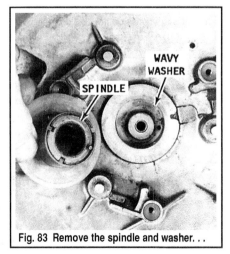

Fig. 83 Remove the spindle and washer. . .

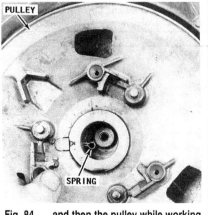

Fig. 84 . . .and then the pulley while working the spring end loose

Fig. 85 A good look at the installed spring

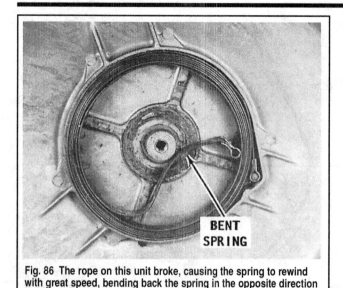

Fig. 86 The rope on this unit broke, causing the spring to rewind with great speed, bending back the spring in the opposite direction

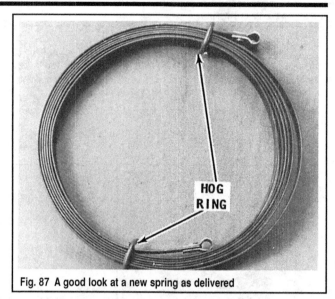

Fig. 87 A good look at a new spring as delivered

ASSEMBLY & INSTALLATION

◆ See Figures 88 thru 99

A new starter spring is held wound with "hog rings". The spring MUST be released to its full extended position before it can be installed. Therefore, use care and remove the "hog rings" and allow the spring to unwind until it is a straight piece of spring steel.

✳✳ WARNING

Wear a good pair of gloves while unwinding and installing the spring. The spring will develop tension and the edges of the spring steel are sharp. The gloves will prevent cuts on your hands and fingers.

A safe method is for one person to remove the hog rings while an assistant holds the spring. After the rings have been removed, both persons work to unwrap the spring, one coil at a time.

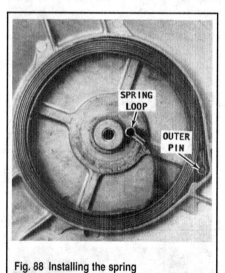

Fig. 88 Installing the spring

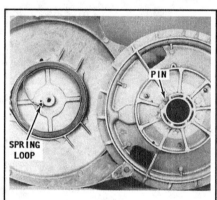

Fig. 89 Housing (left) with the spring properly installed and the spring end bent toward the center. The pin in the pulley (right) must index into the loop of the spring end when installed

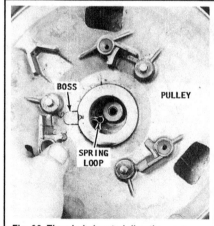

Fig. 90 The pin is located directly underneath the boss on the outside of the pulley

Fig. 91 Slide the wavy washer over the spindle. . .

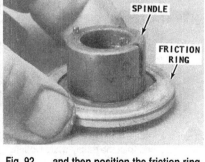

Fig. 92 . . .and then position the friction ring

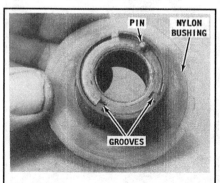

Fig. 93 Proper installation of the bushings

1. Slide the spring onto the outer pin and then start the spring from the outside edge of the housing and insert it into the housing COUNTERCLOCKWISE.

2. Notice the small hump in the housing. This hump prevents the spring from being wound in the wrong direction. Work the first turn into the housing, and then hold the spring down with one hand and continue to wind the spring into the housing. Patience and time are required to work the spring completely into the housing. After the last portion is in place, bend the end of the spring towards the center of the housing. This position will allow the pulley pin to align with the loop in the end of the spring, when the pulley is installed.

3. Lower the pulley down over the top of the spring with the pulley pin indexing into the loop in the end of the spring. Notice the boss on the backside of the pulley. The pin is located directly under the boss. The boss can, therefore, be a guide during pulley installation.

4. Coat the spindle with a thin film of Johnson/Evinrude Type A lubricant. Place the wavy washer onto the spindle.

5. Slide the friction ring onto the spindle with the flats in the washer indexed with the flats on the spindle.

6. Install the nylon bushing onto the spindle with the protrusions on the bushing indexed into the slots in the spindle. Notice the pin protruding from the bottom of the spindle. This pin must drop into the hole in the starter housing. Observe into the housing and visually locate this hole.

7. Lower the spindle assembly down through pulley and index the pin into the hole in the housing.

8. Place the washer inside the spindle housing and then install the bolt through the washer into the housing. Tighten the bolt securely.

9. Install the 3 pawls into their retainers with the tip of each pawl lying over the top of the nylon bushing.

10. Install the retainers securing the pawls to the pulley.

11. Turn the starter over and install the retaining nut onto the thru-bolt (if a nut is used). Tighten the nut securely. Check to be sure the pulley will rotate smoothly and does not bind on the spindle. Rotate the pulley slightly COUNTERCLOCKWISE, then release it to be sure the spring is properly engaged with the pulley and that the pulley has good spring tension.

To Install:

12. Position the starter over the flywheel with the 3 legs aligned over the holes in the powerhead for the retaining bolts.

13. Install the retaining bolts and tighten them securely.

14. Connect the carburetor linkage.

ROPE INSTALLATION

◆ See Figures 100 thru 105

■ The length and diameter of the starter rope required will vary depending on the horsepower size of the model being serviced. Therefore, check the Hand Starter Rope Specifications chart, and then purchase a quality nylon piece of the proper length and diameter size. Only with the proper rope, will you be assured of efficient operation following installation.

1. Each end of the nylon rope should be "fused" by burning them slightly with a very small flame (a match flame will do) to melt the fibers together. After the end fibers have been "fused" and while they are still hot, use a piece of cloth as protection and pull the end out flat to prevent a "glob" from forming.

2. Tie a figure 8 knot in one end of the rope. Set the rope aside, but handy, to be picked up with one hand.

3. Hold the housing with one hand and rotate the pulley three complete turns COUNTERCLOCKWISE with the other hand. After 3 complete turns

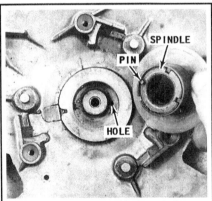

Fig. 94 Lower the assembly through the pulley so the pin lines up with the hole. . .

Fig. 95 . . .install the bolt and washer. . .

Fig. 96 . . .and then the pawls

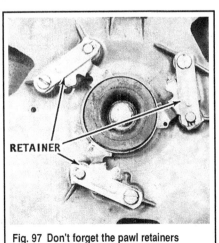

Fig. 97 Don't forget the pawl retainers

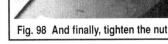

Fig. 98 And finally, tighten the nut

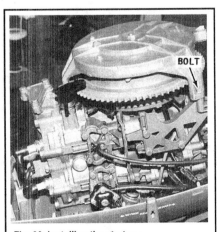

Fig. 99 Installing the starter

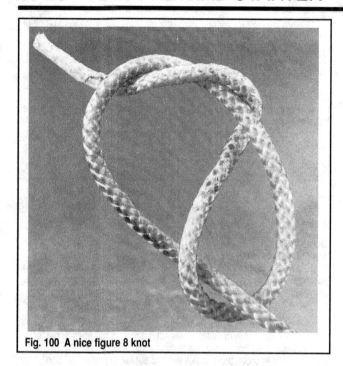

Fig. 100 A nice figure 8 knot

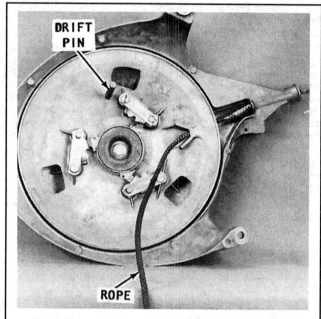

Fig. 101 Rotate the pulley counterclockwise and install a drift pin

have been made, align the rope outlet in the pulley with the outlet in the housing. Insert a drift pin or other suitable tool through the hole in the pulley and the hole in the housing to hold the pulley in the desired position.

4. Feed the rope through the pulley and the housing and out the other side of the housing. Pull the rope tight until the knot is seated against the pulley.

5. A special tool is manufactured by Johnson/Evinrude to install the handle onto the rope. This tool has 3 prongs and is inserted through the handle and then attached to the rope and pulled back through the handle. If the special tool is not available take a stiff piece of wire; insert it through the handle; thread it through the rope; apply just a little oil to the rope; then pull the wire and rope through the handle.

6. Work the end of the rope into the handle anchor. Secure the rope in place by pushing the anchor into the rubber handle.

7. Lightly pull on the rope to relieve tension on the pin installed through the pulley and housing earlier. Maintain some tension on the rope, remove the pin, and allow the spring to SLOWLY wind the rope onto the pulley. Check the bolt through the spindle to be sure it is tight.

8. Lay the starter on its back and pull the rope with quick movements, and at the same time check the pawls to be sure they move towards the center of the pulley. Release the rope slowly and check to be sure the pawls return to their original position under the retainers.

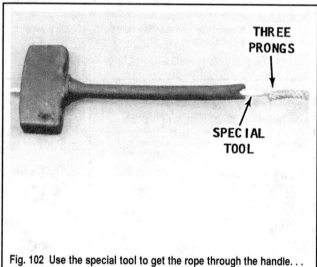

Fig. 102 Use the special tool to get the rope through the handle. . .

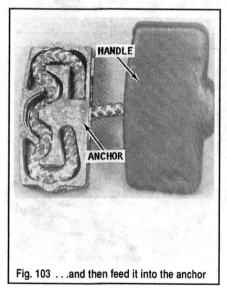

Fig. 103 . . .and then feed it into the anchor

Fig. 104 Remove the drift pin. . .

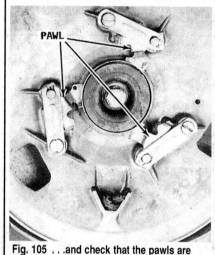

Fig. 105 . . .and check that the pawls are working properly

TYPE V STARTER - MOUNTED ATOP FLYWHEEL W/RETURN SPRINGS

The Type V hand rewind starter is used on:
- 1962-63 28 hp
- 1958-59 35 hp
- 1960-63 40 hp models

■ This starter is virtually identical to the Type IV starter. In fact, the only difference is that this uses 3 pawl return springs and the Type IV does not. To be safe, we have included all procedures here again, but please refer to the Type IV section for most artwork.

This type starter is installed as original equipment by the manufacturer. However, if the starter pulley was damaged sometime in the past, and the hub replaced, the replacement kit would contain parts modifying the unit. The most noticeable change is the absence of the pawl return springs.

However, it is possible, and has happened quite often that the previous owner may have replaced the starter unit with a unit having the pawl return springs. Possibly this was the only unit available from the local dealer.

✳✳ WARNING

As with other types of hand starters, the rewind spring is a potential hazard. The spring is under tremendous tension when it is wound - a real tiger in a cage. If the spring should accidentally be released, severe personal injury could result from being struck by the spring with force. Therefore, the service instructions must, and we say again MUST, be followed closely to prevent release of the spring at the wrong time. Such action would be a bad scene, because serious personal injury could result.

✳✳ CAUTION

The starter rope should NEVER be released from the extended position. Such action would allow the spring to wind with incredible speed, resulting in serious damage to the starter mechanism.

Any time the rope is broken, the starter spring will rewind with incredible speed. Such action will cause the spring to rewind past its normal travel and the end of the spring will be bent back out of shape. Therefore, if the rope has been broken, the starter must be completely disassembled and the spring repaired or replaced.

Starter

REMOVAL & DISASSEMBLY

 ② ◀MODERATE

◆ See Figures 78 thru 81, 83 thru 85 and 106 thru 108

1. Disconnect any linkage between the starter and the carburetor. Move the linkage out of the way.
2. Remove the starter leg retaining bolts and then lift the complete starter from the powerhead.

3. Pull the rope out far enough, and then tie a knot in the rope. Allow the rope to rewind to the knot.

4. Work the rope anchor out of the rubber covered handle, and then remove the rope from the anchor. Remove the handle from the rope.

5. Untie the knot in the rope, and then hold the disc pulley, but permit it to turn and thus allow the rope to wind back on to the pulley SLOWLY. Continue to allow the spring in the pulley to unwind *slowly* until all tension has been released.

6. Remove the center nut from the top side of the starter. Some models do not have a center nut. On other models, the nut may have vibrated loose, but if the center bolt has threads showing, a nut MUST be installed during assembling.

7. Lay the starter on its back on a work surface. Remove the 3 screws securing the pawl retainers to the pulley.

8. Remove the 3 pawl return springs as shown in the illustration.

9. Remove the 3 pawls from their retainers.

10. Remove the center bolt from the hub.

11. Hold the pulley, and at the same time, remove the spindle, the wavy washer, friction ring, and the nylon bushing from the center of the pulley.

✳✳ WARNING

The rewind spring is a potential hazard. The spring is under tremendous tension when it is wound - a real tiger in a cage. If the spring should accidentally be released, severe personal injury could result from being struck by the spring with force. Therefore, the following step must be performed with care to prevent personal injury to self and others in the area. If the spring should be accidentally released at the wrong time, such action would be a bad scene, because serious personal injury could result.

12. Lift the pulley straight up and at the same time work the spring free of the pulley. The spring has a small loop hooked into the pulley.

13. An alternate and safe method is to hold the pulley and the housing together tightly and turn the complete assembly so the legs are facing downward. Now, lower the complete assembly to the floor. When the legs make contact with the floor, release the grip on the pulley. The pulley will fall and the spring will be released from the housing, but the three legs will contain the spring and prevent it from traveling across the room. If the spring was not released from the housing, the only safe method is to jar the three legs on the floor to release the spring. Unwind the rope out of the pulley groove, and then pull it free.

✳✳ WARNING

If the spring was not released from the housing, the only safe method is to jar the three legs on the floor a second or third time to release the spring. Check to be sure ALL of the spring has been released. If some of the spring is hung up in the housing, tap on the top of the housing with a mallet while the legs are still on the floor.

Fig. 106 Removing the starter

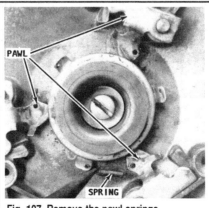

Fig. 107 Remove the pawl springs. . .

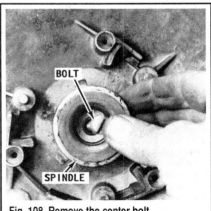

Fig. 108 Remove the center bolt

CLEANING & INSPECTION

◆ **See Figures 86, 87 and 109**

■ **If the rope was broken and the spring is bent backward, it is a simple matter to bend the spring end back to its normal position.**

1. Wash all parts except the rope in solvent and then blow them dry with compressed air.

2. Remove any trace of corrosion and wipe all metal parts with an oil-dampened cloth.

3. Inspect the rope. Replace the rope if it appears to be weak or frayed. If the rope is frayed, check the hole through which the rope passes for rough edges or burrs. Remove the rough edges or burrs with a file, and polish the surface until it is smooth.

4. Inspect the starter spring end loops. Replace the spring if it is weak, corroded or cracked. Check the spring pin located at the back side of the pulley to be sure it is straight and solid.

5. Check the inside surface of the housing and remove any burrs.

6. Check the condition of the pawl springs to be sure they are not stretched out of shape. The end of each spring should be bent back toward the coil of the spring. Inspect the pawls for wear and that the edges are not rounded.

7. Inspect the hub center locating pin to be sure it is straight and tight.

ASSEMBLY & INSTALLATION

◆ **See Figures 88 thru 97**

A new starter spring is held wound with "hog rings". The spring MUST be released to its full extended position before it can be installed. Therefore, use care and remove the "hog rings" and allow the spring to unwind until it is a straight piece of spring steel.

✳✳ WARNING

Wear a good pair of gloves while unwinding and installing the spring. The spring will develop tension and the edges of the spring steel are sharp. The gloves will prevent cuts on your hands and fingers.

A safe method is for one person to remove the hog rings while an assistant holds the spring. After the rings have been removed, both persons work to unwrap the spring, one coil at a time.

1. Slide the spring onto the outer pin and then start the spring from the outside edge of the housing and insert it into the housing COUNTERCLOCKWISE.

2. Notice the small hump in the housing. This hump prevents the spring from being wound in the wrong direction. Work the first turn into the housing, and then hold the spring down with one hand and continue to wind the spring into the housing. Patience and time are required to work the spring completely into the housing. After the last portion is in place, bend the end of the spring towards the center of the housing. This position will allow the pulley pin to align with the loop in the end of the spring, when the pulley is installed.

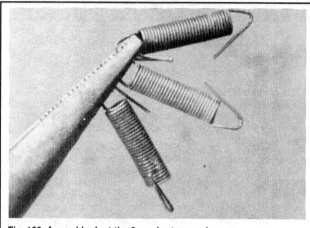

Fig. 109 A good look at the 3 pawl return springs

3. Lower the pulley down over the top of the spring with the pulley pin indexing into the loop in the end of the spring. Notice the boss on the backside of the pulley. The pin is located directly under the boss. The boss can, therefore, be a guide during pulley installation.

4. Coat the spindle with a thin film of Johnson/Evinrude Type A lubricant. Place the wavy washer onto the spindle.

5. Slide the friction ring onto the spindle with the flats in the washer indexed with the flats on the spindle.

6. Install the nylon bushing onto the spindle with the protrusions on the bushing indexed into the slots in the spindle. Notice the pin protruding from the bottom of the spindle. This pin must drop into the hole in the starter housing. Observe into the housing and visually locate this hole.

7. Lower the spindle assembly down through pulley and index the pin into the hole in the housing.

8. Place the washer inside the spindle housing and then install the bolt through the washer into the housing. Tighten the bolt securely.

9. Install the 3 pawls into their retainers with the tip of each pawl lying over the top of the nylon bushing. Connect the return springs.

10. Install the retainers securing the pawls to the pulley.

11. Turn the starter over and install the retaining nut onto the thru-bolt (if a nut is used). Tighten the nut securely. Check to be sure the pulley will rotate smoothly and does not bind on the spindle. Rotate the pulley slightly COUNTERCLOCKWISE, then release it to be sure the spring is properly engaged with the pulley and that the pulley has good spring tension.

To Install:

12. Position the starter over the flywheel with the 3 legs aligned over the holes in the powerhead for the retaining bolts.

13. Install the retaining bolts and tighten them securely.

14. Connect the carburetor linkage.

ROPE INSTALLATION

■ **Please refer to the Type IV procedures for rope installation.**

TYPE VI STARTER - MOUNTED ATOP FLYWHEEL WITH 1 OR 2 NYLON PAWLS

◆ **See Figures 110 and 111**

The Type VI hand rewind starter is used on the following models:

- 1971-72 2 hp
- 1958-68 3 hp
- 1958-64 5.5 hp
- 1958 7.5 hp
- 1958-63 10 hp
- 1958-72 18 hp
- 1966-72 20 hp
- 1969-72 25 hp

✳✳ WARNING

As with other types of hand starters, the rewind spring is a potential hazard. The spring is under tremendous tension when it is wound - a real tiger in a cage. If the spring should accidentally be released, severe personal injury could result from being struck by the spring with force. Therefore, the service instructions must, and we say again MUST, be followed closely to prevent release of the spring at the wrong time. Such action would be a bad scene, because serious personal injury could result.

✳✳ CAUTION

The starter rope should NEVER be released from the extended position. Such action would allow the spring to wind with incredible speed, resulting in serious damage to the starter mechanism.

Any time the rope is broken, the starter spring will rewind with incredible speed. Such action will cause the spring to rewind past its normal travel and the end of the spring will be bent back out of shape. Therefore, if the rope has been broken, the starter must be completely disassembled and the spring repaired or replaced.

Fig. 110 A close up view of the pulley and housing with the arrow on the housing aligned with the marks on the pulley

Arrow

Pull rope

Fig. 111 Notice how far this rope has been pulled before the flywheel has been rotated - this starter has not been timed properly

Hand Starter Timing

Surprising as it may sound, this starter, mounted on top of the powerhead over the flywheel, can actually time to the powerhead. This timing can best be described by using an example.

If 2 marks were made on the flywheel 180° apart, and matching marks made on the powerhead, then each time the powerhead was shut down, one set of marks on the flywheel would align very closely with one of the marks on the powerhead. What is actually happening is the powerhead is stopping with either the top piston at TDC (top dead center) or the bottom piston at the TDC position.

Now, assume the powerhead has been operating at idle speed and then suddenly stops for any number of reasons. The problem is corrected and the powerhead is once again ready to be started.

Two notches are manufactured into the inside diameter of the flywheel. A single dog on the pulley engages with one of these dogs when the rope is pulled.

Now, if it is necessary to pull an excessive amount of rope before the dog is able to engage the flywheel, full starting rotation power would not be available in the rope.

Therefore, the starter is timed to engage the starter with the flywheel after the rope has been pulled exactly the same amount each time. This distance is very short to allow as much rope pull as possible to rotate the crankshaft for fast start.

This "timing" will assure an adequate amount available for starting *and* that one of the pistons will return to TDC when the pull is completed, if the powerhead fails to start. If an excessive amount of pull is necessary before the flywheel begins to rotate, the starter was not assembled properly - the arrow on the pulley was not aligned with the 2 marks on the starter housing when the spring is relaxed and the rope handle is retracted.

Starter

REMOVAL & DISASSEMBLY

◆ **See Figures 77 and 112 thru 117**

1. On most models:

a. Remove the attaching bolts securing the 3 legs of the starter housing to the powerhead. On some smaller horsepower powerheads, the starter housing is attached to the fuel tank with screws.

b. Remove the hand starter and lay it on the bench with the pulley facing toward you.

2. On certain later models:

a. Remove the screw retaining the throttle cable trunnion to the powerhead.

b. Pull the throttle cable down to release it from the armature plate.

c. Remove the nut and 2 bolts securing the starter housing to the powerhead.

d. To remove the throttle cable from the starter housing, remove the 2 screws on the small retaining bracket across the throttle cable and then twist the cable COUNTERCLOCKWISE to release the cable from the throttle knob.

e. Remove the hand starter and lay it on the bench with the pulley facing toward you.

To disassemble:

3. Pull the rope out enough to tie a knot in the rope. Tie a knot, and then allow the rope to rewind to the knot.

4. Work the rope anchor out of the rubber covered handle, and then remove the rope from the anchor.

5. Remove the handle from the rope. Untie the knot in the rope, and then hold the disc pulley, but permit it to turn and thus allow the rope to wind back onto the pulley SLOWLY. Continue to allow the spring in the pulley to unwind SLOWLY until all tension has been released.

■ **The accompanying illustrations depict a hand rewind starter with a single pawl. If servicing a starter with 2 pawls, simply repeat the steps necessary to remove the other pawl. All other procedural steps are valid for both types of starters.**

6. Remove the E-clip from the nylon pawl. Lift the pawl from the stud.

7. Remove the friction spring and friction link from the pawl.

8. Remove the bolt, lock washer and washer from the center of the pulley spindle. Lift the spindle out of the pulley, and at the same time hold the pulley firmly together with the housing.

■ **A starter spring shield is installed between the pulley and the spring two pawl starters.**

9. Lift the pulley straight up and at the same time work the spring free of the pulley. The spring has a small loop hooked into the pulley.

10. An alternate and safe method is to hold the pulley and the housing together tightly and turn the complete assembly with the legs extending downward in the normal manner. Now, lower the complete assembly to the floor. When the legs make contact with the floor, release your grip. The pulley will fall and the spring will be released from the housing, but the three legs will contain the spring and prevent it from traveling across the room. If the spring was not released from the housing, the only safe method is to again make contact with the three legs on the floor and jar the spring free.

11. Unwind the rope out of the pulley groove. Notice the pin next to the knot in the rope and how the rope feeds BEHIND the pin.

12. Pull the knot and the rope out far enough to untie the knot, and then pull the rope free of the pulley.

CLEANING & INSPECTION

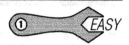

◆ **See Figures 118, 119 and 120**

If the rope was broken and the spring is bent backward, it is a simple matter to bend the spring end back to its normal position.

1. Wash all parts except the rope in solvent and then blow them dry with compressed air.

2. Remove any trace of corrosion and wipe all metal parts with an oil-dampened cloth.

3. Inspect the rope. Replace the rope if it appears to be weak or frayed. If the rope is frayed, check the hole through which the rope passes for rough edges or burrs. Remove the rough edges or burrs with a file, and polish the surface until it is smooth.

4. Inspect the starter spring end loops. Replace the spring if it is weak, corroded or cracked. Check the spring pin located at the back side of the pulley to be sure it is straight and solid.

5. Check the inside surface of the housing and remove any burrs.

6. Check the condition of the pawl spring to be sure it is not stretched out of shape. The end of the spring should be bent back toward the coil of the spring. Inspect the pawl for wear and that the edges are not rounded.

Fig. 112 Remove the handle from the rope, but don't forget to tie a knot downstream of the handle

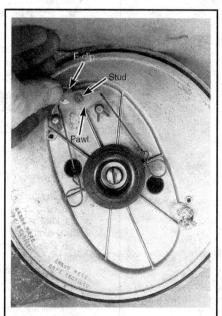

Fig. 113 Remove the E-clip and lift off the pawl

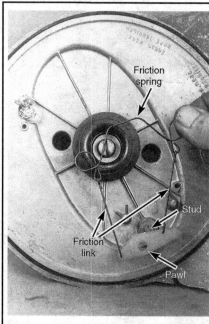

Fig. 114 Remove the friction spring and link...

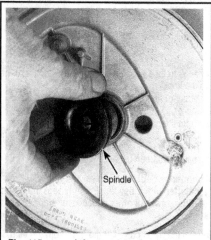

Fig. 115 ...and then remove the spindle

Fig. 116 Lift the pulley straight up while working the spring free

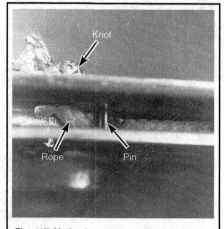

Fig. 117 Notice how the rope feeds behind the pin

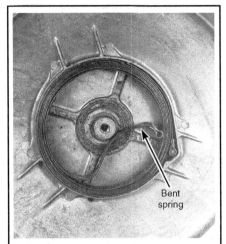

Fig. 118 The rope on this unit broke, causing the spring to rewind with great speed, bending back the spring in the opposite direction

Fig. 119 This pawl has been damaged and is unfit for service

Fig. 120 A nice figure 8 knot

7. Check the friction spring and link to be sure they are not distorted.

8. Inspect the spindle. The spindle must be straight and tight.

ASSEMBLY & INSTALLATION

◆ See Figures 121 thru 129

■ The accompanying illustrations depict a hand starter with a single pawl. If servicing a unit with 2 pawls, simply repeat the necessary steps to install the other pawl. All other procedural steps are valid for both starters.

■ The length and diameter of the starter rope required will vary depending on the horsepower size of the model being serviced. Therefore, check the Hand Starter Rope Specifications chart, and then purchase a quality nylon piece of the proper length and diameter size. Only with the proper rope, will you be assured of efficient operation following installation.

1. Each end of the nylon rope should be "fused" by burning them slightly with a very small flame (a match flame will do) to melt the fibers together. After the end fibers have been "fused" and while they are still hot, use a piece of cloth as protection and pull the end out flat to prevent a "glob" from forming.

2. Tie a figure 8 knot in the end of the rope. Insert one end of the new rope through the hole in the pulley and housing and on the back side of the

pin. Continue to wrap the remainder of the rope COUNTERCLOCKWISE around the pulley.

✳✳ WARNING

Wear a good pair of gloves while installing the spring. The spring will develop tension and the edges of the spring steel are sharp. The gloves will prevent cuts on your hands and fingers.

3. Slide the spring onto the outer pin and then start the spring from the outside edge of the housing and insert it into the housing COUNTERCLOCKWISE.

4. Notice the small hump in the housing. This hump prevents the spring from being wound in the wrong direction. Work the first turn into the housing, and then hold the spring down with one hand and continue to wind the spring into the housing. Patience and time are required to work the spring completely into the housing. After the last portion is in place, bend the end of the spring towards the center of the housing. This position will allow the pulley pin to align with the loop in the end of the spring, when the pulley is installed.

Alternate Method
◆ See Figures 130 thru 132

5. An alternate method of installing a NEW spring into the housing with less risk of personal injury is as follows:

a. Remove only the hog ring next to the end of the outside wrap of the spring.

b. Pull on the outside end of the spring. As the spring is pulled, the inside diameter will get smaller and smaller.

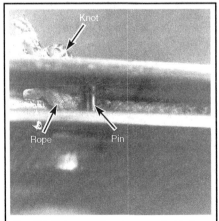

Fig. 121 Make sure the rope goes behind the pin

Fig. 122 Slide the spring onto the outer pin first

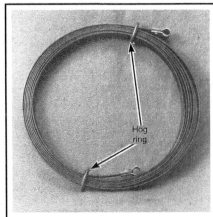

Fig. 123 The pulley pin is located directly under the boss on the outside of the pulley

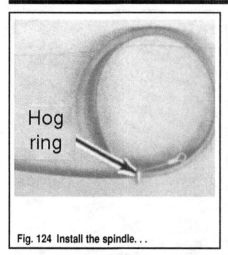

Fig. 124 Install the spindle. . .

Fig. 125 . . .and then the spring, link and pawl

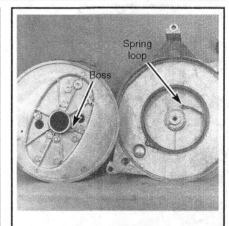

Fig. 126 Now pop in the E-clip

Fig. 127 Installing the handle

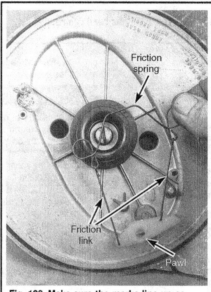
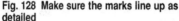

Fig. 128 Make sure the marks line up as detailed

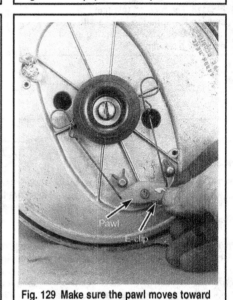

Fig. 129 Make sure the pawl moves toward the center of the pulley when pulling the rope

c. When the diameter is a bit smaller than the inside diameter of the starter housing, wrap the entire free end of the spring around the coiled portion.

d. Carefully lower the coiled spring into the starter housing with the loop on the free end of the spring indexed over the peg in the housing and the spring feeding COUNTERCLOCKWISE.

e. Remove the second hog ring *without* allowing the spring to escape from the housing. An easy and safe method is to cut the hog ring with a pair of "dikes".

f. Bend the inside end of the spring toward the center of the housing to permit the pin in the pulley to index into the loop.

■ **On 2 pawl starters, a starter spring shield is installed between the pulley and the spring on these units. Place the shield against the underside of the pulley aligning the holes.**

6. Lower the pulley down over the top of the spring with the pulley pin indexing into the loop in the end of the spring. Notice the boss on the backside of the pulley. The pin is located directly under the boss. The boss can, therefore, be a guide during pulley installation.

7. Lower the spindle assembly down through pulley.

8. Place the washer and lock washer inside the spindle housing and then install the bolt through the washer into the housing. Tighten the bolt securely. Check to be sure the pulley will rotate smoothly and does not bind on the spindle. Rotate the pulley slightly COUNTERCLOCKWISE and then release it to be sure there is proper engagement with the spring and the pulley has good spring tension.

9. Install the friction spring and link and the nylon pawl onto the starter hub. The friction spring fits into a groove in the spindle. Pull just a little on the pawl and set it over the stud on the flywheel pulley.

10. Snap the E-clip over the top of the pawl to secure it in place.

11. Rotate the pulley 3 complete revolutions, and then work the rope out through the hole in the pulley and the housing. Pull on the rope until a couple of feet are exposed. Tie a knot in the rope and allow the rope to rewind until the knot is tight against the housing.

12. Work the end of the rope into the handle anchor. Secure the rope in place by pushing the anchor into the rubber handle.

13. Pull on the rope enough to untie the knot, and then allow the rope to slowly recoil into the pulley.

14. Lay the starter on its back with the pulley facing toward you. Notice the imprint on the pulley: **ARROW HERE ROPE RECOILED**. Also notice the arrow on the housing. On Johnson powerheads, when the rope is fully coiled (the starter pulley completely wound) the arrow must fall between the marks on the pulley. Further up on the pulley you will notice the letter **E** (for Evinrude). On the Evinrude powerheads, the arrow must fall between the two marks on the pulley. If the arrow is not properly aligned the starter rope is not the proper length. It is either too long or too short. The arrow must align properly for the starter to be timed with the powerhead. For more details, please refer to General Information in this section.

15. With the starter still on its back, pull the rope with quick movements, and at the same time check the pawl to be sure it moves toward the center of the pulley.

16. Release the rope slowly and check to be sure the pawl returns to its original position.

Fig. 130 If using the alternate method, remove only one of the hog rings. . .

Fig. 131 . . .and then pull on the outside of the spring

Fig. 132 Lower the coiled spring into the housing

To Install:

17. Position the starter over the flywheel with the 3 legs aligned over the holes in the powerhead for the retaining bolts.

18. Install the retaining bolts and tighten them to 6.8 ft. lbs. (8 Nm).

Throttle Cable Installation

19. Insert the throttle cable through the cable guide in the housing.

20. Rotate the cable CLOCKWISE to engage the cable with the throttle knob.

21. Place the small retaining bracket across the cable and secure it to the housing with two screws.

22. To adjust the throttle cable, back off the toothed idle stop wheel away from the throttle knob until the throttle cable can be snapped back into place on the armature plate. Push the throttle knob toward the starter housing while pushing the armature plate CLOCKWISE as far as possible. Adjust the position of the throttle cable trunnion to allow the cable to slide into the side retaining bracket and then tighten the bracket retaining screw.

TYPE VII STARTER - MOUNTED ATOP FLYWHEEL W/ADJUSTABLE PAWL

Description & Operation

◆ See Figures 133, 134 and 135

The following engines with manual shift and magneto ignition may be equipped with a hand rewind starter:
- 1958-59 50 hp
- 1960-65 75 hp

The manual starter consists of a spring loaded pulley, a spring loaded pawl and a starter rope.

The pulley has a groove to accept the rope and the spring loaded pawl engages ratchet segments cast into the flywheel when the starter rope is pulled. This pawl automatically disengages the flywheel when force on the rope is released.

The tension built up in the pulley spring as the rope is pulled, causes the pulley to be restored to its original position at the same time the starter rope is rewound around the pulley, reference illustrations.

SAFETY PAWL

◆ See Figures 136 and 137

A safety pawl arrangement is incorporated to prevent any attempt to manually start the engine when it is in gear, either Forward or Reverse. The arrangement consists of a pawl which engages and disengages stop lugs cast onto the underside of the starter pulley and linked with the shift control lever.

When the shift lever is in the Neutral position, the pawl disengages the stop lugs on the starter pulley and hand cranking is permitted.

If the shift lever is in either gear, the stop lugs on the starter pulley engage the pawl and prevent the pulley from rotating. In this manner, hand cranking is not possible with the shift lever in gear.

✳✳ WARNING

As with other types of hand starters, the rewind spring is a potential hazard. The spring is under tremendous tension when it is wound - a real tiger in a cage. If the spring should accidentally be released, severe personal injury could result from being struck by the spring with force. Therefore, the service instructions must, and we say again MUST, be followed closely to prevent release of the spring at the wrong time.

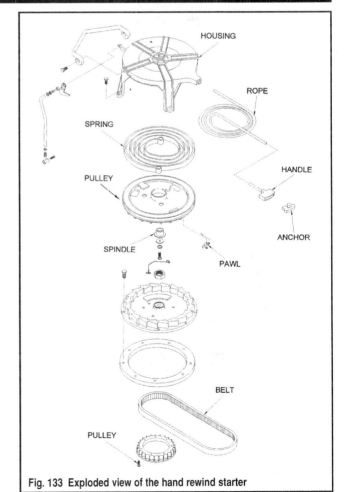

Fig. 133 Exploded view of the hand rewind starter

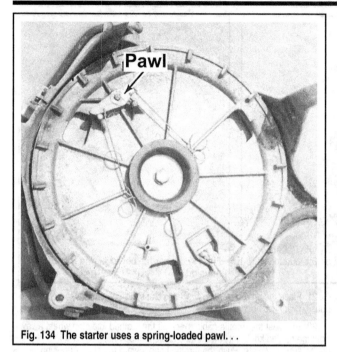

Fig. 134 The starter uses a spring-loaded pawl. . .

Fig. 135 . . .and ratchet segments

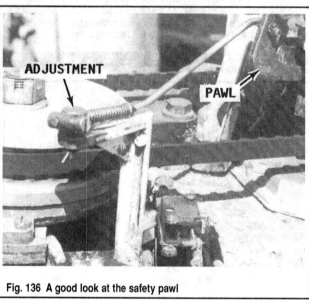

Fig. 136 A good look at the safety pawl

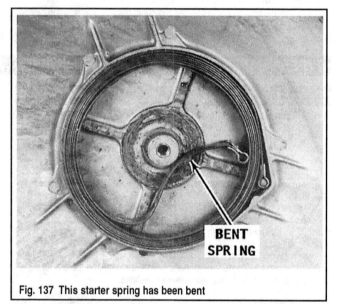

Fig. 137 This starter spring has been bent

The starter rope should NEVER be released from the extended position. Such action would allow the spring to wind with incredible speed, resulting in serious damage to the starter mechanism.

Any time the rope is broken, the starter spring will rewind with incredible speed. Such action will cause the spring to rewind past its normal travel and the end of the spring will be bent back out of shape. If the rope has been broken, the starter should be completely disassembled and the spring repaired or replaced.

Hand Starter

REMOVAL & INSTALLATION

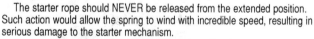

◆ See Figures 138 and 139

1. Remove the neutral latch screw from the starter housing.
2. Remove the attaching bolts securing the 3 legs of the starter housing to the powerhead.

Fig. 138 Remove the neutral latch screw at the housing

3. Remove the hand starter and lay it on the bench with the pulley facing towards you.

4. Position the starter over the flywheel with the 3 legs aligned over the holes in the power head for the retaining bolts.

5. Install the retaining bolts and tighten them securely.

6. Place the neutral latch in position and secure it to the housing with the attaching screw. Tighten the screw securely.

■ **The Neutral lock-out pawl can be adjusted. This pawl is a safety feature for the operator to prevent the hand starter from being operated if the shift lever is in gear - Forward or Reverse.**

Fig. 139 Removing the starter housing

DISASSEMBLY & ASSEMBLY

◆ **See Figures 133 and 140 thru 146**

1. Remove the starter.

2. Pull the rope out enough to tie a knot in the rope. Tie a knot, and then allow the rope to rewind to the knot.

3. Work the rope anchor out of the rubber covered handle, and then remove the rope from the anchor. Remove the handle from the rope.

4. Untie the knot in the rope and then hold the disc pulley, but permit it to turn and thus allow the rope to wind back onto the pulley SLOWLY. Continue to allow the spring in the pulley to unwind slowly until all tension has been released.

5. Remove the E-clip from the nylon pawl.

6. Remove the friction spring and the pawl from the starter housing.

7. Unwind the rope out of the pulley groove. Pull the knot and the rope out far enough to untie the knot, and then pull the rope free of the pulley.

8. Set the starter on the bench in its normal position on its 3 legs. Remove the nut from the center of the starter housing.

9. Remove the bolt, lockwasher and washer from the center of the pulley spindle. Lift the spindle out of the pulley, and at the same time hold the pulley firmly together with the housing.

10. Lift the pulley straight up and at the same time work the spring free of the pulley. The spring has a small loop hooked into the pulley.

An alternate and safer method is to hold the pulley and the housing together tightly and turn the complete assembly with the legs extending downward in the normal manner.

Now, lower the complete assembly to the floor. When the legs make contact with the floor, release your grip. The pulley will fall and the spring will be released from the housing, but the 3 legs will contain the spring and prevent it from traveling across the room. If the spring was not released from the housing, the only safe method is to again make contact with the 3 legs on the floor and jar the spring free.

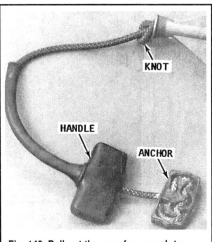

Fig. 140 Pull out the rope far enough to make a knot

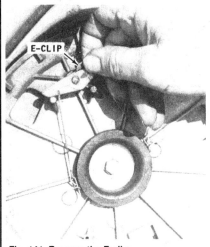

Fig. 141 Remove the E-clip. . .

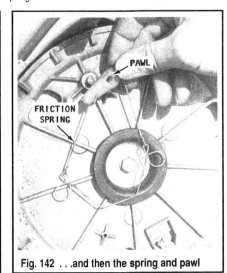

Fig. 142 . . .and then the spring and pawl

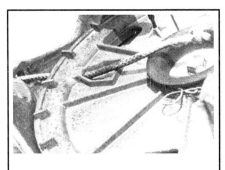

Fig. 143 Pull the rope out

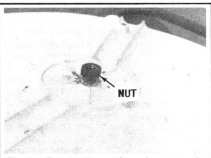

Fig. 144 Remove the nut from the center of the housing

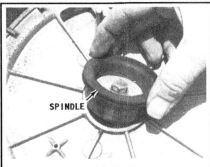

Fig. 145 Remove the spindle. . .

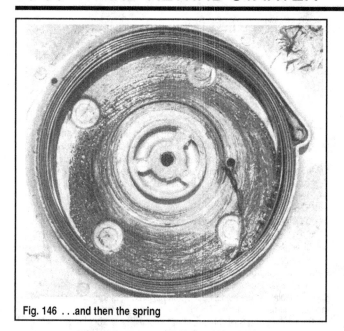

Fig. 146 . . .and then the spring

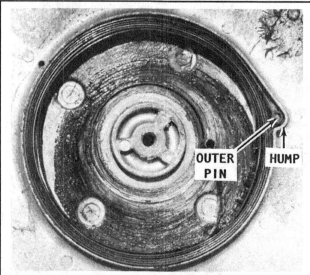

Fig. 147 Make sure that the outer loop goes over the pin in the housing

To assemble:

 See Figures 147 thru 152

The diameter and length of starter rope must be a definite size for maximum efficiency and rope life. The size for all engines covered here is 7/32 in. (5.56mm) diameter and 69 3/4 in. (1.77m) in length. Purchase a quality nylon piece of the proper length and diameter size. Only with the proper rope, will you be assured of efficient operation following installation.

Each end of the nylon rope should be fused by burning them slightly with a very small flame (a match flame will do) to melt the fibers together. After the end fibers have been fused and while they are still hot, use a piece of cloth as protection and pull the end out flat to prevent a glob from forming.

✳✳ CAUTION

Wear a good pair of gloves while installing the spring. The spring will develop tension and the edges of the spring steel material are sharp. The gloves will prevent cuts on your hands and fingers.

11. Slide the spring onto/over the outer pin and then start the spring from the outside edge of the housing and insert it into the housing counterclockwise, as shown in the accompanying illustration. Notice the small hump in the housing which prevents the spring from being wound in the wrong direction.

12. Work the first turn into the housing, and then hold the spring down with one hand and continue to wind the spring into the housing. Patience and time are required to work the spring completely into the housing.

13. After the last portion is in place, bend the end of the spring towards the center of the housing. This position will allow the pulley pin to align with the loop in the end of the spring when the pulley is installed.

Alternate Method

An alternate method to install a new spring into the housing, with less risk of personal injury, is presented with accompanying illustrations.

14. Remove ONLY the hog ring next to the end of the outside wrap of the spring.

15. Pull on the outside end of the spring. As the spring is pulled, the inside diameter will get smaller and smaller, as shown. When the diameter is a bit smaller than the inside diameter of the starter housing, wrap the entire free end of the spring around the coiled portion.

16. Carefully lower the coiled spring into the starter housing with the loop on the free end of the spring indexed over the peg in the housing and the spring feeding counterclockwise, as shown.

17. Remove the second hog ring without allowing the spring to escape from the housing. An easy and safe method is to cut the hog ring with a pair of "dikes". Bend the inside end of the spring toward the center of the housing to permit the pin in the pulley to index into the loop.

18. Lower the pulley down over the top of the spring with the pulley pin indexing into the loop in the end of the spring. In the accompanying illustration, notice the callout for the boss on the backside of the pulley. The

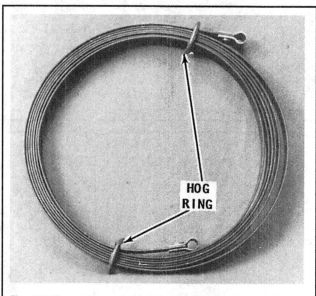

Fig. 148 Alternately, remove only the hog ring near the outer loop. . .

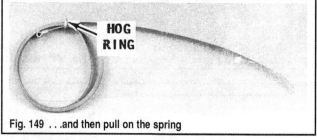

Fig. 149 . . .and then pull on the spring

pin is located directly under the boss. The boss can, therefore, be a guide during pulley installation.

19. Lower the spindle assembly down through the pulley. Place the washer and lockwasher inside the spindle housing and then install the bolt through the washer into the housing. Tighten the bolt securely. Check to be sure the pulley will rotate smoothly and does not bind on the spindle. Rotate the pulley slightly counterclockwise, and then release it to be sure there is proper engagement with the spring and the pulley has good spring tension.

20. Set the starter on the bench in its normal position on its 3 legs. Tighten the nut in the center of the housing.

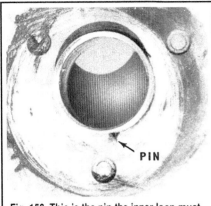

Fig. 150 This is the pin the inner loop must connect to

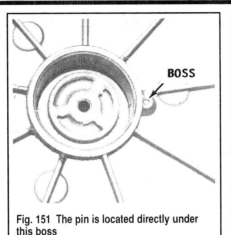

Fig. 151 The pin is located directly under this boss

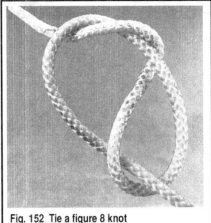

Fig. 152 Tie a figure 8 knot

21. Install the friction spring and link and the nylon pawl onto the starter hub. The friction spring fits into a groove in the spindle. Pull just a little on the pawl and set it over the stud on the flywheel pulley.

22. Snap the E-clip over the top of the pawl to secure it in place.

23. Rotate the hub counterclockwise until it is tight. From this position, back off the pulley about 1/2 turn until the outlet hole in the pulley aligns with the hole in the housing.

24. Once the holes are aligned, insert a drift pin through the pulley and into the housing to prevent the pulley from rewinding.

25. Tie a figure 8 knot in the end of the rope and then feed the rope through the recess in the pulley and out through the housing.

26. Install the pull handle by working the rope through the rubber part of the handle. Secure the rope to the anchor. Work the anchor into the handle. Hold the rope tight and at the same time remove the drift pin from the pulley. Allow the pulley to SLOWLY rewind and at the same time feed the rope into the pulley.

27. With the starter still on its back, pull the rope with quick movements, and at the same time check the pawl to be sure it moves toward the center of the pulley. Release the rope slowly and check to be sure the pawl returns to its original position.

28. Install the starter.

CLEANING & INSPECTION

◆ See Figures 133 and 153

② MODERATE

If the rope was broken and the spring is bent backward, it is a simple matter to bend the spring end back to its normal position.

Wash all parts, except the rope, in solvent and then blow them dry with compressed air.

Remove any trace of corrosion and wipe all metal parts with an oil-dampened cloth. Inspect the rope. Replace the rope if it appears to be weak or frayed. If the rope is frayed, check the hole through which the rope passes for rough edges or burrs. Remove the rough edges or burrs with a file, and polish the surface until it is smooth.

Inspect the starter spring end loops. Replace the spring if it is weak, corroded or cracked. Check the spring pin located at the back side of the pulley to be sure it is straight and solid.

Check the inside surface of the housing and remove any burrs.

Check the condition of the pawl spring to be sure it is not stretched out of shape. The end of the spring should be bent back towards the coil of the spring. Inspect the pawl for wear and make sure the edges are not rounded.

Check the friction spring and link to be sure they are not distorted.

Inspect the spindle. The spindle must be straight and tight.

SAFETY PAWL ADJUSTMENT

◆ See Figure 154

① EASY

1. Disconnect the shift cable at the engine.

2. Manually shift the engine into Forward gear and at the same time have an assistant rotate the propeller. Check with the propeller to be sure the engine is in gear.

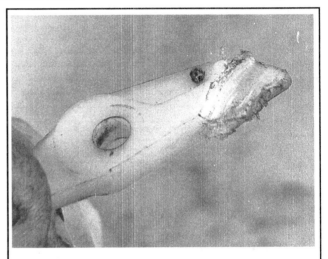

Fig. 153 This damaged pawl can no longer be used

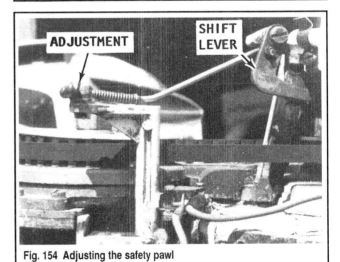

Fig. 154 Adjusting the safety pawl

3. Attempt to pull on the starter rope. The attempt should fail - the starter should NOT rotate.

4. Shift the engine into Reverse gear and again attempt to pull on the starter rope. Again, the attempt should fail - the starter should not rotate.

5. If the starter should rotate during either one of these tests, the linkage to the tower shaft at the rear of the engine must be adjusted to move the pawl up closer to the bottom of the pulley. If the starter cannot be rotated when the shift lever is in the Neutral position, the pawl is too close. Release the eyelet on the tower shaft and lower the pawl away from the pulley a bit.

Hand Rewind Starter Applications

Year	Model (Hp)	No. of Cyl	Displace cu. in. (cc)	Starter Type ①
1958	3	2	5.28 (85.5)	VI
	5.5	2	8.84 (145)	VI
	7.5	2	12.4 (203)	VI
	10	2	16.6 (281)	VI
	18	2	22 (361)	VI
	35	2	40.5 (664)	V
	50	V4	70.7 (1159)	VII
1959	3	2	5.28 (85.5)	VI
	5.5	2	8.84 (145)	VI
	10	2	16.6 (281)	VI
	18	2	22 (361)	VI
	35	2	40.5 (664)	V
	50	V4	70.7 (1159)	VII
1960	3	2	5.28 (85.5)	VI
	5.5	2	8.84 (145)	VI
	10	2	16.6 (281)	VI
	18	2	22 (361)	VI
	40	2	43.9 (719)	V
	75	V4	89.5 (1467)	VII
1961	3	2	5.28 (85.5)	VI
	5.5	2	8.84 (145)	VI
	10	2	16.6 (281)	VI
	18	2	22 (361)	VI
	40	2	43.9 (719)	V
	75	V4	89.5 (1467)	VII
1962	3	2	5.28 (85.5)	VI
	5.5	2	8.84 (145)	VI
	10	2	16.6 (281)	VI
	18	2	22 (361)	VI
	28	2	35.7 (585)	V
	40	2	43.9 (719)	V
	75	V4	89.5 (1467)	VII
1963	3	2	5.28 (85.5)	VI
	5.5	2	8.84 (145)	VI
	10	2	16.6 (281)	VI
	18	2	22 (361)	VI
	28	2	35.7 (585)	V
	40	2	43.9 (719)	V
	75	V4	89.5 (1467)	VII
1964	3	2	5.28 (85.5)	VI
	5.5	2	8.84 (145)	VI
	9.5	2	15.2 (249)	II
	18	2	22 (361)	VI
	28	2	35.7 (585)	IV
	40	2	43.9 (719)	IV
	60	V4	70.7 (1159)	NA
	75	V4	89.5 (1467)	VII
	90	V4	89.5 (1467)	NA
1965	3	2	5.28 (85.5)	VI
	5	2	8.84 (145)	I
	6	2	8.84 (145)	I
	9.5	2	15.2 (249)	II
	18	2	22 (361)	VI

Year	Model (Hp)	No. of Cyl	Displace cu. in. (cc)	Starter Type ①
1965 (cont'd)	33	2	40.5 (664)	IV
	40	2	43.9 (719)	IV
	60	V4	70.7 (1159)	NA
	75	V4	89.5 (1467)	VII
	90	V4	89.5 (1467)	NA
1966	3	2	5.28 (85.5)	VI
	5	2	8.84 (145)	I
	6	2	8.84 (145)	I
	9.5	2	15.2 (249)	II
	18	2	22 (361)	VI
	20	2	22 (361)	VI
	33	2	40.5 (664)	IV
	40	2	43.9 (719)	IV
	60	V4	70.7 (1159)	NA
	80	V4	89.5 (1467)	NA
	100	V4	89.5 (1467)	NA
1967	3	2	5.28 (85.5)	VI
	5	2	8.84 (145)	I
	6	2	8.84 (145)	I
	9.5	2	15.2 (249)	II
	18	2	22 (361)	VI
	20	2	22 (361)	VI
	33	2	40.5 (664)	IV
	40	2	43.9 (719)	IV
	60	V4	70.7 (1159)	NA
	80	V4	89.5 (1467)	NA
	100	V4	89.5 (1467)	NA
1968	1.5	1	2.64 (43.3)	
	3	2	5.28 (85.5)	III, VI
	5	2	8.84 (145)	I
	6	2	8.84 (145)	I
	9.5	2	15.2 (249)	II
	18	2	22 (361)	VI
	20	2	22 (361)	VI
	33	2	40.5 (664)	IV
	40	2	43.9 (719)	IV
	55	3	49.7 (814)	NA
	65	V4	70.7 (1159)	NA
	85	V4	89.5 (1467)	NA
	100	V4	89.5 (1467)	NA
1969	1.5	1	2.64 (43.3)	III, VI
	4	2	5.28 (85.5)	III
	6	2	8.84 (145)	I
	9.5	2	15.2 (249)	II
	18	2	22 (361)	VI
	20	2	22 (361)	VI
	25	2	22 (361)	VI
	33	2	40.5 (664)	IV
	40	2	43.9 (719)	IV
	55	3	49.7 (814)	NA
	85	V4	92.6 (1517)	NA
	115	V4	96.1 (1575)	NA

Hand Rewind Starter Applications

Year	Model (Hp)	No. of Cyl	Displace cu. in. (cc)	Starter Type ①
1970	1.5	1	2.64 (43.3)	III, VI
	4	2	5.28 (85.5)	III
	6	2	8.84 (145)	I
	9.5	2	15.2 (249)	II
	18	2	22 (361)	VI
	20	2	22 (361)	VI
	25	2	22 (361)	VI
	33	2	40.5 (664)	IV
	40	2	43.9 (719)	IV
	60	3	49.7 (814)	NA
	85	V4	92.6 (1517)	NA
	115	V4	96.1 (1575)	NA
1971	2	1	2.64 (43.3)	VI
	4	2	5.28 (85.5)	III
	6	2	8.84 (145)	I
	9.5	2	15.2 (249)	II
	18	2	22 (361)	VI
	20	2	22 (361)	VI
	25	2	22 (361)	VI

Year	Model (Hp)	No. of Cyl	Displace cu. in. (cc)	Starter Type ①
1971 (cont'd)	40	2	43.9 (719)	IV
	50	2	41.5 (680)	NA
	60	3	49.7 (814)	NA
	85	V4	92.6 (1517)	NA
	100	V4	92.6 (1517)	NA
	125	V4	99.6 (1632)	NA
1972	2	1	2.64 (43.3)	VI
	4	2	5.28 (85.5)	III
	6	2	8.84 (145)	I
	9.5	2	15.2 (249)	II
	18	2	22 (361)	VI
	20	2	22 (361)	VI
	25	2	22 (361)	VI
	40	2	43.9 (719)	IV
	50	2	41.5 (680)	NA
	65	3	49.7 (814)	NA
	85	V4	92.6 (1517)	NA
	100	V4	92.6 (1517)	NA
	125	V4	99.6 (1632)	NA

NA Not applicable
① Your starter may differ, please review other starter types if you feel yours is different than listed.

Starter Rope Specifications

Engine Size	Model	Diameter In.	Length In.
1.5 Hp	1968-70	①	①
2 Hp	1971-72	1/8	64
3 Hp	1958-61	5/32	65-1/4
	1962-67	5/32	71-1/2
	1968	1/8	56
4 Hp	1969-72	1/8	64
5 Hp	1965-67	1/8	64
	1968	1/8	56
5.5 Hp	1958-61	5/32	70
	1962-64	5/32	71-1/2
6 Hp	1965-67	1/8	64
	1968-72	1/8	56
7.5 Hp	1958	5/32	70
9.5 Hp	1964-70	5/32	71-1/2
	1971-72	5/32	65-1/4
10 Hp	1958-61	7/32	70-3/16
	1962-63	7/32	75-3/4
18 Hp	1958-61	7/32	70-3/16
	1962-70	7/32	75-3/4
	1971-72	7/32	72-1/4
20 Hp	1966-70	7/32	75-3/4
	1971-72	7/32	72-1/4
25 Hp	1969-70	7/32	75-3/4
	1971-72	7/32	72-1/4
28 Hp	1962-64	7/32	75-3/4
33 Hp	1965-70	7/32	75-3/4
35 Hp	1958-59	7/32	73-3/4
40 Hp	1960-70	7/32	75-3/4
	1971-72	7/32	73-3/4
50 Hp	1958-59 ②	7/32	75-3/4
	1971-72 ③	7/32	69-3/4
75 Hp	1960-65	7/32	69-3/4

NOTE: Always purchase a quality grade nylon rope.

① Order part #382712, rope includes handle
② V4 engine
③ 2 cyl engine

SelocOnLine

SelocOnLine is a maintenance and repair database accessed via the Internet.
Always up-to-date, skill level and special tool icons, quick access buttons to wiring diagrams,
specification charts, maintenance charts, and a parts database.
Contact your local marine dealer or see our demo at www.seloconline.com